THE HUTCHISON SERIES IN MAT

BASIC MATHEMATICAL SKILLS with Geometry

Ninth Edition

Stefan Baratto
Clackamas Community College

Barry Bergman
Clackamas Community College

Don Hutchison
Clackamas Community College

Mc Graw Hill
Connect Learn Succeed

The McGraw·Hill Companies

BASIC MATHEMATICAL SKILLS WITH GEOMETRY, NINTH EDITION
Published by McGraw-Hill, a business unit of The McGraw-Hill Companies, Inc., 1221 Avenue of the Americas, New York, NY 10020.

This book is printed on acid-free paper.

1 2 3 4 5 6 7 8 9 0 QDB/QDB 1 0 9 8 7 6 5 4 3

ISBN 978–0–07–338444–3

MHID 0–07–338444–5

ISBN 978–0–07–757397–3 (Annotated Instructor's Edition)

MHID 0–07–757397–8

Senior Vice President, Products & Markets: *Kurt L. Strand*
Vice President, General Manager, Products & Markets: *Marty Lange*
Vice President, Content Production & Technology Services: *Kimberly Meriwether David*
Managing Director: *Ryan Blankenship*
Director: *Dawn R. Bercier*
Brand Manager: *Mary Ellen Rahn*
Director of Development: *Rose Koos*
Director of Digital Content: *Nicole Lloyd*
Senior Project Manager: *Vicki Krug*
Senior Buyer: *Sandy Ludovissy*
Senior Media Project Manager: *Sandra M. Schnee*
Designer: *Tara McDermott*
Cover Image: © *Leuntje 2012/Getty Images/RF*
Senior Content Licensing Specialist: *Lori Hancock*
Compositor: *MPS Limited*
Typeface: *10/12 Times New Roman*
Printer: Quad/Graphics

About the Cover Photo
A flower symbolizes transformation and growth—a change from the ordinary to the spectacular! The Hutchison Series helps students in an arithmetic/basic math course grow their math skills ***from the ground up*** with a proven approach that motivates them to become stronger math students and to use their mathematical knowledge in everyday life.

Library of Congress Cataloging-in-Publication Data

Baratto, Stefan.
Basic mathematical skills with geometry : the Hutchison series in mathematics. – Ninth edition / Stefan Baratto, Clackamas Community College, Barry Bergman, Clackamas Community College, Donald Hutchison, Clackamas Community College.
pages cm
Includes bibliographical references and index.
ISBN 978–0–07–338444–3 — ISBN 0–07–338444–5 (hard copy : alk. paper) 1. Arithmetic. 2. Algebra. 3. Geometry. I. Bergman, Barry. II. Title.
QA107.2.B37 2014
513'.14–dc23 2012027281

CONTENTS

ABOUT THE AUTHORS & DEDICATION

Stefan Baratto

Stefan began teaching math and science in New York City middle schools. He also ta at the University of Oregon, Southeast Missouri State University, and York County College. Currently, Stefan is a member of the mathematics faculty at Clackamas Co College where he has found a niche, delighting in the CCC faculty, staff, and students. own education includes the University of Michigan (BGS, 1988), Brooklyn College (CUN the University of Oregon (MS, 1996).

Stefan is currently serving on the AMATYC Executive Board as the organization's Nort Vice President. He has been involved with ORMATYC, NEMATYC, NCTM, NADE, an State of Oregon Math Chairs group, as well as other local organizations. He has applied his kn edge of math to various fields, using statistics, technology, and web design. More persona Stefan and his wife, Peggy, try to spend time enjoying the wonders of Oregon and the Pac Northwest. Their activities include scuba diving and hiking.

Barry Bergman

Barry has enjoyed teaching mathematics to a wide variety of students over the years. He began in the field of adult basic education and moved into the teaching of high school mathematics in 1977. He taught high school math for 11 years, at which point he served as a K-12 mathematics specialist for his county. This work allowed him the opportunity to help promote the emerging NCTM standards in his region.

In 1990, Barry began the next portion of his career, having been hired to teach at Clackamas Community College. He maintains a strong interest in the appropriate use of technology and visual models in the learning of mathematics.

Throughout the past 35 years, Barry has played an active role in professional organizations. As a member of OCTM, he contributed several articles and activities to the group's journal. He has presented at AMATYC, OCTM, NCTM, ORMATYC, and ICTCM conferences. Barry also served 4 years as an officer of ORMATYC and participated on an AMATYC committee to provide feedback to revisions of NCTM's standards.

Don Hutchison

Don began teaching in a preschool while he was an undergraduate. He subsequently taught children with disabilities, adults with disabilities, high school mathematics, and college mathematics. Although each position offered different challenges, it was always breaking a challenging lesson into teachable components that he most enjoyed.

It was at Clackamas Community College that he found his professional niche. The community college allowed him to focus on teaching within a department that constantly challenged faculty and students to expect more. Under the guidance of Jim Streeter, Don learned to present his approach to teaching in the form of a textbook. Don has also been an active member of many professional organizations. He has been president of ORMATYC, AMATYC committee chair, and ACM curriculum committee member. He has presented at AMATYC, ORMATYC, AACC, MAA, ICTCM, and a variety of other conferences.

Above all, he encourages you to be involved, whether as a teacher or as a learner. Whether discussing curricula at a professional meeting or homework in a cafeteria, it is the process of communicating an idea that helps one to clarify it.

Dedication

We dedicate this text to the thousands of students who have helped us become better teachers, better communicators, better writers, and even better people. We read and respond to every suggestion we get—every one is invaluable. If you have any thoughts or suggestions, please contact us at

Stefan Baratto: sbaratto@clackamas.edu
Barry Bergman: bfbergman@gmail.com
Don Hutchison: donh@collegemathtext.com

Thank you all.

PREFACE

Welcome Students

Learning math can be an exciting adventure! As you embark on this journey, we are here with you. Whether this material is new to you or you are trying to master topics that previously eluded you, this text is designed to make it easier to learn the essential mathematics that you will need.

Learning math, learning to be a student (again), and growing as a person are all intertwined as part of your experience. We hope that you continue to grow as a student and as a person while you ***grow your math skills from the ground up***.

We have seen many students succeed in our math classes and we offer you some guidance that may help you to be one of the successful students. Through the first half of this text, we offer a series of *Tips for Student Success* features. These cover many of those skills and actions that successful college students exhibit. You can find a complete listing of the *Tips for Student Success* in the *Index*. We would like to highlight two of the more important items in these Tips.

Learning math takes time. Students are expected to study 2 to 3 hours per week, outside of class, for every credit hour in a math course. In order to learn the math necessary to succeed in your course, you will need to make this time commitment. Expect to spend an hour or more every day outside of class learning math. Because of the size of this commitment, you need to schedule your day and week to include these hours.

Second, while your instructor and this text are your primary resources, many other resources are available to you. Your college may offer study skills or new student experience courses. Take such a course if it's been a while since you've been a student. Your college may also offer tutoring or other helpful services.

You may also want to take advantage of the *Student's Solutions Manual* for this text as well as the online resources available to you. McGraw-Hill Higher Education offers the online course management system Connect Math Hosted by ALEKS Corp. and access to the ALEKS platform. These online resources offer the opportunity to practice as much as you need to and provide immediate feedback on your progress.

Practice is one key to learning to do math. Reading mathematics in a text teaches you to read math in a text. Watching instructors do mathematics teaches you to watch instructors do math. In order to learn to do mathematics, you must do math!

In our text, every example is followed by a *Check Yourself* exercise. It is imperative that you complete these exercises. Please do not move on to the next example until you can complete the *Check Yourself* exercise successfully. By being an active learner, you will learn to do math!

There are numerous other features in this text, each designed to help you succeed. Learning to think critically about the math and learning to read and communicate technical information are essential to becoming effective problem solvers. Our many applications and examples will help you achieve this by teaching you to apply the math you are learning to real-life situations.

At the end of each chapter, you will find materials such as a *Summary*, *Chapter Test*, and *Cumulative Review* that will help you coalesce your learning and maintain your knowledge base. They will help you succeed in ways you may have never thought possible.

We wish you fun and success as you continue your journey as a student and as a person.

To Our Colleagues

Over three careers, the author team has learned much about teaching and learning mathematics. Perhaps the biggest item we have learned is how little we know. For example, we can spot the successful students on day one, except when we can't. Just as often, we are pleasantly surprised by the success of a student who struggled at the beginning of the term.

There are some things we feel we do know. We are certain that one important key to learning mathematics is active participation. In our *Welcome* to the students, we write

> Practice is one key to learning to do math. Reading mathematics in a text teaches you to read math in a text. Watching instructors do mathematics teaches you to watch instructors do math. In order to learn to do mathematics, you must do math!

We feel that this may be the most important aspect of becoming a successful math student. We implore you to advise your students that being successful requires that they spend the appropriate amount of time engaged in the process of doing and learning math.

Often, students are tempted to utilize classroom time to complete their homework or engage in other activities that are best completed outside of class. We feel it is important to reinforce the idea that students achieve success by spending their time doing math outside of class. We stress this in our own classrooms and encourage you to do the same.

While we actively encourage students to imitate the habits of successful students, this text is a passive resource. You are their most important resource. We strive to provide you with assistance reiterating those things in print that we all say in the classroom. Between us, your students can ***grow their math skills from the ground up***.

Another key for us is helping students to see the relevance of mathematics to their daily experience. Our students are like your students. They don't always see how their math classes help them in their lives. With your help and ours, they have a world of growth ahead of them.

We offer many real-world applications for you and your students. We've included application examples in the exposition to assist newer instructors demonstrate how to approach these while in the classroom.

We feel that it is very important that students work with these applications. In addition to helping them see the relevance of what they learn, they gain invaluable training in critical thinking and problem solving.

Even more important, we see these applications as providing the student with true transferable skills. Being able to read, comprehend, and communicate technical information encompasses a set of skills that students can use throughout their college careers and their lives. This may be the most important thing they gain from their math classes.

Many people helped us to revise these texts. Our own classroom experiences, our students, and our colleagues at Clackamas Community College were our first line of resources, obviously. But, numerous instructors, users, and reviewers from around the country contributed their thoughts and ideas. Their input provided the impetus for many of the changes you will find. We believe that this helped us to write our strongest text yet and hope you agree. Please don't hesitate to contact the author team if you would like to provide your own input.

Major Changes to the Ninth Edition

Our revisions are based on our own experiences, as well as the comments and feedback that we receive from the many students, instructors, reviewers, and editorial personnel who have engaged with these materials.

Global Changes

Writing, instructions, and other materials edited for clarity

- Revised student learning objectives
- Improved and expanded instruction of reading, interpreting, and solving word problems and applications
- Increased emphasis placed on checking answers for accuracy and reasonableness
- Added and improved thought-provoking exercises

Applications integrated throughout

- Added more application examples to the exposition
- Updated data, current events, and applications

Reorganized and reformatted the exercise sets

- Over 1,000 new exercises; 650 more exercises than the eighth edition

Revised the end-of-chapter materials

- Reorganized the chapter tests to better reflect actual exams

Added a Brief Glossary of key vocabulary terms

Chapter 1

Added instruction and exercises on reading and interpreting word problems and applications (1.2)

Expanded division instruction to include three-digit divisors (1.6)

Expanded grouping symbol discussion and instruction (1.7)

Chapter 2

Improved introduction to fractions material

Chapter 3

Improved discussion of multiplying/dividing by 1 (3.2)

Added fraction number line (3.2)

Major revamp of sections (3.4, 3.5)

Improved grouping symbol instruction (3.5)

> "This series is simple in its explanations, very readable, has very good organization and contains very appropriate problem sets."
>
> —**Sandi Tannen, *Camden County College***

Chapter 4

Resequenced sections

Improved rounding content and expanded rounding with 9s (4.1)

Chapter 5

Improved interpretation of rates (5.2)

Chapter 6

Added fractions of a percent content (6.1)

Improved two methods of solving problems (6.2)

Added repeating decimals content (6.2)

Chapter 7

Two new activities

Revisions based on user feedback (7.4)

Chapter 8

Improved naming of figures materials (8.1)

Expanded Pythagorean materials (8.5)

New Solid Geometry section (8.6)

Chapter 9

Resequenced sections

Emphasis placed on critical thinking and interpretations

Chapter 10

Added mixed number arithmetic (10.3)

Chapter 11

Major revamp with an emphasis on modeling included

Improved applications (11.1)

Improved calculator instruction (11.2, 11.4, 11.5)

Improved solve content; increased emphasis on checking solutions (11.6)

> "An excellent textbook for students who have never taken higher levels of math in high school or have been out of school for a number of years. It would be a great reference book for topics covered in algebra and as a review of basic algebraic and arithmetic concepts needed in all areas of math and science. The text is student-friendly with many worked examples and great summary notes at the end of the chapters."
>
> —Linda Faraone, *Georgia Military College*

GROW YOUR MATHEMATICAL SKILLS

***"Make the Connection"**—Chapter-opening vignettes* provide interesting, relevant scenarios that engage students in the upcoming material. Exercises and *Activities* related to the vignette revisit its themes to more effectively drive mathematical comprehension (marked with a 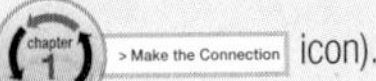 icon).

(a) How many firms in total are located in Washington, Philadelphia, and New York?

(b) What is the total number of employees in all 10 of the areas listed?

(c) What are the total sales for firms in Houston and Dallas?

(d) How many firms in total are located in Chicago and Detroit? 

65. Number Problem These sequences of numbers are called *arithmetic sequences*. Determine the pattern an the next four numbers in each sequence.

(a) 5, 12, 19, 26, ____, ____, ____, ____

(b) 8, 14, 20, 26, ____, ____, ____, ____

(c) 7, 13, 19, 25, ____, ____, ____, _

(d) 9, 17, 25, 33, ____, ____, ____, _

66. Number Prob

1, 1, 2, 3, 5, 8

This sequenc

two precedin

Find the next

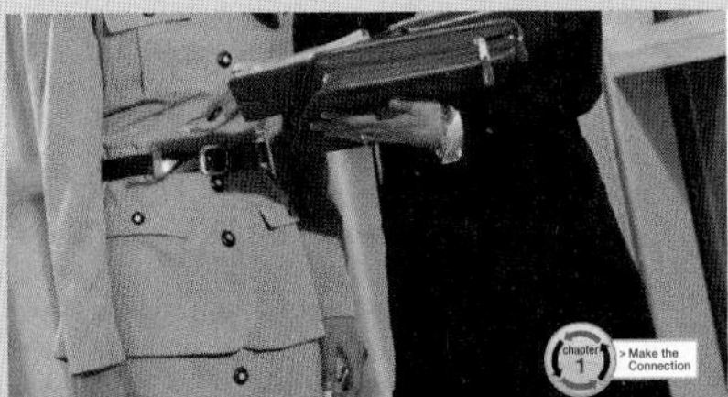

INTRODUCTION

How many? How much? We asked these questions even before we had civilizations. Think about how often you count and use numbers.

In the United States, counting the number of people in the country is one of the more important current examples of this. Article I, Section 2, of the U.S. Constitution directs the government to

Operations on Whole Numbers

Activities promote active learning by requiring students to find, interpret, and manipulate real-world data. The activities tie the chapter together with questions that sharpen mathematical and conceptual understanding of the chapter material. Students can complete the activities on their own or in small groups.

Activity 15 ::

Burning Calories

Many people are interested in losing weight through exercise. An important fact to consider is that a person needs to burn off 3,500 calories more than he or she takes in to lose 1 lb, according to the American Dietetic Association.

The table shows the number of calories burned per hour (cal/hr) for a variety of activities, where the figures are based on a 150-lb person.

Activity	Cal/hr	Activity	Cal/hr
Bicycling 6 mi/hr	240	Running 10 mi/hr	1,280
Bicycling 12 mi/hr	410	Swimming 25 yd/min	275
Cross-country skiing	700	Swimming 50 yd/min	500
Jogging $5\frac{1}{2}$ mi/hr	740	Tennis (singles)	400
Jogging 7 mi/hr	920	Walking 2 mi/hr	240
Jumping rope	750	Walking 3 mi/hr	320
Running in place	650	Walking $4\frac{1}{2}$ mi/hr	440

Reading Your Text offers a brief set of exercises at the end of each section to assess students' knowledge of key vocabulary terms, encourage careful reading, and reinforce understanding of core mathematical concepts. Answers are provided at the end of the book.

Reading Your Text

These fill-in-the-blank exercises will help you understand some of the key vocabulary used in this section. The answers to these exercises are in the Answers Appendix at the back of the text.

(a) Ratios are used to compare ________ quantities.

(b) When we compare measurements with different types of units, we get a ________.

(c) We usually use whole, mixed, or decimal numbers rather than ________ fractions when reporting a unit rate.

(d) ________ prices are used to compare the costs of items in different-sized packages.

THROUGH MORE CAREFUL PRACTICE

Chapter Tests let students check their progress and review important concepts so they can prepare for exams with confidence and proper guidance. Answers are given at the end of the book, with section references provided to help students review important material.

CHAPTER 2 **chapter test 2**

Use this chapter test to assess your progress and to review for your next exam. Allow yourself about an hour to take this test. The answers to these exercises are in the Answers Appendix at the back of the text.

1. Which of the numbers 5, 9, 13, 17, 22, 27, 31, and 45 are prime numbers? Which are composite numbers?

2. Find the prime factorization for 264.

3. Use the divisibility tests to determine which, if any, of the numbers 2, 3, and 5 are factors of 54,204.

Find the greatest common factor (GCF) of each set of numbers.

4. 36 and 84

5. 16, 24, and 72

What fraction names the shaded part of each diagram? Identify the numerator and denominator.

6.

7.

8.

Prerequisite Checks, included starting with Chapter 2, review skills from previous chapters needed for success in the upcoming chapter. These exercises offer students crucial help in identifying which concepts they may need to give extra review before proceeding with the next chapter. Answers are provided at the end of the book.

2 prerequisite check CHAPTER 2

This Prerequisite Check highlights the skills you will need in order to be successful in this chapter. The answers to these exercises are in the Answers Appendix at the back of the text.

1. Does 4 divide exactly into 30 (that is, with no remainder)?

2. Does 5 divide exactly into 29?

3. Does 6 divide exactly into 72?

4. Does 3 divide exactly into 412?

5. Does 2 divide exactly into 238?

6. List all the whole numbers that can divide exactly into 9.

7. List all the whole numbers that can divide exactly into 10.

8. List all the whole numbers that can divide exactly into 17.

9. List all the whole numbers that can divide exactly into 48.

10. List all the whole numbers that can divide exactly into 60.

Cumulative Reviews, included starting with Chapter 2, follow the *Chapter Test* and reinforce previously covered material to help students retain knowledge throughout the course and identify skills necessary for them to review when preparing for exams. Answers are provided at the end of the book, along with section references.

36. CONSTRUCTION A $31\frac{1}{3}$-acre piece of land is subdivided into home lots. Each home lot is to be $\frac{2}{3}$ acre. How many homes can be built?

37. CONSTRUCTION A room measures $5\frac{1}{3}$ yd by $3\frac{3}{4}$ yd. How many square yards of linoleum must be purchased to cover the floor?

38. SOCIAL SCIENCE The scale on a map is 1 in. = 80 mi. If two towns are $2\frac{3}{8}$ in. apart on the map, what is the actual distance in miles between the towns?

cumulative review chapters 1–2

Use this exercise set to review concepts from earlier chapters. While it is not a comprehensive exam, it will help you identify any material that you need to review before moving on to the next chapter. In addition to the answers, you will find section references for these exercises in the Answers Appendix in the back of the text.

1.1

1. Give the place value of 7 in 3,738,500.

2. Give the word name for 302,525.

3. Write two million, four hundred thirty thousand as a numeral.

1.2 *Name the property of addition illustrated.*

4. $5 + 12 = 12 + 5$

5. $9 + 0 = 9$

6. $(7 + 3) + 8 = 7 + (3 + 8)$

GROW YOUR MATHEMATICAL SKILLS

"Check Yourself" Exercises actively involve students in the learning process. Every worked example in the book is followed by an exercise encouraging students to solve a problem similar to the one just presented and check, through practice, what they have just learned. Answers are provided at the end of the section for immediate feedback.

< Objective 3 >

Find the unit price for each item. Round your results to the nearest cent per unit.

(a) 8 oz of cream cost \$2.49.

$\frac{\$2.49}{8 \text{ oz}} = \frac{249 \text{ cents}}{8 \text{ oz}} = \frac{249}{8}\frac{\text{cents}}{\text{oz}}$

$\approx 31 \frac{\text{cents}}{\text{oz}}$

(b) 20 lb of potatoes cost \$9.98.

$\frac{\$9.98}{20 \text{ lb}} = \frac{998 \text{ cents}}{20 \text{ lb}} = \frac{998}{20}\frac{\text{cents}}{\text{lb}}$

$\approx 50 \frac{\text{cents}}{\text{lb}}$

Check Yourself 5

Find the unit price for each item. Round your results to the nearest cent per unit.

(a) 12 soda cans cost \$4.98.
(b) 25 lb of dog food cost \$14.99.

As with ratios, rates are most often used for comparisons. For instance, unit pricing allows people to compare the cost of different-sized items.

In Example 6, we use unit prices to determine whether a glass of milk is less expensive when poured from a 128-oz container (gallon) or a 32-oz container (quart).

End-of-Section Exercises help students evaluate their conceptual mastery of the section through practice. These comprehensive exercise sets are structured to highlight the progression in level, and organized by category and section learning objective to make it easier for instructors to plan assignments. Answers to odd-numbered exercises are provided at the end of the exercise set.

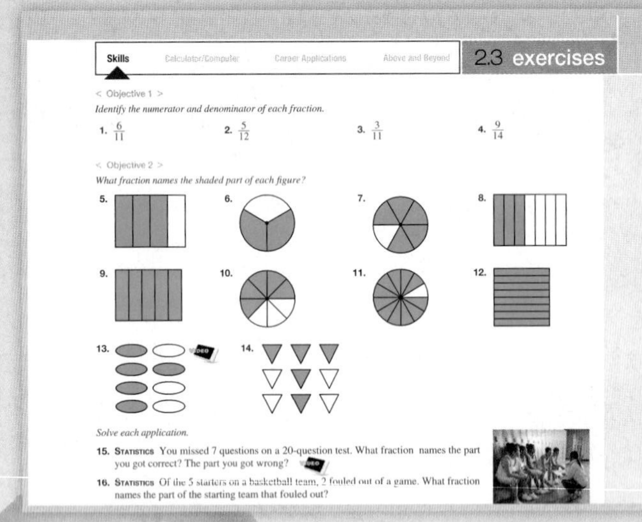

Skills | Calculator/Computer | Career Applications | Above and Beyond | **2.3 exercises**

< Objective 1 >

Identify the numerator and denominator of each fraction.

1. $\frac{6}{11}$ 2. $\frac{5}{12}$ 3. $\frac{3}{11}$ 4. $\frac{9}{14}$

< Objective 2 >

What fraction names the shaded part of each figure?

5. 6. 7. 8.

9. 10. 11. 12.

13. 14.

Solve each application.

15. STATISTICS You missed 7 questions on a 20-question test. What fraction names the part you got correct? The part you got wrong?

16. STATISTICS Of the 5 starters on a basketball team, 2 fouled out of a game. What fraction names the part of the starting team that fouled out?

Summary and Summary Exercises—*Summaries* at the end of each chapter show students key concepts from the chapter that they need to review, and provide page references to where each concept is introduced. *Summary Exercises* give students practice on these important concepts, with section references showing where they can go back to review relevant worked examples. Answers to odd-numbered *Summary Exercises* are provided at the end of the book.

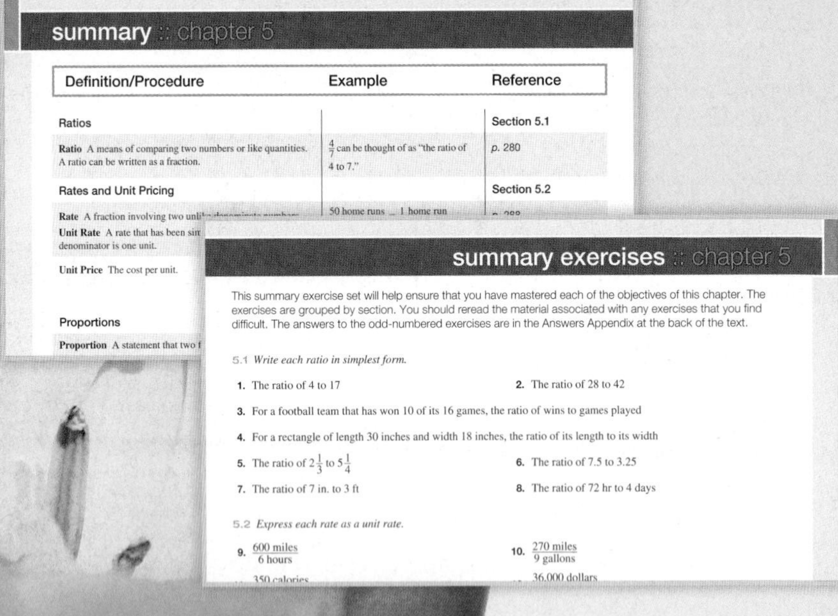

summary :: chapter 5

Definition/Procedure	Example	Reference
Ratios		Section 5.1
Ratio A means of comparing two numbers or like quantities. A ratio can be written as a fraction.	$\frac{4}{7}$ can be thought of as "the ratio of 4 to 7."	p. 280
Rates and Unit Pricing		Section 5.2
Rate A fraction involving two unli... **Unit Rate** A rate that has been sim... denominator is one unit.	50 home runs ... 1 home run	
Unit Price The cost per unit.		
Proportions		
Proportion A statement that two f...		

summary exercises :: chapter 5

This summary exercise set will help ensure that you have mastered each of the objectives of this chapter. The exercises are grouped by section. You should reread the material associated with any exercises that you find difficult. The answers to the odd-numbered exercises are in the Answers Appendix at the back of the text.

5.1 *Write each ratio in simplest form.*

1. The ratio of 4 to 17
2. The ratio of 28 to 42
3. For a football team that has won 10 of its 16 games, the ratio of wins to games played
4. For a rectangle of length 30 inches and width 18 inches, the ratio of its length to its width
5. The ratio of $2\frac{1}{3}$ to $5\frac{1}{4}$
6. The ratio of 7.5 to 3.25
7. The ratio of 7 in. to 3 ft
8. The ratio of 72 hr to 4 days

5.2 *Express each rate as a unit rate.*

9. $\frac{600 \text{ miles}}{6 \text{ hours}}$
10. $\frac{270 \text{ miles}}{9 \text{ gallons}}$

THROUGH BETTER ACTIVE LEARNING TOOLS

Tips for Student Success boxes offer valuable advice and resources to help students new to collegiate mathematics develop the study skills that will help them succeed. These class-tested suggestions provide students with extra direction on preparing for class, studying for exams, and familiarizing themselves with additional resources available outside of class.

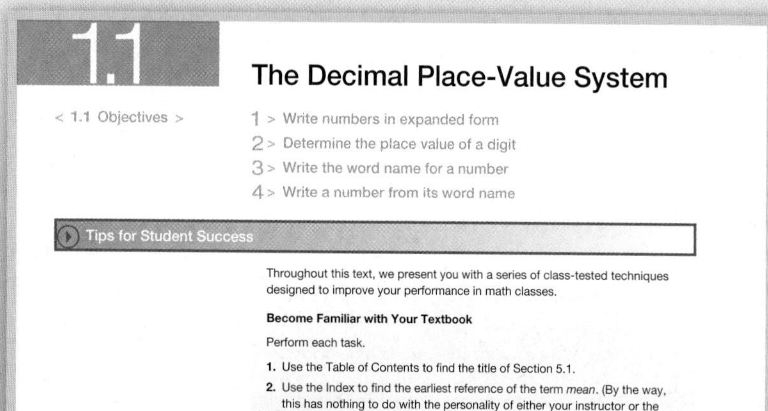

1.1

The Decimal Place-Value System

< 1.1 Objectives >

1 > Write numbers in expanded form
2 > Determine the place value of a digit
3 > Write the word name for a number
4 > Write a number from its word name

Tips for Student Success

Throughout this text, we present you with a series of class-tested techniques designed to improve your performance in math classes.

Become Familiar with Your Textbook

Perform each task.

1. Use the Table of Contents to find the title of Section 5.1.
2. Use the Index to find the earliest reference of the term *mean*. (By the way, this has nothing to do with the personality of either your instructor or the textbook authors!)

Notes and Recalls accompany the worked examples, providing just-in-time reminders that reinforce previously learned material and help students focus on information critical to their success.

NOTE

Ratios are *rarely* written as whole or mixed numbers. We usually write a ratio as a fraction, in simplest terms.

RECALL

In Section 4.3, you learned to multiply a decimal by a power of 10 by moving the decimal point.

Cautions are integrated throughout the textbook to alert students to common mistakes and how to avoid them.

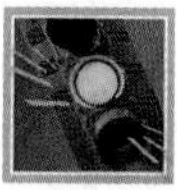

> CAUTION

Solving for x does not give us an answer directly. x represents the total volume, which includes both water and serum. We still need to subtract 8.5 from x to get a final answer.

In preparing
The dilution r
volume, also
a $\frac{1}{20}$ dilution
We set u
that should b

$$\frac{1}{20} = \frac{8.5}{x}$$

Video Exercises offer guided video solutions to selected exercises in each section marked with a VIDEO icon.

Available through *Connect Math Hosted by ALEKS,* these videos feature a presenter working through the exercises just like an instructor would, following the solution methodology from the text. The videos are available closed-captioned for the hearing-impaired or subtitled in Spanish, and meet the Americans with Disabilities Act Standards for Accessible Design.

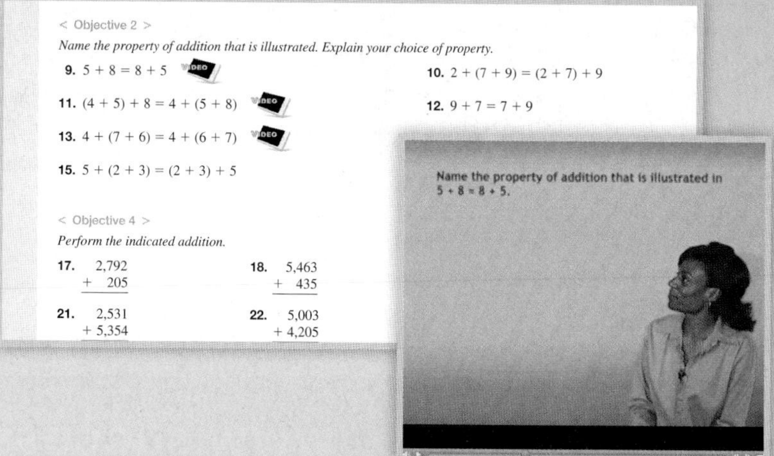

< Objective 2 >

Name the property of addition that is illustrated. Explain your choice of property.

9. $5 + 8 = 8 + 5$
10. $2 + (7 + 9) = (2 + 7) + 9$
11. $(4 + 5) + 8 = 4 + (5 + 8)$
12. $9 + 7 = 7 + 9$
13. $4 + (7 + 6) = 4 + (6 + 7)$
15. $5 + (2 + 3) = (2 + 3) + 5$

< Objective 4 >

Perform the indicated addition.

17. 2,792 + 205
18. 5,463 + 435
21. 2,531 + 5,354
22. 5,003 + 4,205

Quality Content For Today's Online Learners

Online Exercises were carefully selected and developed to provide a seamless transition from textbook to technology.

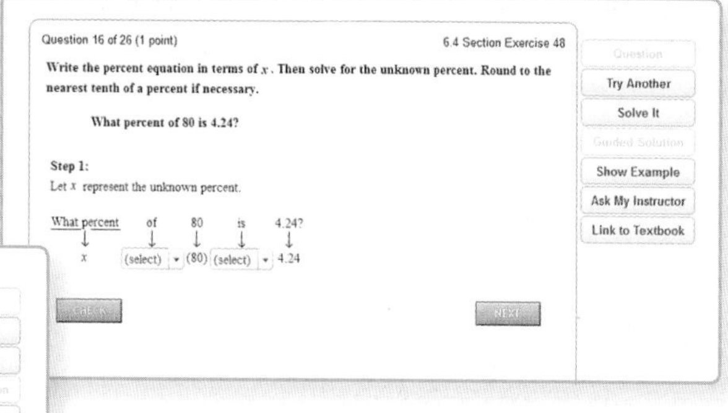

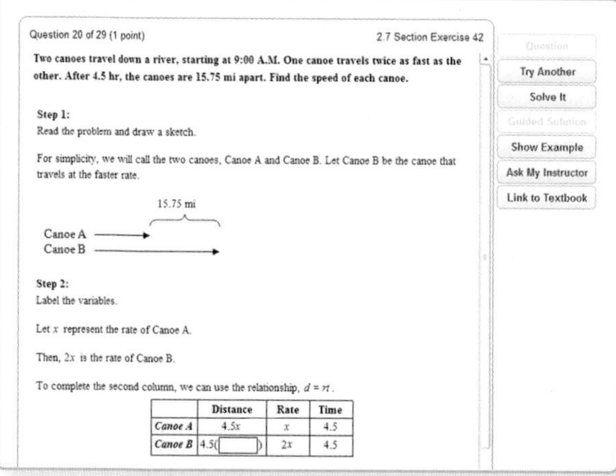

For consistency, the guided solutions match the style and voice of the original text as though the author is guiding the students through the problems.

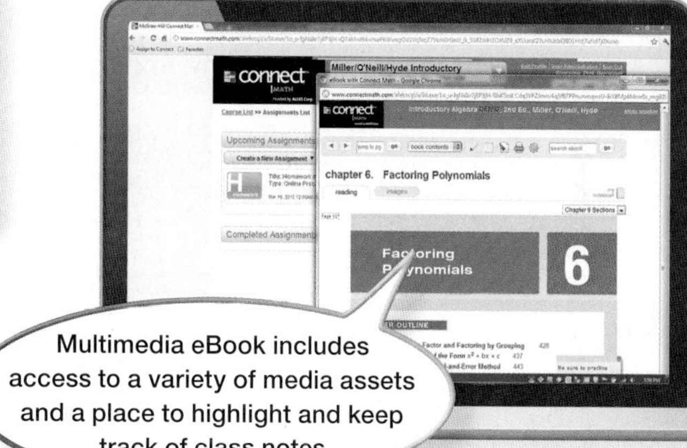

Gradebook - ALEKS Initial Assessment #1 - Goulet, Robert

Assignment: ALEKS Initial Assessment #1
Completion Date: 04/18/11 (time spent: 0 minutes)
Gradebook Score: 100%

ALEKS Assessment Report for Goulet, Robert

Course Mastery (209 of 300 Topics)
Radicals and Rational Exponents (5 of 20)
Complex Numbers and Quadratic Equations (6 of 16)
Rational Expressions and Proportions (13 of 35)
Arithmetic Readiness (63 of 63)
Real Numbers and Variables (29 of 33)
Integer Exponents and Polynomials (27 of 45)

Student Readiness by Topic

This ALEKS Assessment report shows the percentage of students that have mastered the following topics:

Ch.4-Linear Equations in Two Variables

Section 4.1
- Reading a point in the coordinate plane
- Plotting a point in the coordinate plane
- Finding a solution to a linear equation in two variables
- Identifying solutions to linear equations in two variables

Section 4.2
- Graphing a line given the x- and y-intercepts
- Graphing a line given its equation in slope-intercept form
- Graphing a line given its equation in standard form
- Graphing a vertical or horizontal line
- Finding x- and y-intercepts of a line given the equation in standard form

Section 4.3
- Graphing a line through a given point with a given slope
- Finding slope given the graph of a line on a grid
- Finding slope given two points on the line

Section 4.4
- Y-intercept of a line
- Finding the slope of a line given its equation

Section 4.5
- Writing the equations of vertical and horizontal lines through a given point
- Writing equations and drawing graphs to fit a narrative

Section 4.6
- Function tables
- Vertical line test

ALEKS Corporation's experience with algorithm development ensures a commitment to accuracy and a meaningful experience for students to demonstrate their understanding with a focus towards online learning.

The ALEKS® Initial Assessment is an artificially intelligent (AI), diagnostic assessment that identifies precisely what a student knows. Instructors can then use this information to make more informed decisions on what topics to cover in more detail with the class.

ALEKS is a registered trademark of ALEKS Corporation.

www.successinmath.com

ALEKS is a unique, online program that significantly raises student proficiency and success rates in mathematics, while reducing faculty workload and office-hour lines. ALEKS uses artificial intelligence and adaptive questioning to assess precisely a student's knowledge, and deliver individualized learning tailored to the student's needs. With a comprehensive library of math courses, ALEKS delivers an unparalleled adaptive learning system that has helped millions of students achieve math success.

ALEKS Delivers a Unique Math Experience:

- **Research-Based, Artificial Intelligence** precisely measures each student's knowledge
- **Individualized Learning** presents the exact topics each student is most **ready to learn**
- **Adaptive, Open-Response Environment** includes comprehensive tutorials and resources
- **Detailed, Automated Reports** track student and class progress toward course mastery
- **Course Management Tools** include textbook integration, custom features, and more

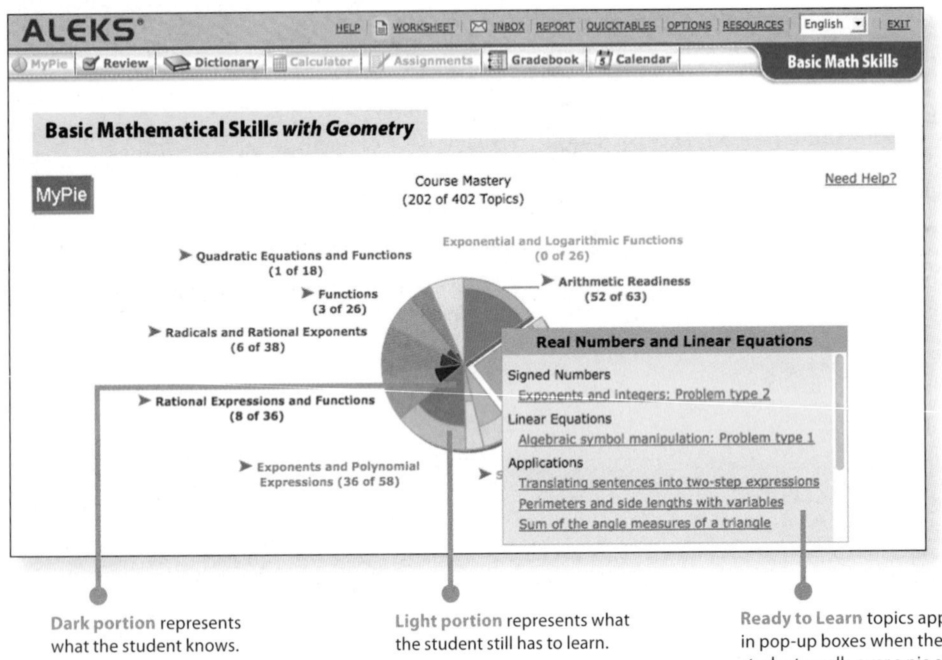

The ALEKS Pie summarizes a student's current knowledge, then delivers an individualized learning path with the exact topics the student is most ready to learn.

Dark portion represents what the student knows.

Light portion represents what the student still has to learn.

Ready to Learn topics appear in pop-up boxes when the student scrolls over a pie slice.

"My experience with ALEKS has been effective, efficient, and eloquent. **Our students' pass rates improved from 49 percent to 82 percent with ALEKS.** We also saw student retention rates increase by 12% in the next course. Students feel empowered as they guide their own learning through ALEKS."

—**Professor Eden Donahou, *Seminole State College of Florida***

To learn more about ALEKS, please visit: **www.aleks.com/highered/math**

ALEKS® Prep Products

ALEKS Prep products focus on prerequisite and introductory material, and can be used during the first six weeks of the term to ensure student success in math courses ranging from Beginning Algebra through Calculus. ALEKS Prep quickly fills gaps in prerequisite knowledge by assessing precisely each student's preparedness and delivering individualized instruction on the exact topics students are most ready to learn. As a result, instructors can focus on core course concepts and see improved student performance with fewer drops.

> "ALEKS is wonderful. It is a professional product that takes very little time as an instructor to administer. Many of our students have taken Calculus in high school, but they have forgotten important algebra skills. ALEKS gives our students an opportunity to review these important skills."
>
> —**Professor Edward E. Allen,** ***Wake Forest University***

A Total Course Solution

A cost-effective total course solution: fully integrated, interactive eBook combined with the power of ALEKS adaptive learning and assessment.

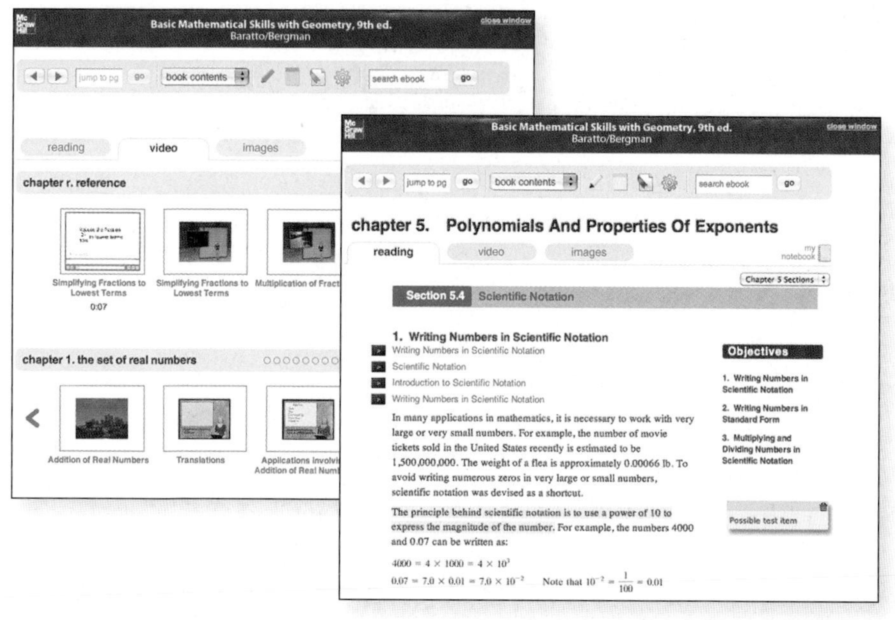

Students can easily access the full eBook content, multimedia resources, and their notes from within their ALEKS Student Accounts.

To learn more about ALEKS, please visit: **www.aleks.com/highered/math**

Our Commitment to Market Development and Accuracy

McGraw-Hill's Development Process is an ongoing, never-ending, market-oriented approach to building accurate and innovative print and digital products. We begin developing a series by partnering with authors that desire to make an impact within their discipline to help students succeed. Next, we share these ideas and manuscript with instructors for review for feedback and to ensure that the authors' ideas represent the needs within that discipline. Throughout multiple drafts, we help our authors adapt to incorporate ideas and suggestions from reviewers to ensure that the series carries the same pulse as today's classrooms. With any new series, we commit to accuracy across the series and its supplements. In addition to involving instructors as we develop our content, we also utilize accuracy checks through our various stages of development and production. The following is a summary of our commitment to market development and accuracy:

1. 3 drafts of author manuscript
2. 5 rounds of manuscript review
3. 2 focus groups
4. 1 consultative, expert review
5. 3 accuracy checks
6. 3 rounds of proofreading and copyediting
7. Towards the final stages of production, we are able to incorporate additional rounds of quality assurance from instructors as they help contribute towards our digital content and print supplements

This process then will start again immediately upon publication in anticipation of the next edition. With our commitment to this process, we are confident that our series has the most developed content the industry has to offer, thus pushing our desire for quality and accurate content that meets the needs of today's students and instructors.

Acknowledgments and Reviewers

The development of this textbook series would never have been possible without the creative ideas and feedback offered by many reviewers. We are especially thankful to the following instructors for their careful review of the manuscript.

Manuscript Review Panels

Over 150 teachers and academics from across the country reviewed the various drafts of the manuscript to give feedback on content, design, pedagogy, and organization. This feedback was summarized by the book team and used to guide the direction of the text.

Reviewers

Paul Ahad, *Antelope Valley College*
Robin Anderson, *Southwestern Illinois College*
Nieves Angulo, *Hostos Community College*
Arlene Atchison, *South Seattle Community College*
Haimd Attarzadeh, *Kentucky Jefferson Community and Technical College*
Jody Balzer, *Milwaukee Area Technical College*
Rebecca Baranowski, *Estrella Mountain Community College*
Wayne Barber, *Chemeketa Community College*
Bob Barmack, *Baruch College*
Chris Bendixen, *Lake Michigan College*
Karen Blount, *Hood College*
Donna Boccio, *Queensborough Community College*
Steve Boettcher, *Estrella Mountain Community College*
Elena Bogardus, *Camden County College*
Karen Bond, *Pearl River Community College, Poplarville*
Laurie Braga Jordan, *Loyola University-Chicago*
Kelly Brooks, *Pierce College*
Dorothy Brown, *Camden County College*
Michael Brozinsky, *Queensborough Community College*
Amy Canavan, *Century Community and Technical College*
Faye Childress, *Central Piedmont Community College*
Kathleen Ciszewski, *University of Akron*
Bill Clarke, *Pikes Peak Community College*
Lois Colpo, *Harrisburg Area Community College*
Christine Copple, *Northwest State Community College*
Jonathan Cornick, *Queensborough Community College*
Julane Crabtree, *Johnson County Community College*
Carol Curtis, *Fresno City College*
Sima Dabir, *Western Iowa Tech Community College*
Reza Dai, *Oakton Community College*
Karen Day, *Elizabethtown Technical and Community College*
Mary Deas, *Johnson County Community College*
Anthony DePass, *St. Petersburg College-Ns*
Shreyas Desai, *Atlanta Metropolitan College*
Deborah Detrick, *Lincoln College*
Robert Diaz, *Fullerton College*
Michaelle Downey, *Ivy Tech Community College*
Ginger Eaves, *Bossier Parish Community College*
Azzam El Shihabi, *Long Beach City College*
Joe Edwards, *Bevill State Community College–Hamilton Campus*
Kristy Erickson, *Cecil College*
Steven Fairgrieve, *Allegany College of Maryland*
Linda Faraone, *Georgia Military College*
Jacqui Fields, *Wake Technical Community College*

Bonnie Filer-Tubaugh, *University of Akron*
Rhoderick Fleming, *Wake Tech Community College*
Matt Foss, *North Hennepin Community College*
Catherine Frank, *Polk Community College*
Ellen Freedman, *Camden County College*
Heather Gallacher, *Cleveland State University*
Matt Gardner, *North Hennepin Community College*
Jeremiah Gilbert, *San Bernardino Valley College*
Judy Godwin, *Collin County Community College-Plano*
Lori Grady, *University of Wisconsin-Whitewater*
Brad Griffith, *Colby Community College*
Robert Grondahl, *Johnson County Community College*
Shelly Hansen, *Mesa State College*
Kristen Hathcock, *Barton County Community College*
Mary Beth Headlee, *Manatee Community College*
Bill Heider, *Hibbing Community College*
Kristy Hill, *Hinds Community College*
Mark Hills, *Johnson County Community College*
Sherrie Holland, *Piedmont Technical College*
Diane Hollister, *Reading Area Community College*
Denise Hum, *Canada College*
Byron D. Hunter, *College of Lake County*
Kelly Jackson, *Camden County College*
Patricia R. Jaquith, *Landmark College*
Nancy Johnson, *Manatee Community College-Bradenton*
Joe Jordan, *John Tyler Community College-Chester*
Sandra Ketcham, *Berkshire Community College*
Lynette King, *Gadsden State Community College*
Kelly Kohlmetz, *University of Wisconsin–Milwaukee*
Chris Kolaczewski-Ferris, *University of Akron*
Jeff Koleno, *Lorain County Community College*
Donna Krichiver, *Johnson County Community College*
Nancy Krueger, *Central Community College*
Indra B. Kshattry, *Colorado Northwestern Community College*
Patricia Labonne, *Cumberland County College*
Ted Lai, *Hudson County Community College*
Krynn Larsen, *Central Community College*
Pat Lazzarino, *Northern Virginia Community College*
Richard Leedy, *Polk Community College*
Jeanine Lewis, *Aims Community College-Main Campus*
Michelle Christina Mages, *Johnson County Community College*
Igor Marder, *Antelope Valley College*
Amina Mathias, *Cecil College*
Donna Martin, *Florida Community College-North Campus*
Jean McArthur, *Joliet Junior College*
Carlea (Carol) McAvoy, *South Puget Sound Community College*
Tim McBride, *Spartanburg Community College*
Joan McNeil, *Quinebaug Valley Community College*
Sonya McQueen, *Hinds Community College*
Maria Luisa Mendez, *Laredo Community College*
Madhu Motha, *Butler County Community College*
Shauna Mullins, *Murray State University*
Julie Muniz, *Southwestern Illinois College*
Kathy Nabours, *Riverside Community College*
Michael Neill, *Carl Sandburg College*
Nicole Newman, *Kalamazoo Valley Community College*
Said Ngobi, *Victor Valley College*
Denise Nunley, *Glendale Community College*
Deanna Oles, *Stark State College of Technology*
Staci Osborn, *Cuyahoga Community College-Eastern Campus*
Linda Padilla, *Joliet Junior College*
Karen D. Pain, *Palm Beach Community College*
George Pate, *Robeson Community College*
Renee Patterson, *Cumberland County College*
Margaret Payerle, *Cleveland State University-Ohio*
Jim Pierce, *Lincoln Land Community College*
Tian Ren, *Queensborough Community College*
Nancy Ressler, *Oakton Community College*
Bob Rhea, *J. Sargeant Reynolds Community College*
Mary Richardson, *Bevill State Community College*
Minnie M. Riley, *Hinds Community College*
Matthew Robinson, *Tallahassee Community College*
Melissa Rossi, *Southwestern Illinois College*
Anna Roth, *Gloucester County College*
Alan Saleski, *Loyola University-Chicago*
Carol Saltsgaver, *University of Illinois at Springfield*
Sheri Sanchez, *Great Basin College*
Lisa Sheppard, *Lorain County Community College*
Mark A. Shore, *Allegany College of Maryland*
Mark Sigfrids, *Kalamazoo Valley Community College*
Amber Smith, *Johnson County Community College*
Leonora Smook, *Suffolk County Community College-Brentwood*
Renee Starr, *Arcadia University*
Larry Stoneburner, *Indiana Institute of Technology*
Jennifer Strehler, *Oakton Community College*
Renee Sundrud, *Harrisburg Area Community College*
Sandi Tannen, *Camden County College*
Harriet Thompson, *Albany State University*
Janet Thompson, *University of Akron*
John Thoo, *Yuba College*
Fred Toxopeus, *Kalamazoo Valley Community College*
Sara Van Asten, *North Hennepin Community College*
Felix Van Leeuwen, *Johnson County Community College*
Josefino Villanueva, *Florida Memorial University*
Howard Wachtel, *Community College of Philadelphia*
Dottie Walton, *Cuyahoga Community College Eastern Campus*
Charles Wang, *California National University*
Walter Wang, *Baruch College*
Brock Wenciker, *Johnson County Community College*
Kevin Wheeler, *Three Rivers Community College*
Latrica Williams, *St. Petersburg College*
Paul Wozniak, *El Camino College*
Christopher Yarrish, *Harrisburg Area Community College*
Steve Zuro, *Joliet Junior College*

Finally, we are truly grateful to the many people at McGraw-Hill Higher Education who have made our text a reality. The exceptional professionals who contributed to this text made it possible for us to reach page 2. We express our heartfelt thanks to the editorial personnel, project managers, proofreaders, accuracy checkers, art professionals, and everyone else who worked with us to better enable our students to succeed.

Supplements for the Student

Student's Solutions Manual (ISBN: 978-0-07-757401-7)

The *Student's Solutions Manual* provides comprehensive, worked-out solutions to the odd-numbered exercises in the Prerequisite Check, Section Exercises, Summary Exercises, Chapter-Test, and the Cumulative Reviews. The steps shown in the solutions match the style of solved examples in the textbook.

Supplements for the Instructor

Instructor's Solutions Manual

The *Instructor's Solutions Manual,* available online to adopting instructors, provides comprehensive, worked-out solutions to all exercises in the Prerequisite Check, Section Exercises, Summary Exercises, Chapter Test, and the Cumulative Reviews. The methods used to solve the problems in the manual are the same as those used to solve the examples in the textbook.

Annotated Instructor's Edition

In the *Annotated Instructor's Edition (AIE),* answers to exercises and tests appear adjacent to each exercise set, in a color used *only* for annotations. Complete answers to all Prerequisite Checks, Reading Your Text, Summary Exercises, Chapter Tests, and Cumulative Reviews are also found at the back of the book.

Instructor's Testing and Resource Online

This computerized test bank, available online to adopting instructors, utilizes Wimba Diploma® algorithm-based testing software to create customized exams quickly. This user-friendly program enables instructors to search for questions by topic, format, or difficulty level; to edit existing questions or to add new ones; and to scramble questions and answer keys for multiple versions of a single test. Hundreds of text-specific, open-ended, and multiple-choice questions are included in the question bank. Sample chapter tests are also provided. CD version is available upon request.

"The text is well suited to our curriculum and is an easy text to use. There are plenty of practice exercises for the student and the check yourself exercises and summary exercises are very good additional practice for the student."

—Bonnie Filer-Tubaugh, *University of Akron*

applications index

BUSINESS AND MANUFACTURING

CONSTRUCTION AND HOME IMPROVEMENT

CONSUMER CONCERNS

CRAFTS AND HOBBIES

GAMES

GEOGRAPHY

GEOMETRY

HEALTH AND MEDICINE

INFORMATION TECHNOLOGY

CHAPTER

1

Operations on Whole Numbers

INTRODUCTION

How many? How much? We asked these questions even before we had civilizations. Think about how often you count and use numbers.

In the United States, counting the number of people in the country is one of the more important current examples of this. Article I, Section 2, of the U.S. Constitution directs the government to conduct a census or count of the population every 10 years.

While the math needed is often no more advanced than counting and whole-number arithmetic, the results of the census have many and far-reaching effects. Some of the ways we use the census include determining the number of congresspersons for each state and aiding in the distribution of transportation funding.

You will have the opportunity to learn about both the census and the U.S. Congress when you work through the activities throughout this text. In Activities 1 and 3, you will work with population results from the U.S. Census Bureau.

CHAPTER 1 OUTLINE

1.1 The Decimal Place-Value System

< 1.1 Objectives >

1 > Write numbers in expanded form

2 > Determine the place value of a digit

3 > Write the word name for a number

4 > Write a number from its word name

Tips for Student Success

Throughout this text, we present you with a series of class-tested techniques designed to improve your performance in math classes.

Become Familiar with Your Textbook

Perform each task.

1. Use the Table of Contents to find the title of Section 5.1.
2. Use the Index to find the earliest reference of the term *mean*. (By the way, this has nothing to do with the personality of either your instructor or the textbook authors!)
3. Use the Selected Definitions list at the end of the text to define the term *mean*.
4. Find the answer to the first Check Yourself exercise in Section 1.1.
5. Find the answers to the odd-numbered end-of-section exercises in Section 1.1.
6. Find the origin of the term *digit* in the margin notes of Section 1.1.

Now you know where to find some of the more important features of the text. When you feel confused, think about using one of these features to clear up your confusion.

You may also be interested in the solutions manual for your text. Many students find it helpful.

Every civilization developed some type of number system, starting with simple *tally* systems to count and keep track of possessions and moving on to more complex systems. We still use the Roman system today when we see Roman numerals. The Egyptians used a set of picture-like symbols called **hieroglyphics.**

Some examples of number systems in addition to our own are shown next.

Numerals	Egyptian	Greek	Roman	Mayan	Traditional Chinese
1	|	Ι	I	•	一
10	∩	Δ	X	═	十
100	୧	H	C	[illegible]	百

NOTE

The prefix *deci* means 10. Our word *digit* comes from the Latin word *digitus*, which means finger.

Any number system provides a way of naming numbers. The system we use is described as a **decimal place-value system.** This system is based on the number 10 and uses symbols called **digits.** (Other numbers have been used as bases. For example, the Mayans used 20 and the Babylonians used 60.)

NOTE

Any number, no matter how large, can be represented using the 10 digits of our system.

The basic symbols of our system are the digits 0, 1, 2, 3, 4, 5, 6, 7, 8, 9.

These basic symbols, or digits, were first used in India and then adopted by Arabic peoples. For this reason, our system is called the Hindu-Arabic numeration system.

Numbers may consist of more than one *digit.* The numbers 3, 45, 567, and 2,359 are examples of numbers in **standard form.** We say that 45 is a two-digit number, 567 is a three-digit number, and so on.

As we said, our decimal system uses a *place-value* concept based on the number 10. Understanding how this system works will help you see the reasons for the rules and methods of arithmetic that you will learn.

Example 1 Writing a Number in Expanded Form

< Objective 1 >

NOTES

Each digit in a number has its own place value.

Here the parentheses are used for emphasis. (4 × 100) means 4 is multiplied by 100. (3 × 10) means 3 is multiplied by 10. (8 × 1) means 8 is multiplied by 1.

Look at the number 438.

We call 8 the *ones digit.* As we move to the left, the digit 3 is the *tens digit.* Again as we move to the left, 4 is the *hundreds digit.*

438

4 hundreds 8 ones

3 tens

When we rewrite a number such that each digit is written with its units, we call it the **expanded form** for the number.

Because 438 = 400 + 30 + 8, we write it in expanded form as

$(4 \times 100) + (3 \times 10) + (8 \times 1)$

Check Yourself 1

Write 593 in expanded form.

This place-value diagram shows the place values of digits as we write larger numbers. For the number 3,156,024,798, we have

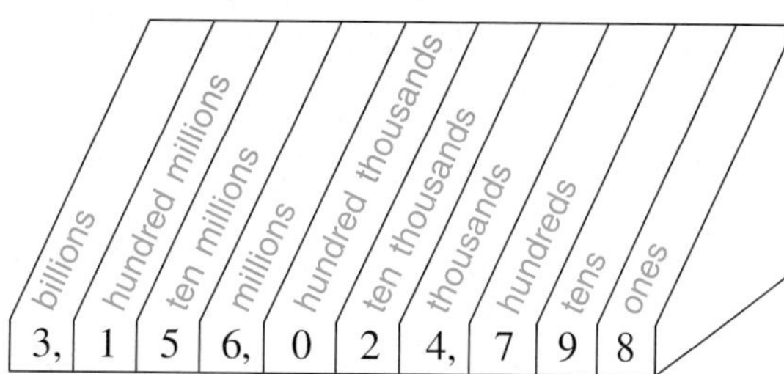

Of course, the naming of place values continues beyond the chart for larger numbers.

For the number 3,156,024,798, the place value of the digit 4 is thousands. As we move to the left, each place value is 10 times the value of the previous place. The place value of 2 is ten thousands, the place value of 0 is hundred thousands, and so on.

Example 2 Identifying Place Value

< Objective 2 >

Identify the place value of each digit in the number 418,295.

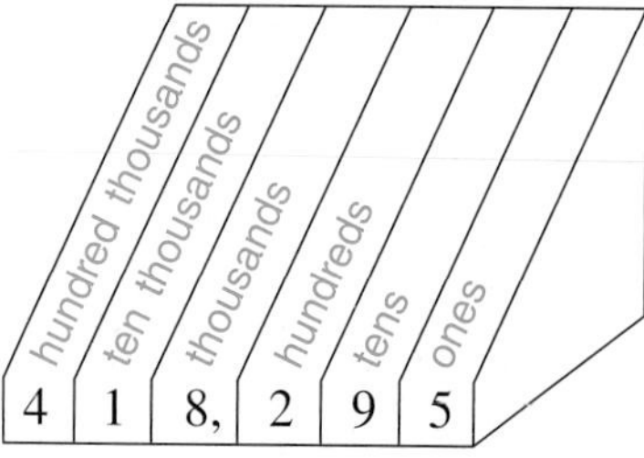

Check Yourself 2

Use a place-value diagram to answer each question for the number 6,831,425,097.

(a) What is the place value of 2?
(b) What is the place value of 4?
(c) What is the place value of 3?
(d) What is the place value of 6?

Understanding place value helps you read and write numbers. Look at the number

7 2,	3 5 8,	6 9 4
Millions	Thousands	Ones

NOTE

A four-digit number, such as 3,456, can be written with or without a comma. We write them with a comma in this text.

Commas are used to set off groups of three digits in the number. The name of each group—millions, thousands, ones, and so on—is then used as we write the number in words. To write a word name for a number, we work from left to right, writing the number in each group, followed by the group name. Here are some group names.

Billions			Millions			Thousands			Ones		
Hundreds	Tens	Ones	Hundreds	Tens	Ones	Hundreds	Tens	Ones	Hundreds	Tens	Ones

Example 3 Writing Numbers in Words

< Objective 3 >

NOTE

The commas in the word statements are in the same places as the commas in the number.

Write the word name of each number.

27,345 is written in words as twenty-seven *thousand,* three hundred forty-five.

2,305,273 is two *million,* three hundred five *thousand,* two hundred seventy-three.

Note: We do *not* write the name of the ones group. Also, the word *and* is not used when a number is written in words. It will have a special meaning later.

Check Yourself 3

Write the word name for each number.

(a) 658,942
(b) 2,305

Reverse the process to write the standard form for numbers given in word form.

Example 4 Translating Words into Numbers

< Objective 4 >

Forty-eight thousand, five hundred seventy-nine in standard form is

48,579

Five hundred three thousand, two hundred thirty-eight in standard form is

503,238

We use 0 as a placeholder when writing the number.

Check Yourself 4

Write twenty-three thousand, seven hundred nine in standard form.

Often, tables are used to present several numbers that are related in some way. When the numbers are large, they might be abbreviated. Abbreviating a number requires us to *round* the number properly. You will learn to estimate and round in Section 1.4. Reading an abbreviated number merely requires us to use what we have learned about the place values of the digits.

Example 5 Reading Numbers from a Table

Use the approximate starting salaries (in thousands) for the set of entry-level positions shown to answer each question.

Profession	Starting Salary (thousands)
Administrative Assistant	$36
Correctional Officer	$39
Web Designer	$51
Interior Designer	$38
Registered Nurse	$52

(a) How much would an interior designer expect to earn in the first year?

Begin by looking at the row containing *interior designer.*

Profession	Starting Salary (thousands)
Administrative Assistant	$36
Correctional Officer	$39
Web Designer	$51
Interior Designer	$38
Registered Nurse	$52

The salary is listed as $38. Because the salary is given in thousands, the starting salary of interior designers is $38,000.

(b) Which profession listed earns the highest starting salary?

The largest salary in the table is earned by registered nurses, who start at $52,000.

Check Yourself 5

Use the approximate starting salaries (in thousands) for the set of entry-level positions shown to answer each question.

Profession	Starting Salary (thousands)
Architectural Drafter	$39
Electrician	$43
Human Resources Generalist	$49
Loan Review Officer	$38
Technical Support Analyst	$49

(a) How much would an electrician expect to earn in the first year?
(b) Which profession(s) listed earn the highest starting salary? What is that salary?

Check Yourself ANSWERS

1. (5 × 100) + (9 × 10) + (3 × 1) **2.** **(a)** Ten thousands; **(b)** hundred thousands; **(c)** ten millions; **(d)** billions **3.** **(a)** Six hundred fifty-eight thousand, nine hundred forty-two; **(b)** two thousand, three hundred five **4.** 23,709 **5.** **(a)** $43,000; **(b)** Human Resources Generalist and Technical Support Analyst; $49,000

Reading Your Text

These fill-in-the-blank exercises will help you understand some of the key vocabulary used in this section. The answers to these exercises are in the Answers Appendix at the back of the text.

(a) The system we use for naming numbers is described as a ________ place-value system.

(b) We say the number 567 is a three-________ number.

(c) A four-digit number can be written with or without a ________.

(d) In words, the number 2,000,000 is written as two ________.

1.1 exercises

Skills | Calculator/Computer | Career Applications | Above and Beyond

< Objective 1 >

Write each number in expanded form.

1. 456

2. 637

3. 5,073 

4. 20,721

5. 1,500

6. 32,005,860

< Objective 2 >

Give the place values for the indicated digits.

7. 4 in the number 416

8. 3 in the number 38,615

9. 6 in the number 56,489

10. 4 in the number 427,083

11. 0 in the number 3,052

12. 1 in the number 5,224,031

13. Consider the number 43,729.

(a) What digit tells the number of thousands?

(b) What digit tells the number of tens?

14. Consider the number 456,719.

(a) What digit tells the number of ten thousands?

(b) What digit tells the number of hundreds?

15. Consider the number 1,403,602.

(a) What digit tells the number of hundred thousands?

(b) What digit tells the number of ones? VIDEO

16. Consider the number 324,678,903.

(a) What digit tells the number of millions?

(b) What digit tells the number of ten thousands?

< Objective 3 >

Write the word name of each number.

17. 5,618 

18. 21,812

19. 1,532,657

20. 3,491,930

21. 200,304 

22. 103,900

< Objective 4 >

Write each number in standard form.

23. Two hundred fifty-three thousand, four hundred eighty-three

24. Three hundred fifty thousand, three hundred fifty-nine

25. Two million, three hundred eight thousand, forty-seven

26. Fourteen million, twenty-six thousand, seven hundred eighty

27. Five hundred two million, seventy-eight thousand 

28. Four billion, two hundred thirty million

Write the number in each sentence in standard form.

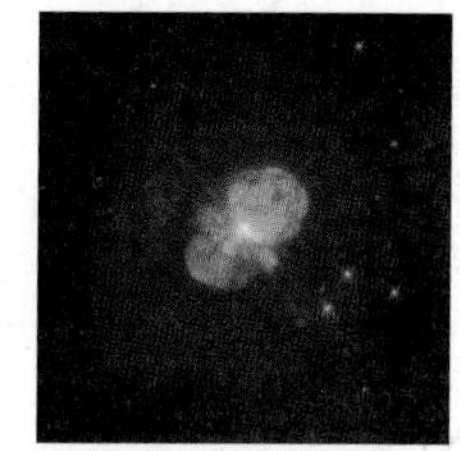

29. **Statistics** Rory McIlroy earned one million, four hundred forty thousand dollars when he won the 2011 U.S. Open golf tournament.

30. **Science and Medicine** Scientists speculate that the universe originated in the explosion of a primordial fireball approximately fourteen billion years ago.

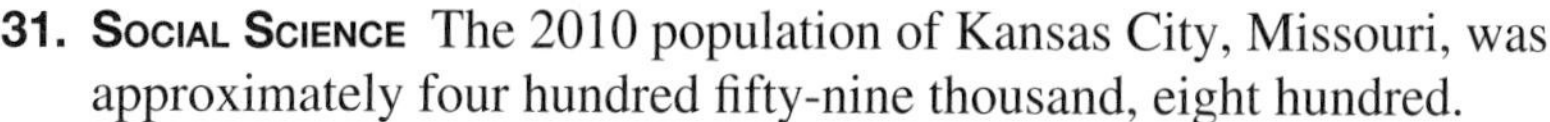

31. **Social Science** The 2010 population of Kansas City, Missouri, was approximately four hundred fifty-nine thousand, eight hundred.

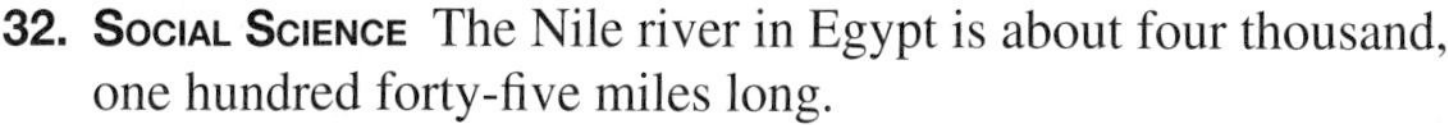

32. **Social Science** The Nile river in Egypt is about four thousand, one hundred forty-five miles long.

The 10 largest U.S. cities, according to the 2010 census are shown. Note that population numbers are given in thousands.

Name	Rank	Population (thousands)
New York City, NY	1	8,175
Los Angeles, CA	2	3,793
Chicago, IL	3	2,696
Houston, TX	4	2,099
Philadelphia, PA	5	1,526
Phoenix, AZ	6	1,446
San Antonio, TX	7	1,327
San Diego, CA	8	1,307
Dallas, TX	9	1,198
San Jose, CA	10	946

Source: U.S. Census Bureau

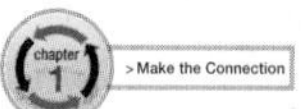

Social Science Use the population table to complete exercises 33–36. Report your results in standard form.

33. What was the population of San Diego in 2010?

34. What was the population of Chicago in 2010?

35. What was the population of Philadelphia in 2010?

36. What was the population of Dallas in 2010?

37. **Business and Finance** Inci had to write a check for $2,565. There is a space on the check to write out the amount of the check in words. What should she write in this space?

38. **Business and Finance** In a rental agreement, the amount of the initial deposit required is two thousand, five hundred forty-five dollars. Write this amount as a number.

Skills | Calculator/Computer | **Career Applications** | Above and Beyond

39. **Allied Health** Doctor Edwards prescribes four hundred eighty thousand units of penicillin G benzathine to treat a 3-year-old child with a streptococcal infection. Write this amount as a number.

40. **Allied Health** Doctor Hill prescribes one thousand, one hundred eighty-three milligrams (mg) of amifostine to be administered together with an adult patient's chemotherapy to reduce the adverse effects of the treatment. Write this amount as a number.

41. **Allied Health** Use the table to answer each question. Report your results in standard form.

Child's Weight (lb)	Dose of Penicillin G Potassium (thousands of units)
30	680
35	795
40	910
45	1,020
50	1,135

(a) Carla weighs 35 pounds (lb). What dose of penicillin G potassium should her doctor prescribe?

(b) Nelson weighs 50 lb. What dose of penicillin G potassium should his doctor prescribe?

(c) What dose of penicillin G potassium is required for a child weighing 40 lb?

42. **Electronics** Use the table to answer each question. Report your results in standard form.

Appliance	Estimated Power Consumption [thousands of watts/hour (W/hr)]
Drip coffeemaker	301
Electric blanket	120
Laser printer	466
Personal computer	25
Video game system	49

(a) What is the estimated power consumption of a laser printer?

(b) What is the estimated power consumption of a video game system?

(c) What is the estimated power consumption of an electric blanket?

Determine the number represented by the scrambled place values.

43. 4 thousands
1 tens
3 ten thousands
5 ones
2 hundreds

44. 7 hundreds
4 ten thousands
9 ones
8 tens
6 thousands

Assume that you have alphabetized the word names for every number from one to one thousand.

45. Which number would appear first in the list?

46. Which number would appear last?

47. **Number Problem** Write the largest five-digit number that can be made using the digits 6, 3, and 9 if each digit is to be used at least once.

48. What are some advantages of a place-value system of numeration?

49. **Social Science** The number 0 was not used initially by the Hindus in our number system (about 250 B.C.). Go to your library (or do a search), and determine when a symbol for zero was introduced. What do you think is the importance of the role of 0 in a numeration system?

50. A *googol* is a very large number. Do some research to find out how big it is. Also, try to find out where the name of this number comes from.

Answers

We provide the answers for the odd-numbered exercises at the end of each exercise set. The answers for the even-numbered exercises may be found in the instructor's resource manual.

1. (4 × 100) + (5 × 10) + (6 × 1) **3.** (5 × 1,000) + (7 × 10) + (3 × 1) **5.** (1 × 1,000) + (5 × 100) **7.** Hundreds **9.** Thousands **11.** Hundreds **13.** (a) 3; (b) 2 **15.** (a) 4; (b) 2 **17.** Five thousand, six hundred eighteen **19.** One million, five hundred thirty-two thousand, six hundred fifty-seven **21.** Two hundred thousand, three hundred four **23.** 253,483 **25.** 2,308,047 **27.** 502,078,000 **29.** $1,440,000 **31.** 459,800 **33.** 1,307,000 **35.** 1,526,000 **37.** Two thousand, five hundred sixty-five **39.** 480,000 **41.** (a) 795,000 units; (b) 1,135,000 units; (c) 910,000 units **43.** 34,215 **45.** Eight **47.** 99,963 **49.** Above and Beyond

1.2 Addition

< 1.2 Objectives >

1 > Add single-digit numbers
2 > Identify the properties of addition
3 > Use addition to solve applications
4 > Add any group of numbers
5 > Find the perimeter of a figure

Tips for Student Success

Become Familiar with Your Syllabus

You probably received a syllabus in your first class meeting. You should add the important information to your calendar and address book.

1. Put all important dates into your *calendar*. These include homework due dates; quiz dates; test dates; and the date, time, and location of the final exam. Never be surprised by a deadline!
2. Put your instructor's name, contact number, and office number into your *address book.* Include your instructor's office hours, phone number, and email address. Make it a point to see your instructor early in the term. Although this is not the only person who can help you, your instructor is one of the most important resources available to you.
3. Familiarize yourself with other resources available to you. You can take advantage of the Student Solutions Manual for your text, as well as the online resources offered through Connect Math Hosted by ALEKS Corp. and the ALEKS platform. Your college may also offer tutoring and other helpful services.

Given these resources, you have no reason to let confusion or frustration mount. If you can't "get it" from the text, try another resource. Take advantage of the resources available to you.

NOTE

The series of three dots (. . .) is called an ellipsis; it means that the obvious pattern continues.

The *natural* or *counting numbers* are the numbers we use to count objects.

The natural numbers are 1, 2, 3, . . .

When we include the number 0, we have the set of *whole numbers*.

The whole numbers are 0, 1, 2, 3, . . .

Let's look at the operation of *addition* on the whole numbers.

Definition

Addition

Addition is the combining of two or more groups of the same kind of objects.

This concept is extremely important, as we see in our later work with fractions. We can only combine or add numbers that represent the same kind of objects.

From your first encounter with arithmetic, you were taught to add "3 apples plus 2 apples."

On the other hand, you have probably encountered a phrase such as "that's like combining apples and oranges." That is to say, what do you get when you add 3 apples and 2 oranges?

You could answer "5 fruits," or "5 objects," but you cannot combine the apples and the oranges.

What if you walked 3 miles and then walked 2 more miles? Clearly, you have now walked 3 miles + 2 miles = 5 miles. The addition is possible because the groups are of the same kind.

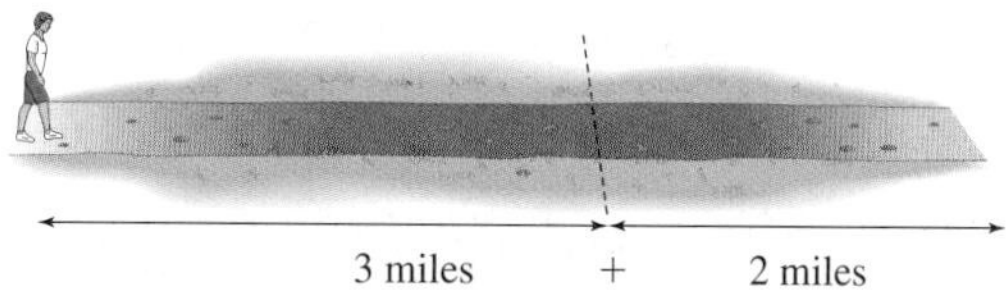

NOTES

The first printed use of the symbol + dates back to 1526 (Smith, *History of Math,* Vol. II).

The point labeled 0 is called the **origin** of the number line.

Each operation of arithmetic has its own special terms and symbols. The addition symbol + is read **plus.** When we write 3 + 4, 3 and 4 are called the **addends.** The answer to an addition problem is called the **sum.**

We can use a number line to illustrate the addition process. To construct a number line, we pick a point on the line and label it 0. We then mark off evenly-spaced units to the right, naming each marked point with a successively larger whole number.

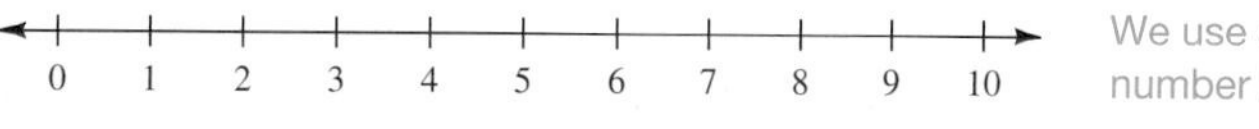

We use arrowheads to show the number line continues.

Example 1 **Representing Addition on a Number Line**

< Objective 1 >

Represent 3 + 4 on a number line.

To represent an addition, such as 3 + 4, on a number line, start by moving 3 spaces to the right of the origin. Then move 4 more spaces to the right to arrive at 7. The number 7 is called the sum of the addends.

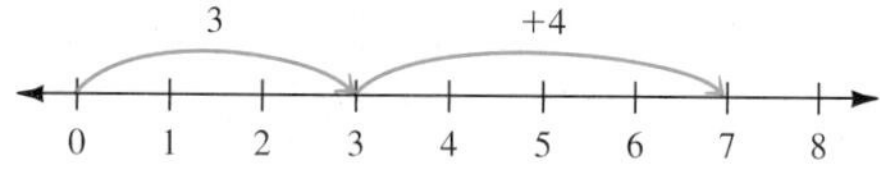

We can write 3 + 4 = 7

Addend (3) Addend (4) Sum (7)

NOTE

Again, addition corresponds to combining groups of the same kind of objects.

3 objects + 4 objects

Addend Addend

7 objects

= Sum

Check Yourself 1

Represent 5 + 6 on a number line.

A statement such as 3 + 4 = 7 is one of the **basic addition facts.** These facts include the sum of every pair of digits. Before you can add larger numbers correctly, you must memorize these basic facts.

Basic Addition Facts

+	0	1	2	3	4	5	6	7	8	9
0	0	1	2	3	4	5	6	7	8	9
1	1	2	3	4	5	6	7	8	9	10
2	2	3	4	5	6	7	8	9	10	11
3	3	4	5	6	7	8	9	10	11	12
4	4	5	6	7	8	9	10	11	12	13
5	5	6	7	8	9	10	11	12	13	14
6	6	7	8	9	10	11	12	13	14	15
7	7	8	9	10	11	12	13	14	15	16
8	8	9	10	11	12	13	14	15	16	17
9	9	10	11	12	13	14	15	16	17	18

NOTES

To find the sum 5 + 8, start with the row labeled 5. Move along that row to the column headed 8 to find the sum, 13.

Commute means to move back and forth, as to school or work.

Examining the basic addition facts leads us to several important properties of addition on whole numbers. For instance, we know that the sum 3 + 4 is 7. What about the sum 4 + 3? It is also 7. This is an illustration of the fact that addition is a **commutative** operation.

Property

The Commutative Property of Addition

The order of two numbers around an addition sign *does not* affect the sum.

Example 2 Using the Commutative Property

< Objective 2 >

$8 + 5 = 13 = 5 + 8$

$6 + 9 = 15 = 9 + 6$

Check Yourself 2

Show that the sum on the left equals the sum on the right.

$7 + 8 = 8 + 7$

NOTE

The *order* does not affect the sum.

If we wish to add *more* than two numbers, we can group them and then add. In mathematics this grouping is indicated by a set of parentheses ( ). This symbol tells us to perform the operation inside the parentheses first.

Example 3 Using the Associative Property

NOTES

We add 3 and 4 as the first step and then add 5.

Here we add 4 and 5 as the first step and then add 3. Again the final sum is 12.

$(3 + 4) + 5 = 7 + 5 = 12$

We also have

$3 + (4 + 5) = 3 + 9 = 12$

This example suggests the following property of whole numbers.

Property

The Associative Property of Addition

The way in which several whole numbers are grouped *does not* affect the final sum when they are added.

Check Yourself 3

Find

$(4 + 8) + 3$ and $4 + (8 + 3)$

The number 0 has a special property in addition.

Property

The Additive Identity Property

The sum of 0 and any whole number is just that whole number.

Because of this property, we call 0 the **identity** for the addition operation.

Example 4 **Adding Zero**

Find the sum of **(a)** $3 + 0$ and **(b)** $0 + 8$.

(a) $3 + 0 = 3$ **(b)** $0 + 8 = 8$

Check Yourself 4

Find each sum.

(a) $4 + 0$ **(b)** $0 + 7$

Next, we turn to the process of adding larger numbers.

Property

Adding Digits of the Same Place Value

We can add the digits of the same place value because they represent the same types of quantities.

NOTE

Remember that 25 means 2 tens and 5 ones; 34 means 3 tens and 4 ones.

Adding two numbers, such as $25 + 34$, can be done in expanded form. Here we write out the place value for each digit.

$$\begin{array}{rl} 25 = & 2 \text{ tens} + 5 \text{ ones} \\ +\ 34 = & 3 \text{ tens} + 4 \text{ ones} \\ \hline = & 5 \text{ tens} + 9 \text{ ones} \\ = & 59 \end{array}$$

Add down.

In actual practice, we use a more convenient short form to perform the addition.

Example 5 **Adding Two Numbers**

NOTE

In using the short form, be very careful to line up the numbers correctly so that each column contains digits of the same place value.

Add $352 + 546$.

Step 1 Add in the ones column.

$$\begin{array}{r} 352 \\ +\ 546 \\ \hline 8 \end{array}$$

Step 2 Add in the tens column.

$$\begin{array}{r} 352 \\ +\ 546 \\ \hline 98 \end{array}$$

Step 3 Add in the hundreds column.

$$\begin{array}{r} 352 \\ +\ 546 \\ \hline 898 \end{array}$$

Check Yourself 5

Add.

$$\begin{array}{r} 245 \\ +\ 632 \\ \hline \end{array}$$

You have seen that the word *sum* indicates addition. There are other words that tell you to add.

The *total* of 12 and 5 is written as

12 + 5 or 17

8 *more than* 10 is written as

10 + 8 or 18

12 *increased by* 3 is written as

12 + 3 or 15

Example 6 Translating Words That Indicate Addition

Find each sum.

(a) 36 increased by 12.

36 increased by 12 is written as 36 + 12 = 48.

(b) The total of 18 and 31.

The total of 18 and 31 is written as 18 + 31 = 49.

NOTE

Get into the habit of writing down *all* your work, rather than just an answer.

Check Yourself 6

Find each sum.

(a) 43 increased by 25 **(b)** The total of 22 and 73

Now we consider applications, or word problems, that require addition. An organized approach is the key to successful problem solving—we suggest the following strategy.

Step by Step

Solving Addition Applications

Step 1 Read the problem carefully to determine the given information and what you are being asked to find.

Step 2 Decide upon the operation (in this case, addition) to be used.

Step 3 Write down the complete statement necessary to solve the problem and do the calculations.

Step 4 Write your answer as a complete sentence. Check to make sure you have answered the question and that your answer seems reasonable.

The first step in solving any word problem or *application* is to read the problem carefully. This sounds easy but often poses problems. The purpose of reading the problem is to determine what we want for a solution and what information we are given.

Example 7 Reading a Problem

(a) Orah Nurseries sold 200 plants on Tuesday. Of those, 16 were returned by their customers. What percent of the plants were sold?

(i) What are we asked to find?

We want to know the percent of the plants sold.

(ii) What information have we been given?

We know:

- 200 plants were sold;
- 16 were returned.

(b) Emma bought a car with a list price of $18,250. She added the air-conditioner option for $445. What was the total price of her car?

(i) What are we asked to find?

We want to know the total price of the car.

(ii) What information have we been given?

The list price of the car is $18,250; the options cost an additional $445.

Check Yourself 7

(a) Four sections of algebra were offered in the fall quarter. The sections had enrollments of 33, 24, 20, and 22 students, respectively. What was the total number of students taking algebra?
(i) What are you asked to find?
(ii) What information have you been given?

(b) Elva won an election for city council with 3,110 votes. Her two opponents received 1,022 and 1,211 votes, respectively. How many votes were cast in the election?
(i) What are you asked to find?
(ii) What information have you been given?

Now we are ready to use the steps to work through an application.

Example 8 Solving an Application

< Objective 3 >

Four sections of algebra were offered in the fall quarter. The sections had enrollments of 33, 24, 20, and 22 students, respectively. What was the total number of students taking algebra?

NOTE

Remember to attach the proper unit (here "students") to your answer.

Step 1 The given information is the number of students in each of the four algebra sections. We want the total number.

Step 2 Since we are looking for a total, we use addition.

Step 3 Write $33 + 24 + 20 + 22 = 99$ students.

Step 4 There were 99 students taking algebra.

Check Yourself 8

Elva won an election for city council with 3,110 votes. Her two opponents received 1,022 and 1,211 votes, respectively. How many votes were cast in the election?

In the previous examples and exercises, the digits in each column added to 9 or less. We now look at the situation in which a column has a two-digit sum. This will involve the process of **carrying.** Look at the process in expanded form.

Example 9 Adding When Regrouping Is Needed

< Objective 4 >

NOTES

Regrouping in addition is also called *carrying*. Of course, the name makes no difference as long as you understand the process.

This is true for any sized number. The place value thousands is 10 times the place value hundreds, and so on.

$$\begin{array}{r} 67 = 60 + 7 \\ +\ 28 = 20 + 8 \\ \hline 80 + 15 \end{array}$$

We have written 15 ones as 1 ten and 5 ones.

or $\underbrace{80 + 10}_{} + 5$ The 1 ten is then combined with the 8 tens.

or $90 + 5$

or 95

The more convenient short form carries the excess units from one column to the next column to the left. Recall that the place value of the next column to the left is 10 times the value of the original column. It is this property of our decimal place-value system that makes carrying work. We work this problem again, this time using the short, or "carrying," form.

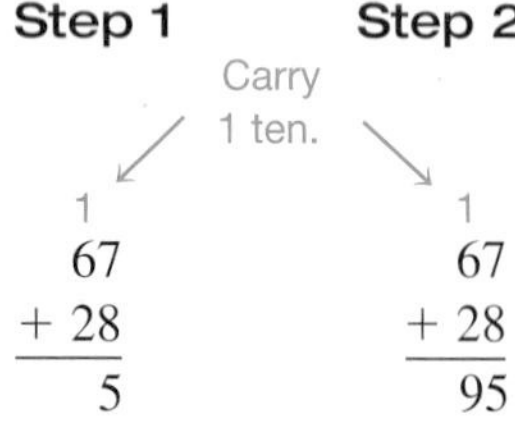

$$\begin{array}{r} ^{1} \\ 67 \\ +\ 28 \\ \hline 5 \end{array} \qquad \begin{array}{r} ^{1} \\ 67 \\ +\ 28 \\ \hline 95 \end{array}$$

Step 1: The sum of the digits in the ones column is 15, so write 5 and make the 10 ones a 1 in the tens column.
Step 2: Now add in the tens column, being sure to include the carried 1.

Check Yourself 9

Add.

(a) $\begin{array}{r} 58 \\ +\ 36 \\ \hline \end{array}$ **(b)** $\begin{array}{r} 73 \\ +\ 18 \\ \hline \end{array}$ **(c)** $\begin{array}{r} 68 \\ +\ 25 \\ \hline \end{array}$

Adding often requires more than one regrouping step, as we see in Example 9.

Example 10 Adding When Regrouping Is Needed

Add 285 and 378.

$$\begin{array}{r} ^{1} \\ 285 \\ +\ 378 \\ \hline 3 \end{array}$$

← Write the 10 as 1 ten.

The sum of the digits in the ones column is 13, so write 3 and carry 1 to the tens column.

Carry → 1 hundred.

$$\begin{array}{r} ^{11} \\ 285 \\ +\ 378 \\ \hline 63 \end{array}$$

Now add in the tens column, being sure to include the carry. We have 16 tens, so write 6 in the tens place and carry 1 to the hundreds column.

$$\begin{array}{r} ^{11} \\ 285 \\ +\ 378 \\ \hline 663 \end{array}$$

Finally, add in the hundreds column.

Check Yourself 10

Add.

(a) $\begin{array}{r} 479 \\ +\ 287 \\ \hline \end{array}$ (b) $\begin{array}{r} 585 \\ +\ 368 \\ \hline \end{array}$

The regrouping process is the same if we want to add more than two numbers.

Example 11 Adding with Multiple Regrouping Steps

Add 53, 2,678, 587, and 27,009.

$$\begin{array}{r} {\scriptstyle 1\,1\,2\,2} \\ 53 \\ 2{,}678 \\ 587 \\ +\ 27{,}009 \\ \hline 30{,}327 \end{array}$$

← Carries

Add in the ones column: 3 + 8 + 7 + 9 = 27. Write 7 in the sum and carry 2 to the tens column.

Now add in the tens column, being sure to include the carry. The sum is 22. Write 2 tens and carry 2 to the hundreds column. Complete the addition by adding in the hundreds column, the thousands column, and the ten thousands column.

Check Yourself 11

Add 46, 365, 7,254, and 24,006.

Finding the *perimeter* of a figure is one application of addition.

Definition

Perimeter — **Perimeter** is the distance around a closed figure.

If the figure has straight sides, the perimeter is the sum of the lengths of its sides.

Example 12 Finding the Perimeter

< Objective 5 >

NOTE

Make sure to include the unit with each number.

We wish to fence in the field shown in the figure. How much fencing, in feet (ft), do we need?

The fencing needed is the perimeter of (or the distance around) the field. We must add the lengths of the five sides.

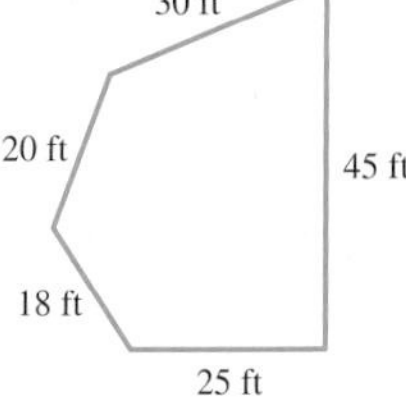

20 ft + 30 ft + 45 ft + 25 ft + 18 ft = 138 ft

So the perimeter is 138 ft.

Check Yourself 12

What is the perimeter of the region shown?

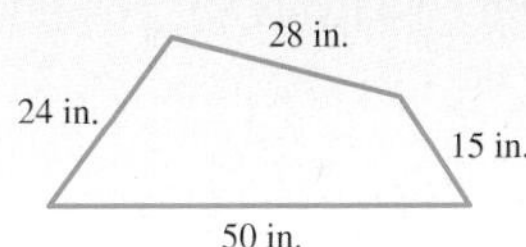

NOTE

We define parallel more carefully at a later point. For now, think that opposite sides have the same direction.

A **rectangle** is a four-sided figure. It has equal-sized (square) corners. Opposite sides of a rectangle are *parallel* and have the same length. Sheets of paper and pages in this text are examples of rectangles.

We can find the perimeter of a rectangle by adding the lengths of the four sides.

Example 13 Finding the Perimeter of a Rectangle

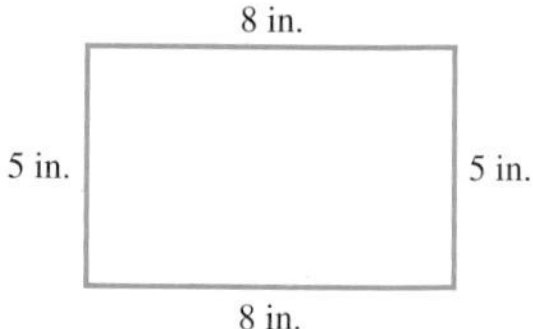

Find the perimeter in inches of the rectangle pictured here.

The perimeter is the sum of the lengths 8 in., 5 in., 8 in., and 5 in.

8 in. + 5 in. + 8 in. + 5 in. = 26 in.

The perimeter of the rectangle is 26 in.

Check Yourself 13

Find the perimeter of the rectangle pictured here.

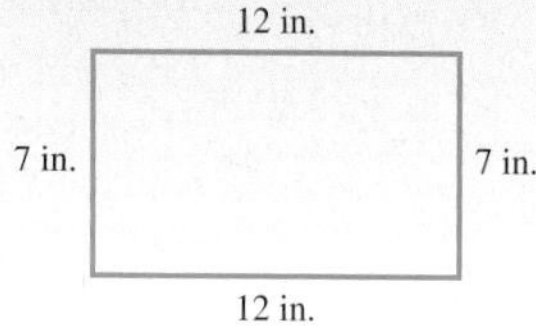

In general, we can find the perimeter of a rectangle by using a *formula*. A **formula** is a set of symbols that describe a general solution to a problem.

Look at a picture of a rectangle.

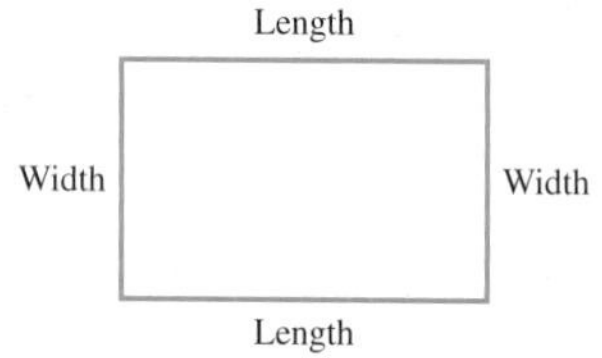

The perimeter can be found by adding the distances, so

Perimeter = length + width + length + width

To make this formula a little more readable, we use only the first letter of each word.

Property

Formula for the Perimeter of a Rectangle $P = L + W + L + W$

Because we add the length (L) twice, we could write that as $2 \cdot L$. Because we add the width (W) twice, we could write that as $2 \cdot W$. This gives us another version of the formula.

Property

Formula for the Perimeter of a Rectangle $P = 2 \cdot L + 2 \cdot W$

In words, we say that the perimeter of a rectangle is twice its length plus twice its width. We use the first version of the formula in Example 14.

Example 14 Finding the Perimeter of a Rectangle

NOTE

We say the rectangle is 8 in. by 11 in.

A rectangle has length 11 in. and width 8 in. What is its perimeter?

Start by drawing a picture of the problem.

Now use the formula.

$P = 11 \text{ in.} + 8 \text{ in.} + 11 \text{ in.} + 8 \text{ in.}$

$= 38 \text{ in.}$

The perimeter is 38 in.

Check Yourself 14

A bedroom is 9 ft by 12 ft. What is its perimeter?

Check Yourself ANSWERS

1.

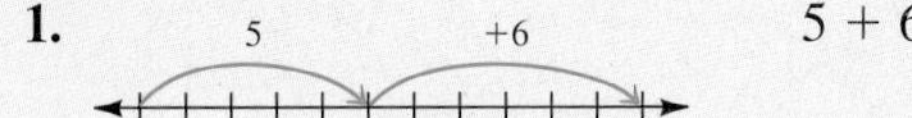

$5 + 6 = 11$ **2.** $7 + 8 = 15$ and $8 + 7 = 15$

3. $(4 + 8) + 3 = 12 + 3 = 15$; $4 + (8 + 3) = 4 + 11 = 15$ **4. (a)** 4; **(b)** 7 **5.** 877

6. (a) 68; **(b)** 95

7. (a) *(i)* The total number of students taking algebra.

(ii) Four sections of algebra are being offered; enrollments are 33, 24, 20, and 22 students.

(b) *(i)* The total number of votes cast.

(ii) Elva received 3,110 votes and her two opponents received 1,022 and 1,211 votes, respectively.

8. 5,343 votes **9. (a)** 94; **(b)** 91; **(c)** 93 **10. (a)** 766; **(b)** 953 **11.** 31,671 **12.** 117 in.

13. 38 in. **14.** 42 ft

Reading Your Text

These fill-in-the-blank exercises will help you understand some of the key vocabulary used in this section. The answers to these exercises are in the Answers Appendix at the back of the text.

(a) The ________ or counting numbers are the numbers used to count objects.

(b) A statement such as $3 + 4 = 7$ is one of the basic ________ facts.

(c) The ________ of two numbers around an addition sign does not affect the sum.

(d) The first step in solving an addition application is to ________ the problem carefully.

1.2 exercises

Skills | Calculator/Computer | Career Applications | Above and Beyond

< Objective 1 >

Add each pair of numbers.

1. $5 + 1$ **2.** $2 + 7$ **3.** $8 + 4$ **4.** $5 + 9$

5. $8 + 8$ **6.** $3 + 3$ **7.** $7 + 3$ **8.** $1 + 9$

< Objective 2 >

Name the property of addition that is illustrated. Explain your choice of property.

9. $5 + 8 = 8 + 5$

10. $2 + (7 + 9) = (2 + 7) + 9$

11. $(4 + 5) + 8 = 4 + (5 + 8)$

12. $9 + 7 = 7 + 9$

13. $4 + (7 + 6) = 4 + (6 + 7)$

14. $5 + 0 = 5$

15. $5 + (2 + 3) = (2 + 3) + 5$

16. $3 + (0 + 6) = (3 + 0) + 6$

< Objective 4 >

Perform the indicated addition.

17. $\begin{array}{r} 2{,}792 \\ +\ 205 \\ \hline \end{array}$ **18.** $\begin{array}{r} 5{,}463 \\ +\ 435 \\ \hline \end{array}$ **19.** $\begin{array}{r} 2{,}345 \\ +\ 6{,}053 \\ \hline \end{array}$ **20.** $\begin{array}{r} 3{,}271 \\ +\ 4{,}715 \\ \hline \end{array}$

21. $\begin{array}{r} 2{,}531 \\ +\ 5{,}354 \\ \hline \end{array}$ **22.** $\begin{array}{r} 5{,}003 \\ +\ 4{,}205 \\ \hline \end{array}$ **23.** $\begin{array}{r} 21{,}314 \\ +\ 43{,}042 \\ \hline \end{array}$ **24.** $\begin{array}{r} 12{,}325 \\ +\ 35{,}403 \\ \hline \end{array}$

25. $\begin{array}{r} 3{,}490 \\ 548 \\ +\ 25 \\ \hline \end{array}$

26. $\begin{array}{r} 678 \\ 4{,}533 \\ +\ 70 \\ \hline \end{array}$ **27.** $\begin{array}{r} 2{,}289 \\ 38 \\ 578 \\ +\ 3{,}489 \\ \hline \end{array}$ **28.** $\begin{array}{r} 3{,}678 \\ 259 \\ 27 \\ +\ 2{,}356 \\ \hline \end{array}$

29. $\begin{array}{r} 23{,}458 \\ +\ 32{,}623 \\ \hline \end{array}$ **30.** $\begin{array}{r} 52{,}591 \\ +\ 59{,}739 \\ \hline \end{array}$ **31.** $46 + 32$ **32.** $123 + 655$

33. $4{,}032 + 2{,}289$ **34.** $12{,}600 + 8{,}905$ **35.** $32 + 867 + 42{,}085$

36. $2{,}940 + 329 + 11 + 71{,}594$

37. In the statement $5 + 4 = 9$
5 is called an ________.
4 is called an ________.
9 is called the ________.

38. In the statement $7 + 8 = 15$
7 is called an ________.
8 is called an ________.
15 is called the ________.

< Objective 5 >

Find the perimeter of each figure.

39.

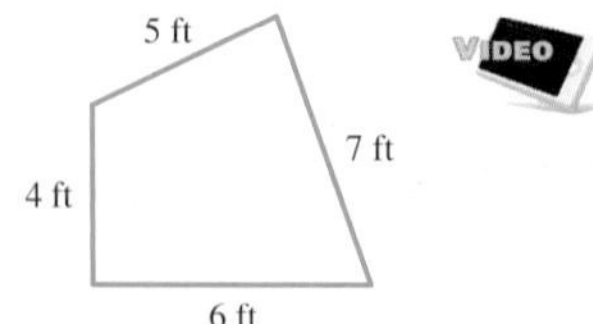

40.

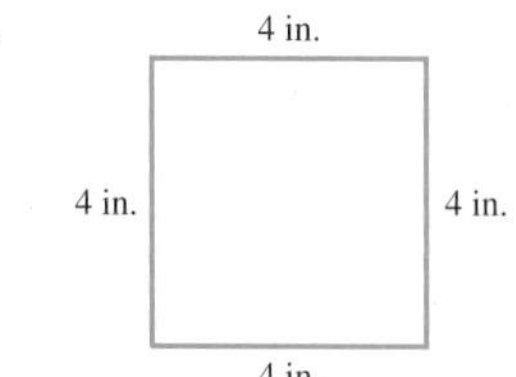

41. 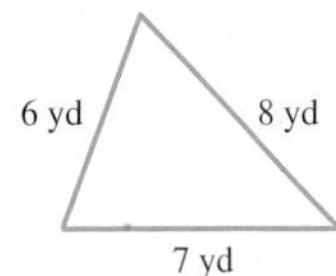

42. 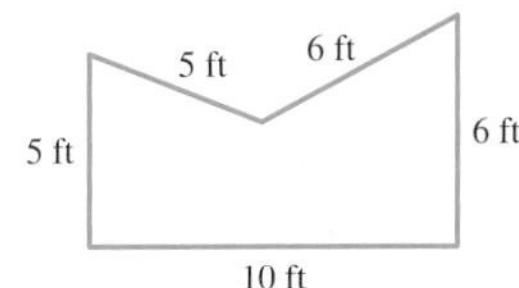

43. 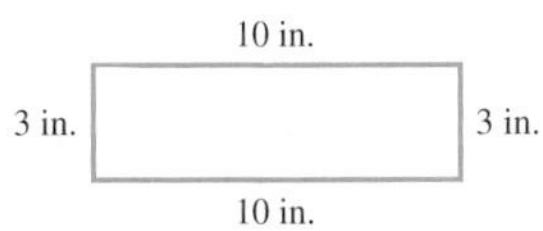

44. 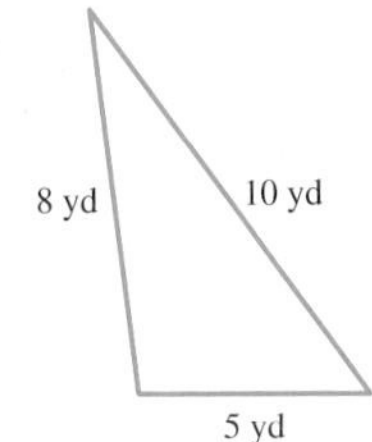

45. In each exercise, find the appropriate sum.
   (a) Find the number that is 356 more than 1,213. 
   (b) Add 23, 2,845, 5, and 589.
   (c) What is the total of the five numbers 2,195, 348, 640, 59, and 23,785?
   (d) Find the number that is 34 more than 125.
   (e) What is 457 increased by 96?

46. In each exercise, find the appropriate sum.
   (a) Find the number that is 567 more than 2,322.
   (b) Add 5,637, 78, 690, 28, and 35,589.
   (c) What is the total of the five numbers 3,295, 9, 427, 56, and 11,100?
   (d) Find the number that is 124 more than 2,351.
   (e) What is 926 increased by 86?

Read each application carefully, and then answer the questions that follow. Do not solve the problems.

47. **Business and Finance** Greg bought a high-end laptop for $2,120 along with an all-in-one printer for $379. After adding $589 worth of software, how much was his total bill?
   (a) What are you asked to find?
   (b) What information have you been given?

48. **Statistics** A bowler scored 201, 153, and 215 in three games. What was his total score for the three games?

   (a) What are you asked to find?
   (b) What information have you been given?

49. **Business and Finance** Angelo's vineyard shipped 4,200 lb of grapes in August, 5,970 lb in September, and 4,850 lb in October. How many pounds of grapes did they ship in the 3-month period?
   (a) What are you asked to find?
   (b) What information have you been given?

50. **Business and Finance** A sales associate drove 68 miles (mi) on Tuesday, 114 mi on Thursday, and 79 mi on Friday for business. How many miles did he drive?
   (a) What are you asked to find?
   (b) What information have you been given?

< Objective 3 >

51. **Business and Finance** Tral bought a 1931 Model A for $26,895, a 1964 Thunderbird convertible for $54,200, and a 1959 Austin Healy Mark I for $69,950. How much did he invest in the three cars?

52. **Business and Finance** Greg bought a high-end laptop for $2,120 along with an all-in-one printer for $379. After adding $589 worth of software, how much was his total bill?

53. **Statistics** A golfer shot a score of 42 on the first nine holes and 46 on the second nine holes. What was her total score for the round?

54. **Statistics** A bowler scored 201, 153, and 215 in three games. What was his total score for the three games?

55. **Business and Finance** Angelo's vineyard shipped 4,200 lb of grapes in August, 5,970 lb in September, and 4,850 lb in October. How many pounds of grapes did they ship in the 3-month period? VIDEO

56. **Business and Finance** A sales associate drove 68 mi on Tuesday, 114 mi on Thursday, and 79 mi on Friday for business. How many miles did he drive?

57. **Statistics** The Torres family drove 325 mi on the first day of a vacation and 273 mi on the second day. How far did they drive in the 2 days?

58. **Statistics** Four performances of a play by a local theater company had attendance figures of 230, 312, 244, and 213. What was the 4-day attendance total?

59. **Business and Finance** Emma bought a car with a list price of $18,250. She added the air-conditioner option for $445. What was the total price of her car?

60. **Business and Finance** An airline ran four shuttle flights between Los Angeles and San Francisco one Monday. The flights had 133, 115, 120, and 111 passengers, respectively. What was the total number of shuttle passengers that Monday?

61. **Business and Finance** Regina's Dress Shop has detailed expenses for the last 3 months of the year. Complete the totals.

Department	Oct.	Nov.	Dec.	Department Totals
Office	$31,714	$32,512	$30,826	______
Production	85,146	87,479	81,234	______
Sales	34,568	37,612	33,455	______
Warehouse	16,588	11,368	13,567	______
Monthly Totals				______

62. **Business and Finance** This chart shows Family Video's monthly rentals for the first 3 months of 2013 by category of film. Complete the totals.

Category of Film	Jan.	Feb.	Mar.	Category Totals
Comedy	4,568	3,269	2,189	______
Drama	5,612	4,129	3,879	______
Action/Adventure	2,654	3,178	1,984	______
Musical	897	623	528	______
Monthly Totals				______

63. Business and Finance The five states with the largest number of organic farms in 2008 are shown in the table, along with the total number of organic farm acres and their total revenue (in thousands of dollars).

State	Farms	Acres	Sales
California	2,691	470,903	$1,148,650
Wisconsin	1,202	195,603	$132,764
Washington	886	82,216	$281,970
New York	819	168,428	$105,133
Oregon	657	105,605	$155,613

Source: United States Department of Agriculture

(a) What is the total number of organic farms in California, Oregon, and Washington?

(b) How many total acres do Wisconsin and New York devote to organic farming?

(c) What were the combined sales from these five states in 2008?

64. Social Science This table ranks the top 10 areas for women-owned firms in the United States.

Metro Area	Number of Firms	Employment	Sales (thousands of dollars)
Los Angeles–Long Beach, CA	360,300	1,056,600	$181,455,900
New York	282,000	1,077,900	193,572,200
Chicago	260,200	1,108,800	161,200,900
Washington, D.C.	193,600	440,000	56,644,000
Philadelphia	144,600	695,900	90,231,000
Atlanta	138,700	331,800	50,206,800
Houston	136,400	560,100	78,180,300
Dallas	123,900	431,900	63,114,900
Detroit	123,600	371,400	50,060,700
Minneapolis–St. Paul, MN	119,600	337,400	51,063,400

(a) How many firms in total are located in Washington, Philadelphia, and New York?

(b) What is the total number of employees in all 10 of the areas listed?

(c) What are the total sales for firms in Houston and Dallas?

(d) How many firms in total are located in Chicago and Detroit? 

65. Number Problem These sequences of numbers are called *arithmetic sequences*. Determine the pattern and write the next four numbers in each sequence.

(a) 5, 12, 19, 26, _____, _____, _____, _____

(b) 8, 14, 20, 26, _____, _____, _____, _____

(c) 7, 13, 19, 25, _____, _____, _____, _____

(d) 9, 17, 25, 33, _____, _____, _____, _____

66. Number Problem Fibonacci numbers occur in the sequence

1, 1, 2, 3, 5, 8, 13, 21, 34, 55, . . .

This sequence begins with the numbers 1 and 1 again, and each subsequent number is obtained by adding the two preceding numbers.

Find the next four numbers in the sequence.

To perform the addition

2,473 + 258 + 35 + 5,823

Note: Do not key in the comma when entering a number into a calculator.

Step 1	Press the clear key.	[C]
Step 2	Enter the first number.	2473
Step 3	Enter the plus key followed by the next number.	[+] 258
Step 4	Continue with the addition until the last number is entered.	[+] 35
		[+] 5823
Step 5	Press the equal sign or enter key.	[=]
The desired sum should now be in the display.		8589

Use your calculator to find each sum.

67. 3,295,153 + 573,128 + 21,257 + 2,586,241 + 5,291

68. 23,563 + 5,638,487 + 385,005 + 27,345

Use your calculator to solve each application.

69. Business and Finance The table shows the number of customers using three branches of a bank during one week. Complete the table by finding the daily, weekly, and grand totals.

Branch	Mon.	Tues.	Wed.	Thurs.	Fri.	Weekly Totals
Downtown	487	356	429	278	834	
Suburban	236	255	254	198	423	
Westside	345	278	323	257	563	
Daily Totals						

70. Statistics The table lists the number of possible types of poker hands. What is the total number of hands possible?

Royal flush	4
Straight flush	36
Four of a kind	624
Full house	3,744
Flush	5,108
Straight	10,200
Three of a kind	54,912
Two pairs	123,552
One pair	1,098,240
Nothing	1,302,540
Total Possible Hands	

71. Manufacturing Technology An inventory of steel round stock shows 248 ft of $\frac{1}{4}$ in., 124 ft of $\frac{3}{8}$ in., 428 ft of $\frac{1}{2}$ in., and 162 ft of $\frac{5}{8}$ in. How many total feet of steel round stock are in inventory?

72. Manufacturing Technology B & L Industries produces three different products. Orders for today are for 351 of product A, 187 of product B, and 94 of product C. How many total products need to be produced today to fill the orders?

73. Allied Health The source-image receptor distance (SID) for radiographic images is the sum of the object-film distance (OFD) and the focus-object distance (FOD). Determine the SID if the distance from the object to the film is 8 in., and the distance from the object to the focus is 48 in.

74. Allied Health Total lung capacity, measured in milliliters (mL), is the sum of the vital capacity and the residual volume. Determine the total lung capacity for a patient whose vital capacity is 4,500 mL and whose residual volume is 1,800 mL.

Skills | Calculator/Computer | Career Applications | **Above and Beyond**

75. Number Problem A magic square is a square in which the sum along any row, column, or diagonal is the same. For example,

35	10	15
0	20	40
25	30	5

Use the numbers 1 to 9 to form a magic square.

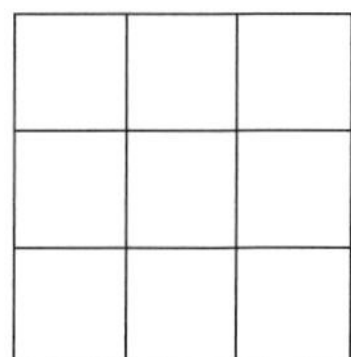

76. This puzzle gives you a chance to practice some of your addition skills.

Across

1. 23 + 22
3. 103 + 42
6. 29 + 58 + 19
8. 3 + 3 + 4
9. 1,480 + 1,624
11. 568 + 730
13. 25 + 25
14. 131 + 132
16. The total of 121, 146, 119, and 132
17. The perimeter of a 4 by 6 rug

Down

1. The sum of 224,000, 155, and 186,000
2. 20 + 30
4. 210 + 200
5. 500,000 + 4,730
7. 130 + 509
10. 90 + 92
12. 100 + 101
15. The perimeter of a 15 by 16 room

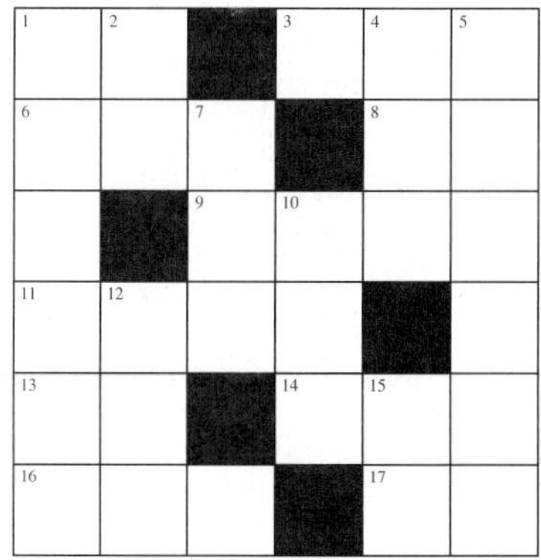

Answers

1. 6 **3.** 12 **5.** 16 **7.** 10 **9.** Commutative property of addition **11.** Associative property of addition **13.** Commutative property of addition **15.** Commutative property of addition **17.** 2,997 **19.** 8,398 **21.** 7,885 **23.** 64,356 **25.** 4,063 **27.** 6,394 **29.** 56,081 **31.** 78 **33.** 6,321 **35.** 42,984 **37.** 5 is an addend, 4 is an addend, 9 is the sum **39.** 22 ft **41.** 21 yd **43.** 26 in. **45.** **(a)** 1,569; **(b)** 3,462; **(c)** 27,027; **(d)** 159; **(e)** 553 **47.** **(a)** The total cost of the purchase. **(b)** He bought 3 items. The laptop cost $2,120, the printer cost $379, and the software cost $589. **49.** **(a)** The total amount of grapes shipped. **(b)** They shipped 4,200 lb, 5,970 lb, and 4,850 lb over the 3-month period. **51.** $151,045 **53.** 88 **55.** 15,020 lb **57.** 598 mi **59.** $18,695

61.

Department	Oct.	Nov.	Dec.	Department Totals
Office	$31,714	$32,512	$30,826	$95,052
Production	85,146	87,479	81,234	$253,859
Sales	34,568	37,612	33,455	$105,635
Warehouse	16,588	11,368	13,567	$41,523
Monthly Totals	$168,016	$168,971	$159,082	$496,069

63. **(a)** 4,234 farms; **(b)** 364,031 acres; **(c)** $1,824,130,000 **65.** **(a)** 33, 40, 47, 54; **(b)** 32, 38, 44, 50; **(c)** 31, 37, 43, 49; **(d)** 41, 49, 57, 65 **67.** 6,481,070

69.

Branch	Mon.	Tues.	Wed.	Thurs.	Fri.	Weekly Totals
Downtown	487	356	429	278	834	2,384
Suburban	236	255	254	198	423	1,366
Westside	345	278	323	257	563	1,766
Daily Totals	1,068	889	1,006	733	1,820	5,516

71. 962 ft **73.** 56 in. **75.**

8	3	4
1	5	9
6	7	2

Answers will vary.

1.3 Subtraction

< 1.3 Objectives >

1 > Subtract whole numbers without borrowing

2 > Use subtraction to solve applications

3 > Use borrowing to subtract whole numbers

Tips for Student Success

Don't Procrastinate!

1. Complete your math homework while you are still fresh. Late at night your mind is tired, making it difficult to understand new concepts.
2. Complete your homework the day it is assigned. The more recent the explanation, the easier it is to recall.
3. When you finish your homework, try reading the next section in the text. This will give you a sense of direction the next time you have class.

Remember, in a typical math class, you are expected to do two or three hours of homework for each hour of class time. This means two or three hours per day. Schedule the time and stay on schedule.

NOTE

By *opposite operation* we mean that subtracting a number "undoes" an addition of that same number. Start with 1. Add 5 and then subtract 5. Where are you?

We are now ready to consider a second operation of arithmetic—subtraction. In Section 1.2, we described addition as the process of combining two or more groups of the same kind of objects. Subtraction can be thought of as the *opposite operation* to addition. Every arithmetic operation has its own notation. The symbol for subtraction, $-$, is called a **minus sign.**

When we write $8 - 5$, we wish to subtract 5 from 8. We call 5 the **subtrahend.** This is the number being subtracted. And 8 is the **minuend.** This is the number we are subtracting from. The **difference** is the result of the subtraction.

To find the *difference* of two numbers, we assume that we wish to subtract the smaller number from the larger. Then we look for a number that, when added to the smaller number, gives us the larger number. For example,

$8 - 5 = 3$ because $3 + 5 = 8$

This special relationship between addition and subtraction provides a method of checking subtraction.

Property

Relationship Between Addition and Subtraction

The sum of the difference and the subtrahend must be equal to the minuend.

Example 1 — Subtraction

< Objective 1 >

$18 - 5 = 13$

Check: $13 + 5 = 18$

Difference ↗ Subtrahend ↖ Minuend ↖

Our check works because $18 - 5$ asks for the number that must be added to 5 to get 18.

Check Yourself 1

Subtract and check your work.

26 − 4

The procedure for subtracting larger whole numbers is similar to the procedure for addition. We subtract digits of the same place value.

Example 2 Subtracting a Larger Number

Step 1	Step 2	Step 3
$\begin{array}{r} 789 \\ -\ 246 \\ \hline 3 \end{array}$	$\begin{array}{r} 789 \\ -\ 246 \\ \hline 43 \end{array}$	$\begin{array}{r} 789 \\ -\ 246 \\ \hline 543 \end{array}$

We subtract in the ones column, then in the tens column, and finally in the hundreds column.

To check: $\begin{array}{r} 789 \\ -\ 246 \\ \hline 543 \end{array}$ Add 543 + 246 = 789

The sum of the difference and the subtrahend must be the minuend.

Check Yourself 2

Subtract and check your work.

(a) $\begin{array}{r} 3{,}468 \\ -\ 2{,}248 \\ \hline \end{array}$ **(b)** $\begin{array}{r} 4{,}984 \\ -\ 1{,}081 \\ \hline \end{array}$

You know that the word *difference* indicates subtraction. There are other words that also tell you to use the subtraction operation. For instance, 5 *less than* 12 is written as

12 − 5 or 7

29 *decreased* by 8 is written as

29 − 8 or 21

Example 3 Translating Words That Indicate Subtraction

Find each of the following.

(a) 4 less than 11

4 less than 11 is written 11 − 4 = 7.

(b) 27 decreased by 6

27 decreased by 6 is written 27 − 6 = 21.

Check Yourself 3

Find each of the following.

(a) 6 less than 19 **(b)** 18 decreased by 3

Units Analysis

This is the first in a series of essays designed to help you solve applications using mathematics. Questions in the exercise sets require the skills that you build by reading these essays.

A number with a unit attached (such as 7 ft or 26 mi/gal) is called a **denominate number.** Any genuine application of mathematics involves denominate numbers.

When adding or subtracting denominate numbers, the units must be identical for both numbers. The sum or difference has the same units.

EXAMPLES:

\$4 + \$9 = \$13 Notice that, although we write the dollar sign first, we read it after the quantity, as in "four dollars."

7 ft + 9 ft = 16 ft

39 degrees − 12 degrees = 27 degrees

7 ft + 12 degrees yields no meaningful answer!

3 ft + 9 in. yields a meaningful result if 3 ft is converted into 36 in. We discuss unit conversion later.

Now we consider subtraction word problems. The strategy is the same one presented in Section 1.2 for addition word problems. It is summarized with the following four basic steps.

Step by Step

Solving Subtraction Applications

Step 1 Read the problem carefully to determine the given information and what you are asked to find.

Step 2 Decide upon the operation(s) to be used.

Step 3 Write down the complete statement necessary to solve the problem and do the calculations.

Step 4 Write your answer as a complete sentence. Check to make sure you have answered the question and that your answer seems reasonable.

Here is an example using these steps.

Example 4 Setting Up a Subtraction Word Problem

< Objective 2 >

Tory has \$37 in his wallet. He is thinking about buying a \$24 pair of pants and a \$10 shirt. If he buys them both, how much money will he have left?

First, we must add the cost of the pants and the shirt.

\$24 + \$10 = \$34

Now, that amount must be subtracted from the \$37.

\$37 − \$34 = \$3

He will have \$3 left.

Check Yourself 4

Sonya has \$97 left in her checking account. If she writes checks for \$12, \$32, and \$21, how much will she have in the account?

Difficulties can arise in subtraction if one or more of the digits of the subtrahend are larger than the corresponding digits in the minuend. We solve this problem by using another version of the regrouping process called **borrowing.**

First, we look at an example in expanded form.

Example 5 Subtracting When Regrouping Is Needed

< Objective 3 >

$$\begin{array}{r} 52 = 50 + 2 \\ -\ 27 = 20 + 7 \\ \hline \end{array}$$

Do you see that we cannot subtract in the ones column?

We regroup by borrowing 1 ten in the minuend and writing that ten as 10 ones:

$$\begin{array}{lrl} & 50 & +\ 2 \\ \text{becomes} & \overbrace{40 + 10} & \underbrace{+\ 2} \\ \text{or} & 40 + & 12 \end{array}$$

We now have

$$\begin{array}{r} 52 = 40 + 12 \\ -\ 27 = 20 + 7 \\ \hline 20 + 5 \\ \text{or} \quad 25 \end{array}$$

We can now subtract as before.

In practice, we use a more convenient short form for the subtraction.

$$\begin{array}{r} 52 \\ -\ 27 \\ \hline \end{array} \qquad \begin{array}{r} {}^{4}\,{}_{1} \\ \not{5}2 \\ -\ 27 \\ \hline 25 \end{array}$$

We indicate the fact that we have borrowed 1 ten by putting a slash through the 5 and then writing 4 tens. Add 10 ones to the original 2 ones to get 12 ones. We can then subtract.

Check: 25 + 27 = 52

Check Yourself 5

Subtract and check your work.

$$\begin{array}{r} 64 \\ -\ 38 \\ \hline \end{array}$$

In Example 6, we work through a subtraction example that requires a number of regrouping steps. Here, zero appears as a digit in the minuend.

Example 6 Subtracting When Regrouping Is Needed

NOTE

Here we borrow 1 thousand; this is written as 10 hundreds.

Step 1

$$\begin{array}{r} {}^{4}\,{}_{1} \\ 4{,}0\not{5}3 \\ -\ 2{,}365 \\ \hline 8 \end{array}$$

In this first step we regroup by borrowing 1 ten. This is written as 10 ones and combined with the original 3 ones. We can then subtract in the ones column.

Step 2

$$\begin{array}{r} 3\ \ \ 4 \\ 10\ 1 \\ \not{4}{,}\not{0}\not{5}3 \\ -\ 2{,}365 \\ \hline 8 \end{array}$$

We must regroup again to subtract in the tens column. There are no hundreds, and so we move to the thousands column.

NOTE

We now borrow 1 hundred; this is written as 10 tens and combined with the remaining 4 tens.

Step 3

```
   9 14
  3 1̸0̸ 4̸1
   4̸,0̸5̸3
 − 2,365
 -------
       8
```

The minuend is now renamed as 3 thousands, 9 hundreds, 14 tens, and 13 ones.

Step 4

```
   9 14
  3 1̸0̸ 4̸1
   4̸,0̸5̸3
 − 2,365
 -------
   1,688
```

The subtraction can now be completed.

To check our subtraction: 1,688 + 2,365 = 4,053

Check Yourself 6

Subtract and check your work.

```
   5,024
 − 1,656
```

You need to use both addition and subtraction to solve some problems, as Example 7 illustrates.

Example 7 Solving a Subtraction Application

Bernard wants to buy a new piece of stereo equipment. He has \$142 and can trade in his old amplifier for \$135. How much more does he need if the new equipment costs \$449?

First, we must add to find out how much money Bernard has available. Then we subtract to find out how much more money he needs.

\$142 + \$135 = \$277 The money available to Bernard

\$449 − \$277 = \$172 The money Bernard still needs

Bernard needs \$172.

Check Yourself 7

Martina spent \$239 in airfare, \$174 for lodging, and \$108 for food on a business trip. Her company allowed her \$375 for the expenses. How much of these expenses will she have to pay herself?

Check Yourself ANSWERS

1. 22 Check: 22 + 4 = 26 **2.** **(a)** 1,220; **(b)** 3,903 **3.** **(a)** 13; **(b)** 15 **4.** Sonya will have \$32 left.

5.

```
   5 1
   6̸4
 − 38
 ----
   26
```

To check 26 + 38 = 64

6. 3,368 Check: 3,368 + 1,656 = 5,024

7.

```
   $239
    174
 +  108
 ------
   $521  ← Total expenses
```

```
   $521  ← Total expenses
 −  375  ← Amount allowed
 ------
   $146
```

Martina will have to pay \$146.

Reading Your Text

These fill-in-the-blank exercises will help you understand some of the key vocabulary used in this section. The answers to these exercises are in the Answers Appendix at the back of the text.

(a) The ________ is the result of subtraction.

(b) 5 ________ than 12 is written as 12 − 5.

(c) The first step in solving a subtraction application is to ________ the problem carefully.

(d) The regrouping process used in subtraction is called ________.

1.3 exercises

Skills | Calculator/Computer | Career Applications | Above and Beyond

< Objectives 1 and 3 >

Find each difference and check your results.

1. 347 − 201

2. 575 − 302

3. 689 − 245

4. 598 − 278

5. 3,446 − 2,326 

6. 5,896 − 3,862

7. 64 − 27

8. 73 − 36

9. 627 − 358

10. 642 − 367

11. 6,423 − 3,678

12. 5,352 − 2,577

13. 6,034 − 2,569 

14. 5,206 − 1,748

15. 4,000 − 2,345

16. 6,000 − 4,349

17. 33,486 − 14,047

18. 53,487 − 25,649

19. 29,400 − 17,900

20. 53,500 − 28,700

21. 58 − 5

22. 72 − 8

23. 148 − 23

24. 352 − 23

25. 127 − 69

26. 1,051 − 920

27. 32,871 − 976

28. 82,723 − 987

29. 19 − 7

30. 281 − 138

31. 4,032 − 2,289

32. 92,087 − 85,879

33. 2,301 − 98

34. 4,582 − 3,683

35. 8,516 − 609

36. 1,290,531 − 439,648

37. In the statement $9 - 6 = 3$
9 is called the ____________.
6 is called the ____________.
3 is called the ____________.
Write the related addition statement ________.

38. In the statement $7 - 5 = 2$
5 is called the ____________.
2 is called the ____________.
7 is called the ____________.
Write the related addition statement ________.

39. Find the number that is 25 less than 76. VIDEO

40. Find the number that results when 58 is decreased by 23.

41. Find the number that is the difference between 97 and 43.

42. Find the number that is 125 less than 265.

43. Find the number that results when 298 is decreased by 47.

44. Find the number that is the difference between 167 and 57.

Based on units, determine if each operation produces a meaningful result.

45. 8 mi − 4 mi

46. \$560 + \$314

47. 7 lb + 11 meters

48. 18°F − \$6

49. 17 yards − 10 yards

50. 4 mi/hr + 6 mi/gal

< Objective 2 >

In exercises 51 to 54, for various treks by a hiker in a mountainous region, the starting elevations and changes are given. Determine the final elevation of the hiker in each case.

51. Starting elevation 1,053 ft, increase of 123 ft, decrease of 98 ft, increase of 63 ft.

52. Starting elevation 1,231 ft, increase of 213 ft, decrease of 112 ft, increase of 78 ft.

53. Starting elevation 7,302 ft, decrease of 623 ft, decrease of 123 ft, increase of 307 ft.

54. Starting elevation 6,907 ft, decrease of 511 ft, decrease of 203 ft, increase of 419 ft.

Solve each application.

55. **Social Science** Shaka's score on a math test was 87. Tony's score was 23 points less than Shaka's. What was Tony's score on the test?

56. **Business and Finance** Duardo's weekly pay of \$879 was decreased by \$175 for withholding. What was he paid for the week? VIDEO

57. **Construction** The Willis Tower in Chicago is 1,454 ft tall. The Empire State Building is 1,250 ft tall. How much taller is the Willis Tower than the Empire State Building?

58. **Statistics** 2,479 students attended Scarlett College in 2012. By 2013, enrollment had increased to 2,653. What was the increase in enrollment?

59. **Number Problem** The difference between two numbers is 134. If the larger number is 655, what is the smaller number?

60. **Number Problem** The sum of two numbers is 850. If the smaller number is 278, what is the larger number?

61. **Business and Finance** In one week, Margaret earned \$480 in regular pay and \$108 for overtime work. \$153 was deducted from her paycheck for income taxes and \$36 for Social Security. What was her take-home pay?

62. **Business and Finance** Rafael opened a checking account and made deposits of \$85 and \$272. He wrote checks during the month for \$35, \$27, \$89, and \$178. What was his balance at the end of the month? VIDEO

63. **Business and Finance** Carmen's frequent-flyer program requires 30,000 mi for a free flight. During 2011 she accumulated 13,850 mi. In 2012 she took three more flights of 2,800, 1,475, and 4,280 mi. How much farther must she fly for her free trip?

64. **Statistics** A professor gives four 100-point exams and a 200-point final exam for a total of 600 points in the course. Students must earn a total of 540 points to receive an A. Find the minimum number of points you would need to score on the final exam if your first four test grades were 95, 84, 82, and 89.

65. **Business and Finance** Complete the monthly expense account.

Monthly income	\$3,240
House payment	1,343
Balance	
Car payment	283
Balance	
Food	512
Balance	
Clothing	189
Amount remaining	

66. **Business and Finance** To keep track of a checking account, you must subtract the amount of each check from the current balance. Complete the statement shown.

Beginning balance	\$351
Check #1	29
Balance	
Check #2	139
Balance	
Check #3	75
Ending balance	

67. **Business and Finance** The value of all crops in the Salinas Valley in 2010 was over \$4 billion. The top four crops are listed in the table.

(a) How much greater is the combined value of both types of lettuce than broccoli?

(b) How much greater is the value of the lettuces and broccoli combined than that of the strawberries?

Crop	Crop value, in millions
Head lettuce	\$512
Leaf lettuce	\$725
Strawberries	\$751
Broccoli	\$297

Source: Agricultural Commissioner, Monterey County, CA

68. **Business and Finance** Karen prices kitchen appliances at two stores and online. She makes a table comparing the prices of the models she wants.

Appliance	Corner Appliances	Big Buys Store	Online
Refrigerator	\$859	\$945	\$829
Range	\$1,049	\$995	\$1,025
Microwave	\$279	\$295	\$349

(a) How much more does it cost to purchase a refrigerator at Big Buys Store compared to an online purchase?

(b) If Karen wants to buy all three appliances from the same supplier, which option offers the lowest total price?

Now that you have reviewed the process of subtracting by hand, look at how a calculator performs this operation.

Find

23 − 13 + 56 − 29

Enter the numbers and the operation signs exactly as they appear in the expression.

23 [−] 13 [+] 56 [−] 29 [ENTER]

Display [37]

An alternative approach would be to add 23 and 56 first and then subtract 13 and 29. The result is the same in either case.

Use a calculator to perform each computation.

69. 5,830 − 3,987

70. 15,280 − 7,595

71. Subtract 235 from the sum of 534 and 678.

72. Subtract 476 from the sum of 306 and 572.

Solve each application.

73. **Business and Finance** Readings from Fast Service Station's storage tanks were taken at the beginning and end of a month. How much of each type of gas was sold? What was the total sold?

	Diesel	Unleaded	Super Unleaded	Total
Beginning reading	73,255	82,349	81,258	
End reading	28,387	19,653	8,654	
Gallons Used				

74. **Business and Finance** A local utility company places a meter on each residence in its area. Each meter keeps a running total of the number of kilowatt hours (kWh) of electricity used. To bill a residence properly, the utility takes beginning- and end-of-month readings and subtracts to find that month's total usage. Meter readings for three months are shown in the table. Determine each month's kWh usage and the three-month total.

	March	April	May	Total
Beginning reading	10,099	10,982	11,795	
End reading	10,982	11,795	12,592	
kWh Used				

Social Science *The land areas, in square miles (mi^2), of three Pacific coast states are California, 155,959 mi^2; Oregon, 95,997 mi^2; and Washington, 66,544 mi^2.*

75. How much larger is California than Oregon?

76. How much larger is California than Washington?

77. How much larger is Oregon than Washington?

Social Science *The land areas, in mi^2, for three of the Gulf Coast states are Alabama, 52,419 mi^2; Louisiana, 51,843 mi^2; and Mississippi, 48,430 mi^2.*

78. How much larger is Alabama than Mississippi?

79. How much larger is Alabama than Louisiana?

80. How much larger is Louisiana than Mississippi?

Skills | Calculator/Computer | **Career Applications** | Above and Beyond

81. Allied Health A patient's anion gap is used to help evaluate her electrolyte balance. The anion gap is equal to the difference between the serum concentration [in milliequivalents per liter (mEq/L)] of sodium and the sum of the serum concentrations of chloride and bicarbonate. Determine the patient's anion gap if the concentration of sodium is 140 mEq/L, chloride is 93 mEq/L, and bicarbonate is 24 mEq/L.

82. Allied Health To increase the sharpness of a radiographic image, we set the focus-object distance (FOD), which is the difference between the source-image receptor distance (SID) and the object-film distance (OFD), to its maximum value. What is the maximum FOD possible if the OFD is fixed at 8 in. and the maximum SID is 72 in.?

83. Information Technology Sally's department needs a new printer, so she bought one for $150. However, the printer is not appropriate for her growing department. Instead, she needs an upgrade, which costs $500. If she returns the printer, how much extra money will she need to buy the upgrade?

84. Information Technology Max has a 20-ft roll of cable, and he needs to run the cable from a wiring closet to an outlet in a room that is adjacent to the closet. The distance is about 25 ft. How much cable will Max need to buy to be able to run the cable from the wiring closet to the outlet?

85. Electronics Solder looks like flexible wire and typically comes wrapped on spools. When heated with a soldering iron or other heat source, solder melts. It is used to connect an electronic component to wires, other components, or conductive traces. If a certain spool holds 10 lb of solder, yet the shipping weight for the spool is 14 lb, how much does the empty spool and shipping materials weigh, in pounds?

86. Manufacturing Technology Kinetics, Inc., sells one model of engine blocks for $840 each. The cost of each block is $360 in materials, $290 in labor, and $95 for shipping and handling. How much is left as profit on each engine block?

Skills | Calculator/Computer | Career Applications | **Above and Beyond**

Number Problem Complete the magic squares.

87.

	7	2
	5	
8		

88.

4	3	
	5	
		6

89.

16	3		13
	10	11	
9	6	7	
			1

90.

7			14
2	13	8	11
16			
	6	15	

91. **Social Science** Use the Internet to find the populations of Arizona, California, Oklahoma, and Pennsylvania in each of the last three censuses.

(a) Find the total change in each state's population over this period.

chapter 1 > Make the Connection

(b) Which state shows the greatest change over the past three censuses?

(c) Write a brief essay describing the changes and any trends you see in these data. List any implications that they might have for future planning.

92. Think of any whole number.

Add 5.
Subtract 3.
Subtract 2 less than the original number.
What number do you end up with?
Check with other people. Does everyone have the same answer? Can you explain the results?

Answers

1. 146 **3.** 444 **5.** 1,120 **7.** 37 **9.** 269 **11.** 2,745 **13.** 3,465 **15.** 1,655 **17.** 19,439 **19.** 11,500 **21.** 53 **23.** 125 **25.** 58 **27.** 31,895 **29.** 12 **31.** 1,743 **33.** 2,203 **35.** 7,907 **37.** 9 is the minuend, 6 is the subtrahend, and 3 is the difference; $3 + 6 = 9$ **39.** 51 **41.** 54 **43.** 251 **45.** Yes **47.** No **49.** Yes **51.** 1,141 ft **53.** 6,863 ft **55.** 64 **57.** 204 ft **59.** $521 **61.** $399 **63.** 7,595 mi **65.** $1,897; $1,614; $1,102; $913 **67.** (a) 940,000,000; (b) 783,000,000 **69.** 1,843 **71.** 977 **73.** Diesel, 44,868 gal; unleaded, 62,696 gal; super unleaded, 72,604 gal; total, 180,168 gal **75.** 59,962 mi^2 **77.** 29,453 mi^2 **79.** 576 mi^2 **81.** 23 mEq/L **83.** $350 **85.** 4 lb

87.

6	7	2
1	5	9
8	3	4

89.

16	3	2	13
5	10	11	8
9	6	7	12
4	15	14	1

91. Above and Beyond

Activity 1 ::

Population Changes

The U.S. Census Bureau compiles population data for the United States. Population figures for the United States as well as the six most populous states according to the 2000 and 2010 censuses are shown. Use this table to complete each exercise.

	2000 Population	2010 Population
United States	281,421,906	308,745,538
California	33,871,648	37,253,956
Texas	20,851,820	25,145,561
New York	18,976,457	19,378,102
Florida	15,982,378	18,801,310
Illinois	12,419,293	12,830,632
Pennsylvania	12,281,054	12,702,379

1. Did any states switch population ranking between the two censuses?
2. By how much did the population of the United States increase between 2000 and 2010?
3. Which state had the greatest increase in population from 2000 to 2010? What was that difference?
4. Which state had the smallest increase in population from 2000 to 2010? What was that difference?
5. What was the total population of the six largest states in 2000?
6. How many people living in the United States did not live in one of the six largest states in 2000?
7. What was the total population of the six largest states in 2010?
8. How many people living in the United States did not live in one of the six largest states in 2010?
9. What regional trends might be true based on what you see in this table?

1.4 Rounding, Estimation, and Order

< 1.4 Objectives >

1 > Round a whole number to a given place value

2 > Estimate sums and differences by rounding

3 > Use inequality symbols

It is a common practice to express numbers to the nearest hundred, thousand, and so on. For instance, the distance from Los Angeles to New York along one route is 2,833 mi. We might say that the distance is 2,800 mi. This is called **rounding,** because we have rounded the distance to the nearest hundred miles.

One way to picture this rounding process is with the use of a number line.

Example 1 Rounding to the Nearest Hundred

< Objective 1 >

To round 2,833 to the nearest hundred, we mark notations on a number line counting by hundreds: 2,800 and 2,900. (We include only those that "surround" 2,833.) Then we mark the spot halfway between these: 2,850. When we also mark 2,833 on the line, estimating its location, we must place the mark to the left of 2,850. This makes it clear that 2,833 is closer to 2,800 than it is to 2,900. So we round *down* to 2,800.

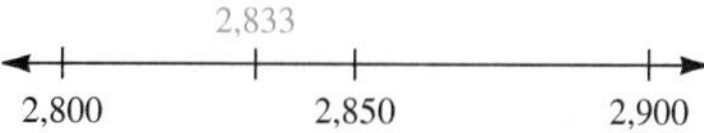

Check Yourself 1

Round 587 to the nearest hundred.

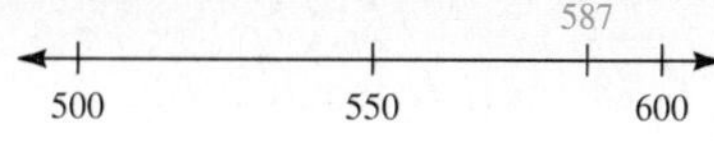

Example 2 Rounding to the Nearest Thousand

To round 28,734 to the nearest thousand:

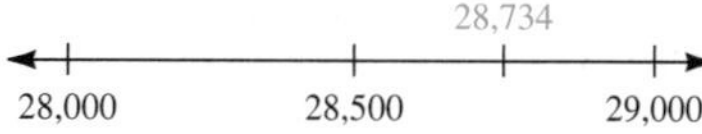

Because 28,734 is closer to 29,000 than it is to 28,000, we round *up* to 29,000.

Check Yourself 2

Locate 1,375 and round to the nearest hundred.

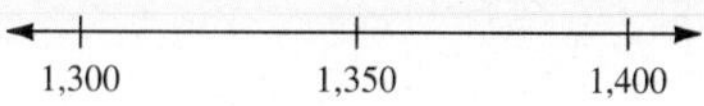

Instead of using a number line, we can apply rounding rules.

Step by Step

Rounding Whole Numbers

Step 1 Identify the place of the digit to be rounded.

Step 2 Look at the digit to the right of that place.

Step 3 a. If that digit is 5 or more, that digit and all digits to the right become 0. The digit in the place you are rounding to is increased by 1.

b. If that digit is less than 5, that digit and all digits to the right become 0. The digit in the place you are rounding to remains the same.

Example 3 Rounding to the Nearest Ten

Round 587 to the nearest ten.

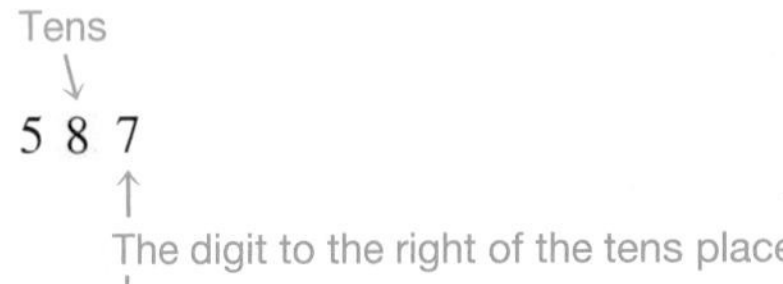

5 8 7 is rounded to 590.

We identify the tens digit. The digit to the right of the tens place, 7, is 5 or more. So round up.

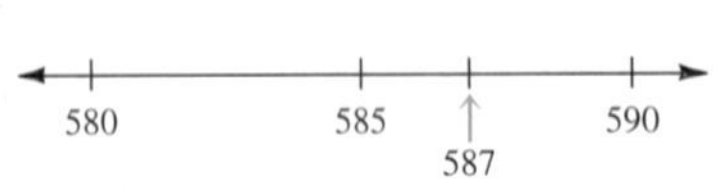

NOTE

587 is between 580 and 590. It is closer to 590, so it makes sense to round up.

Check Yourself 3

Round 847 to the nearest ten.

Example 4 Rounding to the Nearest Hundred

Round 2,638 to the nearest hundred.

2, 6 38 is rounded to 2,600.

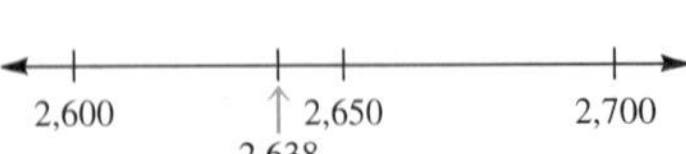

We identify the hundreds digit. The digit to the right, 3, is less than 5. So round down.

NOTE

2,638 is closer to 2,600 than to 2,700. So it makes sense to round down.

Check Yourself 4

Round 3,482 to the nearest hundred.

Examples 5 and 6 use the rounding rules.

Example 5 Rounding Whole Numbers

(a) Round 2,378 to the nearest hundred.

2, 3 78 is rounded to 2,400.

We identified the hundreds digit. The digit to the right is 7. Because this is 5 or more, the hundreds digit is increased by 1. The 7 and all digits to the right of 7 become 0.

(b) Round 53,258 to the nearest thousand.

5 3 ,258 is rounded to 53,000.

We identified the thousands digit. Because the digit to the right is less than 5, the thousands digit remains the same. The 2 and all digits to its right become 0.

(c) Round 685 to the nearest ten.

6 8 5 is rounded to 690.

The digit to the right of the tens place is 5 or more. Round up by our rule.

(d) Round 52,813,212 to the nearest million.

5 2 ,813,212 is rounded to 53,000,000.

Check Yourself 5

(a) Round 568 to the nearest ten.
(b) Round 5,446 to the nearest hundred.

Now, look at a case in which we round up a 9.

Example 6 Rounding to the Nearest Ten

NOTES

Which number is 397 closer to?

390 397 400

An estimate is basically a good guess. If your answer is close to your estimate, then your answer is reasonable.

Suppose we want to round 397 to the nearest ten. We identify the tens digit and look at the next digit to the right.

3 9 7

The digit to the right is 5 or more. If this digit is 9, and it must be increased by 1, replace the 9 with 0 and increase the next digit to the *left* by 1.

So 397 is rounded to 400.

Check Yourself 6

Round 4,961 to the nearest hundred.

When you are doing arithmetic, rounding numbers gives you a handy way of deciding whether an answer seems reasonable. The process is called **estimating,** which we illustrate with an example.

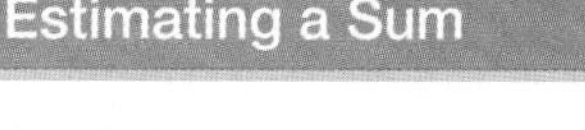

Example 7 Estimating a Sum

< Objective 2 >

NOTE

Placing an arrow above the column to be rounded can be helpful.

Begin by rounding to the nearest hundred.

456	500
235	200
976	1,000
+ 344	+ 300
2,011	2,000 ⟵ Estimate

By rounding to the nearest hundred and adding quickly, we get an estimate or guess of 2,000. Because this is close to the sum calculated, 2,011, the sum seems reasonable.

Check Yourself 7

Round each addend to the nearest hundred and estimate the sum. Then find the actual sum.

$287 + 526 + 311 + 378$

Estimation is a wonderful tool to use when shopping. Every time you go to the store, you should try to estimate the total bill by rounding the price of each item. If you do this regularly, both your addition and rounding skills will improve. The same holds true when you eat in a restaurant. It is always a good idea to know approximately how much you are spending.

Example 8 Estimating a Sum in a Word Problem

Samantha took her family out to dinner, and is ready to pay the bill. The dinner check has no total, only the individual entries.

Soup	\$2.95
Soup	2.95
Salad	1.95
Salad	1.95
Salad	1.95
Lasagne	7.25
Spaghetti	4.95
Ravioli	5.95

What is the approximate cost of the dinner?

Rounding each entry to the nearest whole dollar, we can estimate the total by finding the sum

$3 + 3 + 2 + 2 + 2 + 7 + 5 + 6 = \$30$

Check Yourself 8

Jason is doing the weekly food shopping at FoodWay. So far his basket has items that cost \$3.99, \$7.98, \$2.95, \$1.15, \$2.99, and \$1.95. Approximate the total cost of these items.

Earlier in this section, we used a number line to illustrate the idea of rounding numbers. A number line also gives us an excellent way to picture the concept of **order** for whole numbers, which means that numbers become larger as we move from left to right on a number line.

For instance, we know that 3 is less than 5. On the number line

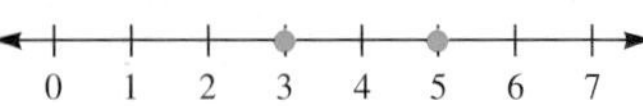

we see that 3 lies *to the left* of 5.

We also know that 4 is greater than 2. On the number line

NOTE

The inequality symbols always "point to" the smaller number.

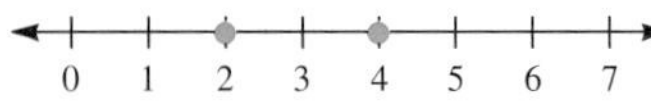

we see that 4 lies *to the right* of 2.

Two symbols, $<$ for "less than" and $>$ for "greater than," are used to indicate these relationships.

Definition

Inequalities

For whole numbers, we can write

1. $2 < 5$ (read "2 is less than 5") because 2 is *to the left* of 5 on a number line.
2. $8 > 3$ (read "8 is greater than 3") because 8 is *to the right* of 3 on a number line.

Example 9 illustrates this notation.

Example 9 Indicating Order with $<$ or $>$

$<$ Objective 3 $>$

Use the symbol $<$ or $>$ to complete each statement.

(a) 7 _____ 10 **(b)** 25 _____ 20
(c) 200 _____ 300 **(d)** 8 _____ 0

(a) $7 < 10$ 7 lies to the left of 10 on a number line.
(b) $25 > 20$ 25 lies to the right of 20 on a number line.
(c) $200 < 300$
(d) $8 > 0$

Check Yourself 9

Use either $<$ or $>$ to complete each statement.

(a) 35 ___ 25 **(b)** 0 ___ 4
(c) 12 ___ 18 **(d)** 1,000 ___ 100

Check Yourself ANSWERS

1. 600 **2.** 

Round 1,375 to 1,400. **3.** 850 **4.** 3,500
5. **(a)** 570; **(b)** 5,400 **6.** 5,000 **7.** 1,500; 1,502 **8.** \$21
9. **(a)** $35 > 25$; **(b)** $0 < 4$; **(c)** $12 < 18$; **(d)** $1,000 > 100$

Reading Your Text

These fill-in-the-blank exercises will help you understand some of the key vocabulary used in this section. The answers to these exercises are in the Answers Appendix at the back of the text.

(a) The practice of expressing numbers to the nearest hundred, thousand, and so on is called _________.

(b) The first step in rounding is to identify the _________ _________ of the digit to be rounded.

(c) A number line gives us an excellent way to picture the concept of _________ for whole numbers.

(d) The symbol $<$ is read as "_________ than."

1.4 exercises

Skills | Calculator/Computer | Career Applications | Above and Beyond

< Objective 1 >

Round each number to the indicated place.

1. 38, the nearest ten 

2. 72, the nearest ten

3. 253, the nearest ten

4. 578, the nearest ten

5. 696, the nearest ten

6. 391, the nearest ten

7. 2,493, the nearest ten

8. 6,497, the nearest ten

9. 683, the nearest hundred

10. 239, the nearest hundred

11. 6,741, the nearest hundred

12. 3,482, the nearest hundred 

13. 5,962, the nearest hundred

14. 8,948, the nearest hundred

15. 12,908, the nearest hundred

16. 34,990, the nearest hundred

17. 4,352, the nearest thousand

18. 8,512, the nearest thousand

19. 4,927, the nearest thousand

20. 1,975, the nearest thousand

21. 23,429, the nearest thousand

22. 39,589, the nearest thousand

23. 9,206, the nearest thousand

24. 9,751, the nearest thousand

25. 129,816, the nearest thousand

26. 599,669, the nearest thousand

27. 787,000, the nearest ten thousand

28. 931,000, the nearest ten thousand

29. 21,800,000, the nearest million

30. 9,120,000, the nearest million

31. 12, the nearest hundred

32. 80, the nearest hundred

33. 741, the nearest thousand

34. 489,000, the nearest million

< Objective 2 >

In exercises 35 to 46, estimate each sum or difference by rounding to the indicated place. Then do the addition or subtraction and use your estimate to see if your actual sum or difference seems reasonable.

Round to the nearest ten.

35. $\begin{array}{r} 58 \\ 27 \\ +\ 33 \\ \hline \end{array}$

36. $\begin{array}{r} 92 \\ 37 \\ 85 \\ +\ 64 \\ \hline \end{array}$

37. $\begin{array}{r} 83 \\ -\ 27 \\ \hline \end{array}$ 

38. $\begin{array}{r} 97 \\ -\ 31 \\ \hline \end{array}$

Round to the nearest hundred.

39. $\begin{array}{r} 379 \\ 1{,}215 \\ +\ 528 \\ \hline \end{array}$

40. $\begin{array}{r} 967 \\ 2{,}365 \\ 544 \\ +\ 738 \\ \hline \end{array}$

41. $\begin{array}{r} 915 \\ -\ 411 \\ \hline \end{array}$

42. $\begin{array}{r} 697 \\ -\ 539 \\ \hline \end{array}$

Round to the nearest thousand.

43. $\begin{array}{r} 2{,}238 \\ 3{,}925 \\ +\ 5{,}217 \\ \hline \end{array}$

44. $\begin{array}{r} 3{,}678 \\ 4{,}215 \\ +\ 2{,}032 \\ \hline \end{array}$

45. $\begin{array}{r} 4{,}822 \\ -\ 2{,}134 \\ \hline \end{array}$

46. $\begin{array}{r} 6{,}120 \\ -\ 4{,}890 \\ \hline \end{array}$

Solve each application.

47. **BUSINESS AND FINANCE** Ed and Sharon go to lunch. The lunch check has no total but lists individual items.

Soup \$1.95	Soup \$1.95
Salad \$1.80	Salad \$1.80
Salmon \$8.95	Flounder \$6.95
Pecan pie \$3.25	Vanilla ice cream \$2.25

Estimate the total amount of the lunch check. 

48. **BUSINESS AND FINANCE** Olivia will purchase several items at the stationery store. Thus far, the items she has collected cost \$2.99, \$6.97, \$3.90, \$2.15, \$9.95, and \$1.10. Approximate the total cost of these items.

49. **STATISTICS** Oscar scored 78, 91, 79, 67, and 100 on his arithmetic tests. Round each score to the nearest ten to estimate his total score.

50. **BUSINESS AND FINANCE** Luigi's pizza parlor makes 293 pizzas on an average day. Estimate (to the nearest hundred) how many pizzas were made on a three-day holiday weekend.

51. **BUSINESS AND FINANCE** Mrs. Gonzalez went shopping for clothes. She bought a sweater for \$32.95, a scarf for \$9.99, boots for \$68.29, a coat for \$125.90, and socks for \$18.15. Estimate the total amount of Mrs. Gonzalez's purchases.

52. **BUSINESS AND FINANCE** Amir bought several items from the bargain bin at the hardware store: hammer, \$8.95; screwdriver, \$3.15; pliers, \$6.90; wire cutters, \$4.25; and sandpaper; \$1.89. Estimate the total cost of Amir's bill.

< Objective 3 >

Use the symbol $<$ or $>$ to complete each statement.

53. 500 _____ 400 

54. 20 _____ 15

55. 100 _____ 1,000 

56. 3,000 _____ 2,000

Skills | Calculator/Computer | **Career Applications** | Above and Beyond

Use the energy consumption chart to complete exercises 57 and 58.

Appliance	Power Required (in watts/hour [W/hr])
Clock radio	10
Electric blanket	100
Clothes washer	500
Toaster oven	1,225
Laptop	50
Hair dryer	1,875
DVD player	25

57. **ELECTRONICS** Assuming all the appliances listed in the table are "on," estimate the total power required to the nearest hundred watts.

58. **ELECTRONICS** Which combination uses more power, the toaster oven and clothes washer or the hair dryer and DVD player?

59. **Manufacturing Technology** An inventory of machine screws shows that bin 1 contains 378 screws, bin 2 contains 192 screws, and bin 3 contains 267 screws. Estimate the total number of screws in the bins.

60. **Manufacturing Technology** A delivery truck must be loaded with the heaviest crates starting in the front to the lightest crates in the back. On Monday, crates weighing 378 lb, 221 lb, 413 lb, 231 lb, 208 lb, 911 lb, 97 lb, 188 lb, and 109 lb need to be shipped. In what order should the crates be loaded?

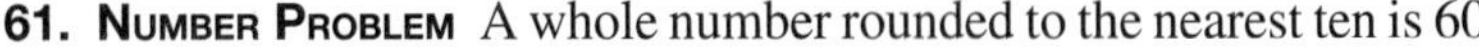

Skills | Calculator/Computer | Career Applications | **Above and Beyond**

61. **Number Problem** A whole number rounded to the nearest ten is 60.

(a) What is the smallest possible number?

(b) What is the largest possible number?

62. **Number Problem** A whole number rounded to the nearest hundred is 7,700.

(a) What is the smallest possible number?

(b) What is the largest possible number?

63. **Statistics** A bag contains 60 marbles. The number of blue marbles, rounded to the nearest 10, is 40, and the number of green marbles in the bag, rounded to the nearest 10, is 20. How many blue marbles are in the bag? (List all answers that satisfy the conditions of the problem.)

64. **Social Science** Describe some situations in which estimating and rounding would not produce a result that would be suitable or acceptable. Review the instructions for filing your federal income tax. What rounding rules are used in the preparation of your tax returns? Do the same rules apply to the filing of your state tax returns? If not, what are these rules?

65. The U.S. Census Bureau listed the population of the United States on September 7, 2011, at 312,160,918 people. Round this number to the nearest ten million. (chapter 1 > Make the Connection)

66. According to the U.S. Census Bureau, the population of the world was believed to be 6,960,550,584 on September 7, 2011. Round this number to the nearest million. (chapter 1 > Make the Connection)

Answers

1. 40 **3.** 250 **5.** 700 **7.** 2,490 **9.** 700 **11.** 6,700 **13.** 6,000 **15.** 12,900 **17.** 4,000 **19.** 5,000 **21.** 23,000 **23.** 9,000 **25.** 130,000 **27.** 790,000 **29.** 22,000,000 **31.** 0 **33.** 1,000 **35.** Estimate: 120, actual sum: 118 **37.** Estimate: 50, actual difference: 56 **39.** Estimate: 2,100, actual sum: 2,122 **41.** Estimate: 500; actual difference: 504 **43.** Estimate: 11,000, actual sum: 11,380 **45.** Estimate: 3,000, actual difference: 2,688 **47.** \$29 **49.** 420 **51.** \$255 **53.** > **55.** < **57.** 3,800 W/hr **59.** 900 screws **61.** (a) 55; (b) 64 **63.** 36, 37, 38, 39, 40, 41, 42, 43, 44 **65.** 310,000,000 people

1.5 Multiplication

< 1.5 Objectives >

1 > Multiply whole numbers
2 > Use the properties of multiplication
3 > Use multiplication to solve applications
4 > Estimate products
5 > Find area and volume

NOTES

The use of the symbol × dates back to the 1600s.

A centered dot is used the same as the times sign (×). We use the centered dot when we are using letters to represent numbers, as we have done with *a* and *b* here.

Our work in this section deals with multiplication, another basic operation of arithmetic. Multiplication is closely related to addition. In fact, we can think of multiplication as a shorthand method for repeated addition. The symbol × is used to indicate multiplication.

3×4 can be interpreted as 3 rows of 4 objects. By counting we see that $3 \times 4 = 12$. Similarly, 4 rows of 3 means $4 \times 3 = 12$.

	4			
3	★	★	★	★
	★	★	★	★
	★	★	★	★

	3		
4	★	★	★
	★	★	★
	★	★	★
	★	★	★

The fact that $3 \times 4 = 4 \times 3$ is an example of the **commutative property of multiplication,** which is given here.

Property

The Commutative Property of Multiplication

Given any two numbers, we can multiply them in either order and we get the same result.

In symbols, we say $a \cdot b = b \cdot a$.

Example 1 Multiplying Single-Digit Numbers

< Objective 1 >

3×5 means 5 multiplied by 3. It is read 3 *times* 5. To find 3×5, we can add 5 three times.

$3 \times 5 = 5 + 5 + 5 = 15$

In a multiplication problem such as $3 \times 5 = 15$, we call 3 and 5 the **factors.** The answer, 15, is the **product** of the factors, 3 and 5.

$3 \times 5 = 15$

Factor ↗ Factor ↖ Product ↖

Check Yourself 1

Name the factors and the product in the equation.

$2 \times 9 = 18$

Statements such as $3 \times 4 = 12$ and $3 \times 5 = 15$ are called **basic multiplication facts.** If you find multiplication difficult, it may be that you do not know some of these facts. The table will help you review before you go on. Notice that, because of the commutative property, you only need to memorize about half of these facts!

Basic Multiplication Facts Table

×	0	1	2	3	4	5	6	7	8	9
0	0	0	0	0	0	0	0	0	0	0
1	0	1	2	3	4	5	6	7	8	9
2	0	2	4	6	8	10	12	14	16	18
3	0	3	6	9	12	15	18	21	24	27
4	0	4	8	12	16	20	24	28	32	36
5	0	5	10	15	20	25	30	35	40	45
6	0	6	12	18	24	30	36	42	48	54
7	0	7	14	21	28	35	42	49	56	63
8	0	8	16	24	32	40	48	56	64	72
9	0	9	18	27	36	45	54	63	72	81

NOTE

To use the table to find the product of 7×6: Find the row labeled 7, and then move to the right in this row until you are in the column labeled 6 at the top. We see that 7×6 is 42.

Armed with these facts, you can become a better, faster problem solver. Take a look at Example 2.

Example 2 Multiplying Instead of Counting

NOTE

This checkerboard is an example of a rectangular array, a series of rows or columns that form a rectangle. When you see such an arrangement, multiply to find the total number of units.

Find the total number of squares on the checkerboard.

You could find the number of squares by counting them. If you counted one per second, it would take you just over a minute. You could make the job a little easier by simply counting the squares in one row (8), and then adding $8 + 8 + 8 + 8 + 8 + 8 + 8 + 8$. Multiplication, which is simply repeated addition, allows you to find the total number of squares by multiplying 8×8. How long that takes depends on how well you know the basic multiplication facts! By now, you know that there are 64 squares on the checkerboard.

Check Yourself 2

Find the number of windows on the displayed side of the building.

In Section 1.2, we called 0 the additive identity because adding 0 to any number leaves the number unchanged. We know that the product of any number and 1 is just that number, so 1 has the same property in multiplication that 0 does in addition.

Property

The Multiplicative Identity Property

The product of 1 and any number is just that number. In symbols, $a \cdot 1 = a$ and $1 \cdot a = a$ for any number a.

Zero is another number with special properties in multiplication. Multiplying any number by 0 results in the product 0.

Property

The Multiplicative Property of Zero

The product of 0 and any number is 0. In symbols, $a \cdot 0 = 0$ and $0 \cdot a = 0$ for any number a.

Example 3 Multiplying by 0 and 1

< Objective 2 >

Find each product.

(a) 8×1
$8 \times 1 = 8$ Any number times 1 is itself.

(b) $(1)(32)$
$(1)(32) = 32$

(c) $5 \cdot 0$
$5 \cdot 0 = 0$ Any number times 0 is 0.

(d) $(0)(32)$
$(0)(32) = 0$

Check Yourself 3

Find each product.

(a) $(1)(6)$ **(b)** 74×0 **(c)** 1×1 **(d)** $74 \cdot 1$ **(e)** $0 \cdot 16$

The next property involves *both* multiplication and addition.

Example 4 Using the Distributive Property

NOTE

Multiplication can also be indicated by using parentheses. A number followed by parentheses or back-to-back parentheses indicate multiplication.

$2 \times (3 + 4)$ could be written as $2(3 + 4)$ or $(2)(3 + 4)$.

$2 \times (3 + 4) = 2 \times 7 = 14$ We added 3 + 4 and then multiplied.

Also,

$$2 \times (3 + 4) = (2 \times 3) + (2 \times 4)$$ We multiplied 2 × 3 and 2 × 4 as the first step.

$$= 6 + 8$$

$$= 14$$ The result is the same.

We see that $2 \times (3 + 4) = (2 \times 3) + (2 \times 4)$. This is an example of the **distributive property of multiplication over addition** because we distributed the multiplication (in this case by 2) over the "plus" sign.

Property

The Distributive Property of Multiplication over Addition

To multiply a factor by a sum of numbers, multiply the factor by each number inside the parentheses. Then add the products. (The result will be the same if we find the sum and then multiply.)
In symbols, we say $a \cdot (b + c) = a \cdot b + a \cdot c$

Check Yourself 4

Show that

$3 \times (5 + 2) = (3 \times 5) + (3 \times 2)$

Regrouping must often be used to multiply larger numbers. We see how regrouping works in multiplication by looking at an example in the expanded form. When regrouping results in changing a digit to the left we sometimes say we "carry" the units.

Example 5 Multiplying by a Single-Digit Number

> **NOTE**
>
> 3 × 25
>
> 3 · 25
>
> (3)(25)
>
> all mean the same thing.

$$
\begin{aligned}
3 \times 25 &= 3 \times (20 + 5) && \text{We use the distributive property.}\\
&= 3 \times 20 + 3 \times 5 \\
&= 60 \quad + 15 && \text{Write the 15 as 10 + 5.}\\
&= 60 + 10 + 5 && \text{Carry 10 ones or 1 ten to the tens place.}\\
&= 70 + 5 \\
&= 75
\end{aligned}
$$

Here is the same multiplication problem using the short form.

Step 1

$$
\begin{array}{r}
\scriptstyle 1 \leftarrow \text{Carry} \\
25 \\
\times\ 3 \\
\hline
5
\end{array}
$$

Multiplying 3 × 5 gives us 15 ones. Write 5 ones and carry 1 ten.

Step 2

$$
\begin{array}{r}
\scriptstyle 1 \\
25 \\
\times\ 3 \\
\hline
75
\end{array}
$$

Now multiply 3 × 2 tens and add the carry to get 7, the tens digit of the product.

Check Yourself 5

Multiply.

(a) $\begin{array}{r} 34 \\ \times\ 6 \\ \hline \end{array}$ **(b)** $\begin{array}{r} 43 \\ \times\ 7 \\ \hline \end{array}$

Units Analysis

When you multiply a denominate number, such as 6 ft, by an abstract number, such as 5, the result has the same units as the denominate number. Some examples are

5 × 6 ft = 30 ft

3 × \$7 = \$21

9 × 4 A's = 36 A's

When you multiply two different denominate numbers, the units must also be multiplied. We discuss this when we look at the area of geometric figures.

We briefly review our discussion of applications, or word problems.

The process of solving applications is the same no matter which operations are required for the solution. In fact, the four-step procedure we suggested in Section 1.2 can be applied here.

Step by Step

Solving Applications

RECALL

It is best to write down the complete statement necessary for the solution of an application.

Step 1 Read the problem carefully to determine the given information and what you are asked to find.

Step 2 Decide upon the operation or operations to be used.

Step 3 Write down the complete statement necessary to solve the problem and do the calculations.

Step 4 Check to make sure you have answered the question and that your answer seems reasonable.

Example 6 Solving an Application Involving Multiplication

< Objective 3 >

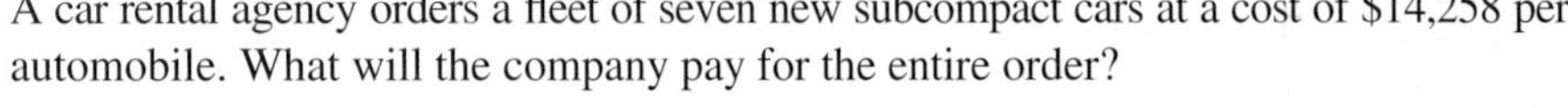

A car rental agency orders a fleet of seven new subcompact cars at a cost of $14,258 per automobile. What will the company pay for the entire order?

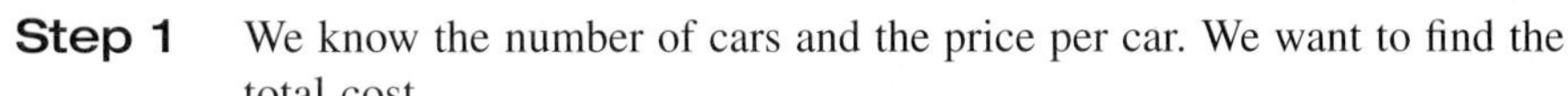

Step 1 We know the number of cars and the price per car. We want to find the total cost.

Step 2 Multiplication is the best approach to the solution.

Step 3 Write

$7 \times \$14,258 = \$99,806$

We could, of course, *add* $14,258 7 times, but multiplication is certainly more efficient.

Step 4 The total cost of the order is $99,806.

Check Yourself 6

Tires sell for $87 apiece. What is the total cost for five tires?

To multiply by numbers with more than one digit, we must multiply each digit of the first factor by each digit of the second. To do this, we form a series of *partial products* and then add them to arrive at the final product.

Example 7 Multiplying by a Two-Digit Number

Multiply 56 × 47.

Step 1

$$\begin{array}{r} {\scriptstyle 4} \\ 56 \\ \times\ 47 \\ \hline 392 \end{array}$$

The first partial product is 7 × 56, or 392. Note that we had to carry 4 to the tens column.

Step 2

$$\begin{array}{r} {\scriptstyle 2} \\ {\scriptstyle \not{4}} \\ 56 \\ \times\ 47 \\ \hline 392 \\ 2240 \\ \hline \end{array}$$

The second partial product is 40 × 56, or 2,240. We must carry 2 during the process.

Step 3

```
    2
    4̸
   56
 × 47
  392
 2240
2,632
```

We add the partial products for our final result.

Check Yourself 7

Multiply.

```
  38
× 76
```

If multiplication involves two three-digit numbers, another step is necessary. In this case, we form three partial products. This ensures that each digit of the first factor is multiplied by each digit of the second.

Example 8 Multiplying Two Three-Digit Numbers

Multiply.

```
   22
   33
   22
  278
× 343
  834
11120
83400
95,354
```

In forming the third partial product, we must multiply by 300.

NOTE

The three partial products are formed when we multiply by the ones, tens, and then the hundreds digits.

Check Yourself 8

Multiply.

```
  352
× 249
```

Now consider an example in which we multiply by a number that has 0 as a digit. There are several ways to arrange the work, as Example 9 shows.

Example 9 Multiplying Larger Numbers

Multiply 573×205.

Method 1

```
    1
   31
   573
 × 205
  2865
  0000
114600
117,465
```

← We can write the second partial product as 0000 to indicate the multiplication by 0 in the tens place.

Here is a second approach to the problem.

Method 2

$$\begin{array}{r} {\scriptstyle 1} \\ {\scriptstyle 31} \\ 573 \\ \times\ 205 \\ \hline 2865 \\ 114600 \\ \hline 117{,}465 \end{array}$$

← We can write a double 0 as our second step. If we place the third partial product on the same line, that product will be shifted *two* places left, indicating that we are multiplying by 200.

We usually use the second method as it is more compact.

Check Yourself 9

Multiply.

(a) 489 × 304 **(b)** 3,408 × 712 **(c)** 6,049 × 208

Example 10 leads us to another property of multiplication.

Example 10 Using the Associative Property

NOTE

You will learn more about the order that you should perform calculations when you study the *order of operations* in Section 1.7.

(2 × 3) × 4 = 6 × 4 = 24 We do the multiplication in the parentheses first, 2 × 3 = 6. Then we multiply 6 × 4.

Also,

2 × (3 × 4) = 2 × 12 = 24 Here we multiply 3 × 4 as the first step. Then we multiply 2 × 12.

We see that

(2 × 3) × 4 = 2 × (3 × 4)

The product is the same no matter which way we *group* the factors. This is called the **associative property** of multiplication.

Property

The Associative Property of Multiplication

Multiplication is an *associative* operation. The way in which you group numbers in multiplication does not affect the final product.

Check Yourself 10

Evaluate each expression by multiplying within the parentheses first and then multiplying that product by the factor outside the parentheses.

(a) (5 × 3) × 6 **(b)** 5 × (3 × 6)

Some shortcuts let you simplify your work when multiplying by a number that ends in 0. Let's see what we can discover in Examples 11 and 12.

Example 11 Multiplying by 10

First we multiply by 10.

$$\begin{array}{r} 67 \\ \times\ 10 \\ \hline 670 \end{array} \quad 10 \times 67 = 670$$

Next we multiply by 100.

$$\begin{array}{r} 537 \\ \times\ 100 \\ \hline 53{,}700 \end{array} \quad 100 \times 537 = 53{,}700$$

Finally, we multiply by 1,000.

$$\begin{array}{r} 489 \\ \times\ 1{,}000 \\ \hline 489{,}000 \end{array} \quad 1{,}000 \times 489 = 489{,}000$$

NOTE

You will learn more about *powers of 10* in greater detail in Section 1.7.

Check Yourself 11

Multiply.

(a) $\begin{array}{r} 257 \\ \times\ 100 \\ \hline \end{array}$ **(b)** $\begin{array}{r} 2{,}436 \\ +\ 1{,}000 \\ \hline \end{array}$

Do you see a pattern? Rather than writing out the multiplication, there is an easier way! We call the numbers 10, 100, 1,000, and so on **powers of 10.**

Property

Multiplying by Powers of 10

When a natural number is multiplied by a power of 10, the product is just that number followed by as many zeros as there are in the power of 10.

Example 12 Multiplying by Numbers That End in Zero

Multiply 400 × 678.

Write

$$\begin{array}{r} 678 \\ \times\ \ 400 \\ \hline \end{array}$$

Shift 400 so that the two zeros are *to the right* of the digits above.

$$\begin{array}{r} {\scriptstyle 33} \\ 678 \\ \times\ \ 400 \\ \hline 271{,}200 \end{array}$$

Bring down the two zeros, then multiply 4 × 678 to find the product.

There is no mystery about why this works. We know that 400 is 4 × 100. In this method, we multiply 678 by 4 and then by 100, adding two zeros to the product by our earlier rule.

Check Yourself 12

Multiply.

300 × 574

Your work in this section, together with our earlier rounding techniques, provides a convenient way to use estimation to check the reasonableness of a product, as Example 13 illustrates.

Example 13 Estimating a Product by Rounding

< Objective 4 >

Estimate the product by rounding each factor to the nearest hundred.

		Rounded
512	⟶	500
$\times$ 289	⟶	$\times$ 300
		150,000

You might want to find the *actual* product and use our estimate to see if your result seems reasonable.

Check Yourself 13

Estimate the product by rounding each factor to the nearest hundred.

$$\begin{array}{r} 689 \\ \times\ 425 \\ \hline \end{array}$$

Rounding the factors can be a very useful way of estimating the solution to an application problem.

Example 14 Estimating the Solution to a Multiplication Application

Bart is thinking of running an ad in the local newspaper for an entire year. The ad costs \$19.95 per week. Approximate the annual cost of the ad.

Rounding the charge to \$20 and rounding the number of weeks in a year to 50, we get

$50 \times 20 = 1{,}000$

The ad would cost approximately \$1,000.

Check Yourself 14

Phyllis is debating whether to join the health club for \$400 per year or just pay \$9 per visit. If she goes about once a week, approximately how much would she spend at \$9 per visit?

Units Analysis

What happens when we multiply two denominate numbers? The units of the result turn out to be the product of the units. This makes sense when we look at an example from geometry.

The area of a square is the square of one side. As a formula, we write that as

$A = s^2$

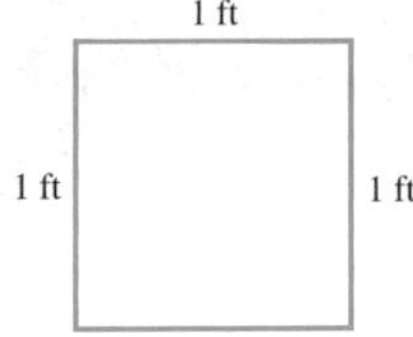

This tile is 1 ft by 1 ft.

$A = s^2 = (1 \text{ ft})^2 = 1 \text{ ft} \times 1 \text{ ft} = 1 (\text{ft}) \times (\text{ft}) = 1 \text{ ft}^2$

In other words, its area is one square foot (1 ft^2).

If we want to find the area of a room, we are actually finding how many of these square feet can be placed in the room.

NOTE

The small, raised 2 is called an *exponent* or *power.* This will be studied in Section 1.7. So in. $\times$ in. can be written in.^{2} and is read "square inches."

Now look at the idea of **area.** Area is a measure we give to a surface. It is measured in terms of **square units.** The area is the number of square units that are needed to cover the surface.

One standard unit of area measure is the **square inch** (written in.^{2}). This is the measure of the space contained by a square with sides of 1 in.

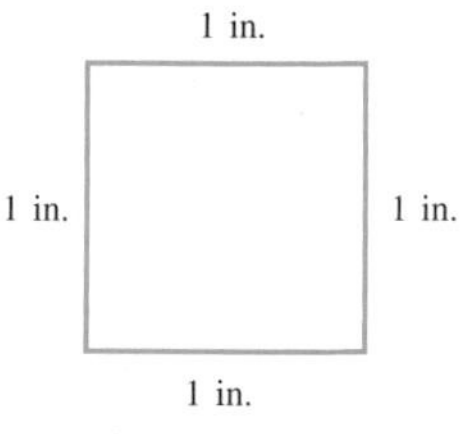

One square inch

Other units of area measure are the square foot (ft^2), the square yard (yd^2), the square centimeter (cm^2), and the square meter (m^2).

Finding the area of a figure means finding the number of square units it contains. One simple case is a rectangle.

The figure shows a rectangle with a length of 4 in. and a 3-in. width. The area of the rectangle is measured in terms of square inches. We can simply count to find the area, 12 in.^{2}. However, because each of the four vertical strips contains 3 in.^{2}, we can multiply:

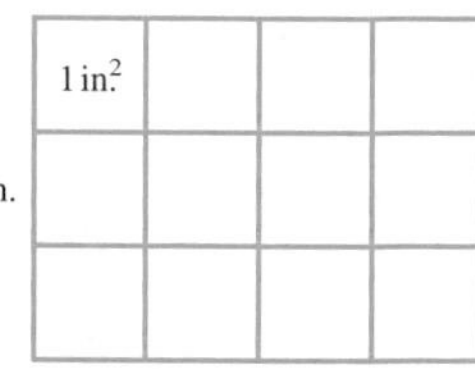

Area = 4 in. × 3 in. = 12 in.^{2}

Property

Formula for the Area of a Rectangle

In general, we can write the formula for the **area of a rectangle:** If the length of a rectangle is L units and the width is W units, then the formula for the area A of the rectangle can be written as

$A = L \cdot W$ (square units)

Example 15 **Find the Area of a Rectangle**

< Objective 5 >

A room has dimensions 12 ft by 15 ft. Find its area.

Use the area formula, with $L = 15$ ft and $W = 12$ ft.

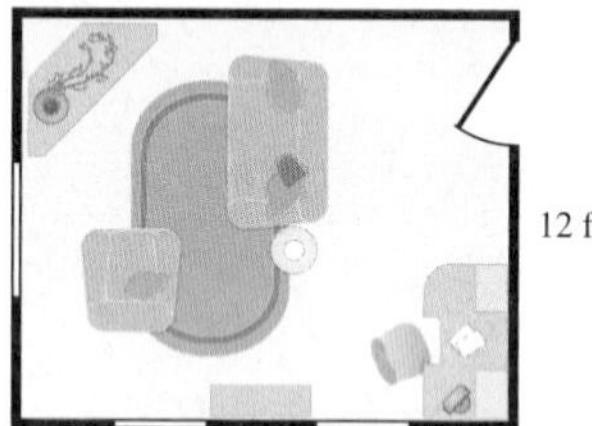

$A = L \cdot W$

$= 15 \text{ ft} \times 12 \text{ ft} = 180 \text{ ft}^2$

The area of the room is 180 ft^2.

Check Yourself 15

A desktop has dimensions 50 in. by 25 in. What is the area of its surface?

We can also write a convenient formula for the area of a square. If the sides of the square have length s, we can write

Property

Formula for the Area of a Square

$A = s \cdot s = s^2$ $\quad$ s^2 is read "s squared."

Example 16 **Finding the Area**

You wish to cover a square table with a plastic laminate that costs 60¢ a square foot. If each side of the table measures 3 ft, what does it cost to cover the table?

We first must find the area of the table. Use the area formula with $s = 3$ ft.

$$A = s^2$$
$$= (3 \text{ ft})^2 = 3 \text{ ft} \times 3 \text{ ft} = 9 \text{ ft}^2$$

Now, multiply by the cost per square foot.

Cost $= 9 \times 60¢ = 540¢ = \$5.40$

Check Yourself 16

You wish to carpet a 4-yd by 4-yd square room, with carpet that costs \$12 per square yard. What is the cost to carpet the room?

Sometimes the total area of an oddly shaped figure is found by adding smaller areas. Example 17 shows how this is done.

Example 17 Finding the Area of an Oddly Shaped Figure

Find the area of the figure.

The area of the figure is found by adding the areas of regions 1 and 2. Region 1 is a 4-in. by 3-in. rectangle; the area of region 1 is 4 in. $\times$ 3 in. = 12 in.^{2}. Region 2 is a 2-in. by 1-in. rectangle; the area of region 2 is 2 in. $\times$ 1 in. = 2 in.^{2}.

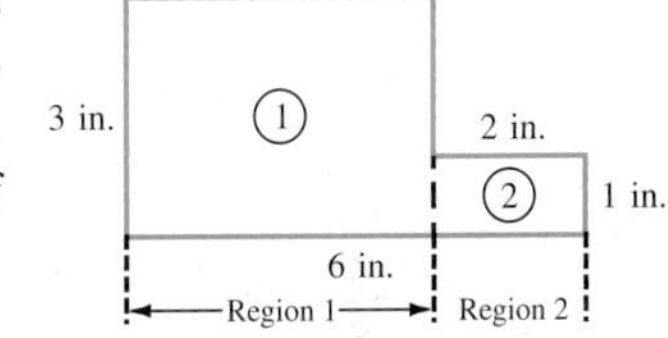

The total area is the sum of the two areas:

Total area = 12 in.^{2} + 2 in.^{2} = 14 in.^{2}

Check Yourself 17

Find the area of the figure.

Hint: You can find the area by adding the areas of three rectangles, or by subtracting the area of the "missing" rectangle from the area of the "completed" larger rectangle.

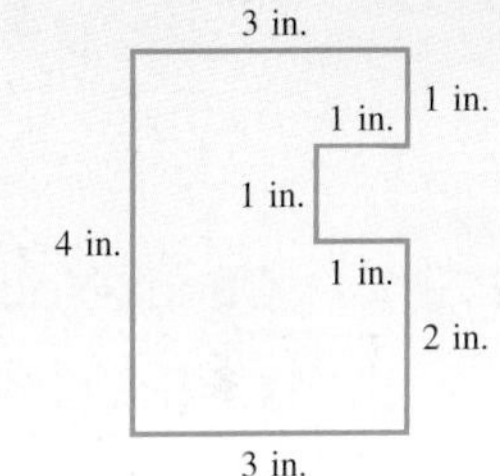

Next, we find the **volume of a solid.** The volume of a solid measures the amount of space contained by the solid.

Definition

Solid A **solid** is a three-dimensional figure. It has length, width, and height.

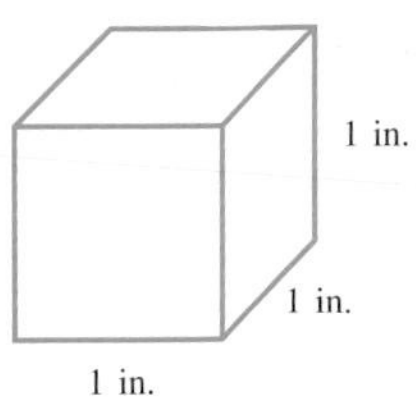

1 cubic inch

Volume is measured in **cubic units.** Examples include cubic inches (in.^{3}), cubic feet (ft^3), and cubic centimeters (cm^3). A cubic inch, for instance, is the measure of the space contained in a cube that is 1 in. on each edge. See the figure to the left.

To find the volume of a figure, we want to know how many cubic units are contained in that figure. Let's start with a simple example, a **rectangular solid.** A rectangular solid is a very familiar figure. A box, a crate, and most rooms are rectangular solids. Say that the dimensions of the solid are 5 in. by 3 in. by 2 in. as pictured.

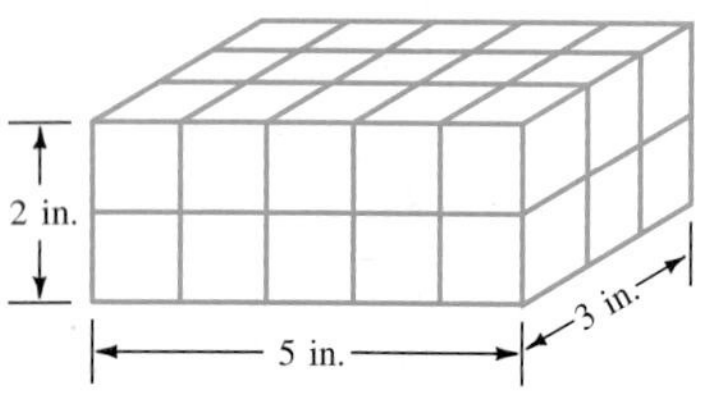

If we divide the solid into units of 1 in.^{3}, we have two layers, each containing 3 units by 5 units, or 15 in.^{3} Because there are two layers, the volume is 30 in.^{3}

In general, the volume of a rectangular solid is the product of its length, width, and height.

Property

Formula for the Volume of a Rectangular Solid

The volume V of a rectangular solid with length L, width W, and height H is given by the product of the length, width, and height.

$V = L \cdot W \cdot H$

Volume is measured in cubic units.

Example 18 Finding the Volume

A crate has dimensions 4 ft by 2 ft by 3 ft. Find its volume.

We use the volume formula with $L = 4$ ft, $W = 2$ ft, and $H = 3$ ft.

$$V = L \cdot W \cdot H$$
$$= 4 \text{ ft} \times 2 \text{ ft} \times 3 \text{ ft}$$
$$= 24 \text{ ft}^3$$

The volume of the crate is 24 cubic feet.

Check Yourself 18

A room is 15 ft long, 10 ft wide, and 8 ft high. What is its volume?

Check Yourself ANSWERS

1. Factors 2, 9; product 18 **2.** 24 **3. (a)** 6; **(b)** 0; **(c)** 1; **(d)** 74; **(e)** 0 **4.** $3 \times 7 = 21$; $15 + 6 = 21$ **5. (a)** 204; **(b)** 301 **6.** $435 **7.** 2,888 **8.** 87,648 **9. (a)** 148,656; **(b)** 2,426,496; **(c)** 1,258,192 **10. (a)** 90; **(b)** 90 **11. (a)** 25,700; **(b)** 2,436,000 **12.** 172,200 **13.** 280,000 **14.** $500 **15.** 1,250 in.^{2} **16.** $192 **17.** 11 in.^{2} **18.** 1,200 ft^3

Reading Your Text

These fill-in-the-blank exercises will help you understand some of the key vocabulary used in this section. The answers to these exercises are in the Answers Appendix at the back of the text.

(a) The final step in solving an application is to make certain that the answer is ________.

(b) The way in which you group numbers in multiplication does not affect the final ____________.

(c) The numbers 10, 100, and 1,000 are called ____________ of 10.

(d) We can write a formula for the area of a ____________ as $A = L \cdot W$.

Skills | Calculator/Computer | Career Applications | Above and Beyond

1.5 exercises

< Objective 1 >

Multiply.

1. $\begin{array}{r} 5 \\ \times 3 \\ \hline \end{array}$
2. $\begin{array}{r} 7 \\ \times 4 \\ \hline \end{array}$
3. $6 \cdot 0$
4. $6 \cdot 6$
5. 4×48
6. 5×53
7. $\begin{array}{r} 508 \\ \times \quad 6 \\ \hline \end{array}$
8. $\begin{array}{r} 903 \\ \times \quad 9 \\ \hline \end{array}$
9. 75(68)
10. 235(49)
11. 327×59
12. $2{,}364 \cdot 67$
13. $4{,}075 \cdot 84$
14. $\begin{array}{r} 315 \\ \times 243 \\ \hline \end{array}$
15. $\begin{array}{r} 124 \\ \times 225 \\ \hline \end{array}$
16. 345×267
17. 639×358
18. $547 \cdot 203$
19. $668 \cdot 305$
20. $\begin{array}{r} 2{,}458 \\ \times \quad 135 \\ \hline \end{array}$
21. $\begin{array}{r} 3{,}219 \\ \times \quad 207 \\ \hline \end{array}$
22. (2,534)(3,106)
23. (3,158)(2,034)
24. $\begin{array}{r} 43 \\ \times 70 \\ \hline \end{array}$
25. $\begin{array}{r} 58 \\ \times 40 \\ \hline \end{array}$
26. $\begin{array}{r} 562 \\ \times 400 \\ \hline \end{array}$
27. $\begin{array}{r} 907 \\ \times 900 \\ \hline \end{array}$
28. $\begin{array}{r} 345 \\ \times 230 \\ \hline \end{array}$
29. $\begin{array}{r} 362 \\ \times 310 \\ \hline \end{array}$
30. $\begin{array}{r} 157 \\ \times 3{,}200 \\ \hline \end{array}$
31. (18)(1)
32. 1×132
33. 64×0
34. (0)(1)
35. Find the product of 304 and 7. 
36. Find the product of 8 and 5,679.
37. What is 21 multiplied by 551?
38. What is 135 multiplied by 507?

< Objective 2 >

Name the property of addition and/or multiplication that is illustrated.

39. $5 \times 8 = 8 \times 5$ 
40. $3 \times (4 + 9) = (3 \times 4) + (3 \times 9)$ 
41. $2 \times (3 \times 5) = (2 \times 3) \times 5$
42. $5 \times (6 + 2) = 5 \times (2 + 6)$
43. $5 \times 1 = 5$
44. $(8)(0) = 0$
45. $0 \times 21 = 0$
46. $1 \times 3{,}296 = 3{,}296$

Use the given property to complete each statement.

47. $7 + (3 \times 8) =$ Commutative property of multiplication
48. $3 \times (2 + 7) =$ Distributive property 

< Objective 3 >

Solve each application.

49. **Business and Finance** A convoy company can transport 8 new cars on one of its trucks. If 34 truck shipments were made in one week, how many cars were shipped?

50. **Business and Finance** A label writer can print 71 labels per minute. How many labels can be printed in 1 hour?

51. **Social Science** A rectangular parking lot has 14 rows of parking spaces, and each row contains 24 spaces. How many cars can be parked in the lot?

52. **Statistics** A petition to get Tom on the ballot for treasurer of student council has 28 signatures on each of 43 pages. How many signatures were collected?

53. **Business and Finance** A manufacturer of wood-burning stoves can make 15 stoves in 1 day. How many stoves can be made in 28 days?

54. **Business and Finance** Each bundle of newspapers contains 25 papers. If 43 bundles are delivered to Jose's house, how many papers are delivered?

55. **Science and Medicine** Sound waves travel at approximately 1,088 feet per second. If you hear thunder 15 seconds after seeing a lightning flash, how far away is the lightning flash?

56. **Science and Medicine** Sound waves travel at approximately 1,088 feet per second. If you hear thunder 23 seconds after seeing a lightning flash, how far away is the lightning flash?

57. **Business and Finance** A ream of paper contains 500 sheets. If an office copy machine goes through 29 reams in 1 week, how many sheets of paper did it use?

58. **Business and Finance** Celeste harvested 34 bushels of corn per acre from 32 acres in June and 43 bushels per acre from 36 acres in July. How many bushels did she harvest over the 2-month period?

Round each factor to the nearest ten and estimate the product.

59. $\begin{array}{r} 38 \\ \times\ 21 \\ \hline \end{array}$

60. $\begin{array}{r} 122 \\ \times\ 76 \\ \hline \end{array}$

Round each factor to the nearest hundred and estimate the product.

61. $\begin{array}{r} 391 \\ \times\ 531 \\ \hline \end{array}$

62. $\begin{array}{r} 729 \\ \times\ 481 \\ \hline \end{array}$

Solve each application.

63. **Social Science** A movie theater has its seats arranged so that there are 42 seats per row. The theater has 48 rows. Estimate the number of seats in the theater.

64. **Statistics** There are 52 mathematics classes with 28 students in each class. Estimate the total number of students in the mathematics classes.

65. **Business and Finance** A company can manufacture 45 sleds per day. Approximately how many can this company make in 128 days?

66. **Business and Finance** The attendance at a basketball game was 2,345. The cost of admission was $12 per person. Estimate the total gate receipts for the game.

< Objective 5 >

Find the area of each figure.

67.

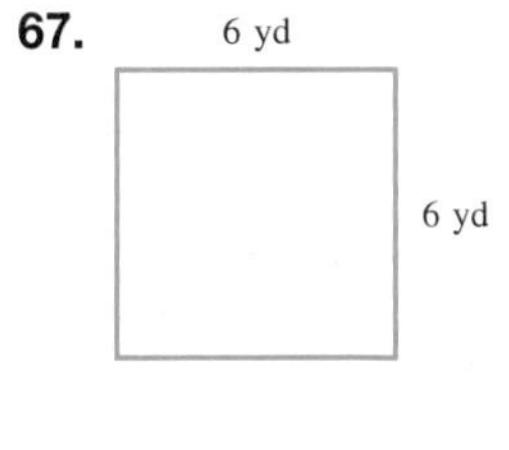

68.

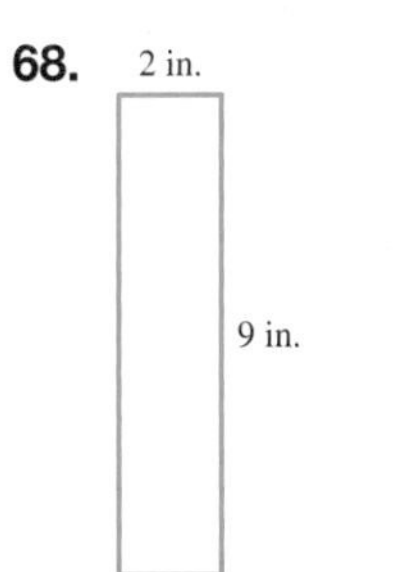

69.

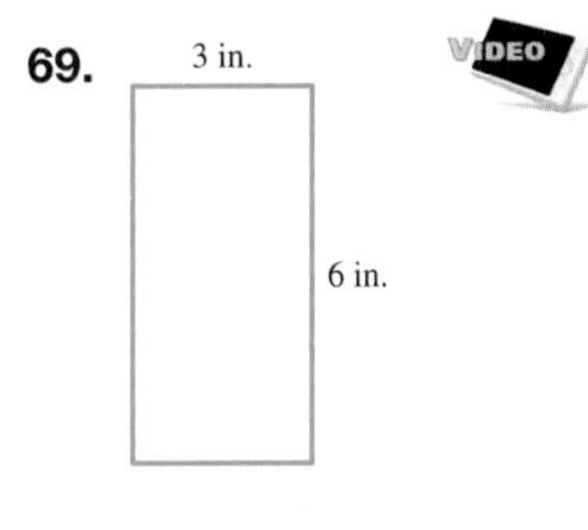

70.

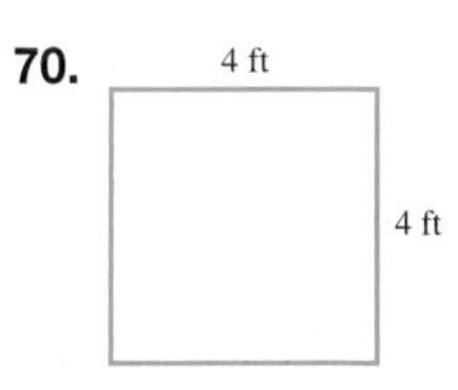

71.

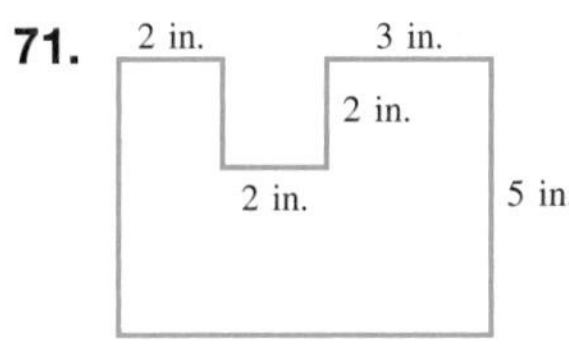

72.

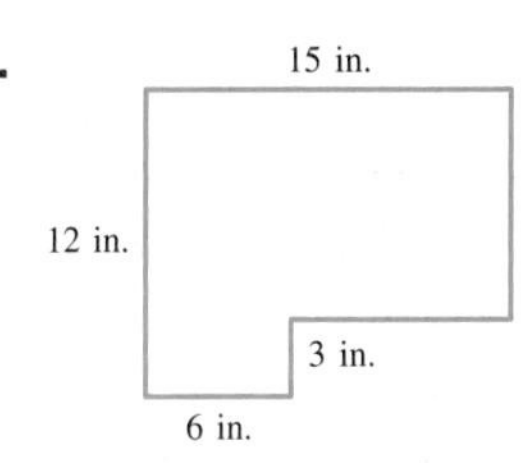

Find the volume of each figure.

73.

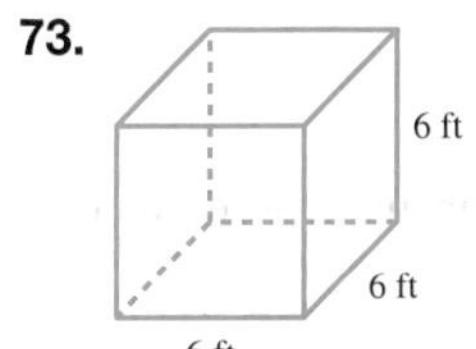

74.

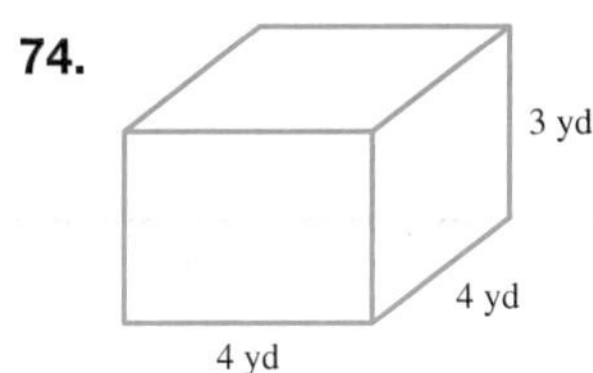

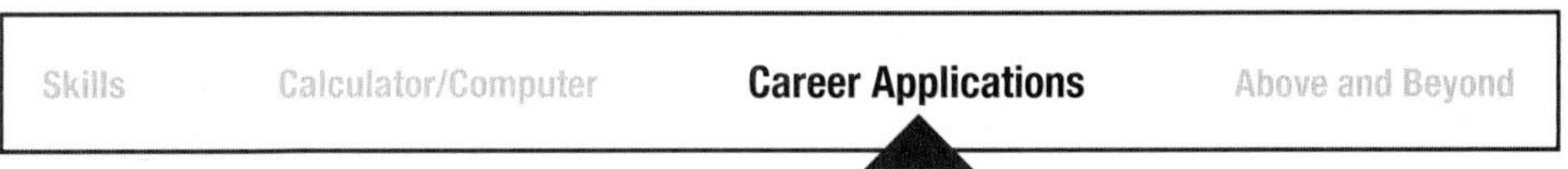

75. **Allied Health** A young male patient is to be administered an intravenous (IV) solution running on an infusion pump set for 125 mL per hour for 6 hours. What is the total volume of solution to be infused?

76. **Allied Health** To help assess breathing efficiency, respiratory therapists calculate the patient's alveolar minute ventilation, in milliliters per minute (mL/min), by taking the product of the patient's respiratory rate, in breaths per minute, and the difference between the patient's tidal volume and dead-space volume, both in milliliters. Calculate the alveolar minute ventilation for a male patient with lung disease given that his respiratory rate is 10 breaths per minute, his tidal volume is 575 mL, and his dead-space volume is 200 mL.

77. **Information Technology** Jack needs a thousand feet of twisted pair cable for a network installation project. He goes to a local electronics store. The store sells cable for 10¢ a foot. How much (in cents) will one thousand feet of cable cost? How much is that in dollars?

78. **Information Technology** Amber needs to buy 100 new computers for new employees that have been hired by ABC consulting. She finds the cost of a decent computer to be $655. How much will she spend to buy 100 new computers?

79. **Electronics** An electronics-component distributor sells resistors, small components that "resist" the flow of electric current, in presealed bags. Each bag contains 50 resistors. If you purchased 25 bags of resistors, how many resistors would you have?

80. **Electronics** Assume that it takes 4 hr to solder all the components on a given printed circuit board. If you are given 36 boards to solder, how many hours of work will the project take?

81. **Manufacturing Technology** A small shop has six machinists each earning $21 per hour, three assembly workers earning $12 per hour, and one supervisor-maintenance person earning $28 per hour. What is the shop's payroll for a 40-hr week?

82. **Manufacturing Technology** What is the distance from the center of hole A to the center of hole B in the diagram?

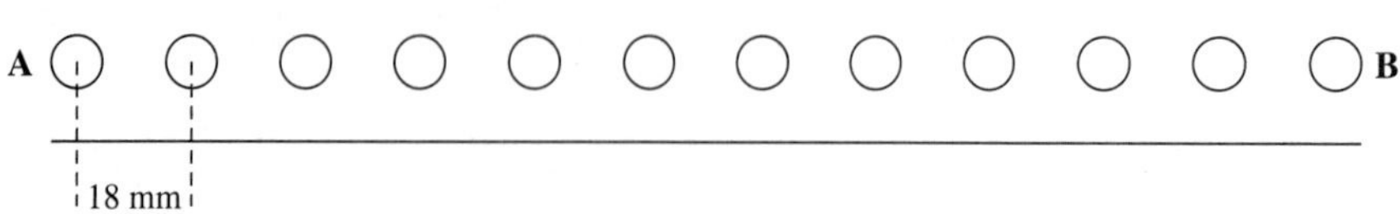

Skills | Calculator/Computer | Career Applications | **Above and Beyond**

83. **Social Science** We have seen that addition and multiplication are commutative operations. Decide which of these pairs of activities is commutative.
   (a) Taking a shower and eating breakfast
   (b) Getting dressed and taking a shower
   (c) Putting on your shoes and your socks
   (d) Brushing your teeth and combing your hair
   (e) Putting your key in the ignition and starting your car

84. **Social Science** The associative properties of addition and multiplication indicate that the result of the operation is the same regardless of where the grouping symbol is placed. This is not always the case in the use of the English language. Many phrases can have different meanings based on how the words are grouped. In each case, explain why the associative property would not hold.
   (a) Cat fearing dog
   (b) Hard test question
   (c) Defective parts department
   (d) Man eating animal

   Write some phrases in which the associative property is satisfied.

85. **Construction** Suppose you wish to build a small, rectangular pen, and you have enough fencing for the pen's perimeter to be 36 ft. Assuming that the length and width are to be whole numbers, complete each exercise.
   (a) List the possible dimensions that the pen could have. (**Note:** A square is a type of rectangle.)
   (b) For each set of dimensions (length and width), find the area that the pen would enclose.
   (c) Which dimensions give the greatest area?
   (d) What is the greatest area?

86. **Construction** Suppose you wish to build a rectangular kennel that encloses 100 ft^2. Assuming that the length and width are to be whole numbers, complete each exercise.
   (a) List the possible dimensions that the kennel could have. (**Note:** A square is a type of rectangle.)
   (b) For each set of dimensions (length and width), find the perimeter that would surround the kennel.
   (c) Which dimensions give the least perimeter?
   (d) What is the least perimeter?

87. **Social Science** Most maps contain legends that allow you to convert the distance between two points on the map to actual miles. For instance, if a map uses a legend that equates 1 in. to 5 mi and the distance between two towns is 4 in. on the map, then the towns are actually 20 mi apart.
   (a) Obtain a map of your state and determine the shortest distance between any two major cities.
   (b) Could you actually travel the route you measured in part (a)?
   (c) Plan a trip between the two cities you selected in part (a) over established roads. Determine the distance that you actually travel using this route.

88. NUMBER PROBLEM Complete the following number cross.

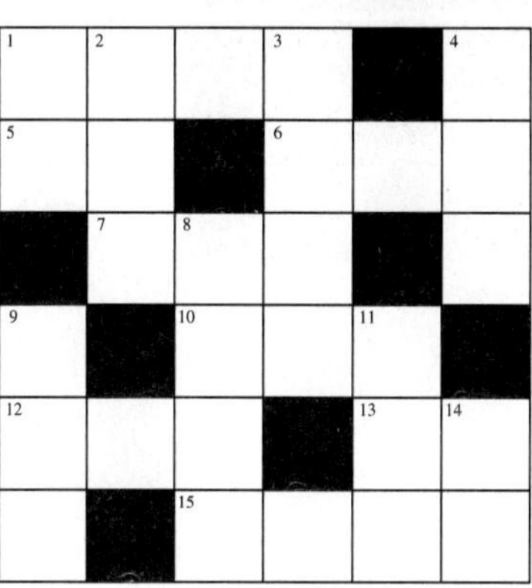

Across

1. 6×551
5. 7×8
6. 27×27
7. 19×50
10. 3×67
12. 6×25
13. 9×8
15. 16×303

Down

1. 5×7
2. 9×41
3. 67×100
4. $2 \times (49 + 100)$
8. $4 \times 1{,}301$
9. $100 + 10 + 1$
11. 2×87
14. $25 + 3$

Answers

1. 15 **3.** 0 **5.** 192 **7.** 3,048 **9.** 5,100 **11.** 19,293 **13.** 342,300 **15.** 27,900 **17.** 228,762 **19.** 203,740 **21.** 666,333 **23.** 6,423,372 **25.** 2,320 **27.** 816,300 **29.** 112,220 **31.** 18 **33.** 0 **35.** 2,128 **37.** 11,571 **39.** Commutative property of multiplication **41.** Associative property of multiplication **43.** Multiplicative identity property **45.** Multiplicative property of 0 **47.** $7 + (8 \times 3)$ **49.** 272 cars **51.** 336 cars **53.** 420 stoves **55.** 16,320 ft **57.** 14,500 sheets **59.** 800 **61.** 200,000 **63.** 2,000 seats **65.** 6,500 sleds **67.** 36 yd^2 **69.** 18 $in.^2$ **71.** 31 $in.^2$ **73.** 216 ft^3 **75.** 750 mL **77.** 10,000¢; $100 **79.** 1,250 resistors **81.** $7,600 **83.** Above and Beyond **85.** Above and Beyond **87.** Above and Beyond

1.6 Division

< 1.6 Objectives >

1 > Write a division problem as repeated subtraction

2 > Divide whole numbers

3 > Estimate quotients

4 > Use division to solve applications

Now we examine a fourth arithmetic operation, division. Just as multiplication is repeated addition, division is repeated subtraction. Division asks *how many times* one number is contained in another.

Example 1 Dividing by Using Subtraction

< Objective 1 >

NOTE

We subtracted 8 six times.

Joel needs to set up 48 chairs in the student union for a concert. If there is room for 8 chairs per row, how many rows will it take to set up all 48 chairs?

This problem can be solved by subtraction. Each row subtracts another 8 chairs.

48	40	32	24	16	8
−8	−8	−8	−8	−8	−8
40	32	24	16	8	0

Because 8 can be subtracted from 48 six times, there will be 6 rows.

This can also be seen as a division problem. Three different notations are commonly used to indicate division. All three are shown here.

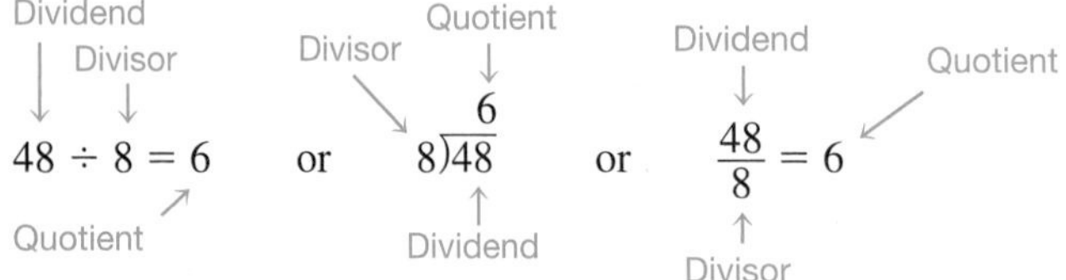

No matter which notation we use, we call 48 the **dividend,** 8 the **divisor,** and 6 the **quotient.**

Check Yourself 1

Carlotta is creating a garden path made of bricks. She has 72 bricks. Each row will have 6 bricks in it. How many rows can she make?

Units Analysis

When you divide a denominate number by an abstract number, the result has the units of the denominate number. Here are a couple of examples:

76 trombones ÷ 4 = 19 trombones

$55 ÷ 11 = $5

When one denominate number is divided by another, the result has the units of the dividend over the units of the divisor.

144 mi ÷ 6 gal = 24 mi/gal (which we read as "miles per gallon")

$120 ÷ 8 hr = 15 dollars/hr ("dollars per hour")

To solve a problem requiring division, first set up the problem as a division statement. Example 2 illustrates this idea.

Example 2 Writing a Division Statement

Write a division statement that corresponds to the following situation. You need not do the division.

The staff at the Wok Inn Restaurant splits all tips at the end of each shift. Yesterday's evening shift collected a total of $224. How much should each of the seven employees get in tips?

$224 ÷ 7 employees The units of the answer are "dollars per employee."

Check Yourself 2

Write a division statement that corresponds to the following situation. You need not do the division.

All nine sections of basic math skills at SCC (Sum Community College) are full. There are a total of 315 students in the classes. How many students are in each class? What are the units for the answer?

In Section 1.5, we used a rectangular array of stars to represent multiplication. These same arrays can represent division. Just as $3 \times 4 = 12$ and $4 \times 3 = 12$, so is it true that $12 \div 3 = 4$ and $12 \div 4 = 3$.

```
* * *
* * *           * * * *
* * *  4 × 3 = 12     * * * *  3 × 4 = 12
* * *           * * * *
  or              or
12 ÷ 3 = 4        12 ÷ 4 = 3
```

This relationship allows us to check our division results by doing multiplication.

Example 3 — Checking Division by Using Multiplication

< Objective 2 >

(a) $\begin{array}{r}3\\7\overline{)21}\end{array}$ Check: $7 \times 3 = 21$

(b) $48 \div 6 = 8$ Check: $6 \times 8 = 48$

NOTES

For a division problem to check, the *product* of the divisor and the quotient *must equal the dividend.*

Because $36 \div 9 = 4$, we say that 36 is *exactly divisible* by 9.

Check Yourself 3

Complete the division statements and check your results.

(a) $9\overline{)45}$ (b) $28 \div 7$

In our examples so far, the product of the divisor and the quotient has been equal to the dividend. This means that the dividend is *exactly divisible* by the divisor. That is not always the case. Look at another example of repeated subtraction.

Example 4 — Dividing by Using Subtraction, Leaving a Remainder

How many times is 5 contained in 23?

$$\begin{array}{r}23\\-\ 5\\\hline 18\end{array}\qquad\begin{array}{r}18\\-\ 5\\\hline 13\end{array}\qquad\begin{array}{r}13\\-\ 5\\\hline 8\end{array}\qquad\begin{array}{r}8\\-\ 5\\\hline 3\end{array}$$

We see that 5 is contained four times in 23, but 3 is "left over."

NOTE

The remainder must be smaller than the divisor or we could subtract again.

We see that 23 is not exactly divisible by 5. It contains 5 four times, but there is still 3 left over. This left over is called the **remainder** in division. We write

$23 \div 5 = 4\text{ r}3$

Check Yourself 4

How many times is 7 contained in 38?

We use the next property to check the result of division when there is a remainder.

Property

Remainder	Dividend = divisor × quotient + remainder

Example 5 — Checking Division by a Single-Digit Number

Using the work of Example 4, we can write

$\begin{array}{r}4\\5\overline{)23}\end{array}$ with remainder 3

We write the result as 4 r3.

To check our work, we use the remainder definition.

NOTE

We multiply 5×4 before adding 3 to the result.

$5 \times 4 + 3 = \underbrace{20}_{5\times4} + 3 = 23$

You will learn more about the *order of operations* in Section 1.7.

$$\begin{aligned}\text{Dividend} \rightarrow 23 &= \overset{\text{Divisor}}{5} \times \overset{\text{Quotient}}{4} + 3 \leftarrow \text{Remainder}\\ 23 &= 20 + 3\\ 23 &= 23 \quad \text{The division checks.}\end{aligned}$$

Check Yourself 5

Evaluate $7\overline{)38}$. Check your answer.

We must be careful when 0 is involved in a division problem. There are two special cases.

Property

Division and Zero

1. Zero divided by any whole number (except 0) is 0.
2. Division by 0 is undefined.

The first case involving zero occurs when we are dividing into zero.

Example 6 **Dividing into Zero**

RECALL

There are three forms that are equivalent:

$0 \div 5 = 0$

$5\overline{)0} = 0$

$\frac{0}{5} = 0$

$0 \div 5 = 0$ because $0 = 5 \times 0$.

Check Yourself 6

(a) $0 \div 7$ **(b)** $9\overline{)0}$ (c) $\frac{0}{12}$

Our second case illustrates what happens when 0 is the *divisor.* Here we have a special problem.

Example 7 **Dividing by Zero**

$8 \div 0 = ?$ This means that $8 = 0 \times ?$

Can 0 times some number ever be 8? From our multiplication facts, we know the answer is *no!* There is no answer to this problem, so we say that $8 \div 0$ is undefined.

Check Yourself 7

Decide whether each problem results in 0 or is undefined.

(a) $\frac{9}{0}$ **(b)** $\frac{0}{9}$ **(c)** $15\overline{)0}$ **(d)** $0\overline{)15}$

It is easy to divide when small whole numbers are involved, because much of the work can be done mentally. In working with larger numbers, we turn to a process called **long division.** This is a shorthand method for performing the steps of repeated subtraction.

To start, look at an example in which we subtract multiples of the divisor.

Example 8 **Dividing by a Single-Digit Number**

NOTE

With larger numbers, repeated subtraction is just too time-consuming to be practical.

Divide 176 by 8.

Because 20 eights are 160, we know that there are at least 20 eights in 176.

Step 1 Write

$$\begin{array}{r} 20 \\ 8\overline{)176} \\ \text{20 eights} \longrightarrow \underline{160} \\ 16 \end{array}$$

Subtracting 160 is just a shortcut for subtracting eight 20 times.

After subtracting the 20 eights, or 160, we are left with 16. There are 2 eights in 16, and so we continue.

Step 2

```
     2 }
    20 }  22   Adding 20 and 2 gives us the quotient, 22.
  8)176
    160
     16
2 eights ⟶ 16
      0
```

Subtracting the 2 eights, we have a 0 remainder. So $176 \div 8 = 22$.

Check Yourself 8

Verify the result of Example 8, using multiplication.

The next step is to simplify this repeated-subtraction process a little more. The result is called *long division.*

Example 9 Dividing by a Single-Digit Number

Divide 358 by 6.

The dividend is 358. We look at the first digit, 3. We cannot divide 6 into 3, and so we look at the *first two digits,* 35. There are 5 sixes in 35, and so we write 5 above the tens digit of the dividend.

```
   5      When we place 5 as the tens digit,
 6)358    we really mean 5 tens, or 50.
```

Now multiply 5×6, place the product below 35, and subtract.

```
   5
 6)358
   30     We actually subtracted 50 sixes
    5     (300) from 358.
```

Because the remainder, 5, is smaller than the divisor, 6, we bring down 8, the ones digit of the dividend.

```
   5
 6)358
   30↓
    58
```

Now divide 6 into 58. There are 9 sixes in 58, and so 9 is the ones digit of the quotient. Multiply 9×6 and subtract to complete the process.

```
   59     We now have
 6)358    358 ÷ 6 = 59 r4
   30↓
    58
    54
     4
```

NOTES

Because 4 is smaller than the divisor, we have a remainder of 4.

Verify that this is true and that the division checks.

To check: $358 = 6 \times 59 + 4$

Check Yourself 9

Divide $7\overline{)453}$.

Long division becomes a bit more complicated when we have a two-digit divisor. It is now a matter of trial and error. We round the divisor and dividend to form a *trial divisor and a trial dividend.* We then estimate the proper quotient and determine whether our estimate was correct.

Example 10 Dividing by a Two-Digit Number

< Objective 3 >

NOTE

Think: $4\overline{)29}$ → 7

Divide.

$38\overline{)293}$

Round the divisor and dividend to the nearest ten. So 38 is rounded to 40, and 293 is rounded to 290. The trial divisor is then 40, and the trial dividend is 290.

Now look at the nonzero digits in the trial divisor and dividend. They are 4 and 29. We know that there are 7 fours in 29, and so 7 is our first estimate of the quotient. Now we see if 7 works.

```
      7  ←—— Your estimate
 38)293
    266       Multiply 7 × 38. The product, 266, is less
    ———       than 293, so we can subtract.
     27
```

The remainder, 27, is less than the divisor, 38, so our estimate was correct. In this case, the process is complete.

$293 \div 38 = 7 \text{ r}27$

Check: $293 = 38 \times 7 + 27$ You should verify that this statement is true.

Check Yourself 10

Divide.

$57\overline{)482}$

Because this process is based on estimation, our first guess may be incorrect.

Example 11 Dividing by a Two-Digit Number

NOTE

Think: $5\overline{)43}$ → 8

Divide.

$54\overline{)428}$

Rounding to the nearest ten, we have a trial divisor of 50 and a trial dividend of 430.

If you look at the nonzero digits, how many fives are in 43? There are 8. This is our first estimate.

```
      8
 54)428
    432  ←—— Too large
```

We multiply 8 × 54. Do you see what's wrong? The product, 432, is too large. We can't subtract. Our estimate of the quotient must be adjusted *downward.*

We adjust the quotient downward to 7. We can now complete the division.

```
      7
 54)428
    378
    ———
     50
```

We have

$428 \div 54 = 7 \text{ r}50$

Check: $428 = 54 \times 7 + 50$

Check Yourself 11

Divide.

$63\overline{)557}$

We must be careful when a 0 appears as a digit in the quotient. In Example 12, this happens with a two-digit divisor.

Example 12 Dividing with Large Dividends

NOTE

Our divisor, 32, divides into 98, the first two digits of the dividend.

Divide.

32)9,871

Rounding to the nearest ten, we have a trial divisor of 30 and a trial dividend of 100. Think, "How many threes are in 10?" There are 3, and this is our first estimate of the quotient.

```
     3
32)9,871
   9 6
     2
```

Everything seems fine so far!

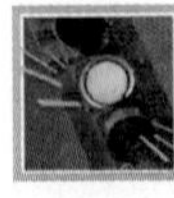

>CAUTION

You must remember to put 0 in the quotient, in the tens place in this case.

Bring down 7, the next digit of the dividend.

```
     30
32)9,871
   9 6↓
     27
```

Now do you see the difficulty? We cannot divide 32 into 27, and so we place 0 in the tens place of the quotient to indicate this fact.

We continue by multiplying by 0. After subtraction, we bring down 1, the last digit of the dividend.

```
     30
32)9,871
   9 6
     27
     00↓
     271
```

Another problem develops here. We round 32 to 30 for our trial divisor, and we round 271 to 270, which is the trial dividend at this point. Our estimate of the last digit of the quotient must be 9.

```
     309
32)9,871
   9 6
     27
     00
     271
     288  ← Too large
```

We cannot subtract because 288 is larger than 271. The trial quotient must be adjusted downward to 8. We can now complete the division.

```
     308
32)9,871
   9 6
     27
     00
     271
     256
      15
```

>CAUTION

Do not forget the 0 in the quotient. The correct answer is 308 r15 *not* 38 r15!

$9{,}871 \div 32 = 308 \text{ r}15$

Check: $9{,}871 = 32 \times 308 + 15$

Check Yourself 12

Divide.

$43\overline{)8{,}857}$

We use the same approach when we have large divisors. In most cases, people use a calculator with three-digit divisors, but occasionally, you need to use long division to perform such a computation by hand.

Example 13 Long Division with Three-Digit Divisors

Divide.

$638{,}976 \div 312$

Begin by putting the problem into long-division form.

$312\overline{)638{,}976}$

Using the rounding technique, we round the divisor to 300. We see that 300 does not go into 6 or 63, so we need to begin with 638. We round this to 600 and estimate that we need two 300s to divide into 600.

```
        2
312)638,976
    624↓
     149
```

Because we cannot divide 149 by 312, we need to place a 0 in the next slot in the quotient and bring down the 7, as well.

```
        2,0
312)638,976
    624  ↓
     1497
```

We round to 300 and 1,500 and estimate that the next number in the quotient is 5.

```
        2,05
312)638,976
    624
     1497
     1560 ←— Too large
```

Because $5 \times 312 = 1{,}560$ is larger than 1,497, we adjust the quotient downward and try 4.

```
        2,04
312)638,976
    624
     1497
     1248
      249
```

That worked. Now we bring down the 6, and complete the division.

```
        2,04
312)638,976
    624
     1497
     1248↓
      2496
```

We estimate that 300 goes into 2,500 eight times.

```
       2,048
312)638,976
    624
     1497
     1248
      2496
      2496
         0
```

There is no remainder, so 638,976 ÷ 312 = 2,048.

Check Yourself 13

Divide.

3,391,365 ÷ 491

Because of the availability of calculators, it is rarely necessary that people find the exact answer when performing long division. On the other hand, it is frequently important to estimate a quotient or confirm that a given answer (particularly from a calculator) is reasonable. As a result, the emphasis in this section is on improving your estimation skills in division.

Let's divide a four-digit number by a two-digit number. We round the divisor to the nearest ten and the dividend to the nearest hundred.

Example 14 Estimating the Result of a Division Application

< Objective 4 >

NOTE

We often see miles per hour written as *mph* and miles per gallon as *mpg*. In this text, we usually use mi/hr and mi/gal to emphasize the relationship between the units used.

The Ramirez family took a trip of 2,394 mi in their new car, using 77 gal of gas. Estimate their gas mileage (mi/gal).

Our estimate is based on dividing 2,400 by 80.

```
    30
80)2,400
```

They got approximately 30 mi/gal.

Check Yourself 14

Troy flew a light plane on a trip of 2,844 mi that took 21 hr. What was his approximate speed in miles per hour (mi/hr)?

As before, we may have to combine operations to solve an application of the mathematics you have learned.

Example 15 Estimating the Result of a Division Application

Charles purchases a used car for \$8,574. The total interest on the loan comes to \$978. He agrees to make payments for 4 years. Approximately what are his monthly payments?

First, we find the amount that Charles owes:

\$8,574 + \$978 = \$9,552

Now, to find the monthly payment, we divide that amount by 48 (months). To estimate the payment, we divide \$9,600 by 50 months.

$$50\overline{)9{,}600}\quad 192$$

The payments are approximately \$192 per month.

Check Yourself 15

One \$10 bag of fertilizer covers 310 ft². Approximately what does it cost to cover 2,145 ft²?

Check Yourself ANSWERS

1. 12 rows **2.** 315 students ÷ 9 classes; students per class **3.** **(a)** 5; 9 × 5 = 45; **(b)** 4; 7 × 4 = 28 **4.** 5 **5.** 5 r3; 38 = 5 × 7 + 3 **6.** **(a)** 0; **(b)** 0; **(c)** 0 **7.** **(a)** Undefined; **(b)** 0; **(c)** 0; **(d)** undefined **8.** 8 × 22 = 176 **9.** 64 r5 **10.** 8 r26 **11.** 8 r53 **12.** 205 r42 **13.** 6,907 r28 **14.** 140 mi/hr **15.** \$70

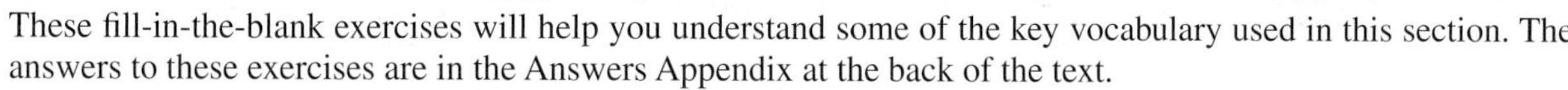

Reading Your Text

These fill-in-the-blank exercises will help you understand some of the key vocabulary used in this section. The answers to these exercises are in the Answers Appendix at the back of the text.

(a) The result from division is called the ______________.

(b) The dividend is equal to the divisor times the quotient plus any ______________.

(c) Zero divided by any whole number (except ______________) is 0.

(d) Division is a shortened form for repeated ______________.

Skills | Calculator/Computer | Career Applications | Above and Beyond

1.6 exercises

1. In 48 ÷ 8 = 6, 8 is the _______, 48 is the _______, and 6 is the _______.

2. In the statement $5\overline{)45}$ with quotient 9, 9 is the ________, 5 is the ________, and 45 is the _______.

< Objective 1 >

3. Find 36 ÷ 9 by repeated subtraction.

4. Find 40 ÷ 8 by repeated subtraction.

5. **Crafts** Stefanie is planting rows of tomato plants. She wants to plant 63 plants with 9 plants per row. How many rows does she need? VIDEO

6. **Construction** Nick is designing a parking lot for a small office building. He must make room for 42 cars with 7 cars per row. How many rows should he plan for?

1.6 exercises

< Objective 2 >

Divide. Include the correct units in your result.

7. 36 pages ÷ 4

8. \$96 ÷ 8

9. 4,900 kilometers (km) ÷ 7

10. 360 gal ÷ 18

11. 160 mi ÷ 4 hr 

12. 264 ft ÷ 3 s

13. 3,720 hr ÷ 5 months (mo)

14. 560 calories (cal) ÷ 7 grams (g)

Divide and check your work.

15. 54 ÷ 9

16. 21 ÷ 3

17. $6\overline{)42}$

18. $7\overline{)63}$

19. $4\overline{)32}$

20. 56 ÷ 8

21. $5\overline{)43}$

22. 40 ÷ 9

23. $9\overline{)65}$

24. $6\overline{)51}$

25. 57 ÷ 8

26. 74 ÷ 8

27. 0 ÷ 5

28. 5 ÷ 0

29. 4 ÷ 0

30. 0 ÷ 12

31. 0 ÷ 6

32. 18 ÷ 0

33. $5\overline{)83}$

34. $9\overline{)78}$

35. 162 ÷ 3

36. 232 ÷ 4

37. $\frac{293}{8}$

38. $\frac{346}{7}$

39. $8\overline{)3{,}136}$

40. $5\overline{)4{,}938}$

41. 5,438 ÷ 8

42. 3,527 ÷ 9

43. $\frac{22{,}153}{8}$

44. $\frac{43{,}287}{5}$

45. $45\overline{)2{,}367}$

46. $53\overline{)3{,}480}$

47. 8,748 ÷ 34

48. 9,335 ÷ 27

49. $\frac{7{,}902}{42}$

50. $\frac{8{,}729}{53}$

51. 1,672 ÷ 8

52. 4,328 ÷ 14

53. $\frac{46{,}653}{53}$

54. $\frac{28{,}132}{26}$

55. 6,720 ÷ 280

56. 11,578 ÷ 642

57. $\frac{125{,}580}{156}$

58. $\frac{1{,}936{,}572}{748}$

59. $245\overline{)857{,}990}$

60. $125\overline{)1{,}510{,}112}$

< Objective 3 >

Estimate the result of each division problem by rounding divisors to the nearest ten and dividends to the nearest hundred.

61. 810 divided by 38

62. 458 divided by 18

63. 4,967 divided by 96

64. 3,971 divided by 39

65. 8,971 divided by 91

66. 3,981 divided by 78

67. 3,879 divided by 126

68. 8,986 divided by 178

69. 3,812 divided by 188

70. 5,245 divided by 255

< Objective 4 >

Solve each application.

71. **Business and Finance** There are 63 candy bars in seven boxes. How many candy bars are in each box?

72. **Business and Finance** A total of 54 printers were shipped to nine stores. How many printers were shipped to each store?

73. **Social Science** Joaquin is putting pictures in an album. He can fit 8 pictures on each page. If he has 77 pictures, how many will be left over after he has filled the last 8-picture page?

74. **Statistics** Kathy is separating a deck of 52 cards into six equal piles. How many cards will be left over?

75. **Social Science** The records of an office show that 1,702 calls were made in 1 day. If there are 37 phones in the office, how many calls were placed per phone?

76. **Business and Finance** A television dealer purchased 23 sets, each the same model, for $5,267. What was the cost of each set?

77. **Business and Finance** A printer can print 340 lines per minute (min). How long will it take to complete a report of 10,880 lines?

78. **Statistics** A train traveled 1,364 mi in 22 hr. What was the speed of the train? *Hint:* Speed is the distance traveled divided by the time.

Skills | **Calculator/Computer** | Career Applications | Above and Beyond

Use your calculator to perform the indicated operations.

79. $583{,}467 \div 129$

80. $464{,}184 \div 189$

81. $6 + 9 \div 3$

82. $18 - 6 \div 3$

83. $24 \div 6 \times 4$

84. $32 \div 8 \times 4$

85. $4{,}368 \div 56 + 726 \div 33$

86. $1{,}176 \div 42 - 1{,}572 \div 524$

87. $3 \times 8 \times 8 \times 8 \div 12$

88. $5 \times 6 \times 6 \div 18$

Skills | Calculator/Computer | **Career Applications** | Above and Beyond

89. **Allied Health** A doctor prescribes 525 milligrams (mg) of bexarotene to be given once daily to an adult female patient with lung cancer. How many pills should the nurse give her if each pill contains 75 mg of bexarotene?

90. **Allied Health** Determine the flow rate, in milliliters per hour (mL/hr), needed for an electronic infusion pump to administer 180 mL of a saline solution via intravenous (IV) infusion over the course of 12 hr.

91. **Information Technology** Marcela is in the process of building a new testing computer lab for ABC software. This lab has five rows of computers and nine computers per row. The distance from each computer to a switch is about 4 ft. She has a 200-ft roll of cable for the job. How many 4-ft cables can she make with a 200-ft roll? Does she have enough cable on the roll for the job?

92. **Information Technology** A computer connection transmits at 56,000 kilobits per second. How long will it take to transmit five hundred sixty thousand kilobits?

93. **Electronics** An electronics component distributor sells resistors in two package sizes. One package contains 500 resistors and the other contains 1,250 resistors. If you need 10,000 resistors, how many packages would you need to buy if you bought all small packages? All large packages?

94. **Electronics** A vendor that makes small-quantity batches of printed circuit boards sells 25 boards for $400. What is the cost per board?

95. **Manufacturing Technology** An order of 24 parts weighs 1,752 lb. Assuming that the parts are identical, how much does each part weigh?

96. **Manufacturing Technology** Triplet Precision Machining has a 3,000-gal liquid petroleum (LP) tank. If the cutting line consumes 125 gal of LP each day, how many days will the LP supply last?

97. **Construction** You are going to recarpet your living room. You have budgeted \$2,000 for the carpet and installation.

(a) Determine how much carpet you need for the job. Draw a sketch to support your measurements.

(b) What is the highest price per square yard you can pay and still stay within budget?

(c) Go to a local store or website and determine the total cost of doing the job for three different grades of carpet. Be sure to include padding, labor costs, and any other expenses.

(d) What considerations (other than cost) would affect your decision about what type of carpet to install?

(e) Write a brief paragraph indicating your final decision and give supporting reasons.

98. **Social Science** Division is the inverse operation of multiplication. Many daily activities have inverses. For each of the following activities, state the inverse activity.

(a) Spending money

(b) Going to sleep

(c) Turning down the volume on your CD player

(d) Getting dressed

99. If you have no money in your pocket and want to divide it equally among your four friends, how much does each person get? Use this situation to explain division of zero by a nonzero number.

100. **Number Problem** Complete the following number cross.

Across

1. $48 \div 4$
3. $1{,}296 \div 8$
6. $2{,}025 \div 5$
8. 4×5
9. 11×11
12. $15 \div 3 \times 111$
14. $144 \div (2 \times 6)$
16. $1{,}404 \div 6$
18. $2{,}500 \div 5$
19. 3×5

Down

1. $(12 + 16) \div 2$
2. 67×3
4. $744 \div 12$
5. $2{,}600 \div 13$
7. $6{,}300 \div 12$
10. $304 \div 2$
11. $5 \times (161 \div 7)$
13. $9{,}027 \div 17$
15. $400 \div 20$
17. 9×5

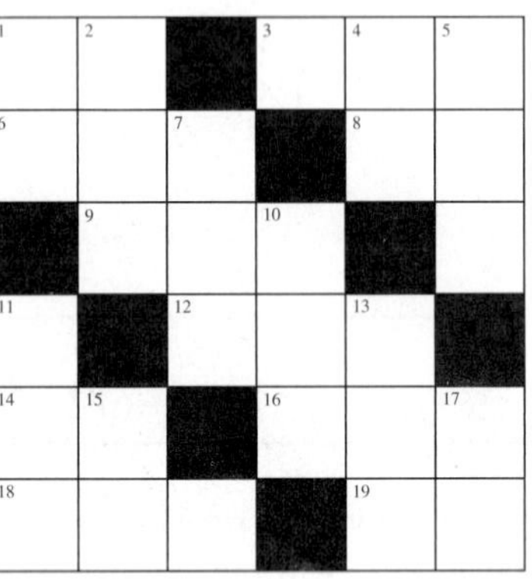

Answers

1. divisor, dividend, quotient **3.** 4 **5.** 7 **7.** 9 pages **9.** 700 km **11.** 40 mi/hr **13.** 744 hr/month **15.** 6 **17.** 7 **19.** 8 **21.** 8 r3 **23.** 7 r2 **25.** 7 r1 **27.** 0 **29.** Undefined **31.** 0 **33.** 16 r3 **35.** 54 **37.** 36 r5 **39.** 392 **41.** 679 r6 **43.** 2,769 r1 **45.** 52 r27 **47.** 257 r10 **49.** 188 r6 **51.** 209 **53.** 880 r13 **55.** 24 **57.** 806 r84 **59.** 3,502 **61.** 20 **63.** 50 **65.** 100 **67.** 30 **69.** 20 **71.** 9 bars **73.** 5 pictures **75.** 46 calls **77.** 32 min **79.** 4,523 **81.** 9 **83.** 16 **85.** 100 **87.** 128 **89.** 7 pills **91.** 50 4-ft cables; yes **93.** 20 packages; 8 packages **95.** 73 lb **97.** Above and Beyond **99.** Above and Beyond

Activity 2 ::

Restaurant Management

There are over 30,000 McDonald's restaurants in the world; approximately 13,000 are in the United States. In total, McDonald's employs over one million people worldwide.

There are 1,525 McDonald's restaurants in Great Britain, just over 600 of which are franchises. We collected data about McDonald's UK and its 1,525 restaurants.

Employees	85,250
Office	935
Management	5,100

Restaurant workers make up the remaining McDonald's UK employees.

Employees under 21 years of age	51,150
Employees between 21 and 29 years old	18,185
Male employees	44,330

1. How many people work in McDonald's restaurants in Great Britain?
2. How many McDonald's UK employees are 30 years old or older?
3. How many women work at McDonald's UK?
4. Divide the total number of McDonald's UK restaurant workers by the number of restaurants to determine approximately how many employees work in each restaurant.
5. If each restaurant were to give each employee a $125 bonus, approximately how much would it cost each restaurant?
6. What would be the total cost of the bonus discussed in exercise 5 (across all McDonald's restaurants in Great Britain)?
7. Based on 52 employees per restaurant and 30,000 McDonald's restaurants in the United States, how many people work in McDonald's restaurants in the United States?
8. If health insurance costs McDonald's restaurants $400 per month for each employee in the United States, how much does it cost for one employee each year?
9. Based on exercises 7 and 8, how much would it cost McDonald's to provide health insurance to all of its U.S. restaurant employees each year?
10. Use the Internet to research the profits earned by McDonald's in the United States and worldwide last year.

1.7 Exponential Notation and the Order of Operations

< 1.7 Objectives >

1 > Use exponent notation

2 > Evaluate expressions containing powers of whole numbers

3 > Use the order of operations to evaluate expressions

Tips for Student Success

Preparing for a Test

Test prep really begins on the first day of class. Everything you do in class and at home is part of that preparation. However, there are a few things that you should focus on in the days before an exam.

1. Plan to complete your test preparation at least 24 hours before the test. The last 24 hours are too late, and besides, you need some rest before the test.
2. Go over your homework and class notes with pencil and paper in hand. Write down all the problem types, formulas, and definitions that you think might give you trouble on the test.
3. The day before the test, take the page(s) of notes from step 2, and transfer the most important ideas to a 3×5 card.
4. Just before the test, review the information on the card. You will be surprised at how much you remember about each concept.
5. If you have been completing your homework assignments successfully, you will do well on your test. This is an obstacle for many students, but it is an obstacle that can be overcome. Truly anxious students are often surprised that they scored as well as they did on a test. They tend to attribute this to blind luck. It is not. It is the first sign that you really do "get it." Enjoy the success.

Earlier, we described multiplication as a shorthand for repeated addition. There is also, a shorthand for repeated multiplication. We use *exponents* or *powers* to indicate repeated multiplication.

Example 1 Writing Repeated Multiplication as a Power

< Objective 1 >

NOTES

Recall that

$3 + 3 + 3 + 3 = 4 \times 3$

We write repeated addition as multiplication.

René Descartes, a French philosopher and mathematician, is generally credited with first introducing our modern exponent notation in about 1637.

$3 \times 3 \times 3 \times 3$ can be written as 3^4. This is read as "3 to the fourth power."

In this case, repeated multiplication is written as the power of a number.

Here, 3 is the **base** of the expression, and the raised number, 4, is the **exponent,** or **power.**

Exponent or power

$$3^4 = \underbrace{3 \times 3 \times 3 \times 3}_{\text{4 factors}}$$

Base

We count the factors and make this the power (or exponent) of the base.

Check Yourself 1

Write $2 \times 2 \times 2 \times 2 \times 2 \times 2$ as a power of 2.

Definition

Exponents — The *exponent* tells us the number of times the base is to be used as a factor.

Example 2 Evaluating a Number Raised to a Power

< Objective 2 >

2^5 is read "2 to the fifth power."

$$2^5 = \underbrace{2 \times 2 \times 2 \times 2 \times 2}_{\text{5 times}} = 32$$

Here 2 is the base, and 5 is the exponent.

2^5 tells us to use 2 as a factor 5 times. The result is 32.

Check Yourself 2

Read and evaluate 3^4.

Example 3 Evaluating a Number Raised to a Power

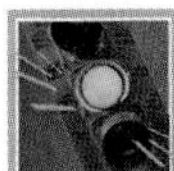

> CAUTION

5^3 is *entirely different* from 5×3. $5^3 = 125$, whereas $5 \times 3 = 15$.

Evaluate 5^3 and 8^2.

$5^3 = 5 \times 5 \times 5 = 125$ Use 3 factors of 5.

5^3 is read "5 to the third power" or "5 cubed."

$8^2 = 8 \times 8 = 64$ Use 2 factors of 8.

And 8^2 is read "8 to the second power" or "8 squared."

Check Yourself 3

Evaluate.

(a) 6^2 **(b)** 2^4

We need two special definitions for powers of whole numbers.

Definition

Raising a Number to the First Power — A number raised to the first power is just that number.

For example, $9^1 = 9$.

Definition

Raising a Number to the Zero Power — Any number, other than 0, raised to the zero power is 1.

For example, $7^0 = 1$.

Example 4 Evaluating Numbers Raised to the Power of 0 or 1

(a) $8^0 = 1$ **(b)** $4^0 = 1$ **(c)** $5^1 = 5$ **(d)** $3^1 = 3$

Check Yourself 4

Evaluate.

(a) 7^0 **(b)** 7^1

NOTES

10^3 is just a 1 followed by *three zeros*.

10^5 is a 1 followed by *five zeros*.

We mentioned *powers of 10* earlier when we multiplied by numbers that end in 0. Because the powers of 10 have special importance, we list some of them here.

$10^0 = 1$

$10^1 = 10$

$10^2 = 10 \times 10 = 100$

$10^3 = 10 \times 10 \times 10 = 1{,}000$

$10^4 = 10 \times 10 \times 10 \times 10 = 10{,}000$

$10^5 = 10 \times 10 \times 10 \times 10 \times 10 = 100{,}000$

Do you see why the powers of 10 are so important?

Property

Powers of 10

The powers of 10 correspond to the place values of our number system—ones, tens, hundreds, thousands, and so on.

NOTE

Archimedes (about 250 B.C.) reportedly estimated the number of grains of sand in the universe to be 10^{63}. This would be a 1 followed by 63 zeros!

This is what we meant earlier when we said that our number system was based on the number 10.

If multiplication is combined with addition or subtraction, you must know which operation to do first in finding the expression's value. We can easily illustrate this problem. How should we simplify the following statement?

$3 + 4 \times 5 = ?$

Both multiplication and addition are involved in this expression, and we must decide which to do first to find the answer.

1. Multiplying first gives us

   $3 + 20 = 23$

2. Adding first gives us

   $7 \times 5 = 35$

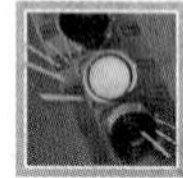

> CAUTION

The answers differ depending on which operation is done first!

Only one of these results can be correct, which is why mathematicians developed some rules to tell us the order in which the operations should be performed.

Step by Step

The Order of Operations

If multiplication, division, addition, and subtraction are involved in the same expression, do the operations in the following order:

Step 1 Do all multiplication and division in order from left to right.

Step 2 Do all addition and subtraction in order from left to right.

Example 5 Using the Order of Operations

< Objective 3 >

(a) $3 \times 4 + 5 = 12 + 5 = 17$ Multiply *first*, then add or subtract.

(b) $5 + 3 \times 6 = 5 + 18 = 23$

(c) $16 - 2 \times 3 = 16 - 6 = 10$

(d) $7 \times 8 - 20 = 56 - 20 = 36$

(e) $5 \times 6 + 4 \times 3 = 30 + 12 = 42$

NOTE

By this rule, we see that the first strategy from before was correct.

NOTE

When learning the order of operations, students sometimes remember this order by relating each step to part of the phrase

"Please Excuse My Dear Aunt Sally."

P
E
MD
AS

Check Yourself 5

Evaluate.

(a) $8 + 3 \times 5$
(b) $15 \times 5 - 3$
(c) $4 \times 3 + 2 \times 6$

We now want to extend our rule for the order of operations to see what happens when parentheses or exponents are involved in an expression.

Step by Step

The Order of Operations

Mixed operations in an expression should be done in the following order:

Step 1 Do any operations inside ***p***arentheses or other grouping symbols.

Step 2 Apply any ***e***xponents.

Step 3 Do all ***m***ultiplication and ***d***ivision in order from left to right.

Step 4 Do all ***a***ddition and ***s***ubtraction in order from left to right.

Parentheses are not the only grouping symbols. As you continue your math studies, you will come across other types of grouping symbols. In every case, you should evaluate expressions inside the grouping symbols as the first step in the order of operations.

Some other examples of grouping symbols include brackets, fraction bars, absolute value bars, and radicals. You will see some of these symbols as you move through this text and many more in your later math classes.

Example 6 Evaluating an Expression

Evaluate 4×2^3.

Step 1 There are no parentheses.

Step 2 Apply exponents.

$4 \times 2^3 = 4 \times 8$

Step 3 Multiply or divide.

$4 \times 8 = 32$

Check Yourself 6

Evaluate.

3×3^2

Example 7 Evaluating an Expression

Evaluate $(2 + 3)^2 + 4 \times 3$.

Step 1 Do operations inside parentheses.

$(2 + 3)^2 + 4 \times 3 = (5)^2 + 4 \times 3$

Step 2 Apply exponents.

$5^2 + 4 \times 3 = 25 + 4 \times 3$

Step 3 Multiply or divide.

$25 + 4 \times 3 = 25 + 12$

Step 4 Add or subtract.

$25 + 12 = 37$

Check Yourself 7

Evaluate.

(a) $4 + (8 - 5)^2$ **(b)** $(6 - 4)^3 + 3 \cdot 2$

Example 8 Using the Order of Operations

Evaluate each expression.

(a) $20 \div 2 \times 5$

$$\underbrace{20 \div 2} \times 5$$
$$= 10 \times 5$$
$$= 50$$

Because the multiplication and division appear next to each other, work in order from left to right. Try it the other way and see what happens!

So $20 \div 2 \times 5 = 50$.

(b) $(5 + 13) \div 6$

$$\underbrace{(5 + 13)} \div 6$$
$$= (18) \div 6$$
$$= 3$$

Do the addition in the parentheses as the first step.

So $(5 + 13) \div 6 = 3$.

(c) $(3 + 4)^2 \div (2^3 - 1)$

$$\underbrace{(3 + 4)^2} \div \underbrace{(2^3 - 1)}$$
$$= (7)^2 \div (8 - 1)$$
$$= 7^2 \div (7)$$
$$= 49 \div 7$$
$$= 7$$

Perform operations inside parentheses first.

So $(3 + 4)^2 \div (2^3 - 1) = 7$.

(d) $3 + [(2 + 5)^2 - 8] - 4^2$

The square brackets are a grouping symbol, so we treat them the same as parentheses. That is, we perform operations inside the brackets first.

Within the brackets, we follow the order of operations, so we do the computation inside the parentheses first, and then apply the exponent before moving to the subtraction.

When that is completed, we move to follow the order of operations outside the grouping symbols.

NOTES

Many people use brackets as a second set of parentheses. It is easier to see parentheses inside brackets than nested parentheses (parentheses inside parentheses).

If you are using a calculator to evaluate an expression, use the parentheses keys when an expression contains brackets.

$3 + [(2 + 5)^2 - 8] - 4^2$

$= 3 + [7^2 - 8] - 4^2$ $\quad 2 + 5 = 7$

$= 3 + [49 - 8] - 4^2$ $\quad 7^2 = 49$

$= 3 + 41 - 4^2$ $\quad 49 - 8 = 41$

$= 3 + 41 - 16$ $\quad$ We now compute $4^2 = 16$.

$= 44 - 16$ $\quad$ Add and subtract, from left to right.

$= 28$

Check Yourself 8

Evaluate.

(a) $36 \div 4 \times 2$ **(b)** $(2 + 4)^2 \div (3^2 - 3)$

(c) $15 \div 3 + (3 - 2)^2 \times 4 - 2$ **(d)** $5^3 - [14 - (8 - 6)^2]^2 + (2 + 3)^2$

Tips for Student Success

Taking a Test

Earlier in this section, we discussed test preparation. Now that you are thoroughly prepared for the test, you must learn how to take it.

There is much to the psychology of anxiety that we can't readily address. There is, however, a physical aspect to anxiety that can be addressed rather easily. When people are in a stressful situation, they frequently start to panic. One symptom of the panic is shallow breathing. In a test situation, this starts a vicious cycle. If you breathe too shallowly, then not enough oxygen reaches your brain. When that happens, you are unable to think clearly. In a test situation, being unable to think clearly can cause you to panic. Hence, we have a vicious cycle.

How do you break that cycle? It's pretty simple. Take a few deep breaths. We have seen students whose performance on math tests improved markedly after they got in the habit of writing "remember to breathe!" at the bottom of every test page. Try breathing; it will almost certainly improve your math test scores!

Check Yourself ANSWERS

1. 2^6 **2.** "Three to the fourth power" is 81 **3.** **(a)** 36; **(b)** 16 **4.** **(a)** 1; **(b)** 7 **5.** **(a)** 23; **(b)** 72; **(c)** 24 **6.** 27 **7.** **(a)** 13; **(b)** 14 **8.** **(a)** 18; **(b)** 6; **(c)** 7; **(d)** 50

Reading Your Text

These fill-in-the-blank exercises will help you understand some of the key vocabulary used in this section. The answers to these exercises are in the Answers Appendix at the back of the text.

(a) Preparation for a test really begins on the ____________ day of class.

(b) Another name for a power is an ____________.

(c) The first step in the order of operations involves doing operations inside ____________ or other grouping symbols.

(d) A whole number (other than zero) raised to the zero power is always equal to ____________.

1.7 exercises

Skills | Calculator/Computer | Career Applications | Above and Beyond

< Objective 1–3 >

Evaluate.

1. 3^2

2. 2^3 

3. 5^1

4. 6^0

5. 10^3

6. 10^6

7. 2×4^3

8. $(2 \times 4)^3$

9. $5 + 2^2$

10. $(5 + 2)^2$

11. $42 - 7 + 9$

12. $27 - 12 + 3$

13. $20 \div 5 \times 2$

14. $48 \div 3 \times 2$

15. $(3 + 2)^3 - 20$

16. $5 + (9 - 5)^2$

17. $(7 - 4)^4 - 30$

18. $(5 + 2)^2 + 20$

19. $8^2 \div 4^2 + 2$

20. $3 \times 5^2 + 2^2$

21. $24 - 6 \div 3$

22. $3 + 9 \div 3$

23. $(24 - 6) \div 3$

24. $(3 + 9) \div 3$

25. $12 + 3 \div (3^2 - 2 \cdot 3)$ 

26. $(8^2 - 2^4) \div 2$

27. $8^2 - 2^4 \div 2$

28. $(5 - 3)^3 + (8 - 6)^2$

29. $30 \div 6 - 12 \div 3$

30. $5 + 8 \div 4 - 3$

31. $16 - 12 \div 3 \cdot 2 + (16 - 12)^2 \cdot 3$

32. $3 \cdot 5 + 3 \cdot 4^2 \div (6 - 4)^2$

33. $6 + 3 \cdot 2^4 - (12 - 7)(10 - 7)$

34. $27 \div (2^2 + 5) - (35 - 33)(24 - 23)$

35. $3 \times [(7 - 5)^3 - 8] + 5 \times 2$

36. $[(3 + 1) + (7 - 2)] \times 4 - 5 \times 7$

Skills | **Calculator/Computer** | Career Applications | Above and Beyond

Use your calculator to evaluate each expression.

37. $4 \times 5 - 7$

38. $3 \times 7 + 8$

39. $9 + 3 \times 7$

40. $6 \times 0 + 3$

41. $4 + 5 \times 0$

42. $23 - 4 \times 5$

43. $5 \times (4 + 7)$

44. $8 \times (6 + 5)$

45. $5 \times 4 + 5 \times 7$

46. $8 \times 6 + 8 \times 5$

Use a calculator to solve each application.

47. **Business and Finance** A car dealer kept the following record of a month's sales. Complete the table.

Model	Number Sold	Profit per Sale	Monthly Profit
Subcompact	38	$528	______
Compact	33	647	______
Standard	19	912	______
		Total Monthly Profit	______

48. **Business and Finance** You take a job paying \$1 the first day. On each following day your pay doubles. That is, on day 2 your pay is \$2, on day 3 the pay is \$4, and so on. Complete the table.

Day	Daily Pay	Total Pay
1	\$1	\$1
2	2	3
3	4	7
4	______	______
5	______	______
6	______	______
7	______	______
8	______	______
9	______	______
10	______	______

Skills | Calculator/Computer | **Career Applications** | Above and Beyond

49. **Electronics** Resistors are commonly identified by colored bands to indicate their approximate resistance, measured in ohms. Each band's color and position corresponds to a specific component of the overall value. In resistors with four colored bands, the third band is typically considered to be the exponent for a base 10. If the first two bands are decoded as 43 and the third band is decoded as 5, what is the total resistance in ohms?

$43 \times 10^5 = ?$

Which of the three bands is most important to read correctly? Why?

50. **Manufacturing Technology** The kinetic energy (KE) of an object (in Joules) is given by the formula.

$$\text{KE} = \frac{mv^2}{2}$$

Find the kinetic energy of an object that has a mass (m) of 46 kg and is moving at a velocity (v) of 16 meters per second.

51. **Manufacturing Technology** The power (P) of a circuit (in Watts) can be given by any of these formulas.

$$P = IV$$

$$P = \frac{V^2}{R}$$

$$P = I^2R$$

Find the power for each circuit.

(a) Voltage (V) = 110 volts (V) and current (I) = 13 amperes (A).

(b) Voltage = 220 V and resistance (R) = 22 ohms.

(c) Current = 25 A and resistance = 9 ohms.

52. Manufacturing Technology A belt is used to connect two pulleys.

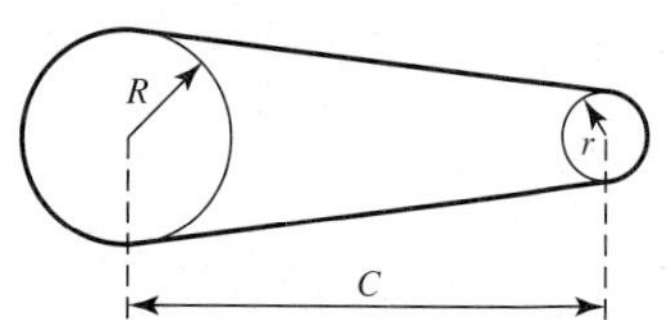

The length of the belt required is given by the formula:

$$\text{Belt length} = 2C + 3(R + r) + \frac{(2R + 2r)^2}{4C}$$

where C is the distance between the centers of the two pulleys.

Find the approximate belt length required to go around a 4-in. radius (r) pulley and a 6-in. radius (R) pulley that are 20 in. apart.

Skills | Calculator/Computer | Career Applications | **Above and Beyond**

Numbers such as 3, 4, and 5 are called ***Pythagorean triples,*** *after the Greek mathematician Pythagoras (sixth century B.C.), because*

$$3^2 + 4^2 = 5^2$$

Determine whether each set of numbers is a Pythagorean triple.

53. 6, 8, 10 **54.** 6, 11, 12 **55.** 5, 12, 13

56. 7, 24, 25 **57.** 8, 16, 18 **58.** 8, 15, 17

59. Is $(a + b)^P$ equal to $a^P + b^P$?

Try a few numbers and decide if you think this is true for all whole numbers, for some whole numbers, or never true. Write an explanation of your findings and give examples.

60. Does $(a \cdot b)^P = a^P \cdot b^P$?

Try a few numbers and decide if you think this is true for all whole numbers, for some whole numbers, or never true. Write an explanation of your findings and give examples.

Answers

1. 9 **3.** 5 **5.** 1,000 **7.** 128 **9.** 9 **11.** 44 **13.** 8 **15.** 105 **17.** 51 **19.** 6 **21.** 22 **23.** 6 **25.** 13 **27.** 56 **29.** 1 **31.** 56 **33.** 39 **35.** 10 **37.** 13 **39.** 30 **41.** 4 **43.** 55 **45.** 55 **47.** \$20,064 21,351 17,328 \$58,743 **49.** 4,300,000 ohms; the third band; misreading it would lead to errors of powers of 10 **51.** **(a)** 1,430 W; **(b)** 2,200 W; **(c)** 5,625 W **53.** Yes **55.** Yes **57.** No **59.** Above and Beyond

Activity 3 ::

Population Growth

According to the U.S. Census Bureau, the U.S. population grew from 281,421,906 in 2000 to 308,745,538 in 2010.

1. By how much did the U.S. population grow during the decade?
2. Round this result to the nearest million.
3. Round the 2000 population to the nearest thousand.
4. Round the 2010 population to the nearest thousand.
5. Use the results of exercises 2 and 3 to approximate the population growth during the decade.
6. How does your answer to exercise 5 compare to your answers to exercises 1 and 2?
7. Every year, the U.S. population grows by approximately 9 people per 1,000 people. Use the result from exercise 4 to predict the population growth in 2011.
8. Based on your answer to exercise 7, how many people would be living in the United States in 2011, to the nearest thousand.
9. Predict the population in the year 2020, rounding each year's population to the nearest thousand.
10. Use the Internet to research factors that contribute to population growth.

summary :: chapter 1

Definition/Procedure	Example	Reference
The Decimal Place-Value System		**Section 1.1**
Digits Digits are the basic symbols of our number system.	0, 1, 2, 3, 4, 5, 6, 7, 8, and 9 are digits.	*p. 2*
Place Value The value of a digit in a number depends on its position or place.	52,589 — 9: Ones; 8: Tens; 5: Hundreds; 2: Thousands; 5: Ten thousands	*p. 2*
The value of a number is the sum of each digit multiplied by its place value.	$2{,}345 = (2 \times 1{,}000) + (3 \times 100) + (4 \times 10) + (5 \times 1)$	*p. 3*
Addition		**Section 1.2**
The Properties of Addition		
The Commutative Property The order in which you add two numbers does not affect the sum.	$5 + 4 = 4 + 5$	*p. 12*
The Associative Property The way in which you group numbers in addition does not affect the final sum.	$(2 + 7) + 8 = 2 + (7 + 8)$	*p. 12*
The Additive Identity The sum of 0 and any number is just that number.	$6 + 0 = 0 + 6 = 6$	*p. 13*
Measuring Perimeter		
The perimeter is the total distance around the outside edge of a shape. We can write a formula for the perimeter of a rectangle in either of two ways. $P = L + W + L + W$ $= 2 \cdot L + 2 \cdot W$	Rectangle: 6 ft, 2 ft, 6 ft, 2 ft $P = L + W + L + W$ $= (6\text{ ft}) + (2\text{ ft}) + (6\text{ ft}) + (2\text{ ft})$ $= 16\text{ ft}$	*p. 17*
Subtraction		**Section 1.3**
Minuend The number we are subtracting from. **Subtrahend** The number that is being subtracted. **Difference** The result of the subtraction.	15 ← Minuend − 9 ← Subtrahend 6 ← Difference	*p. 27*
Rounding, Estimation, and Order		**Section 1.4**
Step 1 Identify the place of the digit to be rounded. **Step 2** Look at the digit to the right of that place. **Step 3** **a.** If that digit is 5 or more, that digit and all digits to the right become 0. The digit in the place you are rounding is increased by 1. **b.** If that digit is less than 5, that digit and all digits to the right become 0. The digit in the place you are rounding remains the same.	To the nearest hundred, 43,578 is rounded to 43,600. To the nearest thousand, 273,212 is rounded to 273,000.	*p. 40*

Continued

Definition/Procedure	Example	Reference
Multiplication		**Section 1.5**
Factors The numbers being multiplied. **Product** The result of the multiplication.	$7 \times 9 = 63$ ⟵ Product ↑ Factors	*p.* 47
The Properties of Multiplication		
The Commutative Property The order in which you multiply two numbers does not affect the product.	$7 \times 9 = 9 \times 7$	*p.* 47
The Distributive Property To multiply a factor by a sum of numbers, multiply the factor by each number inside the parentheses. Then add the products.	$2 \times (3 + 7) = (2 \times 3) + (2 \times 7)$	*p.* 50
The Associative Property The way in which you group numbers in multiplication does not affect the final product.	$(3 \times 5) \times 6 = 3 \times (5 \times 6)$	*p.* 53
Finding Area and Volume		
The area of a rectangle is found using the formula $A = L \cdot W$. The volume of a rectangular solid is found using the formula $V = L \cdot W \cdot H$.	6 ft, 2 ft $A = L \times W = 6\text{ ft} \times 2\text{ ft} = 12\text{ ft}^2$	*p.* 56
Division		**Section 1.6**
Divisor The number we are dividing by. **Dividend** The number being divided. **Quotient** The result of the division. **Remainder** The number "left over" after the division.	Divisor: 7; Quotient: 5 $7\overline{)38}$ ⟵ Dividend 35 3 ⟵ Remainder	*p.* 64
Dividend = divisor × quotient + remainder	$38 = 7 \times 5 + 3$	*p.* 66
Division by 0 is undefined.	$7 \div 0$ is undefined.	*p.* 67
Exponential Notation and the Order of Operations		**Section 1.7**
Using Exponents		
Base The number that is raised to a power. **Exponent** The exponent is written to the right and above the base. The exponent tells the number of times the base is to be used as a factor.	Exponent $5^3 = 5 \times 5 \times 5 = 125$ Base; Three factors	*p.* 78
The Order of Operations		
Mixed operations in an expression should be done in the following order: **Step 1** Do any operations inside parentheses or other grouping symbols. **Step 2** Apply any exponents. **Step 3** Do all multiplication and division in order from left to right. **Step 4** Do all addition and subtraction in order from left to right. Remember *P*lease *E*xcuse *M*y *D*ear *A*unt *S*ally	$4 \times (2 + 3)^2 - 7$ $= 4 \times 5^2 - 7$ $= 4 \times 25 - 7$ $= 100 - 7$ $= 93$	*p.* 81

summary exercises :: chapter 1

This summary exercise set will help ensure that you have mastered each of the objectives of this chapter. The exercises are grouped by section. You should reread the material associated with any exercises that you find difficult. The answers to the odd-numbered exercises are in the Answers Appendix at the back of the text.

1.1 *Give the place value of each of the indicated digits.*

1. 6 in the number 5,674

2. 5 in the number 543,400

Write each number in words.

3. 27,428

4. 200,305

Write each number in standard form.

5. Thirty-seven thousand, five hundred eighty-three

6. Three hundred thousand, four hundred

1.2 *Name the property of addition illustrated.*

7. $4 + 9 = 9 + 4$

8. $(4 + 5) + 9 = 4 + (5 + 9)$

Perform the indicated operations.

9.
$$\begin{array}{r} 784 \\ 385 \\ +\ 247 \\ \hline \end{array}$$

10.
$$\begin{array}{r} 2{,}570 \\ 498 \\ 21{,}456 \\ +\quad 28 \\ \hline \end{array}$$

11.
$$\begin{array}{r} 367 \\ 289 \\ 1{,}463 \\ +\ 2{,}682 \\ \hline \end{array}$$

12.
$$\begin{array}{r} 6{,}389 \\ 1{,}567 \\ 315 \\ +\ 113{,}602 \\ \hline \end{array}$$

Read each application carefully, and then answer the questions that follow. Do not solve the problems.

13. **Business and Finance** An airline had 173, 212, 185, 197, and 202 passengers on five morning flights between Washington, D.C., and New York. What was the total number of passengers on these flights?

(a) What are you asked to find?

(b) What information have you been given?

14. **Business and Finance** Future Stars summer camp employs five junior counselors. Last week, they earned \$432, \$540, \$324, \$540, and \$324. What was the total salary for the junior counselors?

(a) What are you asked to find?

(b) What information have you been given?

Solve each application

15. **Business and Finance** An airline had 173, 212, 185, 197, and 202 passengers on five morning flights between Washington, D.C., and New York. What was the total number of passengers on these flights?

16. **Business and Finance** Future Stars summer camp employs five junior counselors. Last week, they earned \$432, \$540, \$324, \$540, and \$324. What was the total salary for the junior counselors?

1.3 *Find each value.*

17. 34 decreased by 7

18. 7 more than 4

19. The product of 9 and 5 divided by 3.

Perform the indicated operations.

20. $\begin{array}{r} 5{,}325 \\ -\ 847 \\ \hline \end{array}$

21. $\begin{array}{r} 38{,}400 \\ -\ 19{,}600 \\ \hline \end{array}$

22. $\begin{array}{r} 86{,}000 \\ -\ 2{,}169 \\ \hline \end{array}$

23. $\begin{array}{r} 2{,}682 \\ -\ 108 \\ \hline \end{array}$

24. Find the difference of 7,342 and 5,579.

Solve each application.

25. **Business and Finance** Chuck owes $795 on a credit card after a trip. He makes payments of $75, $125, and $90. Interest of $31 is charged. How much remains to be paid on the account?

26. **Business and Finance** Juan bought a new car for $21,985. The manufacturer offers a cash rebate of $1,495. What was the cost after rebate?

1.4 *Round each number to the indicated place.*

27. 6,975 to the nearest hundred

28. 15,897 to the nearest thousand

29. 548,239 to the nearest ten thousand

30. 548,239 to the nearest million

Complete each statement by using the symbol $<$ or $>$.

31. 60 ______ 70

32. 38 ______ 35

Find the perimeter of each figure.

33.

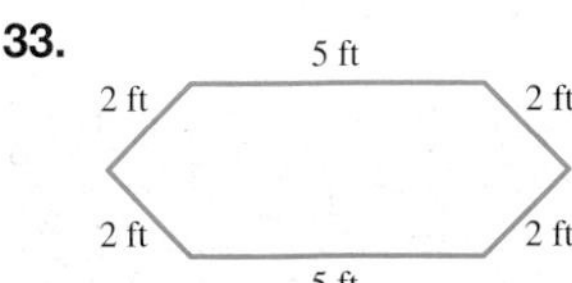

34.

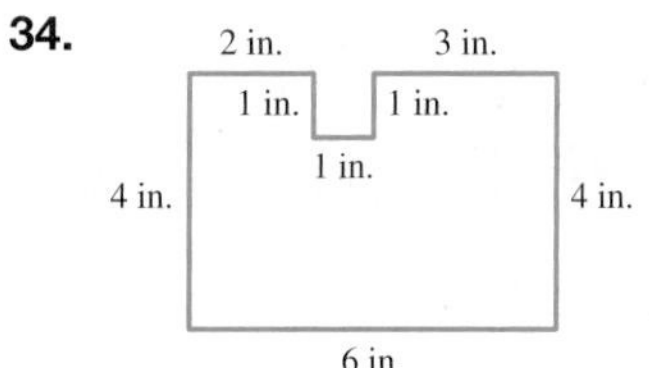

1.5 *Name the property of multiplication illustrated.*

35. $7 \times 8 = 8 \times 7$

36. $3 \times (4 + 7) = 3 \times 4 + 3 \times 7$

37. $(8 \times 9) \times 4 = 8 \times (9 \times 4)$

38. $112 \times 1 = 112$

Perform the indicated operations.

39. $\begin{array}{r} 58 \\ \times\ 32 \\ \hline \end{array}$

40. $\begin{array}{r} 25 \\ \times\ 43 \\ \hline \end{array}$

41. $\begin{array}{r} 378 \\ \times\ 409 \\ \hline \end{array}$

42. $\begin{array}{r} 3{,}412 \\ \times\ \ 641 \\ \hline \end{array}$

43. *Find the area of the figure.*

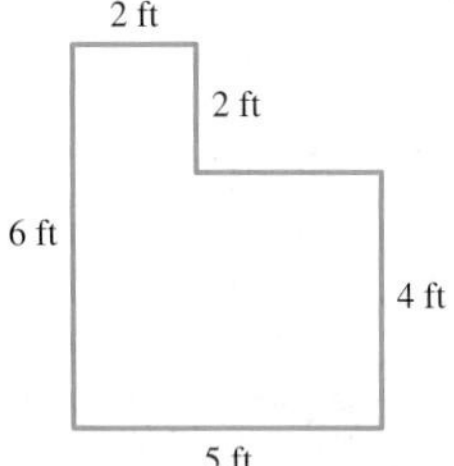

44. *Find the volume of the figure.*

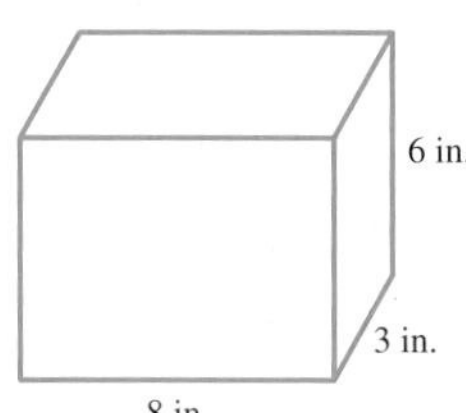

Solve the application.

45. **Crafts** You wish to carpet a room that is 5 yd by 7 yd. The carpet costs $18 per square yard. What is the total cost of the materials?

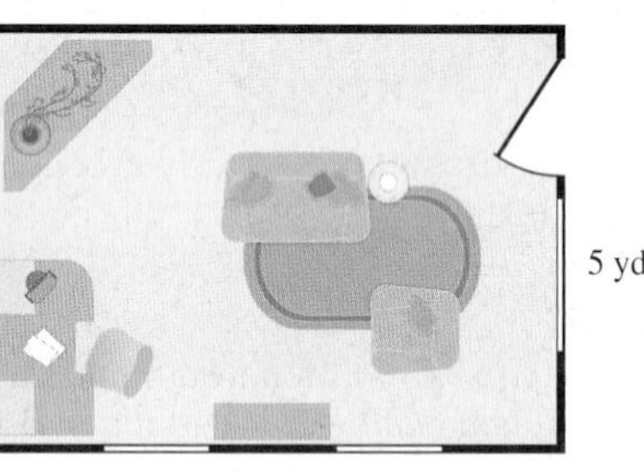

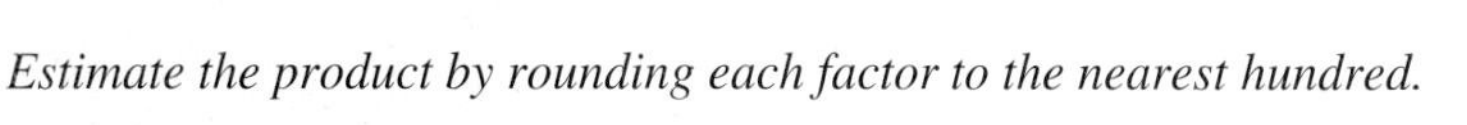

Estimate the product by rounding each factor to the nearest hundred.

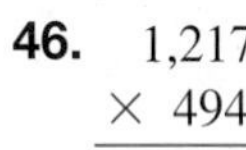

46. $\begin{array}{r} 1{,}217 \\ \times\ \ 494 \\ \hline \end{array}$

1.6 *Divide if possible.*

47. $0 \div 8$

48. $5 \div 0$

Divide.

49. $8\overline{)2{,}469}$

50. $4\overline{)2{,}157}$

51. $64\overline{)31{,}809}$

52. $36\overline{)86{,}915}$

53. **Statistics** Hasina's odometer read 25,235 mi at the beginning of a trip and 26,215 mi at the end. If she used 35 gal of gas for the trip, what was her mileage (mi/gal)?

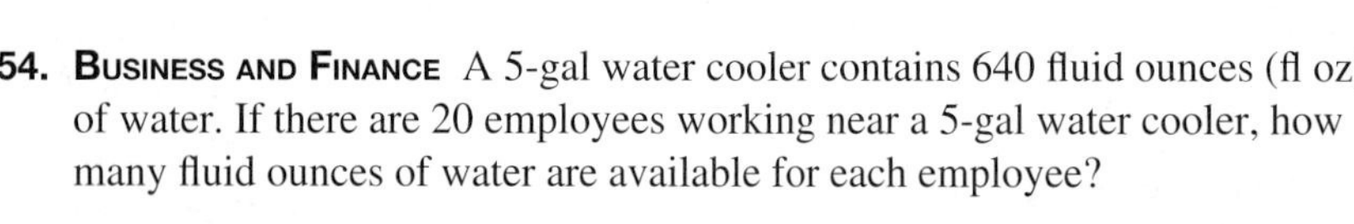

54. **Business and Finance** A 5-gal water cooler contains 640 fluid ounces (fl oz) of water. If there are 20 employees working near a 5-gal water cooler, how many fluid ounces of water are available for each employee?

Estimate.

55. 356 divided by 37

56. 2,125 divided by 123

1.7 *Evaluate each expression.*

57. 5×2^3

58. $(5 \times 2)^3$

59. $4 + 8 \times 3$

60. $48 \div (2^3 + 4)$

61. $(4 + 8) \times 3$

62. $4 \times 3 + 8 \times 3$

63. $8 \div 4 \times 2 - 2 + 1$

64. $63 \times 2 \div 3 - 54 \div (12 \times 2 \div 4)$

65. $(3 \times 4)^2 - 100 \div 5 \times 6$

66. $(16 \times 2) \div 8 - (6 \div 3 \times 2)$

CHAPTER 1

chapter test 1

Use this chapter test to assess your progress and to review for your next exam. Allow yourself about an hour to take this test. The answers to these exercises are in the Answers Appendix at the back of the text.

1. Give the word name for 302,525.

2. Give the place value of 7 in 3,738,500.

3. Write two million, four hundred thirty thousand in number form.

4. What is the total of 392, 95, 9,237, and 11,972?

Evaluate, as indicated.

5. $\begin{array}{r} 489 \\ 562 \\ 613 \\ +\ 254 \\ \hline \end{array}$

6. $\begin{array}{r} 13 \\ 2,543 \\ +10,547 \\ \hline \end{array}$

7. $\begin{array}{r} 89 \\ \times\ 56 \\ \hline \end{array}$

8. $\begin{array}{r} 538 \\ \times\ 103 \\ \hline \end{array}$

9. $289 - 54$

10. $55,342 - 14,787$

11. $32,345 - 1,575$

12. $53,294 - 41,074$

13. $8\overline{)2,135}$

14. $28\overline{)61,382}$

15. $(3 + 4)^2 - (2 + 3^2 - 1)$

16. $15 - 12 \div 2^2 \times 3 + (12 \div 4 \times 3)$

Name the property illustrated.

17. $4 \times (3 + 6) = (4 \times 3) + (4 \times 6)$

18. $(7 + 3) + 8 = 7 + (3 + 8)$

19. $3 \times (2 \times 7) = (3 \times 2) \times 7$

20. $5 + 12 = 12 + 5$

Complete each statement by using the symbol $<$ or $>$.

21. 49 ______ 47

22. 80 ______ 90

Estimate the sum by rounding each addend to the nearest hundred.

23.
```
    943
  3,281
    778
  2,112
+   570
```

Find the perimeter of the figure.

24.

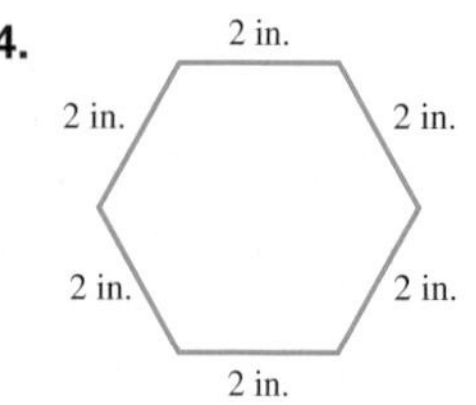

Find the area of the figure.

25.

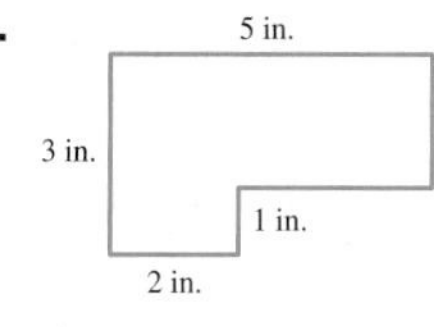

Find the volume of the figure.

26.

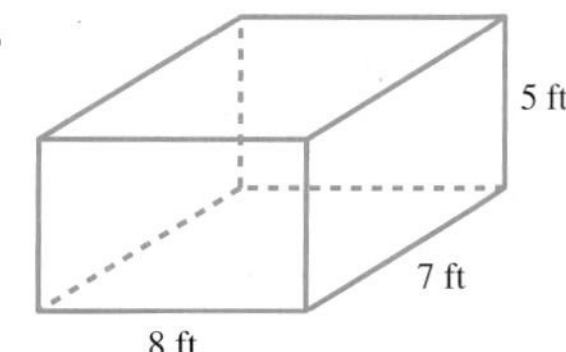

27. Business and Finance A truck rental firm orders 25 new vans at a cost of $22,350 per van. What is the total cost of the order?

28. Business and Finance Eight people estimate that the total expenses for a trip they are planning will be $7,136. If each person pays an equal amount, what is each person's share?

29. Statistics The attendance for the games of a playoff series in basketball was 12,438, 14,325, 14,581, and 14,634. What was the total attendance for the series?

30. Social Science The maximum load for a light plane with full gas tanks is 500 lb. Mr. Whitney weighs 215 lb; his wife, 135 lb; and their daughter, 78 lb. How much luggage can they take on a trip without exceeding the load limit?

An Introduction to Fractions

INTRODUCTION

When was the last time you doubled a recipe? Were you feeding a lot of people or did you want leftovers? What about halving a recipe, perhaps to serve two people instead of four?

Do you use phrases such as "half past three" or "a quarter 'til noon" when giving the time? Do see signs such as "Exit 22 $\frac{1}{4}$ mi" when you drive?

In this chapter, we learn about fractions. We will see how fractions are formed, when they are used, and how to work with them.

To help you relate fractions to everyday life, you will work with nutritional recommendations and even recipes when you complete Activities 4 and 6.

CHAPTER 2 OUTLINE

2 prerequisite check

This Prerequisite Check highlights the skills you will need in order to be successful in this chapter. The answers to these exercises are in the Answers Appendix at the back of the text.

1. Does 4 divide exactly into 30 (that is, with no remainder)?

2. Does 5 divide exactly into 29?

3. Does 6 divide exactly into 72?

4. Does 3 divide exactly into 412?

5. Does 2 divide exactly into 238?

6. List all the whole numbers that can divide exactly into 9.

7. List all the whole numbers that can divide exactly into 10.

8. List all the whole numbers that can divide exactly into 17.

9. List all the whole numbers that can divide exactly into 48.

10. List all the whole numbers that can divide exactly into 60.

2.1 Prime Numbers and Divisibility

< 2.1 Objectives >

1 > Find the factors of a number

2 > Determine whether a number is prime, composite, or neither

3 > Determine whether a number is divisible by 2, 3, 4, 5, 6, 9, or 10

Tips for Student Success

Working Together

How many of your classmates do you know? Whether your nature is outgoing or shy, there are many benefits to getting to know your classmates.

1. It is important to have someone to call when you miss class or if you are unclear on an assignment.
2. Working with another person is almost always beneficial to both people. If you don't understand something, it helps to have someone to ask about it. If you do understand something, nothing cements that understanding better than explaining it to another person.
3. Sometimes we need to commiserate. If an assignment is particularly frustrating, it is reassuring to find that it is also frustrating for other students.
4. Have you ever thought you had the right answer, but it didn't match the answer in the text? Frequently the answers are equivalent, but that's not always easy to see. A different perspective can help you see that. Occasionally there is an error in a textbook. In such cases, it is wonderfully reassuring to find that someone else has the same answer as you do.

NOTE

2 and 5 are called *divisors* of 10 because they divide 10 exactly.

In Section 1.5 we said that because $2 \times 5 = 10$, we call 2 and 5 **factors** of 10.

Definition

Factor A **factor** of a whole number is another whole number that *divides exactly* into the original number. This means that the division has a remainder of 0.

Example 1 Finding Factors

< Objective 1 >

List all the factors of 18.
When finding all the factors for any number, we always start with division by 1 because 1 is a factor of every number.

$1 \times 18 = 18$ Our list of factors starts with 1 and 18.

We continue with division by 2.

$2 \times 9 = 18$ Our list now contains 1, 2, 9, 18.

We check divisibility by each subsequent whole number until we get to a number that is already in the factor list.

$3 \times 6 = 18$ The list is now 1, 2, 3, 6, 9, 18.

NOTE

The factors of 18, except for 18 itself, are *smaller* than 18.

Since 18 is not divisible by either 4 or 5, and 6 is already on the list, our list of factors is complete.

Check Yourself 1

List all the factors of 24.

Listing factors leads us to an important classification of whole numbers. Any whole number larger than 1 is either a *prime* or a *composite* number.

A whole number greater than 1 always has itself and 1 as factors. Sometimes these are the *only* factors. For instance, 1 and 3 are the only factors of 3.

Definition

Prime Number

A **prime number** is any whole number greater than one, that has exactly two factors, 1 and itself.

NOTE

How large can a prime number be? There is no largest prime number. To date, the largest known prime is $2^{43,112,609} - 1$. This is a number with 12,978,189 digits. Of course, computers have to be used to verify that a number of this size is prime. If this number were printed out in 12-point type, it would be 30 mi long. By the time you read this, someone may very well have found an even larger prime number.

As examples, 2, 3, 5, and 7 are prime numbers. Their only factors are 1 and themselves.

To check whether a number is prime, one approach is simply to divide the smaller primes—2, 3, 5, 7, and so on—into the given number. If no factors other than 1 and the given number are found, the number is prime.

Here is the method known as the **sieve of Eratosthenes** for identifying prime numbers.

1. Write down a series of counting numbers, starting with the number 2. In this example, we stop at 50.
2. Start at the number 2. Delete every second number after 2.
3. Move to the number 3. Delete every third number after 3 (some numbers will be deleted twice).
4. Continue this process, deleting every fourth number after 4, every fifth number after 5, and so on.
5. When you have finished, the undeleted numbers are the prime numbers.

	2	3	~~4~~	5	~~6~~	7	~~8~~	~~9~~	~~10~~
11	~~12~~	13	~~14~~	~~15~~	~~16~~	17	~~18~~	19	~~20~~
~~21~~	~~22~~	23	~~24~~	~~25~~	~~26~~	~~27~~	~~28~~	29	~~30~~
31	~~32~~	~~33~~	~~34~~	~~35~~	~~36~~	37	~~38~~	~~39~~	~~40~~
41	~~42~~	43	~~44~~	~~45~~	~~46~~	47	~~48~~	~~49~~	~~50~~

The prime numbers less than 50 are 2, 3, 5, 7, 11, 13, 17, 19, 23, 29, 31, 37, 41, 43, 47.

Example 2 Identifying Prime Numbers

< Objective 2 >

Which of the numbers 17, 29, and 33 are prime?

17 is a prime number. 1 and 17 are the only factors.

29 is a prime number. 1 and 29 are the only factors.

33 is *not* prime. 1, 3, 11, and 33 are all factors of 33.

Note: For two-digit numbers, if the number is *not* a prime, it has one or more of the numbers 2, 3, 5, or 7 as factors.

Check Yourself 2

Identify the prime numbers.

2, 6, 9, 11, 15, 19, 23, 35, 41

We can now define a second class of whole numbers.

Definition

Composite Number

A **composite number** is any whole number greater than 1 that is not prime. Every composite number has more than two factors.

Example 3 Identifying Composite Numbers

NOTE

One helpful memory device (or mnemonic) is to think of composite numbers as composed of other factors.

Which of the numbers 18, 23, 25, and 38 are composite?

18 is a composite number. 1, 2, 3, 6, 9, and 18 are all factors of 18.

23 is *not* a composite number. 1 and 23 are the only factors. This means that 23 is a *prime number.*

25 is a composite number. 1, 5, and 25 are factors.

38 is a composite number. 1, 2, 19, and 38 are factors.

Check Yourself 3

Identify the composite numbers.

2, 6, 10, 13, 16, 17, 22, 27, 31, 35

Definition

Zero and One

The whole numbers 0 and 1 are neither prime nor composite.

NOTE

Divisibility by 2 indicates that a number is *even*.

This is simply a matter of the way in which prime and composite numbers are defined in mathematics. The numbers 0 and 1 are the *only* two whole numbers that cannot be classified as one or the other.

For our work in this text, it is very useful to be able to tell whether a given number is divisible by 2, 3, or 5. The tests that follow will give you some tools to check divisibility without actually having to divide.

Tests for divisibility by other numbers are also available. However, we have limited this section to those tests involving 2, 3, 4, 5, 6, 9, and 10 because they are very easy to use and occur frequently in our work.

Property

Divisibility by 2

A whole number is divisible by 2 if its last digit is 0, 2, 4, 6, or 8.

Example 4 Determining If a Number Is Divisible by 2

< Objective 3 >

Which of the numbers 2,346, 13,254, 23,573, and 57,085 are divisible by 2?

2,346 is divisible by 2. The final digit is 6.

13,254 is divisible by 2. The final digit is 4.

23,573 is *not* divisible by 2. The final digit is not 0, 2, 4, 6, or 8.

57,085 is *not* divisible by 2.

Check Yourself 4

Which numbers are divisible by 2?

274 3,587 7,548 13,593

Property

Divisibility by 3 A whole number is divisible by 3 if the sum of its digits is divisible by 3.

Example 5 Determining If a Number Is Divisible by 3

Which of the numbers 345, 1,243, and 25,368 are divisible by 3?

345 is divisible by 3. The sum of the digits, $3 + 4 + 5$, is 12, and 12 is divisible by 3.

1,243 is *not* divisible by 3. The sum of the digits, $1 + 2 + 4 + 3$, is 10, and 10 is not divisible by 3.

25,368 is divisible by 3. The sum of the digits, $2 + 5 + 3 + 6 + 8$, is 24, and 24 is divisible by 3. Note that 25,368 is also divisible by 2.

Check Yourself 5

(a) Is 372 divisible by 2? By 3? **(b)** Is 5,493 divisible by 2? By 3?

Property

Divisibility by 5 A whole number is divisible by 5 if its last digit is 0 or 5.

Example 6 Determining If a Number Is Divisible by 5

Determine if the numbers 2,435, 23,123, and 123,240 are divisible by 5.

2,435 is divisible by 5. Its last digit is 5.

23,123 is *not* divisible by 5. Its last digit is 3.

123,240 is divisible by 5. Its last digit is 0. Do you see that 123,240 is also divisible by 2 and 3?

Check Yourself 6

(a) Is 12,585 divisible by 5? By 2? By 3?
(b) Is 5,890 divisible by 5? By 2? By 3?

By combining some of these techniques, we can come up with divisibility tests for composite numbers as well.

Property

Divisibility by 4 A whole number is divisible by 4 if the two-digit number formed by its final two digits is divisible by 4.

Example 7 Determining If a Number Is Divisible by 4

Determine if the numbers 1,464 and 2,434 are divisible by 4.

1,464 is divisible by 4. 64 is divisible by 4.

2,434 is *not* divisible by 4. 34 is not divisible by 4.

Check Yourself 7

Determine if each number is divisible by 4.

(a) 6,456 **(b)** 242 **(c)** 22,100

Property

Divisibility by 6 A whole number is divisible by 6 if it is an even number that is divisible by 3.

An even number is divisible by 2, so divisibility by 6 means that the number is divisible by both 2 and 3.

Example 8 Determining If a Number Is Divisible by 6

Determine if the numbers 1,464 and 2,434 are divisible by 6.

1,464 is divisible by 6. It is an even number, and the sum of the digits, 15, is divisible by 3.

2,434 is *not* divisible by 6. Although it is an even number, the sum of the digits, 13, is not divisible by 3.

Check Yourself 8

Determine if each number is divisible by 6.

(a) 6,456 **(b)** 242 **(c)** 22,100

Property

Divisibility by 9 A whole number is divisible by 9 if the sum of its digits is divisible by 9.

Example 9 Determining If a Number Is Divisible by 9

Determine if the numbers 1,494 and 2,634 are divisible by 9.

1,494 is divisible by 9. The sum of the digits, 18, is divisible by 9.

2,634 is *not* divisible by 9. The sum of the digits, 15, is not divisible by 9.

Check Yourself 9

Determine if each number is divisible by 9.

(a) 3,456 **(b)** 243,000 **(c)** 22,200

Property

Divisibility by 10 A whole number is divisible by 10 if it ends with a zero.

Example 10 Determining If a Number Is Divisible by 10

Determine if the numbers 4,390,005 and 6,420 are divisible by 10.

4,390,005 is *not* divisible by 10. The number does not end with a zero.

6,420 is divisible by 10. The number does end with a zero.

Check Yourself 10

Determine whether each number is divisible by 10.

(a) 2,000,020 **(b)** 2,000,002 **(c)** 3,571,110

Check Yourself ANSWERS

1. 1, 2, 3, 4, 6, 8, 12, and 24 **2.** 2, 11, 19, 23, and 41 are prime numbers **3.** 6, 10, 16, 22, 27, and 35 are composite numbers **4.** 274 and 7,548 **5.** **(a)** Yes in both cases; **(b)** only by 3 **6.** **(a)** By 5 and by 3; **(b)** by 5 and by 2 **7.** **(a)** Yes; **(b)** no; **(c)** yes **8.** **(a)** Yes; **(b)** no; **(c)** no **9.** **(a)** Yes; **(b)** yes; **(c)** no **10.** **(a)** Yes; **(b)** no; **(c)** yes

Reading Your Text

These fill-in-the-blank exercises will help you understand some of the key vocabulary used in this section. The answers to these exercises are in the Answers Appendix at the back of the text.

(a) ____________ is a factor of every whole number.

(b) A ____________ number is any whole number that has exactly two factors, 1 and itself.

(c) A __________ number is any whole number greater than 1 that is not prime.

(d) When a whole number is divisible by 2, we call it an ____________ number.

2.1 exercises

Skills Calculator/Computer Career Applications Above and Beyond

< Objective 1 >

List the factors of each number.

1. 4 **2.** 6 **3.** 10 **4.** 12

5. 15 **6.** 21 **7.** 24 VIDEO **8.** 32

9. 64 VIDEO **10.** 66 **11.** 11 VIDEO **12.** 37

13. 135 **14.** 236 **15.** 256 **16.** 512

< Objective 2 >

Use the list of numbers
0, 1, 15, 19, 23, 31, 49, 55, 59, 87, 91, 97, 103, 105 *to complete exercises 17 and 18.*

17. Which numbers are prime? 

18. Which numbers are composite?

19. List all the prime numbers between 30 and 50.

20. List all the prime numbers between 55 and 75.

< Objective 3

Use the list of numbers
45, 72, 158, 260, 378, 569, 570, 585, 3,541, 4,530, 8,300 *to complete exercises 21–27.*

21. Which numbers are divisible by 2? 

22. Which numbers are divisible by 3?

23. Which numbers are divisible by 4?

24. Which numbers are divisible by 5?

25. Which numbers are divisible by 6? 

26. Which numbers are divisible by 9?

27. Which numbers are divisible by 10?

Use the list of numbers
36, 68, 91, 282, 464, 741, 1,840, 6,285, 8,444, 14,320 *to complete exercises 28–34.*

28. Which numbers are divisible by 2?

29. Which numbers are divisible by 3?

30. Which numbers are divisible by 4?

31. Which numbers are divisible by 5?

32. Which numbers are divisible by 6?

33. Which numbers are divisible by 9?

34. Which numbers are divisible by 10?

35. **Number Problem** A school auditorium is to have 350 seats. The principal wants to arrange them in rows with the same number of seats in each row. Use divisibility tests to determine if it is possible to have rows of 10 seats each. Are 15 rows of seats possible?

36. **Social Science** Dr. Mento has a class of 80 students. For a group project, she wants to divide the students into groups of 6, 8, or 10. Is this possible? Explain your answer.

Skills | Calculator/Computer | Career Applications | **Above and Beyond**

37. **Number Problem** Use the *sieve of Eratosthenes* to determine all the prime numbers less than 100.

	2	3	4	5	6	7	8	9	10
11	12	13	14	15	16	17	18	19	20
21	22	23	24	25	26	27	28	29	30
31	32	33	34	35	36	37	38	39	40
41	42	43	44	45	46	47	48	49	50
51	52	53	54	55	56	57	58	59	60
61	62	63	64	65	66	67	68	69	70
71	72	73	74	75	76	77	78	79	80
81	82	83	84	85	86	87	88	89	90
91	92	93	94	95	96	97	98	99	100

38. Why is the statement not a valid divisibility test for 8?

"A number is divisible by 8 if it is divisible by 2 and 4."

Support your answer with an example. Give a valid divisibility test for 8.

39. Prime numbers that differ by 2 are called *twin primes*. Examples are 3 and 5, 17 and 19, and so on. Find one pair of twin primes between 85 and 105.

40. Use the definition of twin primes (exercise 39) to answer these questions.

(a) Search for, and make a list of, several pairs of twin primes in which the primes are greater than 3.

(b) What do you notice about each number that lies *between* a pair of twin primes?

(c) Write an explanation for your observation in part (b).

41. Obtain (or imagine that you have) a quantity of square tiles. Six tiles can be arranged in the shape of a rectangle in two different ways:

(a) Record the dimensions of the rectangles shown.

(b) If you use 7 tiles, how many different rectangles can you form?

(c) If you use 10 tiles, how many different rectangles can you form?

(d) What kind of number (of tiles) permits *only one* arrangement into a rectangle? *More than one* arrangement?

42. The number 10 has 4 factors: 1, 2, 5, and 10. We can say that 10 has an even number of factors. Investigate several numbers to determine which numbers have an *even number* of factors and which numbers have an *odd number* of factors.

43. Number Problem Suppose that a school has 1,000 lockers and that they are all closed. A person passes through, opening every other locker, beginning with locker 2. Then another person passes through, changing every third locker (closing it if it is open, opening it if it is closed), starting with locker 3. Yet another person passes through, changing every fourth locker, beginning with locker 4. This process continues until 1,000 people pass through.

(a) At the end of this process, which locker numbers are closed?

(b) Write an explanation for your answer to part (a).
Hint: It may help to attempt exercise 42 first.

Answers

1. 1, 2, 4 **3.** 1, 2, 5, 10 **5.** 1, 3, 5, 15 **7.** 1, 2, 3, 4, 6, 8, 12, 24 **9.** 1, 2, 4, 8, 16, 32, 64 **11.** 1, 11 **13.** 1, 3, 5, 9, 15, 27, 45, 135 **15.** 1, 2, 4, 8, 16, 32, 64, 128, 256 **17.** 19, 23, 31, 59, 97, 103 **19.** 31, 37, 41, 43, 47 **21.** 72, 158, 260, 378, 570, 4,530, 8,300 **23.** 72, 260, 8,300 **25.** 72, 378, 570, 4,530 **27.** 260, 570, 4,530, 8,300 **29.** 36, 282, 741, 6,285 **31.** 1,840, 6,285, 14,320 **33.** 36 **35.** Yes; no **37.** Above and Beyond **39.** Above and Beyond **41.** Above and Beyond **43.** Above and Beyond

2.2 Factoring Whole Numbers

< 2.2 Objectives >

1 > Find the factors of a whole number

2 > Find the prime factorization for any number

3 > Find the greatest common factor (GCF) of two numbers

4 > Find the GCF for a group of numbers

To **factor a number** means to write the number as a product of its whole-number factors.

Example 1 Factoring a Composite Number

< Objective 1 >

Factor the number 10.

$10 = 2 \times 5$ The order in which you write the factors does not matter, so $10 = 5 \times 2$ is also correct.

Of course, $10 = 10 \times 1$ is also a correct statement. However, in this section we are interested in factors other than 1 and the given number.

Factor the number 21.

$21 = 3 \times 7$

Check Yourself 1

Factor 35.

When writing a composite number as a product of factors, there may be a number of different factorizations possible.

Example 2 Factoring a Composite Number

NOTE

There have to be at least two different factorizations, because a composite number has factors other than 1 and itself.

Find three different factorizations of 72.

$72 = 8 \times 9$

$= 6 \times 12$

$= 3 \times 24$

Check Yourself 2

Find three different factorizations of 42.

Next, we want to write a composite number as the product of its **prime factors.** Look again at the first factored line of Example 2. The process of factoring can be continued until all the factors are prime numbers.

Example 3 Factoring a Composite Number

< Objective 2 >

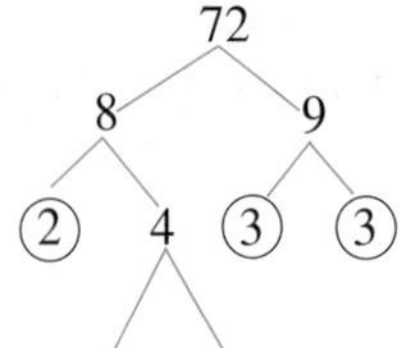

4 is not prime, so we continue by factoring 4.

We can now write 72 as a product of prime factors.

NOTES

This is often called a **factor tree.**

Finding the prime factorization of a number is important when working with fractions.

When we write 72 as $2 \times 2 \times 2 \times 3 \times 3$, no further factorization is possible. This is called the *prime factorization* of 72.

What if we start with a different factored line from the same example, $72 = 6 \times 12$?

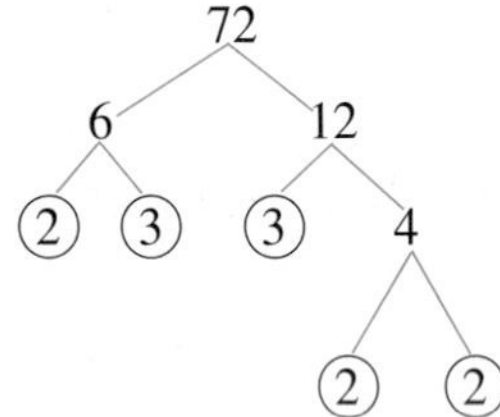

Continue to factor 6 and 12.

Continue again to factor 4. Other choices for the factors of 12 are possible. As we see, the end result is the same.

No matter which pair of factors you start with, you will find the same prime factorization. In this case, there are three factors of 2 and two factors of 3. Because **multiplication is commutative,** the order in which we write the factors does not matter, though we usually write them in size order.

Check Yourself 3

We could also begin

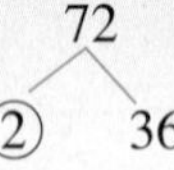

Continue the factorization.

Property

The Fundamental Theorem of Arithmetic There is exactly one prime factorization for any composite number.

The method of Example 3 always works. However, another method for factoring composite numbers exists. This method is particularly useful with large numbers because factor trees become unwieldy.

Property

Factoring by Division To find the prime factorization of a number, divide the number by a series of primes until the final quotient is a prime number.

The prime factorization is then the product of all the prime divisors and the final quotient, as we see in Example 4.

Example 4 Finding the Prime Factorization

NOTE

Do you see how the divisibility tests are used here? 60 is divisible by 2, 30 is divisible by 2, and 15 is divisible by 3.

To write 60 as a product of prime factors, divide 2 into 60 for a quotient of 30. Continue to divide by 2 again for the quotient of 15. Because 2 does not divide evenly into 15, we try 3. Since the quotient 5 is prime, we are done.

$2\overline{)60}$ → 30, $2\overline{)30}$ → 15, $3\overline{)15}$ → 5 Prime

Our factors are the prime divisors and the final quotient. We have

$60 = 2 \times 2 \times 3 \times 5$

Check Yourself 4

Complete the process to find the prime factorization of 90.

$2\overline{)90}$ → 45, $?\overline{)45}$ → ?

Remember to continue until the final quotient is prime.

Writing composite numbers in their completely factored form can be simplified if we use a format called **continued division.**

Example 5 Finding Prime Factors Using Continued Division

NOTE

In each short division, we write the quotient *below* rather than above the dividend. This is just a convenience for the next division.

Use the continued-division method to divide 60 by a series of prime numbers.

$$\begin{array}{r} 2\overline{)60} \\ 2\overline{)30} \\ 3\overline{)15} \\ 5 \end{array}$$

Primes (2, 2, 3)

Stop when the final quotient is prime.

To write the factorization of 60, we list each divisor used and the final prime quotient. In our example, we have

$60 = 2 \times 2 \times 3 \times 5$

Check Yourself 5

Find the prime factorization of 234.

NOTE

The factors of 20 are all less than or equal to 20.

We know that a factor or a divisor of a whole number divides that number exactly. The factors or divisors of 20 are

1, 2, 4, 5, 10, 20

Each of these numbers divides 20 exactly; that is, with no remainder.

Our work in this section involves common factors or divisors. A **common factor** or **divisor** for two numbers is any factor that divides both numbers exactly.

Example 6 Finding Common Factors

Look at the numbers 20 and 30. Is there a common factor for the two numbers?

First, we list the factors. Then we circle the ones that appear in both lists.

Factors

20: (1), (2), 4, (5), (10), 20

30: (1), (2), 3, (5), 6, (10), 15, 30

We see that 1, 2, 5, and 10 are common factors of 20 and 30. Each of these numbers divides both 20 and 30 exactly.

Check Yourself 6

List all common factors of 30 and 36.

Our later work with fractions requires that we find the greatest common factor of a group of numbers.

Definition

Greatest Common Factor

The **greatest common factor** (GCF) of a group of numbers is the *largest* number that divides each of the given numbers exactly.

Example 7 Finding the Greatest Common Factor

< Objective 3 >

In Example 6, we found the common factors of the numbers 20 and 30.

1, 2, 5, 10 Common factors of 20 and 30

The greatest common factor of the two numbers is 10, because 10 is the *largest* of the four common factors.

Check Yourself 7

Find the greatest common factor of 30 and 36.

The method of Example 7 also works in finding the greatest common factor of more than two numbers.

Example 8 Finding the Greatest Common Factor by Listing Factors

< Objective 4 >

NOTES

Looking at the three lists, we see that 1, 2, 3, and 6 are common factors.

If there are no common prime factors, the GCF is 1.

Find the GCF of 24, 30, and 36. We list the factors of each of the three numbers.

24: (1), (2), (3), 4, (6), 8, 12, 24

30: (1), (2), (3), 5, (6), 10, 15, 30

36: (1), (2), (3), 4, (6), 9, 12, 18, 36

So 6 is the greatest common factor of 24, 30, and 36.

Check Yourself 8

Find the greatest common factor of 16, 24, and 32.

The process shown in Example 8 is very time-consuming when larger numbers are involved. A better approach to the problem of finding the GCF of a group of numbers uses the prime factorization of each number. We outline the process here.

Step by Step

Finding the Greatest Common Factor

Step 1 Write the prime factorization for each of the numbers in the group.

Step 2 Locate the prime factors that are *common* to all the numbers.

Step 3 The greatest common factor (GCF) is the *product* of all the common prime factors.

Example 9 Finding the Greatest Common Factor

(a) Find the GCF of 20 and 30.

Step 1 Write the prime factorizations of 20 and 30.

$20 = 2 \times 2 \times 5$

$30 = 2 \times 3 \times 5$

Step 2 Find the prime factors common to each number.

$20 = \textcircled{2} \times 2 \times \textcircled{5}$

$30 = \textcircled{2} \times 3 \times \textcircled{5}$

2 and 5 are the common prime factors.

Step 3 Form the product of the common prime factors.

$2 \times 5 = 10$

So 10 is the greatest common factor.

(b) Find the GCF of 72 and 168.

Step 1 Find the prime factorization of each number.

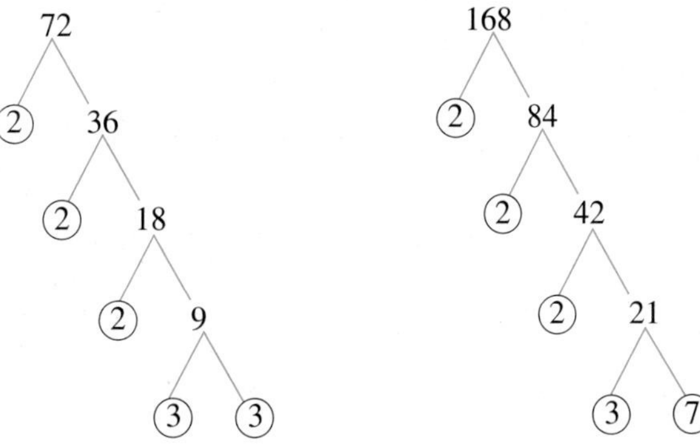

$72 = \textcircled{2} \times \textcircled{2} \times \textcircled{2} \times \textcircled{3} \times 3$

$168 = \textcircled{2} \times \textcircled{2} \times \textcircled{2} \times \textcircled{3} \times 7$

Step 2 Find the common prime factors.

2, 2, 2, and 3 are the common prime factors.

Step 3 Form the product of the common prime factors.

$2 \times 2 \times 2 \times 3 = 24$

24 is the GCF of 72 and 168.

Check Yourself 9

Find the GCF of each pair of numbers.

(a) 36 and 60 **(b)** 147 and 210

To find the greatest common factor of a group of more than two numbers, we use the same process.

Example 10 Finding the Greatest Common Factor

Find the GCF of 24, 30, and 36.

$24 = \textcircled{2} \times 2 \times 2 \times \textcircled{3}$

$30 = \textcircled{2} \times \textcircled{3} \times 5$

$36 = \textcircled{2} \times 2 \times \textcircled{3} \times 3$

So 2 and 3 are the prime factors common to *all three numbers.*

And $2 \times 3 = 6$ is the GCF.

Check Yourself 10

Find the GCF of 15, 30, and 45.

Example 11 Finding the Greatest Common Factor

NOTE

If two numbers, such as 15 and 28, have no common factor other than 1, they are called relatively prime.

Find the greatest common factor of 15 and 28.

$15 = 3 \times 5$

$28 = 2 \times 2 \times 7$

There are no common prime factors listed. But remember that 1 is a factor of every whole number.

The greatest common factor of 15 and 28 is 1.

Check Yourself 11

Find the greatest common factor of 30 and 49.

Check Yourself ANSWERS

1. 5×7 **2.** $2 \times 21, 3 \times 14, 6 \times 7$ **3.** $2 \times 2 \times 2 \times 3 \times 3$ **4.** $2\overline{)90} = 45 \rightarrow 3\overline{)45} = 15 \rightarrow 3\overline{)15} = 5$
$90 = 2 \times 3 \times 3 \times 5$ **5.** $2 \times 3 \times 3 \times 13$ **6.** 1, 2, 3, and 6 **7.** 6 **8.** 8
9. **(a)** 12; **(b)** 21 **10.** 15 **11.** 1 **Note:** 30 and 49 are relatively prime.

Reading Your Text

These fill-in-the-blank exercises will help you understand some of the key vocabulary used in this section. The answers to these exercises are in the Answers Appendix at the back of the text.

(a) Because multiplication is ______________, the order in which we write factors does not matter.

(b) There is exactly one ______________ factorization for any whole number.

(c) A ______________ factor for two numbers is any factor that divides both numbers exactly.

(d) GCF is an abbreviation for ______________ common factor.

Skills | Calculator/Computer | Career Applications | Above and Beyond

2.2 exercises

< Objectives 1 and 2 >

Find the prime factorization of each number.

1. 18 **2.** 22 **3.** 30 **4.** 35

5. 51 **6.** 42 **7.** 66 **8.** 100

9. 130 **10.** 88 **11.** 315 **12.** 400

13. 225 **14.** 132 **15.** 189 **16.** 330

In later mathematics courses, you sometimes want to find factors of a number with a given sum or difference. These exercises use this technique.

17. Find two factors of 48 with a sum of 14.

18. Find two factors of 48 with a sum of 26.

19. Find two factors of 48 with a difference of 8.

20. Find two factors of 48 with a difference of 2.

21. Find two factors of 24 with a sum of 10.

22. Find two factors of 15 with a difference of 2.

23. Find two factors of 30 with a difference of 1.

24. Find two factors of 28 with a sum of 11.

< Objective 3 >

Find the greatest common factor for each pair of numbers.

25. 4 and 6 **26.** 6 and 9 **27.** 10 and 15 **28.** 12 and 14

29. 21 and 24 **30.** 22 and 33 **31.** 20 and 21 **32.** 28 and 42

33. 18 and 24 **34.** 35 and 36 **35.** 18 and 54 **36.** 12 and 48

< Objective 4 >

Find the GCF for each group of numbers.

37. 12, 36, and 60 **38.** 15, 45, and 90 **39.** 105, 140, and 175 **40.** 17, 19, and 31

41. 25, 75, and 150 **42.** 36, 72, and 144

Fill in each blank with either **always, sometimes,** *or* **never.**

43. Factors of a composite number ____________ include 1 and the number itself.

44. Factors of a prime number ____________ include 1 and the number itself.

45. A number with a repeated factor is ____________ a prime number.

46. Factors of an even number are ____________ even numbers.

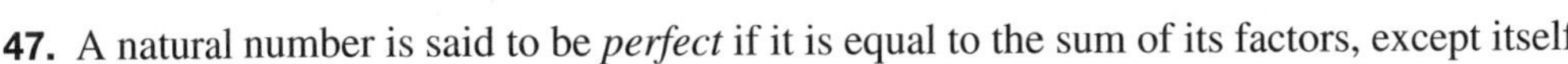

47. A natural number is said to be *perfect* if it is equal to the sum of its factors, except itself.

(a) Show that 28 is a perfect number.

(b) Identify another perfect number less than 28.

48. Find the smallest natural number that is divisible by 2, 3, 4, 6, 8, and 9.

49. Social Science Tom and Dick both work the night shift at the steel mill. Tom has every sixth night off, and Dick has every eighth night off. If they both have August 1 off, when will they both be off together again?

50. **Science and Medicine** Mercury, Venus, and Earth revolve around the sun once every 3, 7, and 12 months, respectively. If the three planets are now in the same straight line, what is the smallest number of months that must pass before they line up again?

Answers

1. $2 \times 3 \times 3$ **3.** $2 \times 3 \times 5$ **5.** 3×17 **7.** $2 \times 3 \times 11$ **9.** $2 \times 5 \times 13$ **11.** $3 \times 3 \times 5 \times 7$ **13.** $3 \times 3 \times 5 \times 5$ **15.** $3 \times 3 \times 3 \times 7$ **17.** 6, 8 **19.** 4, 12 **21.** 4, 6 **23.** 5, 6 **25.** 2 **27.** 5 **29.** 3 **31.** 1 **33.** 6 **35.** 18 **37.** 12 **39.** 35 **41.** 25 **43.** always **45.** never **47.** **(a)** $1 + 2 + 4 + 7 + 14 = 28$; **(b)** 6 **49.** August 25

2.3 Fraction Basics

< 2.3 Objectives >

1 > Identify the numerator and denominator of a fraction

2 > Use fractions to name parts of a whole

3 > Identify proper and improper fractions

4 > Write improper fractions as mixed numbers

5 > Write mixed numbers as improper fractions

Previous sections dealt with whole numbers and the operations that are performed on them. We are now ready to consider a new kind of number, a **fraction.**

Definition

Fraction Whenever a unit or a whole quantity is divided into parts, we call those parts **fractions** of the unit.

NOTES

Our word *fraction* comes from the Latin stem *fractio*, which means "breaking into pieces."

Common fraction is technically the correct term. We just use *fraction* in this text.

In this figure, the whole is divided into five equal parts. We use the symbol $\frac{2}{5}$ to represent the shaded part of the whole.

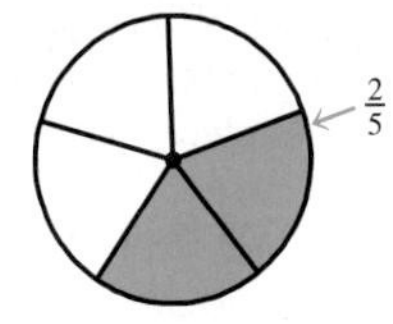

The symbol $\frac{2}{5}$ is called a **common fraction,** or more simply a fraction. A fraction is written in the form $\frac{a}{b}$, in which a and b represent whole numbers and b cannot be equal to 0.

The numbers a and b in the fraction $\frac{a}{b}$ have special names. The number on the bottom b is called the **denominator** and it tells us how many equal parts the unit or whole has been divided into. The number on top a is called the **numerator** and it tells us how many parts of the unit or whole are used.

In the fraction $\frac{2}{5}$, the *denominator* is 5; the unit or whole (the circle) is divided into five equal parts. The *numerator* is 2. Two parts of the unit are shaded.

$\frac{2}{5}$ ⟵ Numerator / ⟵ Denominator

Example 1 Labeling Fraction Components

< Objectives 1 and 2 >

The fraction $\frac{4}{7}$ names the shaded part of the rectangle in the figure.

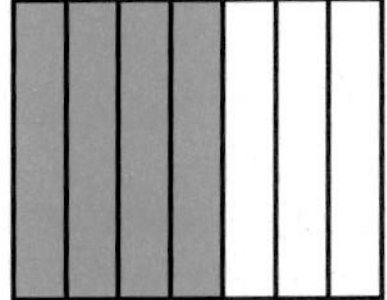

The unit or whole is divided into seven equal parts, so the denominator is 7. Four of those parts are shaded, so we have a numerator of 4.

Check Yourself 1

What fraction names the shaded part of this diagram? Identify the numerator and denominator.

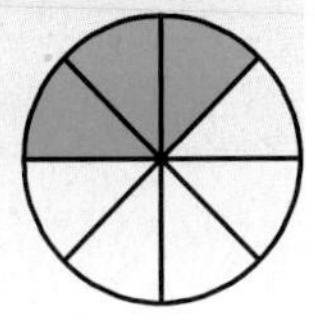

Fractions can also be used to name a part of a collection or a set of identical objects.

Example 2 Naming a Fractional Part

The fraction $\frac{5}{6}$ names the proportion of objects that are shaded in the figure. We have shaded five of the six identical objects.

Check Yourself 2

What fraction names the proportion of squares that are shaded?

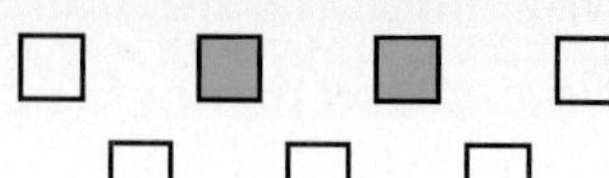

Example 3 Naming a Fractional Part

NOTES

The fraction $\frac{8}{23}$ names the part of the class that is not women.

$\frac{a}{b}$ names the *quotient* when a is divided by b. Of course, b cannot be 0.

In a class of 23 students, 15 are women. We can name the part of the class that is women as $\frac{15}{23}$.

Check Yourself 3

Seven replacement parts out of a shipment of 50 were faulty. What fraction names the portion of the shipment that was faulty?

A fraction also represents division. The symbol $\frac{a}{b}$ also means $a \div b$.

Example 4 Interpreting a Fraction as Division

The fraction $\frac{2}{3}$ names the quotient when 2 is divided by 3. So $\frac{2}{3} = 2 \div 3$.

Note: $\frac{2}{3}$ can be read as "two-thirds" or as "2 divided by 3."

Check Yourself 4

Write $\frac{5}{9}$ using division.

We can use the relative size of the numerator and denominator of a fraction to separate fractions into two different categories.

Definition

Proper Fraction

If the numerator is *less than* the denominator, the fraction names a number less than 1 and is called a **proper fraction.**

Definition

Improper Fraction

If the numerator is *greater than or* equal to the denominator, the fraction names a number greater than or equal to 1 and is called an **improper fraction.**

Example 5 Categorizing Fractions

< Objective 3 >

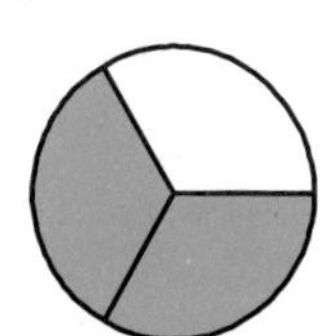

(a) $\frac{2}{3}$ is a proper fraction because the numerator is less than the denominator.

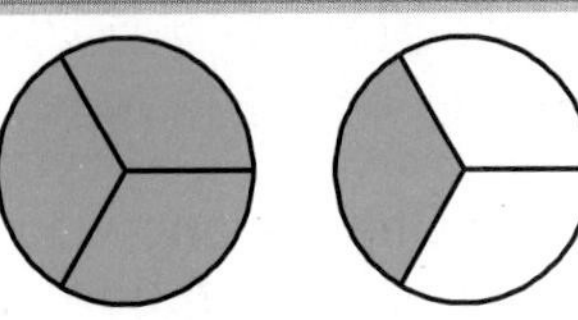

(b) $\frac{4}{3}$ is an improper fraction because the numerator is larger than the denominator.

(c) Also, $\frac{6}{6}$ is an improper fraction because it names exactly 1 unit; the numerator is equal to the denominator.

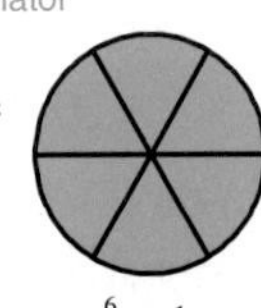

NOTE

In the figure for part (b), the circle on the left is divided into 3 parts and all three parts are shaded, so it represents $\frac{3}{3}$.

Check Yourself 5

List the proper and improper fractions.

$$\frac{5}{4}, \frac{10}{11}, \frac{3}{4}, \frac{8}{5}, \frac{6}{6}, \frac{13}{10}, \frac{7}{8}, \frac{15}{8}$$

Another way to write a fraction that is larger than 1 is as a **mixed number.** Most applications use mixed numbers rather than improper fractions.

Definition

Mixed Number A **mixed number** is the sum of a whole number and a proper fraction.

Example 6 Identifying a Mixed Number

NOTES

$2\frac{3}{4}$ means $2 + \frac{3}{4}$. In fact, we read the mixed number as "two *and* three-fourths." The plus sign is not written.

In subsequent courses, you will find that improper fractions are preferred to mixed numbers for abstract computations.

In most applications, mixed numbers are the preferred form.

The number $2\frac{3}{4}$ is a mixed number. It represents the sum of the whole number 2 and the fraction $\frac{3}{4}$. The figure represents $2\frac{3}{4}$.

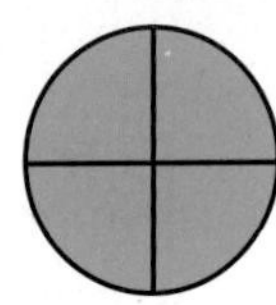

+
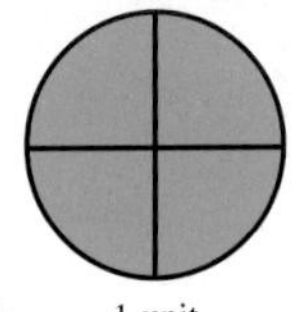

+
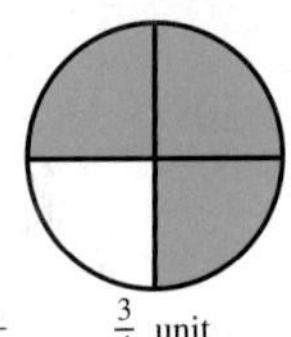

Check Yourself 6

Give the mixed number that names the shaded portion of the diagram.

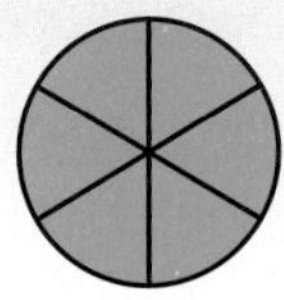
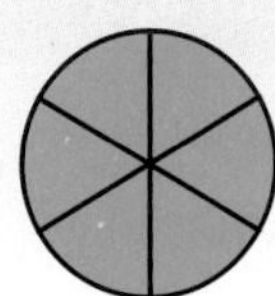
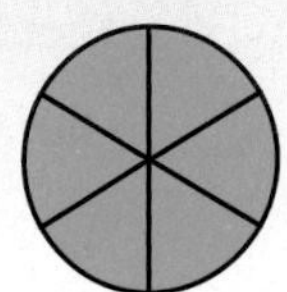
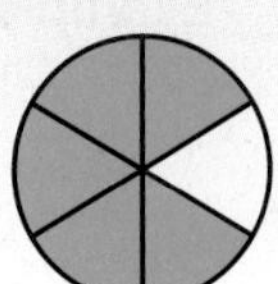

It is important to be able to change back and forth between improper fractions and mixed numbers. Because an improper fraction represents a number that is greater than or equal to 1, we can state a relevant property.

Property

Improper Fractions to Mixed Numbers An improper fraction can always be written as either a mixed number or a whole number.

To do this, remember that a fraction indicates division. The numerator is divided by the denominator. This leads us to a process.

Step by Step

To Change an Improper Fraction to a Mixed Number

Step 1 Divide the numerator by the denominator.

Step 2 If there is a remainder, write the remainder over the original denominator.

Example 7 Converting a Fraction to a Mixed Number

< Objective 4 >

Convert $\frac{17}{5}$ to a mixed number.

Divide 17 by 5.

$$5\overline{)17}\ \text{quotient } 3,\quad 15,\quad \text{remainder } 2 \qquad \frac{17}{5} = 3\frac{2}{5}$$

(Remainder: 2; Original denominator: 5; Quotient: 3)

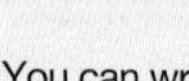

NOTES

You can write the fraction $\frac{17}{5}$ as $17 \div 5$. We divide the numerator by the denominator.

In step 1, the quotient gives the whole-number portion of the mixed number.
Step 2 gives the fraction part of the mixed number.

In diagram form,

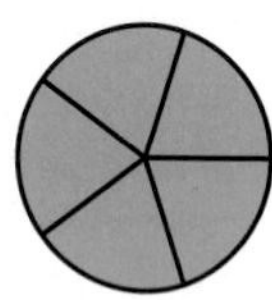 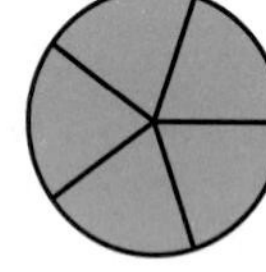 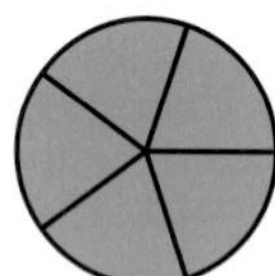 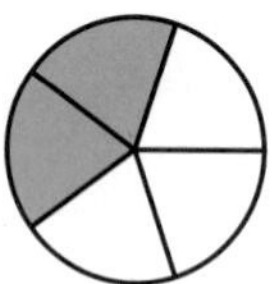

$\frac{17}{5} = 3\frac{2}{5}$

Check Yourself 7

Convert $\frac{32}{5}$ to a mixed number.

Example 8 Converting a Fraction to a Mixed Number

NOTES

If there is *no* remainder, the improper fraction is equal to some whole number, in this case 3.

Any number divided by itself is equal to one.

Any fraction with the same numerator and denominator equals 1.

(a) Convert $\frac{21}{7}$ to a whole or mixed number.

Divide 21 by 7.

$$7\overline{)21}\ \text{quotient } 3,\quad 21,\quad 0 \qquad \frac{21}{7} = 3$$

(b) Convert $\frac{8}{8}$ to a whole or mixed number.

Because a fraction represents division, we have

$$\frac{8}{8} = 8 \div 8$$
$$= 1$$

Check Yourself 8

Convert each improper fraction to a whole or mixed number.

(a) $\frac{48}{6}$ **(b)** $\frac{12}{12}$

You also need to convert mixed numbers to improper fractions. Here is how we do that.

Step by Step

To Change a Mixed Number to an Improper Fraction

Step 1 Multiply the denominator of the fraction by the whole-number part of the mixed number.

Step 2 Add the numerator of the fraction to that product.

Step 3 Write that sum over the original denominator to form the improper fraction.

Example 9 Converting Mixed Numbers to Improper Fractions

< Objective 5 >

(a) Convert $3\frac{2}{5}$ to an improper fraction.

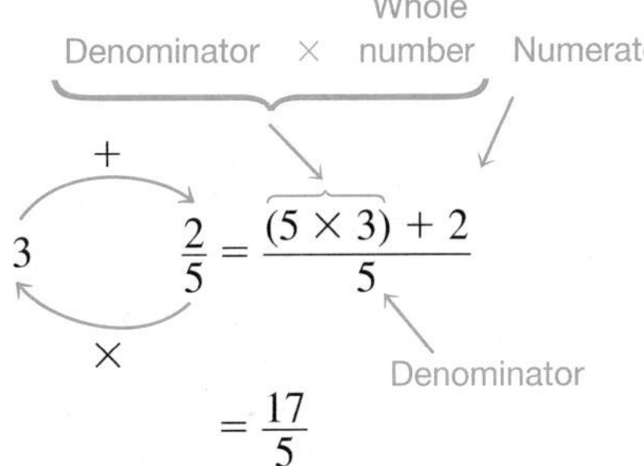

Multiply the denominator by the whole number (5 × 3 = 15). Add the numerator. We now have 17.

$= \frac{17}{5}$

Write 17 over the original denominator.

In diagram form,

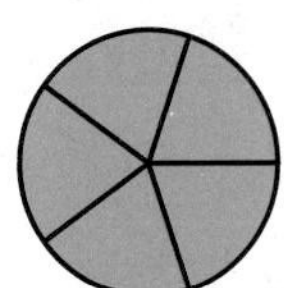 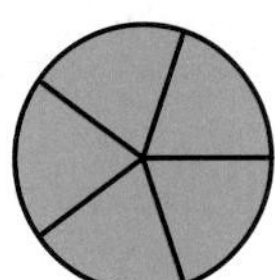 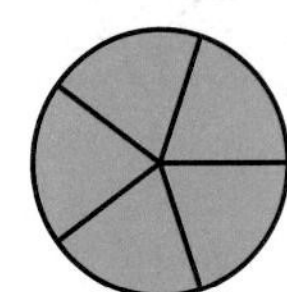 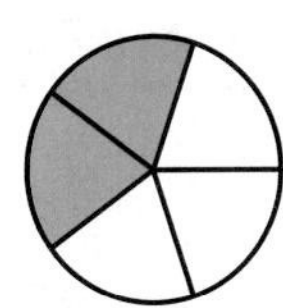

Each of the three units has 5 fifths, so the whole-number part is 5 × 3, or 15, fifths. Then add the $\frac{2}{5}$ from the fraction part for $\frac{17}{5}$.

(b) Convert $4\frac{5}{7}$ to an improper fraction.

$$4\frac{5}{7} = \frac{(7 \times 4) + 5}{7} = \frac{33}{7}$$

NOTE

Multiply the denominator, 7, by the whole number, 4, and add the numerator, 5.

Check Yourself 9

Convert $5\frac{3}{8}$ to an improper fraction.

One special kind of improper fraction should be mentioned at this point: a fraction with a denominator of 1.

Definition

Fractions with a Denominator of 1

Any fraction with a denominator of 1 is equal to the numerator alone. For example,

$\frac{5}{1} = 5$ and $\frac{12}{1} = 12$

This works because any number divided by 1 is equal to itself.

You probably do many conversions between mixed and whole numbers without even thinking about the process, as Example 10 illustrates.

Example 10 Converting Quarter-Dollars to Dollars

Maritza has 53 quarters in her bank. How many dollars does she have?

Because there are 4 quarters in each dollar, 53 quarters can be written as

$\frac{53}{4}$

Converting the amount to dollars is the same as rewriting it as a mixed number.

$\frac{53}{4} = 13\frac{1}{4}$

She has $13\frac{1}{4}$ dollars, which you would probably write as $13.25. (**Note:** We discuss decimals later in this text.)

Check Yourself 10

Kevin is doing the inventory in the convenience store in which he works. He finds there are 11 half-gallons of milk. Write the amount of milk as a mixed number of gallons.

Check Yourself ANSWERS

1. $\frac{3}{8}$ ← Numerator, ← Denominator **2.** $\frac{2}{7}$ **3.** $\frac{7}{50}$ **4.** $5 \div 9$ **5.** Proper fractions: $\frac{10}{11}, \frac{3}{4}, \frac{7}{8}$
Improper fractions: $\frac{5}{4}, \frac{8}{5}, \frac{6}{6}, \frac{13}{10}, \frac{15}{8}$ **6.** $3\frac{5}{6}$ **7.** $6\frac{2}{5}$ **8.** (a) 8; (b) 1 **9.** $\frac{43}{8}$ **10.** $5\frac{1}{2}$ gal

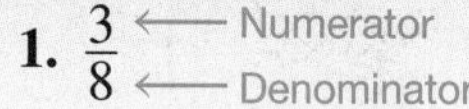

Reading Your Text

These fill-in-the-blank exercises will help you understand some of the key vocabulary used in this section. The answers to these exercises are in the Answers Appendix at the back of the text.

(a) Given a fraction like $\frac{3}{4}$, we call 4 the ______________ of the fraction.

(b) If the numerator is less than the denominator, the fraction names a number less than 1 and is called a ______________ fraction.

(c) An improper fraction can always be written as either a ______________ number or a whole number.

(d) Any fraction with a denominator of 1 is equal to the ______________ alone.

< Objective 1 >

Identify the numerator and denominator of each fraction.

1. $\frac{6}{11}$ **2.** $\frac{5}{12}$ **3.** $\frac{3}{11}$ **4.** $\frac{9}{14}$

< Objective 2 >

What fraction names the shaded part of each figure?

5. 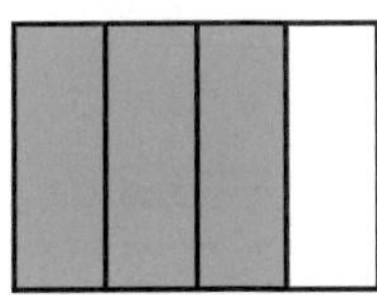

6. 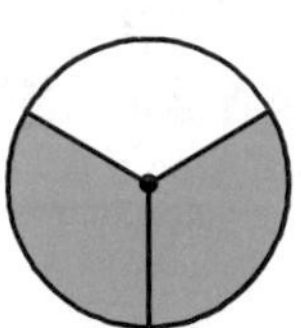

7. 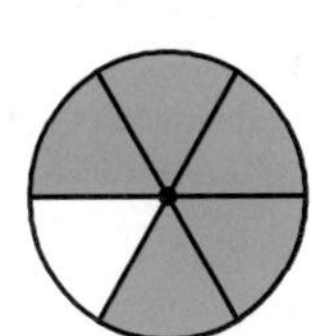

8. 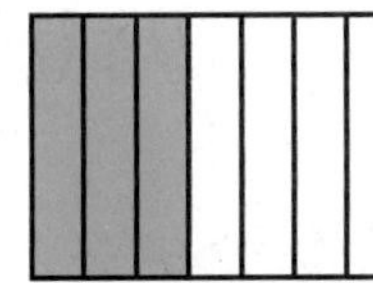

9. 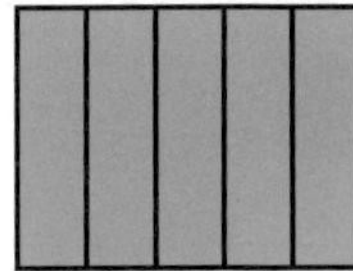

10. 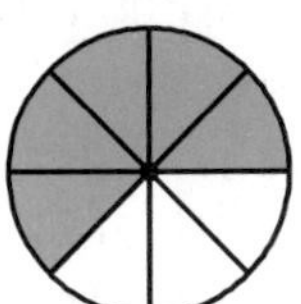

11. 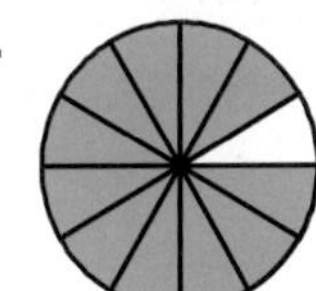

12. 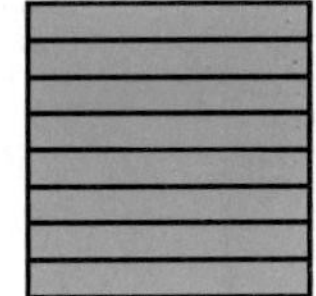

13. 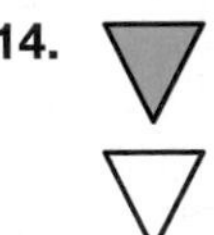

14. 

Solve each application.

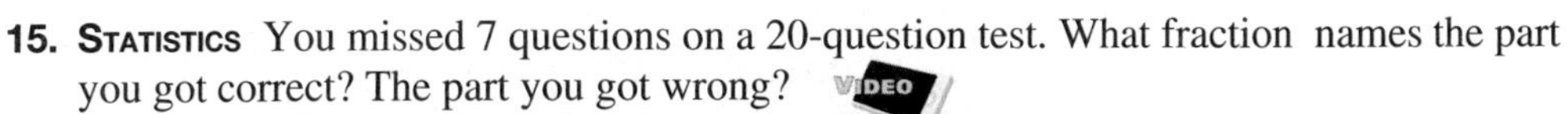

15. **Statistics** You missed 7 questions on a 20-question test. What fraction names the part you got correct? The part you got wrong?

16. **Statistics** Of the 5 starters on a basketball team, 2 fouled out of a game. What fraction names the part of the starting team that fouled out?

17. **Business and Finance** A used-car dealer sold 11 of the 17 cars in stock. What fraction names the portion sold? What fraction names the portion *not* sold?

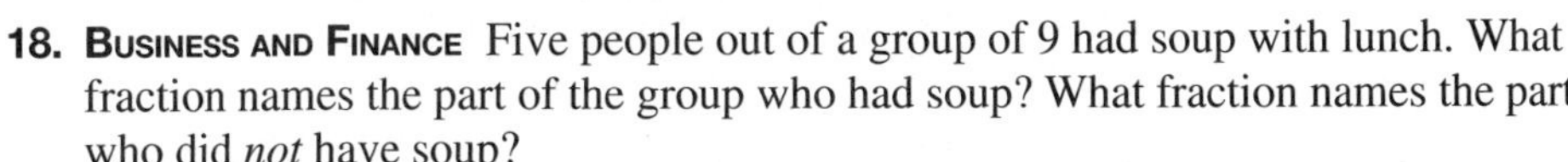

18. **Business and Finance** Five people out of a group of 9 had soup with lunch. What fraction names the part of the group who had soup? What fraction names the part who did *not* have soup?

19. Use division to show another way of writing $\frac{2}{5}$.

20. Use division to show another way of writing $\frac{4}{5}$.

< Objective 3 >

Identify each number as a proper fraction, an improper fraction, or a mixed number.

21. $\frac{3}{5}$ **22.** $\frac{9}{5}$ **23.** $2\frac{3}{5}$

24. $\frac{7}{9}$ **25.** $\frac{6}{6}$ **26.** $1\frac{1}{5}$

27. $\frac{13}{17}$ **28.** $\frac{16}{15}$

Give the mixed number that names the shaded portion of each diagram and write each as an improper fraction.

29.

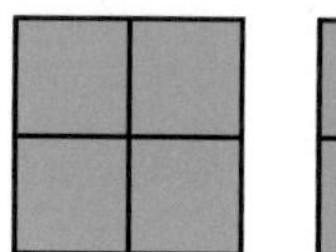

30.

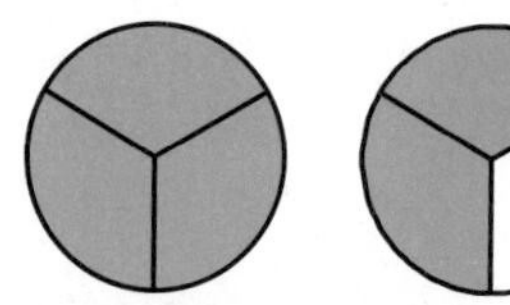

31.

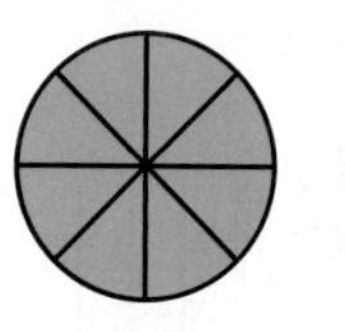

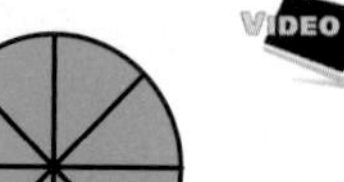

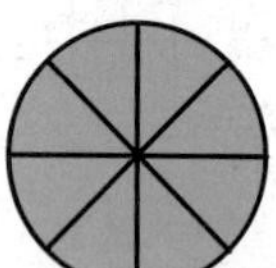

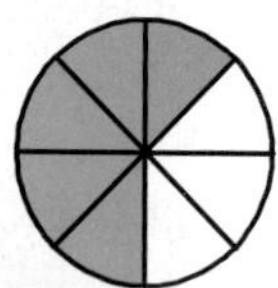

32.

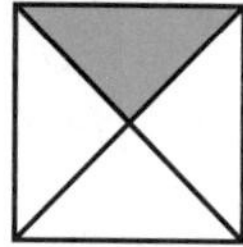

33.

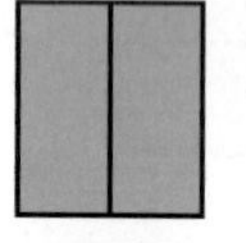

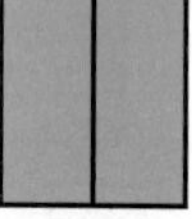

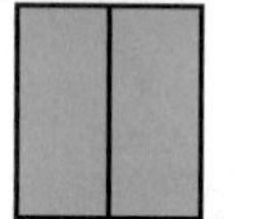

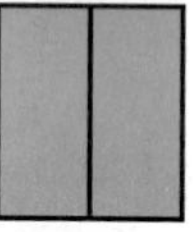

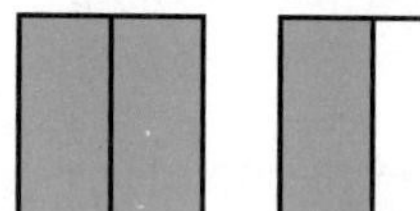

34.

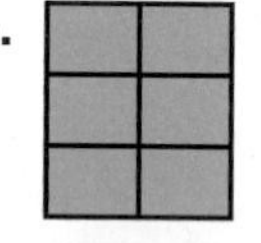

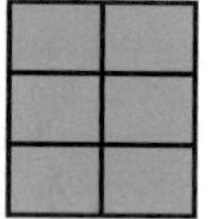

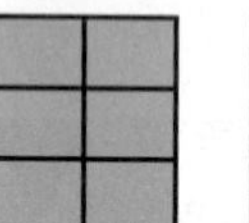

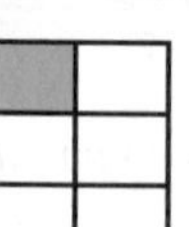

35.

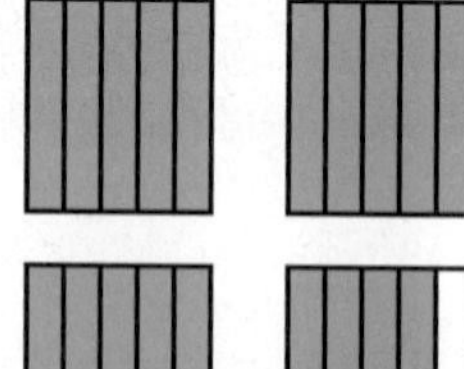

36.

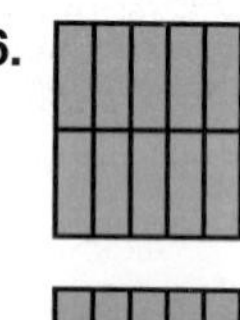

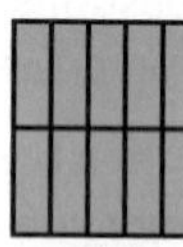

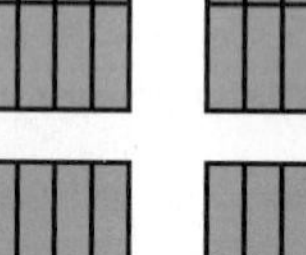

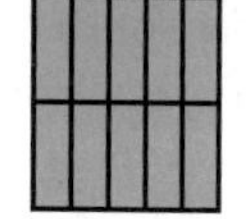

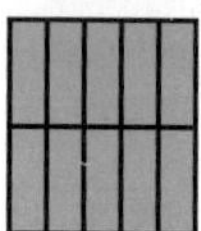

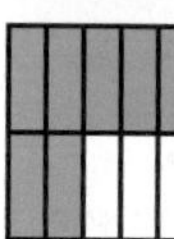

Sketch circles or squares, as in exercises 29–36, to display each fraction or mixed number in diagram form.

37. $\frac{1}{3}$

38. $\frac{2}{5}$

39. $3\frac{1}{4}$

40. $\frac{12}{5}$

41. $\frac{3}{1}$

42. $\frac{4}{2}$

Solve each application.

43. **Business and Finance** Clayton has 64 quarters in his change jar. How many dollars does he have?

44. **Business and Finance** Amy has 19 quarters in her purse. How many dollars does she have?

45. **Business and Finance** Manuel counted 35 half-gallons of orange juice in his store. Write the amount of orange juice as a mixed number of gallons.

46. **Business and Finance** Sarah has 19 half-gallons of turpentine in her paint store. Write the amount of turpentine as a mixed number of gallons.

< Objective 4 >

Convert each fraction to a whole or mixed number.

47. $\frac{22}{5}$ 

48. $\frac{27}{8}$

49. $\frac{34}{5}$

50. $\frac{25}{6}$

51. $\frac{73}{8}$

52. $\frac{151}{12}$

53. $\frac{24}{6}$

54. $\frac{160}{8}$

55. $\frac{9}{1}$

56. $\frac{8}{1}$

< Objective 5 >

Convert each whole or mixed number to an improper fraction.

57. $4\frac{2}{3}$

58. $2\frac{5}{6}$

59. 8

60. $4\frac{5}{8}$

61. $7\frac{6}{13}$

62. $7\frac{3}{10}$

63. $10\frac{2}{5}$

64. $13\frac{2}{5}$

65. $118\frac{3}{4}$

66. $250\frac{3}{4}$

67. 4

68. 10

69. 35

70. 7

Skills	Calculator/Computer	**Career Applications**	Above and Beyond

71. **Allied Health** A dilution contains 3 parts blood serum out of a total of 10 parts. Write this number as a fraction.

72. **Electronics** Write all the fraction values described in the paragraph as fractions.

Betsy ordered electronic components from her favorite supplier. She bought two dozen, one-quarter watt resistors, 10 light-emitting diodes (LEDs) that require ten-thirds of a volt (forward voltage) to illuminate, and one three-eighths henry inductor.

73. **Manufacturing Technology** In the packaging division of Early Enterprises, there are 36 packaging machines. At any given time, five of the machines are shut down for scheduled maintenance and service. What is the fraction of machines that are operating at one time?

74. **Information Technology** On a visit to a wiring closet, Joseph finds a rack of servers that has eight slots. He wants to know what fraction names the two already in the slots. Also, if he buys five more servers, what fraction

names the total servers installed in the slots? If he has to remove two servers later because of failure, what fraction names the total servers installed in the slots?

Skills	Calculator/Computer	Career Applications	**Above and Beyond**

75. **Social Science** U.S. Census information can be found in your library, or on the Web, at www.census.gov. Use the 2010 census to find each fraction.

(a) Fraction of the U.S. population living in your state

(b) Fraction of the U.S. population 65 years of age or older

(c) Fraction of the U.S. population that is female

Answers

1. 6 is the numerator; 11 is the denominator **3.** 3 is the numerator; 11 is the denominator **5.** $\frac{3}{4}$ **7.** $\frac{5}{6}$ **9.** $\frac{5}{5}$ **11.** $\frac{11}{12}$ **13.** $\frac{5}{8}$ **15.** Correct: $\frac{13}{20}$; wrong: $\frac{7}{20}$ **17.** Sold: $\frac{11}{17}$; not sold: $\frac{6}{17}$ **19.** $2 \div 5$ **21.** Proper **23.** Mixed number **25.** Improper **27.** Proper **29.** $1\frac{3}{4}$ or $\frac{7}{4}$ **31.** $3\frac{5}{8}$ or $\frac{29}{8}$ **33.** $5\frac{1}{2}$ or $\frac{11}{2}$ **35.** $3\frac{4}{5}$ or $\frac{19}{5}$ **37.** **39.**

41. 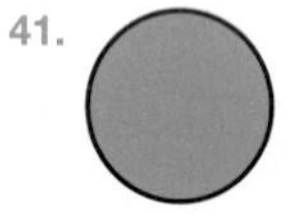   

43. \$16 **45.** $17\frac{1}{2}$ gal **47.** $4\frac{2}{5}$ **49.** $6\frac{4}{5}$ **51.** $9\frac{1}{8}$ **53.** 4 **55.** 9 **57.** $\frac{14}{3}$ **59.** $\frac{8}{1}$ **61.** $\frac{97}{13}$ **63.** $\frac{52}{5}$ **65.** $\frac{475}{4}$ **67.** $\frac{4}{1}$ **69.** $\frac{35}{1}$ **71.** $\frac{3}{10}$ **73.** $\frac{31}{36}$ **75.** Above and Beyond

2.4 Simplifying Fractions

< 2.4 Objectives >

1 > Determine whether two fractions are equivalent

2 > Use the fundamental principle to simplify fractions

It is possible to represent the same portion of a whole by different fractions. Look at the figures representing $\frac{3}{6}$ and $\frac{1}{2}$. The two fractions are simply different names for the same amount, so they are called **equivalent fractions.**

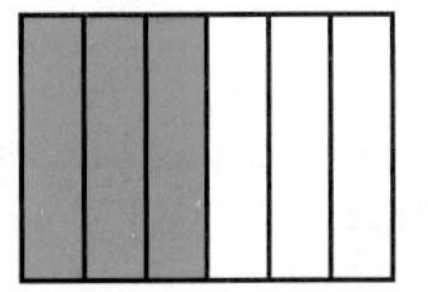

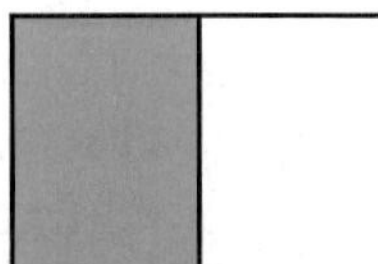

Any fraction has many equivalent fractions. For instance, $\frac{2}{3}$, $\frac{4}{6}$, and $\frac{6}{9}$ are all equivalent fractions because they name the same part of a unit.

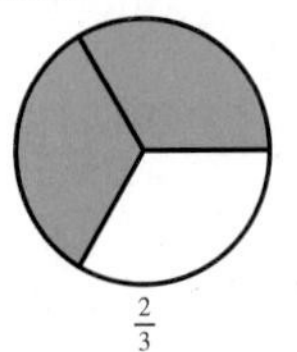

$\frac{2}{3}$

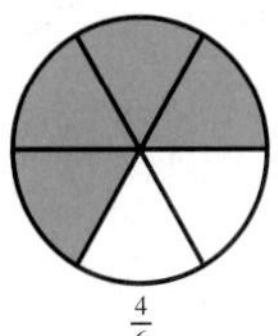

$\frac{4}{6}$

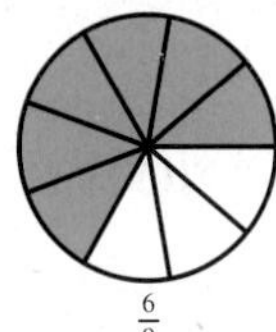

$\frac{6}{9}$

Many more fractions are equivalent to $\frac{2}{3}$. All these fractions can be used interchangeably. An easy way to find out if two fractions are equivalent is to use cross products.

$\frac{a}{b} \times \frac{c}{d}$ We call $a \times d$ and $b \times c$ the **cross products.**

Property

Testing for Equivalence If the **cross products** for two fractions are equal, the two fractions are equivalent.

Example 1 Identifying Equivalent Fractions Using Cross Products

< Objective 1 >

(a) Are $\frac{3}{24}$ and $\frac{4}{32}$ equivalent fractions?

The cross products are 3×32, or 96, and 24×4, or 96. Because the cross products are equal, the fractions are equivalent.

(b) Are $\frac{2}{5}$ and $\frac{3}{7}$ equivalent fractions?

The cross products are 2×7 and 5×3.

$2 \times 7 = 14$ and $5 \times 3 = 15$

Because $14 \neq 15$, the fractions are *not* equivalent.

Check Yourself 1

(a) Are $\frac{3}{8}$ and $\frac{9}{24}$ equivalent fractions?

(b) Are $\frac{7}{8}$ and $\frac{8}{9}$ equivalent fractions?

In writing equivalent fractions, we use an important principle.

Property

The Fundamental Principle of Fractions For the fraction $\frac{a}{b}$ and any nonzero number c,

$$\frac{a}{b} = \frac{a \div c}{b \div c}$$

NOTE

In Sections 2.5 and 2.6, you will see that the fundamental principle of fractions works because we are really multiplying or dividing the entire fraction by 1.

The fundamental principle of fractions tells us that we can divide the numerator and denominator by the same nonzero number. The result is an equivalent fraction. For instance,

$$\frac{2}{4} = \frac{2 \div 2}{4 \div 2} = \frac{1}{2} \qquad \frac{3}{6} = \frac{3 \div 3}{6 \div 3} = \frac{1}{2} \qquad \frac{4}{8} = \frac{4 \div 4}{8 \div 4} = \frac{1}{2}$$

$$\frac{5}{10} = \frac{5 \div 5}{10 \div 5} = \frac{1}{2} \qquad \frac{6}{12} = \frac{6 \div 6}{12 \div 6} = \frac{1}{2} \qquad \frac{7}{14} = \frac{7 \div 7}{14 \div 7} = \frac{1}{2}$$

Simplifying a fraction or *reducing a fraction to lower terms* means finding an equivalent fraction with a *smaller* numerator and denominator than those of the original fraction. Dividing the numerator and denominator by the same nonzero number does exactly that.

Example 2 Simplifying Fractions

< Objective 2 >

Simplify each fraction.

NOTES

We apply the fundamental principle to divide the numerator and denominator by 5.

We divide the numerator and denominator by 2.

(a) $\frac{5}{15} = \frac{5 \div 5}{15 \div 5} = \frac{1}{3}$

$\frac{5}{15}$ and $\frac{1}{3}$ are equivalent fractions. Check this by finding the cross products.

(b) $\frac{4}{8} = \frac{4 \div 2}{8 \div 2} = \frac{2}{4}$

$\frac{4}{8}$ and $\frac{2}{4}$ are equivalent fractions.

Check Yourself 2

Write two fractions that are equivalent to $\frac{30}{45}$.

(a) Divide the numerator and denominator by 5.
(b) Divide the numerator and denominator by 15.

NOTE

In the case of $\frac{2}{4}$, the numerator and denominator are *not* as small as possible. The numerator and denominator have a common factor of 2.

We say that a fraction is in **simplest form,** or in **lowest terms,** if the numerator and denominator have no common factors other than 1. This means that the fraction has the smallest possible numerator and denominator.

In Example 2, $\frac{1}{3}$ is in simplest form because the numerator and denominator have no common factors other than 1. The fraction is in lowest terms.

$\frac{2}{4}$ is *not* in simplest form. Do you see that $\frac{2}{4}$ can also be written as $\frac{1}{2}$?

To write a fraction in simplest form or to *reduce a fraction to lowest terms,* divide the numerator and denominator by their GCF.

Example 3 Simplifying Fractions

Write $\frac{10}{15}$ in simplest form.

From our work earlier in this chapter, we know that the greatest common factor of 10 and 15 is 5. To write $\frac{10}{15}$ in simplest form, divide the numerator and denominator by 5.

$$\frac{10}{15} = \frac{10 \div 5}{15 \div 5} = \frac{2}{3}$$

The resulting fraction, $\frac{2}{3}$, is in lowest terms.

Check Yourself 3

Write $\frac{12}{18}$ in simplest form.

Many students prefer to simplify fractions by using the prime factorizations of the numerator and denominator. Example 4 uses this method.

Example 4 Factoring to Simplify a Fraction

NOTE

Using the prime factorizations of 24 and 42, we divide by the common factors of 2 and 3.

(a) Simplify $\frac{24}{42}$.

To simplify $\frac{24}{42}$, factor.

$$\frac{24}{42} = \frac{\overset{1}{\cancel{2}} \times 2 \times 2 \times \overset{1}{\cancel{3}}}{\underset{1}{\cancel{2}} \times \underset{1}{\cancel{3}} \times 7} = \frac{4}{7}$$

Note: The numerator of the simplified fraction is the *product* of the prime factors remaining in the numerator after dividing by 2 and 3.

(b) Simplify $\frac{120}{180}$.

To reduce $\frac{120}{180}$ to lowest terms, write the prime factorizations of the numerator and denominator. Then divide by any common factors.

$$\frac{120}{180} = \frac{\overset{1}{\cancel{2}} \times \overset{1}{\cancel{2}} \times 2 \times \overset{1}{\cancel{3}} \times \overset{1}{\cancel{5}}}{\underset{1}{\cancel{2}} \times \underset{1}{\cancel{2}} \times \underset{1}{\cancel{3}} \times 3 \times \underset{1}{\cancel{5}}} = \frac{2}{3}$$

Check Yourself 4

Write each fraction in simplest form.

(a) $\frac{60}{75}$ **(b)** $\frac{210}{252}$

Another way to simplify fractions uses the fundamental principle to divide the numerator and denominator by any common factors. We illustrate this with the fractions considered in Example 4.

Example 5 Using Common Factors to Simplify Fractions

(a) Simplify $\frac{24}{42}$.

$$\frac{24}{42} = \frac{\overset{12}{\cancel{24}}}{\underset{21}{\cancel{42}}} = \frac{\overset{4}{\cancel{12}}}{\underset{7}{\cancel{21}}} = \frac{4}{7}$$

Divide by the common factor of 2. Divide by the common factor of 3.

The original numerator and denominator are both divisible by 2, and so we divide by that factor to arrive at $\frac{12}{21}$. Our divisibility tests tell us that a common factor of 3 still exists. (Do you remember why?) Divide again for the result $\frac{4}{7}$, which is in lowest terms.

Note: If we had seen the GCF of 6 at first, we could have divided by 6 and arrived at the same result in one step.

(b) Simplify $\frac{120}{180}$.

$$\frac{120}{180} = \frac{\overset{\overset{2}{\cancel{20}}}{\cancel{120}}}{\underset{\underset{3}{\cancel{30}}}{\cancel{180}}} = \frac{2}{3}$$

Our first step is to divide by the common factor of 6. We then have $\frac{20}{30}$. There is still a common factor of 10, so we again divide.

Again, we could have divided by the GCF of 60 in one step if we had recognized it.

Check Yourself 5

Using the method of Example 5, write each fraction in simplest form.

(a) $\frac{60}{75}$ **(b)** $\frac{84}{196}$

Check Yourself ANSWERS

1. (a) Yes; **(b)** no **2. (a)** $\frac{6}{9}$; **(b)** $\frac{2}{3}$ **3.** $\frac{2}{3}$ **4. (a)** $\frac{4}{5}$; **(b)** $\frac{5}{6}$ **5. (a)** Divide by the common factors of 3 and 5, $\frac{60}{75} = \frac{4}{5}$; **(b)** Divide by the common factors of 4 and 7, $\frac{84}{196} = \frac{3}{7}$

Reading Your Text

These fill-in-the-blank exercises will help you understand some of the key vocabulary used in this section. The answers to these exercises are in the Answers Appendix at the back of the text.

(a) If the ____________ products of two fractions are equal, the two fractions are equivalent.

(b) Two fractions that are simply different names for the same fraction are called ____________ fractions.

(c) In writing equivalent fractions, we use the ____________ principle of fractions.

(d) We say that a fraction is in simplest form if the numerator and denominator have no ____________ factors other than 1.

Skills | Calculator/Computer | Career Applications | Above and Beyond

2.4 exercises

< Objective 1 >

Are the pairs of fractions equivalent?

1. $\frac{1}{3}, \frac{3}{5}$
2. $\frac{3}{5}, \frac{9}{15}$
3. $\frac{1}{7}, \frac{4}{28}$
4. $\frac{2}{3}, \frac{3}{5}$
5. $\frac{5}{6}, \frac{15}{18}$
6. $\frac{3}{4}, \frac{16}{20}$
7. $\frac{2}{21}, \frac{4}{25}$
8. $\frac{20}{24}, \frac{5}{6}$
9. $\frac{2}{7}, \frac{3}{11}$
10. $\frac{12}{15}, \frac{36}{45}$
11. $\frac{16}{24}, \frac{40}{60}$
12. $\frac{15}{20}, \frac{20}{25}$

< Objective 2 >

Write each fraction in simplest form.

13. $\frac{15}{30}$
14. $\frac{100}{200}$
15. $\frac{8}{12}$
16. $\frac{12}{15}$
17. $\frac{10}{14}$
18. $\frac{15}{50}$
19. $\frac{12}{18}$
20. $\frac{28}{35}$
21. $\frac{35}{40}$
22. $\frac{21}{24}$
23. $\frac{11}{44}$
24. $\frac{10}{25}$
25. $\frac{12}{36}$
26. $\frac{18}{48}$
27. $\frac{24}{27}$
28. $\frac{30}{50}$
29. $\frac{32}{40}$
30. $\frac{17}{51}$
31. $\frac{75}{105}$
32. $\frac{62}{93}$
33. $\frac{48}{60}$
34. $\frac{48}{66}$
35. $\frac{105}{135}$
36. $\frac{54}{126}$
37. $\frac{66}{110}$
38. $\frac{280}{320}$
39. $\frac{16}{21}$
40. $\frac{21}{32}$
41. $\frac{31}{52}$
42. $\frac{42}{55}$
43. $\frac{96}{132}$
44. $\frac{33}{121}$
45. $\frac{85}{102}$
46. $\frac{133}{152}$

Solve each application.

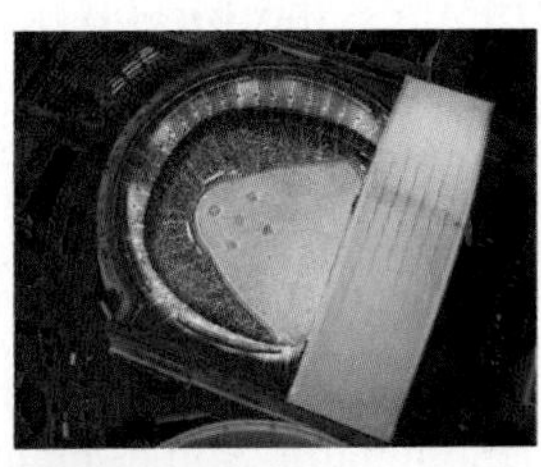

47. **STATISTICS** On a test of 72 questions, Sam answered 54 correctly. On another test, Sam answered 66 correctly out of 88. Did Sam get the same portion of each test correct?

48. **STATISTICS** At one point in a season, Ryan Braun of the Milwaukee Brewers had 164 hits in 498 times at bat. At the same point, Joey Votto of the Cincinnati Reds had 166 hits in 523 at bats. Did they have the same batting average?

    Hint: Batting average is calculated by dividing the number of hits by the number of at bats.

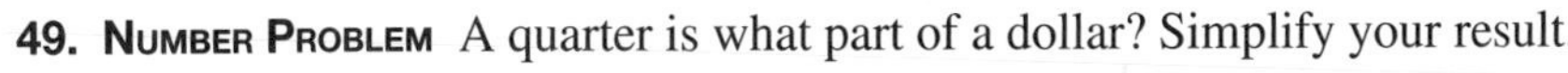

49. **NUMBER PROBLEM** A quarter is what part of a dollar? Simplify your result.

50. **NUMBER PROBLEM** A dime is what part of a dollar? Simplify your result.

51. **STATISTICS** What part of an hour is 15 min? Simplify your result.

52. **STATISTICS** What part of a day is 6 hr? Simplify your result.

53. **Science and Medicine** One meter is equal to 100 cm. What part of a meter is 70 cm? Simplify your result.

54. **Science and Medicine** One kilometer is equal to 1,000 meters (m). What part of a kilometer is 300 m? Simplify your result.

55. **Technology** Susan did a tune-up on her automobile. She found that two of her eight spark plugs were fouled. What simplified fraction represents the number of fouled plugs?

56. **Statistics** Samantha answered 18 of 20 problems correctly on a test. What part did she answer correctly? Simplify your result.

Skills	**Calculator/Computer**	Career Applications	Above and Beyond

Using a Calculator to Simplify Fractions

If you have a calculator that supports fraction arithmetic, you should learn to use it to check your work. Here we look at two different types of these calculators.

Scientific Calculator

Find the **a b/c** key on your calculator. This is the key to use to enter fractions.

Simplify the fraction $\frac{24}{68}$.

There are four steps with a scientific calculator.

(a) Enter the numerator, 24.

(b) Press the **a b/c** key.

(c) Enter the denominator, 68.

(d) Press $\boxed{=}$.

The calculator displays the simplified fraction, $\frac{6}{17}$.

Graphing Calculator

We can simplify the same fraction, $\frac{24}{68}$, using a graphing calculator, such as the TI-84 Plus.

(a) Enter the fraction as a division problem: $24 \div 68$. The calculator displays $\frac{24}{68}$ as 24/68.

(b) Press the **MATH** key.

(c) Select **1: ▶ Frac**.

(d) Press **ENTER**.

The calculator displays the simplified fraction, $\frac{6}{17}$.

A graphing calculator is particularly useful for simplifying fractions with large values in the numerator and denominator. Some scientific calculators cannot handle denominators larger than 999.

Use a calculator to simplify each fraction.

57. $\frac{28}{40}$ **58.** $\frac{121}{132}$ **59.** $\frac{96}{144}$

60. $\frac{445}{623}$ **61.** $\frac{299}{391}$ **62.** $\frac{289}{459}$

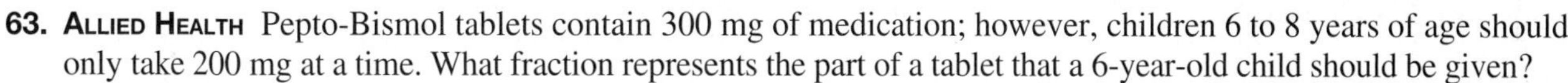

63. ALLIED HEALTH Pepto-Bismol tablets contain 300 mg of medication; however, children 6 to 8 years of age should only take 200 mg at a time. What fraction represents the part of a tablet that a 6-year-old child should be given?

64. ALLIED HEALTH The recommended adult dose of the laxative docusate is 500 mg per day, and the recommended dose for children 3 to 5 years old is 40 mg per day. The dose for a 4-year-old child is what fraction of an adult's dose?

65. MANUFACTURING TECHNOLOGY Express the width of the piece shown in the figure as a simplified fraction of the length.

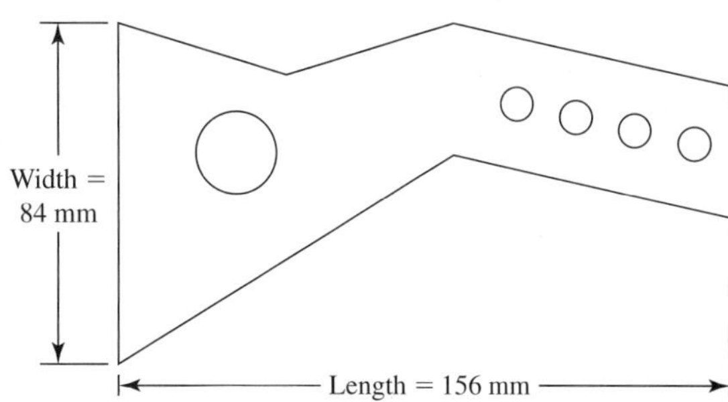

66. MANUFACTURING TECHNOLOGY In the packaging division of Early Enterprises, there are 36 packaging machines. At any given time, 4 of the machines are shut down for scheduled maintenance and service. What is the simplified fraction of machines that are operating at one time?

67. INFORMATION TECHNOLOGY Jo, an executive vice president of information technology, had 10 people on staff before hiring two more people. What part of the total staff are new? Simplify your answer.

68. INFORMATION TECHNOLOGY Jason is responsible for the administration of the servers at his company. He measures that the average arrival rate of requests to the server is 50,000 requests per second, and he also finds out the servers can service 100,000 requests per second. The intensity of the traffic is the quotient of the average arrival rate and the service rate. The intensity shows how busy the servers are. What fraction gives the intensity? Simplify your answer.

Skills | Calculator/Computer | Career Applications | **Above and Beyond**

69. Can any of these fractions be simplified?

(a) $\frac{824}{73}$ **(b)** $\frac{59}{11}$ **(c)** $\frac{135}{17}$

What characteristic do you notice about the denominator of each fraction? What rule would you make up based on your observations?

70. Consider the figures to the right.

(a) Give the fraction that represents the shaded region.

(b) Draw a horizontal line through the figure, as shown. Now give the fraction representing the shaded region.

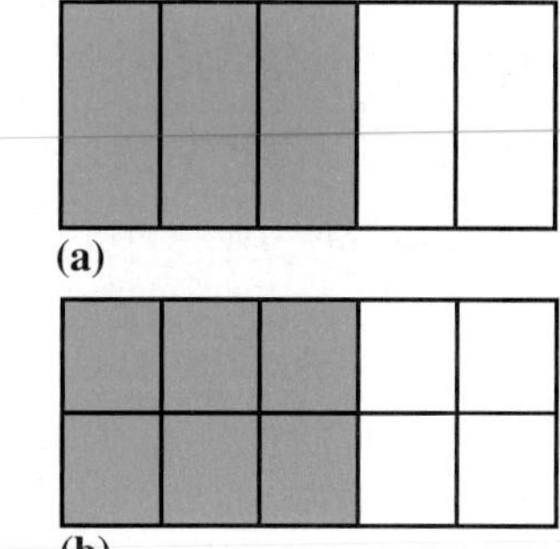

71. Repeat exercise 70, using these figures.

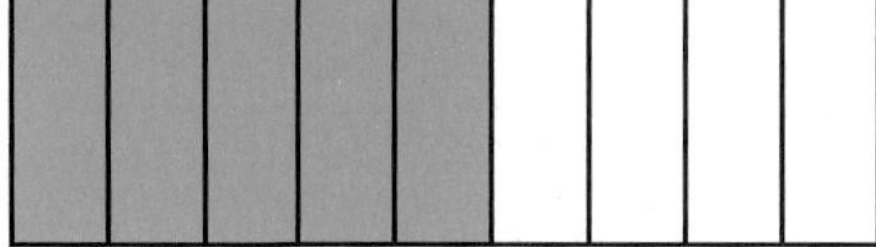

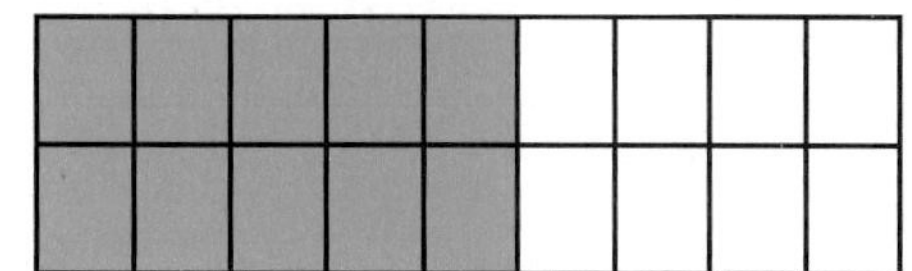

72. A student is attempting to reduce the fraction $\frac{8}{12}$ to lowest terms. He produces the following argument:

$$\frac{8}{12} = \frac{4+4}{8+4} = \frac{4}{8} = \frac{1}{2}$$

What is wrong with this argument? What is the correct answer?

Answers

1. No **3.** Yes **5.** Yes **7.** No **9.** No **11.** Yes **13.** $\frac{1}{2}$ **15.** $\frac{2}{3}$ **17.** $\frac{5}{7}$ **19.** $\frac{2}{3}$ **21.** $\frac{7}{8}$ **23.** $\frac{1}{4}$ **25.** $\frac{1}{3}$ **27.** $\frac{8}{9}$ **29.** $\frac{4}{5}$ **31.** $\frac{5}{7}$ **33.** $\frac{4}{5}$ **35.** $\frac{7}{9}$ **37.** $\frac{3}{5}$ **39.** $\frac{16}{21}$ **41.** $\frac{31}{52}$ **43.** $\frac{8}{11}$ **45.** $\frac{5}{6}$ **47.** Yes **49.** $\frac{1}{4}$ **51.** $\frac{1}{4}$ **53.** $\frac{7}{10}$ **55.** $\frac{1}{4}$ **57.** $\frac{7}{10}$ **59.** $\frac{2}{3}$ **61.** $\frac{13}{17}$ **63.** $\frac{2}{3}$ **65.** $\frac{7}{13}$ **67.** $\frac{1}{6}$ **69.** Above and Beyond **71.** (a) $\frac{5}{9}$; (b) $\frac{10}{18}$

Daily Reference Values

According to the Food and Drug Administration (FDA) in 2003, the following table represents the daily reference values (DRV) for each dietary element, based on a 2,000-calorie diet.

	DRV
Total fat	65 grams
Saturated fat	20 grams
Cholesterol	300 milligrams
Sodium	2,400 milligrams
Carbohydrates	300 grams
Dietary fiber	25 grams (minimum)

A high-performance energy bar made by PowerBar has the following amounts of each food type.

	DRV
Total fat	3 grams
Saturated fat	1 gram
Cholesterol	0 milligrams
Sodium	1,000 milligrams
Carbohydrates	45 grams
Dietary fiber	3 grams

Use the tables to answer each of the following questions.

1. What fraction of the DRV for sodium is contained in the PowerBar?
2. What fraction of the DRV for carbohydrates is contained in the PowerBar?
3. What fraction of the DRV for dietary fiber is contained in the PowerBar?
4. What fraction of the DRV for total fat is contained in the PowerBar?
5. What fraction of the DRV for saturated fat is contained in the PowerBar?

2.5 Multiplying Fractions

< 2.5 Objectives >

1 > Multiply two fractions

2 > Multiply mixed numbers and fractions

3 > Estimate products by rounding

4 > Use multiplication to solve applications

Multiplication is the easiest of the four operations with fractions. We can illustrate multiplication by picturing fractions as parts of a whole. Consider the fractions $\frac{4}{5}$ and $\frac{2}{3}$.

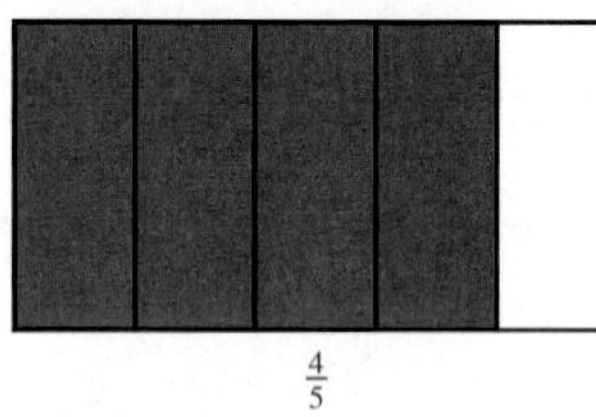

$\frac{4}{5}$

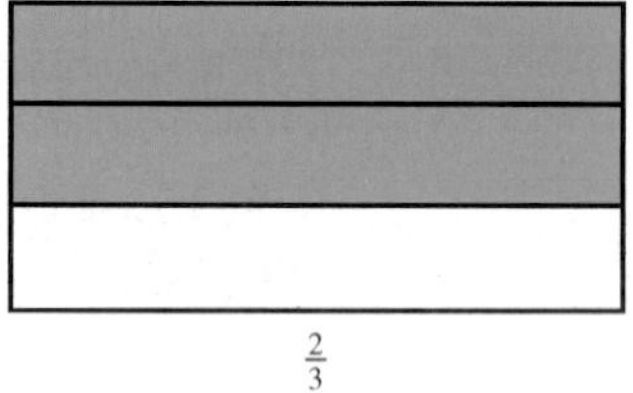

$\frac{2}{3}$

NOTE

A fraction followed by the word *of* means that we want to multiply by that fraction.

Suppose now that we wish to find $\frac{2}{3}$ of $\frac{4}{5}$. We can combine the diagrams as shown below. The part of the whole representing the product $\frac{2}{3} \times \frac{4}{5}$ is the purple region. The unit has been divided into 15 parts, and 8 of those parts are purple, so $\frac{2}{3} \times \frac{4}{5}$ must be $\frac{8}{15}$.

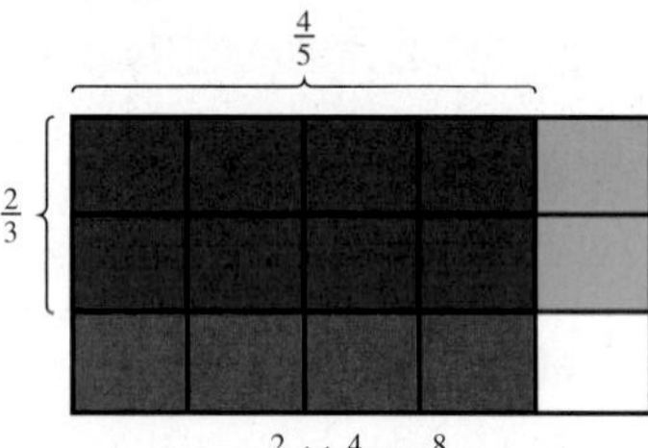

$\frac{2}{3} \times \frac{4}{5} = \frac{8}{15}$

The purple parts represent $\frac{2}{3}$ of the red area, or $\frac{2}{3} \times \frac{4}{5}$.

Based on this, we offer a process to multiply fractions.

Step by Step

Multiplying Fractions

Step 1 Multiply the numerators to find the numerator of the product.

Step 2 Multiply the denominators to find the denominator of the product.

Step 3 Simplify the resulting fraction, if possible.

We need only use the first two steps in Example 1.

Example 1 Multiplying Fractions

< Objective 1 >

Multiply.

(a) $\frac{2}{3} \times \frac{4}{5} = \frac{2 \times 4}{3 \times 5} = \frac{8}{15}$

(b) $\frac{5}{8} \times \frac{7}{9} = \frac{5 \times 7}{8 \times 9} = \frac{35}{72}$

(c) $\left(\frac{2}{3}\right)^2 = \frac{2}{3} \times \frac{2}{3} = \frac{2 \times 2}{3 \times 3} = \frac{4}{9}$

Check Yourself 1

Multiply.

(a) $\frac{7}{8} \times \frac{3}{10}$ (b) $\frac{5}{7} \times \frac{3}{4}$ (c) $\left(\frac{3}{4}\right)^2$

Step 3 indicates that the product of fractions should always be simplified to lowest terms. Consider Example 2.

Example 2 Multiplying Fractions

Multiply and write the result in lowest terms.

$$\frac{3}{4} \times \frac{2}{9} = \frac{3 \times 2}{4 \times 9} = \frac{6}{36} = \frac{1}{6}$$

Noting that $\frac{6}{36}$ is not in simplest form, we divide numerator and denominator by 6 to write the product in simplest terms.

Check Yourself 2

Multiply and write the result in simplest terms.

$\frac{5}{7} \times \frac{3}{10}$

To find the product of a fraction and a whole number, write the whole number as a fraction and apply the multiplication rule. Example 3 illustrates this approach.

Example 3 Multiplying a Whole Number and a Fraction

NOTES

Because any number divided by one is itself, and a fraction indicates division, we can always write a whole number as an improper fraction with 1 as its denominator. Consider,

$5 \div 1 = 5$

Therefore,

$5 = \frac{5}{1}$

Improper fractions and mixed numbers are both correct and equivalently simplified. Your instructor will tell you which form to use for your answer.

Do the indicated multiplication.

Remember that $5 = \frac{5}{1}$.

(a) $5 \times \frac{3}{4} = \frac{5}{1} \times \frac{3}{4} = \frac{5 \times 3}{1 \times 4}$

$= \frac{15}{4}$ or $3\frac{3}{4}$

(b) $\frac{5}{12} \times 6 = \frac{5}{12} \times \frac{6}{1}$

$= \frac{5 \times 6}{12 \times 1}$

$= \frac{30}{12}$

$= \frac{5}{2}$ or $2\frac{1}{2}$

Check Yourself 3

Multiply.

(a) $\frac{3}{16} \times 8$ (b) $4 \times \frac{5}{7}$

When mixed numbers are involved in multiplication, there is an additional step. First change any mixed numbers to improper fractions. Then, multiply the fractions.

Example 4 Multiplying a Mixed Number and a Fraction

< Objective 2 >

$1\frac{1}{2} \times \frac{3}{4} = \frac{3}{2} \times \frac{3}{4}$ Change the mixed number to an improper fraction.

Here $1\frac{1}{2} = \frac{3}{2}$.

$= \frac{3 \times 3}{2 \times 4}$ Multiply as before.

$= \frac{9}{8} = 1\frac{1}{8}$

Check Yourself 4

Multiply.

$\frac{5}{8} \times 3\frac{1}{2}$

If two mixed numbers are involved, change both of the mixed numbers to improper fractions.

Example 5 Multiplying Mixed Numbers

Multiply.

$3\frac{2}{3} \times 2\frac{1}{2} = \frac{11}{3} \times \frac{5}{2}$ Change the mixed numbers to improper fractions.

$= \frac{11 \times 5}{3 \times 2} = \frac{55}{6} = 9\frac{1}{6}$

> CAUTION

Be Careful! Students sometimes think of

$3\frac{2}{3} \times 2\frac{1}{2}$ as $(3 \times 2) + \left(\frac{2}{3} \times \frac{1}{2}\right)$

This is *not* the correct multiplication pattern. You must first change the mixed numbers to improper fractions.

Check Yourself 5

Multiply.

$2\frac{1}{3} \times 3\frac{1}{2}$

When you multiply fractions, it is usually easier to simplify, that is, remove any common factors in the numerator and denominator, *before multiplying*. Remember that to simplify means to *divide* by the same common factor.

Example 6 Simplifying Before Multiplying Fractions

NOTE

Once again we are applying the fundamental principle to divide the numerator and denominator by 3.

Simplify and then multiply.

$\frac{3}{5} \times \frac{4}{9} = \frac{\overset{1}{\cancel{3}} \times 4}{5 \times \underset{3}{\cancel{9}}}$ To simplify, we divide the *numerator* and *denominator* by the common factor 3. Remember that $\overset{1}{\cancel{3}}$ means $3 \div 3 = 1$, and $\underset{3}{\cancel{9}}$ means $9 \div 3 = 3$.

$= \frac{1 \times 4}{5 \times 3}$

$= \frac{4}{15}$ Because we divide by any common factors before we multiply, the resulting product is *in simplest form*.

Check Yourself 6

Simplify and then multiply.

$\frac{7}{8} \times \frac{5}{21}$

Our work in Example 6 leads to a general rule for simplifying fractions in multiplication.

Property

Simplifying Fractions Before Multiplying

In multiplying two or more fractions, we can divide any factor of the numerator and any factor of the denominator by the same nonzero number to simplify the product.

When mixed numbers are involved, the process is similar.

Example 7 Multiplying Mixed Numbers

Multiply.

$2\frac{2}{3} \times 2\frac{1}{4} = \frac{8}{3} \times \frac{9}{4}$ First, convert the mixed numbers to improper fractions.

$= \frac{\overset{2}{\cancel{8}} \times \overset{3}{\cancel{9}}}{\underset{1}{\cancel{3}} \times \underset{1}{\cancel{4}}}$ To simplify, divide by the common factors of 3 and 4.

$= \frac{2 \times 3}{1 \times 1}$ Multiply as before.

$= \frac{6}{1} = 6$

Check Yourself 7

Simplify and then multiply.

$3\frac{1}{3} \times 2\frac{2}{5}$

We use the same process to find the product of more than two fractions.

Example 8 Multiplying Three Numbers

RECALL

We can divide *any* factor of the numerator and *any* factor of the denominator by the same nonzero number.

Simplify and then multiply.

$\frac{2}{3} \times 1\frac{4}{5} \times \frac{5}{8} = \frac{2}{3} \times \frac{9}{5} \times \frac{5}{8}$ Write any mixed or whole numbers as improper fractions.

$= \frac{\overset{1}{\cancel{2}} \times \overset{3}{\cancel{9}} \times \overset{1}{\cancel{5}}}{\underset{1}{\cancel{3}} \times \underset{1}{\cancel{5}} \times \underset{4}{\cancel{8}}}$ To simplify, divide by the common factors in the numerator and denominator.

$= \frac{3}{4}$

Check Yourself 8

Simplify and then multiply.

$\frac{5}{8} \times 4\frac{4}{5} \times \frac{1}{6}$

Now that you can multiply fractions, you are ready to gain a deeper understanding of the fundamental principle of fractions. Consider Example 9.

Example 9 The Fundamental Principle of Fractions

Multiply 20 by $\frac{1}{2}$.

$20 \times \frac{1}{2} = \frac{20}{1} \times \frac{1}{2}$ Rewrite 20 as a fraction.

$= \frac{\overset{10}{\cancel{20}}}{1} \times \frac{1}{\underset{1}{\cancel{2}}}$ Simplify.

$= \frac{10 \times 1}{1 \times 1}$ Multiply.

$= \frac{10}{1}$

$= 10$

Now divide 20 by 2.

$20 \div 2 = 10$

We see that

$20 \times \frac{1}{2} = 20 \div 2$

In fact, there is nothing special about 20 or 2 in this example. We could choose other numbers and get the same result.

Check Yourself 9

(a) Multiply 15 by $\frac{1}{5}$.
(b) Divide 15 by 5.
(c) What do you notice about your answers to (a) and (b)?

Dividing by a number and multiplying by 1 over that number is the same operation. When we apply the fundamental principle of fractions to divide the numerator and denominator by the same number, we are really multiplying the fraction by 1.

Example 10 The Fundamental Principle of Fractions

Simplify $\frac{20}{2}$.

We use the fundamental principle of fractions and divide both the numerator and denominator by 2.

$\frac{20}{2} = \frac{20 \div 2}{2 \div 2}$

$= \frac{10}{1} = 10$

We know that any number divided by itself is one. Therefore, any fraction with the same number in the numerator and denominator equals one. This is true, even if the numerator and denominator are "ugly."

$\frac{1}{2} \div \frac{1}{2} = 1,$

Therefore,

$$\frac{\frac{1}{2}}{\frac{1}{2}} = 1$$

So,

$$\frac{20}{2} = \frac{20}{2} \times 1$$ Multiplying by 1 leaves a number unchanged.

$$= \frac{20}{2} \times \frac{\frac{1}{2}}{\frac{1}{2}}$$ We can write 1 as $\frac{\frac{1}{2}}{\frac{1}{2}}$.

$$= \frac{20 \times \frac{1}{2}}{2 \times \frac{1}{2}}$$ Multiply, as fractions.

$$= \frac{20 \div 2}{2 \div 2} = \frac{10}{1} = 10$$

Thus, multiplication gives us the fundamental principle of fractions.

Check Yourself 10

Multiply $\frac{15}{5}$ by $\frac{\frac{1}{5}}{\frac{1}{5}}$.

We rounded in order to estimate results in our work with whole numbers. Estimation can also be used to check the "reasonableness" of an answer when working with fractions or mixed numbers.

Example 11 **Estimating a Product**

< Objective 3 >

Estimate the product of

$$3\frac{1}{8} \times 5\frac{5}{6}$$

Round each mixed number to the nearest whole number.

$$3\frac{1}{8} \rightarrow 3$$

$$5\frac{5}{6} \rightarrow 6$$

Our estimate of the product is then

$$3 \times 6 = 18$$

Note: The actual product in this case is $18\frac{11}{48}$, which certainly seems reasonable in view of our estimate.

Check Yourself 11

Estimate the product.

$$2\frac{7}{8} \times 8\frac{1}{3}$$

Units Analysis

When you divide two denominate numbers, the units are also divided. This yields a unit in fraction form.

EXAMPLES:

$$250 \text{ mi} \div 10 \text{ gal} = \frac{250 \text{ mi}}{10 \text{ gal}} = \frac{25 \text{ mi}}{1 \text{ gal}} = 25 \text{ mi/gal}$$

$$360 \text{ ft} \div 30 \text{ s} = \frac{360 \text{ ft}}{30 \text{ s}} = 12 \text{ ft/s}$$

When we multiply denominate numbers that have these units in fraction form, they behave just as fractions do.

EXAMPLES:

$$25 \text{ mi/gal} \times 12 \text{ gal} = \frac{25 \text{ mi}}{1 \cancel{\text{gal}}} \times \frac{12 \cancel{\text{gal}}}{1} = 300 \text{ mi}$$

(If we look at the units, we see that the gallons "cancel" when one is in the numerator and the other in the denominator.)

$$12\text{ft/s} \times 60 \text{ s/min} = \frac{12 \text{ ft}}{1 \cancel{\text{s}}} \times \frac{60 \cancel{\text{s}}}{1 \text{ min}} = \frac{720 \text{ ft}}{1 \text{ min}} = 720 \text{ ft/min}$$

(Again, the seconds cancel, leaving feet in the numerator and minutes in the denominator.)

Now we can look at some applications of fractions that involve multiplication. In solving these word problems, we use the same approach we used earlier with whole numbers. Let's review the four-step process introduced in Section 1.2.

Step by Step

Solving Applications Involving the Multiplication of Fractions

Step 1 Read the problem carefully to determine the given information and what you are asked to find.

Step 2 Decide upon the operation or operations to be used.

Step 3 Write down the complete statement necessary to solve the problem and do the calculations.

Step 4 State the answer as a complete sentence and check to make sure that you have answered the question and that your answer seems reasonable.

We can work through some examples, using these steps.

Example 12 An Application Involving Multiplication

< Objective 4 >

Lisa worked $10\frac{1}{4}\,\frac{\text{hr}}{\text{day}}$ for 5 days. How many hours did she work?

Step 1 We are looking for the total hours Lisa worked.

Step 2 We will multiply the hours per day by the days.

Step 3 $10\frac{1}{4}\,\frac{\text{hr}}{\text{day}} \times 5 \text{ days} = \frac{41}{4}\,\frac{\text{hr}}{\text{day}} \times 5 \text{ days} = \frac{205}{4} \text{ hr} = 51\frac{1}{4} \text{ hr}$

Step 4 Note the days cancel, leaving only the unit of hours. The units should always be compared to the desired units from step 1. The answer also seems reasonable. An answer such as 5 hr or 500 hr does not seem reasonable.

Check Yourself 12

Carlos gets 30 mi/gal in his Miata. How far should he be able to drive with an 11-gal tank of gas?

In Example 13, we follow the four steps for solving applications, but do not label the steps. You should still think about these steps as we solve the problem.

Example 13 An Application Involving Mixed Numbers

RECALL

The area of a rectangle is the product of its length and its width.

A sheet of notepaper is $6\frac{3}{4}$ in. wide and $8\frac{2}{3}$ in. long. Find the area of the paper.

Multiply the given length by the width. This gives the desired area. First, we estimate the area.

$9 \text{ in.} \times 7 \text{ in.} = 63 \text{ in.}^2$

Now, we find the exact area.

$$\begin{aligned} 8\tfrac{2}{3} \text{ in.} \times 6\tfrac{3}{4} \text{ in.} &= \frac{26}{3} \text{ in.} \times \frac{27}{4} \text{ in.} \\ &= \frac{117}{2} \text{ in.}^2 \\ &= 58\tfrac{1}{2} \text{ in.}^2 \end{aligned}$$

The units (square inches) are units of area. Note that from our estimate the result is reasonable.

Check Yourself 13

A window is $4\frac{1}{2}$ ft high by $2\frac{1}{3}$ ft wide. What is its area?

Example 14 reminds us that an abstract number multiplied by a denominate number yields the units of the denominate number.

Example 14 An Application Involving Multiplication

A state park contains $38\frac{2}{3}$ acres. According to the plan for the park, $\frac{3}{4}$ of the park is to be left as a wildlife preserve. How many acres is this?

We want to find $\frac{3}{4}$ of $38\frac{2}{3}$ acres, so we multiply.

RECALL

The word *of* usually indicates multiplication.

$$\frac{3}{4} \times 38\frac{2}{3} \text{ acres} = \frac{3}{4} \times \frac{116}{3} \text{ acres} = \frac{\overset{1}{\cancel{3}} \times \overset{29}{\cancel{116}}}{\underset{1}{\cancel{4}} \times \underset{1}{\cancel{3}}} \text{ acres} = 29 \text{ acres}$$

Check Yourself 14

A backyard has $25\frac{3}{4}$ square yards (yd^2) of open space. If Patrick wants to plant a vegetable garden covering $\frac{2}{3}$ of the open space, how many square yards is this?

We see that the word *of* usually indicates multiplication. You should also note that it indicates that the fraction preceding it is an abstract number (it has no units attached). There are even occasions, as in Example 15, when we are looking at the product of two abstract numbers.

Example 15 An Application Involving Fractions

A grocery store survey shows that $\frac{2}{3}$ of the customers buy meat. Of these, $\frac{3}{4}$ will buy at least one package of beef. What portion of the store's customers buy beef?

NOTE

In this problem, *of* means to multiply.

Step 1 We know that $\frac{2}{3}$ of the customers buy meat and that $\frac{3}{4}$ of these customers buy beef.

Step 2 We wish to know $\frac{3}{4}$ of $\frac{2}{3}$. The operation here is multiplication.

Step 3 Multiplying, we have

$$\frac{3}{4} \times \frac{2}{3} = \frac{\overset{1}{\cancel{3}} \times \overset{1}{\cancel{2}}}{\underset{2}{\cancel{4}} \times \underset{1}{\cancel{3}}} = \frac{1}{2}$$

Step 4 From step 3 we have the result: $\frac{1}{2}$ of the store's customers buy beef.

Check Yourself 15

A supermarket survey shows that $\frac{2}{5}$ of the customers buy breakfast foods. Of these, $\frac{3}{4}$ buy cereal. What portion of the store's customers buy cereal?

Example 16 An Application Involving Mixed Numbers

Shirley drives at an average speed of 52 miles per hour (mi/hr) for $3\frac{1}{4}$ hr. How far has she traveled at the end of $3\frac{1}{4}$ hr?

NOTE

Distance is the product of speed and time.

$$\underset{\text{Speed}}{52\,\frac{\text{mi}}{\text{hr}}} \times \underset{\text{Time}}{3\frac{1}{4}\text{ hr}} = \frac{52}{1}\,\frac{\text{mi}}{\cancel{\text{hr}}} \times \frac{13}{4}\,\cancel{\text{hr}}$$

$$= \frac{\overset{13}{\cancel{52}} \times 13}{1 \times \underset{1}{\cancel{4}}}\text{ mi}$$

$$= 169 \text{ mi}$$

Check Yourself 16

(a) The scale on a map is 1 in. = 60 mi. What is the distance in miles between two towns that are $3\frac{1}{2}$ in. apart on the map?

(b) Maria is ordering concrete for a new sidewalk that is to be $\frac{1}{9}$ yd thick, $22\frac{1}{2}$ yd long, and $1\frac{1}{3}$ yd wide. How much concrete should she order if she must order a whole number of cubic yards?

Check Yourself ANSWERS

1. **(a)** $\frac{21}{80}$; **(b)** $\frac{15}{28}$; **(c)** $\frac{9}{16}$ **2.** $\frac{3}{14}$ **3.** **(a)** $1\frac{1}{2}$; **(b)** $2\frac{6}{7}$ **4.** $2\frac{3}{16}$ **5.** $8\frac{1}{6}$ **6.** $\frac{5}{24}$ **7.** 8

8. $\frac{1}{2}$ **9.** **(a)** 3; **(b)** 3; **(c)** They are the same: $15 \times \frac{1}{5} = 15 \div 5$. **10.** $\frac{15}{5} \times \frac{\frac{1}{5}}{\frac{1}{5}} = \frac{15 \times \frac{1}{5}}{5 \times \frac{1}{5}} = \frac{3}{1} = 3$

11. 24 **12.** 330 mi **13.** $10\frac{1}{2}$ ft^2 **14.** $17\frac{1}{6}$ yd^2 **15.** $\frac{3}{10}$

16. **(a)** 210 mi; **(b)** The answer, $3\frac{1}{3}$ yd^3, means she needs 4 yd^3.

Reading Your Text 

These fill-in-the-blank exercises will help you understand some of the key vocabulary used in this section. The answers to these exercises are in the Answers Appendix at the back of the text.

(a) The product of fractions should always be expressed in ________ terms.

(b) When multiplying fractions, it is usually easier to ________ before multiplying.

(c) Estimation can be used to check the ________ of an answer when working with fractions or mixed numbers.

(d) The final step in solving an application is to make sure that your answer seems ________.

Skills Calculator/Computer Career Applications Above and Beyond

2.5 exercises

< Objective 1 >

Multiply. Write each answer in simplest form.

1. $\frac{3}{4} \times \frac{5}{11}$ **2.** $\frac{2}{7} \times \frac{5}{9}$ **3.** $\frac{3}{4} \times \frac{7}{11}$  **4.** $\frac{2}{5} \times \frac{3}{7}$

5. $\frac{3}{5} \cdot \frac{5}{7}$ **6.** $\frac{6}{11} \cdot \frac{8}{6}$ **7.** $\left(\frac{4}{9}\right)^2$ **8.** $\left(\frac{5}{6}\right)^2$

9. $\frac{3}{11} \cdot \frac{7}{9}$ **10.** $\frac{7}{9} \cdot \frac{3}{5}$ **11.** $\frac{3}{10} \times \frac{5}{9}$ **12.** $\frac{5}{21} \times \frac{14}{25}$

13. $\frac{7}{9} \cdot \frac{6}{5}$ **14.** $\frac{48}{63} \cdot \frac{81}{60}$ **15.** $\frac{24}{33} \cdot \frac{55}{40}$ 

< Objective 2 >

16. $3\frac{1}{3} \times \frac{9}{11}$ **17.** $\left(\frac{2}{3}\right)\left(2\frac{2}{5}\right)$  **18.** $\left(3\frac{1}{3}\right)\left(\frac{3}{7}\right)$ **19.** $\frac{2}{5} \times 3\frac{1}{4}$

20. $2\frac{1}{3} \times 2\frac{1}{6}$ **21.** $2\frac{1}{3} \cdot 2\frac{1}{2}$ **22.** $\frac{3}{7} \cdot 14$ **23.** $\left(\frac{5}{6}\right)^3$ VIDEO

24. $\left(\frac{4}{5}\right)^3$ **25.** $\frac{12}{25} \times \frac{11}{18}$ **26.** $\frac{10}{12} \times \frac{16}{25}$ **27.** $\frac{14}{15} \cdot \frac{10}{21}$

28. $\frac{21}{25} \cdot \frac{30}{7}$ **29.** $\left(\frac{18}{28}\right)\left(\frac{35}{22}\right)$ **30.** $\left(3\frac{2}{3}\right)\left(\frac{9}{10}\right)$

Multiply and simplify.

31. $\frac{4}{9} \times 3\frac{3}{5}$ 

32. $5\frac{1}{3} \times \frac{7}{8}$

33. $\frac{10}{27} \times 3\frac{3}{5}$

34. $1\frac{1}{3} \times 1\frac{1}{5}$

35. $2\frac{2}{5} \cdot 3\frac{3}{4}$

36. $2\frac{2}{7} \cdot 2\frac{1}{3}$

37. $4\frac{1}{5} \times \frac{10}{21} \times \frac{9}{20}$ VIDEO

38. $\frac{7}{8} \times 5\frac{1}{3} \times \frac{5}{14}$

39. $3\frac{1}{3} \cdot \frac{4}{5} \cdot 1\frac{1}{8}$

40. $4\frac{1}{2} \cdot 5\frac{5}{6} \cdot \frac{8}{15}$

41. Find $\frac{2}{3}$ of $\frac{3}{7}$

42. What is $\frac{5}{6}$ of $\frac{9}{10}$?

< Objective 3 >

Estimate each product.

43. $3\frac{1}{5} \times 4\frac{2}{3}$

44. $5\frac{1}{7} \times 2\frac{2}{13}$

45. $11\frac{3}{4} \times 5\frac{1}{4}$

46. $3\frac{4}{5} \times 5\frac{6}{7}$

47. $8\frac{2}{9} \cdot 7\frac{11}{12}$

48. $\frac{9}{10} \cdot 2\frac{2}{7}$

Evaluate. Include proper units in your results.

49. 36 mi/hr × 4 hr

50. 80 cal/g × 5 g

51. 55 joules/s × 11 s

52. 5 lb/ft × 3 ft

53. 88 ft/s × 1 mi/5,280 ft × 3,600 s/hr

54. 24 hr/day × 3,600 s/hr × 365 days/yr

< Objective 4 >

Solve each application.

55. **Business and Finance** Maria earns $11 per hour. Last week, she worked 9 hr/day for 6 days. What was her gross pay?

56. **Statistics** The gas tank in Luigi's Toyota Camry holds 17 gal when full. The car gets 32 mi/gal. How far can he travel on 3 full tanks?

57. **Crafts** A recipe calls for $\frac{2}{3}$ cup of sugar per serving. How much sugar is needed for 6 servings?

58. **Crafts** Mom-Mom's French toast requires $\frac{3}{4}$ cup of batter per serving. If five people are expected for breakfast, how much batter is needed?

59. **Science and Medicine** A jet flew at an average speed of 540 mi/hr on a $4\frac{2}{3}$ hr flight. How far did it fly?

60. **Construction** An $11\frac{2}{3}$-acre plot of land is being subdivided for home lots. It is estimated that $\frac{5}{7}$ of the area will be used for lots. What amount will be used for lots?

61. **Geometry** Find the volume of a box that measures $2\frac{1}{4}$ in. by $3\frac{7}{8}$ in. by $4\frac{5}{6}$ in.

62. **Construction** Nico wishes to purchase mulch to cover his garden. The garden measures $7\frac{7}{8}$ ft by $10\frac{1}{8}$ ft. He wants the mulch to be $\frac{1}{3}$ ft deep. How much mulch should Nico order if he must order a whole number of cubic feet?

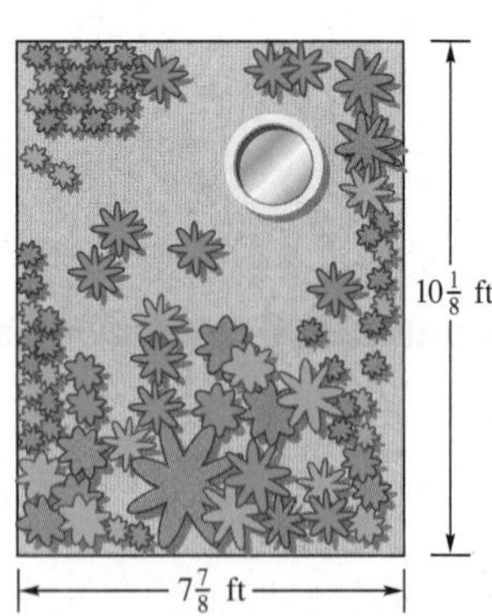

Skills | **Calculator/Computer** | Career Applications | Above and Beyond

Using a Calculator to Multiply Fractions

Scientific Calculator

To multiply fractions on a scientific calculator, you enter the first fraction, using the [a b/c] key, then press the multiplication sign, next enter the second fraction, and then press the equal sign. It is always a good idea to separate the fractions by using parentheses. Note that we do that in the example below.

Graphing Calculator

When using a graphing calculator, you must choose the fraction option [1: ▶ Frac] from the [MATH] menu before pressing [ENTER].

For the fraction problem $\frac{7}{15} \times \frac{5}{21}$, the keystroke sequence is

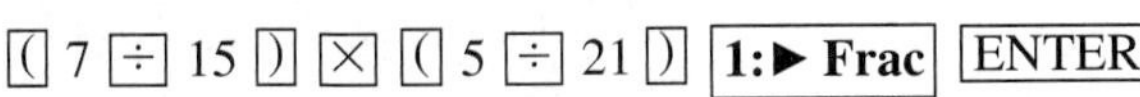

```
(7/15)*(5/21)▸Fr
ac
             1/9
```

The result is $\frac{1}{9}$.

Use a calculator to find each product.

63. $\frac{7}{12} \times \frac{36}{63}$

64. $\frac{8}{27} \times \frac{45}{64}$

65. $\frac{12}{45} \times \frac{27}{72}$

66. $\frac{18}{132} \times \frac{36}{63}$

67. $\frac{27}{72} \cdot \frac{24}{45}$

68. $\frac{81}{136} \cdot \frac{84}{135}$

Skills | Calculator/Computer | **Career Applications** | Above and Beyond

69. Manufacturing Technology Calculate the distance from the center of hole A to the center of hole B.

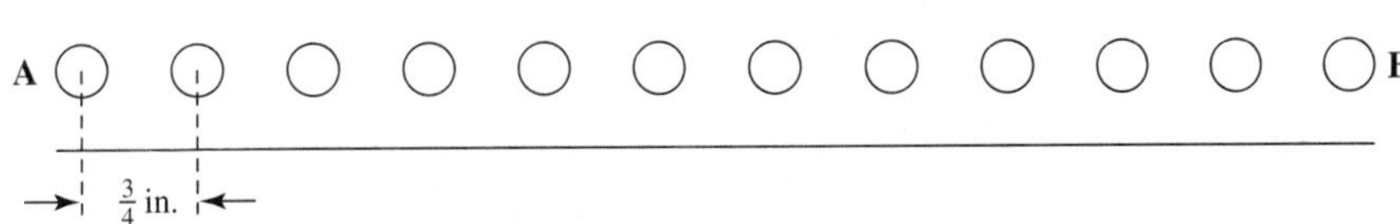

70. Manufacturing Technology A $3\frac{3}{8}$-in. long cut needs to be made in a piece of material. The cut rate is $\frac{3}{4}$ min per in. How many minutes does it take to make the cut?

71. Manufacturing Technology An order requires $14\frac{3}{4}$ ounces of material that costs $7\frac{1}{2}$¢ per ounce. Find the total cost of the material.

72. Manufacturing Technology An order requires $8\frac{5}{8}$ ounces of material that costs $10\frac{1}{2}$¢ per ounce. Find the total cost of the material.

Answers

1. $\frac{15}{44}$ **3.** $\frac{21}{44}$ **5.** $\frac{3}{7}$ **7.** $\frac{16}{81}$ **9.** $\frac{7}{33}$ **11.** $\frac{1}{6}$ **13.** $\frac{14}{15}$ **15.** 1 **17.** $1\frac{3}{5}$ **19.** $1\frac{3}{10}$ **21.** $5\frac{5}{6}$ **23.** $\frac{125}{216}$ **25.** $\frac{22}{75}$ **27.** $\frac{4}{9}$ **29.** $1\frac{1}{44}$ **31.** $1\frac{3}{5}$ **33.** $1\frac{1}{3}$ **35.** 9 **37.** $\frac{9}{10}$ **39.** 3 **41.** $\frac{2}{7}$ **43.** 15 **45.** 60 **47.** 64 **49.** 144 mi **51.** 605 joules **53.** 60 mi/hr **55.** \$594 **57.** 4 cups **59.** 2,520 mi **61.** $42\frac{9}{64}$ in.^{3} **63.** $\frac{1}{3}$ **65.** $\frac{1}{10}$ **67.** $\frac{1}{5}$ **69.** $8\frac{1}{4}$ in. **71.** $110\frac{5}{8}$¢ or \$1.11

Activity 5 ::

Overriding a Presidential Veto

A bill is sent to the president of the United States when it has passed both houses of Congress. A majority vote in both the House of Representatives (218 of 435 members) and the Senate (51 of 100 members) is needed for the bill to be passed on to the president. The majority vote is a majority of the members present, as long as more than one half of all the members are present. More than one half of the members makes up what is called a quorum.

Once a bill passes Congress, the president may sign the bill or veto it. Signing a bill makes it law, whereas vetoing a bill sends it back to Congress.

Congress can still make the bill a law by overriding the presidential veto. To override the veto, $\frac{2}{3}$ of the members of each legislative body must vote to override it. Again, this is $\frac{2}{3}$ of a quorum of members.

1. Assume that 420 members of the House are available to vote. How many votes are necessary to make up a majority, which is anything over $\frac{1}{2}$?
2. If 90 members of the Senate are available to vote, how many votes would constitute a majority?
3. How many votes from a group of 420 members of the House would be necessary to overturn a veto?
4. How many votes from a group of 90 members of the Senate would be necessary to overturn a veto?

2.6 Dividing Fractions

< 2.6 Objectives >

1 > Find the reciprocal of a fraction

2 > Divide fractions

3 > Divide mixed numbers

4 > Use division to solve applications

To divide fractions, it helps to have a new concept, the **reciprocal** of a fraction.

Property

The Reciprocal of a Fraction

Invert, or flip, a nonzero fraction to write its **reciprocal.**

Example 1 Finding the Reciprocal of a Fraction

< Objective 1 >

Find the reciprocal of **(a)** $\frac{3}{4}$, **(b)** 5, and **(c)** $1\frac{2}{3}$.

(a) The reciprocal of $\frac{3}{4}$ is $\frac{4}{3}$. — Just invert, or flip, the fraction.

(b) The reciprocal of 5, or $\frac{5}{1}$, is $\frac{1}{5}$. — Write 5 as $\frac{5}{1}$ and then turn over the fraction.

(c) The reciprocal of $1\frac{2}{3}$, or $\frac{5}{3}$, is $\frac{3}{5}$. — Write $1\frac{2}{3}$ as $\frac{5}{3}$ and then invert.

NOTE

In general, the reciprocal of the fraction $\frac{a}{b}$ is $\frac{b}{a}$. Neither a nor b can be 0.

Check Yourself 1

Find the reciprocal of **(a)** $\frac{5}{8}$ and **(b)** $3\frac{1}{4}$.

There is an important property relating a number and its reciprocal.

Property

Reciprocal Products

The product of any number and its reciprocal is 1. (Every number except zero has a reciprocal.)

We use the reciprocal to find a rule for dividing fractions. Recall that we can represent the operation of division in several ways. We used the symbol ÷ earlier, but remember that a fraction also indicates division. For instance,

$$3 \div 5 = \frac{3}{5}$$

In this statement, 5 is called the *divisor*. It follows the division sign ÷ and is written *below* the fraction bar.

NOTE

$3 \div 5$ and $\frac{3}{5}$ both mean "3 divided by 5."

Therefore, we can write a statement involving fractions and division as a *complex fraction,* which has a fraction as its numerator, its denominator, or both.

Example 2 Writing a Quotient as a Complex Fraction

Write $\frac{2}{3} \div \frac{4}{5}$ as a complex fraction.

$$\frac{\frac{2}{3}}{\frac{4}{5}}$$

The numerator is $\frac{2}{3}$.

A *complex fraction* is written by placing the dividend in the numerator and the divisor in the denominator.

The denominator is $\frac{4}{5}$.

Check Yourself 2

Write $\frac{2}{5} \div \frac{3}{4}$ as a complex fraction.

Let's continue with the same division problem you looked at in Check Yourself 2.

Example 3 Rewriting a Division Problem

$$\frac{2}{5} \div \frac{3}{4} = \frac{\frac{2}{5}}{\frac{3}{4}}$$ Write the original quotient as a complex fraction.

$$= \frac{\frac{2}{5} \times \frac{4}{3}}{\frac{3}{4} \times \frac{4}{3}}$$ Multiply the numerator and denominator by $\frac{4}{3}$, the reciprocal of the denominator. This does *not* change the value of the fraction.

$$= \frac{\frac{2}{5} \times \frac{4}{3}}{1}$$ The denominator becomes 1.

$$= \frac{2}{5} \times \frac{4}{3}$$ Recall that a number divided by 1 is just that number.

We see that

$$\frac{2}{5} \div \frac{3}{4} = \frac{2}{5} \times \frac{4}{3}$$

NOTES

Do you see a rule suggested?

To divide fractions, flip the second fraction and multiply.

One helpful mnemonic (memory device) states, "If you want to flip the right fraction, then flip the RIGHT fraction. The one on the right is right."

Check Yourself 3

Write $\frac{3}{5} \div \frac{7}{8}$ as a multiplication problem.

We would certainly like to be able to divide fractions easily without all the work of this example. Looking carefully at the calculations, we see a good rule to follow.

Property

To Divide Fractions

To divide one fraction by another, multiply the dividend by the reciprocal of the divisor. That is, invert the divisor (the second fraction) and multiply.

In symbols, we write

$$\frac{a}{b} \div \frac{c}{d} = \frac{a}{b} \times \frac{d}{c}$$

In Example 4, we apply this rule to divide fractions.

Example 4 Dividing Fractions

< Objective 2 >

RECALL

The number inverted is the divisor. It *follows* the division sign.

Divide.

(a) $\frac{1}{3} \div \frac{4}{7} = \frac{1}{3} \times \frac{7}{4}$ We invert the divisor, $\frac{4}{7}$, and then multiply.

$= \frac{1 \times 7}{3 \times 4} = \frac{7}{12}$

(b) $\frac{\frac{3}{4}}{6}$

$\frac{\frac{3}{4}}{6} = \frac{3}{4} \div 6 = \frac{\overset{1}{\cancel{3}}}{4} \times \frac{1}{\underset{2}{\cancel{6}}}$

$= \frac{1}{8}$

Check Yourself 4

Divide.

(a) $\frac{\frac{2}{5}}{\frac{3}{4}}$ (b) $\frac{2}{7} \div 4$ (c) $8 \div \frac{2}{3}$

Simplifying can also be useful when dividing fractions.

Example 5 Dividing Fractions

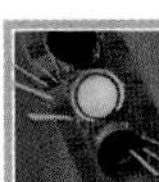

> CAUTION

Be careful! We must invert the divisor *before simplifying.*

Divide.

$\frac{3}{5} \div \frac{6}{7} = \frac{3}{5} \times \frac{7}{6}$ Invert the divisor *first!* Then you can divide by the common factor of 3.

$= \frac{\overset{1}{\cancel{3}} \times 7}{5 \times \underset{2}{\cancel{6}}} = \frac{7}{10}$

Check Yourself 5

Divide.

$\frac{4}{9} \div \frac{8}{15}$

When mixed or whole numbers are involved, the process is similar. Simply change the mixed or whole numbers to improper fractions as the first step. Then proceed with the division rule. Example 6 illustrates this approach.

Example 6 Dividing Mixed Numbers

< Objective 3 >

Divide.

$2\frac{3}{8} \div 1\frac{3}{4} = \frac{19}{8} \div \frac{7}{4}$ Write the mixed numbers as improper fractions.

$= \frac{19}{8} \times \frac{4}{7} = \frac{19 \times \overset{1}{\cancel{4}}}{\underset{2}{\cancel{8}} \times 7}$ Invert the divisor and multiply as before.

$= \frac{19}{14} = 1\frac{5}{14}$

Check Yourself 6

Divide.

$3\frac{1}{5} \div 2\frac{2}{5}$

Example 7 illustrates the division process when whole and mixed numbers are involved.

Example 7 Dividing a Mixed Number and a Whole Number

NOTE

Write the whole number 6 as $\frac{6}{1}$.

Divide and simplify.

$$1\frac{4}{5} \div 6 = \frac{9}{5} \div \frac{6}{1}$$

$$= \frac{9}{5} \times \frac{1}{6} = \frac{\overset{3}{\cancel{9}} \times 1}{5 \times \underset{2}{\cancel{6}}}$$

Invert the divisor, and then divide by the common factor of 3.

$$= \frac{3}{10}$$

Check Yourself 7

Divide.

$8 \div 4\frac{4}{5}$

Units Analysis

When dividing by denominate numbers with units that are fractions, multiply by the reciprocal of the number and its units. That is, we "flip" the units as well.

EXAMPLES:

$$500 \text{ mi} \div \frac{25 \text{ mi}}{1 \text{ gal}} = \frac{500 \text{ mi}}{1} \times \frac{1 \text{ gal}}{25 \text{ mi}} = 20 \text{ gal}$$

$$\$24{,}000 \div \frac{\$400}{1 \text{ yr}} = \frac{24{,}000 \text{ dollars}}{1} \div \frac{400 \text{ dollars}}{1 \text{ yr}} = \frac{24{,}000 \cancel{\text{ dollars}}}{1} \times \frac{1 \text{ yr}}{400 \cancel{\text{ dollars}}} = 60 \text{ yr}$$

In each case, arithmetic on the units produces the final units.

As was the case with multiplication, dividing fractions can be used to solve many applications. The steps of the problem-solving process remain the same.

Example 8 An Application Involving Mixed Numbers

< Objective 4 >

NOTES

One kilometer, abbreviated km, is a metric unit of distance. It is about $\frac{3}{5}$ mi.

The important formula is Speed = distance ÷ time.

Jack traveled 140 kilometers (km) in $2\frac{1}{3}$ hr. What was his average speed?

$$\text{Speed} = \underbrace{140 \text{ km}}_{\text{Distance}} \div \underbrace{2\frac{1}{3} \text{ hr}}_{\text{Time}}$$

$$= \frac{140 \text{ km}}{1} \div \frac{7 \text{ hr}}{3}$$

We know the distance traveled and the time for that travel. To find the *average* speed, we use division. Do you remember why?

$$= \frac{140 \text{ km}}{1} \times \frac{3}{7 \text{ hr}} = \frac{\overset{20}{\cancel{140}} \times 3}{1 \times \underset{1}{\cancel{7}}} \frac{\text{km}}{\text{hr}}$$

$\frac{\text{km}}{\text{hr}}$ is read "kilometers per hour." This is a unit of speed.

$$= 60 \text{ km/hr}$$

Check Yourself 8

A light plane flew 280 mi in $1\frac{3}{4}$ hr. What was its average speed?

Example 9 An Application Involving Division

NOTE

We divide the length of the longer piece by the desired length of the shorter pieces.

An electrician needs pieces of wire $2\frac{3}{5}$ in. long. If she has a $20\frac{4}{5}$-in. piece of wire, how many of the shorter pieces can she cut?

$$\begin{aligned} 20\frac{4}{5}\text{ in.} \div 2\frac{3}{5}\,\frac{\text{in.}}{\text{piece}} &= \frac{104}{5}\text{ in.} \div \frac{13}{5}\,\frac{\text{in.}}{\text{piece}} \\ &= \frac{104}{5}\text{ in.} \times \frac{5}{13}\,\frac{\text{pieces}}{\text{in.}} \\ &= \frac{\overset{8}{\cancel{104}} \times \overset{1}{\cancel{5}}}{\underset{1}{\cancel{5}} \times \underset{1}{\cancel{13}}}\text{ pieces} \\ &= 8\text{ pieces} \end{aligned}$$

Check Yourself 9

A piece of plastic water pipe 63 in. long is to be cut into lengths of $3\frac{1}{2}$ in. How many of the shorter pieces can be cut?

Units Analysis

When you convert units, it is important, and helpful, to carry out all unit arithmetic.

EXAMPLE:

Convert $1\frac{1}{2}$ years (yr) into minutes.

To accomplish this, we must recall that there are $365\,\frac{\text{days}}{\text{yr}}$, $24\,\frac{\text{hr}}{\text{day}}$, and $60\,\frac{\text{min}}{\text{hr}}$.

This allows us to set up an expression.

$$1\frac{1}{2}\text{ yr} \times 365\,\frac{\text{days}}{\text{yr}} \times 24\,\frac{\text{hr}}{\text{day}} \times 60\,\frac{\text{min}}{\text{hr}}$$

Before we work with the numbers, we should check the units to make certain that our result is in minutes.

Because the years, days, and hours all cancel, we are left with only minutes, so we can go ahead with the computation.

$$\frac{3}{2}\text{ yr} \times 365\,\frac{\text{days}}{\text{yr}} \times 24\,\frac{\text{hr}}{\text{day}} \times 60\,\frac{\text{min}}{\text{hr}} = 788{,}400\text{ min}$$

There are 788,400 min in $1\frac{1}{2}$ yr.

Some applications require both multiplication and division. Example 10 is such an application.

Example 10 An Application Involving Division

A parcel of land that is $2\frac{1}{2}$ mi long and $1\frac{1}{3}$ mi wide is to be divided into tracts that are each $\frac{1}{3}$ square mile (mi²). How many of these tracts will the parcel make?

The area of the parcel is its length times its width:

$$\begin{aligned} \text{Area} &= 2\frac{1}{2} \text{ mi} \times 1\frac{1}{3} \text{ mi} \\ &= \frac{5}{2} \text{ mi} \times \frac{4}{3} \text{mi} \\ &= \frac{10}{3} \text{ mi}^2 \end{aligned}$$

We need to divide the total area of the parcel into $\frac{1}{3}$-mi² tracts.

$$\frac{10}{3} \text{ mi}^2 \div \frac{1}{3} \text{ mi}^2 = \frac{10}{3} \text{ mi}^2 \times \frac{3}{1 \text{ mi}^2} = 10$$

The land will provide 10 tracts, each with an area of $\frac{1}{3}$ mi².

Check Yourself 10

A parcel of land that is $3\frac{1}{3}$ mi long and $2\frac{1}{2}$ mi wide is to be divided into $\frac{1}{3}$-mi² tracts. How many of these tracts will the parcel make?

In our final example, we look at a case in which the units in the divisor are fractions.

Example 11 An Application Involving Mixed Numbers

NOTE

We are using the gardener or contractor definition of a "yard" of mulch. It is 1 yd × 1 yd × 1 yd or 1 yd³.

Jackson has $6\frac{1}{2}$ yd of mulch. His garden needs $\frac{2}{3}$ yd per row. How many rows can he cover with the mulch?

We have $6\frac{1}{2}$ yd and $\frac{2}{3}\frac{\text{yd}}{\text{row}}$. Even if you don't immediately see how to solve the problem, units analysis can help. The units of the answer will be "rows." To get there, we need to have the yards units cancel. That will happen if we divide $6\frac{1}{2}$ by $\frac{2}{3}$.

$$\begin{aligned} 6\frac{1}{2} \text{ yd} \div \frac{2}{3}\frac{\text{yd}}{\text{row}} &= \frac{13}{2} \text{ yd} \div \frac{2}{3}\frac{\text{yd}}{\text{row}} \\ &= \frac{13}{2} \text{ yd} \times \frac{3}{2}\frac{\text{rows}}{\text{yd}} = \frac{39}{4} \text{ rows} = 9\frac{3}{4} \text{ rows} \end{aligned}$$

He can cover all of 9 rows and part $\left(\frac{3}{4}\right)$ of the tenth row.

Check Yourself 11

Tangela has \$4,100 to invest in a certain stock. If the stock is selling at \$$25\frac{5}{8}$ per share, how many shares can she buy?

Check Yourself ANSWERS

1. (a) $\frac{8}{5}$; (b) $3\frac{1}{4}$ is $\frac{13}{4}$, so the reciprocal is $\frac{4}{13}$ 2. $\frac{\frac{2}{5}}{\frac{3}{4}}$ 3. $\frac{3}{5} \times \frac{8}{7}$ 4. (a) $\frac{8}{15}$; (b) $\frac{1}{14}$; (c) 12
5. $\frac{5}{6}$ 6. $1\frac{1}{3}$ 7. $1\frac{2}{3}$ 8. 160 mi/hr 9. 18 pieces 10. 25 tracts 11. 160 shares

Reading Your Text

These fill-in-the-blank exercises will help you understand some of the key vocabulary used in this section. The answers to these exercises are in the Answers Appendix at the back of the text.

(a) The product of any nonzero number and its ____________ is 1.

(b) An expression that has a fraction as both its numerator and denominator is called a ____________ fraction.

(c) To divide one fraction by another, invert the ____________ and multiply.

(d) When dividing by denominate numbers with units that are fractions, multiply by the reciprocal of the number and its ____________.

Skills Calculator/Computer Career Applications Above and Beyond

2.6 exercises

< Objective 1 >

Find the reciprocal of each number.

1. $\frac{7}{8}$ VIDEO
2. $\frac{9}{5}$
3. $\frac{5}{2}$
4. $\frac{3}{4}$
5. $\frac{1}{2}$
6. $\frac{1}{8}$
7. $2\frac{1}{3}$ VIDEO
8. $4\frac{3}{5}$
9. $9\frac{3}{4}$
10. $1\frac{4}{5}$
11. 6 VIDEO
12. 20

< Objective 2 >

Divide. Write each result in simplest form.

13. $\frac{1}{5} \div \frac{3}{4}$ VIDEO
14. $\frac{2}{5} \div \frac{1}{3}$
15. $\frac{\frac{2}{5}}{\frac{3}{4}}$
16. $\frac{\frac{5}{8}}{\frac{3}{4}}$
17. $\frac{8}{9} \div \frac{4}{3}$ VIDEO
18. $\frac{5}{9} \div \frac{8}{11}$
19. $\frac{7}{10} \div \frac{5}{9}$
20. $\frac{8}{9} \div \frac{11}{15}$
21. $\frac{8}{15} \div \frac{2}{5}$
22. $\frac{5}{27} \div \frac{15}{54}$ VIDEO
23. $\frac{\frac{5}{27}}{\frac{25}{36}}$
24. $\frac{\frac{9}{28}}{\frac{27}{35}}$
25. $\frac{4}{5} \div 4$
26. $27 \div \frac{3}{7}$
27. $12 \div \frac{2}{3}$
28. $\frac{5}{8} \div 5$
29. $\frac{\frac{12}{17}}{\frac{6}{7}}$
30. $\frac{\frac{3}{4}}{\frac{9}{10}}$

< Objective 3 >

31. $15 \div 3\frac{1}{3}$

32. $2\frac{4}{7} \div 12$

33. $1\frac{3}{5} \div \frac{4}{15}$

34. $\frac{9}{14} \div 2\frac{4}{7}$

35. $\frac{7}{12} \div 2\frac{1}{3}$

36. $1\frac{3}{8} \div \frac{5}{12}$

37. $5\frac{3}{5} \div \frac{7}{15}$ VIDEO

38. $\frac{7}{18} \div 5\frac{5}{6}$

39. $8 \div 5\frac{1}{4}$

40. $3\frac{1}{2} \div 4$

41. $4\frac{3}{8} \div 6\frac{2}{3}$

42. $6\frac{2}{3} \div 4\frac{3}{8}$

43. $\left(\frac{2}{3}\right)^3 \div 4$

44. $\left(\frac{3}{5}\right)^3 \div 9$

45. $\frac{3}{5} \div \left(\frac{3}{10}\right)^2$ VIDEO

46. $\left(\frac{2}{7}\right)^2 \div \frac{3}{14}$

Divide. Be sure to attach the proper units.

47. $900 \text{ mi} \div 15 \frac{\text{mi}}{\text{gal}}$

48. $1{,}500 \text{ joules} \div 75 \frac{\text{joules}}{\text{s}}$

49. $8{,}750 \text{ watts} \div 350 \frac{\text{watts}}{\text{s}}$

50. $\$75{,}744 \div \frac{\$3{,}156}{\text{month}}$

< Objective 4 >

Solve each application.

51. **Construction** A $5\frac{1}{4}$ ft long wire is to be cut into 7 pieces of the same length. How long is each piece?

52. **Crafts** A potter uses $\frac{2}{3}$ lb of clay in making a bowl. How many bowls can be made from 16 lb of clay?

53. **Statistics** Virginia made a trip of 95 mi in $1\frac{1}{4}$ hr. What was her average speed?

54. **Business and Finance** A piece of land measures $3\frac{3}{4}$ acres and is for sale at \$60,000. What is the price per acre?

55. **Crafts** A roast weighs $3\frac{1}{4}$ lb. How many $\frac{1}{4}$-lb servings will the roast provide?

56. **Construction** A bookshelf is 55 in. long. If the books have an average thickness of $1\frac{1}{4}$ in., how many books can be put on the shelf?

Skills | **Calculator/Computer** | Career Applications | Above and Beyond

Using a Calculator to Divide Fractions

Dividing fractions on a calculator is almost exactly the same as multiplying them. You simply press the [÷] key instead of the [×] key. Again, parentheses are important when using a calculator to work with fractions.

Scientific Calculator

Dividing fractions on a scientific calculator requires only that you enter the problem followed by the equal sign. Recall that fractions are entered using the [**a b/c**] key.

Graphing Calculator

When using a graphing calculator, you must choose the fraction option [**1:▶ Frac**] from the [MATH] menu before pressing [ENTER].

Use a calculator to find each quotient.

57. $\frac{1}{5} \div \frac{2}{15}$

58. $\frac{13}{17} \div \frac{39}{34}$

59. $\frac{5}{7} \div \frac{15}{28}$

60. $\frac{3}{7} \div \frac{9}{28}$

61. $\frac{15}{18} \div \frac{45}{27}$

62. $\frac{19}{63} \div \frac{38}{9}$

63. $\frac{25}{45} \div \frac{100}{135}$

64. $\frac{86}{24} \div \frac{258}{96}$

Skills | Calculator/Computer | **Career Applications** | Above and Beyond

65. MANUFACTURING TECHNOLOGY Calculate the distance from the center of hole A to the center of hole B.

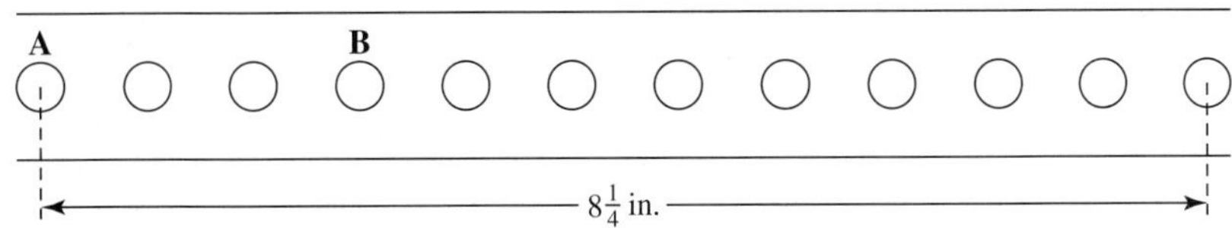

66. MANUFACTURING TECHNOLOGY A staircase is $95\frac{7}{8}$ in. tall and has 13 risers. What is the height of each riser?
Hint: Convert $95\frac{7}{8}$ into an improper fraction and divide by $\frac{13}{1}$.

67. MANUFACTURING TECHNOLOGY A part that is $\frac{3}{4}$ in. wide is to be magnified for a detailed drawing. The scale is 1 in. = $\frac{3}{8}$ in. What is the width of the part in the drawing?
Hint: Divide the width by $\frac{3}{8}$.

68. MANUFACTURING TECHNOLOGY A typical unified threaded bolt has one thread every $\frac{1}{20}$ in. How many threads are in $\frac{1}{4}$ in.?

69. CRAFTS Manuel has $7\frac{1}{2}$ yd of cloth. He wants to cut it into strips $1\frac{3}{4}$ yd long. How many strips will he have? How much cloth remains, if any?

70. CRAFTS Evette has $41\frac{1}{2}$ ft of string. She wants to cut it into pieces $3\frac{3}{4}$ ft long. How many pieces of string will she have? How much string remains, if any?

71. CRAFTS When squeezing oranges for fresh juice, 3 oranges yield about $\frac{1}{3}$ of a cup.

(a) How much juice could you expect to obtain from a bag containing 24 oranges?
(b) If you needed 8 cups of orange juice, how many bags of oranges should you buy?

Skills | Calculator/Computer | Career Applications | **Above and Beyond**

72. NUMBER PROBLEM A farmer left 17 cows to be divided among three employees. The first employee was to receive $\frac{1}{2}$ of the cows, the second employee was to receive $\frac{1}{3}$ of the cows, and the third employee was to receive $\frac{1}{9}$ of the cows. The executor of the farmer's estate realized that 17 cows could not be divided into halves, thirds, or ninths and so added a neighbor's cow to the farmer's. With 18 cows, the executor gave 9 cows to the first employee, 6 cows to the second employee, and 2 cows to the third employee. This accounted for the 17 cows, so the executor returned the borrowed cow to the neighbor. Explain why this works.

73. In general division of fractions is not commutative.

For example, $\frac{3}{4} \div \frac{5}{6} \neq \frac{5}{6} \div \frac{3}{4}$.

There could be an exception. Can you think of a situation in which division of fractions would be commutative?

74. Josephine's boss tells her that her salary is to be divided by $\frac{1}{3}$. Should she quit?

75. Compare the English phrases "divide in half" and "divide by one-half." Do they say the same thing? Create examples to support your answer.

76. **(a)** Compute: $5 \div \frac{1}{10}$; $5 \div \frac{1}{100}$; $5 \div \frac{1}{1{,}000}$; $5 \div \frac{1}{10{,}000}$.

(b) As the divisor gets smaller (approaches 0), what happens to the quotient?

(c) What does this say about the answer to $5 \div 0$?

Answers

1. $\frac{8}{7}$ **3.** $\frac{2}{5}$ **5.** 2 **7.** $\frac{3}{7}$ **9.** $\frac{4}{39}$ **11.** $\frac{1}{6}$ **13.** $\frac{4}{15}$ **15.** $\frac{8}{15}$ **17.** $\frac{2}{3}$ **19.** $1\frac{13}{50}$ **21.** $1\frac{1}{3}$ **23.** $\frac{4}{15}$ **25.** $\frac{1}{5}$ **27.** 18 **29.** $\frac{14}{17}$ **31.** $4\frac{1}{2}$ **33.** 6 **35.** $\frac{1}{4}$ **37.** 12 **39.** $1\frac{11}{12}$ **41.** $\frac{21}{32}$ **43.** $\frac{2}{27}$ **45.** $6\frac{2}{3}$ **47.** 60 gal **49.** 25 s **51.** $\frac{3}{4}$ ft **53.** $76 \frac{\text{mi}}{\text{hr}}$ **55.** 13 servings **57.** $\frac{3}{2}$ or $1\frac{1}{2}$ **59.** $\frac{4}{3}$ or $1\frac{1}{3}$ **61.** $\frac{1}{2}$ **63.** $\frac{3}{4}$ **65.** $2\frac{1}{4}$ in. **67.** 2 in. **69.** 4 strips; $\frac{1}{2}$ yd **71.** $2\frac{2}{3}$ cups; 3 bags **73.** Above and Beyond **75.** Above and Beyond

Adapting a Recipe

Tom and Susan like eating in ethnic restaurants, so they were thrilled when Marco's Cafe, an Indian restaurant, opened in their neighborhood. The first time they ate there, Susan had a bowl of mulligatawny soup and she loved it. She decided that it would be a great soup to serve her friends so she asked Marco for the recipe. Marco said that was no problem. He had already had so many requests for the recipe that he had made up a handout. A copy of it is reproduced here (try it if you are adventurous):

Mulligatawny Soup

This recipe makes 10 gal; recommended serving size is a 12-ounce (12-oz) bowl. Sauté in a steam kettle until the onions are translucent:

10 lb	diced onion
10 lb	diced celery
$\frac{1}{2}$ cup	garlic puree
1 cup	madras curry
2 cups	mild curry

Add and bring to a boil:

4 cups	white wine
$\frac{1}{3}$ cup	sugar
1 #10 can	diced tomato
1 gal	fresh apple juice
$\frac{1}{3}$ cup	lemon juice
2 gal	water
1 #10 can	diced carrots
16 oz	chicken stock

Finish with:

Roux (1 lb butter and 1 lb flour) and 8 quarts cream (temper into hot liquid). Season to taste with salt, pepper, celery seed, basil, and garlic.

How many servings does this recipe make?

Visit local grocery stores to find out how much each item costs. Calculate the total cost for 10 gal of soup. What is the cost for each 12-oz serving? (This is called the *marginal cost*—it does not include the overhead for running the restaurant.)

What is roux?

Susan does want to make this soup for a dinner party she is having. Rewrite the recipe so that it will serve six 12-oz bowls. Use reasonable measures, such as teaspoons and cups. Answering these questions may help. For some items you may have to experiment.

How many ounces in a #10 can?
How many cups in a gallon?
How many ounces in a pound?
How many teaspoons in a cup?
How many cups in a pound of diced onion?

summary :: chapter 2

Definition/Procedure	Example	Reference
Prime Numbers and Divisibility		Section 2.1
Prime Number Any whole number greater than one, that has exactly two factors, 1 and itself.	7, 13, 29, and 73 are prime numbers.	*p.* 98
Composite Number Any whole number greater than 1 that is not prime.	8, 15, 42, and 65 are composite numbers.	*p.* 99
Zero and One Zero and one are neither classified as prime nor composite numbers.		*p.* 99
Divisibility Tests		
By 2 A whole number is divisible by 2 if its last digit is 0, 2, 4, 6, or 8.	932 is divisible by 2; 1,347 is not.	*p.* 99
By 3 A whole number is divisible by 3 if the sum of its digits is divisible by 3.	546 is divisible by 3; 2,357 is not.	*p.* 100
By 5 A whole number is divisible by 5 if its last digit is 0 or 5.	865 is divisible by 5; 23,456 is not.	*p.* 100
Factoring Whole Numbers		Section 2.2
Prime Factorization		
To find the prime factorization of a number, divide the number by a series of primes until the final quotient is a prime number. The prime factors include each prime divisor and the final quotient.	$2\overline{)630}$ $3\overline{)315}$ $3\overline{)105}$ $5\overline{)35}$ 7 So $630 = 2 \times 3 \times 3 \times 5 \times 7$.	*p.* 106
Greatest Common Factor (GCF) The GCF is the *largest* number that is a factor of each of a group of numbers.		*p.* 108
To Find the GCF		
Step 1 Write the prime factorization for each of the numbers in the group. **Step 2** Locate the prime factors that are *common* to all the numbers. **Step 3** The greatest common factor is the *product* of all of the common prime factors. If there are no common prime factors, the GCF is 1.	To find the GCF of 24, 30, and 36: $24 = ② \times 2 \times 2 \times ③$ $30 = ② \times ③ \times 5$ $36 = ② \times 2 \times ③ \times 3$ The GCF is $2 \times 3 = 6$.	*p.* 108
Fraction Basics		Section 2.3
Fraction Fractions name a number of equal parts of a unit or whole. A fraction is written in the form $\frac{a}{b}$, in which a and b are whole numbers and b cannot be zero.		*p.* 113

Continued

Definition/Procedure	Example	Reference
Denominator The number of equal parts into which the whole is divided. **Numerator** The number of parts of the whole that are used.	$\frac{5}{8}$ ← Numerator (5); ← Denominator (8)	*p.* 113
Proper Fraction A fraction whose numerator is *less than* its denominator. It names a number less than 1.	$\frac{2}{3}$ and $\frac{11}{15}$ are proper fractions.	*p.* 114
Improper Fraction A fraction whose numerator *is greater than or equal to* its denominator. It names a number greater than or equal to 1.	$\frac{7}{5}$, $\frac{21}{20}$, and $\frac{8}{8}$ are improper fractions.	*p.* 114
Mixed Number The sum of a whole number and a proper fraction.	$2\frac{1}{3}$ and $5\frac{7}{8}$ are mixed numbers. Note that $2\frac{1}{3}$ means $2 + \frac{1}{3}$.	*p.* 115
To Change an Improper Fraction into a Mixed Number		
Step 1 Divide the numerator by the denominator. The quotient is the whole-number portion of the mixed number. **Step 2** If there is a remainder, write the remainder over the original denominator. This gives the fraction part of the mixed number.	To change $\frac{22}{5}$ to a mixed number: $5\overline{)22}$ gives 4 ← Quotient; 20; 2 ← Remainder $\frac{22}{5} = 4\frac{2}{5}$	*p.* 116
To Change a Mixed Number to an Improper Fraction		
Step 1 Multiply the denominator of the fraction by the whole-number part of the mixed number. **Step 2** Add the numerator of the fraction to that product. **Step 3** Write that sum over the original denominator to form the improper fraction.	Denominator, Whole number, Numerator $5\frac{3}{4} = \frac{(4 \times 5) + 3}{4} = \frac{23}{4}$ Denominator	*p.* 117
Simplifying Fractions		Section 2.4
Equivalent Fractions Two fractions that are equivalent (have equal value) are different names for the same number.		*p.* 123
Cross Products $\frac{a}{b} \times \frac{c}{d}$ $a \times d$ and $b \times c$ are called the *cross products.* If the cross products for two fractions are equal, the two fractions are equivalent.	$\frac{2}{3} = \frac{4}{6}$ because $2 \times 6 = 3 \times 4$	*p.* 123
The Fundamental Principle of Fractions For the fraction $\frac{a}{b}$, and any nonzero number c, $\frac{a}{b} = \frac{a \div c}{b \div c}$ **In words:** We can divide the numerator and denominator of a fraction by the same nonzero number. The result is an equivalent fraction.	$\frac{8}{12} = \frac{8 \div 4}{12 \div 4} = \frac{2}{3}$ $\frac{8}{12}$ and $\frac{2}{3}$ are equivalent fractions.	*p.* 123

Continued

Definition/Procedure	Example	Reference
Simplest Form A fraction is in simplest form, or in lowest terms, if the numerator and denominator have no common factors other than 1. This means that the fraction has the smallest possible numerator and denominator.	$\frac{2}{3}$ is in simplest form. $\frac{12}{18}$ is *not* in simplest form because the numerator and denominator have the common factor 6.	*p.* 124
To Write a Fraction in Simplest Form		
Divide the numerator and denominator by their greatest common factor.	$\frac{10}{15} = \frac{10 \div 5}{15 \div 5} = \frac{2}{3}$	*p.* 124
Multiplying Fractions		**Section 2.5**
To Multiply Two Fractions		
Step 1 Multiply the numerators to find the numerator of the product. **Step 2** Multiply the denominators to find the denominator of the product. **Step 3** Simplify the resulting fraction, if possible. When multiplying fractions it is usually easiest to divide by any common factors in the numerator and denominator *before* multiplying.	$\frac{5}{8} \times \frac{3}{7} = \frac{5 \times 3}{8 \times 7} = \frac{15}{56}$ $\frac{5}{9} \times \frac{3}{10} = \frac{\overset{1}{\cancel{5}} \times \overset{1}{\cancel{3}}}{\underset{3}{\cancel{9}} \times \underset{2}{\cancel{10}}} = \frac{1}{6}$	*p.* 132
Dividing Fractions		**Section 2.6**
To Divide Two Fractions		
Invert the divisor and multiply.	$\frac{3}{7} \div \frac{4}{5} = \frac{3}{7} \times \frac{5}{4} = \frac{15}{28}$	*p.* 146
Multiplying or Dividing Mixed Numbers		
Convert any mixed or whole numbers to improper fractions. Then multiply or divide the fractions as before.	$6\frac{2}{3} \times 3\frac{1}{5} = \frac{\overset{4}{\cancel{20}} \times 16}{3 \times \underset{1}{\cancel{5}}} = \frac{64}{3} = 21\frac{1}{3}$	*p.* 147

summary exercises :: chapter 2

This summary exercise set will help ensure that you have mastered each of the objectives of this chapter. The exercises are grouped by section. You should reread the material associated with any exercises that you find difficult. The answers to the odd-numbered exercises are in the Answers Appendix at the back of the text.

2.1 *List all factors of each number.*

1. 52

2. 41

Use the group of numbers 2, 5, 7, 11, 14, 17, 21, 23, 27, 39, and 43.

3. List the prime numbers; then list the composite numbers.

Use the divisibility tests to determine which, if any, of the numbers 2, 3, and 5 are factors of each number.

4. 2,350 **5.** 33,451

2.2 *Find the prime factorization of each number.*

6. 48 **7.** 420 **8.** 2,640 **9.** 2,250

Find the greatest common factor (GCF).

10. 15 and 20 **11.** 30 and 31 **12.** 24 and 40

13. 39 and 65 **14.** 49, 84, and 119 **15.** 77, 121, and 253

2.3 *Identify the numerator and denominator of each fraction.*

16. $\frac{5}{9}$ **17.** $\frac{17}{23}$

Give the fractions that name the shaded portions in each diagram. Identify the numerator and the denominator.

18.

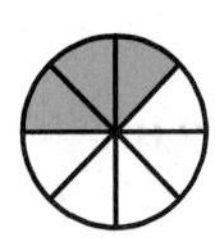

Fraction _________

Numerator _________

Denominator _________

19.

Fraction: _________

Numerator: _________

Denominator: _________

20. From the group of numbers

$\frac{2}{3}, \frac{5}{4}, 2\frac{3}{7}, \frac{45}{8}, \frac{7}{7}, 3\frac{4}{5}, \frac{9}{1}, \frac{7}{10}, \frac{12}{5}, 5\frac{2}{9}$

List the proper fractions. _________

List the improper fractions. _________

List the mixed numbers. _________

Convert to mixed or whole numbers.

21. $\frac{41}{6}$ **22.** $\frac{32}{8}$ **23.** $\frac{23}{3}$ **24.** $\frac{47}{4}$

Convert to improper fractions.

25. $7\frac{5}{8}$ **26.** $4\frac{3}{10}$ **27.** $5\frac{2}{7}$ **28.** $12\frac{8}{13}$

2.4 *Determine whether each pair of fractions is equivalent.*

29. $\frac{5}{8}, \frac{7}{12}$ **30.** $\frac{8}{15}, \frac{32}{60}$

Write each fraction in simplest form.

31. $\frac{24}{36}$

32. $\frac{45}{75}$

33. $\frac{140}{180}$

34. $\frac{16}{21}$

Decide whether each is a true statement.

35. $\frac{15}{25} = \frac{3}{5}$

36. $\frac{36}{40} = \frac{4}{5}$

2.5 *Multiply.*

37. $\frac{7}{15} \times \frac{5}{21}$

38. $\frac{10}{27} \times \frac{9}{20}$

39. $4 \cdot \frac{3}{8}$

40. $3\frac{2}{5} \cdot \frac{5}{8}$

41. $5\frac{1}{3} \times 1\frac{4}{5}$

42. $1\frac{5}{12} \times 8$

43. $3\frac{1}{5} \times \frac{7}{8} \times 2\frac{6}{7}$

Solve each application.

44. **Social Science** The scale on a map is 1 in. = 80 mi. If two cities are $2\frac{3}{4}$ in. apart on the map, what is the actual distance between the cities?

45. **Construction** A kitchen measures $5\frac{1}{3}$ yards (yd) by $4\frac{1}{4}$ yd. If you purchase linoleum costing \$9 per yd^2, what will it cost to cover the floor?

46. **Construction** Your living room measures $6\frac{2}{3}$ yd by $4\frac{1}{2}$ yd. If you purchase carpeting at \$18 per yd^2, what will it cost to carpet the room?

47. **Business and Finance** Maria earns \$72 per day. If she works $\frac{5}{8}$ of a day, how much will she earn?

48. **Science and Medicine** David drove at an average speed of 65 mi/hr for $2\frac{2}{5}$ hr. How many miles did he travel?

49. **Social Science** The scale on a map is 1 in. = 120 mi. What actual distance, in miles, does $3\frac{2}{5}$ in. on the map represent?

50. **Social Science** At a college, $\frac{2}{5}$ of the students take a science course. Of the students taking science, $\frac{1}{4}$ take biology. What fraction of the students take biology?

51. **Social Science** A student survey found that $\frac{3}{4}$ of the students have jobs while going to school. Of those who have jobs, $\frac{5}{6}$ work more than 20 hr/week. What fraction of those surveyed work more than 20 hr/week?

52. **Construction** A living room has dimensions $5\frac{2}{3}$ yd by $4\frac{1}{2}$ yd. How much carpeting must be purchased to cover the room?

2.6 *Divide.*

53. $\frac{5}{12} \div \frac{5}{8}$

54. $\dfrac{\frac{7}{15}}{\frac{14}{25}}$

55. $\dfrac{\frac{9}{20}}{\frac{12}{5}}$

56. $3\frac{3}{8} \div 2\frac{1}{4}$

57. $3\frac{3}{7} \div 8$

58. $6\frac{1}{7} \div \frac{3}{14}$

Solve each application.

59. **Construction** A piece of wire $3\frac{3}{4}$ ft long is to be cut into 5 pieces of the same length. How long will each piece be?

60. **Crafts** A blouse pattern requires $1\frac{3}{4}$ yd of fabric. How many blouses can be made from a piece of silk that is 28 yd long?

61. **Science and Medicine** If you drive 126 mi in $2\frac{1}{4}$ hr, what is your average speed?

62. **Science and Medicine** If you drive 117 mi in $2\frac{1}{4}$ hr, what is your average speed?

63. **Construction** An 18-acre piece of land is to be subdivided into home lots that are each $\frac{3}{8}$ acre. How many lots can be formed?

CHAPTER 2

chapter test 2

Use this chapter test to assess your progress and to review for your next exam. Allow yourself about an hour to take this test. The answers to these exercises are in the Answers Appendix at the back of the text.

1. Which of the numbers 5, 9, 13, 17, 22, 27, 31, and 45 are prime numbers? Which are composite numbers?

2. Find the prime factorization for 264.

3. Use the divisibility tests to determine which, if any, of the numbers 2, 3, and 5 are factors of 54,204.

Find the greatest common factor (GCF) of each set of numbers.

4. 36 and 84

5. 16, 24, and 72

What fraction names the shaded part of each diagram? Identify the numerator and denominator.

6. 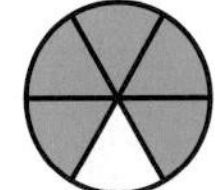

7. 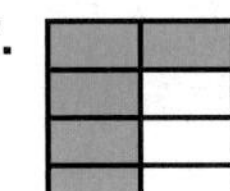

8. 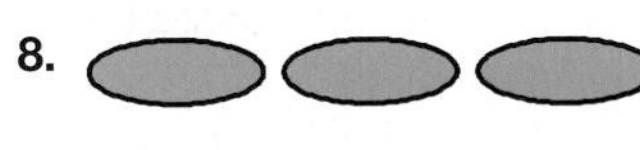

9. Give the mixed number that names the shaded part in the diagram.

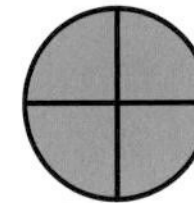 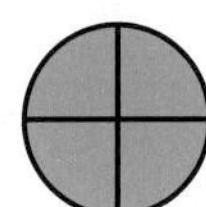 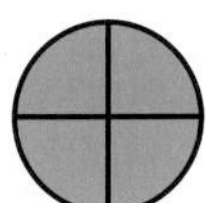 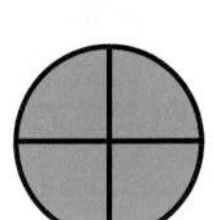 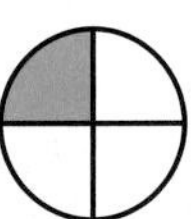

Write each fraction in simplest form.

10. $\frac{21}{27}$

11. $\frac{36}{84}$

12. $\frac{8}{23}$

Use the cross-product method to find out whether each pair of fractions is equivalent.

13. $\frac{2}{7}, \frac{8}{28}$

14. $\frac{8}{20}, \frac{12}{30}$

15. $\frac{3}{20}, \frac{2}{15}$

16. Identify the proper fractions, improper fractions, and mixed numbers in the following group.

$\frac{10}{11}, \frac{9}{5}, \frac{7}{7}, \frac{8}{1}, 2\frac{3}{5}, \frac{1}{8}$

Convert each mixed number to an improper fraction.

17. $5\frac{2}{7}$

18. $4\frac{3}{8}$

19. $8\frac{2}{9}$

Convert each fraction to a mixed or whole number.

20. $\frac{17}{4}$

21. $\frac{15}{1}$

22. $\frac{74}{8}$

23. $\frac{18}{6}$

Evaluate.

24. $\frac{2}{3} \times \frac{5}{7}$

25. $5\frac{1}{3} \times \frac{3}{4}$

26. $\frac{7}{12} \div \frac{14}{15}$

27. $2\frac{2}{3} \times 1\frac{2}{7}$

28. $\frac{16}{35} \times \frac{14}{24}$

29. $5\frac{3}{5} \div 2\frac{1}{10}$

30. $\frac{9}{10} \times \frac{5}{8}$

31. $\frac{6}{7} \div \frac{3}{4}$

32 $3\frac{5}{6} \times 2\frac{2}{5}$

33. $1\frac{3}{4} \div 1\frac{3}{8}$

Solve each application.

34. **Business and Finance** What is the cost of $2\frac{3}{4}$ lb of apples if the price per pound is 48 cents?

35. **Crafts** A bookshelf is 66 in. long. If the thickness of each book on the shelf is $1\frac{3}{8}$ in., how many books can be placed on the shelf?

36. **Construction** A $31\frac{1}{3}$-acre piece of land is subdivided into home lots. Each home lot is to be $\frac{2}{3}$ acre. How many homes can be built?

37. **Construction** A room measures $5\frac{1}{3}$ yd by $3\frac{3}{4}$ yd. How many square yards of linoleum must be purchased to cover the floor?

38. **Social Science** The scale on a map is 1 in. = 80 mi. If two towns are $2\frac{3}{8}$ in. apart on the map, what is the actual distance in miles between the towns?

cumulative review chapters 1–2

Use this exercise set to review concepts from earlier chapters. While it is not a comprehensive exam, it will help you identify any material that you need to review before moving on to the next chapter. In addition to the answers, you will find section references for these exercises in the Answers Appendix in the back of the text.

1.1

1. Give the place value of 7 in 3,738,500.

2. Give the word name for 302,525.

3. Write two million, four hundred thirty thousand as a numeral.

1.2 *Name the property of addition illustrated.*

4. $5 + 12 = 12 + 5$

5. $9 + 0 = 9$

6. $(7 + 3) + 8 = 7 + (3 + 8)$

Perform the indicated operation.

7. $\begin{array}{r} 593 \\ 275 \\ +\ 98 \\ \hline \end{array}$

8. Find the sum of 58, 673, 5,325, and 17,295.

1.4 *Round each number to the indicated place value.*

9. 5,873 to the nearest hundred

10. 953,150 to the nearest ten thousand

Estimate the sum by rounding each addend to the nearest hundred.

11. $\begin{array}{r} 943 \\ 3{,}281 \\ 778 \\ 2{,}112 \\ +\ 570 \\ \hline \end{array}$

Complete each statement by using the symbol < or >.

12. 49________47

13. 80 __________ 90

1.3 *Perform the indicated operation.*

14. $\begin{array}{r} 4{,}834 \\ -\ 973 \\ \hline \end{array}$

15. Find the difference of 25,000 and 7,535.

1.2 *Solve each application.*

16. **Statistics** Attendance for five performances of a play was 172, 153, 205, 193, and 182. How many people in total attended those performances?

1.3

17. **Business and Finance** Alan bought a Volkswagen with a list price of $18,975. He added stereo equipment for $439 and an air conditioner for $615. If he made a down payment of $2,450, what balance remained on the car?

1.5 *Name the property of addition and/or multiplication that is illustrated.*

18. $3 \times (4 \times 7) = (3 \times 4) \times 7$

19. $3 \times 4 = 4 \times 3$

20. $5 \times (2 + 4) = 5 \times 2 + 5 \times 4$

Perform the indicated operation.

21. $\begin{array}{r} 538 \\ \times\ 703 \\ \hline \end{array}$

22. $\begin{array}{r} 1{,}372 \\ \times\ 500 \\ \hline \end{array}$

Solve the application.

23. Construction A classroom is 8 yd wide by 9 yd long. If the room is to be recarpeted with material costing \$14 per square yard, find the cost of the carpeting.

1.6 *Divide, using long division.*

24. $48\overline{)3,259}$

25. $458\overline{)47,350}$

1.7 *Evaluate each expression.*

26. $3 + 5 \times 7$

27. $(3 + 5) \times 7$

28. 4×3^2

29. $2 + 8 \times 3 \div 4$

Solve each application.

30. Business and Finance William bought a washer-dryer combination that, with interest charges, cost \$841. He paid \$145 down and agreed to pay the balance in 12 monthly payments. Find the amount of each payment.

2.1

31. Number Problem Which of the numbers 5, 9, 13, 17, 22, 27, 31, and 45 are prime numbers? Which are composite numbers?

32. Use the divisibility tests to determine which, if any, of the numbers 2, 3, and 5 are factors of 54,204.

2.2

33. Find the prime factorization of 264.

Find the greatest common factor (GCF) of the given numbers.

34. 36 and 96

35. 16, 40, and 72

2.3 *Identify the proper fractions, improper fractions, and mixed numbers.*

$\frac{7}{12}, \frac{10}{8}, 3\frac{1}{5}, \frac{9}{9}, \frac{7}{1}, \frac{3}{7}, 2\frac{2}{3}$

36. Proper fractions:________ Improper fractions:________

Mixed numbers:________

2.4 *Convert to mixed or whole numbers.*

37. $\frac{14}{5}$

38. $\frac{28}{7}$

Convert to improper fractions.

39. $4\frac{1}{3}$

40. $7\frac{7}{8}$

Determine whether each pair of fractions is equivalent.

41. $\frac{7}{21}, \frac{8}{24}$

42. $\frac{7}{12}, \frac{8}{15}$

Write each fraction in simplest form.

43. $\frac{28}{42}$

44. $\frac{36}{96}$

2.5 *Multiply.*

45. $\frac{5}{9} \times \frac{8}{15}$

46. $\frac{20}{21} \cdot \frac{7}{25}$

47. $1\frac{1}{8} \cdot 4\frac{4}{5}$

48. $8 \times 2\frac{5}{6}$

49. $\frac{2}{3} \times 1\frac{4}{5} \times \frac{5}{8}$

2.6 *Divide.*

50. $\frac{5}{8} \div \frac{15}{32}$

51. $2\frac{5}{8} \div \frac{7}{12}$

52. $4\frac{1}{6} \div 5$

53. $2\frac{2}{7} \div 1\frac{11}{21}$

Solve each application.

54. **Construction** Your living room measures $6\frac{2}{3}$ yd by $4\frac{1}{2}$ yd. If you purchase carpeting at \$18 per yd^2, what will it cost to carpet the room?

55. **Construction** If a stack of $\frac{5}{8}$-in. plywood measures 55 in. high, how many sheets of plywood are in the stack?

Adding and Subtracting Fractions

CHAPTER 3 OUTLINE

INTRODUCTION

Carpentry is one of the oldest and most important of all trades. It dates back to ancient times and the earliest use of primitive tools. It includes large-scale work such as architecture and individual pieces such as cabinets and furniture. Carpenters mostly work with wood, but they also use ceramic, metal, and plastic. Some carpenters do roofing, refinishing, remodeling, restoration, and flooring.

Carpenters need to have a vast knowledge of scale drawing and an understanding of blueprints. They use a substantial amount of math for measuring and making their drawings and drafts. They work with models that may be hundreds of times smaller than the actual construction. Sometimes these are actual models, and sometimes they are drawings. Carpenters have to be extremely precise in their measuring, sometimes to tiny fractions of an inch. Errors in measurement can have dire consequences such as warping or cracking.

Carpenters learn their trades from vocational schools or by serving as apprentices to more experienced carpenters. Some carpenters are highly skilled and looked upon as artisans, whereas others do handy work.

3 prerequisite check

This Prerequisite Check highlights the skills you will need in order to be successful in this chapter. The answers to these exercises are in the Answers Appendix at the back of the text.

Write the prime factorization for each number.

1. 24

2. 36

3. 90

Change each mixed number to an improper fraction.

4. $4\frac{2}{3}$

5. $6\frac{2}{5}$

6. $9\frac{1}{10}$

Change each improper fraction to a mixed or whole number.

7. $\frac{29}{4}$

8. $\frac{42}{6}$

9. $\frac{8}{8}$

Evaluate.

10. $7 \times 11 - 2 \times (2 + 3)^2$

Simplify each fraction.

11. $\frac{8}{12}$

12. $\frac{21}{84}$

13. $\frac{21}{35}$

14. $\frac{20}{32}$

3.1 Adding and Subtracting Like Fractions

< 3.1 Objectives >

1 > Add two like fractions

2 > Add a group of like fractions

3 > Subtract two like fractions

We already know to think of adding as combining groups of the *same kinds* of objects. This is also true when we think about adding fractions.

Fractions can be added only if they name the *same parts* of a whole. This means we can add fractions only when they are **like fractions,** that is, when they have the *same* (*common*) denominators. For instance, we can add two nickels and three nickels to get five nickels. We *cannot* directly add two nickels and three dimes!

As long as we are dealing with like fractions, addition is an easy matter.

Step by Step

To Add Like Fractions

Step 1 Add the numerators.

Step 2 Place the sum over the common denominator.

Step 3 Simplify the resulting fraction when necessary.

Example 1 illustrates this rule.

Example 1 Adding Like Fractions

< Objective 1 >

Add.

$\frac{1}{5} + \frac{3}{5}$

Step 1 Add the numerators.

$1 + 3 = 4$

Step 2 Write that sum over the common denominator, 5. We are done at this point because $\frac{4}{5}$ is in simplest form.

Step 1 Step 2

$$\frac{1}{5} + \frac{3}{5} = \frac{1+3}{5} = \frac{4}{5}$$

We illustrate this addition with a diagram.

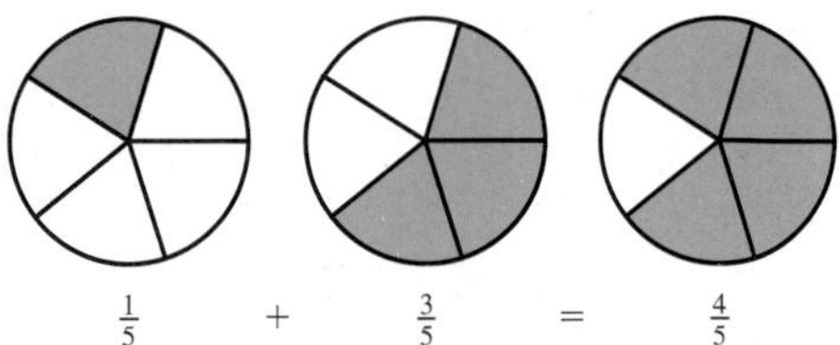

NOTE

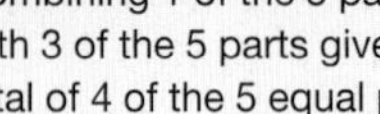

Combining 1 of the 5 parts with 3 of the 5 parts gives a total of 4 of the 5 equal parts.

Check Yourself 1

Add.

$\frac{2}{9} + \frac{5}{9}$

> CAUTION

Be Careful! When adding fractions, *do not* follow the rule for multiplying fractions. To multiply $\frac{1}{5} \times \frac{3}{5}$, we multiply both the numerators and the denominators.

$$\frac{1}{5} \times \frac{3}{5} = \frac{3}{25}$$

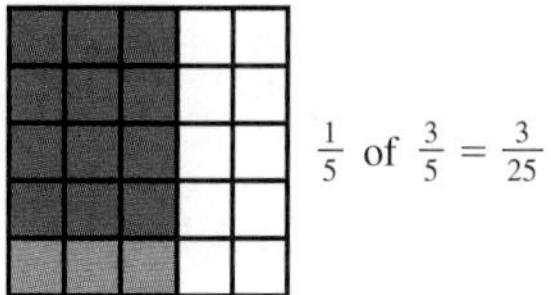

However, when you add like fractions, the sum has the same denominator as the fractions. We *do not* add the denominators.

$$\frac{1}{5} + \frac{3}{5} \neq \frac{4}{10}$$

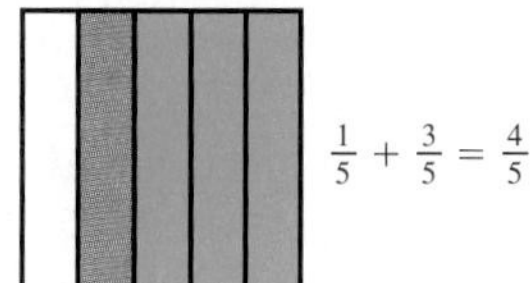

Step 3 of the addition rule for like fractions tells us to *simplify* the sum. Fractions should always be written in lowest terms.

Example 2 Adding Like Fractions and Simplifying

Add and simplify.

$$\frac{3}{12} + \frac{5}{12} = \frac{8}{12} = \frac{2}{3}$$

Step 3 (arrow to $\frac{2}{3}$)

The sum $\frac{8}{12}$ is *not* in lowest terms.

Divide the numerator and denominator by 4 to simplify the result.

Check Yourself 2

Add.

$$\frac{4}{15} + \frac{6}{15}$$

If the sum of two fractions is an improper fraction, we may write that sum as a mixed number.

Example 3 Adding Fractions That Result in Mixed Numbers

NOTE

Add as before. Then convert the sum to a mixed number.

Add.

$$\frac{5}{9} + \frac{8}{9} = \frac{13}{9} = 1\frac{4}{9}$$

Write the sum $\frac{13}{9}$ as a mixed number.

NOTE

Some instructors prefer you to report your results as mixed numbers, whereas others prefer improper fractions. You should ask if your instructor has a preference.

Check Yourself 3

Add.

$$\frac{7}{12} + \frac{10}{12}$$

We can easily extend our addition rule to find the sum of more than two fractions as long as they all have the same denominator, as shown in Example 4.

Example 4 Adding a Group of Like Fractions

< Objective 2 >

Add.

$$\frac{2}{7} + \frac{3}{7} + \frac{6}{7} = \frac{11}{7}$$ Add the numerators: 2 + 3 + 6 = 11.

$$= 1\frac{4}{7}$$

Check Yourself 4

Add.

$$\frac{1}{8} + \frac{3}{8} + \frac{5}{8}$$

Many applications can be solved by adding fractions.

Example 5 An Application Involving Adding Fractions

Noel walked $\frac{9}{10}$ mi to Jensen's house and then walked $\frac{7}{10}$ mi to school. How far did Noel walk?

To find the total distance Noel walked, add the two distances.

$$\frac{9}{10} + \frac{7}{10} = \frac{16}{10} = 1\frac{6}{10} = 1\frac{3}{5}$$

Noel walked $1\frac{3}{5}$ mi.

RECALL

Like fractions have the same denominator.

Check Yourself 5

Emir bought $\frac{7}{16}$ lb of candy at one store and $\frac{11}{16}$ lb at another store. How much candy did Emir buy?

If a problem involves like fractions, then subtraction, like addition, is not difficult.

Step by Step

To Subtract Like Fractions

Step 1 Subtract the numerators.

Step 2 Place the difference over the common denominator.

Step 3 Simplify the resulting fraction when necessary.

Example 6 Subtracting Like Fractions

< Objective 3 >

Subtract.

(a) $\frac{4}{5} - \frac{2}{5} = \frac{4-2}{5} = \frac{2}{5}$ (Step 1, Step 2)

Subtract the numerators: 4 − 2 = 2. Write the difference over the common denominator, 5. Step 3 is not necessary because the difference is in simplest form.

Illustrating with a diagram:

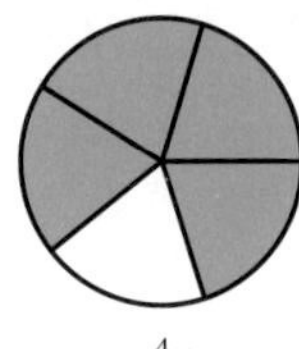
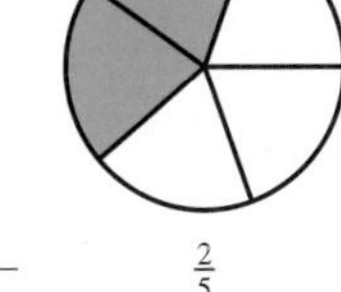
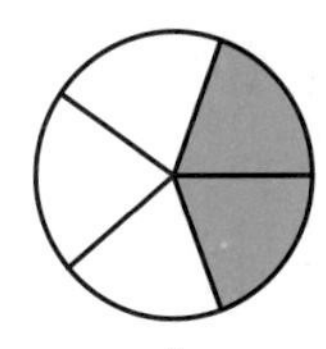

$\frac{4}{5} - \frac{2}{5} = \frac{2}{5}$

NOTES

Subtracting 2 of the 5 parts from 4 of the 5 parts leaves 2 of the 5 parts.

Always write the result in simplest terms.

(b) $\frac{5}{8} - \frac{3}{8} = \frac{5-3}{8} = \frac{2}{8} = \frac{1}{4}$

Check Yourself 6

Subtract.

$\frac{11}{12} - \frac{5}{12}$

Check Yourself ANSWERS

1. $\frac{7}{9}$ 2. $\frac{2}{3}$ 3. $\frac{17}{12}$ or $1\frac{5}{12}$ 4. $\frac{9}{8}$ or $1\frac{1}{8}$ 5. $1\frac{1}{8}$ lb 6. $\frac{1}{2}$

Reading Your Text

These fill-in-the-blank exercises will help you understand some of the key vocabulary used in this section. The answers to these exercises are in the Answers Appendix at the back of the text.

(a) Fractions with the same (common) denominator are called __________ fractions.

(b) When adding like fractions, add the __________.

(c) After adding two fractions, __________ the result when necessary.

(d) The result of subtraction is called the __________.

3.1 exercises

Skills | Calculator/Computer | Career Applications | Above and Beyond

< Objective 1 >

Add. Write all answers in simplest terms.

1. $\frac{3}{5} + \frac{1}{5}$ **2.** $\frac{4}{7} + \frac{1}{7}$ **3.** $\frac{4}{11} + \frac{6}{11}$ **4.** $\frac{5}{16} + \frac{4}{16}$

5. $\frac{2}{10} + \frac{3}{10}$ **6.** $\frac{5}{12} + \frac{1}{12}$ **7.** $\frac{3}{7} + \frac{4}{7}$ **8.** $\frac{13}{20} + \frac{17}{20}$

9. $\frac{29}{30} + \frac{11}{30}$ **10.** $\frac{4}{9} + \frac{5}{9}$ **11.** $\frac{13}{48} + \frac{23}{48}$ **12.** $\frac{17}{60} + \frac{31}{60}$

13. $\frac{3}{7} + \frac{6}{7}$ **14.** $\frac{3}{5} + \frac{4}{5}$ **15.** $\frac{7}{10} + \frac{9}{10}$ **16.** $\frac{5}{8} + \frac{7}{8}$

17. $\frac{11}{12} + \frac{10}{12}$ **18.** $\frac{13}{18} + \frac{11}{18}$

< Objective 2 >

19. $\frac{1}{8} + \frac{1}{8} + \frac{3}{8}$ **20.** $\frac{1}{10} + \frac{3}{10} + \frac{3}{10}$ **21.** $\frac{1}{9} + \frac{4}{9} + \frac{5}{9}$  **22.** $\frac{7}{12} + \frac{11}{12} + \frac{1}{12}$

< Objective 3 >

Subtract. Write all answers in simplest terms.

23. $\frac{3}{5} - \frac{1}{5}$ **24.** $\frac{5}{7} - \frac{2}{7}$ **25.** $\frac{7}{9} - \frac{4}{9}$ **26.** $\frac{7}{10} - \frac{3}{10}$

27. $\frac{13}{20} - \frac{3}{20}$  **28.** $\frac{19}{30} - \frac{17}{30}$ **29.** $\frac{19}{24} - \frac{5}{24}$ **30.** $\frac{25}{36} - \frac{13}{36}$

31. $\frac{11}{12} - \frac{7}{12}$ **32.** $\frac{9}{10} - \frac{6}{10}$ **33.** $\frac{8}{9} - \frac{3}{9}$ **34.** $\frac{5}{8} - \frac{1}{8}$

Solve each application. Write each answer in simplest terms.

35. **Number Problem** You work 7 hr one day, 5 hr the second day, and 6 hr the third day. How long did you work, as a fraction of a 24-hr day?

36. **Number Problem** One task took 7 min, a second task took 12 min, and a third task took 21 min. How long did the three tasks take, as a fraction of an hour?

37. **Geometry** What is the perimeter of a rectangle if the length is $\frac{7}{10}$ in. and the width is $\frac{2}{10}$ in.?

38. **Geometry** Find the perimeter of a rectangular picture if the width is $\frac{7}{9}$ yd and the length is $\frac{5}{9}$ yd.

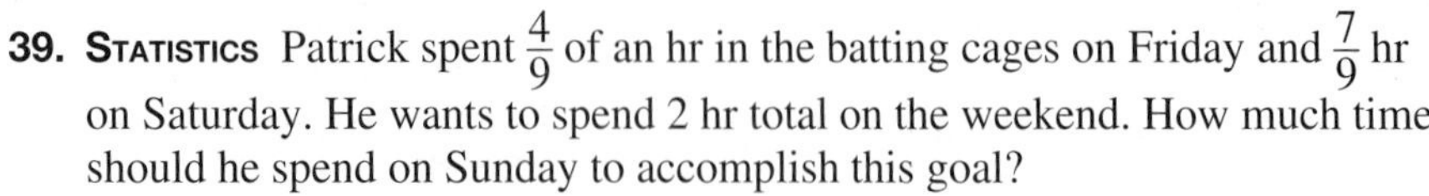

39. **Statistics** Patrick spent $\frac{4}{9}$ of an hr in the batting cages on Friday and $\frac{7}{9}$ hr on Saturday. He wants to spend 2 hr total on the weekend. How much time should he spend on Sunday to accomplish this goal?

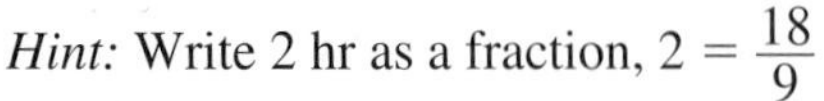

Hint: Write 2 hr as a fraction, $2 = \frac{18}{9}$

40. **Business and Finance** Maria, a road inspector, must inspect $\frac{17}{30}$ of a mile of road. If she has already inspected $\frac{11}{30}$ of a mile, how much more does she need to inspect?

GEOMETRY *Find the perimeter of each triangle.*

41.

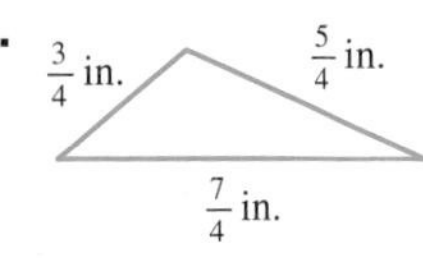

42.

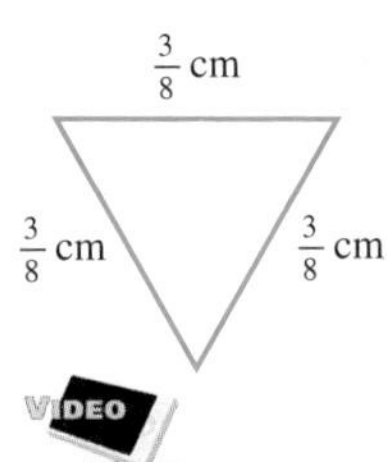

43.

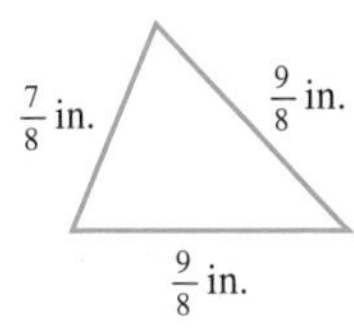

44.

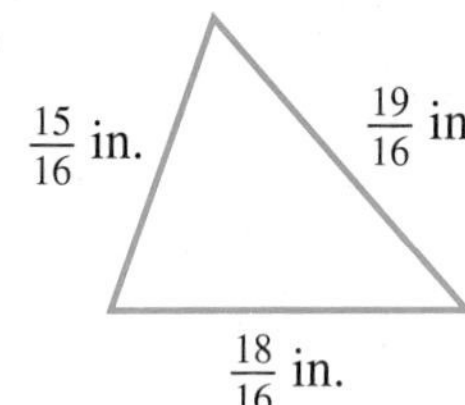

GEOMETRY *Find the perimeter of each polygon.*

45.

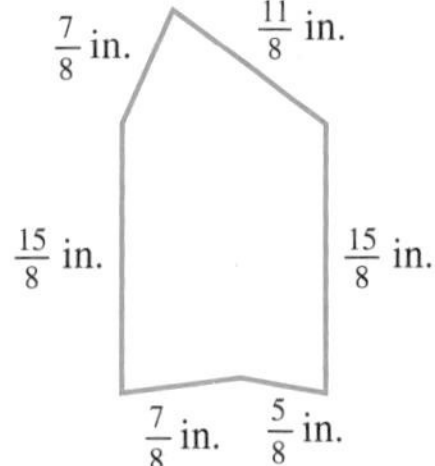

46.

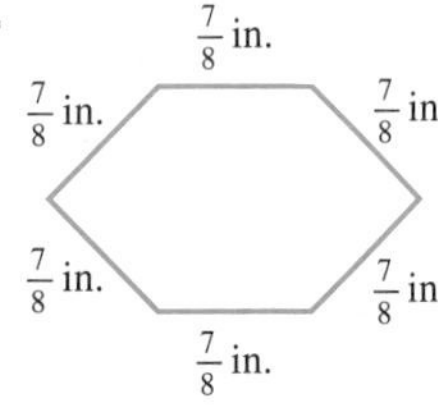

Evaluate. Write all answers in simplest terms.

47. $\frac{7}{12} - \frac{4}{12} + \frac{3}{12}$ 

48. $\frac{8}{9} + \frac{3}{9} - \frac{5}{9}$

49. $\frac{6}{13} - \frac{3}{13} + \frac{11}{13}$

50. $\frac{9}{11} - \frac{3}{11} + \frac{7}{11}$

51. $\frac{18}{23} - \frac{13}{23} - \frac{3}{23}$

52. $\frac{17}{18} - \frac{11}{18} - \frac{5}{18}$

Skills	Calculator/Computer	**Career Applications**	Above and Beyond

53. ALLIED HEALTH Carla is to be given $\frac{1}{8}$ gram (g) of medication in the morning, $\frac{1}{8}$ g at 3 P.M. in the afternoon, and $\frac{3}{8}$ g before she goes to bed. How much medication will she receive in one day?

54. ALLIED HEALTH Prior to chemotherapy, a patient's tumor weighed $\frac{7}{8}$ lb. After the initial round of chemotherapy, the tumor's weight had been reduced by $\frac{3}{8}$ lb. How much did the tumor weigh after the chemotherapy treatment?

55. MANUFACTURING TECHNOLOGY Triplet Precision Machine ordered $\frac{5}{8}$ ton of steel. The first truck arrived with $\frac{1}{8}$ ton of the steel. How much has yet to arrive?

56. MANUFACTURING TECHNOLOGY A sidewalk requires $\frac{15}{16}$ yd of concrete. The concrete mixer is capable of mixing $\frac{3}{16}$ yd at a time. How much concrete is still needed after one mixer load? chapter 3 > Make the Connection

Answers

1. $\frac{4}{5}$ 3. $\frac{10}{11}$ 5. $\frac{1}{2}$ 7. 1 9. $\frac{4}{3} = 1\frac{1}{3}$ 11. $\frac{3}{4}$ 13. $\frac{9}{7} = 1\frac{2}{7}$ 15. $\frac{8}{5} = 1\frac{3}{5}$ 17. $\frac{7}{4} = 1\frac{3}{4}$ 19. $\frac{5}{8}$ 21. $\frac{10}{9} = 1\frac{1}{9}$ 23. $\frac{2}{5}$ 25. $\frac{1}{3}$ 27. $\frac{1}{2}$ 29. $\frac{7}{12}$ 31. $\frac{1}{3}$ 33. $\frac{5}{9}$ 35. $\frac{3}{4}$ day 37. $\frac{9}{5}$ in. $= 1\frac{4}{5}$ in. 39. $\frac{7}{9}$ hr 41. $\frac{15}{4}$ in. $= 3\frac{3}{4}$ in. 43. $\frac{25}{8}$ in. $= 3\frac{1}{8}$ in. 45. $\frac{15}{2}$ in. $= 7\frac{1}{2}$ in. 47. $\frac{1}{2}$ 49. $\frac{14}{13} = 1\frac{1}{13}$ 51. $\frac{2}{23}$ 53. $\frac{5}{8}$ g 55. $\frac{1}{2}$ ton

3.2 Common Multiples

< 3.2 Objectives >

1 > Find the least common multiple (LCM) of two numbers

2 > Find the LCM of a group of numbers

3 > Compare the size of two fractions

In this chapter, we discuss the process used for adding or subtracting fractions. One of the most important concepts we use when we add or subtract fractions is that of **multiples.**

Definition

Multiples

The *multiples* of a number are the products of that number with the natural numbers 1, 2, 3, 4, 5, . . .

Example 1 **Listing Multiples**

List the multiples of 3.

The multiples of 3 are

$3 \times 1, 3 \times 2, 3 \times 3, 3 \times 4, \ldots$

or

3, 6, 9, 12, . . . The three dots indicate that the list continues without stopping.

NOTE

The multiples of 3, except for 3 itself, are all *larger* than 3.

An easy way of listing the multiples of 3 is to think of *counting by threes.*

Check Yourself 1

List the first seven multiples of 4.

Sometimes we need to find common multiples of two or more numbers.

Definition

Common Multiples

If a number is a multiple of each of a group of numbers, it is called a *common multiple* of the numbers; that is, it is a number that is exactly divisible by each of the numbers in the group.

Example 2 **Finding Common Multiples**

Find four common multiples of 3 and 5.

Some common multiples of 3 and 5 are

15, 30, 45, 60

NOTE

15, 30, 45, and 60 are multiples of *both* 3 and 5.

Check Yourself 2

List the first six multiples of 6. Then look at your list from Check Yourself 1 and list some common multiples of 4 and 6.

In our later work, we will use the *least common multiple* of a group of numbers.

Definition

Least Common Multiple

The **least common multiple** (LCM) of a group of numbers is the *smallest* number that is divisible by each number in the group.

It is possible to simply list the multiples of each number and then find the LCM by inspection.

Example 3 Finding the Least Common Multiple (LCM)

< Objective 1 >

NOTE

48 is also a common multiple of 6 and 8, but we are looking for the *smallest* such number.

Find the least common multiple of 6 and 8.

Multiples

6: 6, 12, 18, (24), 30, 36, 42, 48, . . .

8: 8, 16, (24), 32, 40, 48, . . .

We see that 24 is the smallest number common to both lists. So 24 is the LCM of 6 and 8.

Check Yourself 3

Find the least common multiple of 20 and 30 by listing the multiples of each number.

The technique used in Example 3 works for any group of numbers. However, it becomes tedious for larger numbers. Here is a different approach.

Step by Step

Finding the Least Common Multiple

Step 1 Write the prime factorization for each of the numbers in the group.

Step 2 Find all the prime factors that appear in any one of the prime factorizations.

Step 3 Form the product of those prime factors, using each factor the greatest number of times it occurs in any one factorization.

For instance, if a number appears three times in the factorization of a number, it must be included at least three times in forming the least common multiple.

We demonstrate this method in Example 4.

Example 4 Finding the LCM

NOTE

Line up the *like* factors vertically.

To find the LCM of 10 and 18, factor:

$$\begin{aligned} 10 &= 2 \qquad\qquad \times 5 \\ 18 &= \underline{2 \times 3 \times 3 } \\ & 2 \times 3 \times 3 \times 5 \end{aligned}$$

Bring down the factors.

The numbers 2 and 5 appear at most one time in any one factorization. And 3 appears two times in one factorization.

$2 \times 3 \times 3 \times 5 = 90$

So 90 is the LCM of 10 and 18.

Check Yourself 4

Find the LCM of 24 and 36.

The procedure is the same for a group of more than two numbers.

Example 5 **Finding the LCM**

< Objective 2 >

To find the LCM of 12, 18, and 20, we factor each number.

$$12 = 2 \times 2 \times 3$$

$$18 = 2 \quad\quad \times 3 \times 3$$

$$20 = 2 \times 2 \quad\quad\quad\quad \times 5$$

$$2 \times 2 \times 3 \times 3 \times 5$$

NOTE

The different factors that appear are 2, 3, and 5.

The numbers 2 and 3 appear twice in one factorization, and 5 appears just once.

$$2 \times 2 \times 3 \times 3 \times 5 = 180$$

So 180 is the LCM of 12, 18, and 20.

Check Yourself 5

Find the LCM of 3, 4, and 6.

The process of finding the least common multiple is very useful when we are adding, subtracting, or comparing unlike fractions (fractions with different denominators).

Suppose you are asked to compare the sizes of the fractions $\frac{3}{7}$ and $\frac{4}{7}$. Because each circle in the diagram is divided into seven parts, it is easy to see that $\frac{4}{7}$ is larger than $\frac{3}{7}$.

Four parts of seven are a greater portion than three parts. Now compare the size of the fractions $\frac{2}{5}$ and $\frac{3}{7}$.

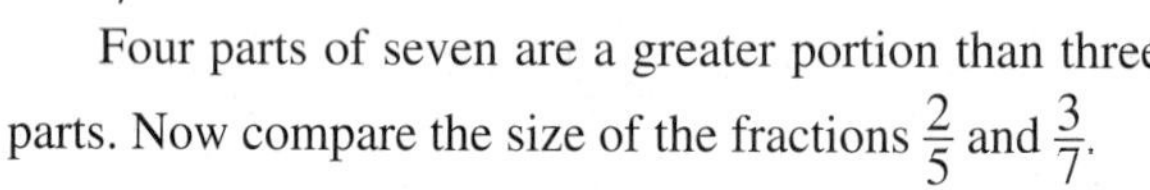

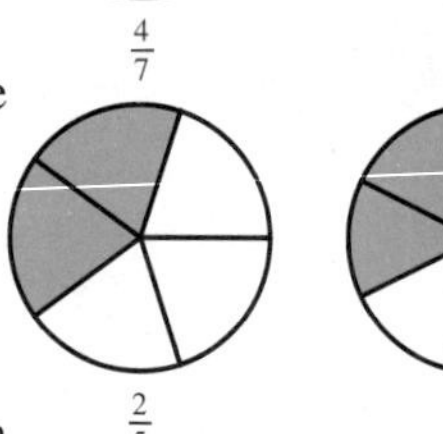

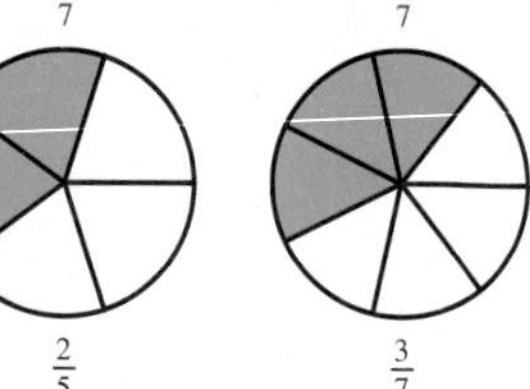

We *cannot* compare fifths with sevenths! The fractions $\frac{2}{5}$ and $\frac{3}{7}$ are *not* like fractions. Because they name different ways of dividing the whole, deciding which fraction is larger is not nearly so easy.

To compare the sizes of fractions, we change them to equivalent fractions having a *common denominator*. This common denominator must be a common multiple of the original denominators.

In Chapter 2, we used the fundamental principle of fractions to simplify fractions. We can also use this property to "go the other way." That is, to find equivalent fractions.

We use the property to take two fractions with different denominators and write them each as equivalent fractions with a common denominator.

Any fraction that has the same numerator and denominator equals 1. Further, when we multiply a number by 1, the number is unchanged. For example, since $3 \div 3 = 1$, we have

$$1 = \frac{3}{3}$$

> **NOTE**
>
> We could have used any form of 1, such as $\frac{5}{5}$.
> Doing so yields a different equivalent fraction.
> $\frac{2}{5} \times \frac{5}{5} = \frac{10}{25}$

Therefore, if we take a fraction, say $\frac{2}{5}$, we can use the fundamental principle of fractions to write it as an equivalent fraction with 15 as its denominator.

$$\frac{2}{5} = \frac{2}{5} \times 1 \qquad \text{Multiplying by 1 does not change a number}$$
$$= \frac{2}{5} \times \frac{3}{3} \qquad 1 = \frac{3}{3}$$
$$= \frac{2 \times 3}{5 \times 3}$$
$$= \frac{6}{15}$$

Therefore, $\frac{2}{5} = \frac{6}{15}$, and we can write $\frac{2}{5}$ as $\frac{6}{15}$ when we find it convenient to do so. When adding fractions with different denominators, we find this approach convenient.

Property

The Fundamental Principle of Fractions

$$\frac{a}{b} = \frac{a \times c}{b \times c}$$

Example 6 Finding Common Denominators

(a) Find the LCM of the denominators of the fractions $\frac{2}{3}$ and $\frac{3}{5}$.

The denominators are 3 and 5. The LCM of 3 and 5 is 15.

(b) Find the LCM of the denominators of the fractions $\frac{3}{4} - \frac{1}{6}$. Do not perform the arithmetic.

The denominators are 4 and 6. The LCM of 4 and 6 is 12.

Check Yourself 6

(a) Find the LCM of the denominators of the fractions $\frac{7}{10}$ and $\frac{5}{6}$.

(b) Find the LCM of the denominators of the fractions $\frac{1}{2} + \frac{2}{5}$. Do not perform the arithmetic.

Finding the LCM of a pair of denominators, coupled with the fundamental principle of fractions, allows us to compare fractions and determine which is larger.

Example 7 Comparing Fractions

< Objective 3 >

> **RECALL**
>
> $\frac{2}{5}$ and $\frac{14}{35}$ are **equivalent fractions.** They name the same part of a whole.

Compare the sizes of $\frac{2}{5}$ and $\frac{3}{7}$.

The original denominators are 5 and 7. Because 35 is a common multiple of 5 and 7, we construct equivalent fractions with 35 as the denominator.

$$\frac{2}{5} \overset{\times 7}{=} \frac{14}{35}$$

Think, "What must we multiply 5 by to get 35?" The answer is 7. Multiply the numerator and denominator by that number.

$$\frac{3}{7} \overset{\times 5}{=} \frac{15}{35}$$

Multiply the numerator and denominator by 5.

> **NOTE**
>
> 15 of 35 parts represents a greater portion of the whole than 14 parts.

Because $\frac{2}{5} = \frac{14}{35}$ and $\frac{3}{7} = \frac{15}{35}$, we see that $\frac{3}{7}$ is larger than $\frac{2}{5}$.

Check Yourself 7

Which is larger, $\frac{5}{9}$ or $\frac{4}{7}$?

Now, consider an example that uses inequality notation.

Example 8 Using an Inequality Symbol

RECALL

The inequality symbol "points" to the smaller quantity.

Fill in the box with either inequality symbol < or >.

$$\frac{5}{8} \square \frac{3}{5}$$

Once again we must compare the sizes of the two fractions, and we do this by converting the fractions to equivalent fractions with a common denominator. Here we use 40 as that denominator.

NOTE

We use the Fundamental Principle of Fractions to rewrite these fractions.

$$\frac{5}{8} = \frac{25}{40} \quad (\times 5) \qquad \frac{3}{5} = \frac{24}{40} \quad (\times 8)$$

Because $\frac{5}{8}\left(\text{or } \frac{25}{40}\right)$ is larger than $\frac{3}{5}\left(\text{or } \frac{24}{40}\right)$, we write

$$\frac{5}{8} > \frac{3}{5}$$

Check Yourself 8

Fill in the box with either inequality symbol < or >.

$$\frac{5}{9} \square \frac{6}{11}$$

You may find it helpful to see some simple fractions on a number line to gauge their proper order. In this diagram, fractions are written in simplest terms, however, you should see that there are equivalent ways to write many of them. For instance,

$\frac{1}{2} = \frac{2}{4} = \frac{3}{6} = \frac{4}{8} = \frac{5}{10}$ and $\frac{2}{5} = \frac{4}{10}$.

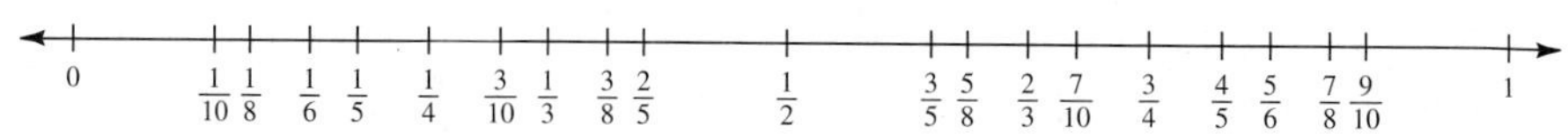

Check Yourself ANSWERS

1. 4, 8, 12, 16, 20, 24, 28 **2.** 6, 12, 18, 24, 30, 36; some common multiples: 12, 24 **3.** 20: 20, 40, 60, 80, 120, . . . ; 30: 30, 60, 90, 120, 150, . . . ; LCM: 60 **4.** 72 **5.** 12 **6.** **(a)** 30; **(b)** 10 **7.** $\frac{4}{7}$ **8.** $\frac{5}{9} > \frac{6}{11}$

Reading Your Text 

These fill-in-the-blank exercises will help you understand some of the key vocabulary used in this section. The answers to these exercises are in the Answers Appendix at the back of the text.

(a) The ______________ of a number are the products of that number with the natural numbers.

(b) The LCM of a group of numbers is the ______________ number that is divisible by each number in that group.

(c) To compare the sizes of fractions, we change them to equivalent fractions having a common ______________.

(d) The statement $\frac{5}{8} > \frac{3}{5}$ is read "$\frac{5}{8}$ is ______________ than $\frac{3}{5}$."

Skills Calculator/Computer Career Applications Above and Beyond

3.2 exercises

< Objective 1 >

Find the least common multiple (LCM) for each group of numbers.

1. 2 and 3 **2.** 3 and 5 **3.** 4 and 6 **4.** 6 and 9

5. 10 and 20 **6.** 12 and 36 **7.** 9 and 12 **8.** 20 and 30

9. 12 and 16 **10.** 10 and 15 **11.** 12 and 15 **12.** 12 and 21

13. 18 and 36  **14.** 25 and 50 **15.** 25 and 40 **16.** 10 and 14

< Objective 2 >

17. 3, 5, and 6 **18.** 2, 8, and 10 **19.** 18, 21, and 28  **20.** 8, 15, and 20

21. 20, 30, and 45 **22.** 12, 20, and 35

Find the LCM of the denominators of each pair of fractions. Do not perform any arithmetic.

23. $\frac{5}{8}$ and $\frac{2}{3}$ **24.** $\frac{3}{4}$ and $\frac{1}{3}$ **25.** $\frac{1}{2}$ and $\frac{1}{8}$ **26.** $\frac{7}{8}$ and $\frac{1}{6}$

27. $\frac{1}{4} + \frac{4}{5}$ **28.** $\frac{1}{2} - \frac{1}{3}$ **29.** $\frac{5}{6} - \frac{2}{9}$ **30.** $\frac{2}{3} + \frac{5}{12}$

Complete each equivalent fraction.

31. $\frac{4}{5} = \frac{\quad}{25}$ **32.** $\frac{6}{13} = \frac{\quad}{26}$ **33.** $\frac{5}{6} = \frac{25}{\quad}$ **34.** $\frac{2}{5} = \frac{14}{\quad}$

35. $\frac{11}{37} = \frac{\quad}{111}$ **36.** $\frac{9}{31} = \frac{\quad}{248}$ **37.** $\frac{1}{2} = \frac{\quad}{16}$ **38.** $\frac{2}{3} = \frac{\quad}{24}$

39. $\frac{5}{12} = \frac{25}{\quad}$ **40.** $\frac{1}{6} = \frac{2}{\quad}$ **41.** $\frac{17}{10} = \frac{\quad}{30}$ **42.** $\frac{6}{5} = \frac{\quad}{15}$

< Objective 3 >

Arrange the fractions from smallest to largest.

43. $\frac{12}{17}, \frac{9}{10}$ **44.** $\frac{4}{9}, \frac{5}{11}$ **45.** $\frac{5}{8}, \frac{3}{5}$ **46.** $\frac{9}{10}, \frac{8}{9}$

47. $\frac{3}{8}, \frac{1}{3}, \frac{1}{4}$ VIDEO **48.** $\frac{7}{12}, \frac{5}{18}, \frac{1}{3}$ **49.** $\frac{11}{12}, \frac{4}{5}, \frac{5}{6}$ **50.** $\frac{5}{8}, \frac{9}{16}, \frac{13}{32}$

Use the symbols < and > to complete each statement.

51. $\frac{5}{6} \square \frac{2}{5}$ 

52. $\frac{3}{4} \square \frac{10}{11}$ **53.** $\frac{4}{9} \square \frac{3}{7}$ **54.** $\frac{7}{10} \square \frac{11}{15}$

55. $\frac{7}{20} \square \frac{9}{25}$ **56.** $\frac{5}{12} \square \frac{7}{18}$

57. $\frac{5}{16} \square \frac{7}{20}$ 

58. $\frac{7}{12} \square \frac{3}{5}$

Solve each application.

59. CRAFTS Three drill bits are marked $\frac{3}{8}, \frac{5}{16}$, and $\frac{11}{32}$. Which drill bit is largest? 

60. CRAFTS Bolts can be purchased with diameters of $\frac{3}{8}, \frac{1}{4}$, or $\frac{3}{16}$ in. Which is smallest? 

61. CONSTRUCTION Plywood comes in thicknesses of $\frac{5}{8}, \frac{3}{4}, \frac{1}{2}$, and $\frac{3}{8}$ in. Which size is thickest? 

62. CONSTRUCTION Dowels are sold with diameters of $\frac{1}{2}, \frac{9}{16}, \frac{5}{8}$, and $\frac{3}{8}$ in. Which size is smallest? 

63. Elian is asked to create a fraction equivalent to $\frac{1}{4}$. His answer is $\frac{4}{7}$. What did he do wrong? Give a correct result.

64. A sign on a busy highway says Exit 5A is $\frac{3}{4}$ mi away and Exit 5B is $\frac{5}{8}$ mi away. Which exit comes first?

Skills | Calculator/Computer | **Career Applications** | Above and Beyond

65. ELECTRONICS Imagine a guitar string vibrating back and forth. Each back-and-forth vibration is called a *cycle.* The number of times that a string vibrates back and forth per second is called the *frequency.* Accordingly, frequency is measured in cycles per second [CPS or hertz (Hz)]. If you had a camera that could zoom in on a vibrating guitar string, you would see that instead of simply vibrating back and forth it actually vibrates differently along different sections of the string. The loudest vibration is called the fundamental frequency; quieter vibrations are called harmonics, which are multiples of the fundamental frequency.

I. If a note's fundamental frequency is 440 Hz, 1,760 Hz is a harmonic.

II. If a note's fundamental frequency is 440 Hz, 771 Hz is a harmonic.

Which conclusion is valid?

(a) I is true and II is true. **(b)** I is true and II is false.

(c) I is false and II is true. **(d)** I is false and II is false.

(e) None of the above.

66. **ELECTRONICS** Music producers know that, when mixing tracks containing instruments and vocals, it is usually necessary to add some reverb and delay (echo) from each track into the mix. For a good mix, every instrument or vocal track usually has different degrees of reverb and delay. Delay, measured in milliseconds (ms), is the time it takes to hear the "echo," which can be quiet or loud. It is usually desirable to have the delay times be "in beat." If we let BPM denote the number of beats per minute a drum machine is set at, then the delay time is a multiple of $\frac{60{,}000}{\text{BPM}}$.

I. If the drum machine is set at 120 BPM, then 1,200 ms will produce an in-beat delay.

II. If the drum machine is set at 120 BPM, then 1,000 ms will produce an in-beat delay.

Which conclusion is valid?

(a) I is true and II is true.
(b) I is true and II is false.
(c) I is false and II is true.
(d) I is false and II is false.
(e) None of the above.

Skills	Calculator/Computer	Career Applications	**Above and Beyond**

67. **BUSINESS AND FINANCE** A company uses two types of boxes, 8 cm and 10 cm long. They are packed in larger cartons to be shipped. What is the shortest length of a container that will accommodate boxes of either size without any room left over? (Each container can contain only boxes of one size—no mixing allowed.)

68. There is an alternate approach to finding the least common multiple of two numbers. The LCM of two numbers can be found by dividing the product of the two numbers by the greatest common factor (GCF) of those two numbers. For example, the GCF of 24 and 36 is 12. If we use this approach, we obtain LCM of 24 and 36 $= \frac{24 \cdot 36}{12} = 72$

(a) Use this approach to find the LCM of 150 and 480.
(b) Verify your result by finding the LCM using the method of prime factorization.

69. Complete the crossword puzzle.

Across

2. The LCM of 11 and 13
4. The GCF of 120 and 300
7. The GCF of 13 and 52
8. The GCF of 360 and 540

Down

1. The LCM of 8, 14, and 21
3. The LCM of 16 and 12
5. The LCM of 2, 5, and 13
6. The GCF of 54 and 90

Answers

1. 6 **3.** 12 **5.** 20 **7.** 36 **9.** 48 **11.** 60 **13.** 36 **15.** 200 **17.** 30 **19.** 252 **21.** 180 **23.** 24 **25.** 8 **27.** 20 **29.** 18 **31.** 20 **33.** 30 **35.** 33 **37.** 8 **39.** 60 **41.** 51 **43.** $\frac{12}{17}, \frac{9}{10}$ **45.** $\frac{3}{5}, \frac{5}{8}$ **47.** $\frac{1}{4}, \frac{1}{3}, \frac{3}{8}$ **49.** $\frac{4}{5}, \frac{5}{6}, \frac{11}{12}$ **51.** > **53.** > **55.** < **57.** < **59.** $\frac{3}{8}$ **61.** $\frac{3}{4}$ in. **63.** He added 3 to both the numerator and denominator; $\frac{1 \times 3}{4 \times 3} = \frac{3}{12}$ **65.** b **67.** Above and Beyond **69.** Above and Beyond

3.3 Adding and Subtracting Unlike Fractions

< 3.3 Objectives >

1 > Add any two fractions

2 > Add any group of fractions

3 > Subtract any two fractions

In Section 3.1, you learned about like fractions (fractions with a common denominator). How should we add **unlike fractions,** such as $\frac{1}{3} + \frac{1}{4}$?

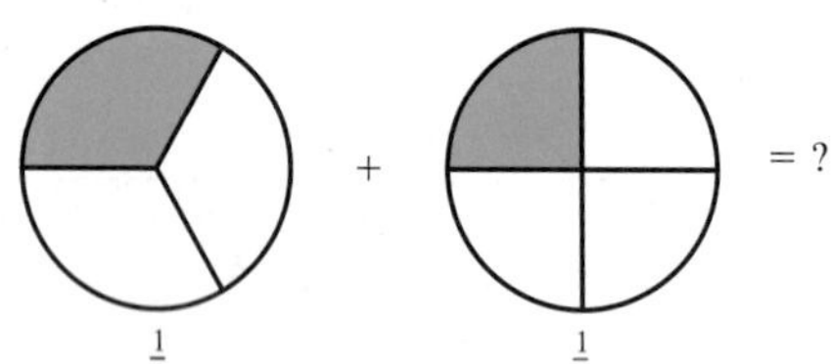

We cannot add unlike fractions because they have different denominators.

To add unlike fractions, write them as equivalent fractions with a common denominator. In this case, we use 12 as the denominator.

NOTES

Only *like* fractions can be added.

We can now add because we have like fractions.

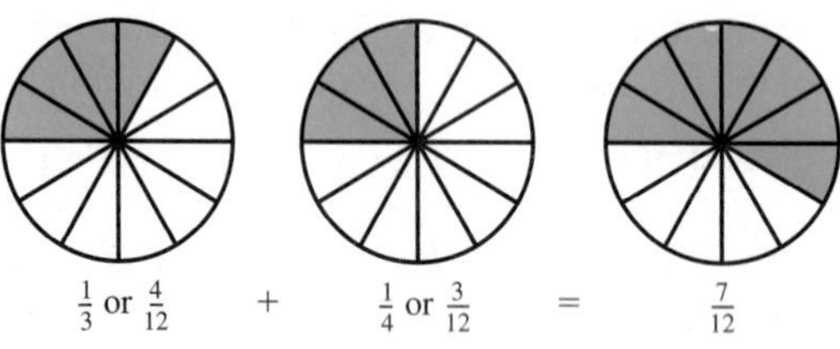

We chose 12 because it is a *multiple* of 3 and 4.

$\frac{1}{3}$ is equivalent to $\frac{4}{12}$.

$\frac{1}{4}$ is equivalent to $\frac{3}{12}$.

Any common multiple of the denominators works when forming equivalent fractions. For instance, we can write $\frac{1}{3}$ as $\frac{8}{24}$ and $\frac{1}{4}$ as $\frac{6}{24}$. Our work is simplest, however, if we use the smallest possible number for the common denominator. This is called the **least common denominator (LCD).**

The LCD is the least common multiple of the denominators of the fractions. This is the *smallest* number that is a multiple of all the denominators. For example, the LCD of $\frac{1}{3}$ and $\frac{1}{4}$ is 12, *not* 24.

Step by Step

To Find the Least Common Denominator

Step 1	Write the prime factorization for each of the denominators.
Step 2	Find all the prime factors that appear in any one of the prime factorizations.
Step 3	Form the product of those prime factors, using each factor the greatest number of times it occurs in any one factorization.

We are now ready to add unlike fractions. In this case, the fractions must be written as equivalent fractions that have the same denominator.

Step by Step

To Add Unlike Fractions

Step 1 Find the LCD of the fractions.

Step 2 Change each unlike fraction to an equivalent fraction with the LCD as its denominator.

Step 3 Add the resulting like fractions as before.

Example 1 shows this process.

Example 1 Adding Unlike Fractions

< Objective 1 >

NOTE

The LCD is just the LCM of the denominators of the fractions.

Add the fractions $\frac{1}{6}$ and $\frac{3}{8}$.

Step 1 We find that the LCD for fractions with denominators of 6 and 8 is 24.

Step 2 Convert the fractions so that they have the denominator 24.

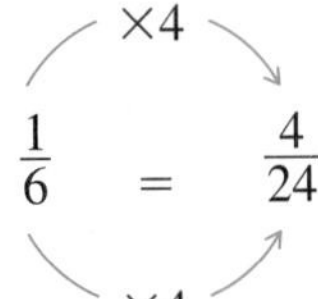

$$\frac{1}{6} = \frac{4}{24}$$

How many sixes are in 24? There are 4. So multiply the numerator and denominator by 4.

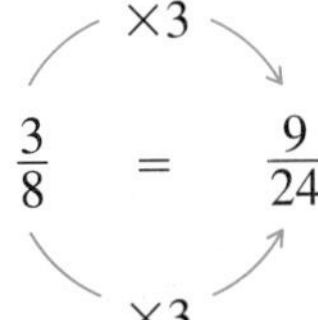

$$\frac{3}{8} = \frac{9}{24}$$

How many eights are in 24? There are 3. So multiply the numerator and denominator by 3.

Step 3 We can now add the equivalent like fractions.

$$\frac{1}{6} + \frac{3}{8} = \frac{4}{24} + \frac{9}{24} = \frac{13}{24}$$

Add the numerators and place that sum over the common denominator.

Check Yourself 1

Add.

$$\frac{3}{5} + \frac{1}{3}$$

Here is a similar example. Remember that the sum should always be written in simplest form.

Example 2 Adding Unlike Fractions and Simplifying

Add the fractions $\frac{7}{10}$ and $\frac{2}{15}$.

Step 1 The LCD for fractions with denominators of 10 and 15 is 30.

Step 2 $\frac{7}{10} = \frac{21}{30}$

$\frac{2}{15} = \frac{4}{30}$

Do you see how the equivalent fractions are formed?

Step 3 $\frac{7}{10} + \frac{2}{15} = \frac{21}{30} + \frac{4}{30}$

$= \frac{25}{30} = \frac{5}{6}$

Add the resulting like fractions. Be sure the sum is in simplest form.

Check Yourself 2

Add.

$\frac{1}{6} + \frac{7}{12}$

We can use the same procedure to add more than two fractions.

Example 3 Adding a Group of Unlike Fractions

< Objective 2 >

Add $\frac{5}{6} + \frac{2}{9} + \frac{4}{15}$.

Step 1 The LCD is 90.

Step 2 $\frac{5}{6} = \frac{75}{90}$ Multiply the numerator and denominator by 15.

$\frac{2}{9} = \frac{20}{90}$ Multiply the numerator and denominator by 10.

$\frac{4}{15} = \frac{24}{90}$ Multiply the numerator and denominator by 6.

Step 3 $\frac{75}{90} + \frac{20}{90} + \frac{24}{90} = \frac{119}{90}$ Now add.

$= 1\frac{29}{90}$ Remember, if the sum is an improper fraction, it may be written as a mixed number.

Check Yourself 3

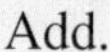

Add.

$\frac{2}{5} + \frac{3}{8} + \frac{7}{20}$

Many of the measurements you deal with in everyday life involve fractions. Following are some typical situations.

Example 4 An Application Involving Unlike Fractions

Jack ran $\frac{1}{2}$ mi on Monday, $\frac{2}{3}$ mi on Wednesday, and $\frac{3}{4}$ mi on Friday. How far did he run during the week?

The three distances that Jack ran are the given information in the problem. We want to find a total distance, so we must add to find the solution.

$$\frac{1}{2} + \frac{2}{3} + \frac{3}{4} = \frac{6}{12} + \frac{8}{12} + \frac{9}{12}$$

$$= \frac{23}{12} = 1\frac{11}{12} \text{ mi}$$

Because we have no common denominator, we must convert to equivalent fractions before we can add.

Jack ran $1\frac{11}{12}$ mi during the week.

Check Yourself 4

Susan is designing an office complex. She needs $\frac{2}{5}$ acre for buildings, $\frac{1}{3}$ acre for driveways and parking, and $\frac{1}{6}$ acre for walks and landscaping. How much land does she need?

Example 5 An Application Involving Unlike Fractions

Sam bought three packages of spices weighing $\frac{1}{4}$, $\frac{5}{8}$, and $\frac{1}{2}$ lb. What was the total weight?

NOTE

The abbreviation for pounds is "lb" from the Latin *libra*, meaning "balance" or "scales."

We need to find the total weight, so we must add.

$$\frac{1}{4} + \frac{5}{8} + \frac{1}{2} = \frac{2}{8} + \frac{5}{8} + \frac{4}{8}$$ Write each fraction with the denominator 8.

$$= \frac{11}{8} = 1\frac{3}{8} \text{ lb}$$

The total weight was $1\frac{3}{8}$ lb.

Check Yourself 5

For three different recipes, Max needs $\frac{3}{8}$, $\frac{1}{2}$, and $\frac{3}{4}$ gallon (gal) of tomato sauce. How many gallons does he need altogether?

Subtracting unlike fractions is similar to adding them.

Step by Step

To Subtract Unlike Fractions

Step 1 Find the LCD of the fractions.

Step 2 Change each unlike fraction to an equivalent fraction with the LCD as its denominator.

Step 3 Subtract the resulting like fractions as before.

Example 6 Subtracting Unlike Fractions

< Objective 3 >

Subtract $\frac{5}{8} - \frac{1}{6}$.

NOTE

You can use a calculator to check your result.

Step 1 The LCD is 24.

Step 2 Convert the fractions so that they have the common denominator 24.

$$\frac{5}{8} = \frac{15}{24}$$

$$\frac{1}{6} = \frac{4}{24}$$

The first two steps are exactly the same as if we were adding.

Step 3 Subtract the equivalent like fractions.

$$\frac{5}{8} - \frac{1}{6} = \frac{15}{24} - \frac{4}{24} = \frac{11}{24}$$

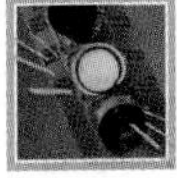

> CAUTION

Be Careful! You *cannot* subtract the numerators and subtract the denominators.

$\frac{5}{8} - \frac{1}{6}$ is *not* $\frac{4}{2}$

Check Yourself 6

Subtract.

$$\frac{7}{10} - \frac{1}{4}$$

We can solve many applications by adding and subtracting fractions.

Example 7 An Application Involving Subtraction

You have $\frac{7}{8}$ yd of handwoven linen. A pattern for a placemat calls for $\frac{1}{2}$ yd. Will you have enough left for two napkins that will use $\frac{1}{3}$ yd?

First, find out how much fabric is left over after the placemat is made.

$$\frac{7}{8}\text{ yd} - \frac{1}{2}\text{ yd} = \frac{7}{8}\text{ yd} - \frac{4}{8}\text{ yd}$$
$$= \frac{3}{8}\text{ yd}$$

Now compare the sizes of $\frac{1}{3}$ and $\frac{3}{8}$.

NOTE

$\frac{3}{8}$ yd is left over but only $\frac{1}{3}$ yd is needed.

$$\frac{3}{8}\text{ yd} = \frac{9}{24}\text{ yd} \quad \text{and} \quad \frac{1}{3}\text{ yd} = \frac{8}{24}\text{ yd}$$

Because $\frac{3}{8}$ yd is *more than* the $\frac{1}{3}$ yd needed, there is enough material for the placemat *and* two napkins.

Check Yourself 7

A concrete walk requires $\frac{3}{4}$ cubic yard (yd^3) of concrete. If you mix $\frac{8}{9}$ yd^3, will enough concrete remain to do a project that uses $\frac{1}{6}$ yd^3?

Our next application involves measurements. Note that on a ruler or yardstick, the marks divide each inch into $\frac{1}{2}$-in., $\frac{1}{4}$-in., and $\frac{1}{8}$-in. sections, and on some rulers, $\frac{1}{16}$-in. sections. We use denominators of 2, 4, 8, and 16 in our measurement applications.

Example 8 An Application Involving Subtraction

Alexei cut two $\frac{3}{16}$-in. slats from a piece of wood that is $\frac{3}{4}$ in. across. How much is left?

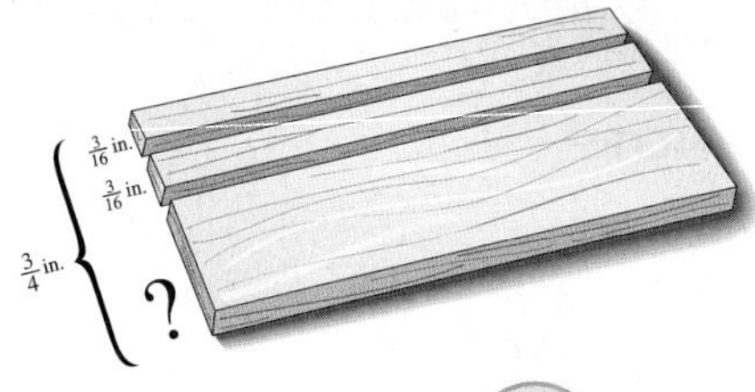

The two $\frac{3}{16}$-in. pieces total

$$2 \times \frac{3}{16} = \frac{6}{16} = \frac{3}{8}\text{ in.}$$
$$\frac{3}{4} = \frac{6}{8}$$
$$\frac{6}{8} - \frac{3}{8} = \frac{3}{8}$$

There is $\frac{3}{8}$ in. of wood left.

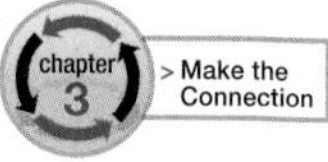

Check Yourself 8

Ricardo cuts three strips from a 1-in. piece of metal. Each strip is $\frac{3}{16}$ in. wide. How much metal remains after the cuts?

Check Yourself ANSWERS

1. $\frac{14}{15}$ 2. $\frac{3}{4}$ 3. $\frac{9}{8} = 1\frac{1}{8}$ 4. $\frac{9}{10}$ acre 5. $1\frac{5}{8}$ gal 6. $\frac{9}{20}$

7. $\frac{5}{36}$ yd^3 remains. You do *not* have enough concrete for both projects. 8. $\frac{7}{16}$ in.

Reading Your Text

These fill-in-the-blank exercises will help you understand some of the key vocabulary used in this section. The answers to these exercises are in the Answers Appendix at the back of the text.

(a) Finding an LCD is nearly identical to finding an ________.

(b) To add ________ fractions, we first find the LCD of the fractions.

(c) Two fractions with the same value but different denominators are called ________ fractions.

(d) When adding fractions with a common denominator, we add the ________ and put that sum over the common denominator.

Skills Calculator/Computer Career Applications Above and Beyond

3.3 exercises

Find the least common denominator (LCD) for fractions with the given denominators.

1. 3 and 4 **2.** 3 and 5 **3.** 4 and 8 **4.** 6 and 12

5. 9 and 27 **6.** 10 and 30 **7.** 8 and 12 **8.** 15 and 40

9. 14 and 21 VIDEO **10.** 15 and 20 **11.** 48 and 80 **12.** 60 and 84

Find the LCD for each set of fractions.

13. $\frac{1}{2}$ and $\frac{2}{3}$ **14.** $\frac{1}{2}$ and $\frac{3}{4}$ **15.** $\frac{2}{3}$ and $\frac{5}{8}$ **16.** $\frac{5}{6}$ and $\frac{3}{10}$

17. $\frac{3}{8}$ and $\frac{7}{12}$ **18.** $\frac{9}{10}$ and $\frac{7}{12}$ **19.** $\frac{5}{12}$ and $\frac{7}{32}$ **20.** $\frac{11}{16}$ and $\frac{4}{15}$

21. $\frac{2}{3}$ and $\frac{4}{9}$ **22.** $\frac{2}{5}$ and $\frac{2}{9}$ **23.** $\frac{5}{9}$ and $\frac{7}{12}$ **24.** $\frac{11}{12}$ and $\frac{5}{18}$

25. $\frac{1}{3}, \frac{3}{4}$, and $\frac{2}{5}$ **26.** $\frac{2}{3}, \frac{3}{4}$, and $\frac{1}{6}$ **27.** $\frac{3}{8}, \frac{3}{10}$, and $\frac{1}{15}$ **28.** $\frac{5}{6}, \frac{5}{22}$, and $\frac{5}{33}$

29. $\frac{1}{5}, \frac{1}{10}$, and $\frac{1}{25}$ **30.** $\frac{7}{8}, \frac{23}{24}$, and $\frac{47}{48}$

< Objectives 1 and 2 >

Add. Report your results in simplest terms.

31. $\frac{2}{3} + \frac{1}{4}$ **32.** $\frac{3}{5} + \frac{1}{3}$ **33.** $\frac{1}{5} + \frac{3}{10}$ **34.** $\frac{1}{3} + \frac{1}{18}$

35. $\frac{3}{4} + \frac{1}{8}$ **36.** $\frac{4}{5} + \frac{1}{10}$ **37.** $\frac{1}{7} + \frac{3}{5}$ VIDEO **38.** $\frac{1}{6} + \frac{2}{15}$

39. $\frac{3}{7} + \frac{3}{14}$ **40.** $\frac{7}{20} + \frac{9}{40}$ **41.** $\frac{7}{15} + \frac{2}{35}$ **42.** $\frac{3}{10} + \frac{3}{8}$

43. $\frac{5}{8} + \frac{1}{12}$

44. $\frac{5}{12} + \frac{3}{10}$

45. $\frac{1}{5} + \frac{7}{10} + \frac{4}{15}$

46. $\frac{2}{3} + \frac{1}{4} + \frac{3}{8}$

47. $\frac{1}{9} + \frac{7}{12} + \frac{5}{8}$

48. $\frac{1}{3} + \frac{5}{12} + \frac{4}{5}$

< Objective 3 >

Subtract.

49. $\frac{4}{5} - \frac{1}{3}$

50. $\frac{7}{9} - \frac{1}{6}$

51. $\frac{11}{15} - \frac{3}{5}$ VIDEO

52. $\frac{5}{6} - \frac{2}{7}$

53. $\frac{3}{8} - \frac{1}{4}$

54. $\frac{9}{10} - \frac{4}{5}$

55. $\frac{5}{12} - \frac{3}{8}$

56. $\frac{13}{15} - \frac{11}{20}$

Evaluate.

57. $\frac{33}{40} - \frac{7}{24} + \frac{11}{30}$ VIDEO

58. $\frac{13}{24} - \frac{5}{16} + \frac{3}{8}$

59. $\frac{15}{16} + \frac{5}{8} - \frac{1}{4}$

60. $\frac{9}{10} - \frac{1}{5} + \frac{1}{2}$

Solve each application.

61. **Number Problem** Paul bought $\frac{1}{2}$ lb of peanuts and $\frac{3}{8}$ lb of cashews. How many pounds of nuts did he buy?

62. **Construction** A countertop consists of a board $\frac{3}{4}$ in. thick and tile $\frac{3}{8}$ in. thick. What is the overall thickness?

63. **Business and Finance** Amy budgets $\frac{2}{5}$ of her income for housing and $\frac{1}{6}$ of her income for food. What fraction of her income is budgeted for these two purposes? What fraction of her income remains? chapter 3 > Make the Connection

64. **Social Science** A person spends $\frac{3}{8}$ of a day at work and $\frac{1}{3}$ of a day sleeping. What fraction of a day do these two activities use? What fraction of the day remains?

65. **Number Problem** Jose walked $\frac{3}{4}$ mi to the store, $\frac{1}{2}$ mi to a friend's house, and then $\frac{2}{3}$ mi home. How far did he walk?

66. **Geometry** Find the perimeter of the triangle to the right.

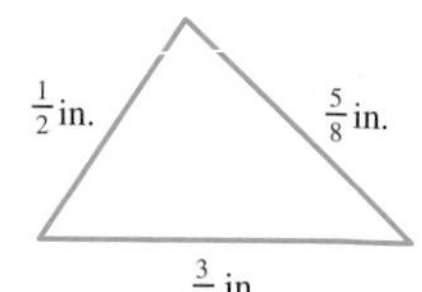

67. **Business and Finance** A budget guide states that you should spend $\frac{1}{4}$ of your salary for housing, $\frac{3}{16}$ for food, $\frac{1}{16}$ for clothing, and $\frac{1}{8}$ for transportation. What total portion of your salary do these four expenses account for?

68. **Business and Finance** Deductions from your paycheck are made roughly as follows: $\frac{1}{8}$ for federal tax, $\frac{1}{20}$ for state tax, $\frac{1}{20}$ for social security, and $\frac{1}{40}$ for a savings withholding plan. What portion of your pay is deducted?

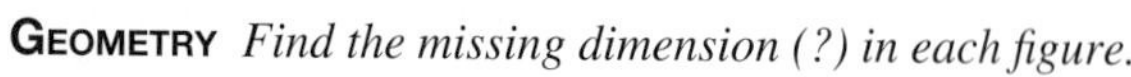

Geometry *Find the missing dimension (?) in each figure.*

69.

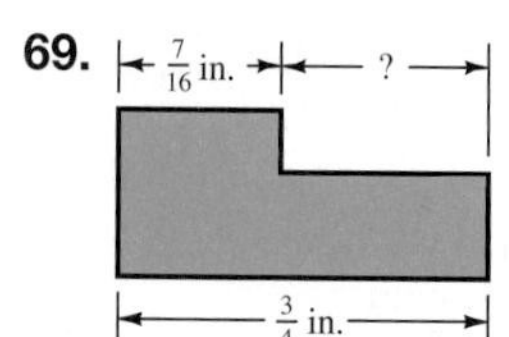

70.

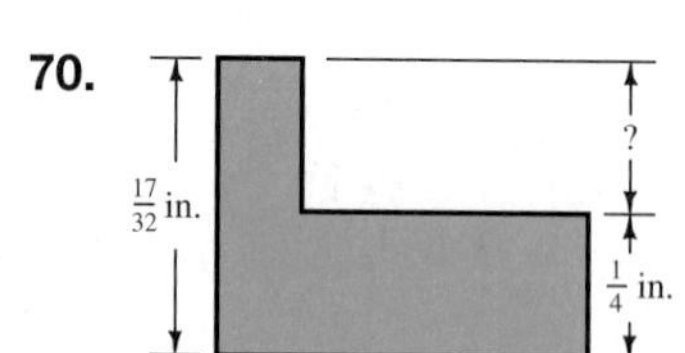

Complete each statement with either **always, sometimes,** *or* **never.**

71. The sum of two like fractions is __________ the sum of the numerators over the common denominator.

72. The LCD for two unlike proper fractions is __________ the same as the GCF of their denominators.

73. The sum of two fractions can __________ be simplified.

74. The difference of two proper fractions is __________ less than either of the two fractions.

Skills | **Calculator/Computer** | Career Applications | Above and Beyond

Using a Calculator to Add and Subtract Fractions

Adding and subtracting fractions on a calculator is very much like multiplication and division. The only thing that changes is the operation. Here's where a calculator is a great tool for checking your work.

Use a calculator to find each sum or difference.

75. $\frac{1}{10} + \frac{7}{12}$

76. $\frac{7}{15} + \frac{17}{24}$

77. $\frac{8}{9} + \frac{6}{7}$

78. $\frac{7}{15} + \frac{2}{5}$

79. $\frac{11}{18} + \frac{5}{12}$

80. $\frac{5}{8} + \frac{4}{9}$

81. $\frac{15}{17} - \frac{9}{11}$

82. $\frac{31}{43} - \frac{18}{53}$

83. $\frac{4}{9} - \frac{2}{5}$

84. $\frac{11}{13} - \frac{2}{3}$

Skills | Calculator/Computer | **Career Applications** | Above and Beyond

85. **ALLIED HEALTH** Prior to chemotherapy, a patient's tumor weighed $\frac{5}{6}$ lb. After a week of chemotherapy, the tumor's weight had been reduced by $\frac{1}{4}$ lb. How much did the tumor weigh at the end of the week?

86. **ALLIED HEALTH** A nurse is treating three children who all need to take the same medication. Charlie needs $\frac{3}{2}$ milliliters (mL), Sharon needs $\frac{2}{3}$ mL, and little Kevin needs $\frac{1}{4}$ mL of medication. How many milliliters of medication will the nurse need in order to give each child a single dose of medicine?

87. **MANUFACTURING TECHNOLOGY** Find the total thickness of this part.

chapter 3 > Make the Connection

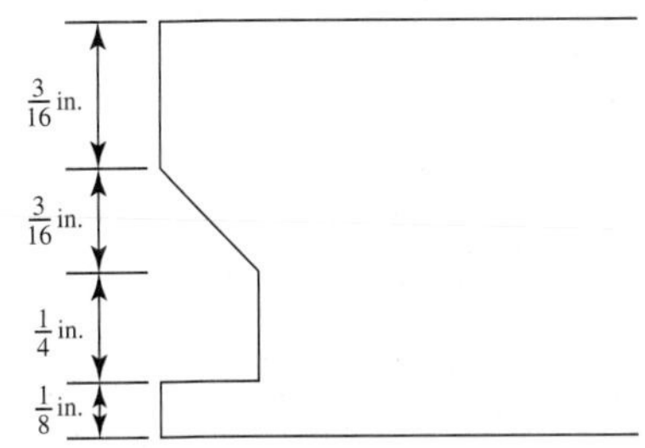

88. **MANUFACTURING TECHNOLOGY** Find the missing dimension (x).

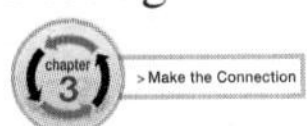

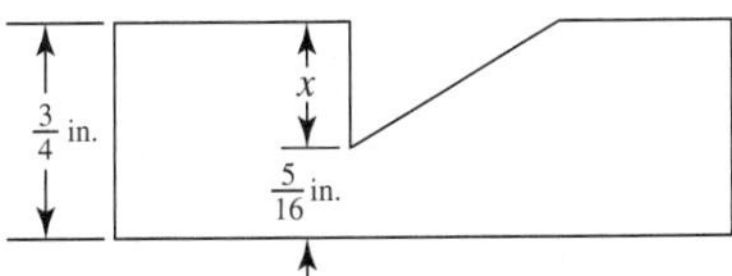

89. ELECTRONICS

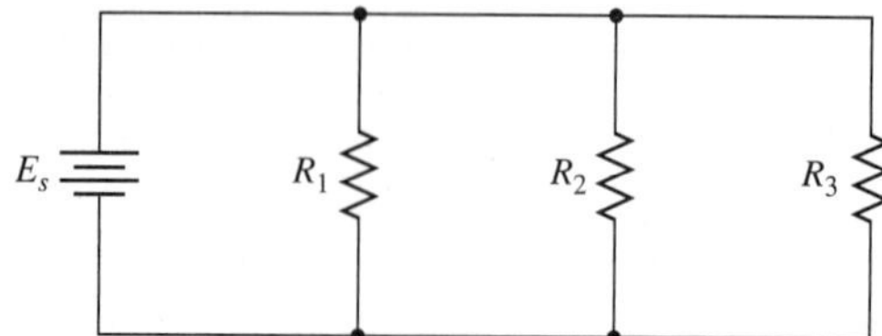

The circuit depicts three resistors (R_1, R_2, R_3) wired in parallel to a source, E_s. The circuit can be simplified by replacing the three separate resistors with a single equivalent resistor, R_{eq}, according to the following equation:

$$\frac{1}{R_{eq}} = \frac{1}{R_1} + \frac{1}{R_2} + \frac{1}{R_3}$$

If $R_1 = 10$ ohms (Ω), $R_2 = 20\ \Omega$, and $R_3 = 40\ \Omega$, what is $\frac{1}{R_{eq}}$? What is R_{eq}?

90. INFORMATION TECHNOLOGY Amin wants to figure out how long it takes to transmit 1 megabyte (MB) over two networks that have transmission rates of 50 and 10 MB per second, respectively. What is the total time of the transmission? Use the fraction of the number of MB to the transmission time to figure out your answer. Represent your answer as a simplified fraction.

Skills	Calculator/Computer	Career Applications	**Above and Beyond**

91. CONSTRUCTION A door is $4\frac{1}{4}$ ft wide. Two hooks are to be attached to the door so that they are $1\frac{1}{2}$ in. apart and the same distance from each edge. How far from the edge of the door should each hook be located? Give your answer in feet. chapter 3 > Make the Connection

92. Complete each exercise.

$\frac{1}{2} + \frac{1}{4} =$ __________.

$\frac{1}{2} + \frac{1}{4} + \frac{1}{8} =$ __________.

$\frac{1}{2} + \frac{1}{4} + \frac{1}{8} + \frac{1}{16} =$ __________.

Based on these results, predict the answer to

$\frac{1}{2} + \frac{1}{4} + \frac{1}{8} + \frac{1}{16} + \frac{1}{32} =$ __________.

Now, do the addition, and check your prediction.

Answers

1. 12 **3.** 8 **5.** 27 **7.** 24 **9.** 42 **11.** 240 **13.** 6 **15.** 24 **17.** 24 **19.** 96 **21.** 9 **23.** 36 **25.** 60 **27.** 120 **29.** 50 **31.** $\frac{11}{12}$ **33.** $\frac{1}{2}$ **35.** $\frac{7}{8}$ **37.** $\frac{26}{35}$ **39.** $\frac{9}{14}$ **41.** $\frac{11}{21}$ **43.** $\frac{17}{24}$ **45.** $\frac{7}{6} = 1\frac{1}{6}$ **47.** $\frac{95}{72} = 1\frac{23}{72}$ **49.** $\frac{7}{15}$ **51.** $\frac{2}{15}$ **53.** $\frac{1}{8}$ **55.** $\frac{1}{24}$ **57.** $\frac{9}{10}$ **59.** $\frac{21}{16} = 1\frac{5}{16}$ **61.** $\frac{7}{8}$ lb **63.** $\frac{17}{30}, \frac{13}{30}$ **65.** $1\frac{11}{12}$ mi **67.** $\frac{5}{8}$ **69.** $\frac{5}{16}$ in. **71.** always **73.** sometimes **75.** $\frac{41}{60}$ **77.** $\frac{110}{63} = 1\frac{47}{63}$ **79.** $\frac{37}{36} = 1\frac{1}{36}$ **81.** $\frac{12}{187}$ **83.** $\frac{2}{45}$ **85.** $\frac{7}{12}$ lb **87.** $\frac{3}{4}$ in. **89.** $\frac{7}{40}; \frac{40}{7} = 5\frac{5}{7}$ **91.** $2\frac{1}{16}$ ft

Kitchen Subflooring

Benjamin and Olivia are putting a new floor in their kitchen. To get the floor up to the desired height, they need to add $1\frac{1}{8}$ in. of subfloor. They can do this in one of two ways. They can put $\frac{1}{2}$-in. sheet on top of $\frac{5}{8}$-in. board (for a total of $\frac{9}{8}$ in. or $1\frac{1}{8}$ in.). They could also put $\frac{3}{8}$-in. board on top of $\frac{3}{4}$-in. sheet.

The table gives the price for sheets of plywood from a construction materials store.

Thickness	Cost for a 4 ft × 8 ft Sheet
$\frac{1}{8}$ in.	\$9.15
$\frac{1}{4}$ in.	13.05
$\frac{3}{8}$ in.	14.99
$\frac{1}{2}$ in.	17.88
$\frac{5}{8}$ in.	19.13
$\frac{3}{4}$ in.	21.36
$\frac{7}{8}$ in.	25.23
1 in.	28.49

1. What is the combined price for a $\frac{1}{2}$-in. sheet and a $\frac{5}{8}$-in. sheet?
2. What is the combined price for a $\frac{3}{8}$-in. sheet and a $\frac{3}{4}$-in. sheet?
3. What other combination of sheets of plywood, using two sheets, yields the needed $1\frac{1}{8}$-in. thickness?
4. Of the four combinations, which is most economical?
5. The kitchen is to be 12 ft × 12 ft. Find the total cost of the plywood you suggested in question 4.

3.4 Adding and Subtracting Mixed Numbers

< 3.4 Objectives >

1 > Add mixed numbers

2 > Subtract mixed numbers

3 > Use mixed numbers to solve applications

Now that we know how to add and subtract fractions, adding and subtracting mixed numbers should be straightforward. Just remember that a mixed number is the sum of a whole number and a proper fraction.

When adding mixed numbers, we use the properties of addition to rewrite the problem so that we are adding the whole number parts separately from the fraction parts. We then simplify before writing the sum as a single number, be it a proper fraction, or a whole or mixed number.

Consider how we use commutativity and associativity when adding mixed numbers.

$$3\frac{1}{5} = 3 + \frac{1}{5}$$

$$4\frac{2}{5} = 4 + \frac{2}{5}$$

Using this, we have

$$3\frac{1}{5} + 4\frac{2}{5} = \left(3 + \frac{1}{5}\right) + \left(4 + \frac{2}{5}\right)$$ The definition of mixed numbers.

$$= 3 + \frac{1}{5} + 4 + \frac{2}{5}$$

$$= 3 + 4 + \frac{1}{5} + \frac{2}{5}$$ Addition is commutative.

$$= (3 + 4) + \left(\frac{1}{5} + \frac{2}{5}\right)$$ Addition is associative.

We continue with this problem in Example 1. First, we summarize the process for adding mixed numbers.

Step by Step

Adding Mixed Numbers

Step 1	Find the LCD of the fraction parts.
Step 2	Rewrite the fraction parts as equivalent fractions with the LCD as their denominator, if necessary.
Step 3	Add the whole number and fraction parts separately.
Step 4	Simplify, if necessary.
Step 5	Rewrite the sum as a proper fraction or as a whole or mixed number.

Example 1 **Adding Mixed Numbers**

< Objective 1 >

$$3\frac{1}{5} + 4\frac{2}{5} = \underbrace{(3 + 4)}_{\text{whole number parts}} + \underbrace{\left(\frac{1}{5} + \frac{2}{5}\right)}_{\text{Fraction parts}}$$

NOTE

Because the fraction parts have the same denominator, we begin with step 3.

$= 7 + \frac{3}{5}$ $\frac{3}{5}$ is already in simplest form.

$= 7\frac{3}{5}$ Write the final sum as a mixed number.

Check Yourself 1

Add $2\frac{3}{10} + 3\frac{4}{10}$.

When the fraction parts of the mixed numbers have different denominators, we must rewrite the fractions as equivalent fractions with the least common denominator to perform the addition in step 3.

Example 2 Adding Mixed Numbers with Different Denominators

Add $3\frac{1}{6} + 2\frac{3}{8}$.

RECALL

$6 = 2 \times 3$

$8 = 2 \times 2 \times 2$

The LCD is

$2 \times 2 \times 2 \times 3 = 24$

The fraction parts have different denominators, so our first step is to write them as equivalent fractions using the LCD. The LCD of 6 and 8 is 24.

$$\frac{1}{6} = \frac{1}{6} \times \frac{4}{4} = \frac{4}{24}$$

$$\frac{3}{8} = \frac{3}{8} \times \frac{3}{3} = \frac{9}{24}$$

Therefore,

$$3\frac{1}{6} + 2\frac{3}{8} = 3\underbrace{\frac{4}{24}}_{\frac{1}{6}} + 2\underbrace{\frac{9}{24}}_{\frac{3}{8}}$$

We can now add.

RECALL

Add the numerators and write the sum over the common denominator.

$$3\frac{1}{6} + 2\frac{3}{8} = 3\frac{4}{24} + 2\frac{9}{24}$$

$$= (3 + 2) + \left(\frac{4}{24} + \frac{9}{24}\right)$$

$$= 5 + \frac{13}{24}$$

$$= 5\frac{13}{24}$$

Check Yourself 2

Add $5\frac{1}{10} + 3\frac{1}{6}$.

Follow the same procedure when there are more than two mixed numbers.

Example 3 Adding Mixed Numbers

Add $2\frac{1}{5} + 3\frac{3}{4} + 4\frac{1}{8}$.

The fraction parts have different denominators, so our first step is to write them as equivalent fractions using the LCD. The LCD of 4, 5, and 8 is 40.

$$\frac{1}{5} = \frac{1}{5} \times \frac{8}{8} = \frac{8}{40}$$

$$\frac{3}{4} = \frac{3}{4} \times \frac{10}{10} = \frac{30}{40}$$

$$\frac{1}{8} = \frac{1}{8} \times \frac{5}{5} = \frac{5}{40}$$

Therefore,

$$2\frac{1}{5} + 3\frac{3}{4} + 4\frac{1}{8} = 2\underbrace{\frac{8}{40}}_{\frac{1}{5}} + 3\underbrace{\frac{30}{40}}_{\frac{3}{4}} + 4\underbrace{\frac{5}{40}}_{\frac{1}{8}}$$

$$= (2 + 3 + 4) + \frac{8}{40} + \frac{30}{40} + \frac{5}{40}$$

$$= 9 + \frac{43}{40}$$

With $9 + \frac{43}{40}$ as an answer, you should see that we are not finished. Because $\frac{43}{40}$ is an improper fraction, we can simplify our result further.

$$\frac{43}{40} = 1\frac{3}{40}$$

So,

$$9\frac{43}{40} = 9 + 1\frac{3}{40}$$

$$= (9 + 1) + \frac{3}{40}$$

$$= 10 + \frac{3}{40}$$

$$= 10\frac{3}{40}$$

Check Yourself 3

Add $5\frac{1}{2} + 4\frac{2}{4} + 3\frac{3}{4}$.

When subtracting mixed numbers, we use the distributive property of multiplication over addition to subtract the whole number and fraction parts separately. Consider,

$$5\frac{7}{12} - 3\frac{5}{12} = \left(5 + \frac{7}{12}\right) - \left(3 + \frac{5}{12}\right)$$

$$= 5 + \frac{7}{12} - 3 - \frac{5}{12}$$ Distribute the negative sign to remove the parentheses.

$$= (5 - 3) + \left(\frac{7}{12} - \frac{5}{12}\right)$$ Now regroup to subtract the parts separately.

Step by Step

Subtracting Mixed Numbers

Step 1	Find the LCD of the fraction parts.
Step 2	Rewrite the fraction parts as equivalent fractions with the LCD as their denominator, if necessary.
Step 3	Subtract the whole number and fraction parts separately.
Step 4	Simplify, if necessary.
Step 5	Rewrite the difference as a proper fraction or as a whole or mixed number.

Example 4 illustrates this process.

Example 4 Subtracting Mixed Numbers

< Objective 2 >

Subtract $5\frac{7}{12} - 3\frac{5}{12}$.

We have

$$5\frac{7}{12} - 3\frac{5}{12} = (5 - 3) + \left(\frac{7}{12} - \frac{5}{12}\right)$$
$$= 2 + \frac{2}{12}$$
$$= 2 + \frac{1}{6} \quad \frac{2}{12} \text{ simplifies, } \frac{2}{12} = \frac{1}{6}.$$
$$= 2\frac{1}{6}$$

Check Yourself 4

Subtract $8\frac{7}{8} - 5\frac{3}{8}$.

As with addition, we must rewrite the fractions when different denominators are involved.

Example 5 Subtracting Mixed Numbers with Different Denominators

Subtract $8\frac{7}{10} - 3\frac{3}{8}$

We need to rewrite the fraction parts as equivalent fractions with the same denominator so we can subtract them. The LCD of 8 and 10 is 40.

$$8\frac{7}{10} - 3\frac{3}{8} = 8\underbrace{\frac{28}{40}}_{\frac{7}{10}} - 3\underbrace{\frac{15}{40}}_{\frac{3}{8}}$$
$$= (8 - 3) + \left(\frac{28}{40} - \frac{15}{40}\right) \quad \text{Subtract the whole number and fraction parts separately.}$$
$$= 5 + \frac{13}{40}$$
$$= 5\frac{13}{40}$$

Check Yourself 5

Subtract $7\frac{11}{12} - 3\frac{5}{8}$.

To subtract a mixed number from a whole number, we must use a form of regrouping, or borrowing.

Example 6 Subtracting Mixed Numbers by Borrowing

RECALL

$6 = 5 + 1$

$= 5 + \frac{4}{4}$

Subtract $6 - 2\frac{3}{4}$.

We write $6 = 5\frac{4}{4}$ so we can subtract the fraction parts.

$$6 - 2\frac{3}{4} = 5\frac{4}{4} - 2\frac{3}{4}$$

$$= (5 - 2) + \left(\frac{4}{4} - \frac{3}{4}\right)$$

$$= 3 + \frac{1}{4}$$

$$= 3\frac{1}{4}$$

Check Yourself 6

Subtract.

(a) $7 - 3\frac{2}{5}$ **(b)** $9 - 2\frac{5}{7}$

We use the same technique when the fraction of the minuend is smaller than the fraction of the subtrahend.

Example 7 Subtracting Mixed Numbers by Borrowing

Subtract $5\frac{3}{8} - 3\frac{3}{4}$.

We begin by rewriting the fraction parts using the LCD, 8.

$$5\frac{3}{8} - 3\frac{3}{4} = 5\frac{3}{8} - 3\frac{6}{8} \qquad \frac{3}{4} = \frac{6}{8}$$

Do you see the problem? We cannot subtract $\frac{6}{8}$ from $\frac{3}{8}$. In this case, we need to *borrow* one from the whole number part of the mixed number.

$$5\frac{3}{8} = 4 + 1 + \frac{3}{8} = 4 + \frac{8}{8} + \frac{3}{8} = 4 + \frac{11}{8}$$

We continue.

$$5\frac{3}{8} - 3\frac{3}{4} = \underbrace{4\frac{11}{8}}_{5\frac{3}{8}} - \underbrace{3\frac{6}{8}}_{3\frac{3}{4}}$$

$$= (4 - 3) + \left(\frac{11}{8} - \frac{6}{8}\right)$$

$$= 1\frac{5}{8}$$

Check Yourself 7

Subtract.

(a) $12\frac{5}{12} - 8\frac{2}{3}$ **(b)** $27\frac{7}{8} - 2\frac{9}{10}$

There is an alternative method that can be used when subtracting fractions. We can rewrite each mixed number as an improper fraction and then subtract. Although this technique requires an extra step, it eliminates the need to regroup.

Example 8 Subtracting by Converting to Improper Fractions

RECALL

$4\frac{5}{9} = \frac{4 \times 9 + 5}{9} = \frac{36 + 5}{9} = \frac{41}{9}$

$2\frac{11}{12} = \frac{2 \times 12 + 11}{12} = \frac{24 + 11}{12} = \frac{35}{12}$

Subtract.

$4\frac{5}{9} - 2\frac{11}{12}$

$\frac{41}{9} - \frac{35}{12}$ First, convert to improper fractions.

$\frac{164}{36} - \frac{105}{36} = \frac{59}{36}$ The LCD is 36. Find the equivalent fractions and subtract.

$\frac{59}{36} = 1\frac{23}{36}$ Rewrite as a mixed number.

Check Yourself 8

Subtract $5\frac{1}{4} - 2\frac{5}{18}$.

Example 9 An Application with Mixed Numbers

< Objective 3 >

Linda was $48\frac{1}{4}$ in. tall on her sixth birthday. By her seventh, she reached $51\frac{5}{8}$ in. How much did she grow during the year?

Because we want the difference in height, we must subtract $48\frac{1}{4}$ from $51\frac{5}{8}$.

$$51\frac{5}{8} - 48\frac{1}{4} = 51\frac{5}{8} - 48\frac{2}{8}$$ Rewrite $\frac{1}{4}$ as $\frac{2}{8}$.

$$= (51 - 48) + \left(\frac{5}{8} - \frac{2}{8}\right)$$

$$= 3\frac{3}{8}$$

Linda grew $3\frac{3}{8}$ in. during the year.

Check Yourself 9

You use $4\frac{3}{4}$ yd of fabric from a 50-yd bolt. How much fabric remains on the bolt?

Often, we have to use more than one operation to solve a problem. Consider Example 10.

Example 10 An Application Involving Mixed Numbers

A rectangular poster is to have a total length of $12\frac{1}{4}$ in. We want a $1\frac{3}{8}$-in. border on the top and a 2-in. border on the bottom. What is the length of the printed part of the poster?

We strongly recommend drawing a sketch when working with geometric problems.

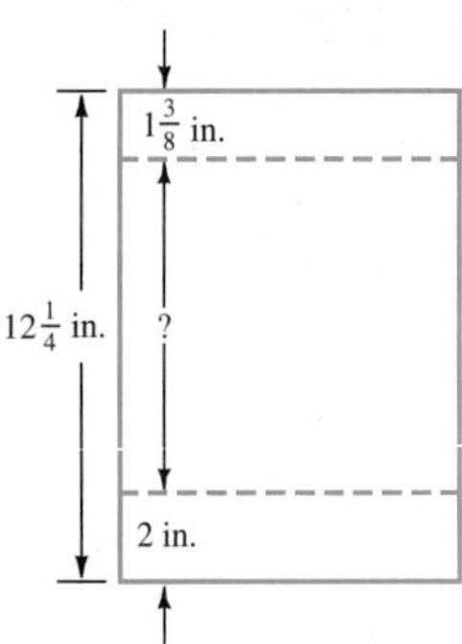

We use this sketch to find the total width of the top and bottom borders.

$$1\frac{3}{8} + 2 = (1 + 2) + \frac{3}{8}$$
$$= 3\frac{3}{8}$$

Now *subtract* that sum (the top and bottom borders) from the total length of the poster.

$$12\frac{1}{4} - 3\frac{3}{8} = 12\frac{2}{8} - 3\frac{3}{8}$$ Rewrite $\frac{1}{4}$ as $\frac{2}{8}$.

$$= 11\frac{10}{8} - 3\frac{3}{8}$$ We need to borrow: $12\frac{2}{8} = 11 + 1 + \frac{2}{8} = 11 + \frac{8}{8} + \frac{2}{8} = 11\frac{10}{8}$.

$$= (11 - 3) + \left(\frac{10}{8} - \frac{3}{8}\right)$$
$$= 8\frac{7}{8}$$

The length of the printed part is $8\frac{7}{8}$ in.

Check Yourself 10

You cut one shelf $3\frac{3}{4}$ ft long and one $4\frac{1}{2}$ ft long from a 12-ft piece of lumber. Can you cut another shelf 4 ft long?

Check Yourself ANSWERS

1. $5\frac{7}{10}$ **2.** $8\frac{4}{15}$ **3.** $13\frac{11}{12}$ **4.** $3\frac{1}{2}$ **5.** $4\frac{7}{24}$ **6.** **(a)** $3\frac{3}{5}$; **(b)** $6\frac{2}{7}$ **7.** **(a)** $3\frac{3}{4}$; **(b)** $24\frac{39}{40}$ **8.** $\frac{107}{36} = 2\frac{35}{36}$ **9.** $45\frac{1}{4}$ yd **10.** No, only $3\frac{3}{4}$ ft remains.

Reading Your Text

These fill-in-the-blank exercises will help you understand some of the key vocabulary used in this section. The answers to these exercises are in the Answers Appendix at the back of the text.

(a) To add mixed numbers with different denominators, we first find the ____________ of the fractions.

(b) To subtract a mixed number from a whole number, we must use a form of ____________, or borrowing.

(c) When the fraction parts of mixed numbers have different denominators, we rewrite the fractions as ____________ fractions.

(d) We always need to borrow when subtracting mixed numbers if the fraction part of the minuend is ____________ than the fraction part of the subtrahend.

3.4 exercises

< Objective 1 >

Evaluate each expression.

1. $4\frac{2}{9} + 5\frac{5}{9}$ **2.** $2\frac{2}{5} + 6\frac{2}{5}$ **3.** $3\frac{3}{8} + 7\frac{3}{8}$ **4.** $8\frac{3}{10} + 12\frac{1}{10}$

5. $8\frac{1}{6} + 8\frac{5}{6}$ **6.** $7\frac{2}{3} + 2\frac{1}{3}$ **7.** $9\frac{5}{8} + 12\frac{7}{8}$ **8.** $1\frac{3}{4} + 5\frac{3}{4}$

9. $3\frac{1}{3} + 6\frac{3}{5}$ **10.** $15\frac{3}{4} + 13\frac{2}{9}$ **11.** $6\frac{1}{2} + 7\frac{1}{8}$ **12.** $7\frac{5}{6} + 12\frac{5}{12}$

13. $11\frac{3}{10} + 4\frac{5}{6}$ **14.** $9\frac{4}{9} + 6\frac{11}{15}$ **15.** $2\frac{7}{12} + 6\frac{7}{9}$ **16.** $1\frac{3}{4} + 3\frac{5}{6}$

17. $9\frac{1}{2} + \frac{3}{4}$ **18.** $\frac{2}{3} + 7\frac{1}{2}$ **19.** $2\frac{1}{4} + 3\frac{5}{8} + 1\frac{1}{6}$ **20.** $3\frac{1}{5} + 2\frac{1}{2} + 5\frac{1}{4}$

21. $3\frac{3}{5} + 4\frac{1}{4} + 5\frac{3}{10}$ **22.** $4\frac{5}{6} + 3\frac{2}{3} + 7\frac{5}{9}$

< Objective 2 >

23. $11\frac{7}{8} - 4\frac{3}{8}$ **24.** $9\frac{5}{6} - 5\frac{1}{6}$ **25.** $6\frac{1}{4} - 1\frac{3}{4}$ **26.** $7\frac{3}{10} - 4\frac{7}{10}$

27. $3\frac{2}{3} - 2\frac{1}{4}$ **28.** $8\frac{7}{9} - 6\frac{7}{12}$ **29.** $7\frac{5}{12} - 3\frac{11}{18}$ **30.** $5\frac{4}{5} - 2\frac{5}{6}$

31. $4\frac{1}{4} - 3\frac{2}{3}$ **32.** $5\frac{3}{10} - 4\frac{5}{6}$ **33.** $1\frac{5}{12} - \frac{11}{18}$ **34.** $8\frac{3}{4} - \frac{9}{10}$

35. $5 - 2\frac{1}{4}$ **36.** $4 - 1\frac{2}{3}$ **37.** $17 - 8\frac{3}{4}$ **38.** $23 - 11\frac{5}{8}$

39. $3\frac{3}{4} + 5\frac{1}{2} - 2\frac{3}{8}$ **40.** $1\frac{5}{6} + 3\frac{5}{12} - 2\frac{1}{4}$ **41.** $2\frac{3}{8} + 2\frac{1}{4} - 1\frac{5}{6}$ **42.** $1\frac{1}{15} + 3\frac{3}{10} - 2\frac{4}{5}$

43. $4\frac{1}{8} + \frac{3}{7} - 2\frac{23}{28}$ **44.** $5\frac{1}{3} + 1\frac{3}{7} - 5\frac{23}{42}$ **45.** $6\frac{1}{11} + \frac{2}{3} - 2\frac{1}{6}$ **46.** $3\frac{1}{5} + 1\frac{7}{8} - 5\frac{1}{20}$

47. $6\frac{1}{11} - \frac{2}{3} + 2\frac{1}{6}$ **48.** $3\frac{1}{5} - 1\frac{7}{8} + 5\frac{1}{20}$ **49.** $\frac{9}{4} + \frac{3}{2}$ **50.** $\frac{7}{3} + \frac{11}{8}$

51. $2\frac{1}{2} + \frac{11}{6}$ **52.** $5\frac{7}{8} + \frac{7}{5}$ **53.** $\frac{9}{4} - \frac{3}{2}$ **54.** $\frac{7}{3} - \frac{11}{8}$

55. $5\frac{2}{3} - \frac{15}{4}$ **56.** $2\frac{1}{2} - \frac{11}{6}$

< Objective 3 >

Solve each application.

57. **Crafts** Senta is working on a project that uses three pieces of fabric with lengths of $\frac{3}{4}$, $1\frac{1}{4}$, and $\frac{5}{8}$ yd. She needs to allow for $\frac{1}{8}$ yd of waste. How much fabric should she buy?

58. **Construction** The framework of a wall is $3\frac{1}{2}$ in. thick. We apply $\frac{5}{8}$-in. wall-board and $\frac{1}{4}$-in. paneling to the inside. $\frac{3}{4}$-in. thick siding is applied to the outside. What is the finished thickness of the wall?

59. **Business and Finance** A stock was listed at $34\frac{3}{8}$ points on Monday. At closing time Friday, it was at $28\frac{3}{4}$. How much did it drop during the week?

60. **Crafts** A roast weighed $4\frac{1}{4}$ lb before cooking and $3\frac{3}{8}$ lb after cooking. How much weight was lost in cooking?

61. **Crafts** A roll of paper contains $30\frac{1}{4}$ yd. If $16\frac{7}{8}$ yd is cut from the roll, how much paper remains?

62. **Geometry** Find the missing dimension.

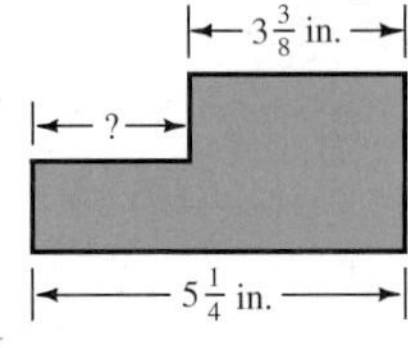

63. **Crafts** A $4\frac{1}{4}$-in. bolt is placed through a board that is $3\frac{1}{2}$ in. thick. How far does the bolt extend beyond the board?

chapter 3 > Make the Connection

64. **Business and Finance** Ben can work 20 hr per week on a part-time job. He works $5\frac{1}{2}$ hr on Monday and $3\frac{3}{4}$ hr on Tuesday. How many more hours can he work during the week?

65. **Geometry** Find the missing dimension.

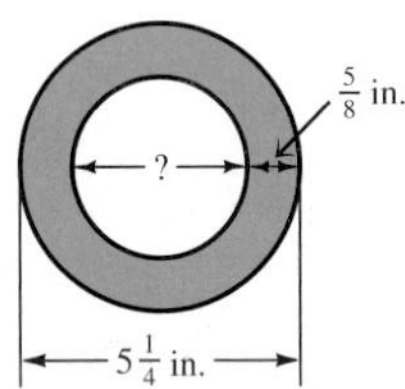

66. **Construction** The Hughes family used $20\frac{3}{4}$ yd² of carpet for their living room, $15\frac{1}{2}$ yd² for the dining room, and $6\frac{1}{4}$ yd² for a hallway. How much remains if they bought a 50-yd² roll of carpet?

Label each statement as **true** *or* **false.**

67. The LCM of 3, 6, and 12 is 24.

68. The GCF of 15, 21, and 300 is 3.

69. The LCD for $\frac{3}{4}$, $\frac{9}{10}$, and $\frac{3}{20}$ is 40.

70. The LCD for $\frac{5}{6}$, $\frac{13}{15}$, and $\frac{1}{21}$ is 3.

Skills | **Calculator/Computer** | Career Applications | Above and Beyond

Using a Calculator to Add and Subtract Mixed Numbers

Scientific Calculator

To enter a mixed number on a scientific calculator, press the fraction key between both the whole number and the numerator and denominator. For example, to enter $3\frac{7}{12}$, press

3 [a b/c] 7 [a b/c] 12

Graphing Calculator

As with multiplying and dividing fractions, when using a graphing calculator, you must choose the fraction option from the math menu before pressing [ENTER].

For the problem $3\frac{7}{12} - 2\frac{11}{16}$, the keystroke sequence is

3 [+] 7 [÷] 12 [−] [(] 2 [+] 11 [÷] 16 [)] [▶ Frac] [ENTER]

Note that the parentheses are very important when doing subtraction. The display will read $\frac{43}{48}$.

Use a calculator to evaluate each expression.

71. $4\frac{7}{9} - 2\frac{11}{18}$

72. $7\frac{8}{11} - 4\frac{13}{22}$

73. $5\frac{11}{16} - 2\frac{5}{12}$

74. $18\frac{5}{24} - 11\frac{3}{40}$

75. $6\frac{2}{3} - 1\frac{5}{6}$

76. $131\frac{43}{45} - 99\frac{27}{60}$

77. $10\frac{2}{3} + 4\frac{1}{5} + 7\frac{2}{15}$

78. $7\frac{1}{5} + 3\frac{2}{3} + 1\frac{1}{5}$

Skills | Calculator/Computer | **Career Applications** | Above and Beyond

79. **Manufacturing Technology** A factory floor is made up of several layers, as shown in the drawing.

What is the total thickness of the floor?

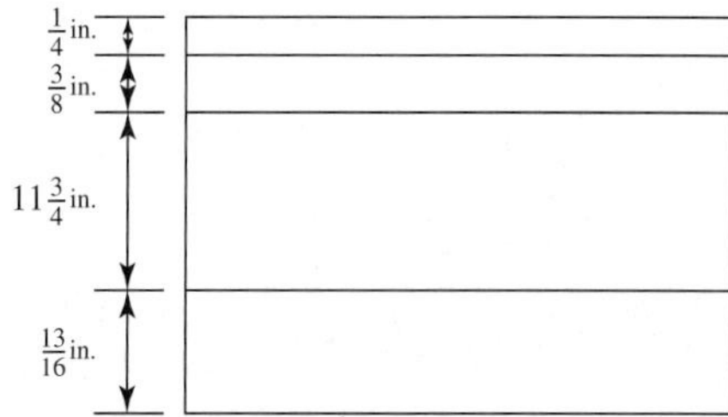

80. **Manufacturing Technology** Pieces that are $2\frac{1}{4}$, $5\frac{1}{16}$, $4\frac{3}{4}$, and $1\frac{7}{8}$ in. long need to be cut from round stock. How long of a piece of round stock is required? $\left(\text{Allow } \frac{3}{32} \text{ in. for each saw kerf.}\right)$

81. **Information Technology** Joy is running cable in a new office building in downtown Washington, D.C. She has $60\frac{1}{4}$ ft of cable but she needs 100 ft. How much more cable does she need? chapter 3 > Make the Connection

82. **Information Technology** Abraham has a part-time technician job while he is going to college. He works the following hours in one week: $3\frac{1}{4}$, $5\frac{3}{4}$, $4\frac{1}{2}$, 4, and $2\frac{1}{4}$. For the week, how many hours did Abraham work?

Answers

1. $9\frac{7}{9}$ **3.** $10\frac{3}{4}$ **5.** 17 **7.** $22\frac{1}{2}$ **9.** $9\frac{14}{15}$ **11.** $13\frac{5}{8}$ **13.** $15\frac{2}{15}$ **15.** $9\frac{13}{36}$ **17.** $10\frac{1}{4}$ **19.** $7\frac{1}{24}$ **21.** $13\frac{3}{20}$ **23.** $7\frac{1}{2}$ **25.** $4\frac{1}{2}$ **27.** $1\frac{5}{12}$ **29.** $3\frac{29}{36}$ **31.** $\frac{7}{12}$ **33.** $\frac{29}{36}$ **35.** $2\frac{3}{4}$ **37.** $8\frac{1}{4}$ **39.** $6\frac{7}{8}$ **41.** $2\frac{19}{24}$ **43.** $1\frac{41}{56}$ **45.** $4\frac{13}{22}$ **47.** $7\frac{13}{22}$ **49.** $\frac{15}{4}$ or $3\frac{3}{4}$ **51.** $\frac{13}{3}$ or $4\frac{1}{3}$ **53.** $\frac{3}{4}$ **55.** $\frac{23}{12}$ or $1\frac{11}{12}$ **57.** $2\frac{3}{4}$ yd **59.** $5\frac{5}{8}$ points **61.** $13\frac{3}{8}$ yd **63.** $\frac{3}{4}$ in. **65.** 4 in. **67.** False **69.** False **71.** $2\frac{1}{6}$ **73.** $3\frac{13}{48}$ **75.** $4\frac{5}{6}$ **77.** 22 **79.** $13\frac{3}{16}$ in. **81.** $39\frac{3}{4}$ ft

Activity 8 ::

Sharing Costs

The Associated Student Government at CCC is sending six students to the national conference in Washington, D.C. Two of the students, Mikaila and Courtney, are in charge of making the room reservations. Looking at the hotels that are either hosting or adjacent to the conference site, they come up with the following information. Accommodations were not available for places on the table that are blank.

Hotel	Price for a Single	Price for a Double	Price for a Triple	Suite (Sleeps 6)
Wyndham	$180	$180	$210	
St. Gregory	168	198	240	$450
Hyatt	190	190	222	
Marriott	159	174		

1. If they get three double rooms at the Marriott and each pays $\frac{1}{6}$ of the bill, what is the cost per person each night?
2. If they stay at the Wyndham in a triple room, they need only two rooms. If they each pay $\frac{1}{6}$ of that total bill, what is the cost per night?
3. If they get the suite at the St. Gregory, what is the per-person cost per night?
4. The Hyatt and the St. Gregory each offer a free breakfast for each person staying there. Is it now cheaper to stay at the Hyatt in a triple room than at the Wyndham? What information would you need to make the decision?
5. What about the St. Gregory suite with a free breakfast? How do you now make the decision?

3.5 Order of Operations with Fractions

< 3.5 Objectives >

1 > Use the order of operations to evaluate expressions

2 > Use expressions to solve applications

In Chapter 1, we introduced the order of operations. As a reminder, we repeat them here.

Step by Step

Evaluating an Expression

Step 1 Do any operations within parentheses or other grouping symbols.

Step 2 Apply any exponents.

Step 3 Do all multiplication and division in order from left to right.

Step 4 Do all addition and subtraction in order from left to right.

We begin with a set of examples demonstrating the order of operations when evaluating expressions containing fractions. You should refer to the Step by Step box as needed.

Example 1 **Evaluating an Expression**

< Objective 1 >

Evaluate.

$$\frac{1}{2} + \frac{2}{3} \times \frac{3}{4}$$

In this expression, there are no parentheses or other grouping symbols and there are no exponents. Therefore, our first step is to complete the multiplication.

$$\frac{1}{2} + \frac{2}{3} \times \frac{3}{4} = \frac{1}{2} + \underbrace{\frac{\overset{1}{\cancel{2}}}{\underset{1}{\cancel{3}}} \times \frac{\overset{1}{\cancel{3}}}{\underset{2}{\cancel{4}}}}_{\text{Multiply first}} \qquad \frac{2}{3} \times \frac{3}{4} = \frac{1}{2}$$

$$= \underbrace{\frac{1}{2} + \frac{1}{2}}_{\text{Now add}}$$

$$= 1$$

RECALL

Rather than reading an expression from left to right, we take a more complete approach. We look at the whole expression and use the order of operations to determine when to do each step.

Check Yourself 1

Evaluate $\frac{2}{3} - \frac{1}{2} \div \frac{3}{2}$.

When evaluating more complicated expressions, simply take it one step at a time. Each step should be straightforward and easy. We do as much as we can in each step, but not so much that we make mistakes.

Example 2 Evaluating an Expression

Evaluate $\frac{14}{15} - \left(\frac{1}{2}\right)^2 \cdot \left(\frac{2}{3} + \frac{4}{5}\right)$.

RECALL

$\frac{2}{3} + \frac{4}{5} = \frac{10}{15} + \frac{12}{15}$

$= \frac{22}{15}$

$\frac{1}{4} \cdot \frac{22}{15} = \frac{11}{30}$

$\frac{14}{15} - \left(\frac{1}{2}\right)^2 \cdot \left(\frac{2}{3} + \frac{4}{5}\right) = \frac{14}{15} - \left(\frac{1}{2}\right)^2 \cdot \left(\frac{22}{15}\right)$ Do the operation inside the parentheses first.

$= \frac{14}{15} - \left(\frac{1}{4}\right) \cdot \left(\frac{22}{15}\right)$ Next, evaluate the exponential expression.

$= \frac{14}{15} - \frac{11}{30}$ Multiply; the centered dot represents multiplication.

$= \frac{17}{30}$ Finally, subtract.

Check Yourself 2

Evaluate $\left(\frac{2}{3}\right)^3 + \left(\frac{1}{3}\right)^2 \cdot \left(\frac{1}{2} + \frac{2}{3}\right)$

Once parentheses are removed and exponents are applied, multiplication and division are always performed left to right.

Example 3 Evaluating an Expression

Evaluate $\left(\frac{1}{13}\right)^2 \cdot \left(\frac{1}{4} + \frac{1}{6}\right) \div \frac{5}{13}$.

RECALL

$\frac{1}{4} + \frac{1}{6} = \frac{3}{12} + \frac{2}{12}$

$= \frac{5}{12}$

$\left(\frac{1}{13}\right)^2 \cdot \left(\frac{1}{4} + \frac{1}{6}\right) \div \frac{5}{13} = \left(\frac{1}{13}\right)^2 \cdot \left(\frac{5}{12}\right) \div \frac{5}{13}$ Do the operation inside the parentheses first.

$= \left(\frac{1}{169}\right) \cdot \left(\frac{5}{12}\right) \div \frac{5}{13}$ Next, evaluate the exponential expression.

$= \frac{1}{169} \cdot \frac{5}{12} \cdot \frac{13}{5}$ Rewrite the division as multiplication.

$= \frac{1}{\cancel{169}_{13}} \cdot \frac{\cancel{5}^{1}}{12} \cdot \frac{\cancel{13}^{1}}{\cancel{5}_{1}}$ Simplify to multiply.

$= \frac{1}{156}$

Check Yourself 3

Evaluate.

$\left(\frac{1}{5}\right)^2 \cdot \left(\frac{3}{10} + \frac{1}{2}\right) \div \frac{2}{5}$

When mixed numbers are involved in a complex expression, it is usually easiest to convert them to improper fractions before continuing.

Example 4 Evaluating an Expression with Mixed Numbers

Evaluate $3\frac{3}{4} + 5 \cdot \left(2\frac{1}{2}\right)$.

RECALL

$3\frac{3}{4} = \frac{3 \times 4 + 3}{4}$

$= \frac{15}{4}$

$3\frac{3}{4} + 5 \cdot \left(2\frac{1}{2}\right) = \frac{15}{4} + 5 \cdot \frac{5}{2}$ Rewrite the mixed numbers as improper fractions.

$= \frac{15}{4} + \frac{25}{2}$ Multiply: $5 \cdot \frac{5}{2} = \frac{5}{1} \cdot \frac{5}{2} = \frac{25}{2}$.

$= \frac{65}{4}$ Add: $\frac{15}{4} + \frac{25}{2} = \frac{15}{4} + \frac{50}{4} = \frac{15 + 50}{4} = \frac{65}{4}$.

$= 16\frac{1}{4}$ Write the final result as a mixed number: $65 \div 4 = 16$ r1.

Check Yourself 4

Evaluate.

$$4\frac{1}{3} - \frac{1}{2} \cdot \left(5\frac{1}{8}\right)$$

Many students are confused by parentheses and grouping symbols. You will see three basic reasons to use parentheses or other grouping symbols in this text.

The first reason is simply clarity. We occasionally enclose some expression in parentheses just to make it easy to see. If you look at Example 4, we placed parentheses around the mixed number $2\frac{1}{2}$. These parentheses were not strictly necessary. Instead, they just make the numbers and operations easier to see.

The second reason we use parentheses and other grouping symbols is to "violate" the order of operations. For instance, consider the expression

$$2 + 3 \times 4$$

The order of operations requires us to multiply first and then add. In this case, we compute

NOTE

A third reason to use grouping symbols is when we need to apply an action to an expression. You will learn one type of action when we look at square roots in Chapter 8.

$$\begin{aligned} 2 + 3 \times 4 &= 2 + 12 \\ &= 14 \end{aligned}$$

If, in fact, we want to add before multiplying, we need to use a grouping symbol so that we add first.

$$\begin{aligned} (2 + 3) \times 4 &= 5 \times 4 \\ &= 20 \end{aligned}$$

Fraction bars are a type of grouping symbol. Since a fraction bar represents division, we should perform any computations in the numerator and denominator first, and then divide.

Example 5 Using a Fraction Bar as a Grouping Symbol

Evaluate $\frac{3 + 2 \times 16}{7}$.

Because the fraction bar is a grouping symbol, we should do the calculations in the numerator before dividing by 7. Within the numerator, we follow the standard order of operations and multiply before adding.

NOTE

Think of this as $(3 + 2 \times 16) \div 7$

$$\begin{aligned} \frac{3 + 2 \times 16}{7} &= \frac{3 + 32}{7} && \text{First multiply in the numerator.} \\ &= \frac{35}{7} && \text{Now add in the numerator.} \\ &= 5 && \text{Finally, divide.} \end{aligned}$$

Check Yourself 5

Evaluate $\frac{30 - 3 \times 2^2}{3}$.

Example 6 illustrates an application of the material in this section.

Example 6 Solving an Application

< Objective 2 >

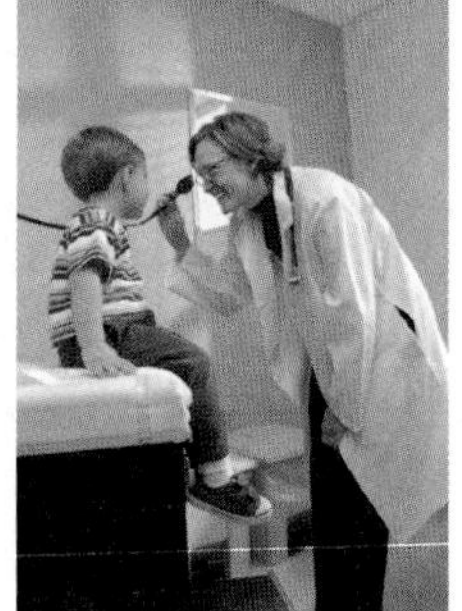

Young's rule is one formula for calculating the children's dosage of a medication. It is based on the adult dosage and the child's age.

$$\text{Child's dose} = \left(\frac{\text{age}}{\text{age} + 12}\right) \times \text{adult dose}$$

According to this rule, the dose prescribed to a 3-year-old child if the recommended adult dose is 24 milligrams (mg) can be found by evaluating the expression

$$\left(\frac{3}{3 + 12}\right) \times 24 \text{ mg}$$

To evaluate this expression, first we do the operations inside the parentheses.

NOTE

The fraction bar is a grouping symbol, so we add 3 and 12 in the denominator first to simplify the expression.

$$\left(\frac{3}{3 + 12}\right) \times 24 \text{ mg}$$

$$= \left(\frac{3}{15}\right) \times 24 \text{ mg}$$

$$= \frac{1}{5} \times \frac{24}{1} \text{ mg} = \frac{24}{5} \text{ mg} = 4\frac{4}{5} \text{ mg}$$

Check Yourself 6

The approximate length of the belt pictured is given by

$$\frac{22}{7}\left(\frac{1}{2} \cdot 15 + \frac{1}{2} \cdot 5\right) + 2 \cdot 21$$

Find the length of the belt.

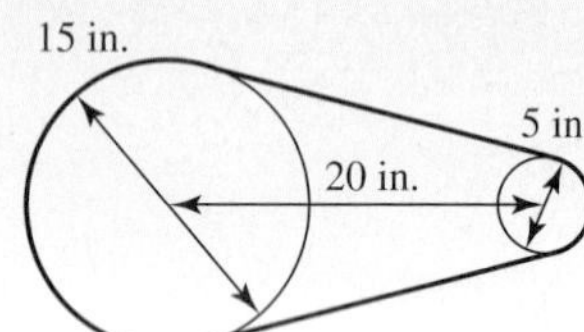

Check Yourself ANSWERS

1. $\frac{1}{3}$ 2. $\frac{23}{54}$ 3. $\frac{2}{25}$ 4. $1\frac{37}{48}$ 5. 6 6. $73\frac{3}{7}$ in.

Reading Your Text

These fill-in-the-blank exercises will help you understand some of the key vocabulary used in this section. The answers to these exercises are in the Answers Appendix at the back of the text.

(a) When evaluating an expression, first do operations inside parentheses or other ______________ symbols.

(b) The second step in evaluating an expression is to evaluate all ______________.

(c) When dividing by a fraction, ______________ and multiply by that fraction.

(d) When mixed numbers are involved in a complex expression, it is almost always best to convert them to __________ fractions before continuing.

Skills | Calculator/Computer | Career Applications | Above and Beyond

3.5 exercises

< Objective 1 >

Evaluate.

1. $\frac{1}{3} - \left(\frac{1}{2} - \frac{1}{4}\right)$

2. $\frac{2}{3} - \left(\frac{3}{4} - \frac{1}{2}\right)$

3. $\frac{3}{4} - \left(\frac{1}{2}\right)\left(\frac{1}{3}\right)$

4. $\frac{5}{6} - \left(\frac{3}{4}\right)\left(\frac{2}{3}\right)$

5. $\left(\frac{1}{2}\right)^2 - \left(\frac{1}{4} - \frac{1}{5}\right)$

6. $\left(\frac{3}{4}\right)^2 - \left(\frac{1}{2} - \frac{1}{3}\right)$

7. $\frac{1}{2} + \frac{2}{3} \cdot \frac{9}{16}$

8. $\frac{2}{3} + \frac{1}{2} \div \frac{3}{4}$

9. $\left(\frac{1}{3}\right)\left(\frac{1}{2} + \frac{1}{4}\right)$

10. $\left(\frac{3}{4}\right)\left(\frac{1}{2} - \frac{1}{3}\right)$

11. $5\left(\frac{2}{3} + \frac{5}{6}\right)$

12. $8\left(\frac{3}{4} - \frac{3}{10}\right)$

13. $\left(4\frac{1}{2}\right)\left(\frac{7}{8} + 2\frac{1}{4}\right)$

14. $\left(3\frac{1}{4}\right)\left(3\frac{1}{3} + 7\frac{3}{8}\right)$

15. $\left(4 - 1\frac{1}{2}\right)\left(2\frac{1}{4} - 1\frac{2}{3}\right)$

16. $\left(9 - 5\frac{3}{5}\right)\left(2\frac{1}{2} + \frac{3}{4}\right)$

17. $\left(\frac{1}{2}\right)^3 + \left(\frac{1}{3}\right) \cdot \left(\frac{1}{2} + \frac{1}{4}\right)$

18. $\left(\frac{1}{3}\right) + \left(\frac{1}{2}\right)^2 \cdot \left(\frac{1}{3} + \frac{1}{6}\right)$

19. $\left(\frac{3}{4}\right)^2 + \left(\frac{1}{2}\right)^3 \cdot \left(\frac{1}{2} + \frac{3}{4}\right)$

20. $\left(\frac{1}{10}\right) + \left(\frac{1}{2}\right)^3 \cdot \left(\frac{1}{5} + \frac{1}{15}\right)$

21. $\left(\frac{2}{3}\right)^2 \cdot \left(\frac{2}{15} + \frac{1}{2}\right) \div \frac{3}{5}$

22. $\left(\frac{1}{3}\right)^3 \cdot \left(\frac{1}{4} + \frac{1}{2}\right) \div \frac{2}{9}$

23. $2\frac{1}{5} - \frac{1}{2} \cdot \frac{1}{4}$

24. $4\frac{3}{8} - \frac{1}{2} \cdot \frac{1}{5}$

25. $11\frac{1}{10} - \frac{1}{2} \cdot \left(6\frac{2}{3}\right)$

26. $12\frac{2}{7} - \frac{1}{3} \cdot \left(3\frac{3}{7}\right)$

27. $\frac{8 + 12}{4}$

28. $\frac{2(9 + 3)}{8}$

29. $\frac{3(8 - 3)}{7}$

30. $\frac{7(12 - 6)}{4}$

31. $6\left(\frac{4 + 9}{2(8 + 5)}\right) - \left(\frac{3}{4} + \frac{3}{4}\right)^2$

32. $\left(1\frac{1}{2}\right) \div \left(\frac{3}{4}\right) \times \left(\frac{4(7 + 2)}{3 + 9}\right)$

33. $\frac{4}{5} + \left(\frac{2 + 3}{1 + 2}\right)^2 - \left(4\frac{1}{2}\right) \div \left(\frac{3(5 - 2)}{2^2}\right)$

34. $\left(6\frac{1}{2}\right)\left(\frac{4 - 1}{2 + 3}\right)^2 - 2\left(\frac{2(3 + 2)}{10}\right)$

< Objective 2 >

35. **Construction** A construction company has bids for paving roads of $1\frac{1}{2}$, $\frac{3}{4}$, and $3\frac{1}{3}$ mi for the month of July. With their present equipment, they can pave 8 mi in 1 month. How much more work can they take on in July?

36. **Statistics** On an 8-hr trip, Jack drives $2\frac{3}{4}$ hr and Pat drives $2\frac{1}{2}$ hr. How many hours are left to drive?

37. **Statistics** A runner told herself that she will run 20 mi each week. She runs $5\frac{1}{2}$ mi on Sunday, $4\frac{1}{4}$ mi on Tuesday, $4\frac{3}{4}$ mi on Wednesday, and $2\frac{1}{8}$ mi on Friday. How far must she run on Saturday to meet her goal?

38. **Science and Medicine** If paper takes up $\frac{1}{2}$ of the space in a landfill and plastic takes up $\frac{1}{10}$ of the space, how much of the landfill is used for other materials?

39. **Science and Medicine** If paper takes up $\frac{1}{2}$ of the space in a landfill and organic waste takes up $\frac{1}{8}$ of the space, how much of the landfill is used for other materials?

40. **Business and Finance** The interest rate on an auto loan in May was $12\frac{3}{8}\%$. By September the rate was up to $14\frac{1}{4}\%$. By how many percentage points did the interest rate increase over the period?

41. **Allied Health** Simone suffers from Gaucher's disease. The recommended dosage of Cerezyme is $2\frac{1}{2}$ units per kilogram (kg) of the patient's weight. How much Cerezyme should the doctor prescribe if Simone weighs $15\frac{1}{3}$ kg?

42. **Information Technology** Kendra is running cable in a new office building in downtown Kansas City. She has $110\frac{1}{4}$ ft of cable, but she needs 150 ft. How much more cable is needed?

43. **Manufacturing Technology** A cut $3\frac{3}{8}$ in. long needs to be made in a piece of material. The cut rate is $\frac{3}{4}$ in. per minute. How many minutes does it take to make the cut?

44. **Manufacturing Technology** Calculate the distance from the center of hole A to the center of hole B.

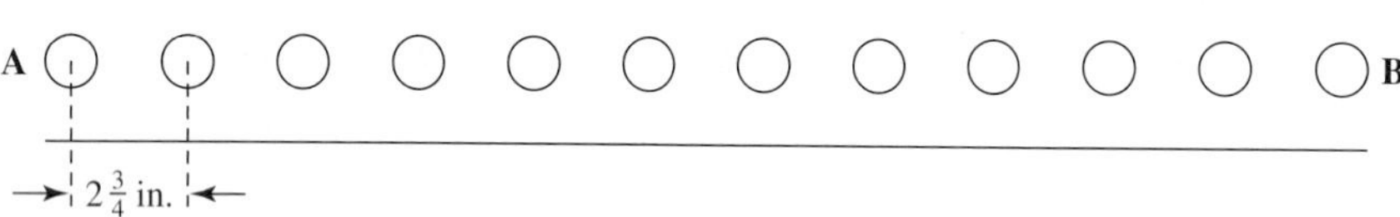

45. **Information Technology** On average, $18\frac{1}{4}$ printed circuit boards can be completely populated with components (with all parts soldered to the board) by Amara in 1 hr. Burt can complete $34\frac{2}{3}$ printed circuit boards in 2 hr. If both workers continue at their respective average paces, how many total printed circuit boards can be populated in 8 hr?

46. **Information Technology** If Carlos joins Amara and Burt in soldering components on circuit boards, the three workers can average 410 complete boards in 8 hr. Assuming the other two workers perform at their respective averages as stated in exercise 45, how many boards can Carlos average in 8 hr? How many boards does Carlos average in 1 hr?

Answers

1. $\frac{1}{12}$ **3.** $\frac{7}{12}$ **5.** $\frac{1}{5}$ **7.** $\frac{7}{8}$ **9.** $\frac{1}{4}$ **11.** $\frac{15}{2}$ or $7\frac{1}{2}$ **13.** $\frac{225}{16}$ or $14\frac{1}{16}$ **15.** $\frac{35}{24}$ or $1\frac{11}{24}$ **17.** $\frac{3}{8}$ **19.** $\frac{23}{32}$ **21.** $\frac{38}{81}$ **23.** $\frac{83}{40} = 2\frac{3}{40}$ **25.** $\frac{233}{30} = 7\frac{23}{30}$ **27.** 5 **29.** $\frac{15}{7}$ or $2\frac{1}{7}$ **31.** $\frac{3}{4}$ **33.** $\frac{71}{45}$ or $1\frac{26}{45}$ **35.** $2\frac{5}{12}$ mi **37.** $3\frac{3}{8}$ mi **39.** $\frac{3}{8}$ **41.** $38\frac{1}{3}$ units **43.** $4\frac{1}{2}$ min **45.** 284 completed boards

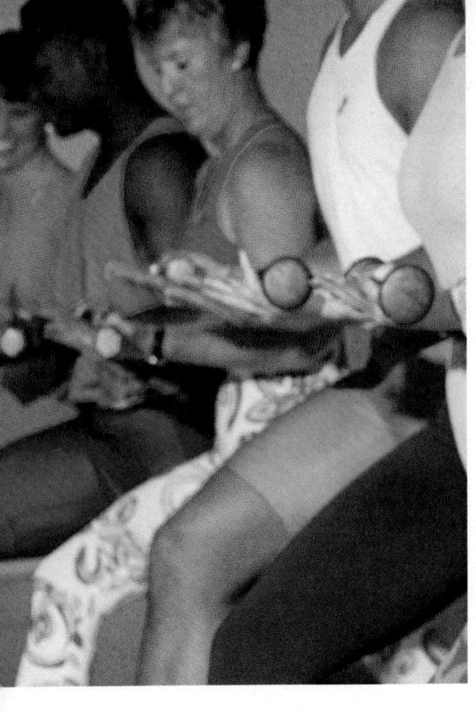

Activity 9 ::

Aerobic Exercise

Some fitness experts believe that there is a *training intensity range* that you can use to improve your level of aerobic fitness without overtaxing your cardiorespiratory system. Your personal training intensity range is a fraction of your *maximal heart rate,* in beats per minute (bpm). Your maximal heart rate can be measured directly, but you can approximate it by subtracting your age from 220.

Multiplying your maximal heart rate by $\frac{3}{5}$ gives the lower limit and multiplying it by $\frac{9}{10}$ gives the upper limit of your aerobic training zone. The best heart rate for a person working out is somewhere in this range. It is primarily determined by their overall physical fitness.

Complete the table by computing the maximal heart rate for each age.

Age	Maximal Heart Rate
20	200
25	
30	
35	
40	
45	
50	
55	

Complete the remainder of the table by calculating the lower and upper limits of each age's aerobic training zone. Anytime your result includes a fraction part, round up to the next whole number.

For instance, the upper limit of the aerobic training zone for 25-year-olds is given by multiplying their maximal heart rate of 195 bpm by $\frac{9}{10}$.

$$195 \times \frac{9}{10} = \frac{351}{2} = 175\frac{1}{2}$$

Because there is a fraction part, the upper limit of their zone is 176 bpm.

Age	Maximal Heart Rate	Lower Limit of Zone $\left(\frac{3}{5}\text{ MHR}\right)$	Upper Limit of Zone $\left(\frac{9}{10}\text{ MHR}\right)$
20	200	120	180
25	195	117	176
30	190		
35	185		
40	180		
45	175		
50	170		
55	165		

Definition/Procedure	Example	Reference
Adding and Subtracting Like Fractions		Section 3.1
To Add Like Fractions		
Step 1 Add the numerators. **Step 2** Place the sum over the common denominator. **Step 3** Simplify the resulting fraction if necessary.	$\frac{5}{18} + \frac{7}{18} = \frac{12}{18} = \frac{2}{3}$	p. 168
To Subtract Like Fractions		
Step 1 Subtract the numerators. **Step 2** Place the difference over the common denominator. **Step 3** Simplify the resulting fraction when necessary.	$\frac{17}{20} - \frac{7}{20} = \frac{10}{20} = \frac{1}{2}$	p. 170
Common Multiples		Section 3.2
Least Common Multiple (LCM) The LCM is the *smallest* number that is a multiple of each of a group of numbers.		p. 175
To Find the LCM		
Step 1 Write the prime factorization for each of the numbers in the group. **Step 2** Find all the prime factors that appear in any one of the prime factorizations. **Step 3** Form the product of those prime factors, using each factor the greatest number of times it occurs in any one factorization.	To find the LCM of 12, 15, and 18: $12 = 2 \times 2 \times 3$ $15 = 3 \times 5$ $18 = 2 \times 3 \times 3$ $2 \times 2 \times 3 \times 3 \times 5$ The LCM is $2 \times 2 \times 3 \times 3 \times 5$, or 180.	p. 175
Adding and Subtracting Unlike Fractions		Section 3.3
To Find the LCD of a Group of Fractions		
Step 1 Write the prime factorization for each of the denominators. **Step 2** Find all the prime factors that appear in any one of the prime factorizations. **Step 3** Form the product of those prime factors, using each factor the greatest number of times it occurs in any one factorization.	To find the LCD of fractions with denominators 4, 6, and 15: $4 = 2 \times 2$ $6 = 2 \times 3$ $15 = 3 \times 5$ $2 \times 2 \times 3 \times 5$ The LCD $= 2 \times 2 \times 3 \times 5$, or 60.	p. 182
To Add Unlike Fractions		
Step 1 Find the LCD of the fractions. **Step 2** Change each unlike fraction to an equivalent fraction with the LCD as its denominator. **Step 3** Add the resulting like fractions as before.	$\frac{3}{4} + \frac{7}{10} = \frac{15}{20} + \frac{14}{20}$ $= \frac{29}{20} = 1\frac{9}{20}$	p. 183
To Subtract Unlike Fractions		
Step 1 Find the LCD of the fractions. **Step 2** Change each unlike fraction to an equivalent fraction with the LCD as its denominator. **Step 3** Subtract the resulting like fractions as before.	$\frac{8}{9} - \frac{5}{6} = \frac{16}{18} - \frac{15}{18} = \frac{1}{18}$	p. 185

Continued

Definition/Procedure	Example	Reference
Adding and Subtracting Mixed Numbers		Section 3.4
To Add or Subtract Mixed Numbers		
Step 1 Find the LCD of the fraction parts. **Step 2** Rewrite the fraction parts as equivalent fractions with the LCD as their denominator, if necessary. **Step 3** Add or subtract the whole number and fraction parts separately. **Step 4** Simplify if necessary. **Step 5** Rewrite the sum or difference as a proper fraction or as a whole or mixed number.	$2\frac{1}{4} + 3\frac{4}{5} = 2\frac{5}{20} + 3\frac{16}{20}$ $= (2 + 3) + \left(\frac{5}{20} + \frac{16}{20}\right)$ $= 5 + \frac{21}{20}$ $= 6\frac{1}{20}$	*p.* 192
Order of Operations with Fractions		Section 3.5
Order of Operations		
Step 1 Do any operations within parentheses or other grouping symbols. **Step 2** Evaluate all powers. **Step 3** Do all multiplication and division in order from left to right. **Step 4** Do all addition and subtraction in order from left to right.	$\frac{2}{3} + \left(\frac{1}{2}\right)^2\left(\frac{1}{3} + \frac{1}{2}\right) = \frac{2}{3} + \left(\frac{1}{2}\right)^2\left(\frac{5}{6}\right)$ $= \frac{2}{3} + \left(\frac{1}{4}\right)\left(\frac{5}{6}\right)$ $= \frac{2}{3} + \frac{5}{24}$ $= \frac{16}{24} + \frac{5}{24}$ $= \frac{21}{24} = \frac{7}{8}$	*p.* 203

summary exercises :: chapter 3

This summary exercise set will help ensure that you have mastered each of the objectives of this chapter. The exercises are grouped by section. You should reread the material associated with any exercises that you find difficult. The answers to the odd-numbered exercises are in the Answers Appendix at the back of the text.

3.1 *Add. Simplify when possible.*

1. $\frac{8}{15} + \frac{2}{15}$

2. $\frac{4}{7} + \frac{3}{7}$

3. $\frac{8}{13} + \frac{7}{13}$

4. $\frac{17}{18} + \frac{5}{18}$

5. $\frac{19}{24} + \frac{13}{24}$

6. $\frac{1}{9} + \frac{2}{9} + \frac{4}{9}$

7. $\frac{2}{9} + \frac{5}{9} + \frac{4}{9}$

8. $\frac{4}{15} + \frac{7}{15} + \frac{7}{15}$

3.2 *Find the least common multiple (LCM) for each group of numbers.*

9. 4 and 12

10. 8 and 16

11. 18 and 24

12. 12 and 18

13. 15 and 20

14. 14 and 21

15. 9, 12, and 24

16. 14, 21, and 28

Arrange the fractions in order from smallest to largest.

17. $\frac{5}{8}, \frac{7}{12}$

18. $\frac{5}{6}, \frac{4}{5}, \frac{7}{10}$

Complete each statement using the symbols <, =, or >.

19. $\frac{5}{12} \square \frac{3}{8}$

20. $\frac{3}{7} \square \frac{9}{21}$

21. $\frac{9}{16} \square \frac{7}{12}$

3.3 *Write as equivalent fractions with the LCD as a common denominator.*

22. $\frac{1}{6}, \frac{7}{8}$

23. $\frac{2}{3}, \frac{4}{5}$

24. $\frac{3}{10}, \frac{5}{8}, \frac{7}{12}$

Find the least common denominator (LCD) for fractions with the given denominators.

25. 12 and 18

26. 20 and 24

27. 25 and 40

28. 6 and 24

29. 3, 4, and 11

30. 2, 5, and 8

31. 3, 6, and 8

32. 4, 5, and 9

Add.

33. $\frac{3}{8} + \frac{5}{12}$

34. $\frac{5}{36} + \frac{7}{24}$

35. $\frac{2}{15} + \frac{9}{20}$

36. $\frac{9}{14} + \frac{10}{21}$

37. $\frac{7}{15} + \frac{13}{18}$

38. $\frac{12}{25} + \frac{19}{30}$

39. $\frac{1}{2} + \frac{1}{4} + \frac{1}{8}$

40. $\frac{1}{3} + \frac{1}{5} + \frac{1}{10}$

41. $\frac{3}{8} + \frac{5}{12} + \frac{7}{18}$

42. $\frac{5}{6} + \frac{8}{15} + \frac{9}{20}$

Subtract.

43. $\frac{8}{9} - \frac{3}{9}$

44. $\frac{9}{10} - \frac{6}{10}$

45. $\frac{5}{8} - \frac{1}{8}$

46. $\frac{11}{12} - \frac{7}{12}$

47. $\frac{7}{8} - \frac{2}{3}$

48. $\frac{5}{6} - \frac{3}{5}$

49. $\frac{11}{18} - \frac{2}{9}$

50. $\frac{5}{6} - \frac{1}{4}$

51. $\frac{5}{8} - \frac{1}{6}$

52. $\frac{13}{18} - \frac{5}{12}$

53. $\frac{8}{21} - \frac{1}{14}$

54. $\frac{13}{18} - \frac{7}{15}$

55. $\frac{11}{12} - \frac{1}{4} - \frac{1}{3}$

56. $\frac{13}{15} + \frac{2}{3} - \frac{3}{5}$

3.4 *Evaluate each expression.*

57. $6\frac{5}{7} + 3\frac{4}{7}$

58. $4\frac{5}{8} - 4\frac{3}{8}$

59. $5\frac{7}{10} + 3\frac{11}{12}$

60. $9\frac{1}{6} - 3\frac{1}{8}$

61. $7\frac{4}{9} - 3\frac{7}{9}$

62. $8\frac{5}{8} - 8\frac{3}{8}$

63. $4\frac{3}{10} - 2\frac{7}{12}$

64. $2\frac{11}{18} - 2\frac{2}{9}$

65. $11\frac{3}{5} - 2\frac{4}{5}$

66. $3\frac{7}{10} - 3\frac{7}{12}$

67. $8 - 4\frac{3}{4}$

68. $6\frac{5}{12} - 4$

69. $2\frac{1}{2} + 3\frac{5}{6} + 3\frac{3}{8}$

70. $2\frac{1}{3} + 5\frac{1}{6} - 2\frac{4}{5}$

Solve each application.

71. **CRAFTS** A recipe calls for $\frac{1}{3}$ cup of milk. You have $\frac{3}{4}$ cup. How much milk will be left over?

72. **CONSTRUCTION** Bradley needs two shelves, one $32\frac{3}{8}$ in. long and the other $36\frac{11}{8}$ in. long. How much total shelving does he need?

73. **GEOMETRY** Find the perimeter of the triangle.

$5\frac{3}{8}$ in. $6\frac{7}{16}$ in. $7\frac{3}{4}$ in.

74. **STATISTICS** At the beginning of one year Miguel was $51\frac{3}{4}$ in. tall. That June, he measured $53\frac{1}{8}$ in. How much did he grow during that period?

75. **CONSTRUCTION** A bookshelf that is $42\frac{5}{16}$ in. long is cut from a board with a length of 8 ft. If $\frac{1}{8}$ in. is wasted in the cut, what length board remains?

76. **CRAFTS** Amelia buys an 8-yd roll of wallpaper on sale. After measuring, she finds that she needs the following amounts of the paper: $2\frac{1}{3}$, $1\frac{1}{2}$, and $3\frac{3}{4}$ yd. Does she have enough for the job? If so, how much will be left over?

77. **CRAFTS** Roberto used $1\frac{3}{4}$ gal of paint in his living room, $1\frac{1}{3}$ gal in the dining room, and $\frac{1}{2}$ gal in a hallway. How much paint did he use?

78. **CONSTRUCTION** A sheet of plywood consists of two outer sections that are $\frac{3}{16}$ in. thick and a center section that is $\frac{3}{8}$ in. thick. How thick is the plywood overall?

3.5 *Evaluate.*

79. $\frac{3}{4} + \left(\frac{1}{2}\right)\left(\frac{1}{3}\right)$

80. $\frac{2}{3} - \left(\frac{3}{4}\right)^2\left(\frac{1}{2}\right)$

81. $\left(4\frac{3}{8}\right)\left(1\frac{1}{2}\right) \div \left(\frac{3}{4} + \frac{1}{2}\right)$

82. $\frac{1}{3} \cdot \left(\frac{3}{4} - \frac{1}{2}\right)^2$

83. $\left(\frac{2}{3}\right)^3 - \frac{1}{2} \cdot \frac{1}{9}$

84. $2\frac{1}{3} - \left[\frac{1}{2} \cdot \left(2 - \frac{1}{3}\right)\right]$

CHAPTER 3

chapter test 3

Use this chapter test to assess your progress and to review for your next exam. Allow yourself about an hour to take this test. The answers to these exercises are in the Answers Appendix at the back of the text.

Find the least common denominator for fractions with the given denominators.

1. 12 and 15

2. 3, 4, and 18

Evaluate each expression.

3. $\frac{2}{5} + \frac{4}{10}$

4. $7\frac{3}{8} - 5\frac{5}{8}$

5. $7 - 5\frac{7}{15}$

6. $\frac{7}{18} - \frac{5}{18}$

7. $\frac{3}{8} + \frac{5}{12}$

8. $7\frac{1}{8} - 3\frac{1}{6}$

9. $\frac{7}{9} - \frac{4}{9}$

10. $\frac{1}{4} + \frac{5}{8} + \frac{7}{10}$

11. $5\frac{3}{10} + 2\frac{4}{10}$

12. $\frac{5}{24} + \frac{3}{8}$

13. $6\frac{3}{8} + 5\frac{7}{10}$

14. $7\frac{3}{8} + 2\frac{7}{8}$

15. $3\frac{5}{6} - 2\frac{2}{9}$

16. $\frac{3}{10} + \frac{6}{10}$

17. $\frac{1}{6} + \frac{3}{7}$

18. $\frac{11}{12} - \frac{3}{20}$

19. $4\frac{2}{7} + 3\frac{3}{7} + 1\frac{3}{7}$

20. $4\frac{1}{6} + 3\frac{3}{4}$

21. $\frac{5}{12} + \frac{3}{12}$

22. $3\frac{1}{2} + 4\frac{3}{4} + 5\frac{3}{10}$

23. $\frac{11}{15} + \frac{9}{20}$

24. $\frac{1}{4} + \left(\frac{1}{2}\right)^2 - \left(\frac{1}{3} + \frac{1}{12}\right)$

25. $\left(1\frac{1}{3} + 2\frac{1}{2}\right) \cdot 3\frac{1}{4}$

26. Find the least common multiple of 18, 24, and 36.

Solve each application.

27. **Statistics** You have $\frac{5}{6}$ hour (hr) to take a three-part test. You use $\frac{1}{3}$ hr for the first section and $\frac{1}{4}$ hr for the second. How much time do you have left to finish the last section of the test?

28. **Crafts** A recipe calls for $\frac{1}{2}$ cup of raisins, $\frac{1}{4}$ cup of walnuts, and $\frac{2}{3}$ cup of rolled oats. What is the total amount of these ingredients?

29. **Statistics** The average coffee drinker has about $3\frac{1}{5}$ cups of coffee per day. If a person works 5 days a week for 50 weeks every year, estimate how many cups of coffee that person will drink in a working lifetime of $51\frac{3}{4}$ years.

30. **Business and Finance** A worker has $2\frac{1}{6}$ hr of overtime on Tuesday, $1\frac{3}{4}$ hr on Wednesday, and $1\frac{5}{6}$ hr on Friday. What is the total overtime for the week?

cumulative review chapters 1–3

Use this exercise set to review concepts from earlier chapters. While it is not a comprehensive exam, it will help you identify any material that you need to review before moving on to the next chapter. In addition to the answers, you will find section references for these exercises in the Answers Appendix in the back of the text.

1.2

Perform the indicated operations.

1. $\begin{array}{r} 1{,}369 \\ +\ 5{,}804 \\ \hline \end{array}$

2. $\begin{array}{r} 489 \\ 562 \\ 613 \\ +254 \\ \hline \end{array}$

3. $\begin{array}{r} 357 \\ 28 \\ +\ 2{,}346 \\ \hline \end{array}$

4. $\begin{array}{r} 13 \\ 2{,}543 \\ +\ 10{,}547 \\ \hline \end{array}$

1.3

5. 289 − 54

6. 53,294 − 41,074

7. 503 − 74

8. 5,731 − 2,492

1.5

9. $\begin{array}{r} 58 \\ \times\ 3 \\ \hline \end{array}$

10. Find the product of 273 and 7.

11. $\begin{array}{r} 89 \\ \times\ 56 \\ \hline \end{array}$

12. $\begin{array}{r} 538 \\ \times\ 103 \\ \hline \end{array}$

1.6

13. $281\overline{)6{,}935}$

14. $571\overline{)12{,}583}$

15. $293\overline{)61{,}382}$

1.7

Evaluate each expression.

16. $12 \div 6 + 3$ **17.** $4 + 12 \div 4$ **18.** $3^3 \div 9$ **19.** $28 \div 7 \times 4$

20. $26 - 2 \times 3$ **21.** $36 \div (3^2 + 3)$

2.3

Identify the proper fractions, improper fractions, and mixed numbers.

$\frac{5}{7}, \frac{15}{9}, 4\frac{5}{6}, \frac{8}{8}, \frac{11}{1}, \frac{2}{5}, 3\frac{5}{6}$

22. Proper fractions: Improper fractions: Mixed numbers:

Rewrite as mixed or whole numbers.

23. $\frac{16}{9}$ **24.** $\frac{36}{5}$

Rewrite as improper fractions.

25. $5\frac{3}{4}$ **26.** $6\frac{1}{9}$

2.4

Determine whether each pair of fractions is equivalent.

27. $\frac{8}{32}, \frac{9}{36}$ **28.** $\frac{6}{11}, \frac{7}{9}$

2.5

Multiply.

29. $\frac{7}{15} \times \frac{5}{21}$ **30.** $\frac{10}{27} \times \frac{9}{20}$ **31.** $4 \times \frac{3}{8}$ **32.** $3\frac{2}{5} \times \frac{5}{8}$

33. $5\frac{1}{3} \times 1\frac{4}{5}$ **34.** $1\frac{5}{12} \times 8$ **35.** $3\frac{1}{5} \times \frac{7}{8} \times 2\frac{6}{7}$

2.6

Divide.

36. $\frac{5}{12} \div \frac{5}{8}$ **37.** $\frac{7}{15} \div \frac{14}{25}$ **38.** $\frac{9}{20} \div 2\frac{2}{5}$

3.1, 3.3

Add.

39. $\frac{4}{15} + \frac{8}{15}$ **40.** $\frac{7}{25} + \frac{8}{15}$ **41.** $\frac{2}{5} + \frac{3}{4} + \frac{5}{8}$

Perform the indicated operations.

42. $\frac{17}{20} - \frac{7}{20}$ **43.** $\frac{5}{9} - \frac{5}{12}$ **44.** $\frac{5}{18} + \frac{4}{9} - \frac{1}{6}$

3.4

Perform the indicated operations.

45. $3\frac{5}{7} + 2\frac{4}{7}$

46. $4\frac{7}{8} + 3\frac{1}{6}$

47. $8\frac{1}{9} - 3\frac{5}{9}$

48. $7\frac{7}{8} - 3\frac{5}{6}$

49. $9 - 5\frac{3}{8}$

50. $3\frac{1}{6} + 3\frac{1}{4} - 2\frac{7}{8}$

Solve each application.

51. **Business and Finance** In his part-time job, Manuel worked $3\frac{5}{6}$ hours (hr) on Monday, $4\frac{3}{10}$ hr on Wednesday, and $6\frac{1}{2}$ hr on Friday. Find the number of hours that he worked during the week.

52. **Crafts** A $6\frac{1}{2}$-in. bolt is placed through a wall that is $5\frac{7}{8}$ in. thick. How far does the bolt extend beyond the wall?

53. **Statistics** On a 6-hr trip, Carlos drove $1\frac{3}{4}$ hr and then Maria drove for another $2\frac{1}{3}$ hr. How many hours remained on the trip?

CHAPTER

4

Decimals

INTRODUCTION

We encounter fractions and mixed numbers written in decimal form every day. Money is the most obvious example of decimals as one quarter of a dollar or 25¢ is usually written as \$0.25 and ten and one half dollars is written as \$10.50.

Because of improvements in our ability to measure time, we use fractions of a second, written as decimals, to give racing times. In fact, decimals are often the form of choice when reporting statistics from sporting events.

You will explore using decimals to report sports statistics when you look at the Tour de France, a famous bicycle race, in Activity 11.

CHAPTER 4 OUTLINE

4 prerequisite check

This Prerequisite Check highlights the skills you will need in order to be successful in this chapter. The answers to these exercises are in the Answers Appendix at the back of the text.

Write the name of each number in words.

1. $3\frac{7}{10}$

2. $6\frac{29}{100}$

3. $17\frac{89}{1,000}$

Evaluate each expression.

4. $183 + 5 + 69$

5. $213 - 49$

6. 426×15

7. $792 \div 10$

8. $792 \div 100$

Find the area of each figure.

9. A rectangle has length 17 ft and width 8 ft.

10. A square has sides of length 15 in.

Round each number, as indicated.

11. 23,456 to the nearest thousand

12. 6,950 to the nearest hundred

Consider the number 4,913,457.

13. Give the place value of 1.

14. Give the place value of 5.

4.1 Place Value and Rounding

< 4.1 Objectives >

1 > Write a number in decimal form

2 > Identify place value in a decimal fraction

3 > Write a decimal as a fraction or mixed number

4 > Compare the sizes of decimals

5 > Round a decimal

In Chapters 2 and 3, we looked at fractions. Now we turn to a special kind of fraction called a **decimal fraction.** A decimal fraction is a fraction whose denominator is a *power of* 10. Some examples of decimal fractions are $\frac{3}{10}$, $\frac{45}{100}$, and $\frac{123}{1,000}$.

When discussing our place-value system, we noted that the place value of each digit increases by a factor of 10 as we move left. For instance, if we consider the number 379, 7 is in the tens place and 9 is in the ones place. The place value of 7 is 10 times the place value of 9 so that we get $(7 \times 10) + (9 \times 1)$. Similarly, the place value of 3 is hundreds, which is 10 times the place value of 7, so 379 is really $(3 \times 100) + (7 \times 10) + (9 \times 1)$.

As we move left, each place is 10 times the place to its right. If we reverse this thinking, we get that each value is *one-tenth* the value of the place to its left. So, in 379, 7 has a place value equal to one-tenth the place value of 3 because tens are one-tenth of 100.

Example 1 — Identifying Place Values

RECALL

The powers of 10 are 1, 10, 100, 1,000, and so on

Label the place values for the number 538.

5	3	8
↑	↑	↑
Hundreds	Tens	Ones

The ones place value is $\frac{1}{10}$ of the tens place value; the tens place value is $\frac{1}{10}$ of the hundreds place value; and so on.

Check Yourself 1

Label the place values for the number 2,793.

NOTE

The decimal point separates the whole-number part and the fraction part of a decimal fraction.

We want to extend this idea *to the right* of the ones place. Write a period to the *right* of the ones place. This is called a **decimal point.** Each digit to the right of that decimal point represents a fraction whose denominator is a power of 10. The first place to the right of the decimal point is the tenths place:

$$0.1 = \frac{1}{10}$$

Example 2 — Writing a Number in Decimal Form

< Objective 1 >

Write the mixed number $3\frac{2}{10}$ in decimal form.

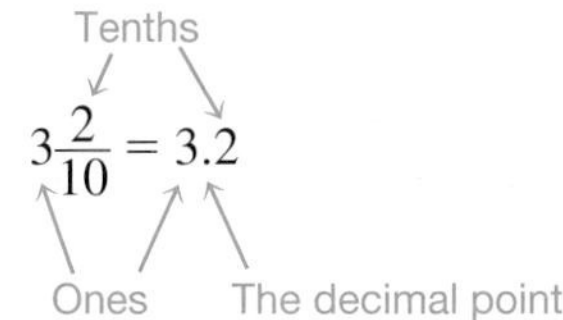

Check Yourself 2

Write $5\frac{3}{10}$ in decimal form.

As you move farther to the *right,* each place value must be $\frac{1}{10}$ of the place value to its left. The second place value is hundredths $\left(0.01 = \frac{1}{100}\right)$. The next place is thousandths, the fourth position is the ten-thousandths place, and so on. The figure illustrates the value of each position as we move to the right of the decimal point.

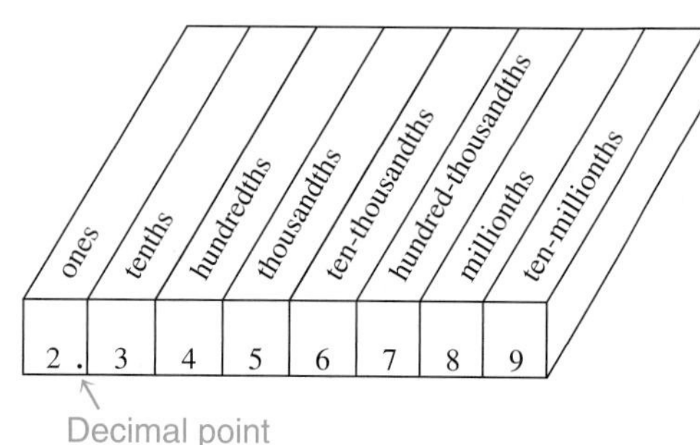

Example 3 Identifying Place Values

< Objective 2 >

What are the place values of the 4 and 6 in the decimal 2.34567?

The place value of 4 is hundredths, and the place value of 6 is ten-thousandths.

NOTE

For convenience we shorten the term *decimal fraction* to *decimal* from this point on.

Check Yourself 3

What is the place value of 5 in the decimal of Example 3?

Understanding place values allows you to read and write decimals. You can use these steps.

Step by Step

Reading and Writing Decimals in Words

Step 1 Read the digits *to the left* of the decimal point as a whole number.

Step 2 Read the decimal point as the word *and*.

Step 3 Read the digits *to the right* of the decimal point as a whole number followed by the place value of the rightmost digit.

Example 4 Writing a Decimal in Words

Write each decimal in words.

5.03 is read "five and three hundredths."

↑ Hundredths — The rightmost digit, 3, is in the hundredths position.

12.057 is read "twelve and fifty-seven thousandths."

↑ Thousandths — The rightmost digit, 7, is in the thousandths position.

0.5321 is read "five thousand, three hundred twenty-one ten-thousandths."

NOTE

If there are *no* nonzero digits to the left of the decimal point, start directly with step 3.

NOTES

An informal way of reading decimals is to simply read the digits in order and use the word *point* to indicate the decimal point. 2.58 can be read "two point five eight." 0.689 can be read "zero point six eight nine."

The number of digits to the right of the decimal point is called the number of **decimal places** in a decimal. So, 0.35 has two decimal places.

When the decimal has no whole-number part, we write a 0 to the left of the decimal point. This helps us avoid careless errors such as missing the decimal point. However, both 0.5321 and .5321 are correct.

Check Yourself 4

Write 2.58 in words.

One quick way to write a decimal as a fraction is to remember that the number of decimal places must be the same as the number of zeros in the denominator of the fraction.

Example 5 Writing a Decimal in Fraction Form

< Objective 3 >

Write each decimal as a fraction or mixed number.

$$0.35 = \frac{35}{100}$$

↑ Two places ↑ Two zeros

Of course, our final answer should be in simplest terms.

$$0.35 = \frac{35}{100}$$

$$= \frac{7}{20} \qquad \text{Divide the numerator and denominator by 5.}$$

NOTE

The 0 to the right of the decimal point is a "placeholder" that is not needed in the fraction form.

The same method can be used with decimals that are greater than 1. Here, the result is a mixed number.

$$2.057 = 2\frac{57}{1{,}000}$$

↑ Three places ↑ Three zeros

Check Yourself 5

Write as fractions or mixed numbers.

(a) 0.527 **(b)** 5.08

RECALL

By the fundamental principle of fractions, multiplying the numerator and denominator of a fraction by the same nonzero number does not change the value of the fraction.

It is often useful to compare the sizes of two decimals. One approach to comparing decimals uses the fundamental principle of fractions.

Adding zeros to the right *does not change* the value of a decimal. The number 0.53 is the same as 0.530.

$$\frac{53}{100} = \frac{530}{1{,}000}$$

The fractions are equivalent because we multiplied both the numerator and denominator by 10.

This allows us to compare decimals as shown in Example 6.

Example 6 Comparing the Sizes of Two Decimals

< Objective 4 >

Which is larger?

0.84 or 0.842

Write 0.84 as 0.840. Then we see that 0.842 (or 842 thousandths) is greater than 0.840 (or 840 thousandths), and we write

$0.842 > 0.84$

Check Yourself 6

Use the symbols < or > to complete the statement.

0.588 ______ 0.59

Whenever a measurement is made, it is not exact. It is correct only to a certain number of places and is called an **approximate number.** Usually, we want to make all decimals in a particular problem precise to a specified decimal place or tolerance. This requires **rounding** the decimals. We can picture the process on a number line.

Example 7 Rounding Decimals

< Objective 5 >

3.74 rounds to the nearest tenth as 3.7.
3.78 rounds to 3.8.

NOTE

3.74 is closer to 3.7 than it is to 3.8, while 3.78 is closer to 3.8.

Check Yourself 7

Round 3.77 to the nearest tenth.

Rather than using a number line, we follow a process to round decimals.

Step by Step

Rounding a Decimal

Step 1 Find the place where the decimal is to be rounded.

Step 2 If the next digit to the right is 5 or more, increase the digit in the place you are rounding to by 1. Discard remaining digits to the right.

Step 3 If the next digit to the right is less than 5, just discard that digit and any remaining digits to the right.

Example 8 Rounding Decimals

Round 34.58 to the nearest tenth.

34.58 Locate the digit you are rounding to. The 5 is in the tenths place.

Because the next digit to the right, 8, is 5 or more, increase the tenths digit by 1. Then discard the remaining digits.

34.58 rounds to 34.6.

NOTE

Many students find it easiest to mark the digit they are rounding to with an arrow.

Check Yourself 8

Round 48.82 to the nearest tenth.

Example 9 Rounding Decimals

Round 5.673 to the nearest hundredth.

5.673 The 7 is in the hundredths place.

The next digit to the right, 3, is less than 5. Leave the hundredths digit as it is and discard the remaining digits to the right.

5.673 rounds to 5.67.

Check Yourself 9

Round 29.247 to the nearest hundredth.

Sometimes, instead of rounding to a place value, we are instructed to round to a certain number of decimal places. For example, instead of "round to the nearest hundredth," we might be told to "round to two decimal places." Do you see that these instructions are the same?

Example 10 Rounding to a Decimal Place

NOTE

The fourth place to the *right* of the decimal point is the ten-thousandths place.

Round 3.14159 to four decimal places.

3.14159 The 5 is in the ten-thousandths place.

The next digit to the right, 9, is 5 or more, so increase the digit you are rounding to by 1. Discard the remaining digits to the right.

3.14159 rounds to 3.1416.

Check Yourself 10

Round 0.8235 to three decimal places.

We need to be careful when there is a 9 in the rounding place. There are no problems if we round *down,* but rounding *up* requires some consideration.

Example 11 Rounding Decimals with Nines

NOTE

Some fields distinguish between 11.05 and 11.050. In such applications, the extra zero indicates the precision of a measurement.

We do not make such a distinction in this text. You should ask if your instructor has a preference.

(a) Round 2.392 to the nearest hundredth.

We see 9 in the hundredths place. The digit to the right is 2, so we drop the digit and leave 9 as it is.

2.392 rounds to 2.39.

(b) Round 11.04961 to three decimal places.

Again, we see 9 in the rounding place. The digit to the right is 6, so we increase 9 by 1 and drop the remaining digits. However, increasing 9 by 1 makes it 10, just like when rounding whole numbers.

11.04961 rounds to 11.050 or 11.05.

Check Yourself 11

(a) Round 15.1992 to the nearest hundredth.
(b) Round 4.918 to one decimal place.

Many applications require us to use decimals.

Example 12 Using Decimals in Applications

A distribution center is putting together a shipment containing items from several sources. Each source provides its own specifications sheet with the weight of its items.

The individual items in the shipment are listed as weighing 5.34 lb, 11.2 lb, 3.071 lb, and 6.96 lb. Because the least precise number is given to the nearest tenth (11.2 lb), it is decided to list all of the weights rounded to the nearest tenth. List the four weights rounded to the nearest tenth.

5.34 lb rounds to 5.3 lb.
11.2 lb is already given to the nearest tenth.
3.071 lb rounds to 3.1 lb.
6.96 lb rounds to 7 lb (or 7.0 lb in this type of application).

Check Yourself 12

Another shipment contains items with weights listed as 1.38 lb, 5.928 lb, 7.97 lb, and 11.421 lb. Round each weight to the nearest tenth.

Check Yourself ANSWERS

1. 2, 7 9 3 (2: Thousands; 7: Hundreds; 9: Tens; 3: Ones) **2.** 5.3 **3.** Thousandths **4.** Two and fifty-eight hundredths

5. **(a)** $\frac{527}{1{,}000}$; **(b)** $5\frac{2}{25}$ **6.** $0.588 < 0.59$ **7.** 3.8 **8.** 48.8 **9.** 29.25 **10.** 0.824

11. **(a)** 15.2; **(b)** 4.9 **12.** 1.4 lb; 5.9 lb; 8 lb or 8.0 lb; 11.4 lb

Reading Your Text

These fill-in-the-blank exercises will help you understand some of the key vocabulary used in this section. The answers to these exercises are in the Answers Appendix at the back of the text.

(a) A ______________ fraction is a fraction whose denominator is a power of 10.

(b) The number of digits to the right of the decimal point is called the number of decimal ______________.

(c) Whenever a decimal represents a measurement made by some instrument, the decimals are not ______________.

(d) The fourth place to the right of the decimal point is called the ___________ place.

4.1 exercises

< Objective 1 >

Write in decimal form.

1. $\frac{23}{100}$

2. $\frac{371}{1{,}000}$

3. $\frac{209}{10{,}000}$ 

4. $3\frac{5}{10}$

5. $23\frac{56}{1{,}000}$

6. $7\frac{431}{10{,}000}$

7. $\frac{2}{10}$

8. $2\frac{8}{100}$

9. $\frac{53}{10}$

10. $\frac{3{,}409}{1{,}000}$

< Objective 2 >

Consider the decimal 8.57932.

11. What is the place value of 7?

12. What is the place value of 5?

13. What is the place value of 3? 

14. What is the place value of 2?

Consider the number 32.06197.

15. What is the place value of 1?

16. What is the place value of 0?

17. What is the place value of 6?

18. What is the place value of 7?

Write in decimal form.

19. Fifty-one thousandths 

20. Two hundred fifty-three ten-thousandths

21. Seven and three tenths

22. Twelve and two hundred forty-five thousandths

23. Eighteen and seven tenths

24. Two hundred forty and twenty-four thousandths

Write in words.

25. 0.23

26. 0.371

27. 0.071 

28. 0.0251

29. 12.07

30. 23.056

< Objective 3 >

Write each number as a fraction or mixed number.

31. 0.65

32. 0.00765

33. 5.231 

34. 4.0171

35. 0.08

36. 0.5

37. 7.25

38. 15.375

< Objective 4 >

Use the symbols <, =, or > to complete each statement.

39. 0.69 ______ 0.689 **40.** 0.75 ______ 0.752 **41.** 1.23 ______ 1.230 **42.** 2.451 ______ 2.45

43. 10 ______ 9.9 **44.** 4.98 ______ 5 **45.** 1.459 ______ 1.46 **46.** 0.235 ______ 0.2350

Arrange in order from smallest to largest.

47. 4.0339, 4.034, $4\frac{3}{10}$, $\frac{432}{100}$, 4.33

48. $\frac{38}{1,000}$, 0.0382, 0.04, 0.37, $\frac{39}{100}$

49. 1, 1.01, $1\frac{1}{10}$, 1.11, $1\frac{11}{1,000}$, $1\frac{111}{1,000}$

50. 0.99, 0.989, $\frac{9}{100}$, $\frac{909}{1,000}$, 0.099, 0.999

51. 0.71, 0.072, $\frac{7}{10}$, 0.007, 0.0069, $\frac{7}{100}$, 0.0701, 0.0619, 0.0712

52. 2.05, $\frac{25}{10}$, 2.0513, 2.059, $\frac{251}{100}$, 2.0515, 2.052, 2.051

< Objective 5 >

Round, as indicated.

53. 21.534 hundredths **54.** 5.842 tenths **55.** 0.342 hundredths **56.** 2.3576 thousandths

57. 2.71828 thousandths **58.** 1.543 tenths **59.** 2.942 tenths **60.** 0.09925 thousandths

61. 0.0475 tenths **62.** 0.85356 ten-thousandths **63.** 4.85344 ten-thousandths

64. 52.8728 thousandths **65.** 2.95 tenths **66.** 0.09625 hundredths

67. 6.734 two decimal places **68.** 12.5467 three decimal places **69.** 6.58739 four decimal places

70. 503.824 two decimal places **71.** 0.59962 thousandths **72.** 16.9951 hundredths

Round to the nearest cent or dollar, as indicated.

73. \$235.1457 cent **74.** \$1,847.9895 cent **75.** \$752.512 dollar

76. \$5,642.4958 dollar **77.** \$49.605 dollar **78.** \$42.696 cent

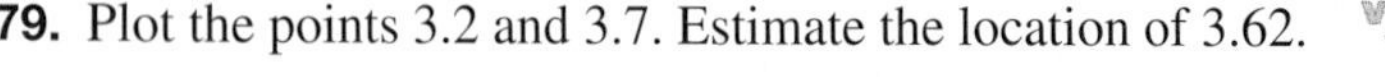

79. Plot the points 3.2 and 3.7. Estimate the location of 3.62. 

80. Plot the points 12.51 and 12.58. Estimate the location of 12.537.

81. Plot the points 7.124 and 7.127. Estimate the location of 7.1253.

82. Plot the points 5.73 and 5.74. Estimate the location of 5.782.

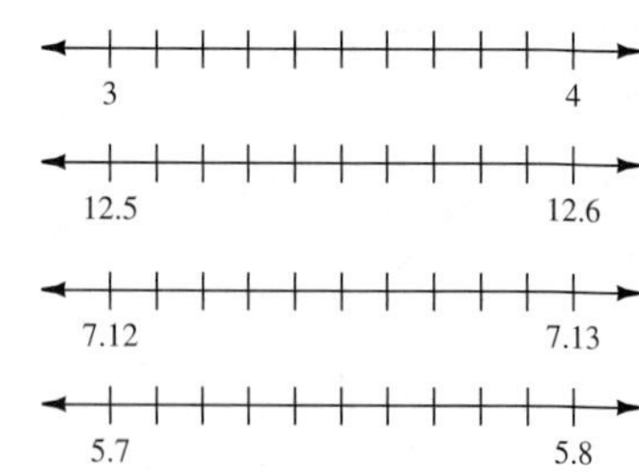

Label each statement as **true** *or* **false.**

83. The only number between 8.6 and 8.8 is 8.7.

84. The smallest number that is greater than 98.6 is 98.7.

85. The place value immediately to the right of the decimal point is tenths.

86. The number 4.586 should always be rounded as 4.6.

Skills	Calculator/Computer	**Career Applications**	Above and Beyond

87. **Allied Health** A nurse calculates a child's dose of Reglan to be 1.53 mg. Round this dose to the nearest tenth of a milligram.

88. **Allied Health** A nurse calculates a young boy's dose of Dilantin to be 23.375 mg every 5 min. Round this dose to the nearest hundredth of a milligram.

89. **Electronics** Write each number as a decimal.

(a) Ten and thirty-five hundredths volts (V)

(b) Forty-seven hundred-thousandths of a farad (F)

(c) One hundred fifty-eight ten-thousandths of a henry (H)

90. **Information Technology** Josie needs to check connectivity of a PC on the network. She uses a tool called ping to see if the PC is configured properly. She receives three readings from ping: 2.1, 2.2, and 2.3 seconds. Convert the decimals to fractions and simplify if needed.

91. **Manufacturing** Put the mill bits in order from smallest to largest.

0.308, 0.297, 0.31, 0.3, 0.311, 0.32

92. **Manufacturing** A drill size is listed as $\frac{372}{1{,}000}$. Express this as a decimal.

Skills	Calculator/Computer	Career Applications	**Above and Beyond**

93. **(a)** What is the difference in the values of 0.120, 0.1200, and 0.12000?

(b) Explain in your own words why placing zeros to the right of a decimal point does not change the value of the number.

94. Lula wants to round 76.24491 to the nearest hundredth. She first rounds 76.24491 to 76.245 and then rounds 76.245 to 76.25 and claims that this is the final answer. What is wrong with this approach?

Answers

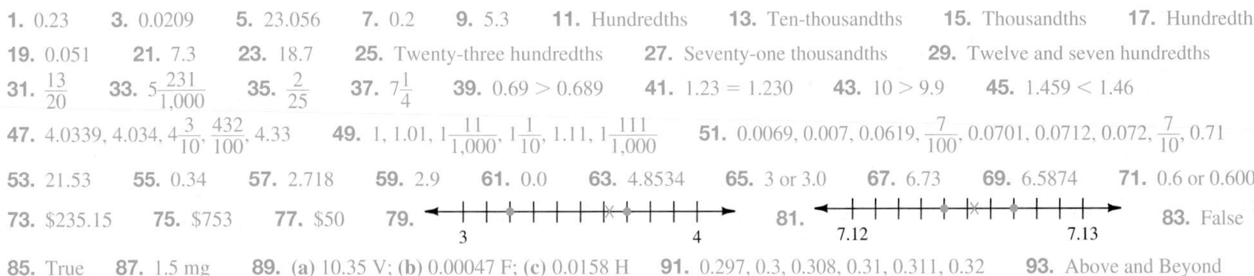

1. 0.23 **3.** 0.0209 **5.** 23.056 **7.** 0.2 **9.** 5.3 **11.** Hundredths **13.** Ten-thousandths **15.** Thousandths **17.** Hundredths **19.** 0.051 **21.** 7.3 **23.** 18.7 **25.** Twenty-three hundredths **27.** Seventy-one thousandths **29.** Twelve and seven hundredths **31.** $\frac{13}{20}$ **33.** $5\frac{231}{1{,}000}$ **35.** $\frac{2}{25}$ **37.** $7\frac{1}{4}$ **39.** $0.69 > 0.689$ **41.** $1.23 = 1.230$ **43.** $10 > 9.9$ **45.** $1.459 < 1.46$ **47.** 4.0339, 4.034, $4\frac{3}{10}$, $\frac{432}{100}$, 4.33 **49.** 1, 1.01, $1\frac{11}{1{,}000}$, $1\frac{1}{10}$, 1.11, $1\frac{111}{1{,}000}$ **51.** 0.0069, 0.007, 0.0619, $\frac{7}{100}$, 0.0701, 0.0712, 0.072, $\frac{7}{10}$, 0.71 **53.** 21.53 **55.** 0.34 **57.** 2.718 **59.** 2.9 **61.** 0.0 **63.** 4.8534 **65.** 3 or 3.0 **67.** 6.73 **69.** 6.5874 **71.** 0.6 or 0.600 **73.** $235.15 **75.** $753 **77.** $50 **79.**

81.

83. False **85.** True **87.** 1.5 mg **89.** **(a)** 10.35 V; **(b)** 0.00047 F; **(c)** 0.0158 H **91.** 0.297, 0.3, 0.308, 0.31, 0.311, 0.32 **93.** Above and Beyond

4.2 Adding and Subtracting Decimals

< 4.2 Objectives >

1 > Add decimals

2 > Use decimals to solve applications

3 > Subtract decimals

Working with decimals rather than fractions makes the basic operations much easier. We begin with addition. One method for adding decimals is to write the decimals as fractions, add, and then change the sum back to a decimal.

$$0.34 + 0.52 = \frac{34}{100} + \frac{52}{100} = \frac{86}{100} = 0.86$$

As you should expect, this is not the most efficient way to add decimals. It would be much better if we could just add them based on the place value of each digit, just like we do with whole numbers.

Take a look at the example we just completed.

$$\frac{34}{100} = \frac{30}{100} + \frac{4}{100} = \underbrace{\frac{3}{10}}_{\frac{30}{100}} + \frac{4}{100} \quad \text{and} \quad \frac{52}{100} = \frac{50}{100} + \frac{2}{100} = \underbrace{\frac{5}{10}}_{\frac{50}{100}} + \frac{2}{100}$$

When we add, we are adding by place value.

NOTE

We can think of adding like terms as adding by place value because the denominators are all powers of ten.

$$\frac{34}{100} + \frac{52}{100} = \underbrace{\frac{3}{10} + \frac{4}{100}}_{\frac{34}{100}} + \underbrace{\frac{5}{10} + \frac{2}{100}}_{\frac{52}{100}}$$

$$= \left(\frac{3}{10} + \frac{5}{10}\right) + \left(\frac{4}{100} + \frac{2}{100}\right)$$ Regroup by common denominators.

$$= \frac{8}{10} + \frac{6}{100}$$ Add like fractions.

$$= \frac{80}{100} + \frac{6}{100}$$ Write the remaining fractions with a common denominator.

$$= \frac{86}{100}$$ Add like fractions.

$$= \frac{43}{50}$$ Simplify.

When adding fractions in which the denominators are powers of 10, we are adding by place value, so we can add decimals in the same way we add whole numbers.

When adding decimals, it is important that we add together the digits in the correct place. When adding vertically, we do this by aligning the decimal points to make certain we are adding by the correct place value.

Step by Step

Adding Decimals

Step 1 Write the numbers being added in column form *with their decimal points in a vertical line.*

Step 2 Add just as you would with whole numbers.

Step 3 Place the decimal point of the sum in line with the decimal points of the addends.

Example 1 illustrates this rule.

Example 1 **Adding Decimals**

< Objective 1 >

NOTE

Placing the decimal points in a vertical line ensures that we are adding digits of the same place value.

Add 0.13, 0.42, and 0.31.

$$\begin{array}{r} 0.13 \\ 0.42 \\ +\ 0.31 \\ \hline 0.86 \end{array}$$

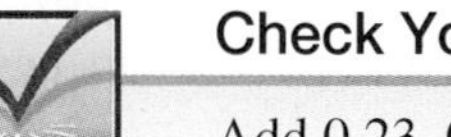

Check Yourself 1

Add 0.23, 0.15, and 0.41.

When adding decimals, we use the *carrying process* just as we did when adding whole numbers.

Example 2 **Adding Decimals**

Add 0.35, 1.58, and 0.67.

$$\begin{array}{r} {\scriptstyle 1\ 2} \\ 0.35 \\ 1.58 \\ +\ 0.67 \\ \hline 2.60 \end{array}$$

1 2 ⟵ Carries

In the hundredths column:
5 + 8 + 7 = 20
Write 0 and carry 2 to the tenths column.
In the tenths column:
2 + 3 + 5 + 6 = 16
Write 6 and carry 1 to the ones column.

The carrying process works with decimals, just as it did with whole numbers, because each place value is 10 times the value of the place to its right.

Check Yourself 2

Add 23.546, 0.489, 2.312, and 6.135.

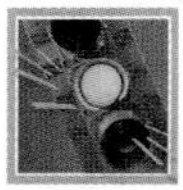

> CAUTION

When adding, we align the decimal points on top of each other. Do not align the numbers on the right.

When adding decimals, the numbers may not have the same number of decimal places. You may find it helpful to fill in as many zeros as needed so that all the numbers added have the same number of decimal places.

Recall that adding zeros to the right *does not change* the value of a decimal. The number 0.53 is the same as 0.530.

Example 3 **Adding Decimals**

NOTE

Be sure that the decimal points are in a vertical line.

Add 0.53, 4, 2.7, and 3.234.

$$\begin{array}{r} 0.53 \\ 4. \\ 2.7 \\ +\ 3.234 \\ \hline \end{array}$$

For a whole number, the decimal is understood to be to its right. So 4 = 4. = 4.0.

Now fill in the missing zeros and add as before.

$$\begin{array}{r} 0.530 \\ 4.000 \\ 2.700 \\ +\ 3.234 \\ \hline 10.464 \end{array}$$

Now all the numbers have *three* decimal places.

Check Yourself 3

Add 6, 2.583, 4.7, and 2.54.

Many applied problems require working with decimals. For instance, filling up at a gas station means reading decimal amounts.

Example 4 An Application with Decimals

< Objective 2 >

On a trip, the Chang family kept track of their gas purchases. They bought 12.3, 14.2, 10.7, and 13.8 gal. How much gas did they buy on the trip?

NOTE

Because we want the total amount, we add.

$$\begin{array}{r} 12.3 \\ 14.2 \\ 10.7 \\ +\ 13.8 \\ \hline 51.0 \text{ gal} \end{array}$$

Check Yourself 4

The Higueras kept track of the gasoline they purchased on a recent trip. If they bought 12.4, 13.6, 9.7, 11.8, and 8.3 gal, how much gas did they buy on the trip?

Because our monetary system is a decimal system, most problems involving money involve decimals.

Example 5 An Application with Decimals

Andre deposits \$3.24, \$15.73, \$50, \$28.79, and \$124.38 during May. How much did he deposit?

$$\begin{array}{r} \$\ \ 3.24 \\ 15.73 \\ 50.00 \\ 28.79 \\ +\ 124.38 \\ \hline \$222.14 \end{array}$$

Add the amounts deposited. We write \$50 as \$50.00.

\$222.14 ⟵ Total deposits for May

Check Yourself 5

Your textbooks for the fall term cost \$63.50, \$78.95, \$43.15, \$82, and \$85.85. What was the total cost of textbooks for the term?

Recall that *perimeter* is the distance around a straight-edged shape. Finding a perimeter often requires that we add decimals.

Example 6 An Application with Decimals

Rachel is going to put a fence around her farm. The figure represents the land, measured in kilometers (km). How much fence does she need to buy?

The perimeter is the sum of the lengths of the sides, so we add those lengths to find the total fencing needed.

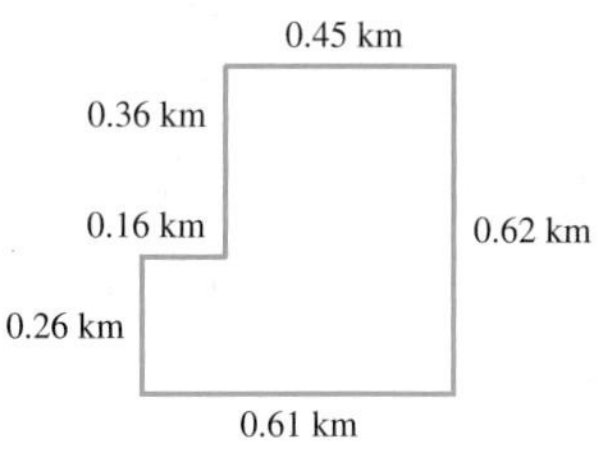

$0.26 + 0.16 + 0.36 + 0.45 + 0.62 + 0.61 = 2.46$

Rachel needs 2.46 km of fence for the perimeter of her farm.

Check Yourself 6

Manuel intends to build a walkway around his garden. What will the total length of the walkway be?

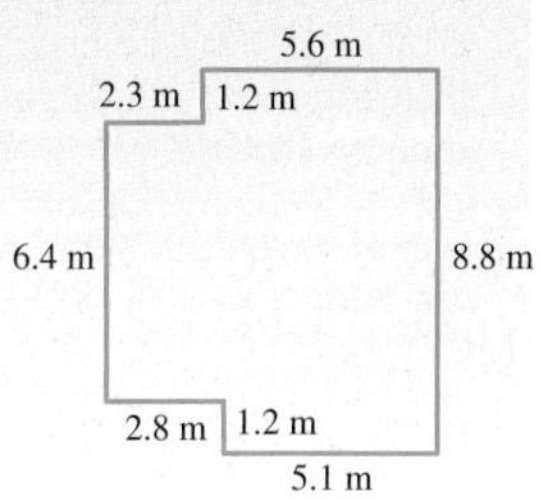

Much of what we have said about adding decimals is also true of subtraction. We can subtract decimals by place value just as we do when adding.

Step by Step

Subtracting Decimals

Step 1 Write the numbers being subtracted in column form *with their decimal points in a vertical line.*

Step 2 Subtract just as you would with whole numbers.

Step 3 Place the decimal point of the difference in line with the decimal points of the numbers being subtracted.

Example 7 illustrates this rule.

Example 7 Subtracting Decimals

< Objective 3 >

RECALL

The number that follows the word *from,* here 3.58, is written first. The number we are subtracting, here 1.23, is then written beneath 3.58.

Subtract 1.23 from 3.58.

$$\begin{array}{r} 3.58 \\ -\ 1.23 \\ \hline 2.35 \end{array}$$

Subtract in the hundredths, the tenths, and then the ones columns.

Check Yourself 7

Subtract $9.87 - 5.45$.

Because each place value is $\frac{1}{10}$ the value of the place to its left, borrowing, when you are subtracting decimals, works just as it did in subtracting whole numbers.

Example 8 Subtracting Decimals

Subtract 1.86 from 6.54.

$$\begin{array}{r} {}^{5}\,{}^{1}4\,{}_{1} \\ \not{6}.\not{5}4 \\ -\ 1.86 \\ \hline 4.68 \end{array}$$

Borrow from the tenths and ones places to do the subtraction.

Check Yourself 8

Subtract $35.35 - 13.89$.

In subtracting decimals, as in adding, we can add zeros to the right of the decimal point so that both decimals have the same number of decimal places.

Example 9 Subtracting Decimals

NOTES

When you are subtracting, align the decimal points, and then add zeros to the right to align the digits.

We rewrite 9 as 9.000.

(a) Subtract 2.36 from 7.5.

$$\begin{array}{r} \overset{4}{7.\cancel{5}}\overset{1}{0} \\ -\ 2.36 \\ \hline 5.14 \end{array}$$

We added a 0 to 7.5. Next, borrow 1 tenth from the 5 tenths in the minuend.

(b) Subtract 3.657 from 9.

$$\begin{array}{r} \overset{8}{\cancel{9}}.\overset{9}{\cancel{0}}\overset{9}{\cancel{0}}\overset{1}{\cancel{0}} \\ -\ 3.657 \\ \hline 5.343 \end{array}$$

In this case, move left to the ones place to begin the borrowing process.

Check Yourself 9

Subtract 5 − 2.345.

We can use subtraction to solve many applications.

Example 10 An Application with Decimals

NOTE

We want to find the difference between the two measurements, so we subtract.

Jonathan was 98.3 cm tall on his sixth birthday. On his seventh birthday he was 104.2 cm. How much did he grow during the year?

$$\begin{array}{r} 104.2\text{ cm} \\ -\ \ 98.3\text{ cm} \\ \hline 5.9\text{ cm} \end{array}$$

Jonathan grew 5.9 cm during the year.

Check Yourself 10

A car's highway mileage before a tune-up was 28.8 miles per gallon (mi/gal). After the tune-up, it measured 30.1 mi/gal. What was the increase in mileage?

The same methods can be used in working with money.

Example 11 An Application with Decimals

NOTE

Sally's change is the *difference* between the price of her purchase and the $20 paid. We use subtraction to find the solution.

Sally buys a package of boxes for $12.37. She pays for her purchase with a $20 bill. How much change should she receive?

$$\begin{array}{r} \$20.00 \\ -\ 12.37 \\ \hline \$\ \ 7.63 \end{array}$$

Add zeros to write $20 as $20.00. Then subtract as before.

Sally should receive $7.63 in change after her purchase.

Check Yourself 11

A stereo system that normally sells for $549.50 is discounted (or marked down) to $499.95 for a sale. What is the savings?

We add and subtract decimals when keeping track of our checking account balance, especially after writing a check or making deposits or withdrawals.

Example 12 An Application with Decimals

Keep a running balance for the check register.

Beginning balance	$234.15
Check # 301	23.88
Balance	______
Check # 302	38.98
Balance	______
Check # 303	114.66
Balance	______
Deposit	175.75
Balance	______
Check # 304	212.55
Ending balance	______

To keep a running balance, we add the deposits and subtract the checks.

Beginning balance	$234.15	
Check # 301	23.88	
Balance	210.27	Subtract.
Check # 302	38.98	
Balance	171.29	Subtract.
Check # 303	114.66	
Balance	56.63	Subtract.
Deposit	175.75	
Balance	232.38	Add.
Check # 304	212.55	
Ending balance	19.83	Subtract.

Check Yourself 12

Keep a running balance for the check register.

	Beginning balance	$398.00
	Check # 401	19.75
(a)	Balance	______
	Check # 402	56.88
(b)	Balance	______
	Check # 403	117.59
(c)	Balance	______
	Deposit	224.67
(d)	Balance	______
	Check # 404	411.48
(e)	Ending balance	______

Check Yourself ANSWERS

1. 0.79 **2.** 32.482 **3.** 15.823 **4.** 55.8 gal **5.** $353.45 **6.** 33.4 m **7.** 4.42 **8.** 21.46 **9.** 2.655 **10.** 1.3 mi/gal **11.** $49.55 **12.** **(a)** $378.25; **(b)** $321.37; **(c)** $203.78; **(d)** $428.45; **(e)** $16.97

Reading Your Text

These fill-in-the-blank exercises will help you understand some of the key vocabulary used in this section. The answers to these exercises are in the Answers Appendix at the back of the text.

(a) To add decimals, write the numbers being added in column form with their ________ ________ in a vertical line.

(b) Adding zeros to the right does not change the ________ of a decimal.

(c) ________ is the distance around a straight-edged shape.

(d) When subtracting one number from another, the number ________ the word *from* is written first.

4.2 exercises

Skills | Calculator/Computer | Career Applications | Above and Beyond

< Objective 1 >

Add.

1. $\begin{array}{r} 0.28 \\ +\ 0.79 \\ \hline \end{array}$ 

2. $\begin{array}{r} 2.59 \\ +\ 0.63 \\ \hline \end{array}$

3. $\begin{array}{r} 13.58 \\ 7.239 \\ +\ \ 1.5 \\ \hline \end{array}$

4. $\begin{array}{r} 8.625 \\ 2.45 \\ +\ 12.6 \\ \hline \end{array}$

5. $\begin{array}{r} 25.3582 \\ 6.5 \\ 1.898 \\ +\ \ 0.69 \\ \hline \end{array}$

6. $\begin{array}{r} 1.336 \\ 15.6857 \\ 7.9 \\ +\ \ 0.85 \\ \hline \end{array}$

7. $0.86 + 5.91$

8. $3.238 + 11.9293$

9. $5 + 0.7$

10. $0.92 + 8$

11. $4.743 + 12$

12. $11 + 9.31$

13. $0.43 + 0.8 + 0.561$

14. $1.25 + 0.7 + 0.259$

15. $42.731 + 1.058 + 103.24$ 

16. $27.4 + 213.321 + 39.38$

< Objective 3 >

Subtract.

17. $\begin{array}{r} 0.85 \\ -\ 0.59 \\ \hline \end{array}$

18. $\begin{array}{r} 5.68 \\ -\ 2.65 \\ \hline \end{array}$

19. $\begin{array}{r} 3.82 \\ -\ 1.565 \\ \hline \end{array}$

20. $\begin{array}{r} 8.59 \\ -\ 5.6 \\ \hline \end{array}$

21. $\begin{array}{r} 7.02 \\ -\ 4.7 \\ \hline \end{array}$

22. $\begin{array}{r} 45.6 \\ -\ 8.75 \\ \hline \end{array}$

23. $\begin{array}{r} 12 \\ -\ 5.35 \\ \hline \end{array}$

24. $\begin{array}{r} 15 \\ -\ 8.85 \\ \hline \end{array}$

25. 5.316 − 2.9

26. 8.1 − 7.9306

27. 7 − 0.5

28. 4.32 − 2

29. 8.1 − 3

30. 8 − 5.999

31. Subtract 2.87 from 6.84.

32. Subtract 3.69 from 10.57.

33. Subtract 7.75 from 9.4.

34. Subtract 5.82 from 12.

35. Subtract 0.24 from 5.

36. Subtract 8.7 from 16.32.

37. Find the difference between 2.4 and 1.4.

38. Find the difference between 8.12 and 3.96.

39. Find the difference between 4.23 and 1.3.

40. Find the difference between 5.4 and 3.271.

< Objective 2 >

Solve each application.

41. **Business and Finance** On a 3-day trip, Dien bought 12.7, 15.9, and 13.8 gal of gas. How many gallons of gas did he buy?

42. **Science and Medicine** Felix ran 2.7 mi on Monday, 1.9 mi on Wednesday, and 3.6 mi on Friday. How far did he run during the week?

43. **Statistics** Rainfall was recorded in centimeters during the winter months, as indicated on the bar graph.

(a) How much rain fell during those months?

(b) How much more rain fell in December than in February?

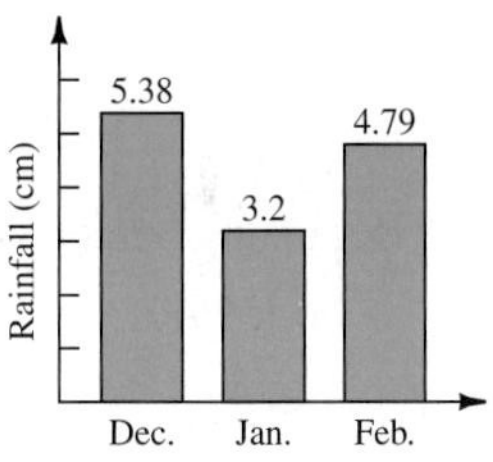

44. **Statistics** During a particularly heavy three-day snowfall, the airport in Bangor, Maine, recorded the daily snowfall, as shown on the chart.

(a) How much snow fell during the three-day period?

(b) How much more snow fell on Friday than on Sunday?

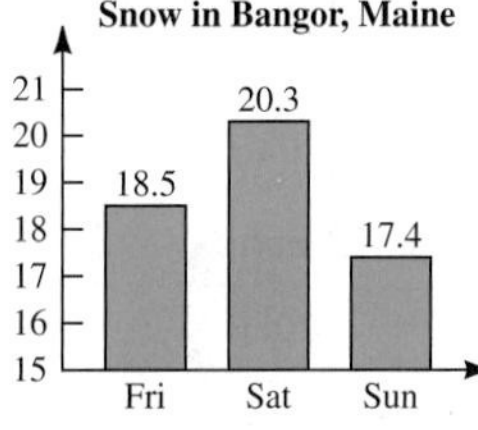

45. **Business and Finance** Nicole recorded her expenses from a business trip.

Expense	Amount
Gas	$78.49
Food	$129.45
Lodging	$149.95
Parking	$8.80

How much did her expenses total?

46. **Construction** A metal fitting has three sections with lengths 2.5, 1.775, and 1.45 in. What is the total length of the fitting?

47. **Construction** Lupe is putting a fence around her yard. Her yard is rectangular and measures 8.16 yd long and 12.68 yd wide. How much fence should Lupe purchase?

48. **Business and Finance** The deposit slip shown indicates the amounts that made up a deposit Peter Rabbit made. What was the total amount of his deposit?

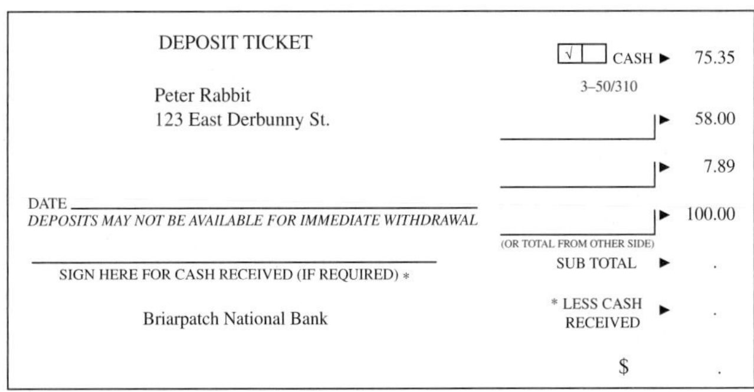

DEPOSIT TICKET

Peter Rabbit
123 East Derbunny St.

√ CASH ► 75.35
3–50/310
► 58.00
► 7.89
► 100.00
(OR TOTAL FROM OTHER SIDE)
SUB TOTAL ► .
* LESS CASH RECEIVED ► .
$.

DATE
DEPOSITS MAY NOT BE AVAILABLE FOR IMMEDIATE WITHDRAWAL

SIGN HERE FOR CASH RECEIVED (IF REQUIRED) *

Briarpatch National Bank

49. **Construction** The figure gives the distances in miles of the boundary sections around a ranch. How much fencing is needed for the property?

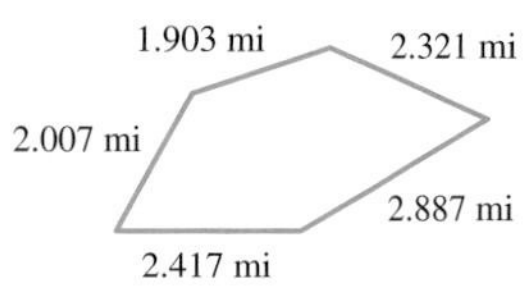

50. **Geometry** Find the perimeter of the figure.

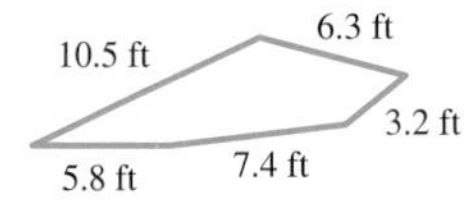

51. **Crafts** The outer radius of a piece of tubing is 2.8325 in. The inner radius is 2.775 in. What is the thickness of the wall of the tubing?

52. **Business and Finance** A television set selling for \$399.50 is discounted (or marked down) to \$365.75. What is the savings?

53. **Business and Finance** You make charges of \$37.25, \$8.78, and \$53.45 on a credit card. If you make a payment of \$73.50, how much do you still owe?

54. **Geometry** Given the figure, find dimension a.

0.65 in.   0.375 in. a 2.000 in.

55. **Business and Finance** Keep a running balance for the check register.

Beginning balance	\$456.00
Check # 601	\$199.29
Balance	
Service charge	\$ 18.00
Balance	
Check # 602	\$ 85.78
Balance	
Deposit	\$250.45
Balance	
Check # 603	\$201.24
Ending balance	

56. **Business and Finance** Keep a running balance for the check register.

Beginning balance	\$896.74
Check # 501	\$425.69
Balance	
Check # 502	\$ 56.34
Balance	
Check # 503	\$ 41.89
Balance	
Deposit	\$123.91
Balance	
Check # 504	\$356.98
Ending balance	

57. **Business and Finance** Keep a running balance for the check register.

Beginning balance	\$1,345.23
Check # 821	\$ 234.99
Balance	
Check # 822	\$ 555.77
Balance	
Deposit	\$ 126.77
Balance	
Check # 823	\$ 53.89
Ending balance	

58. **Business and Finance** Keep a running balance for the check register.

Beginning balance	\$589.21
Check # 678	\$175.63
Balance	
Check # 679	\$ 56.92
Balance	
Deposit	\$121.12
Balance	
Check # 680	\$345.99
Ending balance	

Entering decimals into a calculator is similar to entering whole numbers. There is just one difference: The decimal point key [·] is used to place the decimal point as you enter the number.

Often both addition and subtraction are involved in a calculation. In this case, just enter the decimals and the operation signs, + or −, as they appear in the problem. To find 23.7 − 5.2 + 3.87 − 2.341, enter

23.7 [−] 5.2 [+] 3.87 [−] 2.341 [=]

The display should show 20.029.

Use a calculator to complete each exercise.

59. 10,345.2 + 2,308.35 + 153.58

60. 8.7675 + 2.8 − 3.375 − 6

61. **Business and Finance** Your checking account has a balance of $532.89. You write checks of $50, $27.54, and $134.75 and make a deposit of $50. What is your ending balance?

62. **Business and Finance** A small store makes a profit of $934.20 in the first week of a given month, $1,238.34 in the second week, and $853 in the third week. If the goal is a profit of $4,000 for the month, what profit must the store make during the remainder of the month?

Skills | Calculator/Computer | **Career Applications** | Above and Beyond

63. **Allied Health** A patient is given three capsules of Tc99m sodium pertechnetate containing 79.4, 15.88, and 3.97 millicuries (mCi), respectively. What was the total amount, in millicuries, administered to the patient?

64. **Allied Health** Respiratory therapists calculate the humidity deficit, in milligrams per liter (mg/L), for a patient by subtracting the actual humidity content of inspired air from 43.9 mg/L, which is the maximum humidity content at body temperature. Determine the humidity deficit if the inspired air has a humidity content of 32.7 mg/L.

65. **Electronics** Sandy purchased a number of 12-V batteries from an online auction. There were 10 batteries in the lot. If the batteries were connected in series, the total open-voltage would be the sum of the battery voltages; therefore, she expected an open-voltage of 120 V. Unfortunately, her voltmeter (used to measure voltage) doesn't read above 100 V. So she measured each battery's voltage and recorded it in the table. Calculate the actual open-voltage of the batteries if they are connected in series.

Battery	Measured Voltage (in volts)	Battery	Measured Voltage (in volts)
1	12.20	6	12.82
2	13.84	7	11.93
3	11.42	8	11.01
4	13.00	9	12.77
5	12.45	10	12.03

66. **Manufacturing** A dimension on a computer-aided design (CAD) drawing is given as 3.084 in. ± 0.125 in. What is the minimum and maximum length the feature may be?

67. **NUMBER PROBLEM** Find the next number in the sequence 3.125, 3.375, 3.625, . . .

68. **BUSINESS AND FINANCE** Consider a set of credit card charges.

$8.97, $32.75, $15.95, $67.32, $215.78, $74.95, $83.90, and $257.28

(a) Estimate the total bill for the charges by rounding each number to the nearest dollar and adding the results.

(b) Estimate the total bill by adding the charges and then rounding to the nearest dollar.

(c) What are the advantages and disadvantages of the methods in (a) and (b)?

Recall that a magic square is one in which the sum of every row, column, and diagonal is the same. Complete each magic square.

69.

1.6		1.2
	1	
0.8		

70.

2.4		7.2
10.8		
4.8		

71. Find the next two numbers in each sequence.

(a) 0.75   0.62   0.5   0.39

(b) 1.0   1.5   0.9   3.5   0.8

72. **(a)** Determine the amount of rainfall (to the nearest hundredth of an inch) in your town or city for each of the past 24 months.

(b) Determine the difference in rainfall amounts per month for each month from one year to the next.

Answers

1. 1.07 **3.** 22.319 **5.** 34.4462 **7.** 6.77 **9.** 5.7 **11.** 16.743 **13.** 1.791 **15.** 147.029 **17.** 0.26 **19.** 2.255 **21.** 2.32 **23.** 6.65 **25.** 2.416 **27.** 6.5 **29.** 5.1 **31.** 3.97 **33.** 1.65 **35.** 4.76 **37.** 1 **39.** 2.93 **41.** 42.4 gal **43.** (a) 13.37 cm; (b) 0.59 cm **45.** $366.69 **47.** 41.68 yd **49.** 11.535 mi **51.** 0.0575 in. **53.** $25.98 **55.** $256.71; $238.71; $152.93; $403.38; $202.14 **57.** $1,110.24; $554.47; $681.24; $627.35 **59.** 12,807.13 **61.** $370.60 **63.** 99.25 mCi **65.** 123.47 V **67.** 3.875 **69.**

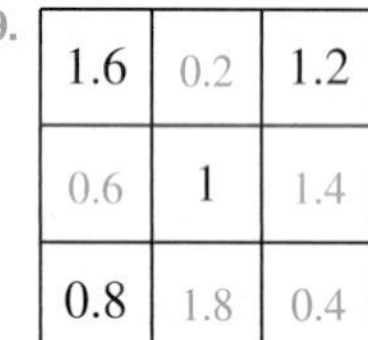

1.6	0.2	1.2
0.6	1	1.4
0.8	1.8	0.4

71. (a) 0.29, 0.2; (b) 5.5, 0.7

4.3

Multiplying Decimals

< 4.3 Objectives >

1 > Multiply decimals

2 > Use multiplication to solve applications

3 > Multiply a decimal by a power of 10

To see how to multiply decimals, we begin by writing a pair in fraction form and multiplying.

Example 1 Multiplying Decimals

< Objective 1 >

$$0.32 \times 0.2 = \frac{32}{100} \times \frac{2}{10} = \frac{64}{1{,}000} = 0.064$$

Here 0.32 has *two* decimal places, and 0.2 has *one* decimal place. The product 0.064 has *three* decimal places.

Check Yourself 1

Find the product and the number of decimal places.

0.14×0.054

Of course, we do not want to rewrite decimals as fractions every time we need to multiply. The preceding example suggests a rule.

When we write the decimals as fractions, the denominators are powers of 10 with the same number of 0s as there are decimal places in original number.

$0.32 = \frac{32}{100}$ Denominator has two zeros.

Two decimal places

$0.2 = \frac{2}{10}$ Denominator has one zero.

One decimal place

When we multiply the fractions, we multiply the denominators. The denominator of the product is a power of 10 with three zeros, so the decimal form of the number has three decimal places.

$$\frac{32}{100} \times \frac{2}{10} = \frac{2 \times 32}{100 \times 10}$$

$= \frac{64}{1{,}000}$ Three zeros in the denominator.

$= 0.064$ Three decimal places.

Step by Step

Multiplying Decimals

Step 1 Multiply the decimals as though they were whole numbers. Temporarily ignore the decimal points.

Step 2 Count the number of decimal places in the numbers being multiplied.

Step 3 Place the decimal point in the product so that the number of decimal places in the product is the sum of the number of decimal places in the factors.

Example 2 illustrates this rule.

Example 2 Multiplying Decimals

Multiply 0.23 by 0.7.

$$
\begin{array}{r l}
0.23 & \longleftarrow \text{Two places} \\
\times\ 0.7 & \longleftarrow \text{One place} \\
\hline
0.161 & \longleftarrow \text{Three places}
\end{array}
$$

Check Yourself 2

Multiply 0.36 × 1.52.

We may have to affix zeros to the left in the product to place the decimal point.

Example 3 Multiplying Decimals

Multiply.

$$
\begin{array}{r l}
0.136 & \longleftarrow \text{Three places} \\
\times\ 0.28 & \longleftarrow \text{Two places} \\
\hline
1088 & \\
272\ & \\
\hline
0.03808 & \longleftarrow \text{Five places}
\end{array}
$$

↑ Insert 0

Insert a 0 to mark off five decimal places.

Check Yourself 3

Multiply 0.234 × 0.24.

Estimation is also helpful when multiplying decimals.

Example 4 Estimating the Product of Decimals

Estimate the product 24.3 × 5.8.

$$
\begin{array}{r c r}
24.3 & \xrightarrow{\text{Round}} & 24 \\
\times\ 5.8 & & \times\ 6 \\
\hline
 & & 144
\end{array}
$$

Multiply to get the estimate.

Check Yourself 4

Estimate the product.

17.95 × 8.17

Many applications require decimal multiplication.

Example 5 An Application of Multiplication

< Objective 2 >

A sheet of paper has dimensions 27.5 by 21.5 cm. What is its area?

We multiply to find the required area.

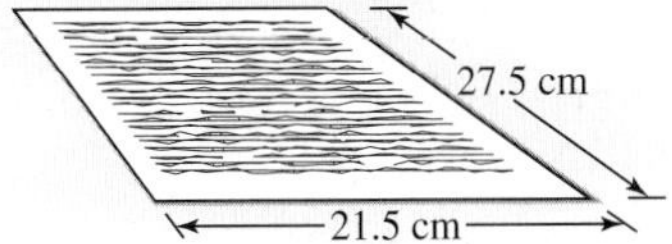

$$\begin{array}{r} 27.5 \text{ cm} \\ \times\, 21.5 \text{ cm} \\ \hline 137\,5 \\ 275 \\ 550 \\ \hline 591.25 \text{ cm}^2 \end{array}$$

RECALL

The area of a rectangle is length times width, so we multiply.

The area of the paper is 591.25 cm^2.

Check Yourself 5

If 1 kg is 2.2 lb, how many pounds equal 5.3 kg?

Example 6 An Application Involving the Multiplication of Decimals

Jack buys 8.7 gal of propane at $\$3.49\frac{9}{10}$ per gallon. Use $\$3.49\frac{9}{10} = \$3.499$ to find the cost of the propane.

Multiply the cost per gallon by the number of gallons, and then round the result to the nearest cent (two decimal places).

Since 3.499 has three decimal places and 8.7 has one, the product has four decimal places.

$$\begin{array}{r} 3.499 \\ \times\quad 8.7 \\ \hline 24493 \\ 279920 \\ \hline 30.4413 \end{array}$$

Four decimal places

NOTE

We usually round money to the nearest cent (hundredth of a dollar).

The propane cost Jack \$30.44.

Check Yourself 6

One liter (L) is approximately 0.265 gal. On a trip to Europe, the Bernards purchased 88.4 L of gas for their rental car. How many gallons of gas did they purchase, to the nearest tenth of a gallon?

Sometimes we have to use more than one operation to solve a problem, as in Example 7.

Example 7 An Application Involving Two Operations

Steve purchased a television set for \$299.50. He agreed to pay for the set by making payments of \$27.70 for 12 months. How much extra does he pay on the installment plan?

First we multiply to find the amount actually paid.

```
  $ 27.70
×      12
   55 40
  277 0
  $332.40 ←— Amount paid
```

Now subtract the listed price. The difference gives the extra amount Steve paid.

```
   $332.40
 −  299.50
   $ 32.90 ←— Extra amount
```

Steve pays an extra $32.90 on the installment plan.

Check Yourself 7

Sandy paid $10,985 for a used car. She paid $1,500 down and will pay $305.35 per month for 36 months on the balance. How much extra will she pay with this loan arrangement?

There are enough applications involving multiplication by the powers of 10 to make it worthwhile to develop a special rule so you can do such operations quickly and easily. Look at the patterns in some of these special multiplications.

```
   0.679                  23.58
×     10               ×     10
  6.790, or 6.79         235.80, or 235.8
```

NOTE

The digits remain the same. Only the *position* of the decimal point changes.

Do you see that multiplying by 10 moves the decimal point *one place to the right?* What happens when we multiply by 100?

```
   0.892                  5.74
×    100               ×   100
  89.200, or 89.2        574.00, or 574
```

Multiplying by 100 shifts the decimal point *two places to the right.* The pattern of these examples gives us a rule.

Property

To Multiply by a Power of 10

Move the decimal point to the right the same number of places as there are zeros in the power of 10.

Example 8 **Multiplying by Powers of 10**

< Objective 3 >

$2.356 \times 10 = 23.56$

↑ One zero — ↑ The decimal point moves one place to the right.

$34.788 \times 100 = 3{,}478.8$

↑ Two zeros — ↑ The decimal point moves two places to the right.

$3.67 \times 1{,}000 = 3{,}670.$

↑ Three zeros — ↑ The decimal point moves three places to the right. Note that we added a 0 to place the decimal point correctly.

NOTE

Multiplying by 10, 100, or any other larger power of 10 makes the number *larger.* Move the decimal point to *the right.*

RECALL

10^5 is just a 1 followed by five zeros.

$0.005672 \times 10^5 = 567.2$

↑ Five zeros — ↖ The decimal point moves five places to the right.

Check Yourself 8

Multiply.

(a) 43.875×100 **(b)** 0.0083×10^3

Example 9 is just one of many applications that require multiplying by a power of 10.

Example 9 An Application Involving Multiplication by a Power of 10

NOTES

There are 1,000 m in 1 km.

If the result is a whole number, there is no need to write the decimal point.

To convert from kilometers to meters, multiply by 1,000. Find the number of meters in 2.45 km.

2.45 km = 2,450. m Just move the decimal point three places right to make the conversion. Note that we added a zero to place the decimal point correctly.

Check Yourself 9

To convert from kilograms to grams, multiply by 1,000. Find the number of grams (g) in 5.23 kilograms (kg).

Check Yourself ANSWERS

1. 0.00756, five decimal places **2.** 0.5472 **3.** 0.05616 **4.** 144 **5.** 11.66 lb **6.** 23.4 gal **7.** $1,507.60 **8.** **(a)** 4,387.5; **(b)** 8.3 **9.** 5,230 g

Reading Your Text

These fill-in-the-blank exercises will help you understand some of the key vocabulary used in this section. The answers to these exercises are in the Answers Appendix at the back of the text.

(a) When multiplying decimals, ________ the number of decimal places in the numbers being multiplied.

(b) The decimal point in the ________ is placed so that the number of places is the sum of the number of decimal places in the factors.

(c) It is sometimes necessary to affix ________ to the left of the product of two decimals to accurately place the decimal point.

(d) Multiplying by 100 shifts the decimal point two places to the ________.

4.3 exercises

Skills | Calculator/Computer | Career Applications | Above and Beyond

< Objective 1 >

Multiply.

1. $\begin{array}{r} 2.3 \\ \times 3.4 \\ \hline \end{array}$

2. $\begin{array}{r} 6.5 \\ \times 4.3 \\ \hline \end{array}$

3. $\begin{array}{r} 8.4 \\ \times 5.2 \\ \hline \end{array}$ 

4. $\begin{array}{r} 9.2 \\ \times 4.6 \\ \hline \end{array}$

5. $\begin{array}{r} 2.56 \\ \times 72 \\ \hline \end{array}$

6. $\begin{array}{r} 56.7 \\ \times 35 \\ \hline \end{array}$

7. (0.78) (2.3)

8. (9.5) (0.45)

9. $\begin{array}{r} 15.7 \\ \times 2.35 \\ \hline \end{array}$

10. $\begin{array}{r} 28.3 \\ \times 0.59 \\ \hline \end{array}$

11. $\begin{array}{r} 0.354 \\ \times 0.8 \\ \hline \end{array}$

12. $\begin{array}{r} 0.624 \\ \times 0.85 \\ \hline \end{array}$

13. $3.28 \cdot 5.07$

14. $0.582 \cdot 6.3$

15. $\begin{array}{r} 5.238 \\ \times 0.48 \\ \hline \end{array}$

16. $\begin{array}{r} 0.372 \\ \times 58 \\ \hline \end{array}$

17. $\begin{array}{r} 1.053 \\ \times 0.552 \\ \hline \end{array}$

18. $\begin{array}{r} 2.375 \\ \times 0.28 \\ \hline \end{array}$

19. $\begin{array}{r} 0.0056 \\ \times 0.082 \\ \hline \end{array}$

20. $\begin{array}{r} 1.008 \\ \times 0.046 \\ \hline \end{array}$

21. 0.8(2.376)

22. 58(3.52)

23. 0.3085×4.5

24. 0.028×0.685

25. 43.8×2.567

26. 18.07×23.123

27. $(2.5)^2$

28. $(1.25)^2$

29. $(3.28)^2$

30. $(11.67)^2$

< Objective 2 >

Solve each application.

31. **Business and Finance** Kurt bought four shirts on sale as pictured. What was the total cost of the purchase?

32. **Business and Finance** Juan makes monthly payments of $245.11 on his car. What will he pay in 1 year?

33. **Science and Medicine** If 1 gal of water weighs 8.34 lb, how much does 2.5 gal weigh?

34. **Business and Finance** Malik worked 37.4 hr in 1 week. If his hourly rate of pay is $12.45, what was his pay for the week?

35. **Business and Finance** To find one year's simple interest on a loan at $9\frac{1}{2}\%$, we multiply the amount of the loan by 0.095. Find the simple interest on a $1,500 loan for 1 year.

36. **Business and Finance** A 5.8-lb chicken costs $3.25 per pound. What is the cost of the chicken?

37. **Business and Finance** Tom's state income tax is found by multiplying his income by 0.054. If Tom's income is $43,640, find the amount of his tax.

38. **Business and Finance** Claudia earns $16.80 per hour. For overtime (each hour over 40 hr) she earns $25.20. If she works 48.5 hr in a week, what pay should she receive?

39. **Geometry** A sheet of copier paper has dimensions as shown. What is its area?

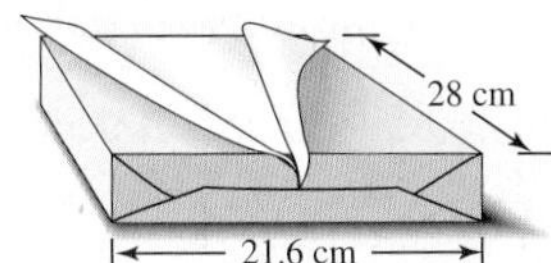

40. **Business and Finance** A rental car costs $68 per day plus 18 cents per mile (mi). If you rent a car for 5 days and drive 785 mi, what will the total car rental bill be?

< Objective 3 >

Multiply.

41. 5.89×10 
42. 0.895×100
43. 23.79×100
44. 2.41×10
45. $10\,(0.045)$
46. $100\,(5.8)$
47. $(0.431)\,(100)$
48. $(0.025)\,(10)$
49. 0.471×100
50. $0.95 \times 10{,}000$
51. $1{,}000 \cdot 0.7125$
52. $23.42 \cdot 1{,}000$
53. 4.25×10^2 
54. 0.36×10^3
55. 3.45×10^4
56. 0.058×10^5

Solve each application.

57. **Business and Finance** A store purchases 100 items for $1.38 each. Find the total cost of the order.
58. **Science and Medicine** To convert from meters to centimeters, multiply by 100. How many centimeters are there in 5.3 m?
59. **Science and Medicine** How many grams are there in 2.2 kg? Multiply by 1,000 to make the conversion. 
60. **Business and Finance** An office purchases 1,000 pens at a cost of 17.8 cents each. What is the cost of the purchase in dollars?

Meyer's Office Supply
371 Maple Dr., Treynor IA 50001

Item	Quantity	Item Price	Total
Pens	1000	$0.178	

Label each statement as **true** *or* **false.**

61. The decimal points must be aligned when finding the sum of two decimals.
62. The decimal points must be aligned when finding the product of two decimals.
63. The number of decimal places in the product of two factors is the product of the number of places in the two factors.
64. When multiplying a decimal by a power of 10, the decimal point is moved to the left the number of places as there are zeros in the power of 10.

The steps for finding the product of decimals on a calculator are similar to the ones we used for multiplying whole numbers. To multiply $2.8 \times 3.45 \times 3.725$, enter

2.8 [×] 3.45 [×] 3.725 [=]

The display should read 35.9835.

You can also easily find powers of decimals with your calculator by using a similar procedure. To find $(2.35)^3$, you can enter

2.35 [×] 2.35 [×] 2.35 [=]

The display should read 12.977875.

Some calculators have keys to find powers more quickly. Look for keys marked [x^2] or [y^x]. Other calculators have a power key marked [∧]. To find $(2.35)^3$, enter

2.35 [∧] 3 [=] or 2.35 [y^x] 3 [=]

Again, the result is 12.977875.

Use a calculator to complete each exercise.

65. $127.85 \times 0.055 \times 15.84$

66. $18.28 \times 143.45 \times 0.075$

67. $(3.95)^3$

68. $(0.521)^2$

69. **Geometry** Find the area of a rectangle with length 3.75 in. and width 2.35 in.

70. **Business and Finance** Mark works 38.4 hr in a given week. If his hourly rate of pay is \$14.85, what will he be paid for the week?

71. **Business and Finance** If fuel oil costs \$3.669 per gallon, what will 150.4 gal cost?

72. **Business and Finance** To find the simple interest on a loan for 1 year at 12.5%, multiply the amount of the loan by 0.125. What simple interest will you pay on a loan of \$1,458 at 12.5% for 1 year?

73. **Business and Finance** You are the office manager for Dr. Rogers. The increasing cost of making photocopies is a concern to Dr. Rogers. She wants to examine alternatives to the current financing plan. The office currently leases a copy machine for \$325 per month and pays \$0.045 per copy. A 3-year payment plan is available that costs \$375 per month and \$0.025 per copy.

(a) If the office expects to run 100,000 copies per year, which is the better plan?

(b) How much money will the better plan save over the other plan?

74. **Business and Finance** In a bottling company, a machine can fill a 2-L bottle in 0.5 second and move the next bottle into place in 0.1 s. How many 2-L bottles can be filled by the machine in 2 hr?

Skills | Calculator/Computer | Career Applications | **Above and Beyond**

What happens when a calculator wants to display an answer that is too big to fit in the display? Suppose you want to evaluate 10^{10}. If you enter 10 [∧] 10 [=], your calculator will probably display 1 E 10 or perhaps 1^{10}, both of which mean 1×10^{10}. Answers that are displayed in this way are said to be in **scientific notation.** This is a topic that you will study later. For now, we can use the calculator to see the relationship between numbers written in scientific notation and decimal notation. For example, 3.485×10^4 is written in scientific notation. To write it in decimal notation, use your calculator to enter

3.485 [×] 10 [y^x] 4 [=] or 3.485 [×] 10 [∧] 4 [=]

The result should be 34,850. Note that the decimal point moves four places (the power of 10) to the right.

Write each number in decimal notation.

75. 3.365×10^3

76. 4.128×10^3

77. 4.316×10^5

78. 8.163×10^6

79. 7.236×10^8

80. 5.234×10^7

Answers

1. 7.82 **3.** 43.68 **5.** 184.32 **7.** 1.794 **9.** 36.895 **11.** 0.2832 **13.** 16.6296 **15.** 2.51424 **17.** 0.581256 **19.** 0.0004592 **21.** 1.9008 **23.** 1.38825 **25.** 112.4346 **27.** 6.25 **29.** 10.7585 **31.** \$39.92 **33.** 20.85 lb **35.** \$142.50 **37.** \$2,356.56 **39.** 604.8 cm² **41.** 58.9 **43.** 2,379 **45.** 0.45 **47.** 43.1 **49.** 47.1 **51.** 712.5 **53.** 425 **55.** 34,500 **57.** \$138 **59.** 2,200 g **61.** True **63.** False **65.** 111.38292 **67.** 61.629875 **69.** 8.8125 in.² **71.** \$551.82 **73.** **(a)** 3-year plan; current plan: \$25,200; 3-year lease: \$21,000; **(b)** \$4,200 **75.** 3,365 **77.** 431,600 **79.** 723,600,000

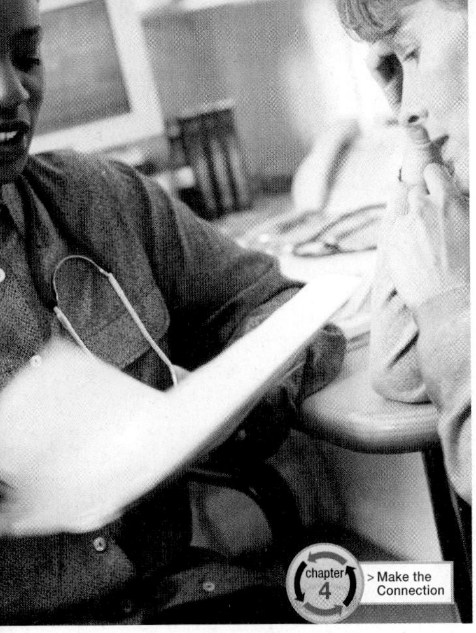

Activity 10 ::

Safe Dosages?

Chemotherapy drug dosages are generally calculated based on a patient's body surface area (BSA) in square meters (m^2). A patient's BSA is calculated using a nomogram and is based on the patient's height and weight. Print out the adult nomogram from the Science Museum of Minnesota's website

www.smm.org/heart/lessons/nomogram_adult.htm

Draw a line connecting the patient's height with his or her weight. The point where this line crosses the middle column denotes the patient's BSA.

Dosages are then calculated using the formula

$$\text{Dose} = \text{recommended dose} \times \text{BSA}$$

In each case, determine if the prescribed dose falls within the recommended dose range for the given patient.

1. The doctor has prescribed Blenoxane to treat an adult male patient with testicular cancer. The patient is 70 in. tall and weighs 260 lb. According to the RxList website (www.rxlist.com), the recommended dose of Blenoxane in the treatment of testicular cancer should be between 10 to 20 units per square meter (units/m^2). The ordered dose is 50 units once per week.

2. The doctor has prescribed BiCNU to treat an adult patient with a brain tumor. The patient is 65 in. tall and weighs 150 lb. According to the RxList website, the recommended dose of BiCNU should be between 150 to 200 milligrams per square meter (mg/m^2). The ordered dose is 300 mg once every 6 weeks.

3. The doctor has prescribed Cisplatin to treat an adult patient with advanced bladder cancer. The patient is 73 in. tall and weighs 275 lb. According to the RxList website, the recommended dose of Cisplatin should be between 50 to 70 mg/m^2. The ordered dose is 150 mg once every 3 to 4 weeks.

4.4 Dividing Decimals

< 4.4 Objectives >

1 > Divide a decimal by a whole number

2 > Use decimals to solve applications

3 > Divide a decimal by a decimal

4 > Divide a decimal by a power of 10

Dividing decimals is very similar to dividing whole numbers. The only difference is in learning to place the decimal point in the quotient. We start by dividing a decimal by a whole number. Here, placing the decimal point is easy.

Step by Step

Dividing a Decimal by a Whole Number

Step 1 Place the decimal point in the quotient *directly above* the decimal point of the dividend.

Step 2 Divide as you would with whole numbers.

Example 1 Dividing a Decimal by a Whole Number

< Objective 1 >

Divide 29.21 by 23.

```
      1.27
23)29.21
   23
    6 2
    4 6
    1 61
    1 61
       0
```

The quotient is 1.27.

NOTE

Do the division just as if you were dealing with whole numbers. Just remember to place the decimal point in the quotient *directly above* the one in the dividend.

Check Yourself 1

Divide 80.24 by 34.

Here is another example of dividing a decimal by a whole number.

Example 2 Dividing a Decimal by a Whole Number

Divide 122.2 by 52.

```
      2.3
52)122.2
   104
    18 2
    15 6
     2 6
```

NOTE

Again place the decimal point of the quotient above that of the dividend.

> **RECALL**
>
> Affixing a zero at the end of a decimal does not change the value of the dividend. It simply allows us to complete the division process in this case.

We normally do not use a remainder when dealing with decimals. Add a 0 to the dividend and continue.

```
      2.35
52)122.20   ← Add a zero.
   104
    18 2
    15 6
     2 60
     2 60
        0
```

So $122.2 \div 52 = 2.35$. The quotient is 2.35.

Check Yourself 2

Divide 234.6 by 68.

Look at Example 2, again. In order to complete the division, we added a 0 to the end of the dividend rather than give a remainder. We were able to do this because adding a zero to the right of a decimal point at the end of a number does not change the number.

$122.2 = 122.20$

This gives us a convenient way to divide any pair of numbers, even whole numbers, without using remainders for any leftover part.

Example 3 Dividing Whole Numbers

> **NOTE**
>
> We study this more in Section 4.5.

Divide.

$3 \div 8$

We write the division problem as $8\overline{)3}$. Immediately, we can see that 8 does not go into 3. We might write this as 0 r3, but that really does nothing for us.

Instead, we can place a decimal point after the whole number 3, and then attach zeros after the decimal point.

$3 = 3. = 3.0 = 3.00 = \cdots$

We can attach as many zeros as is convenient and then divide. We can even attach more, as needed. We begin by attaching one 0 and will attach more if needed.

```
   0.375
8)3.000
  2 4↓     3 × 8 = 24.
    60     Subtract to get 6; bring down another 0 to continue the division.
    56↓    7 × 8 = 56.
     40    We subtract to get 4. We must bring down another 0.
     40    5 × 8 = 40.
      0    This time, the difference is 0, so we are done.
```

So, $3 \div 8 = 0.375$.

Check Yourself 3

Divide $11 \div 4$.

Often, you will be asked to round a quotient to a specific decimal place. This is especially useful if the quotient has many decimal places or even if it does not end at all.

In these cases, continue the division process to *one digit past* the indicated place value. Then round the result back to the desired accuracy.

When working with money, for instance, we normally give the quotient to the nearest hundredth of a dollar (the nearest cent). This means carrying the division out to the thousandths place and then rounding.

Example 4 Rounding a Quotient

(a) Find the quotient of 25.75 ÷ 15 to the nearest hundredth.

```
    1.716
15)25.750
   15        ← Add a zero to carry the division to the thousandths place.
   10 7
   10 5
      25
      15
      100
       90
       10
```

So 25.75 ÷ 15 = 1.72 (to the nearest hundredth).

NOTES

Find the quotient to *one place past* the desired place and then round the result.

We must remember to place a zero in the quotient because 12 does not divide 8.

(b) Round 12.8 ÷ 12 to three decimal places.

```
    1.0666
12)12.8000     Compute four decimal places to round to the nearest thousandth.
   12↓         1 × 12 = 12
    0 8        Subtract to get 0; bring down the 8.
      0↓       12 does not divide 8, so we must place a 0 in the quotient.
      8 0      Subtract to get 8 and bring down a 0.
      7 2↓     6 × 12 = 72
        8 0    Subtract to get 8 and bring down a 0.
        7 2↓
          8 0
          7 2
            8  We have four decimal places, so we can stop and round.
```

To three decimal places, 12.8 ÷ 12 = 1.067.

Check Yourself 4

(a) Round 99.26 ÷ 35 to two decimal places.
(b) Round 91.644 ÷ 15 to the nearest thousandth.

Problems similar to the one in Example 4 often occur when we are working with money. Example 5 is one of the many applications of this type of division.

Example 5 An Application Involving the Division of a Decimal

< Objective 2 >

A carton of 144 key rings costs $56.10. What is the price per key ring, to the nearest cent?

To find the price for each ring, divide the total price by 144.

RECALL

The rules for rounding decimals are given in Section 4.1.

$$\begin{array}{r} 0.389 \\ 144\overline{)56.100} \\ \underline{43\ 2} \\ 12\ 90 \\ \underline{11\ 52} \\ 1\ 380 \\ \underline{1\ 296} \\ 84 \end{array}$$

Carry the division to the thousandths place and then round.

The cost per key ring is rounded to \$0.39, or 39¢.

Check Yourself 5

An office paid \$26.55 for 72 pens. What was the cost per pen, to the nearest cent?

We now know how to divide a number by a whole number. How do we divide by a decimal, though? One method is to rewrite the problem so that we are dividing by a whole number. Since we know how to divide by a whole number, this allows us to divide by a decimal.

Example 6 Rewriting Division by a Decimal

< Objective 3 >

NOTE

We want to divide by a whole number because dividing by a whole number makes it easy to place the decimal point in the quotient.

Divide.

$2.57 \div 3.4 = \frac{2.57}{3.4}$ Write the division as a fraction.

$= \frac{2.57 \times 10}{3.4 \times 10}$ We multiply the numerator and denominator by 10 so the divisor is a whole number. This *does not change* the value of the fraction.

$= \frac{25.7}{34}$ Multiplying by 10 shifts the decimal point in the numerator and denominator *one place to the right*.

$= 25.7 \div 34$ Our division problem is rewritten so that the divisor is a whole number.

So,

$2.57 \div 3.4 = 25.7 \div 34$ After we multiply the numerator and denominator by 10, we see that $2.57 \div 3.4$ is the same as $25.7 \div 34$.

RECALL

Multiplying by a power of 10 is just a matter of shifting the decimal point.

Check Yourself 6

Rewrite the division problem so that the divisor is a whole number.

$3.42 \div 2.5$

In Example 6, we multiplied both the divisor and dividend by the same number. When we write it in fraction form, it is easy to see that we are only multiplying by one, so the quotient remains the same.

$2.57 \div 3.4 = \frac{2.57}{3.4}$

$= \frac{2.57}{3.4} \times 1$ Multiplying by 1 does not change a number.

$= \frac{2.57}{3.4} \times \frac{10}{10}$ $\frac{10}{10} = 1$

$= \frac{25.7}{34}$

$= 25.7 \div 34$

NOTES

Multiplying by 1 is a very common technique in math. Often, we write 1 in some useful way such as $\frac{10}{10}$.

This method does not hold for nonterminating or repeating decimals, which we study in Section 4.5. Nonterminating or repeating decimals do not have powers of 10 as denominators.

When we are dividing by a decimal, we can always multiply both the divisor and dividend by the same power of 10 to ensure that we are dividing by a whole number.

Step by Step

Dividing by a Decimal

Step 1 Move the decimal point in the divisor to the *right,* making the divisor a whole number.

Step 2 Move the decimal point in the dividend to the right *the same number of places.* Add zeros if necessary.

Step 3 Place the decimal point in the quotient directly above the new decimal point in the dividend.

Step 4 Divide. You are dividing by a whole number.

Now look at an example of the use of our division rule.

Example 7 Rounding a Quotient

Divide 1.573 by 0.48 and give the quotient to the nearest tenth.

Write

$0.48\overline{)1.573}$ Shift the decimal points two places to the right to make the divisor a whole number.

Now divide:

```
      3.27
48 )157.30
    144
     13 3
      9 6
      3 70
      3 36
        34
```

We add a zero to carry the division to the hundredths place because we want to find the quotient to the nearest tenth.

NOTE

Once the division statement is rewritten, place the decimal point in the quotient above the new decimal point in the dividend.

Round 3.27 to 3.3. So,

$1.573 \div 0.48 = 3.3$ (to the nearest tenth)

Check Yourself 7

Divide, rounding the quotient to the nearest tenth.

$3.4 \div 1.24$

Many applications involve decimal division.

Example 8 An Application of Decimal Division

Andrea worked 41.5 hr in a week and earned \$488.87 (without overtime). What is her hourly rate of pay?

To find her hourly rate of pay we use division. We divide the number of hours worked into the total pay.

```
          1 1.78
41.5 )488.8 70
       415
        73 8
        41 5
        32 3 7
        29 0 5
         3 3 20
         3 3 20
              0
```

NOTE

We add a zero to the dividend to complete the division process.

Andrea's hourly rate of pay is \$11.78.

Check Yourself 8

A developer wants to subdivide a 12.6-acre piece of land into 0.45-acre lots. How many lots are possible?

Example 9 **An Application of Decimal Division**

At the start of a trip the odometer read 34,563. At the end of the trip, it read 36,235. If 86.7 gal of gas were used, find the number of miles per gallon (to the nearest tenth).

First, find the number of miles traveled by subtracting the initial reading from the final reading.

$$\begin{array}{rl} 36{,}235 & \text{Final reading} \\ -\ 34{,}563 & \text{Initial reading} \\ \hline 1{,}672 & \text{Miles traveled} \end{array}$$

Next, divide the miles traveled by the number of gallons used. This gives us the miles per gallon.

$$\begin{array}{r} 19.28 \\ 86.7_\wedge\overline{)1{,}672.0_\wedge00} \\ \underline{867} \\ 805\ 0 \\ \underline{780\ 3} \\ 24\ 7\ 0 \\ \underline{17\ 3\ 4} \\ 7\ 3\ 60 \\ \underline{6\ 9\ 36} \\ 4\ 24 \end{array}$$

Round 19.28 to 19.3 mi/gal.

Check Yourself 9

John starts his trip with an odometer reading of 15,433 and ends with a reading of 16,238. If he used 25.9 gallons (gal) of gas, find the number of miles per gallon (to the nearest tenth).

Recall that you can multiply decimals by powers of 10 by simply shifting the decimal point to the right. A similar approach works for division by powers of 10.

Example 10 **Dividing by a Power of 10**

< Objective 4 >

(a) Divide.

$$\begin{array}{r} 3.53 \\ 10\overline{)35.30} \\ \underline{30} \\ 5\ 3 \\ \underline{5\ 0} \\ 30 \\ \underline{30} \\ 0 \end{array}$$

The dividend is 35.3. The quotient is 3.53. The decimal point shifts *one place to the left*. Note also that the divisor, 10, has *one* zero.

(b) Divide.

```
       3.785
100)378.500
    300
     78 5
     70 0
      8 50
      8 00
        500
        500
          0
```

Here the dividend is 378.5, whereas the quotient is 3.785. The decimal point now shifts *two places to the left*. In this case the divisor, 100, has *two* zeros.

Check Yourself 10

Divide.

(a) $52.6 \div 10$ **(b)** $267.9 \div 100$

Example 10 suggests a useful rule.

Property

Dividing a Decimal by a Power of 10

Move the decimal point *to the left* the same number of places as there are zeros in the power of 10.

Example 11 **Dividing by a Power of 10**

NOTE

We may add zeros to correctly place the decimal point.

RECALL

10^4 is a 1 followed by *four zeros.*

Divide.

(a) $27.3 \div 10 = 2_\wedge 7.3$ Shift one place to the left.

$= 2.73$

(b) $57.53 \div 100 = 0_\wedge 57.53$ Shift two places to the left.

$= 0.5753$

(c) $39.75 \div 1{,}000 = 0_\wedge 039.75$ Shift three places to the left.

$= 0.03975$

(d) $85 \div 1{,}000 = 0_\wedge 085.$ The decimal after 85 is implied.

$= 0.085$

(e) $235.72 \div 10^4 = 0_\wedge 0235.72$ Shift four places to the left.

$= 0.023572$

Check Yourself 11

Divide.

(a) $3.84 \div 10$ **(b)** $27.3 \div 1{,}000$

Now, look at an application of our work in dividing by powers of 10.

Example 12 — An Application Involving a Power of 10

NOTE

When multiplying by powers of 10, we move the decimal point to the right because our result gets larger.

When dividing, our result gets smaller, so we need to move the decimal to the left.

To convert from millimeters (mm) to meters, we divide by 1,000. How many meters does 3,450 mm equal?

$$3{,}450 \text{ mm} = 3\,450. \text{ m}$$ Shift three places to the left to divide by 1,000.

$$= 3.450 \text{ m}$$

Check Yourself 12

A shipment of 1,000 notebooks cost a stationery store \$658. What was the cost per notebook to the nearest cent?

Recall that the order of operations is always used to simplify a mathematical expression with several operations. Just as we saw with fractions, we always follow the order of operations. Working with decimals does not change that.

Property

The Order of Operations

1. Perform any operations within parentheses or other grouping symbols.
2. Evaluate any exponents.
3. Do any multiplication and division, in order from left to right.
4. Do any addition and subtraction, in order from left to right.

Example 13 — Applying the Order of Operations

Simplify each expression.

(a) $4.6 + (0.5 \times 4.4)^2 - 3.93$

$= 4.6 + (2.2)^2 - 3.93$ Parentheses

$= 4.6 + 4.84 - 3.93$ Exponent

$= 9.44 - 3.93$ Add (left of the subtraction)

$= 5.51$ Subtract

(b) $16.5 - (2.8 + 0.2)^2 + 4.1 \times 2$

$= 16.5 - (3)^2 + 4.1 \times 2$ Parentheses

$= 16.5 - 9 + 4.1 \times 2$ Exponent

$= 16.5 - 9 + 8.2$ Multiply

$= 7.5 + 8.2$ Subtraction (left of the addition)

$= 15.7$ Add

Check Yourself 13

Simplify each expression.

(a) $6.35 + (0.2 \times 8.5)^2 - 3.7$ **(b)** $2.5^2 - (3.57 - 2.14) + 3.2 \times 1.5$

Check Yourself ANSWERS

1. 2.36 **2.** 3.45 **3.** 2.75 **4.** **(a)** 2.84; **(b)** 6.110 or 6.11 **5.** \$0.37, or 37¢
6. 34.2 ÷ 25 **7.** 2.7 **8.** 28 lots **9.** 31.1 mi/gal **10.** **(a)** 5.26; **(b)** 2.679
11. **(a)** 0.384; **(b)** 0.0273 **12.** 66¢ **13.** **(a)** 5.54; **(b)** 9.62

Reading Your Text

These fill-in-the-blank exercises will help you understand some of the key vocabulary used in this section. The answers to these exercises are in the Answers Appendix at the back of the text.

(a) When dividing decimals, place the decimal point in the quotient directly __________ the decimal point of the dividend.

(b) When asked to give a quotient to a certain place value, continue the division process to one digit __________ the indicated place value.

(c) When dividing by a decimal, first move the decimal point in the divisor to the right, making the divisor a __________ __________.

(d) To divide a decimal by a power of 10, move the decimal point to the __________ the same number of places as there are zeros in the power of 10.

4.4 exercises

Skills | Calculator/Computer | Career Applications | Above and Beyond

< Objectives 1–3 >

Divide.

1. 16.68 ÷ 6 
2. 43.92 ÷ 8
3. 1.92 ÷ 4
4. 5.52 ÷ 6
5. 5.48 ÷ 8
6. 2.76 ÷ 8
7. $\frac{13.89}{6}$ 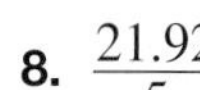
8. $\frac{21.92}{5}$
9. $\frac{185.6}{32}$
10. $\frac{165.6}{36}$
11. 257.6 ÷ 32
12. 33.28 ÷ 16
13. $52\overline{)13.78}$
14. $76\overline{)26.22}$
15. $0.3\overline{)6.24}$
16. $0.5\overline{)32.02}$
17. $\frac{7.22}{3.8}$
18. $\frac{13.34}{2.9}$
19. $\frac{1.717}{3.4}$
20. $\frac{26.156}{6.5}$
21. $0.27\overline{)1.8495}$
22. $0.038\overline{)0.8132}$
23. $0.046\overline{)1.587}$
24. $0.52\overline{)3.2318}$
25. 0.658 ÷ 2.8
26. 0.882 ÷ 0.36
27. 7 ÷ 2
28. 5 ÷ 4
29. 13 ÷ 8
30. 9 ÷ 16

< Objective 4 >

31. 5.8 ÷ 10
32. 5.1 ÷ 10
33. 4.568 ÷ 100
34. 3.817 ÷ 100
35. 24.39 ÷ 1,000
36. 8.41 ÷ 100
37. 6.9 ÷ 1,000
38. 7.2 ÷ 1,000
39. $7.8 \div 10^2$
40. $3.6 \div 10^3$
41. $45.2 \div 10^5$
42. $57.3 \div 10^4$

Divide and round the quotient to the indicated decimal place.

43. $23.8 \div 9$ tenths

44. $5.27 \div 8$ hundredths

45. $38.48 \div 46$ hundredths VIDEO

46. $3.36 \div 36$ thousandths

47. $125.4 \div 52$ tenths

48. $2.563 \div 54$ thousandths

49. $0.7\overline{)1.642}$ hundredths

50. $0.6\overline{)7.695}$ tenths

51. $4.5\overline{)8.415}$ tenths

52. $5.8\overline{)16}$ hundredths

53. $3.4\overline{)27.44}$ hundredths

54. $3.2\overline{)19.526}$ thousandths

55. $4 \div 3$ thousandths

56. $2 \div 9$ hundredths

57. $7 \div 6$ thousandths

58. $8 \div 3$ thousandths

< Objective 2 >

Solve each application.

59. Business and Finance Marv paid \$40.41 for three movies. What was the cost per movie?

60. Business and Finance Seven employees in an office donated a \$172.06 during a charity drive. What was the average donation per person?

61. Business and Finance A shipment of 72 paperback books cost a store \$380.50. What was the average cost per book, to the nearest cent?

62. Business and Finance A restaurant bought 50 glasses at a cost of \$79.80. What was the cost per glass, to the nearest cent?

63. Business and Finance The cost of a box of 48 pens is \$28.20. What is the cost of an individual pen, to the nearest cent?

64. Business and Finance An office bought 18 handheld calculators for \$284. What was the cost per calculator, to the nearest cent?

65. Business and Finance Al purchased a new refrigerator that cost \$736.12 with interest included. He paid \$100 as a down payment and agreed to pay the remainder in 18 monthly payments. What amount will he be paying per month?

66. Business and Finance The cost of a television set with interest is \$490.64. If you make a down payment of \$50 and agree to pay the balance in 12 monthly payments, what is the amount of each monthly payment?

Evaluate each expression.

67. $79 - 28.2 + 13.7$

68. $63.1 - 4.8 + 5.2$

69. $29.64 - (4.2 + 12.39)$

70. $53.6 - (14 + 6.21)$

71. $8.2 \div 0.25 \times 3.6$

72. $7.14 \div 0.3 \times 5.1$

73. $\dfrac{7.8 + 4.2}{9.1 - 6.6}$

74. $\dfrac{6.08 + 3.58}{7.65 - 3.45}$

75. $6.4 + 1.3^2$

76. $(6.4 + 1.3)^2$

77. $15.9 - 4.2 \times 3.5$

78. $23.7 - 8.6 \times 0.8$

79. $6.1 + 2.3 \times (8.08 - 5.9)$

80. $2.09 + 4.5 \times (12.37 - 7.27)$

81. $5.2 - 3.1 \times 1.5 + (3.1 + 0.4)^2$

82. $150 + 4.1 \times 1.5 - (2.5 \times 1.6)^3 \times 2.4$

83. $17.9 \times 1.1 - (2.3 \times 1.1)^2 + (13.4 - 2.1 \times 4.6)$

84. $6.89^2 - 3.14 \times 2.5 + (3.2 \times 1.6 - 4.1)^2$

Skills | **Calculator/Computer** | Career Applications | Above and Beyond

Using a calculator to divide decimals is a good way to check your work and solve applications. However, when using it for applications, we generally round our answers to an appropriate place value.

Use a calculator to divide and round as indicated.

85. 2.546 ÷ 1.38 hundredths

86. 45.8 ÷ 9.4 tenths

87. 0.5782 ÷ 1.236 thousandths

88. 1.25 ÷ 0.785 hundredths

89. 1.34 ÷ 2.63 two decimal places

90. 12.364 ÷ 4.361 three decimal places

Skills | Calculator/Computer | **Career Applications** | Above and Beyond

91. **ALLIED HEALTH** Since people vary in body size, the cardiac index is used to normalize cardiac output measurements. The cardiac index, in liters per minute per square meter L/(min · m^2), is calculated by dividing a patient's cardiac output, in liters per minute (L/min), by his or her body surface area, in m^2. Calculate the cardiac index for a male patient whose cardiac output is 4.8 L/min if his body surface area is 2.03 m^2. Round your answer to the nearest hundredth.

92. **ALLIED HEALTH** The specific concentration of a radioactive drug, or radiopharmaceutical, is defined as the activity, in millicuries (mCi), divided by the volume, in mL. Determine the specific concentration of a vial containing 7.3 mCi of I131 sodium iodide in 0.25 mL. Round your answer to the nearest tenth.

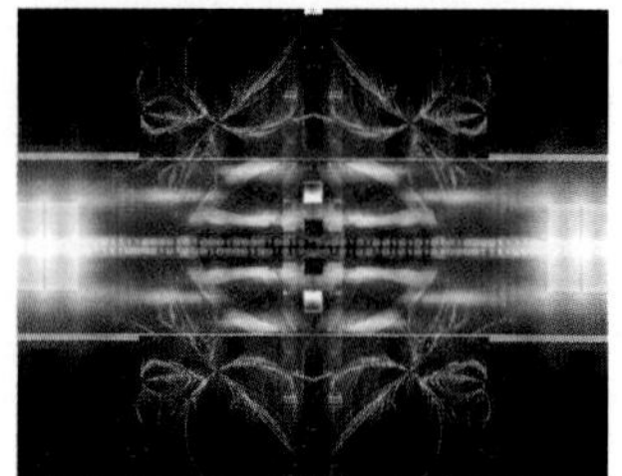

93. **INFORMATION TECHNOLOGY** A Web developer is responsible for designing a Web application for a customer. She uses an FTP program to transmit pages from her local machine to a Web server. She needs to transmit 2.5 megabytes (MB) of data. She notices it takes 1.7 s to transmit the data. How fast is her connection to the Web server in MB/s? Round your answer to the nearest hundredth.

94. **INFORMATION TECHNOLOGY** After creating a presentation for a big customer, Joe sees that the overall size of the files is 16.5 gigabytes (GB). How many 2.9-GB flash drives does he need to store the files?

Skills | Calculator/Computer | Career Applications | **Above and Beyond**

95. The blood alcohol content (BAC) of a person is determined by the Widmark formula. Find that formula using a search engine and use it to solve the next application.

A 125-lb person is driving and is stopped by a police officer on suspicion of driving under the influence (DUI). The driver claims that in the past 2 hr he only consumed six 12-oz bottles of 3.9% beer. If he undergoes a breathalyzer test, what will his BAC be? Will this amount be under the legal limit for your state?

96. Four brands of soap are available in a local store. Compute the unit price (price per ounce) of each brand, rounding to the third decimal place.

Brand	Ounces	Total Price	Unit Price
Squeaky Clean	5.5	$1.95	
Smell Fresh	7.5	$2.29	
Feel Nice	4.5	$1.79	
Look Bright	6.5	$2.49	

97. Sophie is a quality control expert. She inspects boxes of number 2 pencils. Each pencil weighs 4.4 g. The contents of a box of pencils weighs 66.6 g. If a box is labeled CONTENTS: 16 PENCILS, should Sophie approve the box as meeting specifications? Explain your answer.

98. Write a plan to determine the number of miles per gallon your car (or your family car) gets. Use this plan to determine your car's actual mileage.

Answers

1. 2.78 **3.** 0.48 **5.** 0.685 **7.** 2.315 **9.** 5.8 **11.** 8.05 **13.** 0.265 **15.** 20.8 **17.** 1.9 **19.** 0.505 **21.** 6.85 **23.** 34.5 **25.** 0.235 **27.** 3.5 **29.** 1.625 **31.** 0.58 **33.** 0.04568 **35.** 0.02439 **37.** 0.0069 **39.** 0.078 **41.** 0.000452 **43.** 2.6 **45.** 0.84 **47.** 2.4 **49.** 2.35 **51.** 1.9 **53.** 8.07 **55.** 1.333 **57.** 1.167 **59.** \$13.47 **61.** \$5.28 **63.** \$0.59, or 59¢ **65.** \$35.34 **67.** 64.5 **69.** 13.05 **71.** 118.08 **73.** 4.8 **75.** 8.09 **77.** 1.2 **79.** 11.114 **81.** 12.8 **83.** 17.0291 **85.** 1.84 **87.** 0.468 **89.** 0.51 **91.** 2.36 L/(min · m^2) **93.** 1.47 MB/s **95.** Above and Beyond **97.** Above and Beyond

Activity 11 ::

The Tour de France

The Tour de France is perhaps the most grueling of all sporting events. It is a bicycle race that spans 23 days (including 2 rest days), and involves 21 stages of riding in a huge circuit around the country of France. This includes several stages that take the riders through two mountainous regions, the Alps and the Pyrenees. The table presents the winners of each stage in the 2011 race, along with the winning time in hours and the length of the stage expressed in miles. For each stage, compute the winner's average speed by dividing the miles traveled by the winning time. Round your answers to the nearest tenth of a mile per hour.

Stage	Winner	Country	Length (mi)	Time (hr)	Speed (mi/hr)
1	Gilbert	Belgium	119.0	4.69	
2	Hushovd	Norway	14.3	0.41	
3	Farrar	USA	123.0	4.67	
4	Evans	Australia	107.2	4.19	
5	Cavendish	Great Britain	102.2	3.64	
6	Boasson Hagen	Norway	140.7	5.23	
7	Cavendish	Great Britain	135.5	5.65	
8	Costa	Portugal	117.4	4.61	
9	L. Léon Sánchez	Spain	129.2	5.45	
10	Greipel	Germany	98.2	3.52	
11	Cavendish	Great Britain	104.1	3.77	
12	S. Sánchez	Spain	131.1	6.02	
13	Hushovd	Norway	94.8	3.79	
14	Vanendert	Belgium	104.7	5.21	
15	Cavendish	Great Britain	119.6	4.36	
16	Hushovd	Norway	101.0	3.36	
17	Boasson Hagen	Norway	111.2	4.30	
18	Schleck	Luxembourg	124.6	6.13	
19	Rolland	France	68.0	3.21	
20	Martin	Germany	26.4	0.93	
21	Cavendish	Great Britain	59.0	2.45	

Find the total number of miles traveled.

Use the speed of the winner of each stage to determine the stages that were in mountains.

Australian Cadel Evans was the overall winner of the 2011 race. His total time was 86.21 hr. Compute his average speed for the entire race, rounding to the nearest tenth of a mile per hour.

4.5 Converting Between Fractions and Decimals

< 4.5 Objectives >

1 > Convert a fraction to a decimal

2 > Convert a fraction to a repeating decimal

3 > Convert a decimal to a fraction

You have already done most of the work necessary to convert a number between its fraction and decimal forms. In this section, we help you put it all together.

In Example 1, we write a fraction as a decimal. This follows directly from your work in Section 4.4.

Example 1 Converting a Fraction to a Decimal

< Objective 1 >

RECALL

5 can be written as 5.0, 5.00, 5.000, and so on.

Write $\frac{5}{8}$ as a decimal.

To begin, write a decimal point to the right of 5. The decimal point in the quotient is placed directly above this decimal point. We continue the division process by adding zeros to the right of the decimal point in the dividend until a zero remainder is reached.

```
   0.625
8)5.000
  4 8
    20
    16
     40
     40
      0
```

Because $\frac{5}{8}$ means 5 ÷ 8, divide 8 into 5.

We see that $\frac{5}{8} = 0.625$; 0.625 is the decimal equivalent of $\frac{5}{8}$.

Check Yourself 1

Find the decimal equivalent of $\frac{7}{8}$.

Some fractions are used so often that you should know their decimal forms. We list several here as a reference.

NOTE

We stop dividing when the remainder is zero. These are called **terminating decimals.**

Some Common Decimal Equivalents

$\frac{1}{2} = 0.5$	$\frac{1}{4} = 0.25$	$\frac{1}{5} = 0.2$	$\frac{1}{8} = 0.125$
	$\frac{3}{4} = 0.75$	$\frac{2}{5} = 0.4$	$\frac{3}{8} = 0.375$
		$\frac{3}{5} = 0.6$	$\frac{5}{8} = 0.625$
		$\frac{4}{5} = 0.8$	$\frac{7}{8} = 0.875$

If the decimal part of a number does not terminate, you can round the result to approximate the fraction to some specified number of decimal places.

Example 2 Converting a Fraction to a Decimal

Write $\frac{3}{7}$ as a decimal. Round the answer to the nearest thousandth.

$$\begin{array}{r} 0.4285 \\ 7\overline{)3.0000} \\ \underline{2\,8} \\ 20 \\ \underline{14} \\ 60 \\ \underline{56} \\ 40 \\ \underline{35} \\ 5 \end{array}$$

In this example, we are rounding to three decimal places, so we must add enough zeros to carry the division to four decimal places.

So $\frac{3}{7} = 0.429$ (to the nearest thousandth)

Check Yourself 2

Write $\frac{5}{11}$ as a decimal. Round your result to the nearest thousandth.

If the decimal part does *not* terminate, it will *repeat* a sequence of digits. These are called **repeating** or **nonterminating decimals.**

Example 3 Converting a Fraction to a Repeating Decimal

< Objective 2 >

(a) Write $\frac{1}{3}$ in decimal form.

$$\begin{array}{r} 0.333 \\ 3\overline{)1.000} \\ \underline{9} \\ 10 \\ \underline{9} \\ 10 \\ \underline{9} \end{array}$$

The digit 3 repeats indefinitely because each new remainder is 1.

Adding more zeros and going on simply leads to more 3s in the quotient.

We can say $\frac{1}{3} = 0.333\ldots$

The three dots mean "and so on" and tell us that 3 repeats itself indefinitely.

(b) Write $\frac{5}{12}$ as a decimal.

$$\begin{array}{r} 0.4166\ldots \\ 12\overline{)5.0000} \\ \underline{4\,8} \\ 20 \\ \underline{12} \\ 80 \\ \underline{72} \\ 80 \\ \underline{72} \\ 8 \end{array}$$

In this example, the digit 6 repeats itself because the remainder, 8, keeps occurring as we add more zeros and continue the division.

so $\frac{5}{12} = 0.4166\ldots$

Check Yourself 3

Write each fraction as a decimal.

(a) $\frac{2}{3}$ **(b)** $\frac{7}{12}$

Some important decimal equivalents (rounded to the nearest thousandth) are shown here as a reference.

$\frac{1}{3} = 0.333$ $\frac{1}{6} = 0.167$ $\frac{2}{3} = 0.667$ $\frac{5}{6} = 0.833$

Another way to write a repeating decimal is with a bar placed over the digit or digits that repeat. For example, we can write

0.37373737 . . .

as

$0.\overline{37}$

The bar placed over the digits indicates that 37 repeats indefinitely.

Example 4 Converting a Fraction to a Repeating Decimal

Write $\frac{5}{11}$ as a decimal.

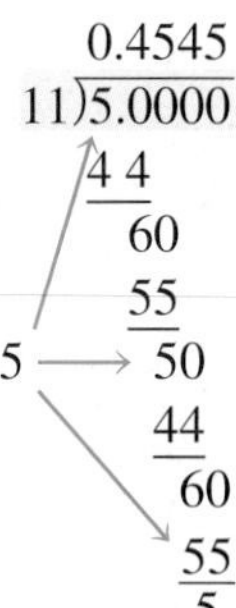

As soon as a remainder repeats itself, as 5 does here, the pattern of digits repeats in the quotient.

Therefore, $\frac{5}{11} = 0.\overline{45} = 0.4545 \ldots.$

Check Yourself 4

Use bar notation to write the decimal equivalent of $\frac{5}{7}$. (Be patient. You have to divide for a while to find the repeating pattern.)

You can write mixed numbers as decimals in a similar way. Find the decimal equivalent of the fraction part of the mixed number and then combine that with the whole-number part.

Example 5 Converting a Mixed Number to a Decimal

Find the decimal equivalent of $3\frac{5}{16}$.

$\frac{5}{16} = 0.3125$ First find the equivalent of $\frac{5}{16}$ by division.

$3\frac{5}{16} = 3.3125$ Add 3 to the result.

Check Yourself 5

Find the decimal equivalent of $2\frac{5}{8}$.

We learned something important in this section. To write a fraction as a decimal, we use division. Because the remainder must be less than the divisor, the remainder must *either repeat or become zero.* Thus, *every fraction* has a *repeating* or a *terminating* decimal as its decimal equivalent.

Next, using what we learned about place values, you can write decimals as fractions.

Step by Step

Converting a Terminating Decimal to a Fraction

Step 1 Write the digits of the decimal without the decimal point. This is the numerator of the fraction.

Step 2 The denominator of the fraction can be written as a 1 followed by as many zeros as there are places in the decimal.

Step 3 Simplify the fraction.

Example 6 Converting a Decimal to a Fraction

< Objective 3 >

$0.7 = \frac{7}{10}$ (One place; One zero)

$0.09 = \frac{9}{100}$ (Two places; Two zeros)

$0.257 = \frac{257}{1{,}000}$ (Three places; Three zeros)

Check Yourself 6

Write as fractions.

(a) 0.3 **(b)** 0.311

When a decimal is converted to a fraction, that resulting fraction should always be simplified.

Example 7 Converting a Decimal to a Fraction

Convert 0.395 to a fraction and write the result in lowest terms.

$$0.395 = \frac{395}{1{,}000} = \frac{79}{200}$$

NOTE

Divide the numerator and denominator by 5.

Check Yourself 7

Write 0.275 as a fraction.

If the decimal has a whole-number part, write the digits to the right of the decimal point as a proper fraction and then form a mixed number for your result.

Example 8 Converting a Decimal to a Mixed Number

Write 12.277 as a mixed number.

As before, we write the decimal part as a fraction.

$$0.277 = \frac{277}{1{,}000}$$

NOTE

You will learn to write a repeating decimal as a fraction in a future math class.

Now we add 12 to construct a mixed number.

$12.277 = 12\frac{277}{1,000}$

Check Yourself 8

Write 32.433 as a mixed number.

The easiest way to compare the size of a fraction to a decimal is to convert the fraction to an equivalent decimal and then compare the decimals.

Example 9 Comparing the Sizes of Fractions and Decimals

RECALL

0.38 can be written as 0.380. Comparing this to 0.375, we see that 0.380 > 0.375.

Which is larger, $\frac{3}{8}$ or 0.38?

Write the decimal equivalent of $\frac{3}{8}$. That decimal is 0.375. Now comparing 0.375 and 0.38, we see that 0.38 is the larger of the numbers.

$0.38 > \frac{3}{8}$

Check Yourself 9

Which is larger, $\frac{3}{4}$ or 0.8?

Check Yourself ANSWERS

1. 0.875 **2.** 0.455 **3.** **(a)** 0.666 . . . ; **(b)** 0.5833 . . . **4.** $\frac{5}{7} = 0.\overline{714285}$ **5.** 2.625 **6.** **(a)** $\frac{3}{10}$; **(b)** $\frac{311}{1,000}$ **7.** $\frac{11}{40}$ **8.** $32\frac{433}{1,000}$ **9.** $0.8 > \frac{3}{4}$

Reading Your Text

These fill-in-the-blank exercises will help you understand some of the key vocabulary used in this section. The answers to these exercises are in the Answers Appendix at the back of the text.

(a) You can __________ the numerator of a fraction by its denominator to convert a fraction to a decimal.

(b) We can write a repeating decimal with a __________ placed over the digit or digits that repeat.

(c) Every fraction has a repeating or a __________ decimal as its equivalent.

(d) The denominator of a fraction equivalent to a given decimal can be written as a 1 followed by as many zeros as there are __________ in the decimal.

4.5 exercises

Skills | Calculator/Computer | Career Applications | Above and Beyond

< Objective 1 >

Write each fraction as a decimal.

1. $\frac{3}{4}$ 
2. $\frac{4}{5}$
3. $\frac{9}{20}$
4. $\frac{3}{10}$

5. $\frac{1}{5}$
6. $\frac{1}{8}$
7. $\frac{5}{16}$
8. $\frac{11}{20}$

9. $\frac{7}{10}$
10. $\frac{7}{16}$
11. $\frac{27}{40}$
12. $\frac{17}{32}$

Write each fraction as a decimal. Round your result as indicated.

13. $\frac{5}{6}$ thousandths
14. $\frac{7}{12}$ hundredths
15. $\frac{4}{15}$ thousandths

16. $\frac{2}{3}$ hundredths
17. $\frac{15}{9}$ two decimal places
18. $\frac{17}{11}$ four decimal places

< Objective 2 >

Use bar notation to write each fraction as a decimal.

19. $\frac{1}{18}$
20. $\frac{4}{9}$
21. $\frac{3}{11}$
22. $\frac{1}{6}$

23. $\frac{1}{12}$
24. $\frac{5}{12}$

Write each number as a decimal.

25. $5\frac{3}{5}$
26. $4\frac{7}{16}$
27. $12\frac{7}{20}$
28. $5\frac{1}{2}$

29. $\frac{9}{4}$
30. $\frac{11}{8}$

< Objective 3 >

Write each number as a fraction or mixed number, in simplest terms.

31. 0.9
32. 0.3
33. 0.8
34. 0.6

35. 0.37
36. 0.97
37. 0.587
38. 0.379

39. 0.48
40. 0.75
41. 18
42. 3

43. 0.425
44. 0.116
45. 0.375
46. 0.225

47. 6.136
48. 11.575
49. 0.059
50. 0.067

51. **Statistics** In a weekend baseball tournament, Joel had 4 hits in 13 times at bat. That is, he hit safely $\frac{4}{13}$ of the time. Write the decimal equivalent for Joel's hitting, rounding to three decimal places. (This is Joel's batting average.)

52. **Statistics** The table gives the wins and losses of the teams in the National League East as of mid-September in a recent season. The winning percentage of each team is calculated by writing the number of wins over the total games played and converting this fraction to a decimal. Convert this fraction to a decimal for every team, rounding to three decimal places.

Team	Wins	Losses
Atlanta	92	56
New York	90	58
Philadelphia	70	77
Washington	61	88
Florida	57	89

53. **Statistics** The table gives the wins and losses of the teams in the Western Division of the National Football Conference for a recent season. Determine the fraction of wins over total games played for every team, rounding to three decimal places for each of the teams.

Team	Wins	Losses
San Francisco	10	6
St. Louis	7	9
Seattle	7	9
Arizona	5	11

54. **Statistics** The table gives the free throws attempted (FTA) and the free throws made (FTM) for the top five players in the NBA for a recent season. Calculate the free throw percentage for each player by writing the FTM over the FTA and converting this fraction to a decimal. Round to three decimal places.

Player	FTM	FTA
Stephen Curry	212	227
Chauncey Billups	384	419
Steve Nash	227	249
D. J. Augustin	269	297
Jodie Meeks	152	170

Use bar notation to write each fraction as a decimal.

55. $\frac{1}{11}$ **56.** $\frac{1}{111}$ **57.** $\frac{1}{1{,}111}$

58. From the pattern of exercises 55 to 57, can you guess the decimal representation for $\frac{1}{11{,}111}$?

Insert > or < to form a true statement.

59. $\frac{31}{34}$ ☐ 0.9118 **60.** $\frac{21}{37}$ ☐ 0.5664 **61.** $\frac{13}{17}$ ☐ 0.7657 **62.** $\frac{7}{8}$ ☐ 0.87 

63. $\frac{5}{16}$ ☐ 0.313 **64.** $\frac{9}{25}$ ☐ 0.4

Find each sum. Give your answers in decimal form, rounded to three decimal places when appropriate.

65. $\frac{1}{2} + 0.385$ **66.** $4.24 + \frac{4}{3}$ **67.** $8.6245 + \frac{18}{11}$ **68.** $\frac{11}{6} + 2.25$

69. $3\frac{3}{5} + 5.608$ **70.** $13.1667 + 9\frac{2}{3}$

Skills | **Calculator/Computer** | Career Applications | Above and Beyond

A calculator is very useful in converting fractions to decimals. Just divide the numerator by the denominator, and the decimal equivalent is displayed. Often, you will have to round the result in the display. For example, to find the decimal equivalent of $\frac{5}{24}$ to the nearest hundredth, enter

5 [÷] 24 [=]

The display may show 0.2083333, and rounding, we have $\frac{5}{24} = 0.21$.

When you are converting a mixed number to a decimal, add the whole-number part. To change $7\frac{5}{8}$ to a decimal, for example, enter

7 [+] 5 [÷] 8 [=]

The result is 7.625.

Use a calculator to write each fraction as a decimal.

71. $\frac{7}{8}$

72. $\frac{9}{16}$

73. $\frac{5}{32}$ to the thousandth

74. $\frac{11}{75}$ to the thousandth

75. $\frac{3}{11}$ using bar notation

76. $\frac{16}{33}$ using bar notation

77. $3\frac{7}{8}$

78. $8\frac{3}{16}$

Skills | Calculator/Computer | **Career Applications** | Above and Beyond

79. Allied Health The internal diameter, in millimeters, of an endotracheal tube for a child is calculated using the formula $\frac{\text{Height}}{20}$, based on the child's height in centimeters. Determine the size of endotracheal tube needed for a girl who is 110 cm tall. Write your answer as a decimal.

80. Allied Health The stroke volume, which measures the average cardiac output per heartbeat (liters/beat), is based on a patient's cardiac output (CO), in liters per minute, and heart rate (HR), in beats per minute (beats/min). It is calculated using the fraction $\frac{CO}{HR}$. Determine the stroke volume for a patient whose cardiac output is 4 L/min and whose heart rate is 80 beats/min. Write your answer as a decimal.

81. Information Technology The propagation delay for a satellite connection is 0.350 seconds (s). Convert to a fraction and simplify.

82. Information Technology From your computer, it takes 0.0021 s to transmit a ping packet to another computer. Convert to a fraction and simplify.

Skills | Calculator/Computer | Career Applications | **Above and Beyond**

83. Use bar notation to write each fraction as a decimal.

$\frac{1}{7}, \frac{2}{7}, \frac{3}{7}, \frac{4}{7}, \frac{5}{7}$

Describe any patterns that you see. Predict the decimal equivalent of $\frac{6}{7}$.

Answers

1. 0.75 **3.** 0.45 **5.** 0.2 **7.** 0.3125 **9.** 0.7 **11.** 0.675 **13.** 0.833 **15.** 0.267 **17.** 1.67 **19.** $0.0\overline{5}$ **21.** $0.\overline{27}$ **23.** $0.08\overline{3}$ **25.** 5.6 **27.** 12.35 **29.** 2.25 **31.** $\frac{9}{10}$ **33.** $\frac{4}{5}$ **35.** $\frac{37}{100}$ **37.** $\frac{587}{1,000}$ **39.** $\frac{12}{25}$ **41.** $\frac{18}{1}$ **43.** $\frac{17}{40}$ **45.** $\frac{3}{8}$ **47.** $6\frac{17}{125}$ **49.** $\frac{59}{1,000}$ **51.** 0.308 **53.** 0.625; 0.438; 0.438; 0.313 **55.** $0.\overline{09}$ **57.** $0.\overline{0009}$ **59.** $<$ **61.** $<$ **63.** $<$ **65.** 0.885 **67.** 10.261 **69.** 9.208 **71.** 0.875 **73.** 0.156 **75.** 0.27 **77.** 3.875 **79.** 5.5 mm **81.** $\frac{7}{20}$ s **83.** Above and Beyond

Terminate or Repeat?

Every fraction can be written as a decimal that either terminates $\left(\text{for example, } \frac{1}{4} = 0.25\right)$ or repeats $\left(\text{for example, } \frac{2}{9} = 0.\overline{2}\right)$. Work with a group to discover which fractions have terminating decimals and which have repeating decimals. You may assume that the numerator of each fraction you consider is one and focus your attention on the denominator. As you complete the table, you will find that the key to this question lies with the prime factorization of the denominator.

Fraction	Decimal Form	Terminate?	Prime Factorization of the Denominator
$\frac{1}{2}$			
$\frac{1}{3}$			
$\frac{1}{4}$			
$\frac{1}{5}$			
$\frac{1}{6}$			
$\frac{1}{7}$			
$\frac{1}{8}$			
$\frac{1}{9}$			
$\frac{1}{10}$			
$\frac{1}{11}$			
$\frac{1}{12}$			

State a general rule describing which fractions have decimal forms that terminate and which have decimal forms that repeat.

Now test your rule on at least three new fractions. That is, be able to predict whether a fraction such as $\frac{1}{25}$ or $\frac{1}{30}$ has a terminating decimal or a repeating decimal. Then confirm your prediction.

summary :: chapter 4

Definition/Procedure	Example	Reference
Place Value and Rounding		**Section 4.1**
Decimal Fraction A fraction whose denominator is a power of 10. We call decimal fractions *decimals*.	$\frac{7}{10}$ and $\frac{47}{100}$ are decimal fractions.	*p.* 219
Decimal Place Each position for a digit to the right of the decimal point. Each decimal place has a place value that is $\frac{1}{10}$ the value of the place to its left.	2.3456 — Tenths (3), Hundredths (4), Thousandths (5), Ten-thousandths (6)	*p.* 220
Reading and Writing Decimals in Words		
Step 1 Read the digits *to the left* of the decimal point as a whole number. **Step 2** Read the decimal point as the word *and*. **Step 3** Read the digits *to the right* of the decimal point as a whole number followed by the place value of the rightmost digit.	Hundredths 8.15 is read "eight and fifteen hundredths."	*p.* 220
Rounding Decimals		
Step 1 Find the place where the decimal is to be rounded. **Step 2** If the next digit to the right is 5 or more, increase the digit in the place you are rounding to by 1. Discard any remaining digits to the right. **Step 3** If the next digit to the right is less than 5, just discard that digit and any remaining digits to the right.	To round 5.87 to the nearest tenth: 5.87 rounds to 5.9 To round 12.3454 to the nearest thousandth: 12.3454 rounds to 12.345.	*p.* 222
Adding and Subtracting Decimals		**Section 4.2**
To Add Decimals		
Step 1 Write the numbers being added in column form *with their decimal points in a vertical line*. **Step 2** Add just as you would with whole numbers. **Step 3** Place the decimal point of the sum in line with the decimal points of the addends.	To add 2.7, 3.15, and 0.48: $\begin{array}{r} 2.7 \\ 3.15 \\ +\ 0.48 \\ \hline 6.33 \end{array}$	*p.* 228
To Subtract Decimals		
Step 1 Write the numbers being subtracted in column form *with their decimal points in a vertical line*. You may have to place zeros to the right of the existing digits. **Step 2** Subtract just as you would with whole numbers. **Step 3** Place the decimal point of the difference in line with the decimal points of the numbers being subtracted.	To subtract 5.875 from 8.5: $\begin{array}{r} 8.500 \\ -\ 5.875 \\ \hline 2.625 \end{array}$	*p.* 231
Multiplying Decimals		**Section 4.3**
To Multiply Decimals		
Step 1 Multiply the decimals as though they were whole numbers. **Step 2** Count the number of decimal places in the factors. **Step 3** Place the decimal point in the product so that the number of decimal places in the product is the sum of the number of decimal places in the factors.	To multiply 2.85×0.045: 2.85 ⟵ Two places $\times$ 0.045 ⟵ Three places 1425 1140 0.12825 ⟵ Five places	*p.* 239

Continued

Definition/Procedure	Example	Reference
Multiplying by Powers of 10		
Move the decimal point to the right the same number of places as there are zeros in the power of 10.	$2.37 \times 10 = 23.7$ $0.567 \times 1{,}000 = 567$	p. 242
Dividing Decimals		**Section 4.4**
To Divide by a Decimal		
Step 1 Move the decimal point in the divisor to the right, making the divisor a whole number. **Step 2** Move the decimal point in the dividend to the right the same number of places. Add zeros if necessary. **Step 3** Place the decimal point in the quotient directly above the new decimal point in the dividend. **Step 4** Divide. You are dividing by a whole number.	To divide 16.5 by 5.5, move the decimal points: $5.5_\wedge\overline{)16.5_\wedge}$ = 3 16 5 0	p. 252
To Divide by a Power of 10		
Move the decimal point to the left the same number of places as there are zeros in the power of 10.	$25.8 \div 10 = 2_\wedge5.8 = 2.58$	p. 254
Converting Between Fractions and Decimals		**Section 4.5**
To Convert a Fraction to a Decimal		
Step 1 Divide the numerator of the fraction by its denominator. **Step 2** The quotient is the decimal equivalent of the fraction.	To convert $\frac{1}{2}$ to a decimal: $2\overline{)1.0}$ = 0.5 1 0 0	p. 261
To Convert a Terminating Decimal to a Fraction		
Step 1 Write the digits of the decimal without the decimal point. This is the numerator of the fraction. **Step 2** The denominator of the fraction can be written as a 1 followed by as many zeros as there are places in the decimal. **Step 3** Simplify the fraction, if necessary.	To convert 0.275 to a fraction: $0.275 = \frac{275}{1{,}000} = \frac{11}{40}$	p. 264

summary exercises :: chapter 4

This summary exercise set will help ensure that you have mastered each of the objectives of this chapter. The exercises are grouped by section. You should reread the material associated with any exercises that you find difficult. The answers to the odd-numbered exercises are in the Answers Appendix at the back of the text.

4.1 *Find the indicated place values.*

1. 7 in 3.5742

2. 3 in 0.5273

Write the fractions in decimal form.

3. $\frac{37}{100}$

4. $\frac{307}{10{,}000}$

Write the decimals in words.

5. 0.071

6. 12.39

Write the fractions in decimal form.

7. Four and five tenths

8. Four hundred and thirty-seven thousandths

Complete each statement, using the symbol <, =, or >.

9. 0.79 ______ 0.785

10. 1.25 ______ 1.250

11. 12.8 ______ 13

12. 0.832 ______ 0.83

Round to the indicated place.

13. 5.837 hundredths

14. 9.5723 thousandths

15. 4.87625 three decimal places

Write each number as a fraction or a mixed number.

16. 0.0067

17. 0.84

18. 21.857

4.2 *Add.*

19. $\begin{array}{r} 2.58 \\ +\ 0.89 \\ \hline \end{array}$

20. $\begin{array}{r} 3.14 \\ 0.8 \\ 2.912 \\ +\ 12 \\ \hline \end{array}$

21. 1.3, 25, 5.27, and 6.158

22. Add eight, forty-three thousandths, five and nineteen hundredths, and seven and three tenths.

Subtract.

23. $\begin{array}{r} 29.21 \\ -\ 5.89 \\ \hline \end{array}$

24. $\begin{array}{r} 6.73 \\ -\ 2.485 \\ \hline \end{array}$

25. Subtract 1.735 from 2.81

26. Subtract 12.38 from 19

Solve each application.

27. GEOMETRY Find the perimeter (to the nearest hundredth of a centimeter) of a rectangle that has dimensions 5.37 cm by 8.64 cm.

28. **Science and Medicine** Janice ran 4.8 mi on Sunday, 5.3 mi on Tuesday, 3.9 mi on Thursday, and 8.2 mi on Saturday. How far did she run during the week?

29. **Geometry** Find dimension a in the figure.

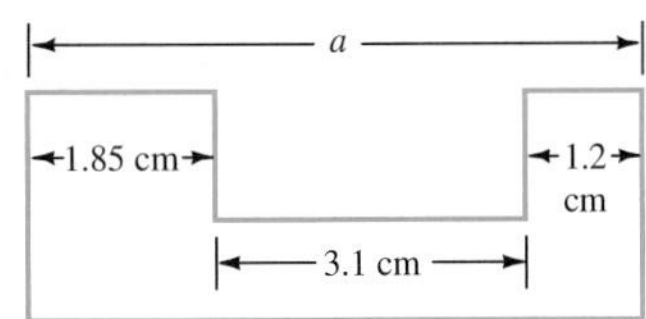

30. **Business and Finance** A stereo system that normally sells for \$499.50 is discounted (or marked down) to \$437.75 for a sale. Find the savings.

4.3 *Multiply.*

31. $\begin{array}{r} 22.8 \\ \times\ 0.72 \\ \hline \end{array}$

32. $\begin{array}{r} 0.0045 \\ \times\ 0.058 \\ \hline \end{array}$

33. 1.24×56

34. 0.0025×0.491

35. $0.052 \times 1{,}000$

36. 0.045×10^4

Solve each application.

37. **Business and Finance** Neal worked for 37.4 hours during a week. If his hourly rate of pay was \$7.25, how much did he earn?

38. **Business and Finance** To find the simple interest on a loan at $11\frac{1}{2}\%$ for 1 year, we must multiply the amount of the loan by 0.115. Find the simple interest on a \$2,500 loan at $11\frac{1}{2}\%$ for 1 year.

39. **Business and Finance** A television set has an advertised price of \$499.50. You buy the set and agree to make payments of \$27.15 per month for 2 years. How much extra are you paying by buying with this installment plan?

40. **Business and Finance** A stereo dealer buys 100 portable radios for a promotion sale. If she pays \$57.42 per radio, what is her total cost?

4.4 *Divide. Round answers to the nearest hundredth.*

41. $58\overline{)269.7}$

42. $55\overline{)17.69}$

Divide. Round answers to the nearest thousandth.

43. $0.7\overline{)1.865}$

44. $3.042 \div 0.37$

45. $5.3\overline{)6.748}$

46. $0.2549 \div 2.87$

Divide.

47. $7.6 \div 10$

48. $80.7 \div 1{,}000$

49. $457 \div 10^4$

50. $322.91 \div 10^2$

Solve each application.

51. **Business and Finance** During a charity fund-raising drive 37 employees of a company donated a total of \$867.65. What was the donation per employee?

52. **Business and Finance** Faith always fills her gas tank as soon as the gauge hits the $\frac{1}{4}$ full mark. In six readings, Faith's gas mileage was 38.9, 35.3, 39.0, 41.2, 40.5, and 40.8 mi/gal. What was the average mileage to the nearest tenth of a mile per gallon?

Hint: First find the sum of the mileages. Then divide the sum by 6, because there are 6 mileage readings.

53. **Construction** A developer is planning to subdivide an 18.5-acre piece of land. She estimates that 5 acres will be used for roads and wants individual lots of 0.25 acre. How many lots are possible?

54. **Business and Finance** Paul drives 949 mi, using 31.8 gal of gas. What is his mileage for the trip (to the nearest tenth of a mile per gallon)?

4.5 *Write each number in decimal form.*

55. $\frac{7}{16}$ **56.** $\frac{3}{7}$ (round to the nearest thousandth) **57.** $\frac{4}{15}$ (use bar notation) **58.** $3\frac{3}{4}$

Write as fractions or mixed numbers. Simplify your answers.

59. 0.21 **60.** 0.084 **61.** 2.03 **62.** 5.28

chapter test 4

CHAPTER 4

Use this chapter test to assess your progress and to review for your next exam. Allow yourself about an hour to take this test. The answers to these exercises are in the Answers Appendix at the back of the text.

1. Find the place value of 8 in 0.5248. **2.** Write 2.53 in words.

3. Write twelve and seventeen thousandths in decimal form.

4. Add: seven and seventy-nine hundredths, and five and thirteen thousandths

Round to the indicated place.

5. 0.5977 thousandths **6.** 23.5724 two decimal places

7. 36,139.0023 thousands **8.** Write $\frac{49}{1{,}000}$ in decimal form.

Evaluate, as indicated.

9. $\begin{array}{r} 3.45 \\ 0.6 \\ +\ 12.59 \\ \hline \end{array}$ **10.** $\begin{array}{r} 18.32 \\ -\ 7.78 \\ \hline \end{array}$ **11.** $4.1\overline{)10.455}$ **12.** 2.75×0.53

13. $27\overline{)63.45}$ **14.** Add: 2.4, 35, 4.73, and 5.123 **15.** $\begin{array}{r} 32.9 \\ \times\ 0.53 \\ \hline \end{array}$

16. $\begin{array}{r} 40 \\ -\ 15.625 \\ \hline \end{array}$ **17.** $4.983 \div 1{,}000$ **18.** $523 \div 10^5$ **19.** $0.735 \times 1{,}000$

20. 1.257×10^4

21. Subtract: 1.742 from 5.63

22. $8\overline{)3.72}$

23. 0.049×0.57

24. $0.6\overline{)1.431}$

25. $3.969 \div 0.54$

26. $2.72 \div 53$ thousandths

27. $0.263 \div 3.91$ three decimal places

Use the symbol < or > to complete each statement.

28. 0.889_______ 0.89

29. 0.531 _______ 0.53

30. 0.168 _______ $\frac{3}{25}$

Write each number in decimal form. When indicated, round to the given place value.

31. $\frac{7}{16}$

32. $\frac{3}{7}$ thousandths

33. $\frac{7}{11}$ use bar notation

Write the decimals as fractions or mixed numbers. Simplify your answers.

34. 0.072

35. 4.44

36. **Business and Finance** A college bookstore purchases 1,000 pens at a cost of 54.3 cents per pen. Find the total cost of the order in dollars.

37. **Business and Finance** On a business trip, Martin bought the following amounts of gasoline: 14.4, 12, 13.8, and 10 gal. How much gasoline did he purchase on the trip?

38. **Construction** A 14-acre piece of land is being developed into home lots. If 2.8 acres of land will be used for roads and each home site is to be 0.35 acre, how many lots can be formed?

39. Find the area of a rectangle with length 3.5 in. and width 2.15 in.

40. **Construction** A street improvement project will cost $57,340, and that cost is to be divided among the 100 families in the area. What will be the cost to each individual family?

41. **Business and Finance** You pay for purchases of $13.99, $18.75, $9.20, and $5 with a $50 bill. How much cash will you have left?

42. A baseball team has a winning percentage of 0.458. Write this as a fraction in simplest form.

cumulative review chapters 1–4

Use this exercise set to review concepts from earlier chapters. While it is not a comprehensive exam, it will help you identify any material that you need to review before moving on to the next chapter. In addition to the answers, you will find section references for these exercises in the Answers Appendix in the back of the text.

1.1

1. Write 286,543 in words.

2. What is the place value of 5 in the number 343,563?

In exercises 3 to 8, perform the indicated operations.

1.2

3. $\begin{array}{r} 2,340 \\ 685 \\ +\ 31,569 \\ \hline \end{array}$

1.3

4. $\begin{array}{r} 75,363 \\ -\ 26,475 \\ \hline \end{array}$

1.5

5. 83×61

6. 231×305

1.6

7. $21\overline{)357}$

8. $463\overline{)16,216}$

1.7

9. Evaluate the expression $18 \div 2 + 4 \times 2^3 - (18 - 6)$.

1.4

10. Round each number to the nearest hundred and find an estimated sum.

$294 + 725 + 2,321 + 689$

1.2 and 1.5

11. Find the perimeter and area of the given figure.

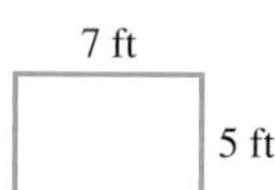

2.4

12. Write the fraction $\frac{15}{51}$ in simplest form.

Perform the indicated operations.

2.5

13. $\frac{2}{3} \times \frac{9}{8}$

14. $1\frac{2}{3} \times 1\frac{5}{7}$

2.6

15. $\frac{3}{4} \div \frac{17}{12}$

3.1

16. $\frac{6}{7} - \frac{3}{7} + \frac{2}{7}$

3.3

17. $\frac{4}{5} - \frac{7}{10} + \frac{4}{30}$

3.4

18. $6\frac{3}{5} - 2\frac{7}{10}$

4.2

19. $35.218 - 22.75$

4.4

20. $2.262 \div 0.58$

21. $523.8 \div 10^5$

4.3

22. 2.53×0.45

4.5

23. 1.53×10^4

24. Write 0.43 as a fraction.

4.5

25. Write the decimal equivalent of each number. Round to the given place value when indicated.

(a) $\frac{5}{8}$

(b) $\frac{9}{23}$ hundredths

4.4

26. Sam has had 15 hits in his last 35 at bats. That is, he has had a hit in $\frac{15}{35}$ of his times at bat. Write this as a decimal, rounding to the nearest thousandth.

Evaluate each expression.

4.4

27. $18.4 - 3.16 \times 2.5 + 6.71$

28. $17.6 \div 2.3 \times 3.4 + 13.812$ (Round to the nearest thousandth.)

Solve each application.

29. GEOMETRY If the perimeter of a square is 19.2 cm, how long is each side?

30. BUSINESS AND FINANCE In 1 week, Tom earned $356.60 by working 36.25 hr. What was his hourly rate of pay to the nearest cent?

31. CONSTRUCTION An 80.5-acre piece of land is being subdivided into 0.35-acre lots. How many lots are possible in the subdivision?

Ratios and Proportions

CHAPTER 5 OUTLINE

INTRODUCTION

People exercise for many reasons. Strength, fitness, health, weight loss, and endurance are all benefits associated with working out and exercising.

Because everyone is physically different, it is difficult or even impossible to provide recommendations that apply to everyone. To help with this, we often use proportions to convert recommendations to appropriate quantities. For instance, we might give a child a fraction of an adult's dosage of medication.

Proportions are also used to recommend exercise routines and determine appropriate levels of breathing and heart rate based on an individual's height, weight, age, and other personal characteristics.

In Activity 15, you will explore proportions in the context of exercise and health.

5 prerequisite check

This Prerequisite Check highlights the skills you will need in order to be successful in this chapter. The answers to these exercises are in the Answers Appendix at the back of the text.

Simplify each fraction.

1. $\frac{24}{32}$

2. $\frac{280}{525}$

3. $\frac{\frac{45}{2}}{30}$

Rewrite the mixed number as an improper fraction.

4. $5\frac{3}{5}$

Perform the indicated operations.

5. $\frac{45}{2} \times \frac{1}{30}$

6. 5.25×100

7. Find the GCF of 280 and 525.

Write each fraction as a decimal. Round to the nearest hundredth.

8. $\frac{50}{9}$

9. $\frac{260}{36}$

Determine whether each pair of fractions is equivalent.

10. $\frac{6}{17}$ and $\frac{4}{11}$

11. $\frac{7}{15}$ and $\frac{84}{180}$

5.1 Ratios

< 5.1 Objectives >

1 > Write the ratio of two quantities

2 > Simplify the ratio of two quantities

In Chapter 2, you learned to think about fractions in two ways.

1. A fraction can name a number of parts of a whole. The fraction $\frac{3}{5}$ names 3 parts of a whole that has been divided into 5 equal parts.
2. A fraction can indicate division. The fraction $\frac{3}{5}$ is the quotient $3 \div 5$.

We now want to turn to a third meaning for a fraction—a fraction can be a *ratio.*

Definition

Ratio A **ratio** is a comparison of two numbers or like quantities.

NOTE

In this text, we write ratios as simplified fractions.

The ratio a to b can also be written as $a:b$ and $\frac{a}{b}$. Ratios are always written in simplest form.

Example 1 Writing a Ratio as a Fraction

< Objective 1 >

Write the ratio 3 to 5 as a fraction.

To compare 3 to 5, we write the ratio of 3 to 5 as $\frac{3}{5}$.

So, $\frac{3}{5}$ also means "the ratio of 3 to 5."

NOTE

Alternatively, the ratio of 3 to 5 can be written as 3:5.

Check Yourself 1

Write the ratio of 7 to 12 as a fraction.

RECALL

Numbers with units attached are called *denominate numbers.*

Ratios are often used to compare *like quantities* such as quarts to quarts, centimeters to centimeters, and apples to apples. In this case, we can simplify the fraction by "canceling" the units. In its simplest form, a ratio is always written without units.

Example 2 Ratios of Denominate Numbers

RECALL

We simplified fractions containing units in Sections 2.5 and 2.6.

A rectangle measures 7 cm wide and 19 cm long.

7 cm

19 cm

(a) Write the ratio of its width to its length, as a fraction.

$$\frac{7 \text{ cm}}{19 \text{ cm}} = \frac{7 \cancel{\text{cm}}}{19 \cancel{\text{cm}}} = \frac{7}{19}$$

We are comparing centimeters to centimeters, so we can simplify the fraction.

NOTE

Ratios are *rarely* written as whole or mixed numbers. We usually write a ratio as a fraction, in simplest terms.

(b) Write the ratio of its length to its width, as a fraction.

We need to write the ratio in the order requested by the example, rather than in the order given in the preceding description.

$$\frac{19 \text{ cm}}{7 \text{ cm}} = \frac{19 \not{\text{cm}}}{7 \not{\text{cm}}} = \frac{19}{7}$$

Check Yourself 2

A basketball team wins 17 of its 29 games in a season.

(a) Write the ratio of wins to games played.
(b) Write the ratio of wins to losses.

We always report a ratio in simplest terms. In practice, this means that we simplify the number part of the fraction. It also means that the units "cancel," so that there are no units in the ratio, only two whole numbers.

Step by Step

Writing a Ratio in Simplest Terms

Step 1 Write the ratio as a fraction.

Step 2 Simplify the units so that you only have numbers in the fraction.

Step 3 Simplify the fraction $\frac{a}{b}$ so that a and b are whole numbers with no common factors other than 1 (that is, write the fraction in simplest terms).

Step 4 Determine whether the application requires the result to be reported as an improper fraction or as a mixed number. Write the simplified ratio in the appropriate form.

Example 3 Simplifying a Ratio

< Objective 2 >

RECALL

This ratio may also be written as 2:3.

Write the ratio of 20 to 30 in simplest terms.

Begin by writing the fraction that represents the ratio: $\frac{20}{30}$. Now, simplify this fraction.

$$\frac{20}{30} = \frac{2}{3}$$ Simplify the fraction by dividing both the numerator and denominator by 10.

Check Yourself 3

Write the ratio of 24 to 32 in simplest terms.

Because ratios are used to compare like quantities, a simplified ratio has no units. In Example 4, we need to simplify both the numbers and the units.

Example 4 Simplifying the Ratio of Denominate Numbers

NOTE

Widescreen movies are often in 16:9 format.

A common size for a movie screen is 32 ft by 18 ft. Write this ratio in simplest form.

$$\frac{32 \text{ ft}}{18 \text{ ft}} = \frac{32 \not{\text{ft}}}{18 \not{\text{ft}}} = \frac{32}{18} = \frac{16}{9}$$ The GCF of 32 and 18 is 2.

Check Yourself 4

A common computer display mode is 640 pixels (picture elements) by 480 pixels. Write this as a ratio in simplest terms.

Often, the quantities in a ratio are given as fractions or decimals. In either of these cases, the ratio should be rewritten as an equivalent ratio comparing whole numbers.

Example 5 Simplifying Ratios

RECALL

In Section 2.3, you learned to rewrite $22\frac{1}{2}$ as $\frac{45}{2}$.

RECALL

In Section 2.6, you learned to simplify complex fractions.

RECALL

In Section 4.3, you learned to multiply a decimal by a power of 10 by moving the decimal point.

(a) Loren sank a $22\frac{1}{2}$-ft putt and Carrie sank a 30-ft putt. Express the ratio of the two distances as a ratio of whole numbers.

30'

$22\frac{1}{2}$'

We begin by writing the ratio of the two distances: $\dfrac{22\frac{1}{2}\text{ ft}}{30\text{ ft}}$. Then, we cancel the units and rewrite the mixed number as an improper fraction.

$$\frac{22\frac{1}{2}\text{ ft}}{30\text{ ft}} = \frac{\frac{45}{2}}{30}$$

To simplify this complex fraction, we rewrite it as a division problem.

$$\frac{\frac{45}{2}}{30} = \frac{45}{2} \div 30$$

$$= \frac{45}{2} \times \frac{1}{30} \quad \text{Flip the second fraction and multiply.}$$

$$= \frac{\overset{3}{\cancel{45}} \times 1}{2 \times \underset{2}{\cancel{30}}} \quad \text{The GCF of 45 and 30 is 15.}$$

$$= \frac{3}{4}$$

The ratio 3 to 4 is equivalent to the ratio $22\frac{1}{2}$ ft to 30 ft.

(b) The diameter of a 20-oz bottle is 2.8 in. The diameter of a 2-L bottle is 5.25 in. Express the ratio of the two diameters as a ratio of whole numbers.

$$\frac{2.8\text{ in.}}{5.25\text{ in.}} = \frac{2.8}{5.25}$$

To simplify this fraction, we need to rewrite it as an equivalent fraction of whole numbers, that is, without the decimals.

If we multiply the numerator by 10, it would be a whole number. However, we need to multiply the denominator by 100 in order to make it a whole number. Because we want to write an equivalent fraction, we multiply it by $1 = \frac{100}{100}$.

$$\frac{2.8}{5.25} \cdot \frac{100}{100} = \frac{280}{525}$$

$$= \frac{8}{15}$$

Divide the numerator and denominator by 5, and then again by 7. Or, we can determine that the GCF of 280 and 525 is 35, and simplify with one division.

The ratio of the bottle diameters is 8 to 15.

Check Yourself 5

(a) One morning Rita jogged $3\frac{1}{2}$ mi, while Yi jogged $4\frac{1}{4}$ mi. Express the ratio of the two distances as a ratio of whole numbers.

(b) A standard newspaper column is 2.625 in. wide and 19.5 in. long. Express the ratio of the two measurements as a ratio of whole numbers.

NOTE

We will learn to convert measurements in more depth in Chapter 7.

Sometimes, we use a ratio to compare the same type of measurement using different units. In Example 6, both quantities are measures of time. To construct and simplify the ratio, we must express both quantities in the same units.

Example 6 Rewriting Denominate Numbers to Find a Ratio

Joe took 2 hr (120 min) to complete his final exam. Jamie finished her exam in 75 min. Write the ratio of the two times in simplest terms.

To find the ratio, both quantities must have the same units. Therefore, we rewrite 2 hr as 120 min. This way, both quantities use minutes as the unit.

$$\frac{2\text{ hr}}{75\text{ min}} = \frac{120\text{ min}}{75\text{ min}} \qquad \text{Rewrite 2 hr as 120 min.}$$

$$= \frac{120}{75} \qquad \text{Simplify the units.}$$

$$= \frac{8}{5} \qquad \text{The GCF of 120 and 75 is 15.}$$

Check Yourself 6

Write the ratio 72 ft to 18 yd.

Note: 1 yd = 3 ft so 18 yd = 54 ft.

Check Yourself ANSWERS

1. $\frac{7}{12}$ **2. (a)** $\frac{17}{29}$; **(b)** $\frac{17}{12}$ **3.** $\frac{3}{4}$ **4.** $\frac{4}{3}$ **5. (a)** $\frac{14}{17}$; **(b)** $\frac{7}{52}$ **6.** $\frac{4}{3}$

Reading Your Text

These fill-in-the-blank exercises will help you understand some of the key vocabulary used in this section. The answers to these exercises are in the Answers Appendix at the back of the text.

(a) A __________ can indicate division.

(b) A ratio is a means of comparing two __________ quantities.

(c) Because a ratio is a fraction, we always write it in __________ terms.

(d) Ratios are never written as __________ numbers.

5.1 exercises

Skills | Calculator/Computer | Career Applications | Above and Beyond

< Objectives 1 and 2 >

Write each ratio in simplest terms.

1. The ratio of 9 to 13
2. The ratio of 5 to 4
3. The ratio of 9 to 4
4. The ratio of 5 to 12
5. The ratio of 10 to 15 VIDEO
6. The ratio of 16 to 12
7. The ratio of 18 to 15
8. The ratio of 42 to 60
9. The ratio of 3 to 21
10. The ratio of 15 to 105
11. The ratio of 24 to 6
12. The ratio of 91 to 13
13. The ratio of $3\frac{1}{2}$ to 14
14. The ratio of 12 to $\frac{5}{3}$
15. The ratio of $1\frac{1}{4}$ to $\frac{3}{2}$
16. The ratio of $5\frac{3}{5}$ to $2\frac{1}{10}$
17. The ratio of 4.5 to 31.5
18. The ratio of 1.4 to 3.6
19. The ratio of 8.7 to 8.4
20. The ratio of 0.6 to 8.1
21. The ratio of 10.5 to 2.7 VIDEO
22. The ratio of 2.2 to 0.6
23. The ratio of 12 mi to 18 mi
24. The ratio of 100 cm to 90 cm
25. The ratio of 40 ft to 65 ft VIDEO
26. The ratio of 12 oz to 18 oz
27. The ratio of \$48 to \$42
28. The ratio of 20 ft to 24 ft
29. The ratio of 75 s to 3 min
    Hint: 3 min = 180 s
30. The ratio of 7 oz to 3 lb
    Hint: 3 lb = 48 oz
31. The ratio of 4 nickels to 3 dimes
    Hint: Convert to cents
32. The ratio of 8 in. to 3 ft
    Hint: 3 ft = 36 in.
33. The ratio of 2 days to 10 hr
    Hint: 2 days = 48 hr VIDEO
34. The ratio of 4 ft to 4 yd
    Hint: 4 yd = 12 ft
35. The ratio of 5 gal to 12 qt
    Hint: 5 gal = 20 qt
36. The ratio of 7 dimes to 3 quarters
    Hint: Convert to cents

Solve each application.

37. **Social Science** An algebra class has 7 men and 13 women. Write the ratio of men to women. Write the ratio of women to men.

38. **Statistics** A football team wins 9 of its 16 games with no ties. Write the ratio of wins to games played. Write the ratio of wins to losses.
39. **Social Science** In a school election 4,500 yes votes were cast, and 3,000 no votes were cast. Write the ratio of yes to no votes. VIDEO
40. **Business and Finance** One car has an $11\frac{1}{2}$-gal tank and another has a $17\frac{3}{4}$-gal tank. Write the ratio of the capacities.
41. **Business and Finance** One compact refrigerator holds $2\frac{2}{3}$ cubic feet (ft^3) of food, and another holds $5\frac{3}{4}$ ft^3 of food. Write the ratio of the capacities. VIDEO
42. **Science and Medicine** The price of an antibiotic in one drugstore is \$37.50 and \$26.25 in another. Write the ratio of the prices.
43. **Social Science** A company employs 24 women and 18 men. Write the ratio of men to women employed by the company.
44. **Geometry** If a room is 30 ft long by 6 yd wide, write the ratio of the length to the width of the room.
45. Canton, OH, experienced 2.4 in. of rain one October. In November, there was 0.4 in. more rain than in October. Write the ratio of rainfall in November to October.

46. St. Cloud, MN, experienced 10.6 in. of snowfall one January. The next month, there was 2.4 in. less snow. Write the ratio of snowfall in February to January.

47. The maximum temperature (°F) for each month in Taos, NM, was collected one summer.

(a) Find the ratio of high temperatures of June to September.

(b) Find the high temperature ratio of August to July.

Taos, NM

Jun	Jul	Aug	Sep
82.1	85.6	83.3	76.5

48. The minimum temperature (°F) for each month in Ann Arbor, MI, was collected one winter.

(a) Find the ratio of low temperatures of January to December.

(b) Find the low temperature ratio of February to March.

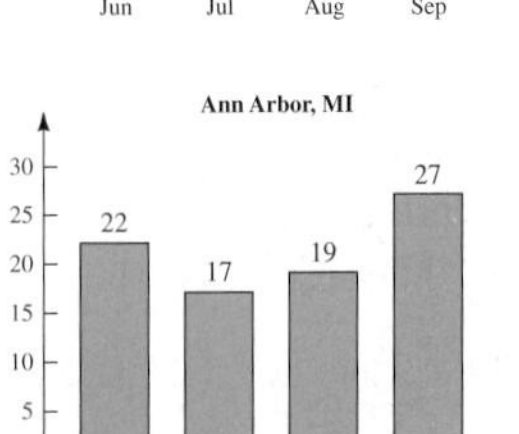

Determine whether each statement is **true** *or* **false.**

49. We use ratios to compare like quantities.

50. We use ratios to compare different types of measurements.

Fill in each blank with **always, sometimes,** *or* **never.**

51. A ratio should __________ be written in simplest form.

52. A ratio is __________ written as a mixed number.

Skills | Calculator/Computer | **Career Applications** | Above and Beyond

53. **Information Technology** Millicent can fix five cell phones per hour. Tyler can fix four cell phones per hour. Express the ratio of the number of cell phones that Millicent can fix to the number that Tyler can fix.

54. **Allied Health** In preparing specimens for testing, it is often necessary to dilute the original solution. The dilution ratio is the ratio of the volume of original solution to the total volume. Each of these volumes is usually measured in milliliters (mL).

Determine the dilution ratio when 2.8 mL of blood serum is diluted with 47.2 mL of water.

55. **Mechanical Engineering** A gear ratio is the ratio of the number of teeth on the driven gear to the number of teeth on the driving gear. In general, the driving gear is attached to the power source or motor. Write the gear ratio for the system shown.

56. **Manufacturing Operations Technology** Of the 384 parts manufactured during a shift, 26 were defective.

(a) Write the ratio of defective parts to total parts.

(b) Write the ratio of defective parts to good parts.

57. **Agricultural Technology** A soil test indicates that a field requires a fertilizer containing 400 lb of nitrogen and 500 lb of phosphorus. Write the ratio of nitrogen to phosphorus needed.

58. FastConnect offers Internet connections speeds of 15 megabytes per second (MB/s) for \$30 per month. BuyNet offers 12 MB/s for \$25 per month.

(a) Write the ratio of connection speeds of FastConnect to BuyNet.

(b) Write the ratio of monthly prices of FastConnect to BuyNet.

Skills | Calculator/Computer | Career Applications | **Above and Beyond**

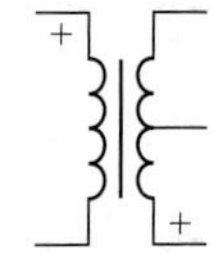

The accompanying image is a common symbol on a schematic (electrical diagram) for a transformer. A transformer uses electromagnetism to change voltage levels. Commonly, two coils of wire (or some conductor) are located in close proximity but kept from directly touching or conducting. Some sort of ferromagnetic core (such as iron) is typically used. When alternating current (AC) is applied to one conductor or coil, referred to as the primary winding, current is induced on the second or secondary winding.

There is a relationship between the voltage supplied to the primary winding and the open-voltage induced in the secondary winding, based on the number of turns in each winding. This relationship is called the turns ratio (a):

$$a = \frac{N_p}{N_s}$$

in which N_p represents the number of turns in the primary winding and N_s represents the number of turns in the secondary winding.

Theoretically, the turns ratio is also equal to the voltage ratio:

$$a = \frac{V_p}{V_s}$$

in which V_p represents the voltage supplied to the primary winding and V_s represents the voltage induced in the secondary winding.

After setting the two ratios equal to one another and performing a multiplication manipulation to isolate V_p, this relationship can be expressed as

$$V_p = \frac{N_p}{N_s}V_s$$

59. Give three combinations of turns of the primary and secondary windings that achieve a turns ratio for a transformer of 3.2.

60. Using a turns ratio of 3.2 and a secondary voltage of 35 volts (V) AC, calculate the voltage supplied to the primary winding.

61. If the turns ratio is 4.5 and the primary voltage is 28 V AC, what is the induced open-voltage on the secondary winding?

62. If the turns ratio from exercise 61 is doubled, how will that affect the voltage on the secondary winding?

63. **(a)** Buy a 1.69-oz (medium-size) bag of M&M's. For each color, determine the ratio of M&M's that are that color to the total number of M&M's in the bag.

(b) Compare your ratios from part (a) to those of a classmate.

(c) Use the information from parts (a) and (b) to estimate the correct ratios for all the different color M&M's in a bag.

(d) Go to the M&M's manufacturer's website (www.mms.com) and see how your ratios compare to their claimed color distribution.

64. Sarah is a field service technician for ABC Networks, Inc. She has been asked to design a wireless home-network for a customer. The customer wants to have the fastest possible throughput for the wireless network in his home.

From your experience, you know that wireless networks come in different varieties: 802.11a, b, and g. The standard for most coffeehouses and restaurants is 802.11b.

If it takes 1 s to transmit a packet on 802.11b, how long does it take to transmit a packet on 802.11a and g? Which technology will you recommend and why?

Answers

1. $\frac{9}{13}$ **3.** $\frac{9}{4}$ **5.** $\frac{2}{3}$ **7.** $\frac{6}{5}$ **9.** $\frac{1}{7}$ **11.** $\frac{4}{1}$ **13.** $\frac{1}{4}$ **15.** $\frac{5}{6}$ **17.** $\frac{1}{7}$ **19.** $\frac{29}{28}$ **21.** $\frac{35}{9}$ **23.** $\frac{2}{3}$ **25.** $\frac{8}{13}$ **27.** $\frac{8}{7}$ **29.** $\frac{5}{12}$ **31.** $\frac{2}{3}$ **33.** $\frac{24}{5}$ **35.** $\frac{5}{3}$ **37.** $\frac{7}{13}$; $\frac{13}{7}$ **39.** $\frac{3}{2}$ **41.** $\frac{32}{69}$ **43.** $\frac{3}{4}$ **45.** $\frac{7}{6}$ **47.** (a) $\frac{821}{765}$; (b) $\frac{833}{856}$ **49.** True **51.** always **53.** $\frac{5}{4}$ **55.** $\frac{5}{8}$ **57.** $\frac{4}{5}$ **59.** Answers will vary. **61.** $6\frac{2}{9}$ V AC **63.** Above and Beyond

Working with Ratios

To solve these problems, we think that you and your group members will find one of these three approaches useful: (1) You may wish to use actual black and white markers; (2) you may wish to make sketches of such markers; or (3) you may wish to simply imagine the necessary markers. Each line in the table is a new (and different) problem, and you are to fill in the missing parts in a given line. Be sure to express a ratio by using the smallest possible whole numbers.

	Ratio of Black to White Markers	**Number of Black Markers**	**Number of White Markers**	**Total Number of Markers**
1.	to		15	20
2.	to	12		30
3.	to		9	21
4.	to	15		33
5.	2 to 5	6		
6.	5 to 3		24	
7.	4 to 1	28		
8.	3 to 7		21	
9.	1 to 3			36
10.	3 to 5			40
11.	7 to 2			360
12.	4 to 7			550
13.				
14.				

For problems 13 and 14, create (and solve!) your own problems of the same sort. You might challenge another group with these.

5.2

Rates and Unit Pricing

< 5.2 Objectives >

1 > Write a rate as a unit rate

2 > Interpret and compare unit rates

3 > Find a unit price

4 > Use unit prices to compare costs

In Section 5.1, we used ratios to compare two like quantities. For instance, the ratio of 9 seconds to 12 seconds is $\frac{3}{4}$.

$$\frac{9 \text{ s}}{12 \text{ s}} = \frac{\overset{3}{\cancel{9}}}{\underset{4}{\cancel{12}}} = \frac{3}{4}$$

RECALL

When simplified, a ratio has no units.

Because the units in the numerator and denominator are the same, we can "cancel" them and simplify the fraction.

We also learned that as long as the two quantities represented the same type of measurement, we could compare them using ratios. For example, if both quantities are measurements of length, then we can convert one of the measurements so that they are like quantities.

In Chapter 7, you will study measurement conversions in more depth. For now, we can do straightforward conversions. For example, we can use a ratio to compare 4 in. and 3 ft by converting 3 ft to 36 in.

$$\frac{4 \text{ in.}}{3 \text{ ft}} = \frac{4 \text{ in.}}{36 \text{ in.}} \qquad 3 \text{ ft} = 36 \text{ in.}$$

$$= \frac{4}{36}$$

$$= \frac{1}{9}$$

RECALL

Denominate numbers have units "attached."

Often, we want to compare denominate numbers with different types of units. For example, we might be interested in the gas mileage that a car gets. In such a case, we are comparing the miles driven (distance) and the gas used (volume). We make this comparison in part (b) of Example 1.

When we compare denominate numbers with different types of units, we get a *rate*.

Definition

Rate A **rate** is a comparison of two denominate numbers with different types of units.

For example, if an animal moves 3 ft in 4 s, we can express the rate as

$$\frac{3 \text{ ft}}{4 \text{ s}}$$

We read this rate as "3 feet per 4 seconds." In general, rates are presented in simplified form as *unit rates*.

Definition

Unit Rate

A **unit rate** is a *rate* that is simplified so that it compares a denominate number with a *single unit* of a different denominate number.

NOTE

Unlike ratios, we sometimes use whole, mixed, or decimal numbers to express unit rates. Different applications have traditionally reported results differently.

A **unit rate** is written so that the numerical value is given followed by the units, written as a fraction. For example, to express the rate $\frac{3 \text{ ft}}{4 \text{ s}}$ as a unit rate, we write it as $\frac{3}{4} \frac{\text{ft}}{\text{s}}$. We read this as "$\frac{3}{4}$ feet per second."

Example 1 Finding a Unit Rate

< Objective 1 >

NOTES

We read the rate $20 \frac{\text{mi}}{\text{gal}}$ as "twenty miles *per* gallon."

In most real-world applications, we use mph and mpg rather than $\frac{\text{mi}}{\text{hr}}$ and $\frac{\text{mi}}{\text{gal}}$. In this text, we use the fraction form to make the comparison more clear and to make it easier to simplify, when necessary.

Express each rate as a unit rate.

(a) $\frac{12 \text{ feet}}{16 \text{ seconds}} = \frac{12}{16} \frac{\text{ft}}{\text{s}} = \frac{3}{4} \frac{\text{ft}}{\text{s}}$

(b) $\frac{200 \text{ miles}}{10 \text{ gallons}} = \frac{200}{10} \frac{\text{mi}}{\text{gal}} = 20 \frac{\text{mi}}{\text{gal}}$

(c) $\frac{10 \text{ gallons}}{200 \text{ miles}} = \frac{10}{200} \frac{\text{gal}}{\text{mi}} = \frac{1}{20} \frac{\text{gal}}{\text{mi}}$

Check Yourself 1

Express each rate as a unit rate.

(a) $\frac{250 \text{ miles}}{10 \text{ hours}}$ **(b)** $\frac{\$60{,}000}{2 \text{ years}}$ **(c)** $\frac{2 \text{ years}}{\$60{,}000}$

Consider part (a) in Example 1. We begin with 12 ft (length) compared to 16 seconds (time). We simplify the rate so that we know the number of feet, $\frac{3}{4}$, per 1 second. In general, we simplify a rate so that we are comparing the quantity of the numerator's units per one of the denominator's units.

In Example 2, we consider parts (b) and (c) of Example 1.

Example 2 Comparing Unit Rates

< Objective 2 >

NOTE

We could also write this rate as a decimal,

$\frac{1}{20} \frac{\text{gal}}{\text{mi}} = 0.05 \frac{\text{gal}}{\text{mi}}$

(a) Write the rate $20 \frac{\text{mi}}{\text{gal}}$ in words.

We write this rate as "twenty miles per gallon."

(b) Write the rate $\frac{1}{20} \frac{\text{gal}}{\text{mi}}$ in words.

We write this rate as "one-twentieth of a gallon per mile."

(c) Describe the difference between the rates in parts (a) and (b).

The rate $20 \frac{\text{mi}}{\text{gal}}$ states that 20 mi can be traveled on a single gallon of gas. The rate $\frac{1}{20} \frac{\text{gal}}{\text{mi}}$ states that $\frac{1}{20}$ of a gallon of gas is used when traveling 1 mi.

We can also interpret these as 20 mi are traveled for each gallon of gas used and $\frac{1}{20}$ of a gallon of gas is used for each mile traveled, respectively.

Check Yourself 2

(a) Write the rate $30{,}000 \dfrac{\text{dollars}}{\text{yr}}$ in words.

(b) Write the rate $\dfrac{1}{30{,}000} \dfrac{\text{yr}}{\text{dollar}}$ in words.

(c) Describe the difference between the rates in parts (a) and (b).

Sometimes, we need to find the appropriate rate within a written statement, as in Example 3.

Example 3 Finding a Unit Rate

RECALL

In Section 5.1 we stated that mixed numbers were inappropriate for ratios. When we write unit rates, mixed numbers and decimals are not only appropriate, depending on the application, they may be preferred.

During a recent season, Milwaukee Brewers pitcher Yovani Gallardo had 200 strikeouts over 185 innings. Find his strikeout per inning rate. Report your result as a decimal, rounded to the nearest hundredth.

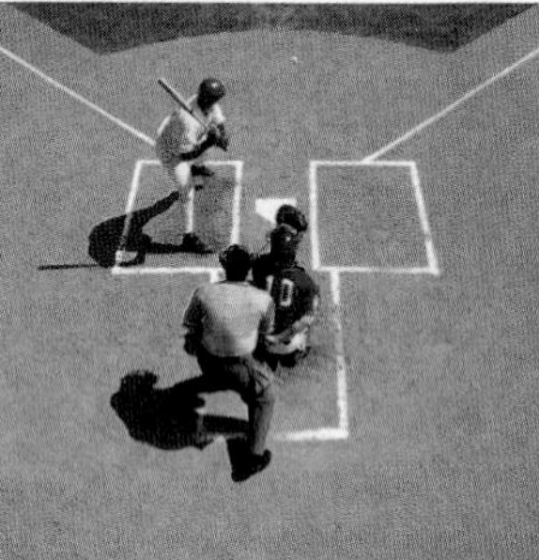

$$\frac{200 \text{ strikeouts}}{185 \text{ innings}} = \frac{200}{185} \frac{\text{strikeouts}}{\text{inning}}$$ Write a unit rate by separating the units.

$$= \frac{40}{37} \frac{\text{strikeouts}}{\text{inning}}$$ Simplify the fraction.

$$\approx 1.08 \frac{\text{strikeouts}}{\text{inning}}$$ Divide to write as a decimal; round to two places.

Check Yourself 3

During a recent season, Phoenix Mercury guard Diana Taurasi scored 702 points over 31 games. Find her points per game rate. Report your result as a decimal, rounded to the nearest tenth.

Rates are often used for comparisons. Report the results based on a field's traditions or on convenience. In Example 4, we use decimals to report our results.

Example 4 Comparing Rates

Player A scores 50 points in 9 games and player B scores 260 points in 36 games. Which player scored at a higher rate?

Player A's rate was $\dfrac{50 \text{ points}}{9 \text{ games}} = \dfrac{50}{9} \dfrac{\text{points}}{\text{game}}$

$$\approx 5.56 \frac{\text{points}}{\text{game}}$$ $\frac{50}{9} = 50 \div 9 \approx 5.56$

Player B's rate was $\dfrac{260 \text{ points}}{36 \text{ games}} = \dfrac{260}{36} \dfrac{\text{points}}{\text{game}}$

$$= \frac{65}{9} \frac{\text{points}}{\text{game}}$$

$$\approx 7.22 \frac{\text{points}}{\text{game}}$$ $65 \div 9 \approx 7.22$

Player B scored at a higher rate.

Check Yourself 4

Hassan scored 25 goals in 8 games and Lee scored 52 goals in 18 games. Which player scored at a higher rate?

Unit pricing represents one of the most common uses of rates. Posted on nearly every item in a supermarket or grocery store is the price of the item as well as its *unit price.*

Definition

Unit Price

The **unit price** relates a price to some common unit.

A unit price is a price *per unit.* The unit used may be ounces, pints, pounds, or some other unit.

Example 5 Finding a Unit Price

< Objective 3 >

Find the unit price for each item. Round your results to the nearest cent per unit.

(a) 8 oz of cream cost \$2.49.

$$\frac{\$2.49}{8 \text{ oz}} = \frac{249 \text{ cents}}{8 \text{ oz}} = \frac{249}{8}\,\frac{\text{cents}}{\text{oz}}$$

$$\approx 31\,\frac{\text{cents}}{\text{oz}}$$

(b) 20 lb of potatoes cost \$9.98.

$$\frac{\$9.98}{20 \text{ lb}} = \frac{998 \text{ cents}}{20 \text{ lb}} = \frac{998}{20}\,\frac{\text{cents}}{\text{lb}}$$

$$\approx 50\,\frac{\text{cents}}{\text{lb}}$$

Check Yourself 5

Find the unit price for each item. Round your results to the nearest cent per unit.

(a) 12 soda cans cost \$4.98.
(b) 25 lb of dog food cost \$14.99.

As with ratios, rates are most often used for comparisons. For instance, unit pricing allows people to compare the cost of different-sized items.

In Example 6, we use unit prices to determine whether a glass of milk is less expensive when poured from a 128-oz container (gallon) or a 32-oz container (quart).

Example 6 Using Unit Prices to Compare Cost

< Objective 4 >

A store sells a 1-gal container (128 oz) of organic whole milk for \$4.89. They sell a 1-quart carton (32 oz) for \$1.29. Which is the better buy?

We begin by determining the unit price of each item. To do this, we compute the cost per ounce for each container of milk. We choose to use cents instead of dollars to make the decimal easier to work with.

NOTE

Usually, we round money to the nearest cent. When comparing unit prices, however, we may round to four decimal places (or more, if necessary).

Gallon

$$\frac{\$4.89}{128 \text{ oz}} = \frac{489 \text{ cents}}{128 \text{ oz}} = \frac{489}{128}\,\frac{\text{cents}}{\text{oz}} \approx 3.8203\frac{\text{cents}}{\text{oz}}$$

Quart

$$\frac{\$1.29}{32 \text{ oz}} = \frac{129 \text{ cents}}{32 \text{ oz}} = \frac{129}{32}\frac{\text{cents}}{\text{oz}} \approx 4.0313\frac{\text{cents}}{\text{oz}}$$

At these prices, the gallon of milk is the better buy.

Check Yourself 6

A store sells a 5-lb bag of Valencia oranges for $4.59. A 12-lb case sells for $11.39. Which is the better buy?

If we compare the two unit prices in Example 6, we see that both items round to 4¢ per ounce. However, milk sold by the gallon is about 0.211¢ cheaper per ounce than milk sold by the quart. We need to consider this small fraction of a cent because we are not buying 1 oz of milk. Rather, we are buying cartons of milk, and for a whole quart of milk, the fraction adds up to nearly 7¢ (it adds up to 27¢ for a gallon).

Check Yourself ANSWERS

1. **(a)** $25 \frac{\text{mi}}{\text{hr}}$; **(b)** $30{,}000 \frac{\text{dollars}}{\text{yr}}$; **(c)** $\frac{1}{30{,}000} \frac{\text{yr}}{\text{dollar}}$ **2.** **(a)** Thirty thousand dollars per year. **(b)** One thirty-thousandth of a year per dollar. **(c)** The rate in part (a) describes the amount of money for each year. The rate in part (b) describes the amount of time per dollar. **3.** $22.6 \frac{\text{points}}{\text{game}}$ **4.** Hassan scored at a higher rate. **5.** **(a)** $\approx 42 \frac{\text{cents}}{\text{can}}$; **(b)** $\approx 60 \frac{\text{cents}}{\text{lb}}$ **6.** The 5-lb bag is the better buy at 91.8¢ per pound compared to approximately 94.9¢ per pound for the case.

Reading Your Text

These fill-in-the-blank exercises will help you understand some of the key vocabulary used in this section. The answers to these exercises are in the Answers Appendix at the back of the text.

(a) Ratios are used to compare ________ quantities.

(b) When we compare measurements with different types of units, we get a ________.

(c) We usually use whole, mixed, or decimal numbers rather than ________ fractions when reporting a unit rate.

(d) ________ prices are used to compare the costs of items in different-sized packages.

< Objective 1 >

Write each rate as a unit rate.

1. $\frac{300 \text{ mi}}{4 \text{ hr}}$ 

2. $\frac{95 \text{ cents}}{5 \text{ pencils}}$

3. $\frac{\$10{,}000}{5 \text{ yr}}$

4. $\frac{680 \text{ ft}}{17 \text{ s}}$ 

5. $\frac{7{,}200 \text{ revolutions}}{16 \text{ mi}}$

6. $\frac{57 \text{ oz}}{3 \text{ cans}}$

7. $\frac{\$2{,}000{,}000}{4 \text{ yr}}$ 

8. $\frac{150 \text{ cal}}{3 \text{ oz}}$

9. $\frac{240 \text{ lb of fertilizer}}{6 \text{ lawns}}$

10. $\frac{192 \text{ diapers}}{32 \text{ babies}}$

Write as unit rates, in decimal form. Round to the nearest tenth.

11. 323 mi on 11 gal of fuel

12. 323 mi in 5 hr

13. 210 ft^2 in 16 hr

14. 410 ft^2 in 12 hr

15. 141 pages in 9 min

16. 12 exercises in 45 min

17. 447 lb on 30 in.^2

18. 24 lb for \$30

Write each as a unit rate. Report each result as a mixed number.

19. 189 points in 42 games

20. 13 mi in 4 hr

21. 72 lengths in 40 min

22. 88 lengths in 50 min

23. 2 cups (c) of water for $1\frac{1}{3}$ c rice

24. 20-lb turkey for 16 people

< Objective 2 >

25. **(a)** Write the rate $32 \frac{\text{ft}}{\text{s}}$ in words.

(b) Write the rate $\frac{1}{32} \frac{\text{s}}{\text{ft}}$ in words.

(c) Describe the difference between the rates given in parts (a) and (b).

26. **(a)** Write the rate $28 \frac{\text{mi}}{\text{gal}}$ in words.

(b) Write the rate $\frac{1}{28} \frac{\text{gal}}{\text{mi}}$ in words.

(c) Describe the difference between the rates given in parts (a) and (b).

27. **(a)** Write the rate $8.9 \frac{\text{bushels}}{\text{dollar}}$ in words.

(b) Write the rate $0.11 \frac{\text{dollars}}{\text{bushel}}$ in words.

(c) Describe the difference between the rates given in parts (a) and (b).

28. **(a)** Write the rate $1.5 \frac{\text{cups of water}}{\text{cup of rice}}$ in words.

(b) Write the rate $\frac{2}{3} \frac{\text{cup of rice}}{\text{cup of water}}$ in words.

(c) Describe the difference between the rates given in parts (a) and (b).

< Objective 3 >

Find the unit price of each item.

29. $57.50 for 5 shirts 

30. $104.93 for 7 DVDs

31. $5.16 for a dozen oranges

32. $10.44 for 18 bottles of water

< Objective 4 >

Find the best buy in each exercise.

33. Dishwashing liquid
- **(a)** 12 fl oz for $3.16
- **(b)** 22 fl oz for $5.16

34. Canned corn
- **(a)** 10 oz for $0.84
- **(b)** 17 oz for $1.56

35. Syrup
- **(a)** 12 fl oz for $3.96
- **(b)** 24 fl oz for $6.36
- **(c)** 36 fl oz for $8.76

36. Shampoo
- **(a)** 4 fl oz for $4.64
- **(b)** 7 fl oz for $6.08
- **(c)** 15 fl oz for $13.56

37. Salad oil (1 qt is 32 fl oz)
- **(a)** 18 fl oz for $3.56
- **(b)** 1 qt for $5.56
- **(c)** 1 qt 16 fl oz for $8.76

38. Tomato juice (1 pt is 16 fl oz)
- **(a)** 8 fl oz for $1.48
- **(b)** 1 pt 10 fl oz for $4.76
- **(c)** 1 qt 14 fl oz for $7.96

39. Peanut butter (1 lb is 16 oz)
- (a) 12 oz for $5.00
- **(b)** 18 oz for $6.88
- **(c)** 1 lb 12 oz for $10.16
- **(d)** 2 lb 8 oz for $15.04

40. Laundry detergent
- **(a)** 1 lb 2 oz for $7.96
- **(b)** 1 lb 12 oz for $11.56
- **(c)** 2 lb 8 oz for $16.76
- **(d)** 5 lb for $31.96

Solve each application.

41. Trac uses 8 gal of gasoline on a 256-mi drive. How many miles per gallon does his car get?

42. Seven pounds of fertilizer covers 1,400 ft^2. How many square feet are covered by 1 lb of fertilizer?

43. A local college has 6,000 registered vehicles for 2,400 campus parking spaces. How many vehicles are there for each parking space?

44. A water pump produces 280 gal in 24 hr. How many gallons per hour is this?

45. It cost $5,992 for 214 shares of stock. What was the cost per share?

46. A printer produces four pages of print in 6 s. How many pages are produced per second?

47. A 12-oz can of tuna costs $4.80. What is the cost of tuna per ounce?

48. The fabric for a dress costs $114.69 for 9 yd. What is the cost per yard?

49. Gerry laid 634 bricks in 35 min. His friend Matt laid 515 bricks in 27 min. Who is the faster bricklayer?

50. Mike drove 135 mi in 2.5 hr. Sam drove 91 mi in 1.75 hr. Who drove faster?

51. Adrian Gonzalez had 198 hits in 585 at bats. Miguel Cabrera had 178 hits in 533 at bats. Who had the higher batting average (hits per at bat)?

52. Which is the better buy: 5 lb of sugar for $4.75 or 20 lb of sugar for $19.92?

Determine whether each statement is **true** *or* **false.**

53. We use rates to compare like quantities.

54. We use rates to compare different types of measurements.

Fill in each blank with **always, sometimes,** *or* **never.**

55. The units __________ cancel in a rate.

56. The units __________ cancel in a ratio.

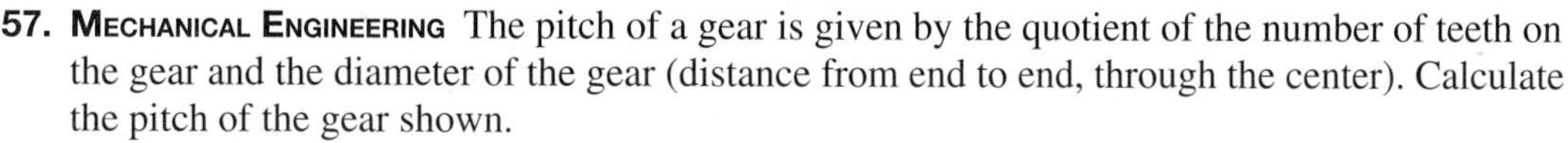

57. **Mechanical Engineering** The pitch of a gear is given by the quotient of the number of teeth on the gear and the diameter of the gear (distance from end to end, through the center). Calculate the pitch of the gear shown.

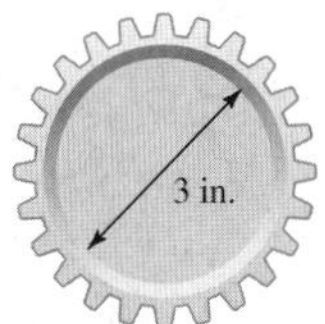

58. **Allied Health** A patient's tidal volume, in milliliters per breath, is the quotient of his or her minute volume (mL/min) and his or her respiratory rate (breaths/min).

Report the tidal volume for an adult, female patient whose minute volume is 7,500 mL/min if her respiratory rate is 12 breaths/min. chapter 5 > Make the Connection

59. **Business and Finance** Determine the unit price of a 1,000-ft cable that costs $99.99.

60. **Electrical Engineering** A 20-volt (V) DC pulse is sent down a 4,000-meter (m) length of conductor (see the figure). Because of resistance, when the pulse reaches the other end, the voltmeter measures the voltage as 4 V. What is the rate of voltage drop per meter of conductor?

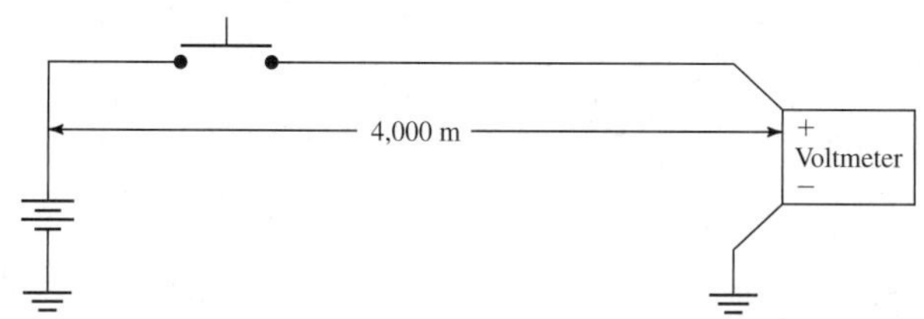

61. **Business and Finance** A 200-bushel load of soybeans sells for $1,780. What is the price per bushel?

62. **Mechanical Engineering** Stress is calculated as the rate of force applied compared to the cross-sectional area of a post. What is the stress on a post that supports 13,475 lb and has a cross-sectional area of 12.25 in.^{2}?

63. Find several real-world examples of ratios and of rates.

64. Explain why unit pricing is useful.

65. Go to a supermarket or grocery store. Choose five items that have price and unit price listed. Check to see if the unit price given for each item is accurate.

Answers

1. $75\,\frac{\text{mi}}{\text{hr}}$ **3.** $2{,}000\,\frac{\text{dollars}}{\text{yr}}$ **5.** $450\,\frac{\text{rev}}{\text{mi}}$ **7.** $500{,}000\,\frac{\text{dollars}}{\text{yr}}$ **9.** $40\,\frac{\text{lb}}{\text{lawn}}$ **11.** $29.4\,\frac{\text{mi}}{\text{gal}}$ **13.** $13.1\,\frac{\text{ft}^2}{\text{hr}}$ **15.** $15.7\,\frac{\text{pages}}{\text{min}}$ **17.** $14.9\,\frac{\text{lb}}{\text{in.}^2}$ **19.** $4\frac{1}{2}\,\frac{\text{points}}{\text{game}}$ **21.** $1\frac{4}{5}\,\frac{\text{lengths}}{\text{min}}$ **23.** $1\frac{1}{2}\,\frac{\text{c of water}}{\text{c of rice}}$ **25.** **(a)** Thirty-two feet per second. **(b)** One thirty-second second per foot. **(c)** Part (a) describes how far you travel in one second. Part (b) describes how long it takes to travel one foot.

27. **(a)** Eight and nine-tenths bushels per dollar. **(b)** Eleven-hundredths dollar per bushel (11¢ per bushel). **(c)** Part (a) describes the number of bushels sold for one dollar. Part (b) describes the cost of one bushel. **29.** $11.50\,\frac{\text{dollars}}{\text{shirt}}$ **31.** $0.43\,\frac{\text{dollar}}{\text{orange}}$

33. (b) **35.** (c) **37.** (b) **39.** (c) **41.** $32\,\frac{\text{mi}}{\text{gal}}$ **43.** $2.5\,\frac{\text{vehicles}}{\text{space}}$ **45.** $28\,\frac{\text{dollars}}{\text{share}}$ **47.** $40\,\frac{\text{cents}}{\text{oz}}$ **49.** Matt

51. Adrian Gonzalez **53.** False **55.** never **57.** $8\,\frac{\text{teeth}}{\text{in.}}$ **59.** $\approx 0.10\,\frac{\text{dollar}}{\text{ft}}$ **61.** $8.90\,\frac{\text{dollars}}{\text{bushel}}$ **63.** Above and Beyond

65. Above and Beyond

Activity 14 ::

Baseball Statistics

There are many statistics in the sport of baseball that are expressed in decimal form. Two of these are batting average and earned run average. Both are actually examples of rates.

A batting average is a rate for which the units are "hits per at bat." To compute the batting average for a hitter, divide the number of hits by the number of times at bat. The result is less than 1 (unless the batter always gets a hit!), and it is always expressed as a decimal to the nearest thousandth. For example, if a hitter has 2 hits in 7 at bats, we divide 2 by 7, getting 0.285714. . . . The batting average is then rounded to 0.286.

Compute the batting average for each major league player.

	Player	Hits	At Bats	Average
1	Cabrera	197	572	
2	Gonzalez	213	630	
3	Young	213	631	
4	Reyes	181	537	
5	Braun	187	563	

The earned run average (ERA) for a pitcher is also a rate; its units are "earned runs per 9 innings." It represents the number of earned runs the pitcher gives up in 9 innings. To compute the ERA for a pitcher, multiply the number of earned runs allowed by the pitcher by 9, and then divide by the number of innings pitched. The result is always rounded to the nearest hundredth.

Compute the earned run average for each major league player.

	Player	Earned Runs	Innings	ERA
1	Kershaw	59	$233\frac{1}{3}$	
2	Halladay	61	$233\frac{2}{3}$	
3	Lee	62	$232\frac{2}{3}$	

Challenge: Suppose a hitter has 54 hits in 200 times at bat. How many hits in a row must the hitter get in order to raise his average to at least 0.300?

5.3 Proportions

< 5.3 Objectives >

1 > Write a proportion

2 > Determine whether two fractions are proportional

3 > Determine whether two rates are proportional

Definition

Proportion A statement that two fractions or rates are equal is called a **proportion.**

NOTES

This is the same as saying the fractions are equivalent. They name the same number.

We call a letter representing an unknown value a *variable*. Here *a*, *b*, *c*, and *d* are variables. We could have chosen other letters.

Because the ratio of 1 to 3 is equal to the ratio of 2 to 6, we can write the proportion

$$\frac{1}{3} = \frac{2}{6}$$

The proportion $\frac{a}{b} = \frac{c}{d}$ is read "*a* is to *b* as *c* is to *d*." We read the proportion $\frac{1}{3} = \frac{2}{6}$ as "one is to three as two is to six."

Example 1 Writing a Proportion

< Objective 1 >

Write the proportion 3 is to 7 as 9 is to 21.

$$\frac{3}{7} = \frac{9}{21}$$

Check Yourself 1

Write the proportion 4 is to 12 as 6 is to 18.

When you write a proportion for two rates, placement of similar units is important.

Example 2 Writing a Proportion with Two Rates

Write a proportion that is equivalent to the statement: If it takes 3 hr to mow 4 acres of grass, it will take 6 hr to mow 8 acres.

$$\frac{3\text{ hr}}{4\text{ acres}} = \frac{6\text{ hr}}{8\text{ acres}}$$

Note that, in both fractions, the hours units are in the numerator and the acres units are in the denominator.

Check Yourself 2

Write a proportion that is equivalent to the statement: If it takes 5 rolls of wallpaper to cover 400 ft^2, it will take 7 rolls to cover 560 ft^2.

If two fractions form a true proportion, we say that they are **proportional.**

Property

The Proportion Rule

If $\frac{a}{b} = \frac{c}{d}$, then $a \cdot d = b \cdot c$.

We say that the fractions $\frac{a}{b}$ and $\frac{c}{d}$ are proportional.

Example 3 Determining Whether Two Fractions Are Proportional

< Objective 2 >

NOTE

We use the centered dot (·) for multiplication rather than the cross (×).

Determine whether each pair of fractions is proportional.

(a) $\frac{5}{6} \stackrel{?}{=} \frac{10}{12}$

Multiply:

$5 \cdot 12 = 60$

$6 \cdot 10 = 60$

Equal

Because $a \cdot d = b \cdot c$, $\frac{5}{6}$ and $\frac{10}{12}$ are proportional.

(b) $\frac{3}{7} \stackrel{?}{=} \frac{4}{9}$

Multiply:

$3 \cdot 9 = 27$

$7 \cdot 4 = 28$

Not equal

The products are not equal, so $\frac{3}{7}$ and $\frac{4}{9}$ are not proportional.

Check Yourself 3

Determine whether each pair of fractions is proportional.

(a) $\frac{5}{8} \stackrel{?}{=} \frac{20}{32}$ **(b)** $\frac{7}{9} \stackrel{?}{=} \frac{3}{4}$

The proportion rule can also be used when fractions or decimals are involved.

Example 4 Verifying a Proportion

Determine whether each pair of fractions is proportional.

(a) $\frac{3}{\frac{1}{2}} \stackrel{?}{=} \frac{30}{5}$

$3 \cdot 5 = 15$

$\frac{1}{2} \cdot 30 = 15$

Because the products are equal, the fractions are proportional.

(b) $\frac{0.4}{20} \stackrel{?}{=} \frac{3}{100}$

$0.4 \cdot 100 = 40$

$20 \cdot 3 = 60$

Because the products are *not* equal, the fractions are not proportional.

Check Yourself 4

Determine whether each pair of fractions is proportional.

(a) $\frac{0.5}{8} \stackrel{?}{=} \frac{3}{48}$ (b) $\frac{\frac{1}{4}}{6} \stackrel{?}{=} \frac{3}{80}$

We can use the proportion rule to verify that rates are proportional.

Example 5 Determining Whether Two Rates are Proportional

< Objective 3 >

NOTE

Colones are the monetary unit of Costa Rica.

Is the rate $\frac{5 \text{ U.S. dollars}}{2{,}500 \text{ colones}}$ proportional to the rate $\frac{27 \text{ U.S. dollars}}{13{,}500 \text{ colones}}$?

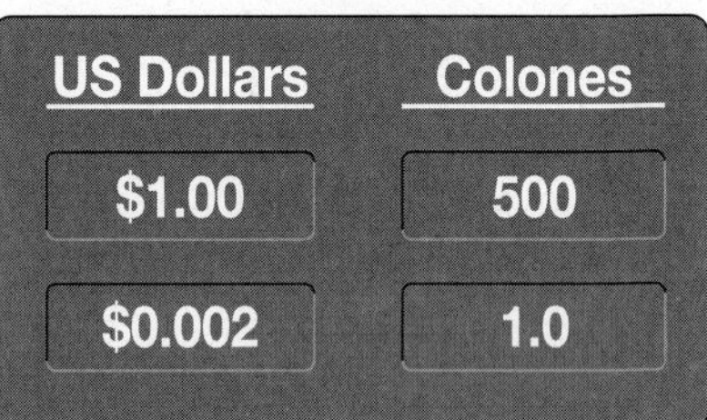

We want to know if the following is true.

$$\frac{5}{2{,}500} \stackrel{?}{=} \frac{27}{13{,}500}$$

$$5 \cdot 13{,}500 = 67{,}500$$

$$27 \cdot 2{,}500 = 67{,}500$$

The rates are proportional.

Check Yourself 5

Is the rate $\frac{50 \text{ pages}}{45 \text{ min}}$ proportional to the rate $\frac{30 \text{ pages}}{25 \text{ min}}$?

In Section 5.4, we use proportions to solve many applications. For instance, if a 12-ft piece of steel stock weighs 27.6 lb, how much would a 25-ft piece weigh? Here, we check the accuracy of such a proportion.

Example 6 An Application of Proportions

NOTE

We need to set up each rate correctly. To do this, we make sure the same units are in the same position in each rate.

Because feet are in the numerator and pounds are in the denominator on the left side, we place feet on top and pounds in the bottom on the right side.

The order that the units appear in the sentences does not matter!

A 12-ft piece of steel stock weighs 27.6 lb. If 57.5 lb is the weight of a 25-ft piece, do the two pieces have the same density?

We check that the two rates are proportional.

$$\frac{12 \text{ ft}}{27.6 \text{ lb}} \stackrel{?}{=} \frac{25 \text{ ft}}{57.5 \text{ lb}}$$

$$12 \cdot 57.5 = 690$$

$$27.6 \cdot 25 = 690$$

Because the two products are equal, we have a true proportion, so the two pieces have the same density.

Check Yourself 6

One supplier sells a 200-lb lot of steel for $522.36. A second supplier charges $789.09 for a 300-lb lot of steel. Determine whether the suppliers are offering steel for the same price per pound.

Check Yourself ANSWERS

1. $\frac{4}{12} = \frac{6}{18}$ **2.** $\frac{5 \text{ rolls}}{400 \text{ ft}^2} = \frac{7 \text{ rolls}}{560 \text{ ft}^2}$ **3.** **(a)** Yes; **(b)** no **4.** **(a)** Yes; **(b)** no **5.** No **6.** No

Reading Your Text 

These fill-in-the-blank exercises will help you understand some of the key vocabulary used in this section. The answers to these exercises are in the Answers Appendix at the back of the text.

(a) A statement that two rates are ______________ is called a proportion.

(b) A letter used to represent an unknown value is called a ______________.

(c) If two fractions are ______________, we say they are proportional.

(d) When writing a proportion for two ______________, corresponding units must be similarly placed.

Skills Calculator/Computer Career Applications Above and Beyond

5.3 exercises

< Objective 1 >

Write each statement as a proportion.

1. 4 is to 9 as 8 is to 18.
2. 6 is to 11 as 18 is to 33.
3. 2 is to 9 as 8 is to 36.
4. 10 is to 15 as 20 is to 30.
5. 3 is to 5 as 15 is to 25.
6. 8 is to 11 as 16 is to 22.
7. 9 is to 13 as 27 is to 39.
8. 15 is to 21 as 60 is to 84.

< Objective 2 >

Determine whether each pair of fractions is proportional.

9. $\frac{3}{4} \stackrel{?}{=} \frac{9}{12}$ 
10. $\frac{6}{7} \stackrel{?}{=} \frac{18}{21}$
11. $\frac{3}{4} \stackrel{?}{=} \frac{15}{20}$
12. $\frac{3}{5} \stackrel{?}{=} \frac{6}{10}$
13. $\frac{11}{15} \stackrel{?}{=} \frac{9}{13}$ 
14. $\frac{9}{10} \stackrel{?}{=} \frac{2}{7}$
15. $\frac{8}{3} \stackrel{?}{=} \frac{24}{9}$
16. $\frac{5}{8} \stackrel{?}{=} \frac{15}{24}$
17. $\frac{6}{17} \stackrel{?}{=} \frac{9}{11}$
18. $\frac{5}{12} \stackrel{?}{=} \frac{8}{20}$
19. $\frac{7}{16} \stackrel{?}{=} \frac{21}{48}$
20. $\frac{2}{5} \stackrel{?}{=} \frac{7}{9}$
21. $\frac{10}{3} \stackrel{?}{=} \frac{150}{50}$
22. $\frac{5}{8} \stackrel{?}{=} \frac{75}{120}$
23. $\frac{3}{7} \stackrel{?}{=} \frac{18}{42}$
24. $\frac{12}{7} \stackrel{?}{=} \frac{96}{50}$
25. $\frac{7}{15} \stackrel{?}{=} \frac{84}{180}$
26. $\frac{76}{24} \stackrel{?}{=} \frac{19}{6}$
27. $\frac{60}{36} \stackrel{?}{=} \frac{25}{15}$
28. $\frac{\frac{1}{2}}{4} \stackrel{?}{=} \frac{5}{40}$
29. $\frac{3}{\frac{1}{5}} \stackrel{?}{=} \frac{30}{6}$
30. $\frac{\frac{2}{3}}{6} \stackrel{?}{=} \frac{1}{12}$
31. $\frac{\frac{3}{4}}{12} \stackrel{?}{=} \frac{1}{16}$
32. $\frac{0.3}{4} \stackrel{?}{=} \frac{1}{20}$
33. $\frac{3}{60} \stackrel{?}{=} \frac{0.3}{6}$
34. $\frac{0.6}{0.12} \stackrel{?}{=} \frac{2}{0.4}$
35. $\frac{0.6}{15} \stackrel{?}{=} \frac{2}{75}$
36. $\frac{5}{8} \stackrel{?}{=} \frac{11}{\frac{5}{2}}$

< Objective 3 >

Determine whether each pair of rates is proportional.

37. $\frac{7 \text{ c of flour}}{4 \text{ loaves of bread}} \stackrel{?}{=} \frac{4 \text{ c of flour}}{3 \text{ loaves of bread}}$ 
38. $\frac{5 \text{ U.S. dollars}}{27 \text{ krone}} \stackrel{?}{=} \frac{15 \text{ U.S. dollars}}{81 \text{ krone}}$
39. $\frac{22 \text{ mi}}{15 \text{ gal}} \stackrel{?}{=} \frac{55 \text{ mi}}{35 \text{ gal}}$
40. $\frac{46 \text{ pages}}{30 \text{ min}} \stackrel{?}{=} \frac{18 \text{ pages}}{8 \text{ min}}$

41. $\frac{9 \text{ in.}}{57 \text{ mi}} \stackrel{?}{=} \frac{6 \text{ in.}}{38 \text{ mi}}$

42. $\frac{12 \text{ yen}}{5 \text{ pesos}} \stackrel{?}{=} \frac{108 \text{ yen}}{45 \text{ pesos}}$

43. $\frac{300 \text{ ft}^2}{18 \text{ hr}} \stackrel{?}{=} \frac{200 \text{ ft}^2}{12 \text{ hr}}$

44. $\frac{12 \text{ gal of paint}}{8{,}329 \text{ ft}^2} \stackrel{?}{=} \frac{9 \text{ gal of paint}}{1{,}240 \text{ ft}^2}$

45. $\frac{12 \text{ in. of snow}}{1.4 \text{ in. of rain}} \stackrel{?}{=} \frac{36 \text{ in. of snow}}{7 \text{ in. of rain}}$

46. $\frac{9 \text{ people}}{2 \text{ cars}} \stackrel{?}{=} \frac{11 \text{ people}}{3 \text{ cars}}$

Write proportions equivalent to each statement.

47. If 15 lb of string beans cost \$20, then 45 lb will cost \$60.

48. If Maria hit 8 home runs in 15 softball games, then she should hit 24 home runs in 45 games.

49. If 3 credits at Berndt Community College cost \$216, then 12 credits cost \$864.

50. If 16 lb of fertilizer cover 1,520 ft^2, then 21 lb should cover 1,995 ft^2.

51. If Audrey travels 180 mi on interstate I-95 in 3 hr, then she should travel 300 mi in 5 hr.

52. If 2 vans can transport 18 people, then 5 vans can transport 45 people.

Determine whether each statement is **true** *or* **false.**

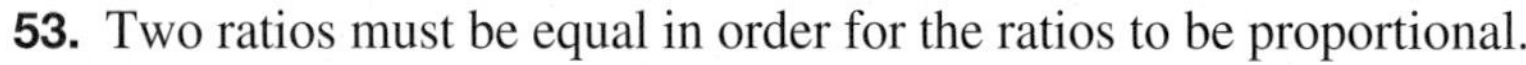

53. Two ratios must be equal in order for the ratios to be proportional.

54. If $\frac{a}{b} = \frac{c}{d}$, then $a \cdot c = b \cdot d$.

Fill in each blank with **always, sometimes,** *or* **never.**

55. Proportions are ____________ used to compare two rates.

56. When writing a proportion for two rates, the placement of units is ____________ important.

57. **Allied Health** Quinidine is an antiarrhythmic heart medication. It is available for injection as an 80 milligrams per milliliter (mg/mL) solution. A patient receives a prescription for 300 mg of quinidine dissolved in 3.75 mL of solution. Are these rates proportional?

58. **Information Technology** A computer transmits 5 Web pages in 2 s to a Web server. A second computer transmits 20 pages in 10 s to the server. Are these two computers transmitting at the same speed?

59. **Mechanical Engineering** A gear has a pitch diameter of 5 in. and 20 teeth. A second gear has a pitch diameter of 18 in. and 68 teeth. In order to mesh, the teeth to diameter rates must be proportional. Will these two gears mesh?

60. **Agricultural Technology** A 13-acre field requires 7,020 lb of fertilizer. Will 11,340 lb of fertilizer cover a 21-acre field?

Answers

1. $\frac{4}{9} = \frac{8}{18}$ 3. $\frac{2}{9} = \frac{8}{36}$ 5. $\frac{3}{5} = \frac{15}{25}$ 7. $\frac{9}{13} = \frac{27}{39}$ 9. Yes 11. Yes 13. No 15. Yes 17. No 19. Yes 21. No 23. Yes 25. Yes 27. Yes 29. No 31. Yes 33. Yes 35. No 37. No 39. No 41. Yes 43. Yes 45. No 47. $\frac{15 \text{ lb}}{\$20} = \frac{45 \text{ lb}}{\$60}$ 49. $\frac{3 \text{ credits}}{\$216} = \frac{12 \text{ credits}}{\$864}$ 51. $\frac{180 \text{ mi}}{3 \text{ hr}} = \frac{300 \text{ mi}}{5 \text{ hr}}$ 53. True 55. sometimes 57. Yes 59. No

5.4 Solving Proportions

< 5.4 Objectives >

1 > Solve a proportion for an unknown value

2 > Use proportions to solve applications

NOTE

$\frac{?}{3} = \frac{10}{15}$ is a proportion in which the first value is unknown. Our work in this section involves learning how to find that unknown value.

A proportion consists of four values. If three of the four values of a proportion are known, you can always find the missing or unknown value.

In the proportion $\frac{a}{3} = \frac{10}{15}$, the first value is unknown. We choose to represent the unknown value with the letter a. In this problem, the letter a is called a variable. Any letter can be used as the variable. We proceed, using the proportion rule.

$$\frac{a}{3} = \frac{10}{15}$$

$$15 \cdot a = 3 \cdot 10 \qquad \text{or} \qquad 15 \cdot a = 30$$

The equal sign tells us that $15 \cdot a$ and 30 are just different names for the same number. This type of statement is called an **equation.**

Definition

Equation An **equation** is a statement that two expressions are equal.

One important property of an equation is that we can divide both sides by the same nonzero number. Here we divide by 15.

NOTE

We always divide by the number multiplying the variable. This is called the *coefficient* of the variable.

$$15 \cdot a = 30$$

$$\frac{15 \cdot a}{15} = \frac{30}{15}$$

$$\frac{\overset{1}{\cancel{15}} \cdot a}{\underset{1}{\cancel{15}}} = \frac{\overset{2}{\cancel{30}}}{\underset{1}{\cancel{15}}}$$

Divide by the coefficient of the variable. Do you see why we divided by 15? It leaves our unknown a by itself on the left side.

$$a = 2$$

You should always check your result. It is easy in this case. We found a value of 2 for a. Replace the unknown a with that value. Then verify that the fractions are proportional. We started with $\frac{a}{3} = \frac{10}{15}$ and found a value of 2 for a. So we write

$$\frac{2}{3} \stackrel{?}{=} \frac{10}{15}$$

$$2 \cdot 15 \stackrel{?}{=} 3 \cdot 10$$

$$30 = 30$$

The value of 2 for a is correct.

We summarize the procedure for solving a proportion.

Step by Step

Solving a Proportion

Step 1 Use the proportion rule to write the equivalent equation $a \cdot d = b \cdot c$.

Step 2 Divide both terms of the equation by the coefficient of the variable.

Step 3 Use the value found to replace the unknown in the original proportion. Check that the ratios or the rates are proportional.

Example 1 Solving Proportions

< Objective 1 >

Find the unknown value.

(a) $\frac{8}{x} = \frac{6}{9}$

NOTES

You are really using algebra to solve these proportions. In algebra, we write the product $6 \cdot x$ as $6x$, omitting the dot. Multiplication of the number and the variable is understood.

This gives us the unknown value. Now check the result.

Step 1 Using the proportion rule, we have

$$6 \cdot x = 8 \cdot 9$$

or $$6x = 72$$

Step 2 Locate the coefficient of the variable, 6, and divide both sides of the equation by that coefficient.

$$\frac{\overset{1}{\cancel{6}}x}{\underset{1}{\cancel{6}}} = \frac{\overset{12}{\cancel{72}}}{\underset{1}{\cancel{6}}}$$

$$x = 12$$

Step 3 To check, replace x with 12 in the original proportion.

$$\frac{8}{12} \stackrel{?}{=} \frac{6}{9}$$

Multiply.

$$12 \cdot 6 \stackrel{?}{=} 8 \cdot 9$$

$$72 = 72$$ The value of 12 checks for x.

(b) $\frac{3}{4} = \frac{c}{25}$

Step 1 Use the proportion rule.

$$4 \cdot c = 3 \cdot 25$$

or $$4c = 75$$

Step 2 Locate the coefficient of the variable, 4, and divide both sides of the equation by that coefficient.

$$\frac{\overset{1}{\cancel{4}}c}{\underset{1}{\cancel{4}}} = \frac{75}{4}$$

$$c = \frac{75}{4}$$ $\frac{75}{4}$ does not simplify.

Step 3 To check, replace c with $\frac{75}{4}$ in the original proportion.

$$\frac{3}{4} \stackrel{?}{=} \frac{\frac{75}{4}}{25}$$

Multiply.

$$3 \cdot 25 = 75$$

$$4 \cdot \frac{75}{4} = 75$$ The products are the same, so the value of $\frac{75}{4}$ checks for c.

RECALL

In the proportion equation $\frac{a}{b} = \frac{c}{d}$, we check to see if $ad = bc$.

Check Yourself 1

Solve the proportions for n. Check your result.

(a) $\frac{4}{5} = \frac{n}{25}$ **(b)** $\frac{5}{9} = \frac{12}{n}$

In solving for a missing term in a proportion, we may find an equation involving fractions or decimals. Example 2 involves finding the unknown value in such cases.

Example 2 Solving Proportions

(a) Solve the proportion for x.

$$\frac{\frac{1}{4}}{3} = \frac{4}{x}$$

$$\frac{1}{4}x = 12$$

> **RECALL**
>
> The coefficient is the number multiplying the variable, in this case $\frac{1}{4}$.

$$\frac{\frac{1}{4}x}{\frac{1}{4}} = \frac{12}{\frac{1}{4}}$$

We divide by the coefficient of x. In this case, the coefficient is $\frac{1}{4}$.

$$x = \frac{12}{\frac{1}{4}}$$

Remember: $\frac{12}{\frac{1}{4}}$ is $12 \div \frac{1}{4}$. Invert the divisor and multiply.

$$x = 48$$

To check, replace x with 48 in the original proportion.

$$\frac{\frac{1}{4}}{3} \stackrel{?}{=} \frac{4}{48}$$

$$\frac{1}{4} \cdot 48 \stackrel{?}{=} 3 \cdot 4$$

$$12 = 12$$

> **RECALL**
>
> We always simplify a ratio or fraction if it has decimals or additional fractions.
>
> Here we must divide 6 by 0.5 to find the unknown value.
>
> $$\begin{array}{r} 12. \\ 0.5\overline{)6.0} \\ 5 \\ \hline 10 \\ 10 \\ \hline 0 \end{array}$$
>
> 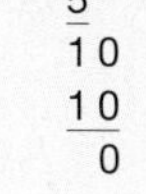

(b) Solve the proportion for d.

$$\frac{0.5}{2} = \frac{3}{d}$$

$$0.5d = 6$$

$$\frac{0.5d}{0.5} = \frac{6}{0.5}$$

Divide by the coefficient, 0.5.

$$d = 12$$

We leave it to you to confirm that $0.5 \cdot 12 = 2 \cdot 3$.

Check Yourself 2

(a) Solve for d.

$$\frac{\frac{1}{2}}{5} = \frac{3}{d}$$

(b) Solve for x.

$$\frac{0.4}{x} = \frac{2}{30}$$

Now that we know how to find an unknown value in a proportion, we can solve many applications.

Step by Step

Solving Applications of Proportions

Step 1 Read the problem carefully to determine the given information.

Step 2 Write the proportion necessary to solve the problem. Use a letter or variable to represent the unknown quantity. Be sure to include the units when writing the proportion.

Step 3 Solve, answer the question of the original problem, and check the proportion.

Example 3 Solving an Application

< Objective 2 >

(a) In a shipment of 400 parts, 14 are found to be defective. How many defective parts should be expected in a shipment of 1,000?
Assume that the ratio of defective parts to the total number remains the same.

$$\frac{14 \text{ defective}}{400 \text{ total}} = \frac{x \text{ defective}}{1{,}000 \text{ total}}$$ We decided to let x be the unknown number of defective parts.

Multiply.

$400x = 14{,}000$

Divide by the coefficient, 400.

$x = 35$

So 35 defective parts should be expected in the shipment.
Checking the original proportion, we get

$14 \cdot 1{,}000 \stackrel{?}{=} 400 \cdot 35$

$14{,}000 = 14{,}000$

(b) Jill works 4.2 hr and receives \$52.50. How much will she get if she works 10 hr? The rate comparing hours worked and the amount of pay remains the same.

$$\frac{4.2 \text{ hr}}{\$52.50} = \frac{10 \text{ hr}}{\$a}$$ Let a be the unknown amount of pay.

$4.2a = 525$

$$\frac{4.2a}{4.2} = \frac{525}{4.2}$$ Divide both sides by 4.2.

$a = \$125$

Jill would receive \$125 for 10 hr of work.

Check Yourself 3

(a) A \$3,000 investment earned \$330 for 1 year. How much will a \$10,000 investment earn at the same rate for 1 year?

(b) A piece of cable 8.5 cm long weighs 68 grams (g). What does a 10-cm length of the same cable weigh?

Example 4 Using Proportions to Solve an Application

The scale on a map is given as $\frac{1}{4}$ in. = 3 mi. The distance between two towns is 4 in. on the map. How far apart are the towns in miles?

To solve this problem, we use the fact that the ratio of inches (on the map) to miles remains the same. We also use another important property of an equation: We can multiply both sides of the equation by the same nonzero number.

$$\frac{\frac{1}{4}\text{ in.}}{3\text{ mi}} = \frac{4\text{ in.}}{x\text{ mi}}$$

$$\frac{1}{4} \cdot x = 3 \cdot 4$$

$$4 \cdot \frac{1}{4} \cdot x = 4 \cdot 3 \cdot 4$$

We multiply both sides of the equation by 4, since $4 \cdot \frac{1}{4} = 1$

$$1 \cdot x = 4 \cdot 3 \cdot 4$$

$$x = 48 \text{ (mi)}$$

The towns are 48 mi apart.

NOTE

We could divide both sides by $\frac{1}{4}$:

$$\frac{\frac{1}{4} \cdot x}{\frac{1}{4}} = \frac{3 \cdot 4}{\frac{1}{4}}$$

$$x = \frac{3 \cdot 4}{\frac{1}{4}}$$

$$x = \frac{12}{\frac{1}{4}}$$

then invert and multiply.

$$x = \frac{12}{1} \cdot \frac{4}{1}$$

$$= 48$$

Check Yourself 4

Jack drives 125 mi in $2\frac{1}{2}$ hr. At the same rate, how far will he be able to travel in 4 hr?

Hint: Write $2\frac{1}{2}$ as an improper fraction.

In Example 5 we must convert the units stated in the problem.

Example 5 Using Proportions to Solve an Application

A machine can produce 15 tin cans in 2 min. At this rate how many cans can it make in an 8-hr period?

In writing a proportion for this problem, we must write the times involved in terms of the same units.

$$\frac{15\text{ cans}}{2\text{ min}} = \frac{x\text{ cans}}{480\text{ min}}$$

Because 1 hr is 60 min, convert 8 hr to 480 min.

$$15 \cdot 480 = 2x$$

or $7{,}200 = 2x$

$$x = 3{,}600 \text{ cans}$$

NOTE

Always check that your units are properly placed.

Check Yourself 5

Instructions on a can of film developer call for 2 oz of concentrate to 1 quart (qt) of water. How much of the concentrate is needed to mix with 1 gal of water? (4 qt = 1 gal.)

Proportions are important when working with *similar* geometric figures. These are figures that have the same shape and whose corresponding sides are proportional. For instance, in the similar triangles shown here, a proportion involving corresponding sides is

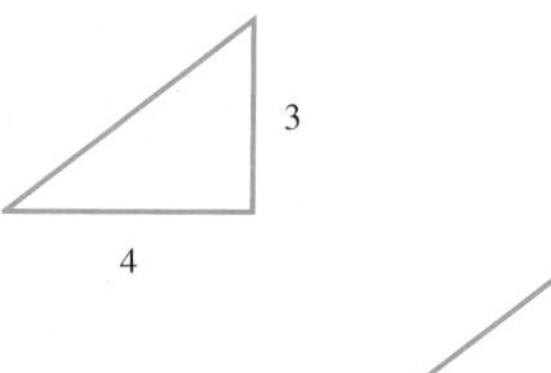

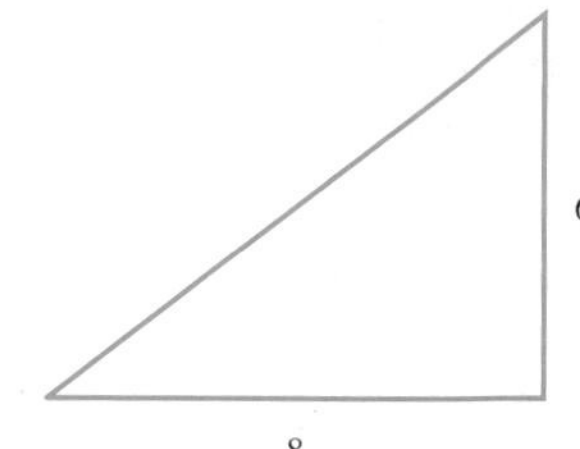

$$\frac{3}{4} = \frac{6}{8}$$

Example 6 Solving an Application Using Similar Triangles

NOTE

Connect the top of the tree to the end of the shadow to create a triangle. Connecting the top of the man to the end of his shadow creates a similar triangle

If a 6-ft-tall man casts a shadow that is 10 ft long, how tall is a tree that casts a shadow that is 140 ft long?

Look at a picture of the two triangles involved.

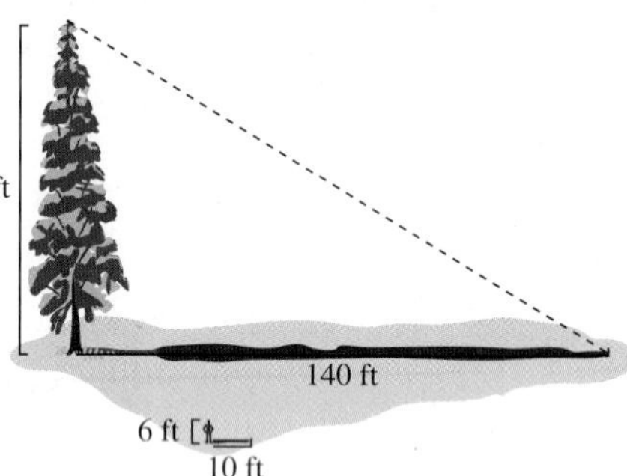

From the similar triangles, we have the proportion

$$\frac{6}{10} = \frac{h}{140}$$

Using the proportion rule, we have $6 \cdot 140 = 10 \cdot h$, so

$$10 \cdot h = 840$$

$$\frac{10 \cdot h}{10} = \frac{840}{10}$$

$$h = 84$$

The tree is 84 ft tall.

Check Yourself 6

If a woman who is $5\frac{1}{2}$ ft tall casts a shadow that is 3 ft long, how tall is a building that casts a 90-ft shadow?

Proportions are used in solving a variety of problems such as the allied health application in Example 7.

Example 7 An Allied Health Application

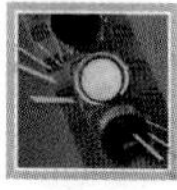

> CAUTION

Solving for x does not give us an answer directly. x represents the total volume, which includes both water and serum. We still need to subtract 8.5 from x to get a final answer.

In preparing specimens for testing, it is often necessary to dilute the original solution. The dilution ratio is the ratio of the volume, in milliliters, of original solution to the total volume, also in milliliters, of the diluted solution. How much water is required to make a $\frac{1}{20}$ dilution from 8.5 mL of serum?

We set up a proportion equation. The original solution is the 8.5 mL of serum, so that should be in the numerator. We name the denominator x.

$$\frac{1}{20} = \frac{8.5}{x}$$

$$1 \cdot x = 20 \cdot 8.5$$

$$x = 170$$

Therefore, the total volume should be 170 mL. Because 8.5 mL of the total solution is serum, we need to add $170 - 8.5 = 161.5$ mL of water.

Check Yourself 7

How much water is required to make a $\frac{3}{50}$ dilution from 11.25 mL of serum?

Check Yourself ANSWERS

1. **(a)** $n = 20$; **(b)** $n = \frac{108}{5}$ **2.** **(a)** $d = 30$; **(b)** $x = 6$ **3.** **(a)** \$1,100; **(b)** 80 g **4.** 200 mi **5.** 8 oz **6.** 165 ft **7.** 176.25 mL

Reading Your Text

These fill-in-the-blank exercises will help you understand some of the key vocabulary used in this section. The answers to these exercises are in the Answers Appendix at the back of the text.

(a) An ________ is a statement that two expressions are equal.

(b) When a number and a variable are multiplied, the number is called a ________.

(c) The first step to solving application problems is to ________ the problem carefully.

(d) Two triangles are similar if corresponding sides are ________.

Skills Calculator/Computer Career Applications Above and Beyond

5.4 exercises

< Objective 1 >

Solve for the unknown in each proportion.

1. $\frac{x}{3} = \frac{6}{9}$ 

2. $\frac{x}{6} = \frac{3}{9}$

3. $\frac{10}{n} = \frac{15}{6}$

4. $\frac{4}{3} = \frac{8}{n}$

5. $\frac{4}{7} = \frac{y}{14}$

6. $\frac{5}{8} = \frac{a}{16}$

7. $\frac{5}{7} = \frac{x}{35}$

8. $\frac{4}{15} = \frac{8}{n}$

9. $\frac{11}{a} = \frac{2}{44}$

10. $\frac{35}{40} = \frac{7}{n}$

11. $\frac{x}{8} = \frac{15}{24}$

12. $\frac{7}{12} = \frac{m}{24}$

Solve each proportion. Write your answers as proper fractions or mixed numbers.

13. $\frac{x}{8} = \frac{7}{16}$

14. $\frac{5}{4} = \frac{x}{5}$

15. $\frac{4}{y} = \frac{6}{1}$

16. $\frac{4}{1} = \frac{1}{y}$

17. $\frac{a}{4} = \frac{9}{14}$

18. $\frac{29}{16} = \frac{n}{2}$

Solve each proportion. Write your answers as decimals, rounded to the nearest hundredth.

19. $\frac{5}{m} = \frac{12}{5}$ **20.** $\frac{3}{8} = \frac{8}{b}$ **21.** $\frac{x}{10} = \frac{9}{7}$ **22.** $\frac{4}{13} = \frac{y}{6}$

23. $\frac{x}{8} = \frac{1}{12}$ **24.** $\frac{1}{11} = \frac{x}{2}$

Solve each proportion.

25. $\frac{\frac{1}{2}}{2} = \frac{3}{a}$ **26.** $\frac{x}{5} = \frac{2}{\frac{1}{3}}$ **27.** $\frac{0.2}{2} = \frac{1.2}{a}$ **28.** $\frac{0.5}{x} = \frac{1.25}{5}$

29. $\frac{\frac{2}{5}}{8} = \frac{1.2}{n}$ **30.** $\frac{\frac{2}{3}}{0.75} = \frac{x}{27}$

Solve each proportion. Write your answers as proper fractions or mixed numbers.

31. $\frac{\frac{1}{4}}{12} = \frac{m}{40}$ **32.** $\frac{1}{6} = \frac{\frac{2}{x}}{18}$ **33.** $\frac{12}{\frac{1}{3}} = \frac{80}{y}$ **34.** $\frac{3}{4} = \frac{4}{\frac{x}{10}}$

Solve each proportion. Write your answers in decimal form. Round to the nearest tenth when appropriate.

35. $\frac{x}{3.3} = \frac{1.1}{6.6}$ **36.** $\frac{2.4}{5.7} = \frac{m}{1.1}$ **37.** $\frac{4}{a} = \frac{\frac{1}{4}}{0.8}$ **38.** $\frac{x}{9} = \frac{\frac{5}{4}}{8.2}$

< Objective 2 >

Solve each application.

39. **Business and Finance** If 12 books are purchased for \$80, how much will you pay for 18 books at the same rate?

40. If 6 ears of corn cost \$2, how much do 21 ears cost?

41. **Business and Finance** A box of 18 tea bags is marked \$4.89. At that price, what should a box of 48 tea bags cost?

42. **Construction** If an 8-foot (ft) two-by-four costs \$4.64, what should a 12-ft two-by-four cost?

43. **Social Science** The ratio of yes to no votes in an election was 3 to 2. How many no votes were cast if there were 2,880 yes votes?

44. **Business and Finance** A worker can complete the assembly of 15 MP3 players in 6 hours (hr). At this rate, how many can the worker complete in a 40-hr workweek?

45. **Crafts** A photograph 5 in. wide by 6 in. high is to be enlarged so that the new width is 15 in. What will the height of the enlargement be?

46. **Business and Finance** Christy can travel 110 mi in her new car on 5 gal of gas. How far can she travel on a full (12-gal) tank?

47. **Business and Finance** The Changs purchased a \$120,000 home, and the property taxes were \$2,100. If they make improvements and the house is now valued at \$150,000, what will the new property tax be?

48. **Business and Finance** A car travels 165 mi in 3 hr. How far will it travel in 8 hr if it continues at the same speed?

Using the given map, find the distances between the cities named in exercises 49 to 52. Measure distances to the nearest sixteenth of an inch.

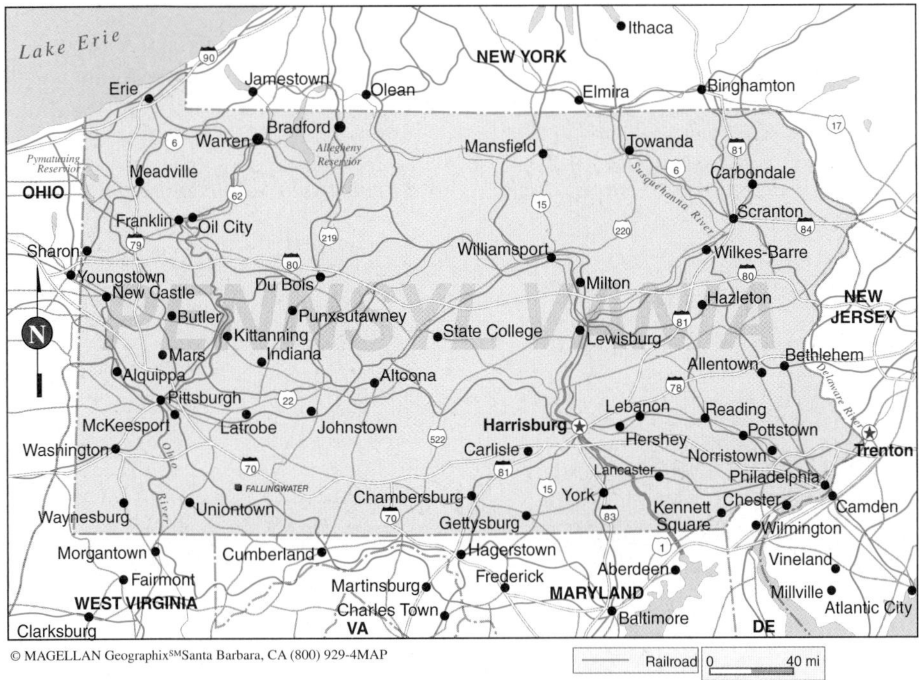

© MAGELLAN Geographix[SM]Santa Barbara, CA (800) 929-4MAP

49. Find the distance from Harrisburg to Philadelphia.

50. Find the distance from Punxsutawney (home of the groundhog) to State College (home of the Nittany Lions).

51. Find the distance from Gettysburg to Meadville.

52. Find the distance from Scranton to Waynesburg.

53. **Business and Finance** An inspection reveals 27 defective parts in a shipment of 500. How many defective parts should be expected in a shipment of 1,200? Report your result to the nearest whole number.

54. **Business and Finance** You invest \$4,000 in a stock that pays a \$180 dividend in 1 year. At the same rate, how much will you need to invest to earn \$250? Round your result to the nearest dollar.

55. **Construction** A 5-ft fence post casts a 9-ft shadow. How tall is a nearby pole that casts a 15-ft shadow? Round your result to the nearest tenth of a foot.

56. **Construction** A 9-ft light pole casts a 15-ft shadow. Find the height of a nearby tree that is casting a 36-ft shadow. Round your result to the nearest tenth.

57. **Construction** On the blueprint of the Wilsons' new home, the scale is 5 in. equals 7 ft. What will the actual length of a bedroom be if it measures 10 in. long on the blueprint?

58. **Social Science** The scale on a map is $\frac{1}{2}$ in. = 50 mi. If the distance between two towns on the map is 6 in., how far apart are they in miles?

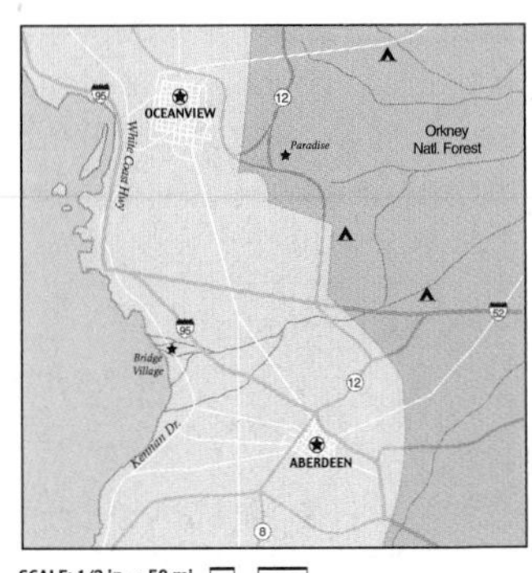

SCALE: 1/2 in. = 50 mi
0 50 100 200

59. **Science and Medicine** A metal bar expands $\frac{1}{4}$ in. for each 12°F rise in temperature. How much will it expand if the temperature rises 44°F? Write your result in fraction form.

60. **Business and Finance** Your car burns $2\frac{1}{2}$ qt of oil on a trip of 5,000 mi. How many quarts should you expect to use when driving 7,200 mi? Write your result as a fraction or mixed number.

61. **Social Science** Approximately 7 out of every 10 people in the United States workforce drive to work alone. During morning rush hour there are 115,000 cars on the streets of a medium-sized city. How many of these cars have one person in them?

62. **Social Science** Approximately 15 out of every 100 people in the United States workforce carpool to work. There are an estimated 320,000 people in the workforce of a given city. How many of these people are in carpools?

Use a proportion to find the unknown side, labeled x, in each pair of similar figures. Report your results as proper fractions or mixed numbers, when appropriate.

63.

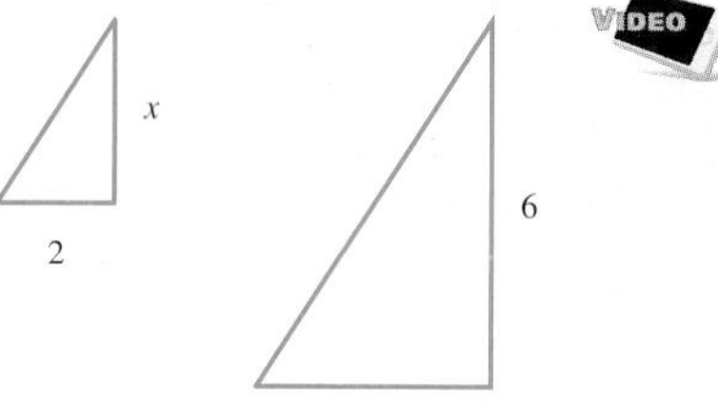

VIDEO

64.

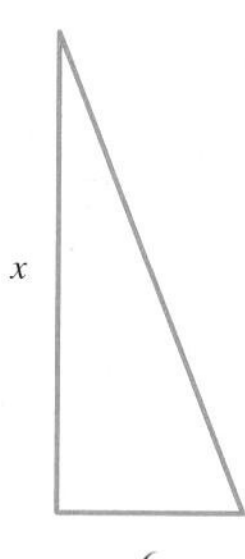

65.

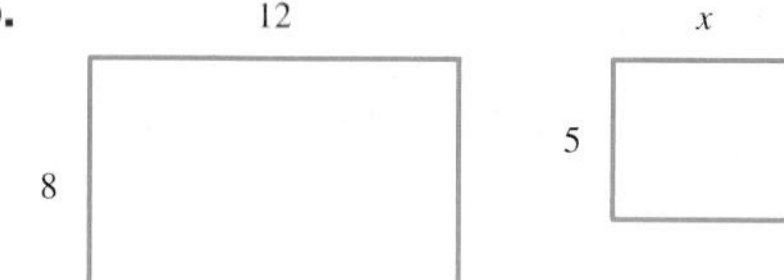

66.

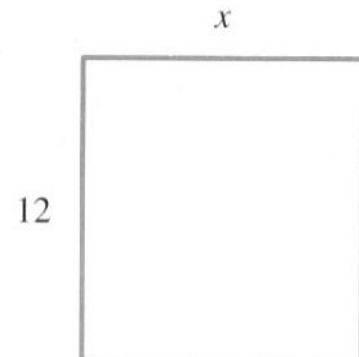

Determine whether each statement is **true** *or* **false.**

67. Given the product of a number and a variable, the variable is called the coefficient.

68. Given the product of a number and a variable, the number is called the coefficient.

Fill in each blank with **always, sometimes,** *or* **never.**

69. In solving a proportion, we can __________ divide by the coefficient of the variable.

70. Two similar triangles ________ have corresponding sides that are proportional.

Skills	**Calculator/Computer**	Career Applications	Above and Beyond

In real-world applications, the numbers involved in a proportion may be large or contain inconvenient decimals. A calculator is likely to be the tool of choice for solving such proportions. Typically, we set up the solution without doing any calculating, and then we put the calculator to work, usually rounding the result to an appropriate place value. For example, suppose you drive 278 mi on 13.6 gal of gas. If the gas tank holds 21 gal, and you want to know how far you can travel on a full tank of gas, you write

$$\frac{278 \text{ mi}}{13.6 \text{ gal}} = \frac{x \text{ mi}}{21 \text{ gal}}$$

Solving for x, you obtain

$$x = \frac{(278)(21)}{13.6}$$

With your calculator, you enter

278 [×] 21 [÷] 13.6 [=]

The display shows 429.26471. Rounding to the nearest mile, you can travel 429 mi.

Use your calculator to solve each proportion.

71. $\frac{630}{1,365} = \frac{15}{a}$

72. $\frac{770}{1,988} = \frac{n}{71}$

73. $\frac{x}{4.7} = \frac{11.8}{16.9}$ (to nearest tenth)

74. $\frac{13.9}{8.4} = \frac{n}{9.2}$ (to nearest hundredth)

75. $\frac{2.7}{3.8} = \frac{5.9}{n}$ (to nearest tenth)

76. $\frac{12.2}{0.042} = \frac{x}{0.08}$ (to nearest hundredth)

Solve each application.

77. **Business and Finance** Bill earns $496.80 for working 34.5 hr. How much will he receive if he works at the same pay rate for 31.75 hr?

78. **Construction** Construction-grade pressure-treated lumber costs $961.25 per 1,000 board-feet. What will be the cost of 686 board-feet? Round your answer to the nearest cent.

79. **Science and Medicine** A speed of 88 feet per second (ft/s) is equal to a speed of 60 miles per hour (mi/hr). If the speed of sound is 750 mi/hr, what is the speed of sound in feet per second?

80. **Business and Finance** A shipment of 75 parts is inspected, and 6 are found to be faulty. At the same rate, how many defective parts should be found in a shipment of 139? Round your result to the nearest whole number.

81. **Information Technology** A computer transmits 5 Web pages in 2 s to a Web server. How many pages can the computer transmit in 1 min?

82. **Automotive Technology** A tire shows $\frac{1}{8}$ in. of tread wear after 32,000 mi. Assuming that the rate of wear remains constant, how much tread wear would you expect the tire to show after 48,000 mi?

83. **Manufacturing Technology** Cutting 7 holes removes 0.322 lb of material from a frame. How much weight would 43 holes remove?

84. **Electrical Engineering** The voltage output V_{out} of a transformer is given by the proportion

$$\frac{N_{out}}{N_{in}} = \frac{V_{out}}{V_{in}}$$

in which N gives the number of turns in the coil (see the figure).

In the system shown, what voltage input is required in order for the output voltage to reach 630 V?

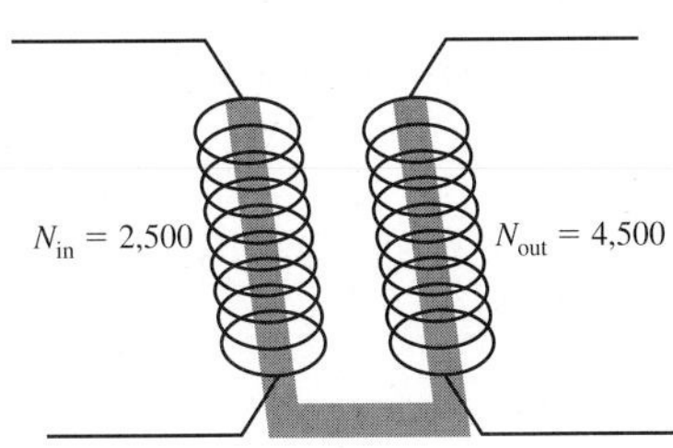

85. A cookbook lists the ingredients for a recipe that serves 12 people.

12 cups ziti	7 cups spaghetti sauce	4 cups ricotta cheese
$\frac{1}{2}$ cup parsley	1 teaspoon garlic powder	$\frac{1}{2}$ teaspoon pepper
4 cups mozzarella cheese	2 tablespoons parmesan cheese	

Determine the amount of each ingredient needed to serve 5 people.

Answers

1. 2 **3.** 4 **5.** 8 **7.** 25 **9.** 242 **11.** 5 **13.** $3\frac{1}{2}$ **15.** $\frac{2}{3}$ **17.** $2\frac{4}{7}$ **19.** 2.08 **21.** 12.86 **23.** 0.67 **25.** 12 **27.** 12 **29.** 24 **31.** $\frac{5}{6}$ **33.** $2\frac{2}{9}$ **35.** 0.6 **37.** 12.8 **39.** $120 **41.** $13.04 **43.** 1,920 no votes **45.** 18 in. **47.** $2,625 **49.** 110 mi **51.** 215 mi **53.** 65 defective parts **55.** 8.3 ft **57.** 14 ft **59.** $\frac{11}{12}$ in. **61.** 80,500 cars with one person **63.** 3 **65.** $7\frac{1}{2}$ **67.** False **69.** always **71.** 32.5 **73.** 3.3 **75.** 8.3 **77.** $457.20 **79.** 1,100 ft/s **81.** 150 Web pages **83.** 1.978 lb **85.** Above and Beyond

Burning Calories

Many people are interested in losing weight through exercise. An important fact to consider is that a person needs to burn off 3,500 calories more than he or she takes in to lose 1 lb, according to the American Dietetic Association.

The table shows the number of calories burned per hour (cal/hr) for a variety of activities, where the figures are based on a 150-lb person.

Activity	Cal/hr	Activity	Cal/hr
Bicycling 6 mi/hr	240	Running 10 mi/hr	1,280
Bicycling 12 mi/hr	410	Swimming 25 yd/min	275
Cross-country skiing	700	Swimming 50 yd/min	500
Jogging $5\frac{1}{2}$ mi/hr	740	Tennis (singles)	400
Jogging 7 mi/hr	920	Walking 2 mi/hr	240
Jumping rope	750	Walking 3 mi/hr	320
Running in place	650	Walking $4\frac{1}{2}$ mi/hr	440

Work with your group members to solve each problem. You may find that setting up proportions is helpful.

For problems 1 through 4, assume a 150-lb person.

1. If a person jogs at a rate of $5\frac{1}{2}$ mi/hr for $3\frac{1}{2}$ hr in a week, how many calories will the person burn?
2. If a person runs in place for 15 min, how many calories will the person burn?
3. If a person cross-country skis for 35 min, how many calories will the person burn?
4. How many hours would a person have to jump rope in order to lose 1 lb? (Assume calorie consumption is just enough to maintain weight, with no activity.)

Heavier people burn more calories (for the same activity), and lighter people burn fewer. In fact, you can calculate similar figures for burning calories by setting up the appropriate proportion.

5. Assuming that the calories burned are proportional, at what rate would a 120-lb person burn calories while bicycling at 12 mi/hr?
6. Assuming that the calories burned are proportional, at what rate would a 180-lb person burn calories while bicycling at 12 mi/hr?
7. Assuming that the calories burned are proportional, how many hours of jogging at $5\frac{1}{2}$ mi/hr would be needed for a 200-lb person to lose 5 lb? (Again, assume calorie consumption is just enough to maintain weight, with no activity.)

summary :: chapter 5

Definition/Procedure	Example	Reference
Ratios		**Section 5.1**
Ratio A means of comparing two numbers or like quantities. A ratio can be written as a fraction.	$\frac{4}{7}$ can be thought of as "the ratio of 4 to 7."	*p.* 280
Rates and Unit Pricing		**Section 5.2**
Rate A fraction involving two unlike denominate numbers. **Unit Rate** A rate that has been simplified so that the denominator is one unit.	$\frac{50 \text{ home runs}}{150 \text{ games}} = \frac{1}{3} \frac{\text{home run}}{\text{game}}$	*p.* 288
Unit Price The cost per unit.	$\frac{\$2}{5 \text{ rolls}} = \$0.40 \text{ per roll}$ $= 40 \frac{\text{cents}}{\text{roll}}$	*p.* 291
Proportions		**Section 5.3**
Proportion A statement that two fractions or rates are equal.	$\frac{3}{5} = \frac{6}{10}$ is a proportion read "three is to five as six is to ten."	*p.* 298
The Proportion Rule If $\frac{a}{b} = \frac{c}{d}$, then $a \cdot d = b \cdot c$.	If $\frac{3}{5} = \frac{6}{10}$, then $3 \cdot 10 = 5 \cdot 6$	*p.* 299
Solving Proportions		**Section 5.4**
To Solve a Proportion		
Step 1 Use the proportion rule to write the equivalent equation $a \cdot d = b \cdot c$. **Step 2** Divide both terms of the equation by the coefficient of the variable. **Step 3** Use the value found to replace the unknown in the original proportion. Check that the ratios or rates are proportional.	To solve: $\frac{x}{5} = \frac{16}{20}$ $20x = 5 \cdot 16$ $20x = 80$ $\frac{\cancel{20}x}{\cancel{20}} = \frac{80}{20}$ $x = 4$ Check: $\frac{4}{5} \stackrel{?}{=} \frac{16}{20}$ $4 \times 20 \stackrel{?}{=} 5 \times 16$ $80 = 80$	*p.* 303
Solving a Problem by Using Proportions		
Step 1 Read the problem carefully to determine the given information. **Step 2** Write the proportion necessary to solve the problem, using a letter or variable to represent the unknown quantity. Be sure to include the units when writing the proportion. **Step 3** Solve, answer the question of the original problem, and check the proportion as before.	A machine can produce 250 units in 5 min. At this rate, how many can it produce in a 12-hr period? $\frac{250 \text{ units}}{5 \text{ min}} = \frac{x \text{ units}}{12 \text{ hr}}$ $\frac{250 \text{ units}}{5 \text{ min}} = \frac{x \text{ units}}{720 \text{ min}}$ $250 \cdot 720 = 5x$ or $5x = 180{,}000$ $x = 36{,}000$ The machine can produce 36,000 units in 12 hr.	*p.* 306

This summary exercise set will help ensure that you have mastered each of the objectives of this chapter. The exercises are grouped by section. You should reread the material associated with any exercises that you find difficult. The answers to the odd-numbered exercises are in the Answers Appendix at the back of the text.

5.1 *Write each ratio in simplest form.*

1. The ratio of 4 to 17

2. The ratio of 28 to 42

3. For a football team that has won 10 of its 16 games, the ratio of wins to games played

4. For a rectangle of length 30 inches and width 18 inches, the ratio of its length to its width

5. The ratio of $2\frac{1}{3}$ to $5\frac{1}{4}$

6. The ratio of 7.5 to 3.25

7. The ratio of 7 in. to 3 ft

8. The ratio of 72 hr to 4 days

5.2 *Express each rate as a unit rate.*

9. $\frac{600 \text{ miles}}{6 \text{ hours}}$

10. $\frac{270 \text{ miles}}{9 \text{ gallons}}$

11. $\frac{350 \text{ calories}}{7 \text{ ounces}}$

12. $\frac{36{,}000 \text{ dollars}}{9 \text{ years}}$

13. $\frac{5{,}000 \text{ feet}}{30 \text{ seconds}}$

14. $\frac{10{,}000 \text{ revolutions}}{3 \text{ minutes}}$

15. A baseball team has had 117 hits in 18 games. Find the team's hits per game rate.

16. A basketball team has scored 216 points in 8 quarters. Find the team's points per quarter rate.

17. Taniko scored 246 points in 20 games. Marisa scored 216 points in 16 games. Which player has the highest points per game rate?

18. One shop will charge $306 for a job that takes $4\frac{1}{2}$ hr. A second shop can do the same job in 4 hr and will charge $290. Which shop has the higher cost per hour rate?

Find the unit price for each item.

19. A 32-oz bottle of dishwashing liquid costs $2.88.

20. A 35-oz box of breakfast cereal costs $5.60.

21. A 24-oz loaf of bread costs $2.28.

22. Five large jars of fruit cost $67.30.

23. Three CDs cost $44.85.

24. Six tickets cost $267.60.

5.3 *Write each proportion.*

25. 4 is to 9 as 20 is to 45.

26. 7 is to 5 as 56 is to 40.

27. If Jorge can travel 110 mi in 2 hr, he can travel 385 mi in 7 hr.

28. If it takes 4 gal of paint to cover 1,000 ft^2, it takes 10 gal of paint to cover 2,500 ft^2.

Determine whether each pair of fractions is proportional.

29. $\frac{4}{13} \stackrel{?}{=} \frac{7}{22}$

30. $\frac{8}{11} \stackrel{?}{=} \frac{24}{33}$

31. $\frac{9}{24} \stackrel{?}{=} \frac{12}{32}$

32. $\frac{7}{18} \stackrel{?}{=} \frac{35}{80}$

33. $\frac{5}{\frac{1}{6}} \stackrel{?}{=} \frac{120}{4}$

34. $\frac{0.8}{4} \stackrel{?}{=} \frac{12}{50}$

35. Is $\frac{35 \text{ Euros}}{40 \text{ dollars}}$ proportional to $\frac{75.25 \text{ Euros}}{86 \text{ dollars}}$?

36. Is $\frac{188 \text{ words}}{8 \text{ minutes}}$ proportional to $\frac{121 \text{ words}}{5 \text{ minutes}}$?

5.4 *Solve each proportion.*

37. $\frac{16}{24} = \frac{m}{3}$

38. $\frac{6}{a} = \frac{27}{18}$

39. $\frac{14}{35} = \frac{t}{10}$

40. $\frac{55}{88} = \frac{10}{p}$

41. $\frac{\frac{1}{2}}{18} = \frac{5}{w}$

42. $\frac{\frac{3}{2}}{9} = \frac{5}{a}$

43. $\frac{5}{x} = \frac{0.6}{12}$

44. $\frac{s}{2.5} = \frac{1.5}{7.5}$

Solve each application.

45. Business and Finance If 4 tickets to a civic theater performance cost $90, what is the price for 6 tickets?

46. Social Science The ratio of first-year to second-year students at a school is 8 to 7. If there are 224 second-year students, how many first-year students are there?

47. Crafts A photograph that is 5 in. wide by 7 in. tall is to be enlarged so that the new height will be 21 in. What will be the width of the enlargement?

48. Business and Finance Marcia assembles disk drives for a computer manufacturer. If she can assemble 11 drives in 2 hr, how many can she assemble in a workweek (40 hr)?

49. Business and Finance A firm finds 14 defective parts in a shipment of 400. How many defective parts can be expected in a shipment of 800 parts?

50. Social Science The scale on a map is $\frac{1}{4}$ in. = 10 mi. How many miles apart are two towns that are 3 in. apart on the map?

51. Business and Finance A piece of tubing that is 16.5 cm long weighs 55 g. What is the weight of a piece of the same tubing that is 42 cm long?

52. Crafts If 1 qt of paint covers 120 ft^2, how many square feet does 2 gal cover? (1 gal = 4 qt.)

chapter test 5

Use this chapter test to assess your progress and to review for your next exam. Allow yourself about an hour to take this test. The answers to these exercises are in the Answers Appendix at the back of the text.

Write each ratio in simplest form.

1. The ratio of 7 to 19

2. The ratio of 75 to 45

3. The ratio of 8 ft to 4 yd

4. The ratio of 6 hr to 3 days

Express each rate as a unit rate.

5. $\frac{840 \text{ mi}}{175 \text{ gal}}$

6. $\frac{132 \text{ dollars}}{16 \text{ hr}}$

Solve for the unknown in each proportion.

7. $\frac{45}{75} = \frac{12}{x}$

8. $\frac{a}{26} = \frac{45}{65}$

9. $\frac{\frac{1}{2}}{p} = \frac{5}{30}$

10. $\frac{3}{m} = \frac{0.9}{4.8}$

Determine whether each pair of fractions is proportional.

11. $\frac{3}{9} \stackrel{?}{=} \frac{27}{81}$

12. $\frac{6}{7} \stackrel{?}{=} \frac{9}{11}$

13. $\frac{9}{10} \stackrel{?}{=} \frac{27}{30}$

14. $\frac{\frac{1}{2}}{5} \stackrel{?}{=} \frac{2}{18}$

15. Find the unit price, if 11 gal of milk cost $28.16.

16. A basketball team wins 26 of its 33 games during a season. What is the ratio of wins to games played? What is the ratio of wins to losses?

Solve each application using a proportion.

17. Business and Finance If ballpoint pens are marked 5 for 95¢, how much does a dozen cost?

18. Business and Finance Your new compact car travels 324 miles on 9 gal of gas. If the tank holds 16 usable gallons, how far can you drive on a tank of gas?

19. Business and Finance An assembly line can install 5 car mufflers in 4 min. At this rate, how many mufflers can be installed in an 8-hr shift?

20. Crafts Instructions on a package of concentrated plant food call for 2 teaspoons (tsp) to 1 qt of water. We wish to use 3 gal of water. How much of the plant food concentrate should be added to the 3 gal of water?

cumulative review chapters 1–5

Use this exercise set to review concepts from earlier chapters. While it is not a comprehensive exam, it will help you identify any material that you need to review before moving on to the next chapter. In addition to the answers, you will find section references for these exercises in the Answers Appendix in the back of the text.

1.1

1. Write 45,789 in words.

2. What is the place value of 2 in the number 621,487?

Perform the indicated operations.

1.2

3. $\begin{array}{r} 2{,}790 \\ 831 \\ +\ 22{,}683 \\ \hline \end{array}$

1.3

4. $\begin{array}{r} 84{,}793 \\ -\ 36{,}987 \\ \hline \end{array}$

1.5

5. 76×58

1.6

6. $72\overline{)5{,}683}$

1.3

7. Luis owes \$815 on a credit card after a trip. He makes payments of \$125, \$80, and \$90. Interest amounting to \$48 is charged. How much does he still owe on the account?

1.7

8. Evaluate the expression: $48 \div 8 \times 2 - 3^2$.

1.2 and 1.5

9. Find the perimeter and area of the figure.

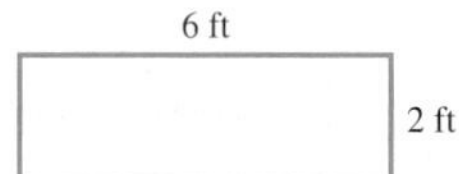

10. A room that measures 6 yd by 8 yd is to be carpeted. The carpet costs \$23 per square yard. What is the cost of the carpet?

2.2

11. Write the prime factorization of 924.

12. Find the greatest common factor (GCF) of 42 and 56.

2.4

13. Write the fraction $\frac{42}{168}$ in simplest form.

Perform the indicated operations.

2.5

14. $\frac{3}{4} \times \frac{24}{15}$

15. $2\frac{2}{3} \times 3\frac{3}{4}$

2.6

16. $\frac{6}{7} \div \frac{4}{21}$

17. $5\frac{1}{2} \div 3\frac{1}{4}$

3.3

18. $\frac{9}{11} - \frac{3}{4} + \frac{1}{2}$

3.4

19. $4\frac{5}{6} + 2\frac{3}{4}$

20. $8\frac{2}{7} - 3\frac{11}{14}$

3.2

21. Find the least common multiple (LCM) of 36 and 60.

2.5

22. Maria drove at an average speed of 55 mi/hr for $3\frac{1}{5}$ hr. How many miles did she travel?

2.6

23. Stefan drove 132 mi in $2\frac{3}{4}$ hr. What was his average speed?

Perform the indicated operations.

4.2
24. $36.169 - 28.341$

4.4
25. $3.1488 \div 2.56$

4.3
26. 4.89×1.35

4.5
27. Write 0.36 as a fraction and simplify.

28. Write $\frac{7}{22}$ as a decimal (to the nearest hundredth).

4.2
29. Find the perimeter of a rectangle that has dimensions 4.23 m by 2.8 m.

4.3
30. Find the area of a rectangle with dimensions 8 cm by 6.28 cm.

5.1
Write each ratio in simplest form.

31. 12 to 26

32. 60 to 18

33. 6 dimes to 3 quarters

5.3
Determine whether each pair of fractions is proportional.

34. $\frac{5}{6} \stackrel{?}{=} \frac{20}{24}$

35. $\frac{3}{7} \stackrel{?}{=} \frac{9}{22}$

5.4
Solve for the unknown.

36. $\frac{x}{3} = \frac{8}{12}$

37. $\frac{5}{x} = \frac{4}{12}$

5.2
38. Give the unit price for an item that weighs 20 oz and costs $4.88.

5.4
39. On a map, 3 cm represents 250 km. How far apart are two cities if the distance between them on the map is 7.2 cm?

40. A company finds 15 defective items in a shipment of 600 items. How many defective items can be expected in a shipment of 2,000 items?

Percents

CHAPTER 6 OUTLINE

INTRODUCTION

A store pays \$15.96 for T-shirts that it sells for \$19.95 each. How do they determine the selling or retail price? Did they simply add \$3.99 to the wholesale price? Would the same store charge \$503.99 for a refrigerator that costs them \$500 wholesale? Would such a store stay in business very long?

Instead of adding \$3.99 to the wholesale price, the store adds 25% of the wholesale cost to figure out how much to charge. With this approach, the T-shirts would still sell for \$19.95, but now the refrigerator sells for \$625 (or even \$629.95).

Of course, the real benefit of increasing the price by a fixed percentage is that every dollar grows by the same amount. This is how they compute the return on their investment. Properly calculating their return is the key to being successful in the business environment.

Determining markup (or percent increase), markdown (or percent decrease), and the return on an investment are just some of the real-world applications of percents. There are many others. We use percents in nearly every aspect of life.

In this chapter, we learn about percents, how to apply them to situations, and how to solve problems that rely on them. You will have the opportunity to look at how money grows and to compute the return on an investment when you complete Activity 17.

6 prerequisite check

This Prerequisite Check highlights the skills you will need in order to be successful in this chapter. The answers to these exercises are in the Answers Appendix at the back of the text.

Simplify each fraction.

1. $\frac{44}{100}$

2. $\frac{125}{100}$

Rewrite each number in decimal form.

3. $\frac{3}{8}$

4. $5\frac{1}{2}$

Rewrite each number as a simplified fraction or mixed number.

5. 0.08

6. 6.25

Evaluate each expression. Report your results in decimal form.

7. $0.04 \times 1{,}040$

8. 1.25×58.95

9. $1\frac{1}{3} \times 62.1$

10. $\frac{2}{3} \times 95$

11. $\frac{5{,}250}{100}$

12. $8 \times 12 \div 100$

Solve each proportion.

13. $\frac{4}{5} = \frac{x}{60}$

14. $\frac{7}{12} = \frac{x}{100}$

15. $\frac{21}{100} = \frac{x}{18}$

16. $\frac{280}{x} = \frac{3.5}{100}$

Read the application carefully and then complete exercises 17 and 18.

Business and Finance A store charges $899.95 for an electric range that costs them $674.96. What is the store's profit margin on this electric range?

17. What does the application ask you to find?

18. What information are you given in the application?

6.1 Writing Percents as Fractions and Decimals

< 6.1 Objectives >

1 > Use percent notation

2 > Write a percent as a fraction or mixed number

3 > Write a percent as a decimal

In earlier chapters, we used fractions and decimals to describe parts of a whole. Another common method is *percents*. Percents are ratios with 100 in the denominator. In fact, the word **percent** means "for each hundred." Consider the figure.

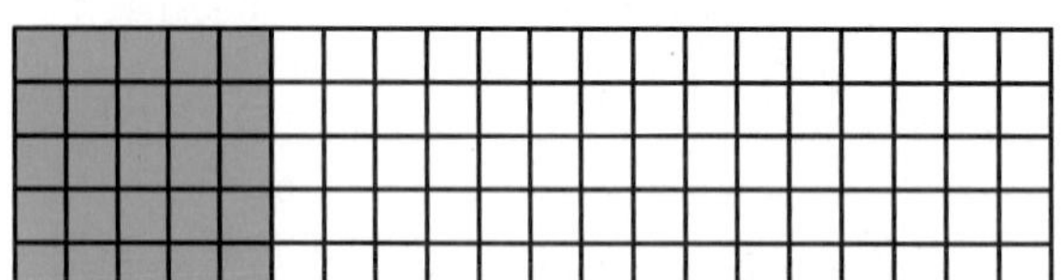

RECALL

Multiplying by $\frac{1}{100}$ is the same as dividing by 100.

The symbol for percent, %, represents multiplication by $\frac{1}{100}$. In the figure, 25 of 100 squares are shaded. We write this in fraction form as $\frac{25}{100}$.

$$\frac{25}{100} = 25\left(\frac{1}{100}\right) = 25\%$$

25 percent of the squares are shaded.

Example 1 Using Percent Notation

< Objective 1 >

(a) Four out of five geography students passed their midterm exams. What percent passed their midterm exams?

The ratio of passing students to all students is $\frac{4}{5}$, which we need to write as an equivalent fraction with a denominator of 100.

RECALL

You learned to solve proportions in Section 5.4.

$$\frac{4}{5} = \frac{x}{100}$$

Using the proportion rule, we have

$$5 \cdot x = 4 \cdot 100$$

or $\quad 5x = 400$

The coefficient of the variable is 5, so we divide both sides of the equation by 5.

$$\frac{5x}{5} = \frac{400}{5}$$

$x = 80 \qquad 400 \div 5 = 80$

Therefore, we write

$$\frac{4}{5} = \frac{80}{100} = 80\left(\frac{1}{100}\right) = 80\%$$

80% of the geography students passed their midterm exams.

NOTE

The ratio of compact cars to all cars is $\frac{35}{50}$.

(b) Of 50 automobiles sold by a dealer in 1 month, 35 were compact cars. What percent of the cars sold were compact cars?

$$\frac{35}{50} = \frac{70}{100} = 70\left(\frac{1}{100}\right) = 70\%$$

70% of the cars sold were compact cars.

Check Yourself 1

Four of the 50 parts in a shipment were defective. What percent of the parts were defective?

Because there are different ways of naming the parts of a whole, you need to know how to change from one of these ways to another. First, we look at writing a percent as a fraction. Because a percent is a fraction or ratio with 100 in the denominator, we can easily write a percent in fraction form.

Property

Writing a Percent as a Fraction

To change a percent to a fraction, replace the percent symbol with $\frac{1}{100}$ and multiply. Simplify the result, if necessary.

Example 2 Writing a Percent as a Fraction

< Objective 2 >

Write each percent as a fraction.

(a) $7\% = 7\left(\frac{1}{100}\right) = \frac{7}{100}$

(b) $25\% = 25\left(\frac{1}{100}\right) = \frac{25}{100} = \frac{1}{4}$

NOTE

You should write $\frac{25}{100}$ in simplest form.

Check Yourself 2

Write 12% as a fraction.

If a percent is *greater than 100,* the equivalent fraction is *greater than 1,* as shown in Example 3.

Example 3 Writing a Percent as a Mixed Number

Write 150% as a mixed number.

$$150\% = 150\left(\frac{1}{100}\right) = \frac{150}{100} = 1\frac{50}{100} = 1\frac{1}{2}$$

Check Yourself 3

Write 125% as a mixed number.

In Examples 2 and 3, we wrote percents as fractions by replacing the percent sign with $\frac{1}{100}$ and multiplying. How do we convert percents to decimal form? Since multiplying by $\frac{1}{100}$ is the same as dividing by 100, we just move the decimal point two places to the left.

Property

Writing a Percent as a Decimal

To write a percent as an equivalent decimal, remove the percent symbol and divide the number by 100.

In practice, this means we move the decimal point *two* places to the *left* and remove the percent symbol.

Example 4 Writing a Percent as a Decimal

< Objective 3 >

NOTE

A percent greater than 100 gives a decimal greater than 1.

Change each percent to its decimal equivalent.

(a) $25\% = 0.25$ The decimal point in 25% is understood to be after the 5.

(b) $8\% = 0.08$ We must add a zero to place the decimal point.

(c) $130\% = 1.30 = 1.3$

Check Yourself 4

Write as decimals.

(a) 5% **(b)** 32% **(c)** 115%

We apply the same approach, even when a percent has a decimal part.

Example 5 Writing a Percent as a Decimal

RECALL

We move the decimal point two places left when removing the percent symbol because we are dividing by 100.

Write as decimals.

(a) $4.5\% = 0.045$

(b) $0.5\% = 0.005$

Check Yourself 5

Write as decimals.

(a) 8.5% **(b)** 0.3%

There are many situations in which common fractions are involved in percents. Example 6 illustrates this situation.

Example 6 Writing a Percent as a Decimal

NOTE

Write the fraction as a decimal. Then remove the percent symbol by our earlier rule.

Write as decimals.

(a) $9\frac{1}{2}\% = 9.5\% = 0.095$

(b) $\frac{3}{4}\% = 0.75\% = 0.0075$

Check Yourself 6

Write as decimals.

(a) $7\frac{1}{2}\%$ **(b)** $\frac{1}{2}\%$

RECALL

You learned to use bar notation to write repeating decimals in Section 4.5.

In Example 6, we were able to write the fraction part of each percent as a terminating decimal. When that is not the case, we are usually better off writing the percent as an equivalent fraction. If we want our final answer to be a decimal, we divide and round, or use bar notation to indicate a repeating decimal.

Example 7 Converting a Percent

Write $18\frac{1}{3}\%$ as an equivalent fraction and decimal.

There are several methods for writing the percent in fraction form. We prefer to write the percent as an improper fraction and then convert the percent to a fraction.

RECALL

You learned to write a mixed number as an improper fraction in Section 2.3.

Step 1 Write $18\frac{1}{3}\%$ as an improper fraction percent.

$$18\frac{1}{3}\% = \frac{18 \times 3 + 1}{3}\%$$

$$= \frac{55}{3}\%$$

Step 2 Convert the percent to a fraction.

$$18\frac{1}{3}\% = \frac{55}{3}\%$$

$$= \frac{55}{3}\left(\frac{1}{100}\right)$$ Remove the % symbol by multiplying by $\frac{1}{100}$.

$$= \frac{55}{300}$$

$$= \frac{11}{60}$$ Always simplify your final result.

We can convert the fraction to a decimal by dividing.

$$\frac{11}{60} = 11 \div 60$$

$$= 0.18\overline{3}$$ Place the bar over the 3 because only that digit repeats.

Check Yourself 7

Write each percent as an equivalent fraction and decimal.

(a) $87\frac{2}{3}\%$ (use bar notation)

(b) $15\frac{1}{6}\%$ (round to the fifth decimal place)

Writing a percent in fraction or decimal form is required in many applications. One such application is presented in Example 8.

Example 8 A Technology Application

A motor has an 86% efficiency rating. Express its efficiency rating as a fraction and as a decimal.

To express 86% as a fraction, we replace the % symbol with $\frac{1}{100}$ and simplify.

$$86\% = \frac{86}{100} = \frac{43}{50}$$

To express 86% as a decimal, we remove the % symbol and move the decimal point two places to the left.

86% = 0.86

Check Yourself 8

An inspection of Carina's hard drive reveals that it is 60% full. Write the amount of hard drive capacity that is full as a fraction and as a decimal.

Check Yourself ANSWERS

1. 8% were defective **2.** $\frac{3}{25}$ **3.** $1\frac{1}{4}$ **4. (a)** 0.05; **(b)** 0.32; **(c)** 1.15 **5. (a)** 0.085; **(b)** 0.003 **6. (a)** 0.075; **(b)** 0.005 **7. (a)** $\frac{263}{300}$; $0.87\overline{6}$; **(b)** $\frac{91}{600}$; 0.15167 **8.** $\frac{3}{5}$; 0.6

Reading Your Text

These fill-in-the-blank exercises will help you understand some of the key vocabulary used in this section. The answers to these exercises are in the Answers Appendix at the back of the text.

(a) The word *percent* means, "for each ________."

(b) To rewrite a percent as a ________, divide the number by 100, remove the percent symbol, and simplify the result.

(c) To write a percent in decimal form, remove the % symbol and move the decimal two places to the ________.

(d) A percent that is larger than 100% is equivalent to a number ________ than 1.

6.1 exercises

Skills Calculator/Computer Career Applications Above and Beyond

< Objective 1 >

Use percents to name the shaded portion of each drawing.

1. 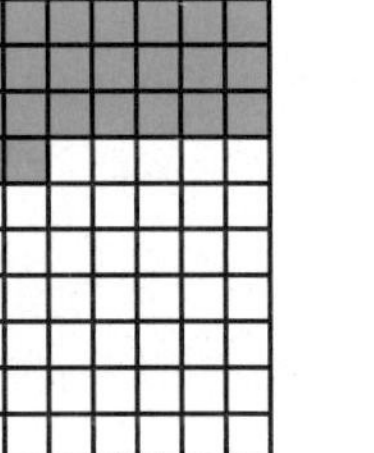

2. 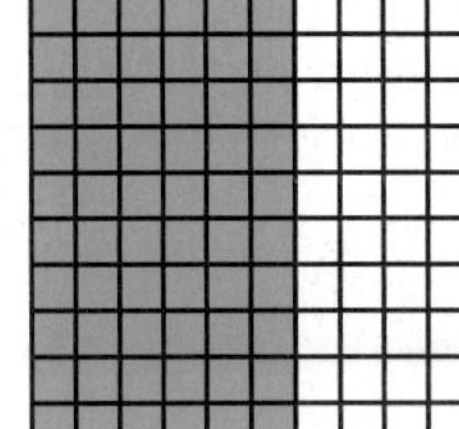

3. 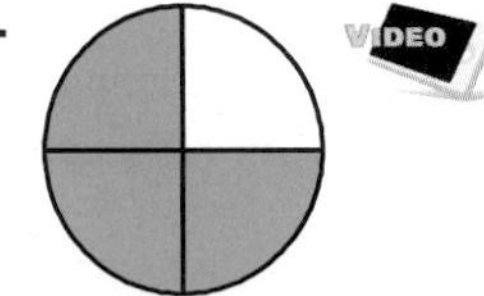

4. 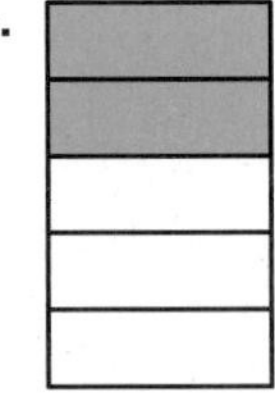

Use percents to describe each statement.

5. Out of every 100 eligible people, 53 voted in a recent election. 

6. You receive $5 in interest for every $100 saved for 1 year.

7. Out of every 100 entering students, 74 register for English composition.
8. Of 100 people surveyed, 29 watched a particular sports event on television.
9. Out of 10 voters in a state, 3 are registered as independents.
10. A dealer sold 9 of the 20 cars available during a 1-day sale.
11. Of 50 houses in a development, 27 are sold.
12. Of the 25 employees of a company, 9 are part-time.
13. Out of 50 people surveyed, 23 prefer decaffeinated coffee.
14. 17 out of 20 college students work at part-time jobs.
15. Of the 20 students in an algebra class, 5 receive a grade of A. VIDEO
16. Of the 50 families in a neighborhood, 31 have children in public schools.

< Objective 2 >

Write each percent as a fraction or mixed number.

17. 6% VIDEO
18. 17%
19. 75%
20. 20%
21. 65% VIDEO
22. 48%
23. 50%
24. 52%
25. 46%
26. 35%
27. 66% VIDEO
28. 4%
29. 150%
30. 140%
31. 225%
32. 450%
33. $166\frac{2}{3}\%$ VIDEO
34. $233\frac{1}{3}\%$
35. $212\frac{1}{2}\%$
36. $116\frac{2}{3}\%$

< Objective 3 >

Write each percent as an equivalent decimal.

37. 20%
38. 70%
39. 35%
40. 75%
41. 39%
42. 27%
43. 5% VIDEO
44. 7%
45. 135%
46. 250%
47. 240%
48. 160%
49. 23.6%
50. 10.5%
51. 6.4% VIDEO
52. 3.5%
53. 0.2%
54. 0.5%
55. 1.05%
56. 0.023%
57. $7\frac{1}{2}\%$ VIDEO
58. $8\frac{1}{4}\%$
59. $87\frac{1}{2}\%$
60. $16\frac{2}{5}\%$
61. $128\frac{3}{4}\%$
62. $220\frac{3}{20}\%$
63. $\frac{1}{2}\%$
64. $\frac{3}{4}\%$

Write each percent as an equivalent fraction and decimal (use bar notation when appropriate).

65. 50%
66. 10%
67. $43\frac{1}{2}\%$
68. $80\frac{3}{4}\%$
69. $33\frac{1}{3}\%$
70. $66\frac{2}{3}\%$
71. $24\frac{1}{6}\%$
72. $76\frac{5}{12}$

73. **Social Science** Automobiles account for 85% of the travel between cities in the United States. What fraction does this percent represent?

74. **Social Science** Automobiles and small trucks account for 84% of the travel to and from work in the United States. What fraction does this percent represent?

75. Convert the discount shown to a decimal and a fraction.

76. Convert the discount shown to a decimal and a fraction.

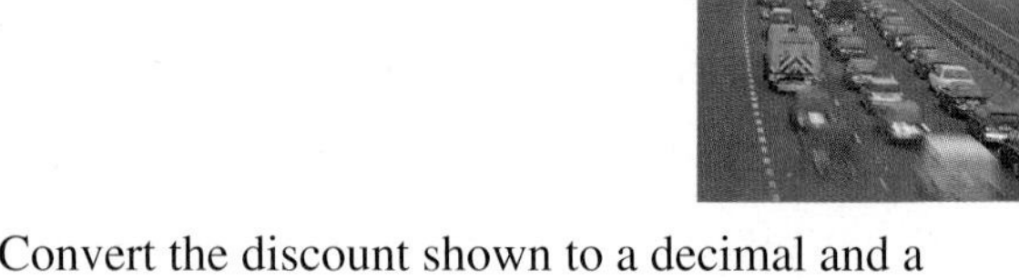

77. The given *percent daily values* (%DV) for some foodstuffs are based on a 2,000-calorie diet. Write each as an equivalent decimal and fraction.

(a) 1 cup of organic whole milk provides 30% of the %DV of calcium.

(b) 1 tablespoon of extra virgin olive oil provides 10% of the %DV of saturated fat.

(c) ¼ cup of long-grain brown rice provides 12% of the %DV of carbohydrates.

(d) ¼ cup of Zante currants provides 8% of the %DV of potassium.

(e) 2 tablespoons of peanut butter provides 7% of the %DV of dietary fiber.

(f) 2 ounces of tuna packed in olive oil provides 2% of the %DV of iron.

78. Complete the table for the percentages given in the bar graph.

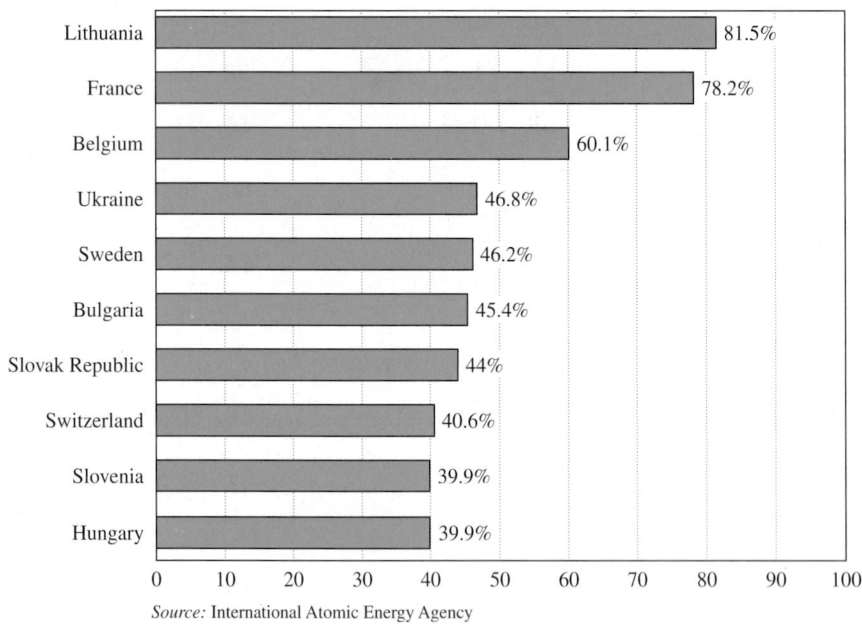

Country	Decimal Equivalent	Fraction Equivalent
Lithuania		
France		
Belgium		
Ukraine		
Sweden		
Bulgaria		
Slovak Republic		
Switzerland		
Slovenia		
Hungary		

Skills | Calculator/Computer | **Career Applications** | Above and Beyond

79. **Information Technology** An information technology project manager has a $49,000 annual budget. She spent 10.2% of her budget in March. What fraction of her annual budget did she spend?

80. **Manufacturing Technology** A packaging company advertises that 87.5% of the machines that it produced in the last 20 years are still in operation. Express the proportion of machines still in service as a fraction.

81. **Manufacturing Technology** On an assembly line, 12% of all products are produced in the color red. Express this as a decimal and a fraction.

82. **Construction Technology** A board that starts with a width of 5.5 in. shrinks to 95% of its original width as it dries. What fraction is the dry width of the original width?

83. **Mechanical Engineering** As a piece of metal cools, it shrinks 3.125%. What fraction of its original size is lost due to shrinkage?

84. **Agricultural Technology** 12.5% of a growing season's 52 in. of rain fell in August. Write the percent of rain that fell in August as a decimal and a fraction.

85. Match each percent in column A with the equivalent fraction in column B.

Column A	Column B
(a) $37\frac{1}{2}\%$	**(1)** $\frac{3}{5}$
(b) 5%	**(2)** $\frac{5}{8}$
(c) $33\frac{1}{3}\%$	**(3)** $\frac{1}{20}$
(d) $83\frac{1}{3}\%$	**(4)** $\frac{3}{8}$
(e) 60%	**(5)** $\frac{5}{6}$
(f) $62\frac{1}{2}\%$	**(6)** $\frac{1}{3}$

86. Explain the difference between $\frac{1}{4}$ of a quantity and $\frac{1}{4}\%$ of a quantity.

Answers

1. 35% **3.** 75% **5.** 53% **7.** 74% **9.** 30% **11.** 54% **13.** 46% **15.** 25% **17.** $\frac{3}{50}$ **19.** $\frac{3}{4}$ **21.** $\frac{13}{20}$ **23.** $\frac{1}{2}$ **25.** $\frac{23}{50}$ **27.** $\frac{33}{50}$ **29.** $1\frac{1}{2}$ **31.** $2\frac{1}{4}$ **33.** $1\frac{2}{3}$ **35.** $2\frac{1}{8}$ **37.** 0.2 **39.** 0.35 **41.** 0.39 **43.** 0.05 **45.** 1.35 **47.** 2.4 **49.** 0.236 **51.** 0.064 **53.** 0.002 **55.** 0.0105 **57.** 0.075 **59.** 0.875 **61.** 1.2875 **63.** 0.005 **65.** $\frac{1}{2}$; 0.5 **67.** $\frac{87}{200}$; 0.435 **69.** $\frac{1}{3}$; $0.\overline{3}$ **71.** $\frac{29}{120}$; $0.241\overline{6}$ **73.** $\frac{17}{20}$ **75.** 0.15; $\frac{3}{20}$ **77.** **(a)** 0.3; $\frac{3}{10}$; **(b)** 0.1; $\frac{1}{10}$; **(c)** 0.12; $\frac{3}{25}$; **(d)** 0.08; $\frac{2}{25}$; **(e)** 0.07; $\frac{7}{100}$; **(f)** 0.02; $\frac{1}{50}$ **79.** $\frac{51}{500}$ **81.** 0.12; $\frac{3}{25}$ **83.** $\frac{1}{32}$ **85.** **(a)** (4); **(b)** (3); **(c)** (6); **(d)** (5); **(e)** (1); **(f)** (2)

6.2 Writing Decimals and Fractions as Percents

< 6.2 Objectives >

1 > Write a decimal as a percent

2 > Write a fraction or mixed number as a percent

When a number is given as a percent, we rewrite it as an equivalent decimal by removing the percent symbol and dividing by 100. In practice, this means that we move the decimal point two places to the left.

We simply reverse this process to rewrite a decimal as a percent.

Property

Writing a Decimal as a Percent

To write a decimal as an equivalent percent, multiply by 100 and attach the percent symbol.

In practice, this means we move the decimal point *two* places to the *right* and attach the percent symbol.

Example 1 — Writing a Decimal as a Percent

< Objective 1 >

(a) Write 0.18 as a percent.

$0.18 = 18\%$

(b) Write 0.03 as a percent.

$0.03 = 3\%$

NOTE

$0.18 = \frac{18}{100} = 18\left(\frac{1}{100}\right) = 18\%$

$0.03 = \frac{3}{100} = 3\left(\frac{1}{100}\right) = 3\%$

Check Yourself 1

Write each number as a percent.

(a) 0.27 (b) 0.05

There are many instances in which we use percents greater than 100%. For example, if a retailer marks goods up 25%, then the retail price is 125% of the wholesale price.

Example 2 — Writing a Decimal as a Percent

Write 1.25 as a percent.

$1.25 = 125\%$

NOTE

$1.25 = \frac{125}{100} = 125\left(\frac{1}{100}\right)$

$= 125\%$

Check Yourself 2

Write 1.3 as a percent.

If a percent includes numbers to the right of the decimal point after the decimal is moved two places to the right, the fraction part can be written as a decimal or as a fraction.

Example 3 Writing a Decimal as a Percent

Write each number as a percent.

(a) 0.045 = 4.5% or $4\frac{1}{2}\%$

(b) 0.003 = 0.3% or $\frac{3}{10}\%$

Check Yourself 3

Write 0.075 as a percent.

RECALL

You learned to use the proportion method in Section 5.4.

You learned to convert fractions to decimals in Section 4.5.

There are two good methods for converting a fraction to a percent. The first method uses a proportions approach. The second method converts the fraction to a decimal, and then converts the decimal to a percent.

The benefit of the proportion method is that your result is an exact value. This is especially useful if the fraction is equivalent to a nonterminating decimal number. The drawback is that it is sometimes more work.

The benefit of the decimal-conversion method is that it is quick and easy, especially if you use a calculator. The drawback is that it is limited in reporting results as exact values.

Property

Writing a Fraction as a Percent

To write a fraction as a percent, use one of these methods.

Method 1: Proportions
Write the given fraction as a proportion equal to $\frac{x}{100}$. Solve the resulting proportion equation for x and attach the percent symbol.

Method 2: Decimal Conversion
Use division to write the fraction as an equivalent decimal. Then multiply by 100 to move the decimal two places to the right and attach the percent symbol.

Example 4 Writing a Fraction as a Percent

< Objective 2 >

Write $\frac{3}{5}$ as a percent.

RECALL

We used this method in Example 1 in Section 6.1.

Method 1: Proportions

Write a proportion in which $\frac{3}{5}$ is equal to a fraction with 100 in the denominator.

$\frac{3}{5} = \frac{x}{100}$ Write a proportion.

$5 \cdot x = 3 \cdot 100$ Use the proportion rule.

$5x = 300$

$\frac{5x}{5} = \frac{300}{5}$ Divide by 5 because it is the coefficient of the variable.

$x = 60$ Simplify.

So, $\frac{3}{5} = 60\%$.

Method 2: Decimal Conversion

First write the decimal equivalent.

$\frac{3}{5} = 0.6$ To find the decimal equivalent, just divide the denominator into the numerator.

Now write the percent.

$\frac{3}{5} = 0.60 = 60\%$ Affix zeros to the right of the decimal if necessary.

Check Yourself 4

Write $\frac{3}{4}$ as a percent.

Again, you will find both decimals and fractions used in writing percents. Consider Example 5.

Example 5 Writing a Fraction as a Percent

Write $\frac{1}{8}$ as a percent.

Method 1: Proportions

Write a proportion in which $\frac{1}{8}$ is equal to a fraction with 100 in the denominator.

$\frac{1}{8} = \frac{x}{100}$ Write a proportion.

$8 \cdot x = 1 \cdot 100$ Use the proportion rule.

$8x = 100$

$\frac{8x}{8} = \frac{100}{8}$ Divide by 8 because it is the coefficient of the variable.

$x = \frac{25}{2}$ Simplify.

$= 12\frac{1}{2} = 12.5$

So, $\frac{1}{8} = 12.5\%$ or $12\frac{1}{2}\%$.

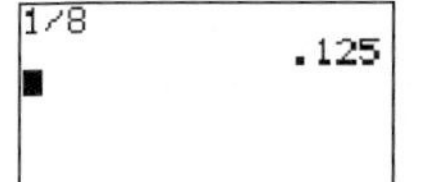

Method 2: Decimal Conversion

$\frac{1}{8} = 0.125 = 12.5\%$ or $12\frac{1}{2}\%$

Check Yourself 5

Write $\frac{3}{8}$ as a percent.

Units Analysis

We rarely express the units when computing percentages. When we say "percent" we are essentially saying, "numerator units per 100 denominator units."

EXAMPLES:

Of 800 students, 200 were boys. What percent of the students were boys?

$\frac{200 \text{ boys}}{800 \text{ students}} = 0.25 = 25\%$

But what happened to our units? At the decimal, the units boys/student (0.25 boy per student) wouldn't make much sense, but we can read the % as

25 "boys per 100 students"

and have a reasonable unit phrase.

Of 500 computers sold, 180 were equipped with scanners. What percent were equipped with scanners?

$\frac{180}{500} = 0.36 = 36\%$

36 computers were equipped with scanners for each 100 computers sold.

To write a mixed number as a percent, we use exactly the same steps. In Example 6, we use the decimal-conversion method because it is easier in this case.

Example 6 Writing a Mixed Number as a Percent

NOTE

The percent must be greater than 100% because the number is greater than 1.

Write $1\frac{1}{4}$ as a percent.

$1\frac{1}{4} = 1.25 = 125\%$

Check Yourself 6

Write $1\frac{2}{5}$ as a percent.

When a fraction is equivalent to a repeating decimal, we have two choices. We can write the percent as an exact value with a fraction part or we can write the fraction part of the percent as a decimal, usually rounded to some specified place.

If we want to write the percent as an exact value, with a fraction part, we prefer to use the proportion method. If we want to write the percent as a rounded decimal, we usually use the decimal-conversion method.

Example 7 illustrates both approaches.

Example 7 Writing a Fraction as a Percent

(a) Write $\frac{1}{3}$ as a percent (report an exact-value result).

We use the proportion method to report our result as an exact value.

$\frac{1}{3} = \frac{x}{100}$ Write a proportion.

$3 \cdot x = 1 \cdot 100$ Use the proportion rule.

$3x = 100$

$\frac{3x}{3} = \frac{100}{3}$ Divide by 3 because it is the coefficient of the variable.

$x = 33\frac{1}{3}$ Simplify.

So, $\frac{1}{3} = 33\frac{1}{3}\%$.

(b) Write $\frac{5}{7}$ as a percent (round to the nearest tenth of a percent).

We use the decimal-conversion method to report a rounded result. The nearest tenth of a percent is equivalent to three decimal places, so we need to compute $\frac{5}{7}$ to four decimal places to round in the appropriate place.

5/7
.7142857143

$\frac{5}{7} \approx 0.7142$ Divide; carry the division to four decimal places.

$= 71.42\%$ Multiply by 100 to move the decimal two places to the right and attach the % symbol.

Rounded to the nearest tenth percent, we have

$\frac{5}{7} = 71.4\%$

Check Yourself 7

(a) Write $\frac{2}{3}$ as a percent (report an exact-value result).

(b) Write $\frac{5}{9}$ as a percent (round to the nearest tenth of a percent).

Percents are used in many applications.

Example 8 An Application of Percents

The hard drive in Emma's computer has a 74.5-gigabyte (GB) capacity. Currently, she is using 10.8 GB. What percent of her hard drive's capacity is Emma using?

We begin by constructing a fraction based on the given information.

$\frac{\text{Used}}{\text{Capacity}} = \frac{10.8 \text{ GB}}{74.5 \text{ GB}}$

$= \frac{10.8 \not{\text{GB}}}{74.5 \not{\text{GB}}}$ As with proportions, we can simply "cancel" the units.

≈ 0.145 Divide and round the result to the nearest thousandth.

$= 14.5\%$

Emma is using 14.5% of her hard drive's capacity.

Check Yourself 8

Three cylinders are not firing properly in an 8-cylinder engine. What percent of the cylinders are firing properly?

Certain percents appear frequently enough that you should memorize their fraction and decimal equivalents.

Conversion Facts

$100\% = 1$	$10\% = 0.1 = \frac{1}{10}$	$12\frac{1}{2}\% = 0.125 = \frac{1}{8}$
$200\% = 2$	$20\% = 0.2 = \frac{1}{5}$	$37\frac{1}{2}\% = 0.375 = \frac{3}{8}$
	$30\% = 0.3 = \frac{3}{10}$	$62\frac{1}{2}\% = 0.625 = \frac{5}{8}$
$25\% = 0.25 = \frac{1}{4}$	$40\% = 0.4 = \frac{2}{5}$	$87\frac{1}{2}\% = 0.875 = \frac{7}{8}$
$50\% = 0.5 = \frac{1}{2}$	$60\% = 0.6 = \frac{3}{5}$	
$75\% = 0.75 = \frac{3}{4}$	$70\% = 0.7 = \frac{7}{10}$	$33\frac{1}{3}\% = 0.\overline{3} = \frac{1}{3}$
	$80\% = 0.8 = \frac{4}{5}$	$66\frac{2}{3}\% = 0.\overline{6} = \frac{2}{3}$
$5\% = 0.05 = \frac{1}{20}$	$90\% = 0.9 = \frac{9}{10}$	

Check Yourself ANSWERS

1. **(a)** 27%; **(b)** 5% **2.** 130% **3.** 7.5% or $7\frac{1}{2}\%$ **4.** 75% **5.** 37.5% or $37\frac{1}{2}\%$
6. 140% **7.** **(a)** $66\frac{2}{3}\%$; **(b)** 55.6% **8.** 62.5%

Reading Your Text

These fill-in-the-blank exercises will help you understand some of the key vocabulary used in this section. The answers to these exercises are in the Answers Appendix at the back of the text.

(a) To write a decimal in percent form, move the decimal point two places to the ____________ and add the % symbol.

(b) To write a fraction as a percent, you can first convert the fraction to a ________.

(c) When writing a repeating decimal as a ____________, we round to some indicated place or we write the percent using an exact fraction.

(d) When writing a decimal as a percent, we may need to add ____________ to the right as placeholders.

6.2 exercises

Skills Calculator/Computer Career Applications Above and Beyond

< Objectives 1 and 2 >

Write each number as a percent.

1. 0.08 
2. 0.09
3. 0.05
4. 0.13
5. 0.18
6. 0.63
7. 0.86
8. 0.45
9. 0.4
10. 0.3
11. 0.7
12. 0.6
13. 1.10 
14. 2.50
15. 4.40
16. 5
17. 0.065 
18. 0.095
19. 0.025
20. 0.085
21. $\frac{1}{4}$
22. $\frac{4}{5}$
23. $\frac{2}{5}$
24. $\frac{1}{2}$
25. $\frac{1}{5}$
26. $\frac{3}{4}$
27. $\frac{5}{8}$
28. $\frac{7}{8}$
29. $\frac{5}{16}$
30. $1\frac{1}{5}$
31. $3\frac{1}{2}$
32. $\frac{2}{3}$
33. 0.002
34. 0.008
35. 0.004
36. 0.001
37. $\frac{1}{6}$ (exact value)
38. $\frac{3}{16}$
39. $\frac{7}{9}$ (to nearest tenth of a percent) 

40. $\frac{5}{11}$ (to nearest tenth of a percent)

41. $\frac{7}{9}$ (exact value)

42. $\frac{5}{11}$ (exact value)

43. $5\frac{1}{4}$

44. $1\frac{3}{4}$

45. $4\frac{1}{3}$ (to the nearest tenth of a percent)

46. $2\frac{11}{12}$ (to the nearest tenth of a percent)

47. $4\frac{1}{3}$ (exact value)

48. $2\frac{11}{12}$ (exact value)

49. $\frac{11}{6}$ (to the nearest tenth of a percent)

50. $\frac{10}{9}$ (to the nearest tenth of a percent)

51. $\frac{11}{6}$ (exact value)

52. $\frac{10}{9}$ (exact value)

53. $\frac{10}{7}$ (to the nearest tenth of a percent)

54. $\frac{19}{12}$ (to the nearest tenth of a percent)

55. $\frac{10}{7}$ (exact value)

56. $\frac{19}{12}$ (exact value)

Express each shaded region as a decimal, fraction, and percent of each whole.

57. 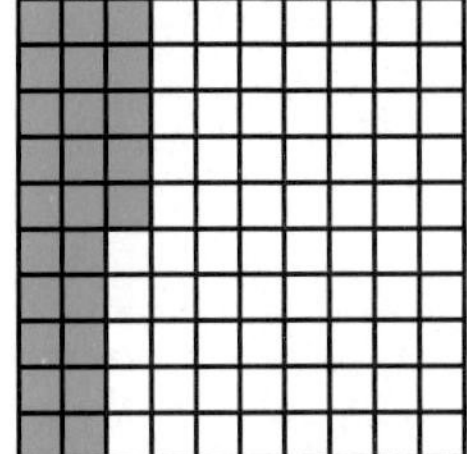

58. 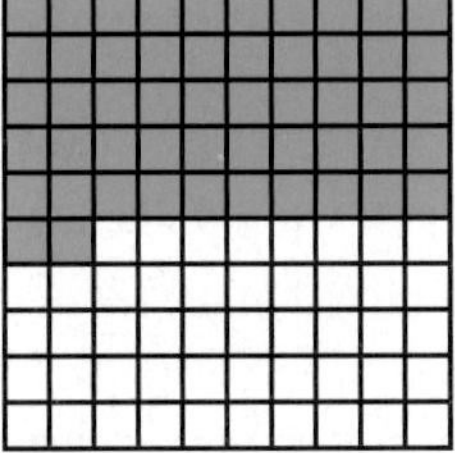

59. 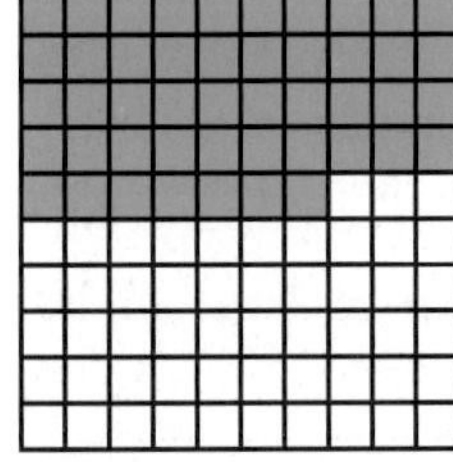

60. 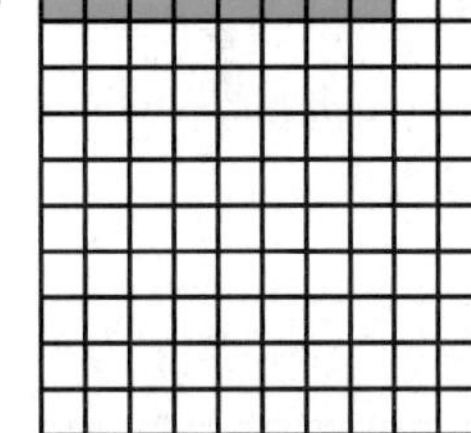

61. 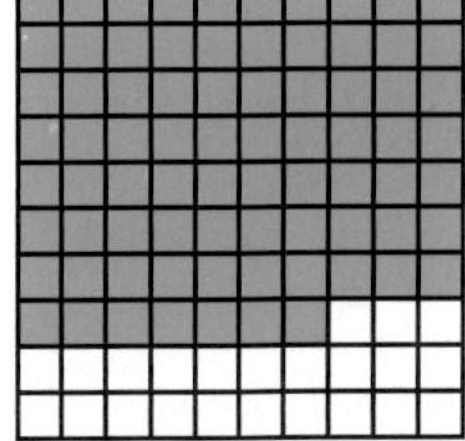

62. 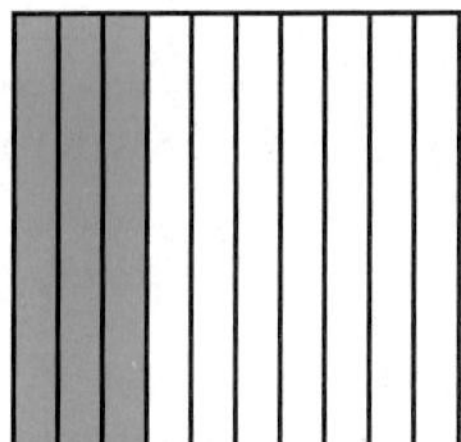

63. 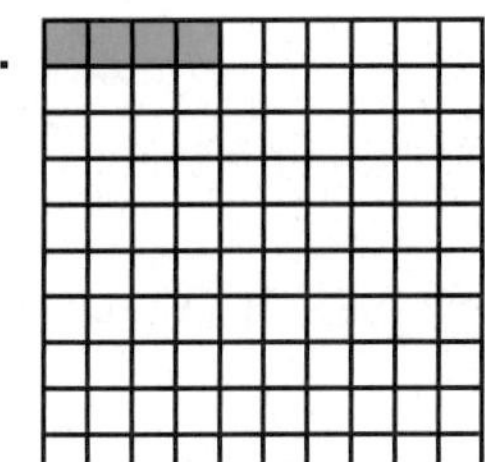

64. 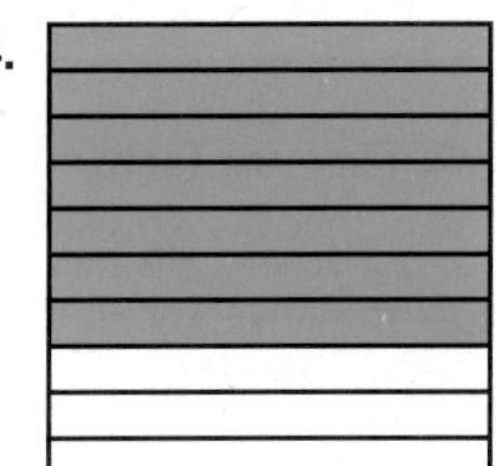

65. **Science and Technology** Between 1990 and 2010, the average fuel efficiency of new domestic U.S. cars increased from 26.9 to 32.9 miles per gallon (mi/gal). We compute the increase in fuel efficiency with the expression

$$\frac{32.9 - 26.9}{26.9}$$

Compute the increase in fuel efficiency and report your result as a percent, rounded to the nearest tenth of a percent.

66. **Science and Technology** Between 1990 and 2010, the average fuel efficiency of new imported U.S. cars increased from 29.9 to 35.1 mi/gal. We compute the increase in fuel efficiency with the expression

$$\frac{35.1 - 29.9}{29.9}$$

Compute the increase in fuel efficiency and report your result as a percent, rounded to the nearest tenth of a percent.

Business and Finance *Business travelers were asked how much they spent on different items during a business trip. The circle graph shows the results for every $1,000 spent. Use this information to complete exercises 67 to 70.*

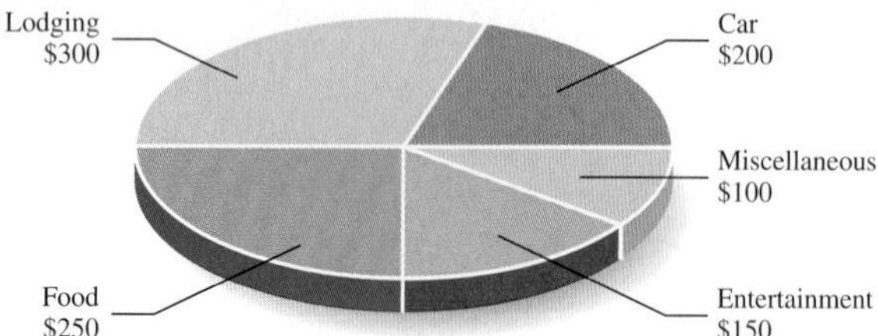

67. What percent was spent on car expenses?

68. What percent was spent on food?

69. Where was the least amount of money spent? What percent was this?

70. What percent was spent on food and lodging together?

Determine whether each statement is **true** *or* **false**.

71. To write a decimal as a percent, move the decimal point two places to the right and add the % symbol.

72. To write a decimal as a percent, move the decimal point two places to the left and add the % symbol.

Fill in each blank with **always, sometimes,** *or* **never.**

73. A decimal greater than 1 is __________ equivalent to a percent greater than 100%.

74. A percent less than 100% can __________ be written as a mixed number.

Skills	Calculator/Computer	**Career Applications**	Above and Beyond

75. **Automotive Technology** When an automobile engine is cold, its fuel efficiency is only $\frac{13}{16}$ of its maximum efficiency. Express this as a percent.

76. **Automotive Technology** Driving on underinflated tires can reduce the life of a set of tires by $\frac{3}{20}$. Write this reduction in tire life as a percent.

77. **Welding Technology** It was found that 165 of a total of 172 welds exceeded the required tensile strength. What percent of the welds exceeded the required tensile strength? Round your answer to the nearest whole percent.

78. **Welding Technology** Of the 220 welds on a jobsite, 209 exceeded the required tensile strength. What percent did not exceed the required tensile strength?

79. **Electrical Engineering** The efficiency of an electric motor is defined as the output power divided by the input power. It is usually written as a percent. For a particular motor, the output power is measured to be 400 watts (W) given an input power level of 435 W. What is the efficiency of this motor? Round your answer to the nearest whole percent.

80. **Agricultural Technology** There were 52 in. of rainfall in one growing season. If 6.5 in. fell in August, what percent of the season's rainfall fell in August?

81. **Manufacturing Technology** A manufacturer determines that 2 out of every 27 products will be returned due to defects. What percent of the products will be returned? Round to the nearest tenth of a percent.

82. **Manufacturing Technology** A manufacturer determines that 299 of every 300 fire extinguishers passes inspection. What percent are defective? Write your result as a fraction of a percent.

83. Complete the table of equivalents. Round decimals to the nearest ten-thousandth. Round percents to the nearest hundredth percent.

Fraction	Decimal	Percent
$\frac{7}{12}$		
	0.08	
		35%
$\frac{11}{18}$		
	0.265	
		$4\frac{3}{8}\%$

84. Complete the table of equivalents. Round decimals to the nearest ten-thousandth. Round percents to the nearest hundredth percent.

Fraction	Decimal	Percent
$\frac{3}{7}$		
	2.6875	
		$83\frac{1}{3}\%$
$\frac{31}{24}$		
	$0.\overline{27}$	
		$14\frac{1}{16}\%$

85. When writing a fraction as a percent, explain when you should convert the fraction to a decimal and then write it as a percent, rather than using the proportion method.

86. When writing a fraction as a percent, explain when you should use the proportion method rather than converting the fraction to a decimal and then writing it as a percent.

Answers

1. 8% **3.** 5% **5.** 18% **7.** 86% **9.** 40% **11.** 70% **13.** 110% **15.** 440% **17.** 6.5% **19.** 2.5% **21.** 25% **23.** 40% **25.** 20% **27.** 62.5% **29.** 31.25% **31.** 350% **33.** 0.2% **35.** 0.4% **37.** $16\frac{2}{3}\%$ **39.** 77.8% **41.** $77\frac{7}{9}\%$ **43.** 525% **45.** 433.3% **47.** $433\frac{1}{3}\%$ **49.** 183.3% **51.** $183\frac{1}{3}\%$ **53.** 142.9% **55.** $142\frac{6}{7}\%$ **57.** 0.25; $\frac{1}{4}$; 25% **59.** 0.47; $\frac{47}{100}$; 47% **61.** 0.77; $\frac{77}{100}$; 77% **63.** 0.04; $\frac{1}{25}$; 4% **65.** 22.3% **67.** 20% **69.** Miscellaneous; 10% **71.** True **73.** always **75.** 81.25% **77.** 96% **79.** 92% **81.** 7.4% **83.**

Fraction	Decimal	Percent
$\frac{7}{12}$	0.5833	58.33%
$\frac{2}{25}$	0.08	8%
$\frac{7}{20}$	0.35	35%
$\frac{11}{18}$	0.6111	61.11%
$\frac{53}{200}$	0.265	26.5%
$\frac{7}{160}$	0.0438	$4\frac{3}{8}\%$

85. Above and Beyond

Activity 16 ::

M&M's

According to the M&M's/Mars company, there are currently six colors used for M&M's. Each package contains approximately the same distribution of colors.

Brown:	30%
Yellow:	20%
Red:	20%
Orange:	10%
Green:	10%
Blue:	10%

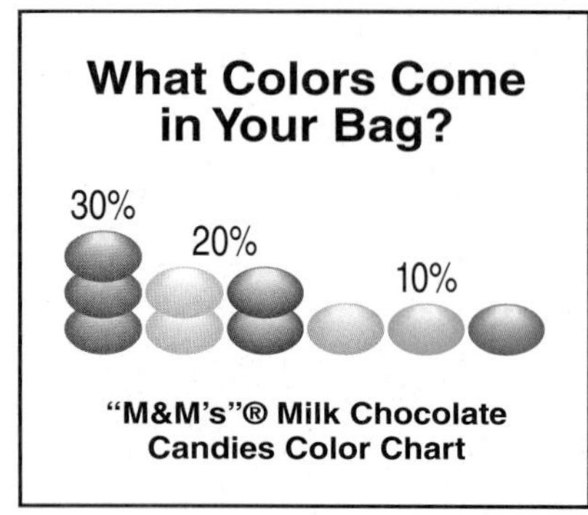

We opened an M&M's package and recorded the number of M&M's of each color.

Brown	Yellow	Red	Orange	Green	Blue
14	8	15	6	6	10

Calculate the percent for each color M&M in this pack. Round to the nearest percent.

Brown	Yellow	Red	Orange	Green	Blue

Do any of these seem to differ markedly from the percents given by the M&M's/MARS company? If so, give reasons why this may have occurred.

Obtain your own package of M&M's and determine the percents for each color. Comment on how closely your percents agree with the percents given by the company.

6.3

Solving Percent Problems

< 6.3 Objectives >

1 > Identify the parts of a percent problem

2 > Solve basic percent problems

There are many practical applications of our work with percents. All of these problems have three basic parts that need to be identified. Here are some definitions that will help with that process.

Definition

Base, Amount, and Rate

The **base,** denoted B, is the whole in a problem. It is the standard used for comparison.

The **amount,** denoted A, is the part of the whole being compared to the base.

The **rate,** denoted R, is the ratio of the amount to the base. It is written as a percent.

The first set of examples provides some practice in determining the parts of a percent problem.

Example 1 **Identifying Rates**

< Objective 1 >

NOTE

The *rate R* is the easiest of the terms to identify. The rate is written with the percent symbol (%) or the word *percent.*

Identify each rate.

(a) What is 15% of 200?
↑
R

Here 15% is the rate because it has the percent symbol attached.

(b) 25% of what number is 50?
↑
R

25% is the rate.

(c) 20 is what percent of 40?
↑
R

Here the rate is unknown.

Check Yourself 1

Identify the rate.

(a) 15% of what number is 75? **(b)** What is 8.5% of 200?
(c) 200 is what percent of 500?

Instructor's note: In this section, we use proportions to solve percent problems. After introducing students to algebra, we take an equations approach in Section 11.5.

The *base B* is the whole, or 100%, in the problem. The base often follows the phrase *percent of,* as shown in Example 2.

Example 2 Identifying Bases

Identify each base.

(a) What is 15% of 200?
↑ B (under 200)

200 is the base.

(b) 25% of what number is 50?
↑ B (under "what number")

Here the base is the unknown.

(c) 20 is what percent of 40?
↑ B (under 40)

40 is the base.

Check Yourself 2

Identify the base.

(a) 70 is what percent of 350? **(b)** What is 25% of 300?
(c) 14% of what number is 280?

The *amount A* is the part of the problem remaining once the rate and the base have been identified.

In many applications, the amount is found with the word *is.*

Example 3 Identifying Amounts

Identify the amount.

(a) What is 15% of 200?
↑ A (under "What")

Here the amount is the unknown part of the problem.

(b) 25% of what number is 50?
↑ A (under 50)

Here the amount, 50, follows the word *is*.

(c) 20 is what percent of 40?
↑ A (under 20)

Again the amount, here 20, can be found with the word *is*.

Check Yourself 3

Identify the amount.

(a) 30 is what percent of 600? **(b)** What is 12% of 5,000?
(c) 24% of what number is 96?

In Example 4, we identify all three parts in a percent problem.

Example 4 Identifying the Rate, Base, and Amount

Determine the rate, base, and amount in the problem.

12% of 800 is what number?

Finding the *rate* is not difficult. Just look for the percent symbol or the word *percent*. In this example, 12% is the rate.

The *base* is the whole. Here it follows the word *of*. 800 is the whole or the base.

The *amount* remains when the rate and the base have been found. Here the amount is the unknown. It follows the word *is*. "What number" asks for the unknown amount.

Check Yourself 4

Find the rate, base, and amount in each statement or question.

(a) 75 is 25% of 300. **(b)** 20% of what number is 50?

We use percents to solve a variety of applied problems. In all these situations, you should identify the three parts of the problem. Example 5 is intended to help you build that skill.

Example 5 Identifying the Rate, Base, and Amount

Identify the rate, base, and amount in each application.

(a) In an algebra class of 35 students, 7 received a grade of A. What percent of the class received an A?

The *base* is the whole in the problem, or the number of students in the class. 35 is the base.

The *amount* is the portion of the base, here the number of students that received an A grade. 7 is the amount.

The *rate* is the unknown in this example. "What percent" asks for the unknown rate.

(b) Doyle borrows $2,000 for 1 year. If the interest rate is 12%, how much interest will he pay?

The *base* is again the whole, the size of the loan in this example. $2,000 is the base.

The *rate* is, of course, the interest rate. 12% is the rate.

The *amount* is the quantity left once the base and rate have been identified. Here the amount is the amount of interest that Doyle must pay. The amount is the unknown in this example.

Check Yourself 5

Identify the rate, base, and amount in each application.

(a) In a shipment of 150 parts, 9 were found to be defective. What percent were defective?

(b) Robert earned $120 interest from an investment paying an 8% return. How much did he invest?

So far, you have learned that statements and problems about percents consist of three parts: the rate, base, and amount. In fact, most percent problems consist of these three parts.

In nearly every percent problem, one of these three parts is missing. Solving a percent problem is a matter of identifying and computing the missing part. To do this, we use the **percent proportion.**

Property

The Percent Proportion

$$\frac{A}{B} = \frac{r}{100}$$

where A is the amount, B is the base, and $\frac{r}{100}$ is the rate.

Previously, we said that the rate, R, is written as a percent. Note that R and r represent different things. For example, if $R = 54\%$ then $r = 54$. This means that r is the number before the percent symbol. These symbols (A, B, r, and R) should become clear through Examples 6 to 10.

Example 6 Finding an Unknown Amount

< Objective 2 >

What is 18% of 300?

We know that the rate R is 18%. This means that $r = 18$. We also know the base B equals 300. The amount A is unknown. Use the percent relationship to write a proportion:

$$\frac{A}{300} = \frac{18}{100}$$

Applying the proportion rule from Section 5.3 gives

$$100 \cdot A = 18 \cdot 300$$

or $$100A = 5{,}400$$

We identify the coefficient of the variable as 100 and divide by this number.

$$\frac{100A}{100} = \frac{5{,}400}{100}$$

$$A = 54$$

So, 54 is 18% of 300.

Check Yourself 6

Find 65% of 200.

The second type of percent problem requires us to find an unknown rate. We solve such a problem in the same way as before—we write the percent proportion and solve it.

Example 7 Finding an Unknown Rate

30 is what percent of 150?

We know that 30 is the amount A and 150 is the base B. The rate R is the unknown.

$$\frac{30}{150} = \frac{r}{100} \qquad \frac{A}{B} = \frac{r}{100}$$

$$150 \cdot r = 30 \cdot 100$$ Rewrite the equation using the proportion rule.

$$150r = 3{,}000$$ The coefficient of the variable is 150.

$$\frac{150r}{150} = \frac{3{,}000}{150}$$ Divide both sides by the coefficient of the variable.

$$r = 20$$ $3{,}000 \div 150 = 20$

Since $r = 20$, the rate R is 20%.

Therefore, 30 is 20% of 150.

Check Yourself 7

75 is what percent of 300?

The final type of percent problem is one with a missing base.

Example 8 **Finding an Unknown Base**

RECALL

$\frac{A}{B} = \frac{r}{100}$

28 is 40% of what number?

The amount A is 28 and the rate R is 40%. This means that $r = 40$. We use these to set up the percent proportion with a missing base.

$$\frac{28}{B} = \frac{40}{100}$$

As before, we use the proportion rule to solve this.

$$40B = 2{,}800 \qquad 28 \cdot 100 = 2{,}800$$

$$\frac{40B}{40} = \frac{2{,}800}{40} \qquad \text{40 is the coefficient of the variable.}$$

$$B = 70 \qquad 2{,}800 \div 40 = 70$$

Therefore, 28 is 40% of 70.

Check Yourself 8

70 is 35% of what number?

Remember that a percent (the rate) can be greater than 100.

Example 9 **Finding an Unknown Amount**

NOTE

The rate is 125%. The base is 300.

What is 125% of 300?

In the percent proportion, we have

$$\frac{A}{300} = \frac{125}{100} \qquad \text{Since } R = 125\%, r = 125.$$

Again, we begin by applying the proportion rule.

$$100 \cdot A = 300 \cdot 125$$

$$100A = 37{,}500 \qquad 300 \cdot 125 = 37{,}500$$

$$A = \frac{37{,}500}{100} = 375$$

NOTE

When the rate is greater than 100%, the amount is *greater than* the base.

So $A = 375$. Therefore, 375 is 125% of 300.

Check Yourself 9

Find 150% of 500.

We next look at an example of a problem with a fraction of a percent.

Example 10 **Finding an Unknown Base**

NOTE

The amount is 34 and the rate is 8.5%. We want to find the base.

34 is 8.5% of what number?

Using the percent proportion yields

$$\frac{34}{B} = \frac{8.5}{100} \qquad \text{Since } R = 8.5\%, r = 8.5.$$

Solving the proportion yields

$8.5 \cdot B = 34 \cdot 100$ Use the proportion rule.

$8.5B = 3{,}400$

$\frac{8.5B}{8.5} = \frac{3{,}400}{8.5}$ The coefficient of the variable is 8.5, so we divide.

$B = 400$ $3{,}400 \div 8.5 = 400$

Therefore, 34 is 8.5% of 400.

Check Yourself 10

12.5% of what number is 75?

Check Yourself ANSWERS

1. **(a)** 15%; **(b)** 8.5%; **(c)** "what percent" (the unknown) **2.** **(a)** 350; **(b)** 300; **(c)** "what number" (the unknown) **3.** **(a)** 30; **(b)** "What" (the unknown); **(c)** 96 **4.** **(a)** $R = 25\%$, $B = 300$, $A = 75$; **(b)** $R = 20\%$, $B =$ "What number," $A = 50$ **5.** **(a)** $R =$ "What percent" (the unknown), $B = 150$, $A = 9$; **(b)** $R = 8\%$, $B =$ "How much," $A = \$120$ **6.** 130 **7.** 25% **8.** 200 **9.** 750 **10.** 600

Reading Your Text

These fill-in-the-blank exercises will help you understand some of the key vocabulary used in this section. The answers to these exercises are in the Answers Appendix at the back of the text.

(a) In percent problems, the ______________ is the standard used for comparison.

(b) In percent problems, the ______________ is the ratio of the amount to the base.

(c) In percent problems, the ______________ is often written as a percent.

(d) In percent problems, the ______________ is the part of the whole being compared to the base.

6.3 exercises

Skills | Calculator/Computer | Career Applications | Above and Beyond

< Objective 1 >

Identify the rate, base, and amount in each statement or question. Do not solve *the exercise at this point.*

1. 23% of 400 is 92. 

2. 150 is 20% of 750.

3. 40% of 600 is 240.

4. 200 is 40% of 500.

5. What is 7% of 325?

6. 80 is what percent of 400?

7. 16% of what number is 56? 

8. What percent of 150 is 30?

9. 480 is 60% of what number?

10. What is 60% of 250?

11. What percent of 120 is 40? 

12. 150 is 75% of what number?

Identify the rate, base, and amount in each application. Do not solve *the application at this point.*

13. **Business and Finance** Jan has a 5% commission rate on all her sales. If she sells $40,000 worth of merchandise in 1 month, what commission will she earn?

14. **Business and Finance** 22% of Shirley's monthly salary is deducted for withholding. If those deductions total $209, what is her salary?

15. **Science and Medicine** In a chemistry class of 30 students, 5 received a grade of A. What percent of the students received A's?

16. **Business and Finance** A can of mixed nuts contains 80% peanuts. If the can holds 16 oz, how many ounces of peanuts does it contain?

17. **Business and Finance** The sales tax rate in a state is 5.5%. If you pay a tax of $3.30 on an item that you purchase, what is its selling price?

18. **Business and Finance** In a shipment of 750 parts, 75 were found to be defective. What percent of the parts were faulty?

19. **Social Science** A college had 9,000 students at the start of a school year. If there is an enrollment increase of 6% by the beginning of the next year, how many additional students are there?

20. **Business and Finance** Paul invested $5,000 in a time deposit. What interest will he earn for 1 year if the interest rate is 3.5%? chapter 6 > Make the Connection

< Objective 2 >

Solve each percent problem.

21. What is 35% of 600? 

22. 20% of 400 is what number?

23. 45% of 200 is what number?

24. What is 40% of 1,200?

25. Find 40% of 2,500.

26. What is 75% of 120?

27. What percent of 50 is 4? 

28. 51 is what percent of 850?

29. What percent of 500 is 45?

30. 14 is what percent of 200?

31. What percent of 200 is 340? 

32. 392 is what percent of 2,800?

33. 46 is 8% of what number? 

34. 7% of what number is 42?

35. Find the base if 11% of the base is 55.

36. 16% of what number is 192?

37. 58.5 is 13% of what number? 

38. 21% of what number is 73.5?

39. Find 110% of 800.

40. What is 115% of 600?

41. What is 108% of 4,000?

42. Find 160% of 2,000.

43. 210 is what percent of 120?

44. What percent of 40 is 52?

45. 360 is what percent of 90?

46. What percent of 15,000 is 18,000?

47. 625 is 125% of what number?

48. 140% of what number is 350?

49. Find the base if 110% of the base is 935.

50. 130% of what number is 1,170?

51. Find 8.5% of 300.

52. $8\frac{1}{4}\%$ of 800 is what number?

53. Find $11\frac{3}{4}\%$ of 6,000. 

54. What is 3.5% of 500?

55. What is 5.25% of 3,000?

56. What is 7.25% of 7,600?

57. 60 is what percent of 800?

58. 500 is what percent of 1,500?

59. What percent of 180 is 120?

60. What percent of 800 is 78?

61. What percent of 1,200 is 750?

62. 68 is what percent of 800?

63. Find 87% of 112.

64. What is 136% of 82?

65. 17 is 30% of what number?

66. 43 is 12% of what number?

67. Find $66\frac{2}{3}\%$ of 180.

68. What is $58\frac{1}{3}\%$ of 276?

69. 75 is $33\frac{1}{3}\%$ of what number?

70. $16\frac{2}{3}\%$ of what number is 72?

71. Find $55\frac{5}{9}\%$ of 120.

72. 8 is $83\frac{1}{3}\%$ of what number?

73. 10.5% of what number is 420? 

74. Find the base if $11\frac{1}{2}\%$ of the base is 46.

75. 58.5 is 13% of what number?

76. 6.5% of what number is 325?

77. 195 is 7.5% of what number?

78. 21% of what number is 73.5?

Determine whether each statement is **true** *or* **false.**

79. The rate, in a percent problem, is never greater than 100%.

80. The base, in a percent problem, often follows the phrase *percent of.*

Identify the rate, base, and amount in each statement or question. Do not solve *the problems at this point.*

81. **Allied Health** How much 25% alcohol solution can be prepared using 225 milliliters (mL) of ethyl alcohol?

82. **Information Technology** When a particular network transmits data packets, 1.7% of the overall size is "overhead." If a network transmission measures 1,500 kilobytes (KB), how much of the transmission is overhead?

83. **Agricultural Technology** Milk that is labeled "3.5%" is made up of 3.5% butterfat. How many grams of butterfat are in 1 liter (938 g) of 3.5% milk?

84. **Environmental Technology** In some communities, "green" laws require that 40% of a lot remains green (covered in grass or other vegetation). How much green space is required in a 12,680-ft^2 lot?

Answers

1. 23% of 400 is 92 **3.** 40% of 600 is 240 **5.** What is 7% of 325 **7.** 16% of what number is 56 **9.** 480 is 60% of what number

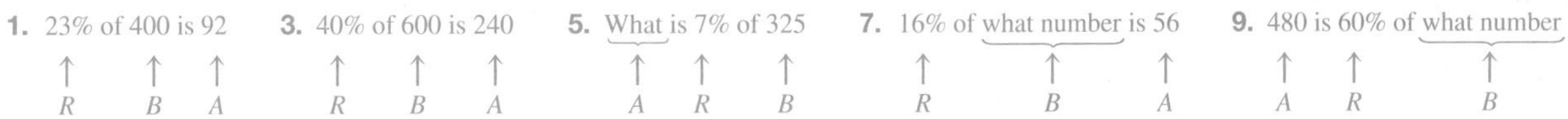

11. What percent of 120 is 40 (What percent: R; 120: B; 40: A) **13.** \$40,000 is the base. 5% is the rate. Her commission, the unknown, is the amount.

15. 30 is the base. 5 is the amount. The unknown percent is the rate.

17. 5.5% is the rate. The tax, \$3.30, is the amount. The unknown selling price is the base.

19. The base is 9,000. The rate is 6%. The unknown number of additional students is the amount.

21. 210 **23.** 90 **25.** 1,000 **27.** 8% **29.** 9% **31.** 170% **33.** 575 **35.** 500 **37.** 450 **39.** 880 **41.** 4,320 **43.** 175% **45.** 400% **47.** 500 **49.** 850 **51.** 25.5 **53.** 705 **55.** 157.5 **57.** 7.5% **59.** $66\frac{2}{3}\%$ **61.** 62.5% **63.** 97.44 **65.** $\frac{170}{3}$ or $56\frac{2}{3}$ **67.** 120 **69.** 225 **71.** $\frac{200}{3}$ or $66\frac{2}{3}$ **73.** 4,000 **75.** 450 **77.** 2,600 **79.** False

81. $R = 25\%$, $B =$ unknown, $A = 225$ mL **83.** $R = 3.5\%$, $B = 938$ g, $A =$ unknown

Activity 17 ::

A Matter of Interest

Many people put their money in some investment vehicle. In general, you invest some money, called the **principal,** and leave it for a period of time to allow it to grow based on the return. Periodically, your investment earns a return based on a percentage of the initial investment.

Many investment vehicles calculate the return annually and add it to the principal, according to the **growth** formula.

Interest = Principal × Rate × Time

or

$I = P \cdot R \cdot T$

To compute the earnings and account balance after 1 year, we use $T = 1$, which simplifies the formula to

$I = P \cdot R$

Suppose that your original investment is \$1,000 and that your money grows 4% (per year). Let us further suppose that the return is computed at the end of each year and then added to the principal. The principal increases, and the return in the next year is greater. To see how this works, complete the table.

Year	Principal	Interest	Investment at End of Year
1	\$1,000	\$40	\$1,040
2	1,040		
3			
4			
5			
6			
7			

After 7 years, how much has your investment grown?

For how many years must the money sit to earn a total return equal to 50% of the original investment?

Hint: Continue the calculations in the table.

6.4 Percent Applications

< 6.4 Objectives >

1 > Solve percent applications

2 > Solve applications of percent increase and decrease

3 > Solve percent applications involving interest

Percents may be the most common real-world application in this text. In this section, we show some of the many applications of percents and the special terms that are used with them.

In Example 1, we solve a percent application in which the amount is the unknown quantity. You should keep in mind some percent basics as you work through this section.

If the rate is *less than* 100%, then the amount is *less than* the base.

20 is 40% of 50 and $20 < 50$

If the rate is *greater than* 100%, then the amount is *greater than* the base.

75 is 150% of 50 and $75 > 50$

Example 1 Solving a Percent Application

< Objective 1 >

A student needs to answer at least 70% of the questions correctly on a 50-question exam in order to pass. How many questions must the student get right?

In this case, the rate is 70%, so $r = 70$, and the base, or total, is 50 questions, so $B = 50$. The amount A is unknown.

We write the percent proportion and solve.

$$\frac{A}{50} = \frac{70}{100}$$

$100A = 70 \cdot 50$ If $\frac{a}{b} = \frac{c}{d}$, then $ad = bc$.

$100A = 3{,}500$ The coefficient of the variable is 100.

$\frac{100A}{100} = \frac{3{,}500}{100}$ Divide both sides by the coefficient of the variable.

$A = 35$ $3{,}500 \div 100 = 35$

The student must answer at least 35 questions correctly in order to pass the exam.

Check Yourself 1

Generally, 72% of the students in a chemistry course pass the class. If there are 150 students in the class, how many are expected to pass?

Now consider an application involving an unknown rate.
If the amount is *less than* the base, then the rate is *less than* 100%.
If the amount is *greater than* the base, then the rate is *greater than* 100%.

Instructor's note: In this section, we use proportions to solve percent problems. After introducing students to algebra, we use an equations approach in Section 11.5.

Example 2 Solving a Percent Application

Simon waits tables at La Catalana, an upscale restaurant. A group left a \$45 tip on a \$250 meal. What percent of the bill did the group leave as a tip?

We are asked to find the rate given a base $B = \$250$ and an amount $A = \$45$. To find the rate, we use the percent proportion.

$$\frac{45}{250} = \frac{r}{100}$$

$$250r = 4{,}500 \qquad 45 \cdot 100 = 4{,}500$$

$$\frac{250r}{250} = \frac{4{,}500}{250}$$

$$r = 18 \qquad 4{,}500 \div 250 = 18$$

NOTE

Remember to write your answer as a percent.

Therefore, the group left an 18% tip.

Check Yourself 2

Last year, Xian reported an income of \$47,500 on her tax return. Of that, she paid \$13,300 in taxes. What percent of her income did she pay in taxes?

Now look at an application with an unknown base.

Example 3 Solving a Percent Application

A computer ran 60% of a scan in 120 seconds (s). How long should it take to complete an entire scan?

In this case, the rate is 60% and the amount is 120 s. We employ the percent proportion again, and solve.

$$\frac{120}{B} = \frac{60}{100}$$

$$60B = 12{,}000$$

$$B = \frac{12{,}000}{60} \qquad \text{Divide by the coefficient of the variable, 60.}$$

$$B = 200$$

The entire scan should take 200 s.

Check Yourself 3

An indexing program takes 4 min to check 30% of the files on a laptop computer. How long should it take to index all the files (to the nearest second)?

Percents are used in too many ways for us to list. Notice the variety in the following examples, which illustrate some additional situations in which you will find percents.

Example 4 Solving a Percent Application: Commission

A sales associate sells a used car for \$9,500. His commission rate is 4%. What is his commission for the sale?

NOTE

A **commission** is the amount that a person is paid for a sale.

The base is the total of the sale, in this problem, \$9,500. The rate is 4%, and we want to find the commission. This is the amount. By the percent proportion,

$$\frac{A}{9{,}500} = \frac{4}{100}$$

$$100A = 4 \cdot 9{,}500$$

$$100A = 38{,}000$$

$$A = \frac{38{,}000}{100}$$

$$A = 380$$

The sales associate's commission is \$380.

Check Yourself 4

Jenny sells a \$76,000 building lot. If her real estate commission rate is 3%, what commission does she earn on the sale?

Example 5 Solving a Percent Application: Commission

A clerk sold \$3,500 in merchandise during 1 week. If he received a commission of \$140, what was the commission rate?

The base is \$3,500, and the amount is the commission of \$140. Using the percent proportion, we have

$$\frac{140}{3{,}500} = \frac{r}{100}$$

$$3{,}500r = 140 \cdot 100$$

$$3{,}500r = 14{,}000$$

$$r = \frac{14{,}000}{3{,}500}$$

$$r = 4$$

So $R = 4\%$. The commission rate is 4%.

Check Yourself 5

On a purchase of \$500 you pay a sales tax of \$21. What is the tax rate?

Example 6 shows how to find the total amount sold.

Example 6 Solving a Percent Application: Commission

A sales associate earns a commission rate of 3.5%. To earn \$280, how much must she sell?

The rate is 3.5%. The amount is the commission, \$280. We want to find the base. In this case, this is the amount that the sales associate needs to sell.

By the percent proportion,

$$\frac{280}{B} = \frac{3.5}{100} \qquad \text{or} \qquad 3.5B = 280 \cdot 100$$

$$B = \frac{28{,}000}{3.5}$$

$$B = 8{,}000$$

The sales associate must sell \$8,000 to earn \$280 in commissions.

Check Yourself 6

Kerri earns a commission rate of 5.5%. If she wants to earn \$825 in commissions, find the total sales that she must make.

Tax rates are another common application of percents.

Example 7 Solving a Percent Application: Tax

A state taxes sales at 7%. If a 160-GB iPod is listed for \$249, what is the total you have to pay?

The tax you pay is the amount A (the part of the whole). Here the base B is the purchase price, \$249. The rate R is 7%, so $r = 7$.

NOTE

In applications involving taxes, the tax paid is usually the amount.

$$\frac{A}{249} = \frac{7}{100} \quad \text{or} \quad 100A = 249 \cdot 7$$

$$A = \frac{1{,}743}{100} = 17.43$$

The sales tax is \$17.43. You will have to pay \$249 + \$17.43 or \$266.43.

Check Yourself 7

Suppose that a state has a sales tax rate of $6\frac{1}{2}\%$. If you buy a used car for \$1,200, what is the total you have to pay?

In applications with discounts or markups, the base B always represents the original price. This might be the price before the discount, or it might be the wholesale price prior to a markup. The amount A is usually the amount of the discount, or the amount of the markup.

Example 8 Solving a Percent Application: Discount

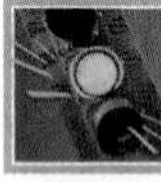

> CAUTION

The amount is not \$92. The amount is the size of the discount.

A kitchen store offers an All-Clad saucepan on sale for \$92. The saucepan normally sells for \$115. What is the discount rate?

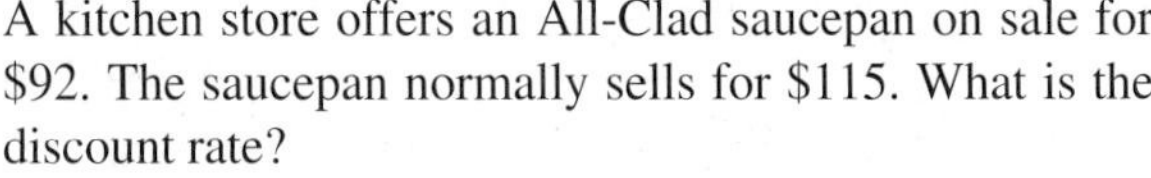

We begin by identifying the base as the original price, \$115, and the rate as the unknown. The amount A is the amount of the discount:

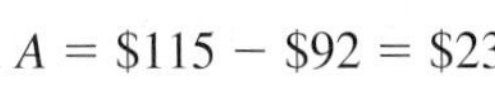

$A = \$115 - \$92 = \$23$

Now we are ready to apply the percent proportion to solve this problem.

NOTE

The *markup* is the amount a store adds to the price of an item to cover expenses and profit. Retailers usually mark an item up by a percentage of its wholesale cost.

$$\frac{23}{115} = \frac{r}{100}$$

$115r = 2{,}300$ $23 \cdot 100 = 2{,}300$

$r = \frac{2{,}300}{115}$ Divide both sides by the coefficient of the variable, 115.

$r = 20$ $2{,}300 \div 115 = 20$.

Therefore, the discount rate is 20%.

Check Yourself 8

An electronics store sells a certain Kicker amplifier for a car stereo system for $250. If the store pays $200 for the amplifier, what is the markup rate?

A common type of application involves *percent increase* or *percent decrease*. *Percent increase* and *percent decrease* always describe the change starting from the base.

Example 9 Solving a Percent Application: Percent Decrease

< Objective 2 >

NOTE

The base is the *original* population.

The population of a town decreased 15% in a 3-year period. If the original population was 12,000, what was the population at the end of the period?

Since the original population was 12,000, this is the base B. The rate of decrease R is 15%. The amount of the decrease A is unknown.

$$\frac{A}{12{,}000} = \frac{15}{100} \quad \text{so} \quad 100A = 15 \cdot 12{,}000$$

$$A = \frac{180{,}000}{100}$$

$$A = 1{,}800$$

Since the amount of the decrease is 1,800, the population at the end of the period must be

$$12{,}000 - 1{,}800 = 10{,}200$$

Original population (12,000), Decrease (1,800), New population (10,200)

Check Yourself 9

A school's enrollment increased by 8% from one year to the next. If the enrollment was 550 students the first year, how many students were enrolled the second year?

Example 10 Solving a Percent Application: Percent Increase

Enrollment at a school increased from 800 to 888 students from one year to the next. What was the rate of increase?

First we must subtract to find the amount of the increase.

Increase: $888 - 800 = 88$ students

NOTE

We use the *original* enrollment, 800, as our base. The size of the increase is the amount.

Now to find the rate, we have

$$\frac{88}{800} = \frac{r}{100} \quad \text{so} \quad 800r = 88 \cdot 100$$

$$r = \frac{8{,}800}{800}$$

$$r = 11$$

The enrollment increased by 11%.

Check Yourself 10

Car sales at a dealership increased from 350 units one year to 378 units the next. What was the rate of increase?

Example 11 Solving a Percent Application: Percent Increase

A company hired 18 new employees in 1 year. If this was a 15% increase, how many employees did the company have before the increase?

The rate is 15%. The amount is 18, the number of new employees. The base in this problem is the number of employees *before the increase.* So

$$\frac{18}{B} = \frac{15}{100}$$

$$15B = 18 \cdot 100$$

$$B = \frac{1{,}800}{15}$$

$$B = 120$$

The company had 120 employees before the increase.

> **NOTE**
>
> The size of the increase is the amount. The original number of employees is the base.

Check Yourself 11

A school had 54 fewer students in one term than in the previous term. If this was a 12% decrease from the previous term, how many students were there before the decrease?

> **NOTE**
>
> The money borrowed or invested is called the principal.

Another common application of percent is interest. When you take out a home loan, you pay interest; when you invest money in a savings account, you earn interest. Interest is a percent of the whole, and the percent is called the **interest rate.**

When we work with interest on a certain amount of money (the principal) for a specific time period, the interest is called **simple interest.**

In the applications here, we confine our study to interest earned or owed after one year. We again put the percent proportion to use. In this case, the *interest* plays the same role as the "amount," and the *principal* takes the part of the "base."

Example 12 Solving a Percent Application: Simple Interest

< Objective 3 >

Find the interest you must pay if you borrow \$2,000 for 1 year at an interest rate of $9\frac{1}{2}\%$.

The principal, \$2,000, is the base B. The interest rate R is 9.5%, so $r = 9.5$. We want the amount of interest A.

$$\frac{A}{2{,}000} = \frac{9.5}{100}$$

$$100 \cdot A = 2{,}000 \cdot 9.5$$

$$\frac{100 \cdot A}{100} = \frac{19{,}000}{100}$$

$$A = 190$$

The amount of interest is \$190.

Check Yourself 12

You invest \$5,000 for 1 year at an annual rate of $8\frac{1}{2}\%$. How much interest will you earn?

As with other percent problems, you can find the principal or the rate using the percent proportion.

Example 13 Solving a Percent Application: Simple Interest

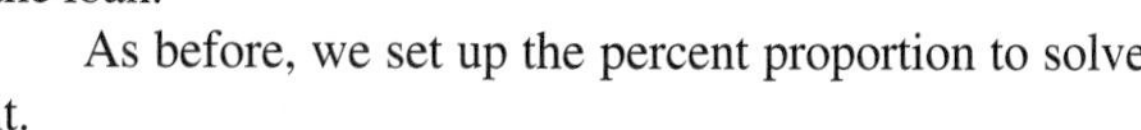

Ms. Hobson agrees to pay 11% interest on a loan for her new car. She is charged \$2,200 interest on the loan for 1 year. How much did she borrow?

The rate is 11% and the interest (or amount) is \$2,200. We need to find the principal (or base), which is the size of the loan.

As before, we set up the percent proportion to solve it.

$$\frac{2{,}200}{B} = \frac{11}{100}$$

$$11B = 220{,}000 \qquad 2{,}200 \cdot 100 = 220{,}000$$

$$\frac{11B}{11} = \frac{220{,}000}{11} \qquad \text{The coefficient of the variable is 11.}$$

$$B = 20{,}000 \qquad 220{,}000 \div 11 = 20{,}000$$

Ms. Hobson borrowed \$20,000 to purchase her car.

Check Yourself 13

Sue pays \$210 in interest for a 1-year loan at 10.5% interest. What is the size of her loan?

The true power of interest to earn money over time comes from the idea of **compound interest.** This means that once you earn (or are charged) interest, you start earning (or paying) interest on the interest. This is what is meant by compounding.

Compound interest is an exceptionally powerful idea. For many people, their retirement plans are not based on the amount of money invested, but rather on the fact that their retirement account grows with time.

The earnings on someone's retirement investments *compound* over decades and grow into much more than the original investments. This is why financial advisors always suggest that you should start saving for retirement as early as possible. The more time an investment has to grow, the larger it gets.

We conclude this section by looking at an example involving the compounding of interest.

Example 14 Solving a Percent Application: Compound Interest

Suppose you invest \$1,000 at 5% (compounded annually) in a savings account for 2 years. How much will you have in the account at the end of the 2-year period?

After the first year, you earn 5% interest on your \$1,000 principal.

$$\frac{A}{1{,}000} = \frac{5}{100}$$

$$100A = 5{,}000$$

$$A = \frac{5{,}000}{100}$$

$$A = 50$$

So, you earn \$50 after 1 year. The compounding comes into play in the second year when your account begins with \$1,000 + \$50 = \$1,050 in it. This is your new principal.

$$\underset{\text{Start}}{\$1,000} \xrightarrow{\text{At } 5\%} \underset{\text{Year 1}}{\$1,050}$$

In the second year, you earn 5% interest on this new principal. This process is called *compound interest.*

$$\frac{A}{1,050} = \frac{5}{100}$$

$$100A = 5,250 \qquad 5 \cdot 1,050 = 5,250$$

$$A = \frac{5,250}{100}$$

$$A = 52.5 \qquad 5,250 \div 100 = 52.5$$

You earn \$52.50 in the second year of your investment, therefore, your account balance will be \$1,050 + \$52.50 = \$1,102.50.

$$\underset{\text{Start}}{\$1,000} \xrightarrow{\text{At } 5\%} \underset{\text{Year 1}}{\$1,050} \xrightarrow{\text{At } 5\%} \underset{\text{Year 2}}{\$1,102.50}$$

Check Yourself 14

If you invest \$6,000 at 4% (compounded annually) for 2 years, how much will you have after 2 years?

Check Yourself ANSWERS

1. 108 students **2.** 28% **3.** 800 s **4.** \$2,280 **5.** 4.2% **6.** \$15,000 **7.** \$1,278 **8.** 25% **9.** 594 students **10.** 8% **11.** 450 students **12.** \$425 **13.** \$2,000 **14.** \$6,489.60

Reading Your Text

These fill-in-the-blank exercises will help you understand some of the key vocabulary used in this section. The answers to these exercises are in the Answers Appendix at the back of the text.

(a) If the rate is greater than 100%, then the amount will be ______________ than the base.

(b) In an application involving taxes, the tax paid is usually the ______________.

(c) The __________ is the amount a store adds to the price of an item to cover expenses and profit.

(d) In an interest application, the money borrowed or invested is called the ______________.

< Objectives 1–3 >

Solve each application.

1. **Business and Finance** What interest will you pay on a $3,400 loan for 1 year if the interest rate is 12%?

2. **Science and Medicine** A chemist has 300 mL of solution that is 18% acid. How many milliliters of acid are in the solution?

3. **Business and Finance** If a sales associate is paid a $140 commission on the sale of a $2,800 sailboat, what is his commission rate?

4. **Business and Finance** Ms. Jordan received a $2,500 loan for 1 year. If the interest charged is $275, what is the interest rate on the loan?

5. **Statistics** On a test, Alice had 80% of the problems right. If she had 20 problems correct, how many questions were on the test? VIDEO

6. **Business and Finance** A state sales tax rate is 3.5%. If the tax on a purchase was $7, what was the price of the purchase?

7. **Business and Finance** A state sales tax is levied at a rate of 6.4%. How much tax would one pay on a purchase of $260? VIDEO

8. **Business and Finance** Clarence must make a $9\frac{1}{2}\%$ down payment on the purchase of a $6,000 motorcycle. What is his down payment?

9. **Social Science** A study has shown that 102 of the 1,200 people in the workforce of a small town are unemployed. What is the town's unemployment rate? VIDEO

10. **Social Science** A survey of 400 people found that 66 were left-handed. What percent of those surveyed were left-handed?

11. **Statistics** In a recent survey, 65% of those responding were in favor of a freeway improvement project. If 780 people were in favor of the project, how many people responded to the survey?

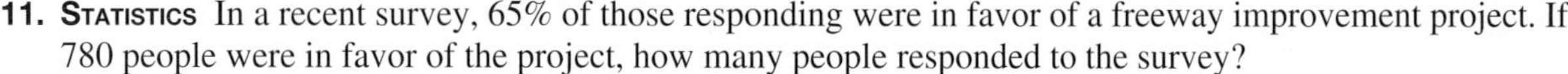

12. **Social Science** A college finds that 42% of the students taking a foreign language are enrolled in Spanish. If 1,512 students are taking Spanish, how many foreign language students are there?

13. **Business and Finance** An appliance dealer marks up refrigerators 22% (based on cost). If the wholesale cost of one model is $1,200, what should its selling price be?

14. **Social Science** A school had 900 students at the start of a school year. If there is an enrollment increase of 7% by the beginning of the next year, what is the new enrollment?

15. **Business and Finance** The price of a new van increases $2,030, which amounts to a 14% increase. What was the price of the van before the increase?

16. **Business and Finance** A television set is marked down $75 for a sale. If this is a 12.5% decrease from the original price, what was the selling price before the sale?

17. **Business and Finance** Carlotta received a monthly raise of $162.50. If this represented a 6.5% increase, what was her monthly salary before the raise?

18. **Business and Finance** Mr. Hernandez buys stock for $15,000. At the end of 6 months, the stock's value has decreased 7.5%. What is the stock worth at the end of the period? chapter 6 > Make the Connection

19. **Social Science** The population of a town increases 14% in 2 years. If the population was 6,000 originally, what is the population after the increase?

20. **Business and Finance** A store marks up merchandise 25% to allow for profit. If an item costs the store $11, what is its selling price?

21. **Information Technology** A virus scanning program is checking every file for viruses. It has completed checking 40% of the files in 300 s. How long should it take to check all the files?

22. **Social Science** In 2010, the United States consumed 138.6 billion gallons of gas. Consumption increased 2.8% for 2011. How much gas did the United States consume in 2011 (to the nearest tenth billion gallon)?

23. **Social Science** Of the 254.2 million registered vehicles in the United States in 2009, 194.0 million were passenger cars. What percentage of the vehicles registered were passenger cars (to the nearest tenth percent)?

24. **Manufacturing Technology** One 42-gal barrel of crude oil yields 44.68 gal of refined oil. What percent of volume gain does this represent (to the nearest tenth percent)?

25. **Science and Medicine** In 2008, the United States emitted 6,924.56 million metric tons (mmt of CO_2 equivalent) of greenhouse gasses. This represented a 13.3% increase over 1990 levels of emissions. What were the total greenhouse gas emissions in the United States in 1990 (round to the nearest hundredth mmt)?

26. **Science and Medicine** Global CO_2 emissions decreased by 0.5 gigatonnes (Gt) or 1.5% from 2008 to 2009. How much CO_2 was emitted globally in 2009 (round to the nearest tenth Gt)?

The chart shows U.S. trade with Mexico from 2006 to 2010. Use this information for exercises 27 to 30.

U.S. Trade with Mexico, 2006–2010

(millions of dollars) MEXICO			
Year	**Exports**	**Imports**	**Trade Balance**
2006	133,722	198,253	−64,531
2007	135,918	210,714	−74,796
2008	151,220	215,942	−64,722
2009	128,892	176,654	−47,762
2010	163,473	229,908	−66,435

Source: U.S. Census Bureau.

27. What was the rate of increase (to the nearest whole percent) of exports from 2006 to 2010?

28. What was the rate of increase (to the nearest whole percent) of imports from 2006 to 2010?

29. By what percent did imports exceed exports in 2006?

30. By what percent did imports exceed exports in 2010?

Assume interest is compounded annually (at the end of each year) and find the amount in an account with the given interest rate and principal.

31. \$4,000, 6%, 2 years **32.** \$3,000, 7%, 2 years **33.** \$4,000, 5%, 3 years **34.** \$5,000, 6%, 3 years

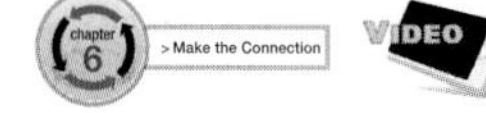

Use the number line to complete exercises 35 to 38.

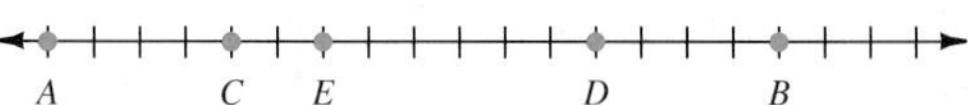

35. Length AC is what percent of length AB?

36. Length AD is what percent of length AB?

37. Length AE is what percent of length AB?

38. Length AE is what percent of length AD?

In many everyday applications of percent, the computations can become quite messy. A calculator can be a great help. Whether we use the proportion approach or an equations approach in solving such an application, we typically set up the problem and isolate the desired variable before doing any calculations.

In some percent increase or percent decrease applications, we can set up the problem so it can be done in one step. (Previously, we did these as two-step problems.) For example, suppose that a store marks up an item 22.5%. If the original cost to the store was \$36.40, we want to know what the selling price will be. Since the selling price is the cost to the store plus the markup, the selling price is 122.5% of the store's cost (100% + 22.5%). We can restate the problem as "What is 122.5% of \$36.40?" The base is \$36.40 and the rate is 122.5%, and we want the amount.

$$\frac{A}{36.40} = \frac{122.5}{100} \quad \text{so} \quad A = \frac{36.40 \times 122.5}{100}$$

We enter

36.40 [×] 122.5 [÷] 100 [=]

The selling price should be \$44.59.

Suppose now that a certain card collection *decreases* 8.2% in value from \$750. Note that 100% − 8.2% = 91.8%. To find the new value, we can restate the problem as: "What is 91.8% of \$750?" We set up the problem accordingly:

$$\frac{A}{750} = \frac{91.8}{100} \quad \text{so} \quad A = \frac{750 \times 91.8}{100}$$

Entering 750 [×] 91.8 [÷] 100 [=], we get \$688.50.

Use your calculator to solve each application.

39. **Social Science** The population of a town increases 4.2% in 1 year. If the original population was 19,500, what is the population after the increase?

40. **Business and Finance** A store marks up items 42.5% to allow for profit. If an item costs a store \$24.40, what will its selling price be?

41. **Business and Finance** A jacket that originally sold for \$98.50 is marked down by 12.5% for a sale. Find its sale price (to the nearest cent).

42. **Business and Finance** Jerry earned \$36,500 one year and then received a 10.5% raise. What is his new salary?

43. **Business and Finance** Carolyn's salary is \$5,220 per month. If deductions average 24.6%, what is her take-home pay?

44. **Business and Finance** Yi Chen made a \$6,400 investment at the beginning of a year. By the end of the year, the value of the investment had decreased by 8.2%. What was its value at the end of the year?

chapter 6 > Make the Connection

Skills | Calculator/Computer | **Career Applications** | Above and Beyond

Solve each application.

45. **Allied Health** How much 25% alcohol solution can be prepared using 225 mL of ethyl alcohol?

46. **Information Technology** When a particular network transmits data packets, 1.7% of the overall size is "overhead." If a network transmission measures 1,500 kilobytes (KB), how much of the transmission is overhead?

47. **Agricultural Technology** Milk that is labeled "3.5%" is made up of 3.5% butterfat. How many grams of butterfat are in 1 liter (938 g) of 3.5% milk?

48. **Environmental Technology** In some communities, "green" laws require that 40% of a lot remains green (covered in grass or other vegetation). How much green space is required in a 12,680-ft^2 lot?

49. **Business and Finance** The two ads pictured appeared last week and this week in the local paper. Is this week's ad accurate? Explain.

50. **Business and Finance** At True Grip hardware, you pay $10 in tax for a barbecue grill, which is 6% of the purchase price. At Loose Fit hardware, you pay $10 in tax for the same grill, but it is 8% of the purchase price. At which store do you get the better buy? Why?

51. It is customary when eating in a restaurant to leave a 15% tip (or more).
   (a) Outline a method to do a quick approximation for the amount of tip to leave.
   (b) Use this method to figure a 15% tip on a bill of $47.76.

52. The dean of enrollment management at a college states, "Last year was not a good year. Our enrollments were down 25%. But this year we increased our enrollment by 30% over last year. I think we have turned the corner." Evaluate the dean's analysis.

Answers

1. $408 **3.** 5% **5.** 25 questions **7.** $16.64 **9.** 8.5% **11.** 1,200 people **13.** $1,464 **15.** $14,500 **17.** $2,500 **19.** 6,840 people **21.** 750 s **23.** 76.3% **25.** 6,111.70 mmt **27.** 22% **29.** 48% **31.** $4,494.40 **33.** $4,630.50 **35.** 25% **37.** 37.5% **39.** 20,319 people **41.** $86.19 **43.** $3,935.88 **45.** 900 mL **47.** 32.83 g **49.** Above and Beyond **51.** Above and Beyond

Activity 18 ::

Population Changes Revisited

The U.S. Census Bureau conducts a census of the U.S. population every 10 years, as required by the U.S. Constitution. The table gives the population of the nation and the six largest states in 2000 and 2010. Use the table to complete the exercises that follow. Round all percents to the nearest tenth percent.

	2000 Population	2010 Population
United States	281,421,906	308,745,538
California	33,871,648	37,253,956
Texas	20,851,820	25,145,561
New York	18,976,457	19,378,102
Florida	15,982,378	18,801,310
Illinois	12,419,293	12,830,632
Pennsylvania	12,281,054	12,702,379

1. Find the percent increase in the U.S. population from 2000 to 2010.
2. By examining the table (no actual calculations yet!), predict which state had the greatest percent increase.
3. Predict which state had the smallest percent increase.
4. Now find the percent increase in population during this period for each state.

   California: Texas:
   New York: Florida:
   Illinois: Pennsylvania:

5. Which state had the greatest percent increase during this period? Which had the smallest?
6. The population of the six largest states combined represented what percent of the U.S. population in 2000?
7. The population of the six largest states combined represented what percent of the U.S. population in 2010?
8. Determine the percent increase in population from 2000 to 2010 for the combined six largest states.

summary :: chapter 6

Definition/Procedure	Example	Reference
Writing Percents as Fractions and Decimals		**Section 6.1**
Percent Another way of naming parts of a whole. *Percent* means per hundred if necessary.	Fractions and decimals are other ways of naming parts of a whole. $21\% = 21\left(\frac{1}{100}\right) = \frac{21}{100} = 0.21$	*p.* 324
To write a percent as a fraction, replace the percent symbol with $\frac{1}{100}$ and then multiply and simplify.	$40\% = 40\left(\frac{1}{100}\right) = \frac{40}{100} = \frac{2}{5}$	*p.* 325
To write a percent as a decimal, remove the percent symbol and move the decimal point two places to the left.	$37\% = 0.37$	*p.* 326
Writing Decimals and Fractions as Percents		**Section 6.2**
To write a decimal as a percent, move the decimal point two places to the right and attach the percent symbol.	$0.581 = 58.1\%$	*p.* 333
There are two methods for writing a fraction as a percent. *Method 1: The Proportion Method* Use proportions to write an equivalent fraction with a denominator of 100. Then, write the fraction as a percent.	$\frac{3}{5} = \frac{x}{100} \Rightarrow 3 \cdot 100 = 5x$ $\frac{300}{5} = \frac{5x}{5} \Rightarrow 60 = x$ Therefore, $\frac{3}{5} = 60\%$.	*p.* 334
Method 2: The Decimal Method Write the decimal equivalent of the fraction, and then write that decimal as a percent.	$\frac{3}{5} = 0.6 = 60\%$	*p.* 334
Solving Percent Problems		**Section 6.3**
Generally, percent problems have three parts. **1.** The *base B.* This is the whole amount or starting amount in the problem. It is the standard used for comparison. **2.** The *amount A.* This is the part of the whole being compared to the base. **3.** The *rate R.* This is the ratio of the amount to the base. The rate is generally written as a percent.	45 is 30% of 150. ↑ (A) ↑ (R) ↑ (B)	*p.* 343
Percent problems are solved by identifying A, B, R, and r. The rate R is related to r as $R = \frac{r}{100}$ The base B, the amount A, and r are related by the **percent proportion** $\frac{A}{B} = \frac{r}{100}$	Suppose that $A = 45$, $B = 150$, $R = 30\%$ If the rate R is 30%, then $r = 30$. The percent proportion is $\frac{45}{150} = \frac{30}{100}$	*p.* 346
Use the percent proportion to solve percent problems. **Step 1** Substitute the two known values into the proportion. **Step 2** Solve the proportion for the unknown value.	What is 24% of 300? $\frac{A}{300} = \frac{24}{100}$ $100A = 7{,}200$ $A = 72$ 72 is 24% of 300.	*p.* 346

Continued

Definition/Procedure	Example	Reference
Percent Applications		**Section 6.4**
Applications involving percentages are varied and include solving problems related to commissions, taxes, discounts, markups, and rates of increase or decrease.	An electronics store normally sells a DVD player for \$130. During a sale, they apply a 15% discount to the price. What is the sale price? $B = \$130$ and $R = 15\%$ (so $r = 15$). $\frac{A}{130} = \frac{15}{100}$ $100A = 15 \cdot 130 = 1{,}950$ $A = \frac{1{,}950}{100}$ $= 19.50$ The sale price is \$130 − \$19.50 = \$110.50	p. 353

summary exercises :: chapter 6

This summary exercise set will help ensure that you have mastered each of the objectives of this chapter. The exercises are grouped by section. You should reread the material associated with any exercises that you find difficult. The answers to the odd-numbered exercises are in the Answers Appendix at the back of the text.

6.1 *Use a percent to describe the shaded portion of each diagram.*

1. 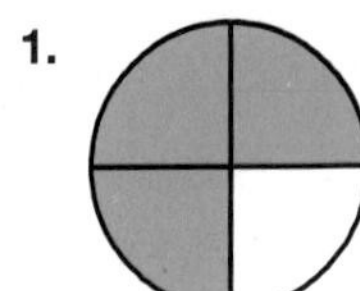

2. 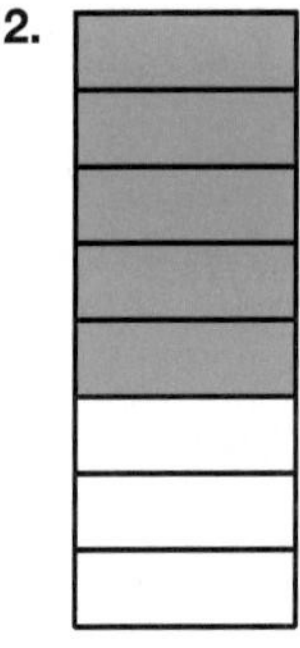

Write each percent as a fraction or a mixed number.

3. 2%

4. 20%

5. 37.5%

6. 150%

7. $233\frac{1}{3}\%$

8. 300%

Write each percent as a decimal.

9. 75%

10. 4%

11. 6.25%

12. 13.5%

13. 0.6%

14. 225%

6.2 *Write each number as a percent.*

15. 0.06

16. 0.375

17. 2.4

18. 7

19. 0.035

20. 0.005

21. $\frac{43}{100}$

22. $\frac{7}{10}$

23. $\frac{2}{5}$

24. $1\frac{1}{4}$

25. $2\frac{2}{3}$ (exact value)

26. $\frac{3}{11}$ (to nearest tenth of a percent)

6.3 *Complete each exercise.*

27. 80 is 4% of what number?

28. 70 is what percent of 50?

29. 11% of 3,000 is what number?

30. 24 is what percent of 192?

31. Find the base if 12.5% of the base is 625.

32. 90 is 120% of what number?

33. What is 9.5% of 700?

34. Find 150% of 50.

35. Find the base if 130% of the base is 780.

36. 350 is what percent of 200?

37. 28.8 is what percent of 960?

38. 18 is $66\frac{2}{3}\%$ of what number?

6.4 *Solve each application.*

39. **Business and Finance** Joan works on a 4% commission basis. She sold $45,000 in merchandise during 1 month. What was the amount of her commission?

40. **Business and Finance** David buys a dishwasher that is marked down $77 from its original price of $350. What is the discount rate?

41. **Science and Medicine** A chemist prepares a 400-milliliter (mL) acid-water solution. If the solution contains 30 mL of acid, what percent of the solution is acid?

42. **Business and Finance** The price of a new compact car has increased $819 over that of the previous year. If this amounts to a 4.5% increase, what was the price of the car before the increase?

43. **Business and Finance** A store advertises, "Buy the red-tagged items at 25% off their listed price." If you buy a coat marked $136, what will you pay for the coat during the sale?

44. **Business and Finance** Tom has 6% of his salary deducted for a retirement plan. If that deduction is $168, what is his monthly salary?

45. **Social Science** A college finds that 35% of its science students take biology. If there are 252 biology students, how many science students are there altogether?

46. **Business and Finance** A company finds that its advertising costs increased from $72,000 to $76,680 in 1 year. What was the rate of increase?

47. **Business and Finance** A bank offers 5.25% interest on 1-year time deposits. If you place $3,000 in an account, how much will you have at the end of the year? (chapter 6 > Make the Connection)

48. **Business and Finance** Maria's company offers her a 4% pay raise. This amounts to a $126 per month increase in her salary. What is her monthly salary before and after the raise?

49. **Business and Finance** A virus scanning program is checking every file for viruses. It has completed 30% of the files in 150 seconds. How long should it take to check all the files?

50. **Business and Finance** Gary is currently using 143 GB of his 232-GB hard drive. What percent of his hard drive is Gary using? Round your result to the nearest tenth of a percent.

chapter test 6

Use this chapter test to assess your progress and to review for your next exam. Allow yourself about an hour to take this test. The answers to these exercises are in the Answers Appendix at the back of the text.

1. Use a percent to name the shaded portion of the following diagram.

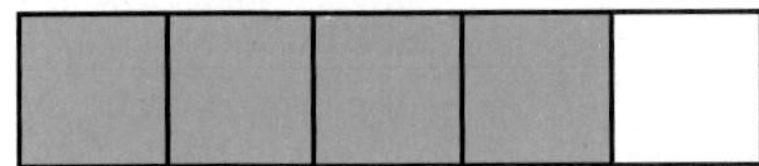

Write each number as a percent.

2. 0.03
3. 0.042
4. $\frac{2}{5}$
5. $\frac{5}{8}$

Write each percent as an equivalent decimal.

6. 42%
7. 6%
8. 160%

Write each percent as an equivalent fraction.

9. 7%
10. 72%

In exercises 11 to 13, identify the rate, base, and amount. Do not solve *at this point.*

11. 50 is 25% of 200.
12. What is 8% of 500?
13. **Business and Finance** A state sales tax rate is 6%. If the tax on a purchase is $30, what is the amount of the purchase?

Solve each percent problem.

14. What is 4.5% of 250?
15. What percent of 300 is 60?
16. $33\frac{1}{3}$% of 1,500 is what number?
17. Find 125% of 600.
18. 875 is what percent of 500?
19. 96 is 12% of what number?
20. 8.5% of what number is 25.5?
21. 4.5 is what percent of 60?

Solve each application.

22. **Business and Finance** A state sales tax rate is 6.2%. What tax will you pay on a sweater that costs $80?

23. **Statistics** You receive a grade of 75% on a test of 80 questions. How many questions did you have correct?

24. **Business and Finance** A shirt that costs a store \$54 is marked up 30% (based on cost). Find its selling price.

25. **Business and Finance** Mrs. Sanford pays \$300 in interest on a \$2,500 loan for 1 year. What is the interest rate for the loan?

26. **Business and Finance** A car is marked down \$1,552 from its original selling price of \$19,400. What is the discount rate?

27. **Business and Finance** Sarah earns \$540 in commissions in one month. If her commission rate is 3%, what were her total sales?

28. **Social Science** A community college has 480 more students in fall 2008 than in fall 2007. If this is a 7.5% increase, what was the fall 2007 enrollment?

29. **Business and Finance** Shawn arranges financing for his new car. The interest rate for the financing plan is 12%, and he will pay \$2,220 interest for 1 year. How much money did he borrow to finance the car?

30. **Business and Finance** Jovita's monthly salary is \$2,200. If the deductions for taxes from her monthly paycheck are \$528, what percent of her salary goes for these deductions?

cumulative review chapters 1–6

Use this exercise set to review concepts from earlier chapters. While it is not a comprehensive exam, it will help you identify any material that you need to review before moving on to the next chapter. In addition to the answers, you will find section references for these exercises in the Answers Appendix in the back of the text.

1.1

1. What is the place value of 4 in the number 234,768?

Evaluate each expression.

1.5

2. 56×203

1.6

3. $3{,}026 \div 34$

1.7

4. $8 - 5 + 2$

5. $15 - 3 \times 2$

6. $6 + 4 \times 3^2$

2.1

7. List the prime numbers between 50 and 70.

2.2

8. Write the prime factorization of 260.

9. Find the greatest common factor (GCF) of 84 and 140.

3.2

10. Find the least common multiple (LCM) of 18, 20, and 30.

Evaluate each expression.

2.5

11. $3\frac{2}{5} \times 2\frac{1}{2}$

2.6

12. $5\frac{1}{3} \div 4$

3.4

13. $4\frac{3}{4} + 3\frac{5}{6}$

14. $7\frac{1}{6} - 2\frac{3}{8}$

2.5

15. A kitchen measures $5\frac{1}{2}$ yd by $3\frac{1}{4}$ yd. If vinyl flooring costs \$16 per yd^2, what will it cost to cover the floor?

2.6

16. If you drive 180 mi in $3\frac{1}{3}$ hr, what is your average speed?

3.4

17. A bookshelf that is $54\frac{5}{8}$ in. long is cut from a board that is 8 ft long. If $\frac{1}{8}$ in. is wasted in the cut, what length board remains?

4.1

Find the indicated place values.

18. 8 in 4.2835

19. 4 in 6.09743

Complete each statement with <, =, or >.

20. 6.28 _______ 6.3

21. 3.75 _______ 3.750

Evaluate each expression.

4.3

22. 2.8×4.03

4.4

23. $54.528 \div 3.2$

4.3

24. A television set has an advertised price of \$599.95. You buy the set and agree to make payments of \$29.50 per month for 2 years. How much extra are you paying on this installment plan?

25. Find the area of a rectangle with length 3.4 m and width 1.85 m.

26. Find the volume of a box with dimensions 5.1 ft $\times$ 3.6 ft $\times$ 2.4 ft.

4.5

Write as a fraction or a mixed number. Simplify.

27. 0.36

28. 5.125

5.1

Write each ratio in simplest form.

29. $8\frac{1}{2}$ to $12\frac{3}{4}$

30. 34 feet to 8 yards

5.4

Solve for the unknown.

31. $\frac{3}{7} = \frac{8}{x}$

32. $\frac{1.9}{y} = \frac{5.7}{1.2}$

33. On a map the scale is $\frac{1}{4}$ in. = 25 mi. How many miles apart are two towns that are $3\frac{1}{2}$ in. apart on the map?

34. Diane worked 23.5 hours on a part-time job and was paid \$131.60. She is asked to work 25 hours the next week at the same pay rate. What salary will she receive?

6.1

35. Write 34% as a decimal and as a fraction.

6.2

36. Write $\frac{11}{20}$ as a decimal and as a percent.

6.3

37. Find 18% of 250.

38. 11% of what number is 55?

6.4

39. A company reduced the number of employees by 8% this year. 10 employees were laid off. How many were there last year?

40. The sales tax on an item priced at $72 is $6.12. What percent is the tax rate?

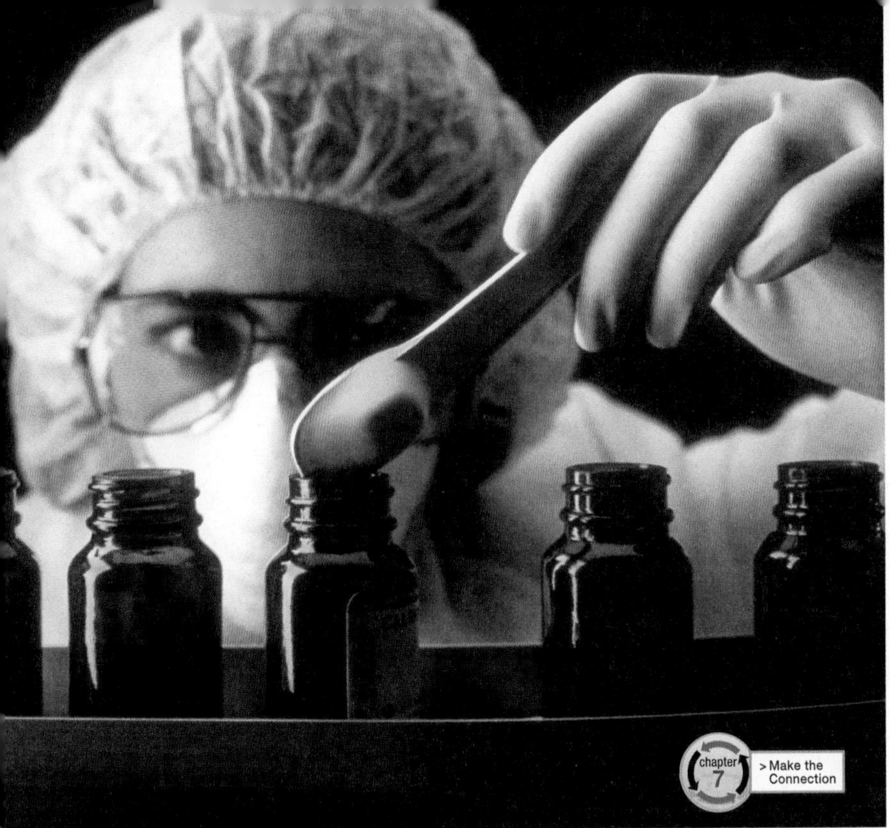

CHAPTER

7

Measurement

INTRODUCTION

A recipe calls for 2 teaspoons (tsp) of cider vinegar, but you are tripling the recipe. Instead of measuring 6 tsp, you could measure 2 tablespoons (Tb). You just need to know that 1 Tb = 3 tsp.

We measure things every day. How cold is it out? How far away is that? What does something weigh? These are typical questions we ask.

Of course, most of the world uses the metric system to answer these questions. In fact, the United States, Liberia, Yemen, and Myanmar are the only nations that have not switched over to the metric system.

In this chapter, we study both measurement systems and learn to move between the systems. You will have the opportunity to consider such conversions by looking at tools and hardware in both systems when you complete Activity 21.

CHAPTER 7 OUTLINE

7 prerequisite check

CHAPTER 7

This Prerequisite Check highlights the skills you will need in order to be successful in this chapter. The answers to these exercises are in the Answers Appendix at the back of the text.

Evaluate each expression.

1. 42.84×10^4

2. $42.84 \div 10^4$

Simplify each fraction.

3. $\frac{144}{108}$

4. $\frac{\frac{9}{28}}{\frac{27}{35}}$

Write each ratio in simplest form.

5. 15 to 6

6. 1.2 to 2

Write each rate as a unit rate.

7. $\frac{\$344}{32 \text{ hr}}$

8. $\frac{175 \text{ mi}}{3.5 \text{ hr}}$

Find the unit price of each item.

9. A 64-fluid-ounce (fl oz) container of orange juice sells for \$3.29 (round your result to the nearest hundredth of a cent).

10. A case of 50 blank DVD-R disks sells for \$22.50.

Find the perimeter of each figure.

11.

12.

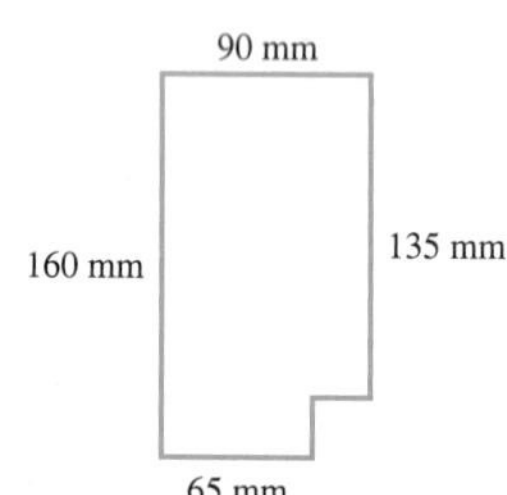

7.1 The U.S. Customary System of Measurement

< 7.1 Objectives >

1 > Convert between two U.S. Customary units of measure

2 > Use denominate numbers

3 > Solve measurement applications

Many problems involve **measurements.** When we measure an object, we describe some property it has by using a number and the appropriate unit. For instance, we might say that a board is 6 feet long to describe its length, or that a package weighs 5 pounds to describe its weight or mass. Feet and pounds are examples of units of measure.

The system you are probably most familiar with is the U.S. Customary system of measurement. The United States is the only industrialized nation in the world that has not switched to the *metric system*, which we introduce in Section 7.2.

In order to work with the U.S. Customary system of measurement, we need to know what the units are and how they relate to each other. The table gives some commonly used units and their equivalents.

> CAUTION

Ounces and fluid ounces are different. An ounce is a measure of weight or mass, whereas fluid ounces measure volume.

U.S. Customary Units of Measure

Length	**Weight/Mass**
1 foot (ft) = 12 inches (in.)	1 pound (lb) = 16 ounces (oz)
1 yard (yd) = 3 ft	1 ton = 2,000 lb
1 mile (mi) = 5,280 ft	
Volume	**Time**
1 cup (c) = 8 fluid ounces (fl oz)	1 minute (min) = 60 seconds (s)
1 pint (pt) = 2 c	1 hour (hr) = 60 min
1 quart (qt) = 2 pt	1 day = 24 hr
1 gallon (gal) = 4 qt	1 week = 7 days

There are two good ways to use the equivalencies in the table to convert between units of measure. We show you the *substitution method* first. It is the easier method and should be used when the problem is not too complex or complicated. It works especially well when converting from larger units to smaller units.

Step by Step

Converting Measures with the Substitution Method

Step 1 Rewrite the given measure, separating the number and the unit.

Step 2 Replace the unit of measure with an equivalent measure.

Step 3 Perform the arithmetic and simplify, if necessary.

Example 1 Converting Measures with the Substitution Method

< Objective 1 >

(a) How many feet are in 8 yards?

Begin by writing 8 yd as $8 \times (1 \text{ yd})$.

We know from the measurement table that 1 yd is equivalent to 3 ft, therefore we can replace 1 yd with 3 ft in the statement without changing the actual length.

$8 \text{ yd} = 8 \times (1 \text{ yd})$
$= 8 \times (3 \text{ ft})$ Replace 1 yd with its equivalent in feet.
$= 24 \text{ ft}$ Perform the arithmetic: $8 \times 3 = 24$.

There are 24 ft in 8 yd.

(b) Find the number of hours in 4 days.

Again, we separate the number from the units (days) and then substitute in the equivalent number of hours in a day.

$4 \text{ days} = 4 \times (1 \text{ day})$
$= 4 \times (24 \text{ hr})$
$= 96 \text{ hr}$

There are 96 hours in 4 days.

Check Yourself 1

Complete each statement.

(a) 4 ft = _______ in. **(b)** 12 qt = _______ pt
(c) 18 lb = _______ oz **(d)** 6.5 min = _______ s

To keep our conversion table from getting too large, we choose not to include some equivalences that we can compute using two or more steps. Some examples include the number of yards in a mile and the number of seconds in an hour. We can use the substitution method to make these conversions, as well.

Example 2 **Using the Substitution Method in Multistep Conversions**

(a) Find the number of seconds in 3 hours.

We do not have a direct conversion between hours and seconds in our table. However, we do have hour-minute and minute-second conversions, so we can still solve this problem.

$3 \text{ hr} = 3 \times (1 \text{ hr})$
$= 3 \times (60 \text{ min})$ Replace 1 hr with its equivalent in minutes.
$= 180 \text{ min}$ Perform the arithmetic: $3 \times 60 = 180$.
$= 180 \times (1 \text{ min})$ Repeat the process to convert minutes to seconds.
$= 180 \times (60 \text{ s})$ Replace 1 min with its equivalent in seconds.
$= 10{,}800 \text{ s}$ Perform the arithmetic: $180 \times 60 = 10{,}800$.

There are 10,800 s in 3 hr.

(b) Convert gallons to cups.

$1 \text{ gal} = 4 \text{ qt}$
$= 4 \times (1 \text{ qt})$
$= 4 \times (2 \text{ pt})$
$= 8 \text{ pt}$
$= 8 \times (2 \text{ c})$
$= 16 \text{ c}$

One gallon is equivalent to 16 cups.

Check Yourself 2

Convert, as indicated.

(a) Find the number of inches in 5 miles.
(b) Find the number of ounces in 3 tons.
(c) Find the number of seconds in 1 hour.

The substitution method is best when making straightforward conversions, especially when converting to smaller units. When we have to make more complex conversions, especially when converting to larger units, we prefer the *unit-ratio method.*

RECALL

A unit ratio is a ratio equal to one.

Multiplying any quantity by 1 leaves it unchanged.

In Example 1(a), we converted yards to feet and found that 8 yd = 24 ft. We look at this example again, with the unit-ratio method. We know that 1 yd = 3 ft. Since these amounts are equal, we can form two *unit ratios* with them.

$$\frac{1 \text{ yd}}{3 \text{ ft}} = 1 \qquad \text{and} \qquad \frac{3 \text{ ft}}{1 \text{ yd}} = 1$$

Since each of these ratios is equal to 1, we can multiply a quantity by one of these ratios and still have the same amount.

$$\begin{aligned} 8 \text{ yd} &= 8 \text{ yd} \times 1 \\ &= 8 \text{ yd} \times \frac{3 \text{ ft}}{1 \text{ yd}} && \frac{3 \text{ ft}}{1 \text{ yd}} = 1 \\ &= \frac{8 \cancel{\text{yd}}}{1} \times \frac{3 \text{ ft}}{1 \cancel{\text{yd}}} && \text{We “cancel” yards because we are multiplying fractions.} \\ &= \frac{8 \times 3 \text{ ft}}{1} && \text{Simplify.} \\ &= 24 \text{ ft} \end{aligned}$$

Do you see why we chose to use $\frac{3 \text{ ft}}{1 \text{ yd}}$ rather than $\frac{1 \text{ yd}}{3 \text{ ft}}$? This allowed us to “cancel” the yards unit so that we were left with feet. When using the unit-ratio method, always select the form that leaves you with the units you want.

Step by Step

Converting Measures with the Unit-Ratio Method

Step 1 Construct a unit ratio so that the original units “cancel” and you are left with either the desired units or an intermediary measure.

Step 2 Multiply your original measure by the unit ratio from step 1 and simplify.

Step 3 If you used an intermediary measure in step 1, repeat the process until you have a measurement with the desired units.

Example 3 Converting Measures with the Unit-Ratio Method

(a) Find the number of yards in 4 miles.

NOTE

We are converting to feet so it goes in the numerator; we are converting from miles, so it goes in the denominator.

We want to convert miles to yards but our table does not have a direct conversion. We can work through feet (miles to feet to yards), but then we would be going from a smaller unit, feet, to a larger unit, yards. The unit-ratio method is the better method in this case.

First, we convert miles to feet. To do so, we construct a unit ratio with feet in the numerator and miles in the denominator.

$$\frac{5{,}280 \text{ ft}}{1 \text{ mi}} = 1$$

NOTE

If it helps, you can think of 4 mi as $\frac{4 \text{ mi}}{1}$.

Because there are 5,280 feet in 1 mile, this ratio is equal to 1. Therefore, the distance described by 4 miles is unchanged if we multiply it by the ratio we just constructed.

$$\begin{aligned}4 \text{ mi} &= 4 \text{ mi} \times \frac{5{,}280 \text{ ft}}{1 \text{ mi}} \\ &= 4 \cancel{\text{mi}} \times \frac{5{,}280 \text{ ft}}{1 \cancel{\text{mi}}} \\ &= 4 \times 5{,}280 \text{ ft} \\ &= 21{,}120 \text{ ft}\end{aligned}$$

The unit ratio allows us to "cancel" miles.

To convert to yards, we construct a unit ratio with yards in the numerator and feet in the denominator.

$$\begin{aligned}21{,}120 \text{ ft} &= 21{,}120 \cancel{\text{ft}} \times \frac{1 \text{ yd}}{3 \cancel{\text{ft}}} \\ &= \frac{21{,}120}{3} \text{ yd} \\ &= 7{,}040 \text{ yd}\end{aligned}$$

In this case, we have to divide.

There are 7,040 yd in 4 mi.

(b) Find the number of minutes in 45 seconds.

We use the unit ratio that contains minutes in the numerator and seconds in the denominator.

$$\begin{aligned}45 \text{ s} &= 45 \cancel{\text{s}} \times \frac{1 \text{ min}}{60 \cancel{\text{s}}} \\ &= \frac{45}{60} \text{ min} \\ &= \frac{3}{4} \text{ min}\end{aligned}$$

45 seconds are three-quarters of a minute.

(c) Find the number of gallons in 160 fl oz.

From Example 2(b), we know that 1 gal = 16 c, so we convert fluid ounces to cups and then cups to gallons. In this case, we use unit ratios to make the conversions.

$$\begin{aligned}160 \text{ fl oz} &= 160 \text{ fl oz} \times \frac{1 \text{ c}}{8 \text{ fl oz}} \\ &= 160 \cancel{\text{fl oz}} \times \frac{1 \text{ c}}{8 \cancel{\text{fl oz}}} \\ &= \frac{160}{8} \text{ c} \\ &= 20 \text{ c}\end{aligned}$$

1 c = 8 fl oz

Now use the unit ratio $\frac{1 \text{ gal}}{16 \text{ c}}$ to convert 20 cups to gallons.

$$\begin{aligned}20 \text{ c} &= 20 \text{ c} \times \frac{1 \text{ gal}}{16 \text{ c}} \\ &= 20 \cancel{\text{c}} \times \frac{1 \text{ gal}}{16 \cancel{\text{c}}} \\ &= \frac{20}{16} \text{ gal} \\ &= 1\frac{1}{4} \text{ gal}\end{aligned}$$

Therefore, 160 fl oz is $1\frac{1}{4}$ gal.

Check Yourself 3

Convert, as indicated.

(a) Find the number of yards in 1 mi.
(b) Find the number of seconds in 4 hr.
(c) Find the number of feet in 8 in.
(d) Find the number of tons in 5,000 oz.

RECALL

You should always consider whether your answer to a problem is reasonable.

Here is a good way of determining whether your answer to a conversion problem is reasonable.

- If you are converting to larger units, your number should get smaller.
- If you are converting to smaller units, your number should get larger.

This is because it takes fewer of a larger unit and more of a smaller unit to measure the same amount.

Consider Example 3(a). We began with 4 miles and converted it to yards. Because yards are smaller than miles, we needed many more than 4 of them to measure a 4-mi distance, so our answer needs to be larger than 4. In this case, our answer was 4 mi = 7,040 yd. Since 7,040 is larger than 4 and yards are smaller than miles, our answer passes this test of reasonableness.

NOTE

Historically, units were associated with various things. A foot was the length of a foot, of course. The yard was the distance from the end of a nose to the fingertips of an outstretched arm. Objects were weighed by comparing them with grains of barley.

We use the phrase *denominate number* to mean a number with units attached. That is, 3 ft, 12 students, and $45.55 are all examples of denominate numbers, whereas 6 is not since there are no units. The number 6 is an abstract number and is not necessarily measuring anything. 3 ft is a measurement of length; 12 students is a count of the number of students. This is what makes them denominate numbers.

A denominate number may involve two or more different units. We regularly combine feet and inches, pounds and ounces, and so on. The measures 5 lb 6 oz and 4 ft 7 in. are examples. When simplifying a denominate number with multiple units, the *largest unit should include as much of the measure as possible.* For example, 3 ft 2 in. is simplified, whereas 2 ft 14 in. is not.

Example 4 shows the steps used to simplify a denominate number with multiple units.

Example 4 Simplifying Denominate Numbers

< Objective 2 >

NOTE

18 in. is larger than 1 ft so it can be simplified.

(a) Simplify 4 ft 18 in.

$$4\text{ ft }18\text{ in.} = \underbrace{4\text{ ft} + \overbrace{1\text{ ft}}^{18\text{ in.}}} \overbrace{+\ 6\text{ in.}} = 5\text{ ft }6\text{ in.}$$

Write 18 in. as 1 ft 6 in. because 12 in. is 1 ft.

(b) Simplify 5 hr 75 min.

$$5\text{ hr }75\text{ min} = \underbrace{5\text{ hr} + \overbrace{1\text{ hr}}^{75\text{ min}}} \overbrace{+\ 15\text{ min}} = 6\text{ hr }15\text{ min}$$

Write 75 min as 1 hr 15 min because 1 hr is 60 min.

Check Yourself 4

(a) Simplify 5 lb 24 oz. **(b)** Simplify 7 ft 20 in.

Denominate numbers with the same units are called *like numbers.* We can always add or subtract denominate numbers that have the same units.

Step by Step

Adding Denominate Numbers

Step 1 Arrange the numbers so that the like units are in the same vertical column.

Step 2 Add in each column.

Step 3 Simplify if necessary.

Example 5 illustrates this rule for adding denominate numbers.

Example 5 Adding Denominate Numbers

Add 5 ft 4 in., 6 ft 7 in., and 7 ft 9 in.

NOTES

The columns here represent inches and feet.

Be sure to simplify the results.

5 ft 4 in.	Arrange in vertical columns.
6 ft 7 in.	
+ 7 ft 9 in.	
18 ft 20 in.	Add in each column.
= 19 ft 8 in.	Simplify as before.

Check Yourself 5

Add 3 hr 15 min, 5 hr 50 min, and 2 hr 40 min.

To subtract denominate numbers, we have a similar rule.

Step by Step

Subtracting Denominate Numbers

Step 1 Arrange the numbers so that the like units are in the same vertical column.

Step 2 Subtract in each column. You may have to borrow from the larger unit at this point.

Step 3 Simplify if necessary.

Example 6 Subtracting Denominate Numbers

Subtract 3 lb 6 oz from 8 lb 13 oz.

8 lb 13 oz	Arrange vertically.
− 3 lb 6 oz	
5 lb 7 oz	Subtract in each column.

Check Yourself 6

Subtract 5 ft 9 in. from 10 ft 11 in.

As Example 7 shows, subtracting denominate numbers may involve borrowing.

Example 7 Subtracting Denominate Numbers

NOTES

Borrowing with denominate numbers is not the same as in the place-value system, in which we always borrow a power of 10.

9 ft becomes 8 ft 12 in. Combine the 12 in. with the original 3 in.

Subtract 5 ft 8 in. from 9 ft 3 in.

$$\begin{array}{r} 9 \text{ ft } 3 \text{ in.} \\ -\ 5 \text{ ft } 8 \text{ in.} \\ \hline \end{array}$$

Do you see the problem? We cannot subtract in the inches column.

To complete the subtraction, we borrow 1 ft and rename. The "borrowed" number will depend on the units involved.

$$\begin{array}{r} 9 \text{ ft } 3 \text{ in.} \\ -\ 5 \text{ ft } 8 \text{ in.} \\ \hline \end{array} \longrightarrow \begin{array}{r} 8 \text{ ft } 15 \text{ in.} \\ -\ 5 \text{ ft } \ \ 8 \text{ in.} \\ \hline 3 \text{ ft } \ \ 7 \text{ in.} \end{array}$$

Check Yourself 7

Subtract 3 lb 9 oz from 8 lb 5 oz.

Certain types of problems involve multiplying or dividing denominate numbers by abstract numbers, that is, numbers without a unit of measure attached.

Step by Step

Multiplying or Dividing by Abstract Numbers

Step 1 Multiply or divide each part of the denominate number by the abstract number.

Step 2 Simplify if necessary.

Example 8 Multiplying Denominate Numbers

NOTE

Multiply each part of the denominate number by 3.

(a) Multiply 4×5 in.

4×5 in. = 20 in. or 1 ft 8 in.

(b) Multiply $3 \times (2$ ft 7 in.).

$$\begin{array}{r} 2 \text{ ft } \ \ 7 \text{ in.} \\ \times \qquad \quad 3 \\ \hline 6 \text{ ft } \ 21 \text{ in.} \end{array}$$

Simplify. The product is 7 ft 9 in.

Check Yourself 8

Multiply 5 lb 8 oz by 4.

Division is illustrated in Example 9.

Example 9 Dividing Denominate Numbers

NOTE

Divide each part of the denominate number by 4.

Divide 8 lb 12 oz by 4.

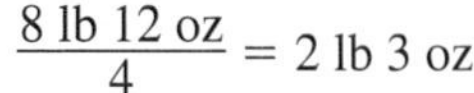

$$\frac{8 \text{ lb } 12 \text{ oz}}{4} = 2 \text{ lb } 3 \text{ oz}$$

Check Yourself 9

Divide 9 ft 6 in. by 3.

We encounter the need to make such calculations in many applications.

Example 10 An Application Involving Denominate Numbers

< Objective 3 >

There were 482 lb 6 oz of steel in stock before a shipment of 219 lb 13 oz arrived. How much steel was in stock after the shipment?

We begin by lining up like numbers.

482 lb	6 oz
219 lb	13 oz
701 lb	19 oz

Then, since 19 oz = 1 lb 3 oz, we simplify our result to 702 lb 3 oz.

Check Yourself 10

There are 6 gal 1 qt of aluminum sealer on hand. A run of parts requires 1 gal 2 qt of sealer. How much sealer is left after the run?

RECALL

We studied rates in Section 5.2.

Earlier in this section, we showed that the unit-ratio method is better for complicated conversions. In many applications in science, engineering, and technology, speed is measured in feet per second (fps) rather than in miles per hour. We need to construct several unit ratios in cases such as these.

Example 11 An Application Using the Unit-Ratio Method

Convert 65 miles per hour to feet per second.

We take the same approach converting miles to feet and hours to seconds. You can do this in one step, but we will use two steps to make it clearer.

RECALL

In Check Yourself 2, you showed that

1 hr = 3,600 s,

so

$\frac{1 \text{ hr}}{3{,}600 \text{ s}}$ is a unit ratio.

$$\frac{65 \text{ mi}}{1 \text{ hr}} = \frac{65 \cancel{\text{mi}}}{1 \text{ hr}} \times \frac{5{,}280 \text{ ft}}{1 \cancel{\text{mi}}}$$ We chose to convert miles to feet first.

$$= \frac{65 \times 5{,}280 \text{ ft}}{1 \text{ hr}}$$ $65 \cdot 5{,}280 = 343{,}200$

$$= \frac{343{,}200 \text{ ft}}{1 \cancel{\text{hr}}} \times \frac{1 \cancel{\text{hr}}}{3{,}600 \text{ s}}$$ Now we convert hours to seconds.

$$= \frac{343{,}200}{3{,}600} \frac{\text{ft}}{\text{s}}$$ Perform the division.

$$= \frac{286}{3} \frac{\text{ft}}{\text{s}}$$ $\frac{286}{3} = 95\frac{1}{3}$

65 miles per hour is equivalent to $95\frac{1}{3}$ feet per second.

In the previous conversion, we wanted seconds in the denominator and we wanted to change from hours, so our unit ratio needed to have hours in the numerator so that it would "cancel."

Check Yourself 11

In an 8-hr shift, a continuous cutting shear cuts 7,000 yd of steel. Calculate the rate of steel cut in feet per minute.

Check Yourself ANSWERS

1. **(a)** 48; **(b)** 24; **(c)** 288; **(d)** 390 2. **(a)** 316,800 in.; **(b)** 96,000 oz; **(c)** 3,600 s
3. **(a)** 1,760 yd; **(b)** 14,400 s; **(c)** $\frac{2}{3}$ ft; **(d)** 0.15625 ton or $\frac{5}{32}$ ton 4. **(a)** 6 lb 8 oz; **(b)** 8 ft 8 in.
5. 11 hr 45 min 6. 5 ft 2 in. 7. 4 lb 12 oz 8. 22 lb 9. 3 ft 2 in. 10. 4 gal 3 qt
11. $\frac{175}{4}\ \frac{\text{ft}}{\text{min}}$ or $43\frac{3}{4}\ \frac{\text{ft}}{\text{min}}$

Reading Your Text

These fill-in-the-blank exercises will help you understand some of the key vocabulary used in this section. The answers to these exercises are in the Answers Appendix at the back of the text.

(a) The United States is the only industrialized nation that has not switched to the ____________ system.

(b) A fluid ounce is a measure of ____________.

(c) A unit ratio is equal to ____________.

(d) Denominate numbers with the same units are called ____________ numbers.

Skills | Calculator/Computer | Career Applications | Above and Beyond

7.1 exercises

< Objective 1 >

Complete each statement.

1. 8 ft = ______ in.
2. 9 gal = ______ qt
3. 3 lb = ______ oz
4. 300 s = ______ min
5. 360 min = ______ hr
6. 5 pt = ______ fl oz
7. 4 days = ______ hr
8. 6 hr = ______ min
9. 16 qt = ______ gal
10. 11 min = ______ s
11. 10,000 lb = ______ tons
12. 5 mi = ______ ft
13. 30 pt = ______ qt
14. 64 fl oz = ______ pt
15. 64 oz = ______ lb
16. 540 min = ______ hr
17. 7 yd = ______ ft
18. 24 qt = ______ gal
19. 39 ft = ______ yd
20. 192 oz = ______ lb
21. 8 min = ______ s
22. 18 qt = ______ pt
23. 192 hr = ______ days
24. 360 hr = ______ days
25. 16 qt = ______ pt
26. 7 days = ______ hr
27. $7\frac{1}{4}$ hr = ______ min
28. 43 pt = ______ qt
29. 56 oz = ______ lb
30. 20 fl oz = ______ pt
31. 225 s = ______ min
32. 44 in. = ______ ft
33. 1.55 lb = ______ oz
34. 4.72 ft = ______ in.
35. 40 in. = ______ yd
36. 12 gal = ______ c
37. 16 gal = ______ fl oz
38. 16 fl oz = ______ gal
39. 1 ton = ______ oz
40. 4 mi = ______ in.
41. 6 weeks = ______ min
42. 15 s = ______ days

< Objective 2 >

Simplify.

43. 4 ft 18 in.

44. 6 lb 20 oz

45. 7 qt 5 pt

46. 7 yd 50 in.

47. 5 gal 9 qt

48. 3 min 110 s

49. 9 min 75 s

50. 9 hr 80 min

Add.

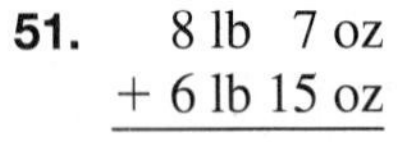

51. 8 lb 7 oz + 6 lb 15 oz

52. 9 ft 7 in. + 3 ft 10 in.

53. 3 hr 20 min, 4 hr 25 min + 5 hr 35 min

54. 5 yd 2 ft, 4 yd + 6 yd 1 ft

55. 4 lb 7 oz, 3 lb 11 oz, and 5 lb 8 oz

56. 7 ft 8 in., 8 ft 5 in., and 9 ft 7 in.

Subtract.

57. 9 lb 15 oz − 5 lb 8 oz

58. 7 ft 11 in. − 4 ft 3 in.

59. 6 hr 30 min − 3 hr 50 min

60. 7 gal 3 qt − 1 gal 3 qt

61. Subtract 2 yd 2 ft from 5 yd 1 ft.

62. Subtract 2 hr 30 min from 7 hr 25 min.

Multiply.

63. 4×13 oz

64. 4×10 in.

65. $3 \times$ (4 ft 5 in.)

66. $5 \times$ (4 min 20 s)

Divide.

67. $\frac{\text{4 ft 6 in.}}{2}$

68. $\frac{\text{12 lb 15 oz}}{3}$

69. $\frac{\text{16 min 28 s}}{4}$

70. $\frac{\text{25 hr 40 min}}{5}$

Evaluate and simplify.

71. 2 gal 3 qt 1 pt + 3 gal 2 qt 1 pt

72. 7 weeks 3 days 15 hours + 3 weeks 9 days 10 hours

73. 13 yd 15 ft 10 in. − 9 yd 16 ft 15 in.

74. 8 gal 3 qt 2 pt − 5 gal 5 qt 3 pt

75. 2 weeks 7 days 18 hr 40 min × 2

76. 4 gal 5 qt 3 pt 10 fl oz × 2

< Objective 3 >

Solve each application.

77. **Science and Medicine** The United States emitted approximately 9 million tons of suspended particulates into the atmosphere in 1 year. How many pounds of (suspended) particulates did the United States emit that year?

78. **Science and Medicine** The United States emitted approximately 20 million tons of volatile organic compounds into the atmosphere in 1 year. How many pounds of volatile organic compounds did the United States emit that year?

79. **Construction** A railing for a deck requires pieces of cedar 4 ft 8 in., 11 ft 7 in., and 9 ft 3 in. long. What is the total length of material needed?

80. **Business and Finance** Ted worked 3 hr 45 min on Monday, 5 hr 30 min on Wednesday, and 4 hr 15 min on Friday. How many hours did he work during the week?

81. **Crafts** A pattern requires a 2-ft 10-in. length of fabric. If a 2-yd length is used, what length remains?

82. **Crafts** You use 2 lb 8 oz of hamburger from a package that weighs 4 lb 5 oz. How much is left over?

83. **Crafts** A picture frame is to be 2 ft 6 in. long and 1 ft 8 in. wide. A 9-ft piece of molding is available for the frame. Will this be enough for the frame?

84. **Construction** A plumber needs two pieces of plastic pipe that are 6 ft 9 in. long and one piece that is 2 ft 11 in. long. He has a 16-ft piece of pipe. Is this enough for the job?

85. **Crafts** Mark uses 1 pt 9 fl oz and then 2 pt 10 fl oz from a container of film developer that holds 3 qt. How much of the developer remains?

86. **Business and Finance** Some flights limit passengers to 44 lb of checked-in luggage. Susan checks three pieces, weighing 20 lb 5 oz, 7 lb 8 oz, and 15 lb 7 oz. By how much is she under or over the limit?

87. **Business and Finance** Six packages weighing 2 lb 9 oz each are to be mailed. What is the total weight of the packages?

88. **Construction** A bookshelf requires four boards 3 ft 8 in. long and two boards 2 ft 10 in. long. How much lumber is needed for the bookshelf?

89. **Business and Finance** You can buy three 12-oz cans of peanuts for \$3 or one large can containing 2 lb 8 oz for the same price. Which is the better buy?

90. **Statistics and Mathematics** Rich, Susan, and Marc agree to share the driving on a 12-hr trip. Rich has driven for 4 hr 45 min, and Susan has driven for 3 hr 30 min. How long must Marc drive to complete the trip?

91. **Statistics** Colette's car has an average fuel efficiency of 30 mi/gal. How many yards can Colette travel on 1 fl oz of fuel?

92. **Statistics** In exercise 91, how many fluid ounces of fuel does Colette need in order to travel 1 yd?

93. **Business and Finance** A half-gallon bottle of organic milk sells for \$3.29. Find the cost per fluid ounce (to the nearest cent).

94. **Construction** 220 gal of water flow through a pipe every 3 hr. Find the flow rate, in fluid ounces per minute (round your result to the nearest whole fluid ounce).

95. **Statistics** A driver travels at a rate of 55 mi/hr. How many seconds does it take the driver to travel 1,000 ft (to the nearest tenth of a second)?

96. **Science and Medicine** A greyhound can run at an average rate of 37 mi/hr for $\frac{5}{16}$ of a mile. How long does it take the greyhound to run $\frac{5}{16}$ of a mile?

Determine whether each statement is **true** *or* **false.**

97. A denominate number may have more than two units of measure attached.

98. It is impossible to divide a denominate number by an abstract number.

In each statement, fill in the blank with **always, sometimes,** *or* **never.**

99. We can __________ add or subtract denominate numbers with the same units if we arrange the numbers so that the like units are in the same vertical column.

100. Subtracting denominate numbers __________ requires borrowing.

Skills | Calculator/Computer | **Career Applications** | Above and Beyond

101. **Mechanical Engineering** A piece of steel that is 13 ft 8 in. long is to be sheared into four equal pieces. How long will each piece be?

102. **Manufacturing Technology** A part weighs 4 lb 11 oz. How much does a lot of 24 parts weigh?

103. **Automotive Technology** An engine block weighs 218 lb 12 oz. Each head weighs 36 lb 3 oz. There are two heads on the engine. What is the total weight of the block with the two heads?

104. **Automotive Technology** Each piston in an engine weighs 4 lb 13 oz. How much do the eight pistons add to the weight of an engine?

105. **Manufacturing Technology** A knee wall is to be constructed from studs that are 4 ft 9 in. long. If the wall will use 13 studs, what is the total length required for the studs?

106. **Allied Health** A premature newborn baby boy weighs 98 oz. Determine his weight in pounds and ounces.

107. **Science and Medicine** On average, a human's heart beats at a rate of 72 beats per minute.

(a) Find the age at which the average person's heart has beaten 1 billion times (to the nearest whole day).

(b) Find the age at which the average person's heart has beaten 1 billion times (to the nearest whole year).

108. **Science and Medicine** Each human heartbeat pumps about 1 fl oz of blood. How many gallons of blood does the human heart pump in a day? (See exercise 107.)

109. The average person takes about 17 breaths per minute. How many breaths have you taken in your lifetime?

110. **Science and Medicine** The average breath takes in about $1\frac{1}{2}$ pints of air. The air is about 20% oxygen. Of the oxygen that we breathe, about 25% makes its way into our bloodstream. How much oxygen does the average person take in every day? (See exercise 109.)

111. What is your age in seconds? (Remember to consider leap years!)

112. **Science and Medicine**

(a) John is traveling at a speed of 60 mi/hr. What is his speed in feet per second?

(b) Use the information in part (a) to develop a method to convert any speed from miles per hour to feet per second.

113. A unit of measurement used in surveying is the **chain.** There are 80 chains in a mile. If you measured the distance from your home to school, how many chains would you have traveled?

114. Refer to several sources and write a brief history of how the units used in the U.S. Customary system of measurement originated. Discuss some units that were previously used but are no longer in use today.

Answers

1. 96 **3.** 48 **5.** 6 **7.** 96 **9.** 4 **11.** 5 **13.** 15 **15.** 4 **17.** 21 **19.** 13 **21.** 480 **23.** 8 **25.** 32 **27.** 435 **29.** 3.5 **31.** 3.75 **33.** 24.8 **35.** $\frac{10}{9}$ **37.** 2,048 **39.** 32,000 **41.** 60,480 **43.** 5 ft 6 in. **45.** 9 qt 1 pt **47.** 7 gal 1 qt **49.** 10 min 15 s **51.** 15 lb 6 oz **53.** 13 hr 20 min **55.** 13 lb 10 oz **57.** 4 lb 7 oz **59.** 2 hr 40 min **61.** 2 yd 2 ft **63.** 3 lb 4 oz **65.** 13 ft 3 in. **67.** 2 ft 3 in. **69.** 4 min 7 s **71.** 6 gal 2 qt **73.** 3 yd 1 ft 7 in. **75.** 6 weeks 1 day 13 hr 20 min **77.** 18 billion lb **79.** 25 ft 6 in. **81.** 3 ft 2 in. **83.** Yes, 8 in. will remain **85.** 1 pt 13 fl oz **87.** 15 lb 6 oz **89.** The 2-lb 8-oz can **91.** 412.5 yd/fl oz **93.** 5¢/fl oz **95.** 12.4 s **97.** True **99.** always **101.** 3 ft 5 in. **103.** 291 lb 2 oz **105.** 61 ft 9 in. **107.** (a) 9,645 days; (b) 26 years **109.** Above and Beyond **111.** Above and Beyond **113.** Above and Beyond

Activity 19 ::

Measurements

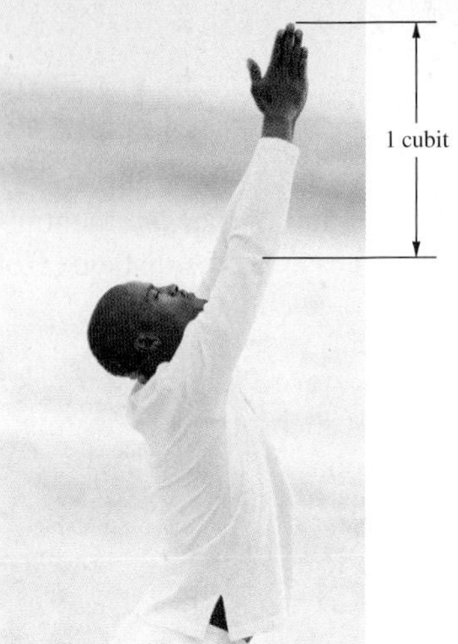

Many units of measure have interesting histories. Supposedly, a yard was defined by the English king, Henry I, in the twelfth century as the distance from the tip of his nose to the tip of his outstretched finger. Of course, this definition makes it difficult to compare a twenty-first century length to such a yard.

In addition to the more familiar units in the English system such as the inch, pound, or cup, history contains examples of numerous other units. Length was measured in cubits and rods, weight was measured by comparing an object with grains of barley, and cordwood was measured using steres.

According to tradition, a **cubit** was the distance between the tip of an elbow and the tip of the middle finger.

1. Find a desk or table. According to your own forearm, what are the dimensions of the desk or table, in cubits?
2. Ask a second person to make the same measurement using their own forearm. What are the dimensions of the desk or table, using this person's cubit?
3. How long is your cubit, in inches?

According to biblical tradition, the cubit was defined according to Noah's forearm. Noah was commanded to build an ark 300 cubits long, 50 cubits wide, and 30 cubits high.

4. Using the cubit defined by your own forearm, what are the dimensions of such an ark, in inches?

At some point in history, the length of a cubit was standardized as 18 inches.

5. How different is your cubit from the standard cubit?
6. What are the dimensions of your desk or table, in standard cubits?
7. Using standard cubits, find the dimensions of the ark discussed in exercise 4, in inches.

Often, common mathematical ideas have long and storied histories.

8. Earlier, we discussed a yard as the distance from the tip of the nose to the tip of the outstretched finger. How does your cubit compare to your yard?
9. How does the standard cubit compare to the standard yard?
10. Research the history of the yard as a unit of measure. Write a brief paragraph of this history.

7.2

Length and the Metric System

< 7.2 Objectives >

1 > Estimate metric units of length

2 > Convert between metric units of length

NOTES

Even in the United States, the metric system is used in science, medicine, the automotive industry, the food industry, and many other areas.

The basic unit of length in the metric system is also spelled *metre* (the British spelling).

The meter is one of the basic units of the International System of Units (abbreviated SI). This is a standardization of the metric system agreed to by scientists in 1960.

There is a standard pattern of abbreviation in the metric system. We will introduce the abbreviation for each term as we go along. The abbreviation for meter is m (no period!).

In Section 7.1, we studied the U.S. Customary system of measurement, which is used in the United States and a few other countries. The rest of the world uses the **metric system** to make measurements.

We begin our study of the metric system by looking at length. Once you understand how to convert units of length, you will apply that knowledge to other types of measurement in the metric system, such as mass or weight and volume, in Section 7.3.

The difficulty with the U.S. Customary system is that each conversion has a different conversion factor. For instance, 12 in. = 1 ft, 3 ft = 1 yd, and 5,280 ft = 1 mi.

This makes for a lot of different conversion factors and even unit names to remember. The metric system is a base-10 system, just like our number system. This makes it very easy to convert between units because we only need to multiply and divide by powers of 10.

The second benefit to the metric system is that each type of measure only has one basic unit. The standard unit of length is the **meter.** Any other unit of length is named by adding a prefix to the word meter. These same prefixes are applied to the basic units of mass and volume, which you will see in Section 7.3.

These advantages and the need for uniformity throughout the world led to legislation promoting the metric system in the United States.

To see how the metric system works, we start with measures of length and compare a basic U.S. Customary unit, the yard, with the meter.

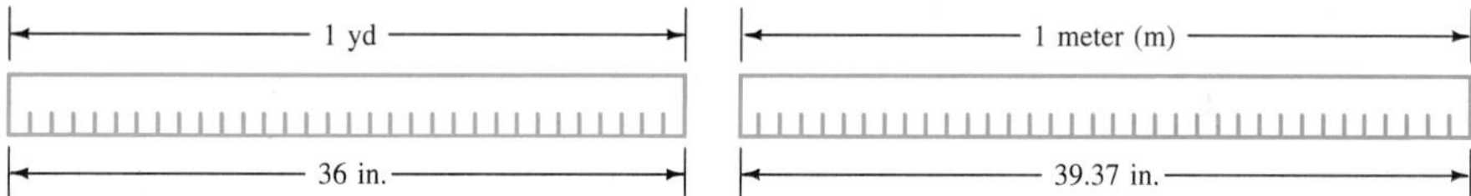

As you can see, a meter is just slightly longer than a yard. It is used for measuring the same things you might measure in feet or yards. Look at Example 1 to get a feel for the size of a meter.

Example 1 **Estimating Metric Length**

< Objective 1 >

A room might be 6 meters (6 m) long.

A building lot could be 30 m wide.

A fence is 2 m tall.

Check Yourself 1

Estimate each length, in meters.

(a) A traffic lane is __________ m wide.
(b) A small car is __________ m long.
(c) You are __________ m tall.

For other units of length, the meter is multiplied or divided by powers of 10. One commonly used unit is the **centimeter (cm).**

Definition

Centimeter (cm)

1 centimeter (cm) = $\frac{1}{100}$ meter (m) or 1 m = 100 cm

NOTE

The prefix *centi* means one-hundredth. This should be no surprise as one cent is one hundredth of a dollar.

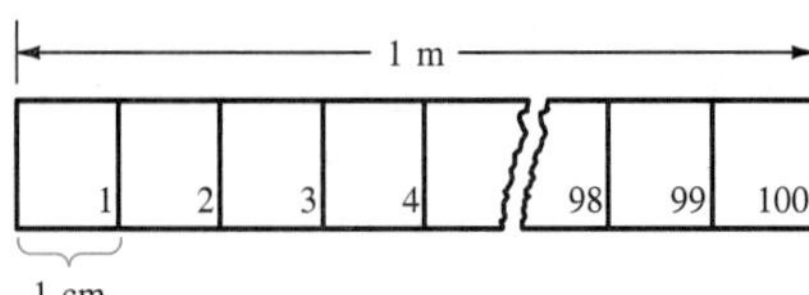

To give you an idea of the size of the centimeter, it is about the width of your little finger. There are about $2\frac{1}{2}$ cm to 1 in., and the unit is used to measure small objects. Look at Example 2 to get a feel for the length of a centimeter.

Example 2 Estimating Metric Length

A small paperback book is 10 cm wide.

A playing card is 8 cm long.

A ballpoint pen is 16 cm long.

Check Yourself 2

Estimate each length. Then use a metric ruler to check your guess.

(a) This page is __________ cm long.
(b) A dollar bill is __________ cm long.
(c) The seat of the chair you are on is __________ cm from the floor.

We use **millimeters (mm)** to describe very small lengths. To give you an idea of its size, a millimeter is about the thickness of a dime.

Definition

Millimeter (mm)

1 millimeter (mm) = $\frac{1}{1,000}$ m or 1 m = 1,000 mm

NOTES

The prefix *milli* means one-thousandth.

There are 10 mm to 1 cm.

To get used to the millimeter, consider Example 3.

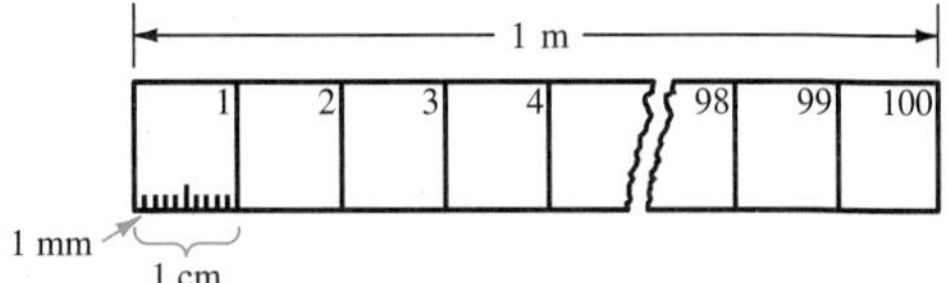

Example 3 Estimating Metric Length

Standard camera film is 35 mm wide.

A small paper clip is 7 mm wide.

A water glass might be 2 mm thick.

NOTE

The prefix *kilo* means 1,000. You are already familiar with this. For instance, 1 kilowatt (kW) = 1,000 watts (W).

Check Yourself 3

Estimate each length. Then use a metric ruler to check your guess.

(a) Your pencil is __________ mm wide.
(b) The tabletop you are working on is __________ mm thick.

The **kilometer (km)** is used to measure long distances. A kilometer is about $\frac{6}{10}$ of a mile.

Definition

Kilometer (km)

1 kilometer (km) = 1,000 m or 1 m = $\frac{1}{1,000}$ km

Example 4 uses kilometers.

Example 4 Estimating Metric Length

The distance from New York to Boston is 338 km.

A popular distance for runners is 5 km.

Check Yourself 4

Choose the most reasonable measure for each statement.

(a) The width of a doorway: 50 mm, 1 m, or 50 cm.
(b) The length of your pencil: 20 m, 20 mm, or 20 cm.
(c) The distance from your house to school: 500 km, 5 km, or 50 m.
(d) The height of a basketball center: 2.2 m, 22 m, or 22 cm.

To convert units of measure within the metric system, we multiply or divide by the appropriate power of 10. To accomplish this, we move the decimal point to the right or left the required number of places. This is the big advantage of the metric system.

Property

Converting Metric Measurements to Smaller Units

To convert to a *smaller* unit of measure, we *multiply* by a power of 10, moving the decimal point *to the right.*

Example 5 Converting Metric Length

< Objective 2 >

5.2 m = 520 cm — The *smaller* the unit, the *more* units it takes, so we multiply by 100 to convert from meters to centimeters.

8 km = 8,000 m — Multiply by 1,000.

6.5 m = 6,500 mm — Multiply by 1,000.

2.5 cm = 25 mm — Multiply by 10.

Check Yourself 5

Complete each statement.

(a) 3 km = __________ m **(b)** 4.5 m = __________ cm
(c) 1.2 m = __________ mm **(d)** 6.5 cm = __________ mm

Property

Converting Metric Measurements to Larger Units

To convert to a *larger* unit of measure, we *divide* by a power of 10, moving the decimal point *to the left.*

Example 6 **Converting Metric Length**

NOTE

The *larger* the unit, the *fewer* units it takes, so *divide.*

43 mm = 4.3 cm   Divide by 10.

3,000 m = 3 km   Divide by 1,000.

450 cm = 4.5 m   Divide by 100.

Check Yourself 6

Complete each statement.

(a) 750 cm = __________ m **(b)** 5,000 m = __________ km

(c) 78 mm = __________ cm **(d)** 3,500 mm = __________ m

We have introduced the commonly used units of length in the metric system. There are other prefixes that are used to form linear measures. The prefix *deci* means $\frac{1}{10}$, *deka* means 10, and *hecto* means 100.

Definition

Metric Prefixes

1 *kilo*meter (km) = 1,000 m

1 *hecto*meter (hm) = 100 m

1 *deka*meter (dam) = 10 m

1 meter (m)

1 *deci*meter (dm) = $\frac{1}{10}$ m

1 *centi*meter (cm) = $\frac{1}{100}$ m

1 *milli*meter (mm) = $\frac{1}{1{,}000}$ m

You may find this chart helpful when converting between metric units. Think of the chart as a set of stairs. Note that the largest unit is at the highest step on the stairs.

1 km
1,000 m

1 hm
100 m

1 dam
10 m

1 m

1 dm
0.1 m

1 cm
0.01 m

1 mm
0.001 m

To move from a smaller unit to a larger unit, you move to the left up the stairs, so move the decimal point to the left.

To move from a larger unit to a smaller unit, you move to the right down the stairs, so move the decimal point to the right.

Example 7 Converting Between Metric Lengths

(a) 800 cm = ? m

To convert from centimeters to meters, you can see from the chart that you must move the decimal point *two places to the left.*

800 cm = 8.00 m = 8 m

(b) 500 m = ? km

To convert from meters to kilometers, move the decimal point *three places to the left.*

500 m = .500 km = 0.5 km

(c) 6 m = ? mm

To convert from meters to millimeters, move the decimal point *three places to the right.*

6 m = 6000. mm = 6,000 mm

RECALL

When converting, if the unit gets larger, the amount must get smaller; if the unit gets smaller, the amount must get larger.

Check Yourself 7

Complete each statement.

(a) 300 cm = ________ m **(b)** 370 mm = ________ m
(c) 4,500 m = ________ km

Example 8 A Manufacturing Application

You remove 372-cm piece from a 6-m length of steel. How much is left? Express your answer in meters.

First, we convert the length 372 cm to meters. Since we move the decimal point two places to the left, we see

372 cm = 3.72 m

Now subtract.

$$\begin{array}{r} 6.00 \text{ m} \\ -3.72 \text{ m} \\ \hline 2.28 \text{ m} \end{array}$$

2.28 m of steel remains.

Check Yourself 8

A 2-m piece of wire is cut into five equal pieces. How long is each piece, in centimeters?

Check Yourself ANSWERS

1. **(a)** About 3 m; **(b)** perhaps 5 m; **(c)** you are probably between 1.5 and 2 m tall.
2. **(a)** About 28 cm; **(b)** almost 16 cm; **(c)** about 45 cm
3. **(a)** About 8 mm; **(b)** probably between 25 and 30 mm
4. **(a)** 1 m; **(b)** 20 cm; **(c)** 5 km; **(d)** 2.2 m
5. **(a)** 3,000 m; **(b)** 450 cm; **(c)** 1,200 mm; **(d)** 65 mm
6. **(a)** 7.5 m; **(b)** 5 km; **(c)** 7.8 cm; **(d)** 3.5 m
7. **(a)** 3 m; **(b)** 0.37 m; **(c)** 4.5 km **8.** 40 cm

Reading Your Text

These fill-in-the-blank exercises will help you understand some of the key vocabulary used in this section. The answers to these exercises are in the Answers Appendix at the back of the text.

(a) The ________ system is based on one unit of length, the meter.

(b) A meter is just slightly longer than a ________.

(c) The prefix ________ means one-hundredth.

(d) ________ are used to measure long distances.

7.2 exercises

Skills | Calculator/Computer | Career Applications | Above and Beyond

< Objective 1 >

Choose the most reasonable measure.

1. The height of a ceiling 

(a) 25 m
(b) 2.5 m
(c) 25 cm

2. The diameter of a quarter
(a) 24 mm
(b) 2.4 mm
(c) 24 cm

3. The height of a kitchen counter
(a) 9 m
(b) 9 cm
(c) 90 cm

4. The diagonal measure of a television screen
(a) 50 mm
(b) 50 cm
(c) 5 m

5. The height of a two-story building
(a) 7 m
(b) 70 m
(c) 70 cm

6. An hour's drive on a freeway
(a) 9 km
(b) 90 m
(c) 90 km

7. The width of a roll of cellophane tape
(a) 1.27 mm
(b) 12.7 mm
(c) 12.7 cm

8. The width of a sheet of printer paper
(a) 21.6 cm
(b) 21.6 mm
(c) 2.16 cm

9. The thickness of window glass
(a) 5 mm
(b) 5 cm
(c) 50 mm

10. The height of a refrigerator
(a) 16 m
(b) 16 cm
(c) 160 cm

11. The length of a ballpoint pen
(a) 16 mm
(b) 16 m
(c) 16 cm

12. The width of a calculator key
(a) 1.2 mm
(b) 12 mm
(c) 12 cm

Use a metric unit of length to complete each statement.

13. A playing card is 6 __________ wide.

14. The diameter of a penny is 19 __________.

15. A doorway is 2 __________ high.

16. A table knife is 22 __________ long.

17. A basketball court is 28 __________ long.

18. A commercial jet flies 800 __________ per hour.

19. The width of a nail file is 12 __________.

20. The distance from New York to Washington, D.C., is 387 __________.

21. A recreation room is 6 __________ long.

22. A ruler is 22 __________ wide.

23. A long-distance run is 35 __________.

24. A paperback book is 11 __________ wide.

< Objective 2 >

Complete each statement.

25. 3,000 mm = __________ m

26. 150 cm = __________ m

27. 8 m = __________ cm

28. 77 mm = __________ cm

29. 250 km = __________ cm

30. 500 cm = __________ m

31. 25 cm = __________ mm

32. 150 mm = __________ m

33. 7,000 m = __________ km

34. 9 m = __________ cm

35. 8 cm = __________ mm

36. 45 cm = __________ mm

37. 5 km = __________ m

38. 4,000 m = __________ km

39. 5 m = __________ mm

40. 7 km = __________ m

Convert, as indicated.

41. 250 cm to m

42. 250 cm to km

43. 12 mm to cm

44. 12 mm to km

45. 3.4 m to mm

46. 3.4 mm to m

47. 132 km to m

48. 132 m to km

49. $90\ \frac{\text{km}}{\text{hr}}$ to $\frac{\text{m}}{\text{s}}$

50. $2{,}500\ \frac{\text{m}}{\text{s}}$ to $\frac{\text{km}}{\text{hr}}$

51. $800\ \frac{\text{km}}{\text{day}}$ to $\frac{\text{m}}{\text{hr}}$

52. $5{,}000\ \frac{\text{m}}{\text{hr}}$ to $\frac{\text{km}}{\text{day}}$

Use a metric ruler to measure the necessary dimensions and complete each statement.

53. The perimeter of the parallelogram is __________ cm.

54. The perimeter of the triangle is __________ mm.

55. The perimeter of the rectangle is __________ cm.

56. The area of the rectangle in exercise 55 is __________ cm^2.

57. The perimeter of the square is __________ mm.

58. The area of the square in exercise 57 is __________ mm^2.

Determine whether each statement is **true** *or* **false.**

59. To convert to a larger measure in the metric system, we divide by a power of 10.

60. To convert kilometers to meters, move the decimal point three decimal places to the left.

In each statement, fill in the blank with **always, sometimes,** *or* **never.**

61. When converting units of measure within the metric system, __________ multiply or divide by a power of 10.

62. If we are converting from a smaller unit to a larger unit, we ________ move the decimal point to the right.

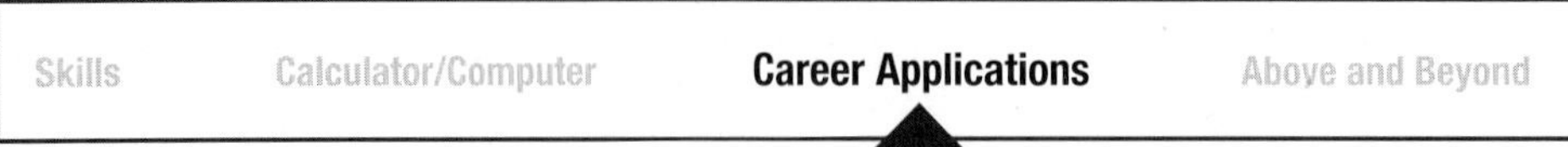

63. **MANUFACTURING TECHNOLOGY** A 2-by-6 plank is 3 m long. If lengths of 86 cm, 9.3 dm, and 29 cm are cut from the plank, how long is the remaining piece? chapter 7 > Make the Connection

64. **MECHANICAL ENGINEERING** A 2.5-m piece of steel stock has lengths of 82 cm, 2.4 dm, and 190 mm cut from it. How long is the remaining portion? chapter 7 > Make the Connection

65. **AGRICULTURE** In a barn, there are 40 stalls over an 81.2-m span. How many centimeters wide is each stall?

66. **ELECTRONICS** A printed circuit board (PCB) measures 45 mm by 67 mm. What is the area of the board in *square centimeters?*

67. **INFORMATION TECHNOLOGY** Radio waves travel at the speed of light, which is 300,000 km per second. What is the rate in meters per second?

68. **INFORMATION TECHNOLOGY** Satellites are located in geostationary orbit at an altitude of approximately 35,786 km above the equator. What is this distance in meters?

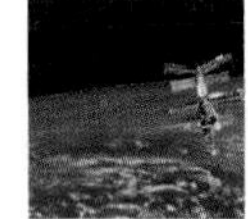

Skills | Calculator/Computer | Career Applications | **Above and Beyond**

69. **(a)** Determine the world record speed for both men and women in meters per second (m/s) for each event: 100-, 400-, 1,500-, and 5,000-m run. The record times can be found at any one of several websites.

(b) Rank all the speeds obtained in order from fastest to slowest.

70. What units in the metric system would you use to measure each quantity?

(a) Distance from Los Angeles to New York

(b) Your waist measurement

(c) Width of a hair

(d) Your height

Answers

1. (b) **3.** (c) **5.** (a) **7.** (b) **9.** (a) **11.** (c) **13.** cm **15.** m **17.** m **19.** mm **21.** m **23.** km **25.** 3 **27.** 800 **29.** 25,000,000 **31.** 250 **33.** 7 **35.** 80 **37.** 5,000 **39.** 5,000 **41.** 2.5 m **43.** 1.2 cm **45.** 3,400 mm **47.** 132,000 m **49.** $25\frac{\text{m}}{\text{s}}$ **51.** $33{,}333\frac{1}{3}\ \frac{\text{m}}{\text{hr}}$ **53.** 4.6 **55.** 7 **57.** 60 **59.** True **61.** always **63.** 92 cm **65.** 203 cm **67.** 300,000,000 m/s **69.** Above and Beyond

Computers and Measurements

Computers give us reasons to take extremely small and large measurements. The amount of time it takes a computer to process a single task might be measured in *milliseconds* (ms), *microseconds* (μs or mcs), or even *nanoseconds* (ns).

1. Write 1 ms as a fraction of a second, in decimal form.
2. Write 1 μs as a fraction of a second, in decimal form.
3. Write 1 ns as a fraction of a second, in decimal form.
4. What do you notice about how these prefixes relate to each other?

NOTE

1 s = 1,000 ms
= 10^3 ms
= 1,000,000 μs
= 10^6 μs
= 10^9 ns

On the other hand, file size might be measured in large units such as *kilobytes* (KB) or *megabytes* (MB), and it is not unusual to measure the amount of memory available in *gigabytes* (GB), *terabytes* (TB), or even *petabytes* (PB).

It gets a little bit trickier with large units. Because computers are based on a binary system, each prefix is not 1,000 times larger than the previous prefix. Rather than using base-10, these units are measured in a base-2 system.

In practice, this means

1 KB = 2^{10} bytes (B)

1 MB = 2^{10} KB

1 GB = 2^{10} MB

1 TB = 2^{10} GB

5. How large is 2^{10}?
6. How many bytes are in 1 MB?
7. How many bytes are in 1 GB?
8. Write the number of bytes in 1 GB as a power of 2.

Historically, a byte was the number of *bits* used to encode a single character of text in a computer. A **bit** (or binary unit) represents the most basic unit of information in computing and telecommunications.

A bit distinguishes between two distinct states, usually represented as 0 and 1, but often meaning true/false, off/on, or some other pair of states.

1 byte (B) = 8 bits (b) = 2^3 b

9. How many bits are in 1 kilobyte?
10. Express the number of bits in 1 kilobyte as a power of 2.

7.3 Metric Units of Weight and Volume

< 7.3 Objectives >

1 > Use appropriate metric units of weight/mass

2 > Convert metric units of weight/mass

3 > Use appropriate metric units of volume

4 > Convert metric units of volume

NOTE

The gram is a unit of mass rather than weight. Weight is a measure of the force of gravity on an object. If you were on the moon, your mass would be the same, but you would weigh less. As long as we are on Earth, we use *mass* and *weight* interchangeably.

The basic unit of weight or mass in the metric system is a very small unit called the **gram.** Think of a paper clip. It weighs roughly 1 gram (g). About 28 g make 1 oz in the U.S. Customary system. Grams are most often used to measure items that are fairly light. For heavier items, a more convenient unit of weight is the **kilogram (kg).** From the prefix *kilo* you should be able to deduce that a kilogram is equal to 1,000 grams.

Definition

Kilogram (kg)

1 kilogram (kg) = 1,000 grams (g) or 1 g = $\frac{1}{1,000}$ kg

Note: 1 kg is a little more than 2 lb.

Example 1 **Using Metric Units**

< Objective 1 >

The weight of a box of breakfast cereal is 320 g.

A woman might weigh 65 kg.

A nickel weighs 5 g.

Check Yourself 1

Choose the most reasonable measure.

(a) A penny: 30 g, 3 g, or 3 kg.
(b) A bar of soap: 120 g, 12 g, or 1.2 kg.
(c) A car: 5,000 kg, 1,000 kg, or 5,000 g.

Milligrams are another common metric unit of mass or weight.

Definition

Milligram (mg)

1 milligram (mg) = $\frac{1}{1,000}$ gram (g) or 1 g = 1,000 mg

Milligrams are very small. In medicine, milligrams are often used to measure drug amounts. For example, an aspirin tablet might be 300 mg.

Just as with units of length, converting metric units of weight or mass is simply a matter of moving the decimal point. Remember to think of the chart as a set of stairs. Recall that the largest unit is at the highest step on the stairs.

NOTES

The prefix *milli* means one-thousandth.

kg, g, and mg are the units in common use.

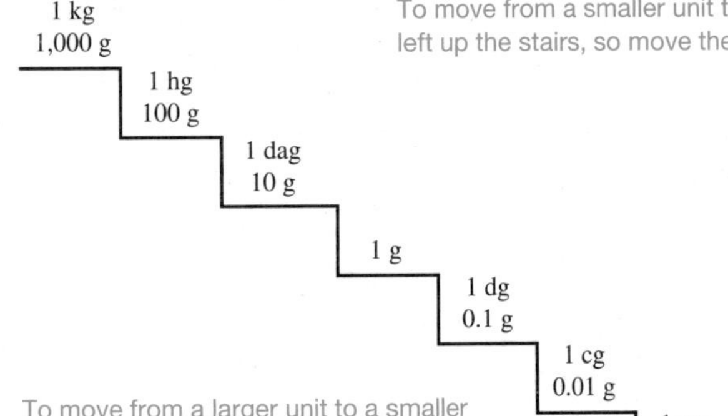

To move from a smaller unit to a larger unit, you move to the left up the stairs, so move the decimal point to the left.

To move from a larger unit to a smaller one, you move to the right down the stairs, so move the decimal point to the right.

Example 2 Converting Metric Weight

< Objective 2 >

NOTES

When converting to a *smaller* unit, the number gets larger.

When converting to a *larger* unit, the number gets smaller.

Complete each statement.

(a) 7 kg = ? g

7 kg = 7000^ g = 7,000 g

Move the decimal point *three places to the right* (to multiply by 1,000).

(b) 5,000 mg = ? g

5,000 mg = 5^000 g = 5 g

Move the decimal point *three places to the left* (to divide by 1,000).

Check Yourself 2

(a) 3,000 g = __________ kg **(b)** 500 cg = __________ g

For massive objects, we use *metric tons* (t).

Definition

Metric Ton

1 metric ton (t) = 1,000 kg so $\frac{1\text{ t}}{1{,}000\text{ kg}} = 1$ and $\frac{1{,}000\text{ kg}}{1\text{ t}} = 1$

We use this measure in Example 3.

Example 3 Converting to Metric Tons

Complete each statement.

(a) 7,500 kg = ? t

$$7{,}500\text{ kg} = \frac{7{,}500\text{ kg}}{1} \times \frac{1\text{ t}}{1{,}000\text{ kg}} = 7.500\text{ t}$$

Move the decimal point three places to the left to divide by 1,000.

(b) 12.25 t = ? kg

$$12.25\text{ t} = \frac{12.25\text{ t}}{1} \times \frac{1{,}000\text{ kg}}{1\text{ t}}$$

$$= 12{,}250\text{ kg}$$

Move the decimal point three places to the right to multiply by 1,000.

NOTE

In both parts of Example 3, we multiply by a unit ratio (which equals 1).

Check Yourself 3

(a) 13,400 kg = __________ t **(b)** 0.76 t = __________ kg

In the metric system, the basic unit of volume is the **liter (L).** A liter is slightly more than a quart and is used for soft drinks, milk, oil, gasoline, and so on.

The metric unit used to measure smaller volumes is the **milliliter (mL).** From the prefix we know that it is one-thousandth of a liter.

Definition

Milliliter (mL)

$$1 \text{ milliliter (mL)} = \frac{1}{1{,}000} \text{ liter (L)} \quad \text{or} \quad 1 \text{ L} = 1{,}000 \text{ mL}$$

NOTE

This unit of volume is also spelled *litre* (the British spelling).

NOTE

The liter is related to the meter. It is defined as the volume of a cube 10 cm on each edge, so

$1 \text{ L} = 1{,}000 \text{ cm}^3$

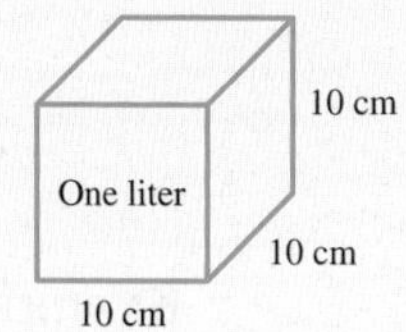

A milliliter contains the volume of a cube 1 cm on each edge. So 1 mL is equal to 1 cm^3. These units can be used interchangeably.

The medical community often uses the cubic centimeter as a unit of volume, though they use the abbreviation *cc* rather than cm^3.

Example 4 will help you get used to the metric units of volume. We explore area and volume more in Chapter 8.

Example 4 **Using Appropriate Units of Metric Volume**

< Objective 3 >

A teaspoon is about 5 mL or 5 cm^3.

A 6-fl-oz cup of coffee is about 180 mL.

A quart of milk is 946 mL (just less than 1 L).

A gallon is just less than 4 L.

Check Yourself 4

Choose the most reasonable measure.

(a) A can of soup: 3 L, 30 mL, or 300 mL.
(b) A pint of cream: 4.73 L, 473 mL, or 47.3 mL.
(c) A home-heating oil tank: 100 L, 1,000 L, or 1,000 mL.
(d) A tablespoon: 150 mL, 1.5 L, or 15 mL.

Converting metric units of volume is again just a matter of moving the decimal point. A chart similar to the ones you saw earlier may be helpful.

NOTE

L, cL, and mL are the most commonly used units. We show the other units simply to indicate that the prefixes and abbreviations are used in a consistent fashion.

1 kL
1,000 L

1 hL
100 L

1 daL
10 L

1 L

1 dL
0.1 L

1 cL
0.01 L

1 mL
0.001 L

To move from a smaller unit to a larger unit, you move to the left up the stairs, so move the decimal point to the left.

To move from a larger unit to a smaller one, you move to the right down the stairs, so move the decimal point to the right.

Example 5 Converting Metric Volume

< Objective 4 >

Complete each statement.

(a) 4 L = ? mL

From the chart, we see that we should move the decimal point three places to the right (to multiply by 1,000).

4 L = 4 000∧ mL = 4,000 mL

NOTES

We are converting to a *smaller* unit.

(b) 3,500 mL = ? L

Move the decimal point three places to the left (to divide by 1,000).

3,500 mL = 3∧500 L = 3.5 L

We are converting to a *larger* unit.

(c) 30 cL = ? mL

Move the decimal point one place to the right (to multiply by 10).

30 cL = 30 0∧ mL = 300 mL

We are converting to a *smaller* unit.

Check Yourself 5

Complete the following statements.

(a) 5 L = _______ mL **(b)** 7,500 mL = _______ L
(c) 550 mL = _______ cL

Consider the next health science application.

Example 6 An Allied Health Application

Cardiac output, measured in liters per minute (L/min), is the product of a patient's stroke volume, in liters per beat, times the heart rate, in beats per minute (beats/min). Determine the cardiac output for a patient with a stroke volume of 45 milliliters per beat (mL/beat) and a heart rate of 80 beats/min.

Since the stroke volume is 45 mL/beat and the patient's heart rate is 80 beats/min, the cardiac output is

$$45\frac{\text{mL}}{\text{beat}} \times 80\frac{\text{beats}}{\text{min}} = 3{,}600\frac{\text{mL}}{\text{min}}$$

Converting 3,600 mL to liters, the cardiac output is 3.6 L/min.

Check Yourself 6

Determine the cardiac output for a patient with a stroke volume of 68 mL/beat and a heart rate of 95 beats/min.

Check Yourself ANSWERS

1. **(a)** 3 g; **(b)** 120 g; **(c)** 1,000 kg **2.** **(a)** 3 kg; **(b)** 5 g
3. **(a)** 13.4 t; **(b)** 760 kg **4.** **(a)** 300 mL; **(b)** 473 mL; **(c)** 1,000 L; **(d)** 15 mL
5. **(a)** 5,000 mL; **(b)** 7.5 L; **(c)** 55 cL **6.** 6.46 L/min

Reading Your Text

These fill-in-the-blank exercises will help you understand some of the key vocabulary used in this section. The answers to these exercises are in the Answers Appendix at the back of the text.

(a) The basic unit of weight or mass in the metric system is called the ________.

(b) The ________ is a small unit of mass that is used, for example, in measuring drug amounts.

(c) In the metric system, the basic unit of volume is the ________.

(d) A ________ contains the volume of a cube 1 cm on each edge.

7.3 exercises

Skills | Calculator/Computer | Career Applications | Above and Beyond

< Objective 1 >

Choose the most reasonable measure of mass/weight.

1. A nickel VIDEO
(a) 5 kg
(b) 5 g
(c) 50 g

2. A portable television set
(a) 8 g
(b) 8 kg
(c) 80 kg

3. A flashlight battery
(a) 8 g
(b) 8 kg
(c) 80 g

4. A 10-year-old boy
(a) 30 kg
(b) 3 kg
(c) 300 g

5. A Toyota Prius
(a) 150 kg
(b) 1,500 kg
(c) 1,500 g

6. A 10-lb bag of flour
(a) 45 kg
(b) 4.5 kg
(c) 45 g

7. A dinner fork
(a) 50 g
(b) 5 g
(c) 5 kg

8. A can of spices
(a) 3 g
(b) 300 g
(c) 30 g

9. A slice of bread
(a) 3 g
(b) 30 g
(c) 3 kg

10. A house paintbrush
(a) 120 g
(b) 12 kg
(c) 12 g

11. A sugar cube
(a) 2 mg
(b) 20 g
(c) 2 g

12. A salt shaker
(a) 10 g
(b) 100 g
(c) 1 g

Use a metric unit of mass to complete each statement.

13. A marshmallow weighs 5 _______.

14. A toaster weighs 2 _______.

15. 1 _______ is $\frac{1}{1{,}000}$ g. 

16. A bag of peanuts weighs 100 _______.

17. An electric razor weighs 250 _______.

18. A soup spoon weighs 50 _______.

19. A heavyweight boxer weighs 98 _______.

20. A vitamin C tablet weighs 500 _______.

21. A legal-sized envelope weighs 30 _______.

22. A clock radio weighs 1.5 _______.

23. A household broom weighs 300 _______.

24. A 60-watt lightbulb weighs 25 _______.

< Objective 2 >

Complete each statement.

25. 8 kg = _______ g 

26. 5,000 mg = _______ g

27. 9,500 kg = _______ t

28. 3 kg = _______ g

29. 1.45 t = _______ kg

30. 12,500 kg = _______ t

31. 3 g = _______ mg

32. 2,000 g = _______ kg

< Objective 3 >

Choose the most reasonable measure of volume.

33. A bottle of wine
(a) 75 mL
(b) 7.5 L
(c) 750 mL

34. A gallon of gasoline
(a) 400 mL
(b) 4 L
(c) 40 L

35. A bottle of perfume
(a) 15 mL
(b) 150 mL
(c) 1.5 L

36. A can of frozen orange juice
(a) 1.5 L
(b) 150 mL
(c) 15 mL

37. A hot-water heater
(a) 200 mL
(b) 50 L
(c) 200 L

38. An oil drum
(a) 220 L
(b) 220 mL
(c) 22 L

39. A bottle of ink
(a) 60 cm^3
(b) 6 cm^3
(c) 600 cm^3

40. A cup of tea
(a) 18 mL
(b) 180 mL
(c) 18 L

41. A jar of mustard
(a) 150 mL
(b) 15 L
(c) 15 mL

42. A bottle of aftershave lotion
(a) 50 mL
(b) 5 L
(c) 5 mL

43. A cream pitcher
(a) 12 mL
(b) 120 mL
(c) 1.2 L

44. One tablespoon
(a) 1.5 mL
(b) 1.5 L
(c) 15 mL

Use a metric unit of volume to complete each statement.

45. A can of tomato soup is 300 _______.

46. 1 _______ is $\frac{1}{100}$ L.

47. A saucepan holds 1.5 _______.

48. A thermos bottle contains 500 _______ of liquid.

49. A coffee pot holds 720 _______.

50. A garbage can holds 120 _______.

51. A car's engine capacity is 2,000 cm^3. It is advertised as a 2.0 _______ model.

52. A bottle of vanilla extract contains 60 _______.

53. 1 _______ is $\frac{1}{10}$ cL.

54. A can of soft drink is 35 _______.

55. A garden sprinkler delivers 8 _______ of water per minute.

56. 1 kL is 1,000 _______.

< Objective 4 >

Complete each statement.

57. 7 L = _______ mL VIDEO

58. 4,000 cm^3 = _______ L

59. 4 hL = _______ L

60. 7 L = _______ cL

61. 8,000 mL = _______ L

62. 12 L = _______ mL

63. 5 L = _______ cm^3

64. 2 L = _______ cL

65. 75 cL = _______ mL

66. 5 kL = _______ L

67. 5 L = _______ cL

68. 400 mL = _______ cL

Solve each application.

69. **Business and Finance** A caterer expects to serve 300-mL portions of soup to 70 people. How many liters of soup does the caterer need to prepare?

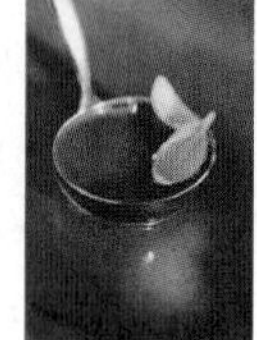

70. **Business and Finance** A produce stand sells local walnuts for $19.96 per kilogram. How much does 750 g cost?

71. **Business and Finance** Dried oregano sells for 8¢ per gram, in bulk, at a local store. How much does 0.15 kg cost?

72. **Construction** If 788 g of mortar are required to mortar a brick into place, how many kilograms do you need to mortar 238 bricks into place (to the nearest whole kilogram)?

73. **Science and Medicine** The United States emitted 67.3 million metric tons (t) of carbon monoxide (CO) into the atmosphere in one year. How many kilograms of CO were emitted to the atmosphere in the United States that year?

74. **Science and Medicine** The United States emitted 19.5 million t of nitrogen oxides (NO) into the atmosphere in one year. How many kilograms of NO were emitted to the atmosphere in the United States that year?

Determine whether each statement is **true** *or* **false.**

75. A milliliter is the same as a cubic centimeter.

76. A kilogram is 1,000 times heavier than a milligram.

77. **Allied Health** Many medications are expressed as percent solutions where the percent indicates how many grams of the active ingredient are dissolved in 100 milliliters (mL) of diluting element (usually water or saline). Consider a 0.6% solution of metaproterenol, an oral inhalation medication used to treat bronchospasm.

    (a) Determine the number of milligrams (mg) of metaproterenol per milliliter of diluting element in a 0.6% solution.

    (b) Determine the volume (in milliliters) of the 0.6% solution to be administered if a dose of 15 mg of metaproterenol is ordered.

78. **Allied Health** Many medications are expressed as percent solutions where the percent indicates how many grams of the active ingredient are dissolved in 100 mL of diluting element (usually water or saline). Consider a 2.5% solution of Demerol, a potent pain medication.

    (a) Determine the number of milligrams of Demerol per milliliter of diluting element in a 2.5% solution.

    (b) Determine the volume (in milliliters) of the 2.5% solution to be administered if a dose of 120 mg of Demerol is ordered.

79. **Manufacturing Technology** There are 312.83 kg of steel in stock. A new part will use 1,600 mg. If 30 of these parts will be produced, how much steel will be left in stock?

80. **Manufacturing Technology** There are 8,370 mL of sealer in stock. A run of parts uses 2.4 L of sealer. How much is left after the run?

81. **Automotive Technology** In a 4-cylinder engine, the displacement of a single cylinder is 575 mL. What is the displacement of the engine?

82. **Automotive Technology** A quart of oil weighs 972 g. What is the weight, in kg, of the oil in a 4-qt oil pan (to the nearest tenth kilogram)?

83. **Manufacturing Technology** A green treated deck board weighs 4.19 kg. As it dries out, it loses 247 g of weight. How much does the board now weigh?

84. **Agriculture** A pallet containing 65 bags of fertilizer weighs 1.482 t. How many kilograms does each bag weigh?

Skills Calculator/Computer Career Applications **Above and Beyond**

85. Mass (weight) and volume are connected in the metric system. The weight of water in a cube 1 cm on a side is 1 g. Does such a relationship exist in the U.S. Customary system of measurement? If so, what is it? If not, why not?

86. (a) Determine how many liters of gasoline your car will hold.

    (b) Using current prices, determine what a liter of gasoline should cost to make it competitive.

    (c) How much would it cost to fill your car?

87. Does each dose seem reasonable or unreasonable?

    (a) Rinse with 5 L of mouthwash every morning.

    (b) Soak your feet in 5 L of epsom salt bath every evening.

    (c) Inject $\frac{3}{4}$ L of insulin every day.

Answers

1. (b) **3.** (c) **5.** (b) **7.** (a) **9.** (b) **11.** (c) **13.** g **15.** mg **17.** g **19.** kg **21.** g **23.** g **25.** 8,000 **27.** 9.5 **29.** 1,450 **31.** 3,000 **33.** (c) **35.** (a) **37.** (c) **39.** (a) **41.** (a) **43.** (b) **45.** mL (or cm^3) **47.** L **49.** mL **51.** L **53.** mL **55.** L **57.** 7,000 **59.** 400 **61.** 8 **63.** 5,000 **65.** 750 **67.** 500 **69.** 21 L **71.** $12 **73.** 67.3 billion kg **75.** True **77.** (a) 6 mg/mL; (b) 2.5 mL **79.** 312.782 kg **81.** 2.3 L **83.** 3.943 kg **85.** Above and Beyond **87.** (a) unreasonable; (b) reasonable; (c) unreasonable

7.4 Converting Between Measurement Systems

< 7.4 Objectives >

1 > Convert units of length

2 > Convert units of mass/weight

3 > Convert units of volume

4 > Convert between Fahrenheit and Celsius temperature scales

Often, we need to convert between the U.S. Customary and metric systems of measurement. We usually do this with the help of conversion tables and a calculator.

The conversion table gives conversion factors between some U.S. Customary and metric units of length. Each factor is rounded to the nearest hundredth (two decimal places). Even though most of these factors are approximate, we use equal signs for convenience.

Property: Converting Length

U.S. to Metric	Metric to U.S.
1 in. = 2.54 cm	1 cm = 0.39 in.
1 ft = 0.30 m	1 m = 39.37 in.
1 yd = 0.91 m	1 m = 1.09 yd
1 mi = 1.61 km	1 km = 0.62 mi

When reading this table, remember that a mile is larger than a kilometer, an inch is larger than a centimeter, and that a yard is smaller than a meter.

In Section 7.1, we presented two methods for converting within the U.S. Customary system—the substitution method and the unit-ratio method. Both methods work well when converting between the U.S. Customary and metric systems.

We usually use the substitution method for straightforward conversions and the unit-ratio method for more complex conversions.

Example 1 Converting Between Measurement Systems—Substitution

< Objective 1 >

Convert, as indicated.

(a) 5 in. to cm

1 in. is 2.54 cm, so we can substitute as we did in Section 7.1.

5 in. = 5 × (1 in.) — We write 5 in. as 5 × (1 in.).
= 5 × (2.54 cm) — We substitute 2.54 cm for 1 in.
= 12.7 cm — Finally, we perform the arithmetic.

NOTE

To use the table, choose the conversion equation that has 1 of your given unit. In (a), we use 1 in. = 2.54 cm.

(b) 12 mi to km

According to the table, 1 mi is 1.61 km.

12 mi = 12 × (1 mi)
= 12 × (1.61 km) — We substitute 1.61 km for 1 mi.
= 19.32 km

(c) 3 m to in.

We begin with a metric measure, so we use the *Metric to U.S.* column in the Property table, with 1 m = 39.37 in.

$$\begin{aligned} 3 \text{ m} &= 3 \times (1 \text{ m}) && \text{We write 3 m as } 3 \times (1 \text{ m}). \\ &= 3 \times (39.37 \text{ in.}) && \text{We substitute 39.37 in. for 1 m.} \\ &= 118.11 \text{ in.} && \text{Finally, we perform the arithmetic.} \end{aligned}$$

(d) 25 km to mi

$$\begin{aligned} 25 \text{ km} &= 25 \times (1 \text{ km}) \\ &= 25 \times (0.62 \text{ mi}) && \text{We substitute 0.62 mi for 1 km.} \\ &= 15.5 \text{ mi} \end{aligned}$$

RECALL

Most conversions between the U.S. Customary and metric systems of measurement are approximations.

Check Yourself 1

Convert, as indicated.

(a) 8 in. to cm **(b)** 14 ft to m **(c)** 90 km to mi **(d)** 5 cm to in.

To prevent substitution tables from growing too large, we omit some conversions. For instance, the Property table does not have a direct meters-to-feet conversion. In this case, we could convert from meters to yards and then from yards to feet or we could use unit ratios to do the conversion directly.

We choose to do the conversions directly with unit ratios in Example 2.

Example 2 **Converting Between Measurement Systems—Unit Ratios**

RECALL

The unit you want to convert to should be in the numerator. The unit you want to "cancel" needs to be in the denominator.

Convert 6 m to ft.

The *U.S. to Metric* column in the Property table lists the conversion 1 ft = 0.30 m. We use this to form a unit ratio with feet in the numerator and meters in the denominator.

$$\begin{aligned} 6 \text{ m} &= \frac{6 \cancel{\text{m}}}{1} \times \frac{1 \text{ ft}}{0.30 \cancel{\text{m}}} && 1 = \frac{1 \text{ ft}}{0.30 \text{ m}} \text{ is a unit ratio.} \\ &= \frac{6}{0.30} \text{ ft} && \text{After simplifying the units, we divide.} \\ &= 20 \text{ ft} && 6 \div 0.30 = 20 \end{aligned}$$

Check Yourself 2

Convert 42 in. to m (round your result to the nearest hundredth meter).

Even with unit ratios, we may need to pass through one measurement to get to another. Unit ratios are still the better method for these conversions, as shown in Example 3.

Example 3 **Converting Between Measurement Systems—Unit Ratios**

NOTE

When doing multiple conversions, it is a good idea to outline a plan before beginning.

Convert 2 km to in.

We can reach inches from kilometers in several ways. We might convert from kilometers to miles, miles to feet, and then feet to inches. Or, we can convert from kilometers to meters, and then meters to inches. We choose this second method in this example.

$$\begin{aligned} 2 \text{ km} &= \frac{2 \cancel{\text{km}}}{1} \times \frac{1{,}000 \cancel{\text{m}}}{1 \cancel{\text{km}}} \times \frac{39.37 \text{ in.}}{1 \cancel{\text{m}}} && 1 \text{ m} = 39.37 \text{ in., so } \frac{39.37 \text{ in.}}{1 \text{ m}} \text{ is a unit ratio.} \\ &= 2 \times 1{,}000 \times 39.37 \text{ in.} && \text{After simplifications, we are left with inches.} \\ &= 78{,}740 \text{ in.} \end{aligned}$$

Check Yourself 3

Convert 1 km to ft.

When we need to convert rates between systems, we rely on the unit-ratio method, as we show in Example 4.

Example 4 Converting Rates Between Systems

> **RECALL**
>
> 1 mi = 1.61 km
> 1 km = 1,000 m
> 1 hr = 3,600 s

Convert 55 $\frac{\text{mi}}{\text{hr}}$ to $\frac{\text{m}}{\text{s}}$. Round the result to two decimal places.

We write 55 $\frac{\text{mi}}{\text{hr}}$ as $\frac{55\text{ mi}}{\text{hr}}$ and use unit ratios to convert miles to meters (miles to kilometers, kilometers to meters) and hours to seconds.

$$\frac{55\text{ }\cancel{\text{mi}}}{1\text{ }\cancel{\text{hr}}} \times \frac{1.61\text{ }\cancel{\text{km}}}{1\text{ }\cancel{\text{mi}}} \times \frac{1{,}000\text{ m}}{1\text{ }\cancel{\text{km}}} \times \frac{1\text{ }\cancel{\text{hr}}}{3{,}600\text{ s}} = \frac{55 \times 1.61 \times 1{,}000}{3{,}600}\ \frac{\text{m}}{\text{s}}$$

$$= 24.60\ \frac{\text{m}}{\text{s}}$$

Check Yourself 4

Convert 85 $\frac{\text{ft}}{\text{s}}$ to $\frac{\text{km}}{\text{hr}}$. Round your result to one decimal place.

> **RECALL**
>
> Weight is the effect of gravity on an object's mass. We generally use the terms interchangeably when on Earth.

We use the same methods to convert between U.S. and metric units of weight/mass or volume.

Property: Converting Weight/Mass

U.S. to Metric	Metric to U.S.
1 oz = 28.35 g	1 g = 0.04 oz
1 lb = 0.45 kg	1 kg = 2.20 lb

As before, we can convert directly with substitution or use unit ratios for more complex conversions.

Example 5 Converting Mass/Weight

< Objective 2 >

(a) A large chicken weighs 3 kg. What is its weight in pounds?

We have the conversion 1 kg = 2.20 lb.

3 kg = 3 × (1 kg) — Write 3 kg as 3 · (1 kg).
= 3 × (2.2 lb) — 1 kg = 2.20 lb
= 6.6 lb — 3 · 2.2 = 6.6

(b) A package weighs 5 oz. Find its weight in grams.

Use 1 oz = 28.35 g.

5 oz = 5 × (1 oz)
= 5 × (28.35 g)
= 141.75 g

RECALL

1 lb = 16 oz

(c) How many ounces are in 3.5 kg?

We use unit ratios to convert from kilograms to pounds to ounces.

$$\begin{aligned} 3.5 \text{ kg} &= \frac{3.5 \cancel{\text{kg}}}{1} \times \frac{2.2 \cancel{\text{lb}}}{1 \cancel{\text{kg}}} \times \frac{16 \text{ oz}}{1 \cancel{\text{lb}}} \\ &= 3.5 \times 2.2 \times 16 \text{ oz} \\ &= 123.2 \text{ oz} \end{aligned}$$

Check Yourself 5

(a) A radio weighs 8 lb. Find its weight in kilograms.
(b) A box of cereal lists its contents at 375 g. Convert this to ounces.
(c) A medium-sized package contains 10.5 oz of candy. How many kilograms does a case of 24 packages contain? Round your result to the nearest hundredth kilogram.

NOTE

We convert square and cubic units in Chapter 8.

Property: Converting Volume

U.S. to Metric	Metric to U.S.
1 qt = 0.95 L	1 L = 1.06 qt
1 fl oz = 29.57 mL	1 mL = 0.03 fl oz

Example 6 Converting Units of Volume

< Objective 3 >

(a) How many liters does a 2-qt bottle of milk contain?

We use the conversion 1 qt = 0.95 L, and substitute.

$$\begin{aligned} 2 \text{ qt} &= 2 \times (1 \text{ qt}) && \text{Write 2 qt as } 2 \cdot (1 \text{ qt}). \\ &= 2 \times (0.95 \text{ L}) && 1 \text{ qt} = 0.95 \text{ L} \\ &= 1.9 \text{ L} && 2 \cdot 0.95 = 1.9 \end{aligned}$$

(b) How many liters of fuel does a 16-gal gas tank hold?

We do not have a direct conversion for gallons to liters, so we use unit ratios to convert gallons to quarts and quarts to liters.

$$\begin{aligned} 16 \text{ gal} &= \frac{16 \cancel{\text{gal}}}{1} \times \frac{4 \cancel{\text{qt}}}{1 \cancel{\text{gal}}} \times \frac{0.95 \text{ L}}{1 \cancel{\text{qt}}} \\ &= 16 \times 4 \times 0.95 \text{ L} \\ &= 60.8 \text{ L} \end{aligned}$$

Check Yourself 6

(a) How many liters are in a 40-gal hot-water tank?
(b) How many quarts does a 3-L bottle contain?

As you would expect, there are more important applications requiring us to convert between the systems.

Example 7 An Application Involving Conversions

An automobile has a 16-gal fuel tank and gets 32 mi/gal. How many kilometers can the car travel on a single tank of gas?

There are a couple of ways to approach this problem. For instance, we could convert all measurements to their metric equivalent, and then compute the number of kilometers that the car could travel.

Alternatively, we can find the total number of miles the car can go, and then convert that to km. In this example, we choose the second method as it requires only a single conversion.

To find the distance, we multiply 16 gal and 32 mi/gal.

$$16\,\cancel{\text{gal}} \times 32\,\frac{\text{mi}}{\cancel{\text{gal}}} = 512\text{ mi}$$

The car can drive 512 miles on a full tank of gas.

$$512\text{ mi} = 512(1.61\text{ km}) = 824.32\text{ km}$$

The car can drive 824.32 km on a single tank of gas.

Check Yourself 7

The load limit of a trailer is listed at 2,500 lb. It is to be loaded with pallets of concrete blocks. Each block weighs 3 kg and there are 50 blocks to the pallet; each pallet weighs 10 kg. What is the maximum number of pallets that can be loaded onto the trailer?

NOTE

Temperature conversions produce exact-value results.

Before moving on to temperature conversions, we should discuss rounding errors that arise when converting between the measurement systems. Most of the conversion factors presented so far are approximations, rounded to two decimal places. As such, you may get different answers to some problems depending on whether you divide or multiply, as we see in Example 8.

Example 8 Converting Measurements and Rounding Errors

Convert 15 in. to cm.

We have two conversion factors between inches and centimeters, which produce two sets of unit ratios.

$$1\text{ in.} = 2.54\text{ cm gives } \frac{1\text{ in.}}{2.54\text{ cm}} = 1 \quad\text{and}\quad \frac{2.54\text{ cm}}{1\text{ in.}} = 1.$$

$$1\text{ cm} = 0.39\text{ in. gives } \frac{1\text{ cm}}{0.39\text{ in.}} = 1 \quad\text{and}\quad \frac{0.39\text{ in.}}{1\text{ cm}} = 1.$$

If we use unit ratios from the first conversion, we have

$$\begin{aligned}15\text{ in.} &= 15\text{ in.} \times \frac{2.54\text{ cm}}{1\text{ in.}} = 15\,\cancel{\text{in.}} \times \frac{2.54\text{ cm}}{1\,\cancel{\text{in.}}}\\ &= 15 \times 2.54\text{ cm}\\ &= 38.1\text{ cm}\end{aligned}$$

Using the second conversion and rounding to two decimal places gives

$$\begin{aligned}15\text{ in.} &= 15\text{ in.} \times \frac{1\text{ cm}}{0.39\text{ in.}} = 15\,\cancel{\text{in.}} \times \frac{1\text{ cm}}{0.39\,\cancel{\text{in.}}}\\ &= \frac{15}{0.39}\text{ cm}\\ &= 38.46\text{ cm}\end{aligned}$$

These answers are not the same!

If we carry more decimal places in the conversion, this discrepancy vanishes.

$$1\text{ cm} = 0.3937\text{ in. gives } \frac{1\text{ cm}}{0.3937\text{ in.}} = 1.$$

$$\begin{aligned}15\text{ in.} &= 15\text{ in.} \times \frac{1\text{ cm}}{0.3937\text{ in.}} = 15\,\cancel{\text{in.}} \times \frac{1\text{ cm}}{0.3937\,\cancel{\text{in.}}}\\ &= \frac{15}{0.3937}\text{ cm}\\ &= 38.10\text{ cm}\end{aligned}$$

We see that with more precision in our conversion factors, we get more precise answers.

Check Yourself 8

(a) Use 1 oz = 28.35 g to write a unit ratio and convert 20 oz to grams.
(b) Use 1 g = 0.04 oz to write a unit ratio and convert 20 oz to grams.
(c) Use 1 g = 0.0353 oz to write a unit ratio and convert 20 oz to grams.
(d) What can you conclude, based on your answers to the previous exercises?

In this text, we provide more than one answer in the answers sections when multiple approaches are equivalently straightforward and rounding errors can lead to more than one result. This should assist you and your instructor when determining if you arrived at an answer through legitimate methods.

While we may not be able to list all possible answers for more complex conversions, if you show your work, you should be able to see if your answers are the result of correct conversions, even if you use different factors than the authors.

Temperature is expressed in **degrees Celsius** in the metric system, whereas **degrees Fahrenheit** is the unit of temperature used in the U.S. Customary system. The boiling point of water (at sea level) is 100 degrees Celsius, written 100°C, while it is 212 degrees Fahrenheit, written 212°F. The freezing point of water (at sea level) is 0°C, which corresponds to 32°F. The temperature on a warm day might be 30°C, corresponding to 86°F.

We use formulas to convert between units of temperature. In the case of temperature conversions, we get exact values rather than approximations.

Property

Converting Temperature Units

To convert from degrees Celsius (°C) to degrees Fahrenheit (°F), multiply by 9, divide by 5, and then add 32. A formula that describes this is

$$F = \frac{9C}{5} + 32$$

To convert from degrees Fahrenheit (°F) to degrees Celsius (°C), subtract 32, multiply by 5, and then divide by 9. A formula that describes this is

$$C = \frac{5(F - 32)}{9}$$

Example 9 Converting Between Fahrenheit and Celsius Temperatures

< Objective 4 >

(a) Convert 18°C to Fahrenheit.
Using the first formula given,

$$F = \frac{9(18)}{5} + 32 = \frac{162}{5} + 32 = 32.4 + 32 = 64.4$$

So 18°C corresponds to 64.4°F.

(b) Convert 77°F to Celsius.
Using the second rule,

$$C = \frac{5(77 - 32)}{9} = \frac{5(45)}{9} = \frac{225}{9} = 25$$

77°F corresponds to 25°C.

Check Yourself 9

Complete each statement.

(a) 12°C = _______ °F **(b)** 83°F = _______ °C

Example 10 A Temperature Conversion Application

The melting point in a cast-iron block is approximately 2,300°F. What is this temperature in degrees Celsius?

Using the second formula given,

$$C = \frac{5(2{,}300 - 32)}{9} = \frac{5(2{,}268)}{9} = \frac{11{,}340}{9} = 1{,}260$$

The corresponding temperature is 1,260°C.

Check Yourself 10

The thermostat in a car engine opens at a temperature of 165°F. Convert this temperature to degrees Celsius. Round your result to the nearest tenth degree Celsius.

We close this section with a brief table of the conversions between units of length, mass/weight, volume, and temperature.

Length	
U.S. to Metric	**Metric to U.S.**
1 in. = 2.54 cm	1 cm = 0.39 in.
1 ft = 0.30 m	1 m = 39.37 in.
1 yd = 0.91 m	1 m = 1.09 yd
1 mi = 1.61 km	1 km = 0.62 mi
Weight/Mass	
U.S. to Metric	**Metric to U.S.**
1 oz = 28.35 g	1 g = 0.04 oz
1 lb = 0.45 kg	1 kg = 2.20 lb
Volume	
U.S. to Metric	**Metric to U.S.**
1 qt = 0.95 L	1 L = 1.06 qt
1 fl oz = 29.57 mL	1 mL = 0.03 fl oz
Temperature	
U.S. to Metric	**Metric to U.S.**
$C = \frac{5}{9}(F - 32)$	$F = \frac{9}{5}C + 32$

Check Yourself ANSWERS

1. **(a)** 20.32 cm; **(b)** 4.2 m; **(c)** 55.8 mi; **(d)** 1.95 in. **2.** 1.07 m **3.** 3,273.6 ft **4.** 91.8 $\frac{\text{km}}{\text{hr}}$ **5.** **(a)** 3.6 kg; **(b)** 15 oz; **(c)** 7.09 kg or 7.14 kg (rounding) **6.** **(a)** 152 L; **(b)** 3.18 qt **7.** 7 pallets **8.** **(a)** 567 g; **(b)** 500 g; **(c)** 566.57 g; **(d)** Answers will vary. **9.** **(a)** 53.6°F; **(b)** $28\frac{1}{3}$°C **10.** 73.9°C

Reading Your Text 

These fill-in-the-blank exercises will help you understand some of the key vocabulary used in this section. The answers to these exercises are in the Answers Appendix at the back of the text.

(a) For straightforward conversions, we usually use the ________ method.

(b) When we need to convert rates between systems, we rely on the ________ method.

(c) Weight is the effect of ________ on an object's mass.

(d) In the metric system, temperature is expressed in degrees ________.

Skills | Calculator/Computer | Career Applications | Above and Beyond

7.4 exercises

< Objectives 1–4 >

Complete each statement. Round to the nearest hundredth, when appropriate.

1. 250 km = ________ mi  

2. 9 cm = ________ in.

3. 150 mi = ________ km

4. 9 yd = ________ m

5. 2.6 m = ________ in. 

6. 72 in. = ________ cm

7. 3 ft = ________ mm

8. 16 km = ________ ft

9. 19.4 mm = ________ yd

10. 45 ft = ________ km

11. $75\ \frac{\text{km}}{\text{hr}} = \_\_\_\_\_\_\_\_\ \frac{\text{ft}}{\text{s}}$

12. $50\ \frac{\text{mi}}{\text{hr}} = \_\_\_\_\_\_\_\_\ \frac{\text{m}}{\text{s}}$ 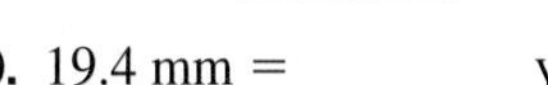

13. 6 lb = ________ kg

14. 8 oz = ________ g

15. 0.25 kg = ________ oz

16. 5 lb = ________ g

17. 12 kg = ________ lb

18. 450 g = ________ oz

19. 8,000 lb = ________ t (metric tons)

20. 4,500 kg = ________ U.S. tons

21. $14.5\ \frac{\text{oz}}{\text{week}} = \_\_\_\_\_\_\_\_\ \frac{\text{g}}{\text{day}}$

22. $2.3\ \frac{\text{kg}}{\text{day}} = \_\_\_\_\_\_\_\_\ \frac{\text{oz}}{\text{hr}}$

23. 4 qt = ________ L

24. 7 L = ________ qt

25. 8 fl oz = ________ mL

26. 15.9 gal = ________ L

27. 760 mL = ________ qt

28. 15 L = ________ gal

29. 72 mL = ________ fl oz

30. 450 fl oz = ________ L

31. $12\ \frac{\text{gal}}{\text{hr}} = \_\_\_\_\_\_\_\_\ \frac{\text{L}}{\text{min}}$

32. $3.8\ \frac{\text{L}}{\text{min}} = \_\_\_\_\_\_\_\_\ \frac{\text{fl oz}}{\text{s}}$

33. 52°F = ________ °C

34. 6°C = ________ °F

35. 24°C = ________ °F

36. 95°F = ________ °C

37. 86°F = ________ °C

38. 10°C = ________ °F

39. 20°C = ________ °F

40. 72°F = ________ °C

41. 100°F = ________ °C

42. 27°C = ________ °F

43. 37°C = ________ °F

44. 98.6°F = ________ °C

Solve each application.

45. **Science and Medicine** A football team's fullback weighs 250 lb. How many kilograms does he weigh?

46. **Science and Medicine** Samantha's speedometer reads in kilometers per hour. If the legal speed limit is 55 mi/hr, how fast can she drive?

Science and Medicine A Boeing 747 can travel 8,336 mi on one 57,285-gal tank of airplane fuel.

47. How many liters of fuel can a Boeing 747 tank hold?

48. How many kilometers can the Boeing 747 travel on one fuel tank (to the nearest kilometer)?

49. Express the fuel efficiency of the Boeing 747 in kilometers per liter (round to three decimal places).

50. How many liters of fuel does the Boeing 747 use per kilometer traveled (round to one decimal place)?

Science and Medicine A Boeing 777 can travel 11,029 km on one 171,835-L tank of airplane fuel.

51. How many gallons of fuel can a Boeing 777 tank hold (to the nearest gallon)?

52. How many miles can the Boeing 777 travel on one tank of fuel (to the nearest mile)?

53. Express the fuel efficiency of the Boeing 777 in miles per gallon (round to three decimal places).

54. How many gallons of fuel does the Boeing 777 use per mile traveled (round to one decimal place)?

Determine whether each statement is **true** *or* **false.**

55. A mile is shorter than a kilometer.

56. An inch is larger than a centimeter.

57. A quart is smaller then a liter.

58. A pound is more than half of a kilogram.

Skills | Calculator/Computer | **Career Applications** | Above and Beyond

59. Manufacturing Technology The label for a primer states that it should not be applied to surfaces at less than 10°C. What is this temperature in degrees Fahrenheit?

60. Manufacturing Technology A construction adhesive should not be exposed to temperatures above 140°F. What is this temperature in degrees Celsius?

61. Agriculture For a corn seed to germinate, a soil temperature of at least 12.5°C is required. What is this temperature in degrees Fahrenheit?

62. Manufacturing Technology A copper-nickel alloy of 40% nickel becomes liquid at 1,280°C and becomes solid at 1,240°C. Convert these temperatures into degrees Fahrenheit.

63. Allied Health An infant measures 51 cm long at birth. Determine the baby's length at birth in inches. Round to the nearest inch. chapter 7 > Make the Connection

64. Allied Health An adult female patient is 5 ft 4 in. tall. Determine her height in centimeters. Round to the nearest centimeter. chapter 7 > Make the Connection

65. Manufacturing Technology A machine is listed with a gross weight of 1,200 kg. The load limit of the trailer is listed at 2,500 lb. Can the machine be hauled on the trailer?

66. Manufacturing Technology A piece of steel rod stock is 5 m long. For a prototype, 6 ft 2 in. is used. How much is left (in meters)? Round to the nearest hundredth meter. chapter 7 > Make the Connection

67. Automotive Technology A diesel engine oil pan calls for 2 gal of oil. How many liters of oil would this be?

68. Manufacturing Technology Which is larger: a metric ton of aluminum or a standard U.S. ton of aluminum?

69. In exercise 49, you expressed the fuel efficiency of the Boeing 747 in terms of kilometers per liter. In exercise 50, you expressed its rate in liters per kilometer. Use complete sentences to describe how these two rates differ.

70. In exercise 53, you expressed the fuel efficiency of the Boeing 777 in miles per gallon. In exercise 54, you expressed its rate in gallons per mile. Use complete sentences to describe how these two rates differ.

71. Complete the puzzle.

Across

1. 6,000 mL
6. 1,760 yd
7. paradise
8. LV + LV
9. extraterrestrial
10. out of whack
13. tome
14. seven thousandths g

Down

1. 600 cm
2. top
3. _____ de France
4. _____ dm in 1 m
5. sixty thousand g
10. presidential nickname
11. didn't lose
12. read-only memory

MATHWORK PUZZLE

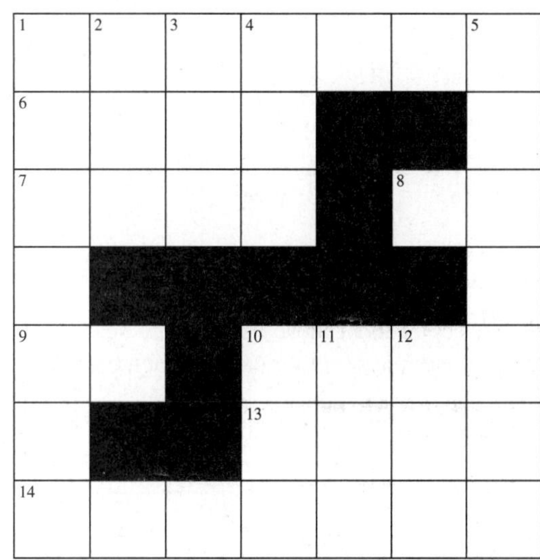

Answers

1. 155 **3.** 241.5 **5.** 102.36 **7.** 900 **9.** 0.02 **11.** 68.2 **13.** 2.7 **15.** 8.8 **17.** 26.4 **19.** 3.6 **21.** 58.73 **23.** 3.8 **25.** 236.56 **27.** 0.81 **29.** 2.16 **31.** 0.76 **33.** 11.11 **35.** 75.2 **37.** 30 **39.** 68 **41.** 37.78 **43.** 98.6 **45.** 112.5 kg **47.** 217,683 L **49.** 0.062 $\frac{\text{km}}{\text{L}}$ **51.** 45,536 gal **53.** 0.150 $\frac{\text{mi}}{\text{gal}}$ **55.** False **57.** True **59.** 50°F **61.** 54.5°F **63.** 20 in. **65.** No **67.** 7.6 L **69.** Above and Beyond **71.**

Activity 21 ::

Tool Sizes

Perhaps you work in construction, remodeling, or automotive repair. Or maybe you simply enjoy working on your own car or doing projects around the house. You have probably found the need to use both U.S. Customary and metric tool sizes.

These drill bits come in a typical set. Convert each bit size to millimeters (mm), rounding to the nearest tenth of a millimeter.

$\frac{1}{16}$ in.	$\frac{5}{64}$ in.	$\frac{3}{32}$ in.	$\frac{7}{64}$ in.	$\frac{1}{8}$ in.
$\frac{9}{64}$ in.	$\frac{5}{32}$ in.	$\frac{3}{16}$ in.	$\frac{7}{32}$ in.	$\frac{1}{4}$ in.

A set of wrenches with metric unit sizes consists of those listed in the next chart. Convert each size to a corresponding U.S. Customary unit wrench. In each case, find a U.S. Customary size rounded to the nearest $\frac{1}{32}$ of an inch.

8 mm	10 mm	12 mm	13 mm	14 mm	17 mm

Locate at least three other tool sizes in your home or apartment and make appropriate conversions, either U.S. Customary to metric or metric to U.S. Customary.

Definition/Procedure	Example	Reference
The U.S. Customary System of Measurement		Section 7.1
U.S. Customary Units of Measure *Length* 1 foot (ft) = 12 inches (in.) 1 yard (yd) = 3 ft 1 mile (mi) = 5,280 ft *Weight/Mass* 1 pound (lb) = 16 ounces (oz) 1 ton = 2,000 lb *Volume* 1 cup (c) = 8 fluid ounces (fl oz) 1 pint (pt) = 2 c 1 quart (qt) = 2 pt 1 gallon (gal) = 4 qt *Time* 1 minute (min) = 60 seconds (s) 1 hour (hr) = 60 min 1 day = 24 hr 1 week = 7 days		p. 375
To Convert Units in the U.S. System *Substitution Method*		
Step 1 Rewrite the given measure, separating the number and the unit. **Step 2** Replace the unit of measure with an equivalent measure. **Step 3** Perform the arithmetic and simplify, if necessary.	8 yd = 8 × (1 yd) = 8 × (3 ft) = 24 ft	p. 375
Unit Ratios A fraction whose value is 1. Unit ratios can be used to convert units.	$\frac{12 \text{ in.}}{1 \text{ ft}}$ and $\frac{60 \text{ min}}{1 \text{ hr}}$ are unit ratios.	p. 377
To Convert Units in the U.S. System *Unit-Ratio Method*		
Step 1 Construct a unit ratio so that the original units "cancel" and you are left with either the desired units or an intermediary measure. **Step 2** Multiply your original measure by the unit ratio from step 1 and simplify. **Step 3** If you used an intermediary measure in step 1, repeat the process until you have a measurement with the desired units.	$12 \text{ pt} = \frac{12 \cancel{\text{pt}}}{1} \times \frac{1 \text{ qt}}{2 \cancel{\text{pt}}}$ $= \frac{12}{2} \text{ qt}$ $= 6 \text{ qt}$	p. 377
Adding or Subtracting Like Denominate Numbers		
Step 1 Arrange the numbers so that the like units are in the same column. **Step 2** Add or subtract in each column. **Step 3** Simplify if necessary.	To add 4 ft 7 in. and 5 ft 10 in.: 4 ft 7 in. + 5 ft 10 in. 9 ft 17 in. = 10 ft 5 in.	p. 380
Multiplying or Dividing Denominate Numbers by Abstract Numbers		
Step 1 Multiply or divide each part of the denominate number by the abstract number. **Step 2** Simplify if necessary.	2 × (3 yd 2 ft) = 6 yd 4 ft = 7 yd 1 ft	p. 381

Continued

Definition/Procedure	Example	Reference
Length and the Metric System		Section 7.2
Metric units of length are the meter (m), centimeter (cm), millimeter (mm), and kilometer (km).		p. 389
Basic Metric Prefixes		
*milli** means $\frac{1}{1{,}000}$ *kilo** means 1,000 *centi** means $\frac{1}{100}$ *hecto* means 100 *deci* means $\frac{1}{10}$ *deka* means 10	*These are the most commonly used and should be memorized.	p. 392
Converting Metric Units		
You can use the following chart. 1 km 1,000 m; 1 hm 100 m; 1 dam 10 m; 1 m; 1 dm 0.1 m; 1 cm 0.01 m; 1 mm 0.001 m To move from a smaller unit to a larger unit, you move to the left up the stairs, so move the decimal point to the left. To move from a larger unit to a smaller unit, you move to the right down the stairs, so move the decimal point to the right. To convert between metric units, just move the decimal point the same number of places to the left or right as indicated by the chart.	500 cm = ? m To convert from centimeters to meters, move the decimal point two places to the *left*. 500 cm = 5∧00. cm = 5 m	p. 391
Metric Units of Weight and Volume		Section 7.3
Conversions between units of volume (liters) or units of weight (grams) work in exactly the same fashion as those between units of length. 1 kg 1,000 g; 1 hg 100 g; 1 dag 10 g; 1 g; 1 dg 0.1 g; 1 cg 0.01 g; 1 mg 0.001 g To move from a smaller unit to a larger unit, you move to the left up the stairs, so move the decimal point to the left. To move from a larger unit to a smaller one, you move to the right down the stairs, so move the decimal point to the right.	3 L = ? mL To convert from liters to milliliters, move the decimal point three places to the *right*. 3 L = 3.000∧L = 3,000 mL	p. 398

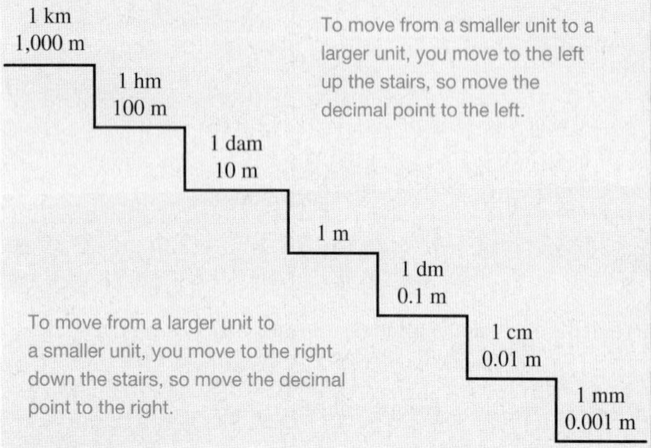

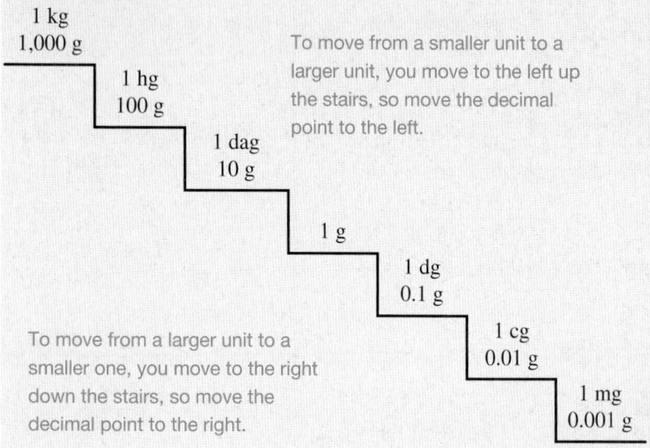

Continued

Definition/Procedure	Example	Reference
1 kL 1,000 L; 1 hL 100 L; 1 daL 10 L; 1 L; 1 dL 0.1 L; 1 cL 0.01 L; 1 mL 0.001 L. To move from a smaller unit to a larger unit, you move to the left up the stairs, so move the decimal point to the left. To move from a larger unit to a smaller one, you move to the right down the stairs, so move the decimal point to the right.		
Converting Between Measurement Systems		Section 7.4
Length		
U.S. to Metric 1 in. = 2.54 cm 1 ft = 0.30 m 1 yd = 0.91 m 1 mi = 1.61 km ***Metric to U.S.*** 1 cm = 0.39 in. 1 m = 39.37 in. 1 m = 1.09 yd 1 km = 0.62 mi	90 km = 90 × (1 km) = 90 × (0.62 mi) = 55.8 mi	*p.* 406
Weight/Mass ***U.S. to Metric*** 1 oz = 28.35 g 1 lb = 0.45 kg ***Metric to U.S.*** 1 g = 0.04 oz 1 kg = 2.20 lb *Volume* ***U.S. to Metric*** 1 qt = 0.95 L 1 fl oz = 29.57 mL ***Metric to U.S.*** 1 L = 1.06 qt 1 mL = 0.03 fl oz	$50\frac{\text{gal}}{\text{hr}} = \frac{50\ \cancel{\text{gal}}}{1\text{ hr}} \times \frac{4\ \cancel{\text{qt}}}{1\ \cancel{\text{gal}}} \times \frac{0.95\text{ L}}{1\ \cancel{\text{qt}}}$ $= 50 \times 4 \times 0.95\frac{\text{L}}{\text{hr}}$ $= 190\frac{\text{L}}{\text{hr}}$	
Temperature Conversions		
From degrees Celsius (°C) to degrees Fahrenheit (°F): $F = \frac{9C}{5} + 32$ From degrees Fahrenheit (°F) to degrees Celsius (°C): $C = \frac{5(F - 32)}{9}$	To convert 34°C, $F = \frac{9(34)}{5} + 32 = 93.2$ 34°C = 93.2°F To convert 75°F, $C = \frac{5(75 - 32)}{9} \approx 23.9$ 75°F = 23.9°C	*p.* 411

summary exercises :: chapter 7

This summary exercise set will help ensure that you have mastered each of the objectives of this chapter. The exercises are grouped by section. You should reread the material associated with any exercises that you find difficult. The answers to the odd-numbered exercises are in the Answers Appendix at the back of the text.

7.1 *Complete each statement.*

1. 11 ft = _______ in.

2. 72 hr = _______ days

3. 6 gal = _______ qt

4. 80 fl oz = _______ pt

5. 4 lb = _______ oz

6. 5 mi = _______ ft

7. 8,000 lb = _______ tons

8. 16 pt = _______ qt

Simplify.

9. 3 ft 23 in.

10. 4 lb 20 oz

Add.

11.
$$\begin{array}{r} 3 \text{ lb } \ 9 \text{ oz} \\ +\ 5 \text{ lb } 10 \text{ oz} \\ \hline \end{array}$$

12.
$$\begin{array}{r} 5 \text{ hr } 20 \text{ min} \\ 3 \text{ hr } 40 \text{ min} \\ +\ 2 \text{ hr } 20 \text{ min} \\ \hline \end{array}$$

Subtract.

13.
$$\begin{array}{r} 7 \text{ ft } 11 \text{ in.} \\ -\ 2 \text{ ft } \ 4 \text{ in.} \\ \hline \end{array}$$

14.
$$\begin{array}{r} 3 \text{ hr } 30 \text{ min} \\ -\ 1 \text{ hr } 50 \text{ min} \\ \hline \end{array}$$

Multiply.

15. 3 × (1 hr 25 min)

16. 8 × (5 ft 2 in.)

Divide.

17. $\frac{10 \text{ lb } 12 \text{ oz}}{2}$

18. $\frac{5 \text{ gal } 2 \text{ qt}}{3}$

19. **Business and Finance** John worked 6 hr 15 min, 8 hr, 5 hr 50 min, 7 hr 30 min, and 6 hr during 1 week. What were the total hours worked?

20. **Construction** A room requires two pieces of floor molding 12 ft 8 in. long, one piece 6 ft 5 in. long, and one piece 10 ft long. Will 42 ft of molding be enough for the job?

7.2 *Choose the most reasonable measure.*

21. A marathon race

(a) 40 km
(b) 400 km
(c) 400 m

22. The distance around your wrist

(a) 15 mm
(b) 15 cm
(c) 1.5 m

23. The diameter of a penny

(a) 19 cm
(b) 1.9 mm
(c) 19 mm

24. The width of a portable television screen

(a) 28 mm
(b) 28 cm
(c) 2.8 m

Use metric units of length to complete each statement.

25. A small Post-it is 39 _______ wide.

26. The distance from San Francisco to Los Angeles is 618 _______.

27. A 1-lb coffee can has a diameter of 10 _______.

28. A doorway is 90 _______ wide.

Complete each statement.

29. 2 km = _______ m

30. 3 cm = _______ mm

31. 3,000 mm = _______ m

32. 8 m = _______ mm

33. 6 cm = _______ m

34. 8 m = _______ km

7.3 *Choose the most reasonable measure of weight/mass.*

35. A quarter

(a) 6 g
(b) 6 kg
(c) 60 g

36. A tube of toothpaste

(a) 20 kg
(b) 200 g
(c) 20 g

37. A refrigerator

(a) 120 kg
(b) 1,200 kg
(c) 12 kg

38. A paperback book

(a) 1.2 kg
(b) 120 g
(c) 12 g

Use metric units of mass to complete each statement.

39. A loaf of bread weighs 500 _______.

40. A compact car weighs 900 _______.

41. A television set weighs 25 _______.

42. A fully loaded pick-up truck weighs 3 _______.

Complete each statement.

43. 5 kg = _______ g

44. 5 t = _______ kg

45. 2,000 g = _______ kg

46. 2,000 mg = _______ g

Choose the most reasonable measure of volume.

47. The gas tank of your car

(a) 500 mL
(b) 5 L
(c) 50 L

48. A bottle of eye drops

(a) 18 cm^3
(b) 180 cm^3
(c) 1.8 L

49. A can of soft drink

(a) 3.5 L
(b) 350 mL
(c) 35 mL

50. A punch bowl

(a) 200 L
(b) 20 L
(c) 200 mL

Use metric units of volume to complete each statement.

51. The crankcase of an automobile takes 5.5 _______ of oil.

52. The correct dosage for a cough medicine is 40 _______.

53. A bottle of iodine holds 20 _______.

54. A large mixing bowl holds 6 _______.

Complete each statement.

55. 5 L = _______ mL

56. 9 L = _______ cm^3

57. 6,000 cm^3 = _______ L

58. 10 mL = _______ L

7.4 *Complete each statement. Round to the nearest hundredth when appropriate.*

59. 8.3 m = _______ in.

60. 42 in. = _______ cm

61. 15 lb = _______ kg

62. 27.5 kg = _______ lb

63. 5.2 L = _______ qt

64. 18 gal = _______ L

65. 65 cm = _______ in.

66. 5 ft = _______ mm

67. 19 yd = _______ cm

68. 12 m = _____ in.

69. 13 lb = _____ g

70. 375 g = _______ lb

71. 750 mL = _____ qt

72. 15 gal = _____ L

73. 17°C = _______ °F

74. 41°F = _______ °C

75. 98.6°F = _______ °C

76. 6°C = _______ °F

77. 59°F = _______ °C

78. 30°C = _______ °F

79. 5°C = _______ °F

80. 35°F = _______ °C

chapter test 7

CHAPTER 7

Use this chapter test to assess your progress and to review for your next exam. Allow yourself about an hour to take this test. The answers to these exercises are in the Answers Appendix at the back of the text.

Complete each statement. Round to the nearest tenth.

1. 8 ft = _______ in.

2. 3 pt = _______ fl oz

3. 5 m = _______ mm

4. 3 kg = _______ g

5. 300 cL = _______ L

6. 40 in. = _______ cm

7. 5.2 km = _______ mi

8. 14.5 gal = _______ L

9. 150 lb = _______ kg

10. 12 in. = _______ cm

11. 58°F = _______ °C

12. 24°C = _______ °F

Simplify.

13. 5 ft 21 in.

14. 2 days 47 hr 72 min

Evaluate as indicated.

15. $\begin{array}{r} 7 \text{ ft } 9 \text{ in.} \\ + \ 3 \text{ ft } 8 \text{ in.} \\ \hline \end{array}$

16. $\begin{array}{r} 7 \text{ lb } \ 3 \text{ oz} \\ - \ 4 \text{ lb } 10 \text{ oz} \\ \hline \end{array}$

17. $4 \times (3 \text{ hr } 50 \text{ min})$

18. $\dfrac{12 \text{ lb } 18 \text{ oz}}{3}$

Choose the most reasonable measure.

19. The width of your hand
 (a) 50 cm
 (b) 10 cm
 (c) 1 m

20. The speed limit on a freeway
 (a) 10 km/hr
 (b) 100 km/hr
 (c) 100 m/hr

21. A football player
 (a) 12 kg
 (b) 120 kg
 (c) 120 g

22. A small can of tomato juice
 (a) 4 L
 (b) 400 mL
 (c) 40 mL

23. **CONSTRUCTION** The Martins are fencing in a rectangular yard that is 110 ft long by 40 ft wide. If the fencing costs \$3.50 per linear foot, what is the total cost of the fencing?

24. **AUTOMOTIVE TECHNOLOGY** A quart of oil weighs 972 g. What is the weight, in pounds, of the oil in a 4-qt oil pan (to the nearest tenth pound)?

25. **BUSINESS AND FINANCE** A gallon of orange juice is on sale for \$4.79. Find the cost per milliliter (to the nearest hundredth cent).

cumulative review chapters 1–7

Use this exercise set to review concepts from earlier chapters. While it is not a comprehensive exam, it will help you identify any material that you need to review before moving on to the next chapter. In addition to the answers, you will find section references for these exercises in the Answers Appendix in the back of the text.

1.5

1. **CONSTRUCTION** A classroom is 7 yd wide by 8 yd long. If the room is to be recarpeted with material costing \$16 per square yard, find the cost of the carpeting.

1.6

2. **BUSINESS AND FINANCE** Michael bought a washer-dryer combination that, with interest, cost \$959. He paid \$215 down and agreed to pay the balance in 12 monthly payments. Find the amount of each payment.

1.7

3. Evaluate. $8 + 16 \div 4 \times 2$

2.2

4. Write the prime factorization for 168.

5. Find the greatest common factor of 12 and 20.

2.4

6. Arrange in order from smallest to largest. $\frac{5}{8}, \frac{3}{5}, \frac{2}{3}$

2.5

7. Multiply. $\frac{2}{3} \times 1\frac{4}{5} \times \frac{5}{8}$

2.7

8. Divide. $4\frac{1}{6} \div 10$

3.2

9. Find the least common multiple of 6, 15, and 20.

3.3

10. Add. $\frac{3}{5} + \frac{1}{6} + \frac{4}{15}$

3.4

11. Subtract. $7\frac{3}{8} - 3\frac{5}{6}$

4.2

12. Business and Finance You pay for purchases of \$14.95, \$18.50, \$11.25, and \$7 with a \$70 gift certificate. How much cash will you have left?

4.3

13. Geometry Find the area of a rectangle that is 6.4 cm long and 4.35 cm wide.

4.2

14. Geometry Find the perimeter of a rectangle that is 6.4 cm long and 4.35 cm wide.

4.5

15. Find the decimal equivalent of $\frac{9}{16}$.

16. Write the decimal form of $\frac{7}{13}$. Round to the nearest thousandth.

5.4

17. Solve the proportion $\frac{15}{x} = \frac{10}{16}$

18. Geometry If the scale on a map is $\frac{1}{4}$ in. equals 20 mi, how far apart are two towns that are 5 in. apart on the map?

19. Science and Medicine Felipe traveled 342 mi using 19 gal of gas. At this rate, how far can he travel on 25 gal?

6.2

20. Write as a percent. 0.375

6.1

21. Write as a simplified fraction. 12.5%

6.3

22. What is 43% of 8,200?

23. 315 is what percent of 140?

24. 120% of what number is 180?

6.4

25. Business and Finance A home that was purchased for \$125,000 increased in value by 14% over a 3-year period. What was its value at the end of that period?

7.1

26. Complete the statement 5 days = ________ hours

27. Subtract:

$$\begin{array}{r} 4 \text{ min } 10 \text{ s} \\ -\ 2 \text{ min } 35 \text{ s} \\ \hline \end{array}$$

28. Find the sum of 8 lb 14 oz and 12 lb 13 oz.

29. Find the difference between 7 ft 2 in. and 4 ft 5 in.

30. Multiply and simplify: $8 \times (2 \text{ hr } 40 \text{ min})$

Complete each statement.

7.2

31. 43 cm = _______ m

7.3

32. 62 kg = _______ g

7.2

33. 740 mm = _______ cm

7.3

34. 14 L = _______ mL

35. 500 mL = _______ L

36. 375 g = _______ kg

7.4

Complete each statement. Round to the nearest tenth.

37. 8.3 mi = _______ km

38. 68 kg = _______ lb

Convert each temperature. Round to the nearest tenth.

39. 85°F = _______ °C

40. 9°C = _______ °F

Geometry

CHAPTER 8 OUTLINE

INTRODUCTION

Norman architecture has been popular since the Norman conquests in Europe. In the eleventh century, constructions such as churches, castles, and universities began including Norman windows in their design. Later designs, such as Gothic architecture, grew out of the Norman designs.

A Norman window is one example of a composite geometric figure. These are figures that incorporate several simpler geometric figures. A racetrack, formed by attaching a semicircle to two ends of a rectangle, is another example of a composite geometric figure.

Composite geometric figures are all around us. We work with their features and properties when we carpet rooms, figure out the amount of fertilizer needed for the grass inside a racetrack, determine the amount of glass needed for a custom window, or calculate the amount of fluid a bottle can hold.

The major properties of geometric figures, such as perimeter, area, and volume, are critical to many careers. Architects, craftspeople, and contractors frequently work with two- and three-dimensional figures.

You will have the opportunity to work with composite geometric figures like Norman windows when you complete Activity 24.

8 prerequisite check

This Prerequisite Check highlights the skills you will need in order to be successful in this chapter. The answers to these exercises are in the Answers Appendix at the back of the text.

Evaluate, as indicated.

1. 5^2

2. 13^2

3. $3.14 \times (5)^2$

4. $\frac{1}{2}(8)(6 + 3)$

5. $2(5.6) + 2(3.1)$

6. $\frac{1}{2}(9)(5)$

7. Evaluate 3.14×2.5^2 (round your result to the nearest hundredth).

8. Convert 18 cm to inches (round your result to the nearest tenth of an inch).

9. Convert 14 ft 8 in. to meters (round your result to the nearest tenth of a meter).

10. Find the perimeter of the figure shown.

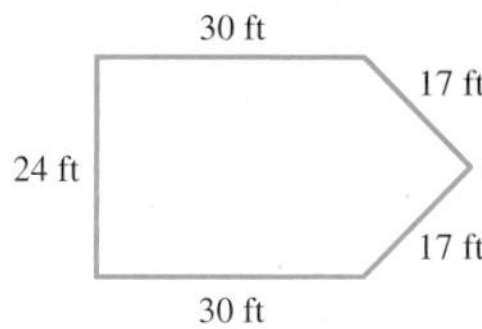

Find the area of each figure.

11.

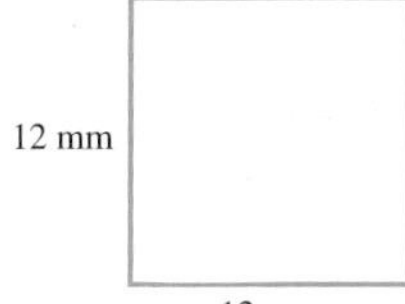

12.

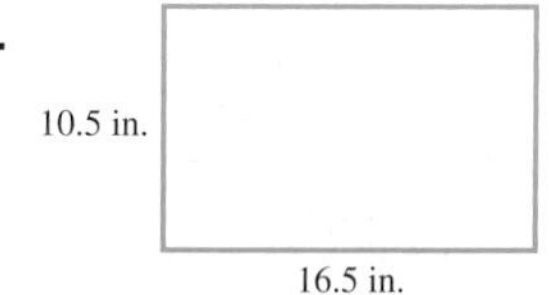

8.1 Lines and Angles

< 8.1 Objectives >

1 > Recognize, name, and relate lines and line segments

2 > Identify parallel and perpendicular lines

3 > Name and classify angles

4 > Use a protractor to measure an angle

5 > Use geometry to measure an angle

NOTE

Geo means earth, just as it does in the words *geography* and *geology.*

Once ancient cultures mastered counting, they became interested in measuring land. This is the foundation of geometry. Literally translated, *geometry* means earth measurement. Many of the topics we consider in geometry (topics such as angles, perimeter, and area) were first studied as part of surveying.

As is usually the case, we start the study of a new topic by learning some vocabulary. Most of the terms we discuss are familiar to you. It is important that you understand what we mean when we use these words in the context of geometry.

We begin with the word *point.* A point is a location; it has no size and covers no area.

When infinitely many points are attached to each other in two directions we have a *curve*. If the curve is "straight," and continues indefinitely in opposite directions, we have a *line*. We use arrowheads to indicate that a line continues "forever."

A piece of a line with endpoints on both sides is called a *line segment.* To help you see the difference, consider Example 1.

Example 1 — Recognizing Lines and Line Segments

< Objective 1 >

NOTE

The capital letters are labels for points.

Label each figure as a line or a line segment.

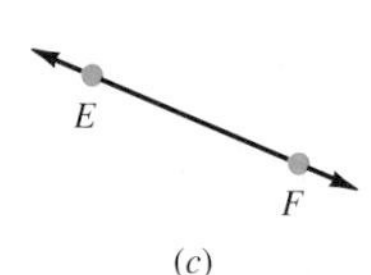

Both (*a*) and (*c*) continue forever in both directions. They are lines. Part (*b*) has two endpoints. It is a line segment.

Check Yourself 1

Label each figure as a line or a line segment.

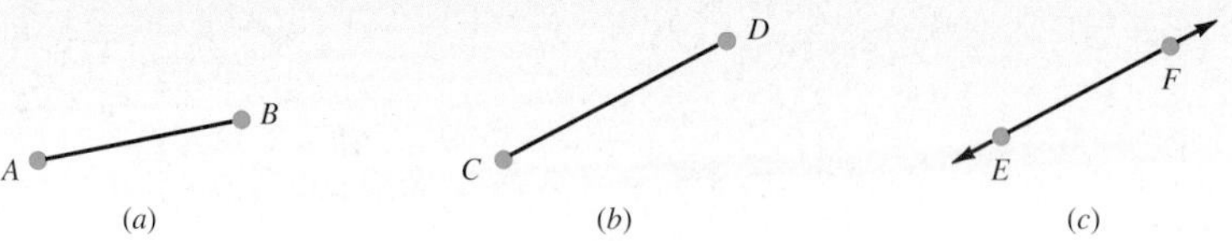

We need to be able to name points, lines, line segments, and other geometric figures so that we can distinguish one from another.

We can name a point anything we like. In this text, we use capital letters to name points. Lines and line segments are defined by two points, so we use two named points on lines and line segments to label them. Often, we use some symbol over the point labels as shown in Example 2.

Example 2 Naming Lines and Line Segments

Name each geometric figure.

(a) We name points with capital letters. Here, we have the point A.

(b) In this case, we have a line segment with two named points. We combine the point labels with an overbar to indicate that we are naming a line segment.

We name this line segment $\overline{BC}$.

(c) The arrows indicate that the curve continues indefinitely, so we have a line rather than a line segment. We add arrows to the overbar to indicate a line.

We name this line $\overleftrightarrow{ED}$. Note that the line $\overleftrightarrow{ED}$ contains the line segment $\overline{ED}$.

Check Yourself 2

Name all points, lines, and line segments in the figure.

The next geometric figure we look at is an *angle*.

Definition

Angle An **angle** is a geometric figure consisting of two line segments that share a common endpoint. The common endpoint is called the **vertex.**

$\overline{OA}$ and $\overline{OB}$ are line segments. O is the vertex of the angle.

Surveyors use an instrument called a *transit.* A transit allows surveyors to measure angles so that they can determine where property lines are.

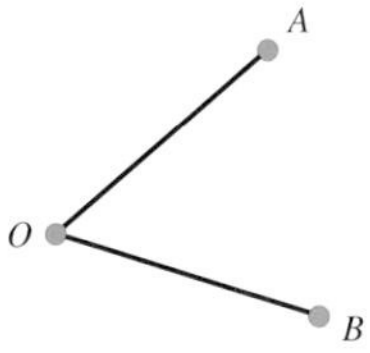

Definition

Perpendicular Lines When two lines cross (or intersect), they form four angles. If the lines intersect such that four equal angles are formed, we say that the two lines are **perpendicular.**

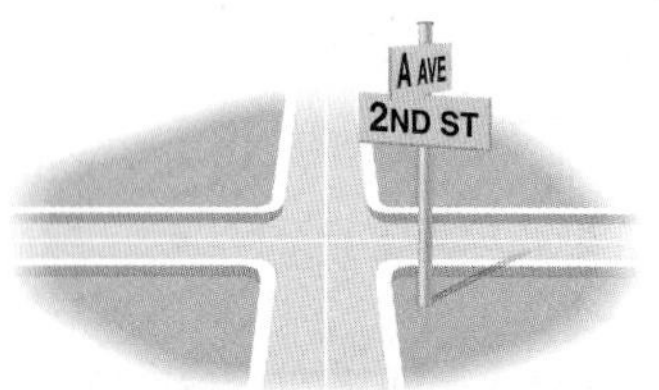

At most intersections, the two roads are perpendicular.

Definition

Parallel Lines

If two lines are drawn so that they never intersect (even if we extend the lines forever), we say that the two lines are **parallel.**

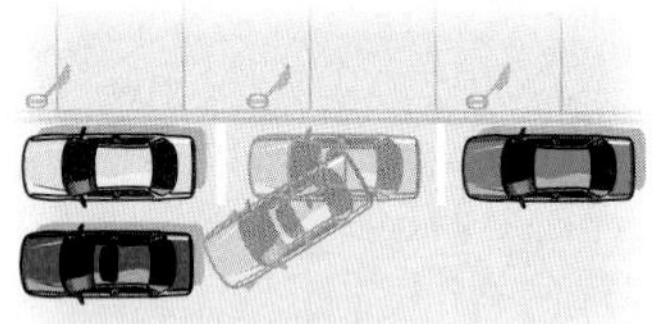

Parallel parking gets its name from the fact that the parking spot is parallel to the traffic lane.

Example 3 Recognizing Parallel and Perpendicular Lines

< Objective 2 >

Label each pair of lines as parallel, perpendicular, or neither.

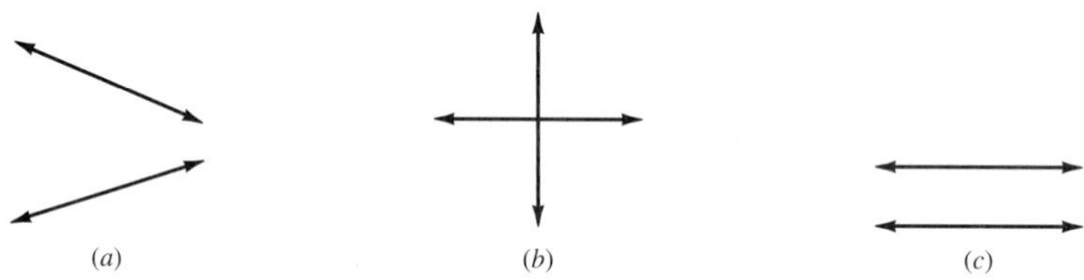

Although part (*a*) does not show the lines intersecting, if they were extended as the arrowheads indicate, they would. The lines of part (*b*) are perpendicular because the four angles formed are equal. The lines in part (*c*) are parallel.

Check Yourself 3

Label each pair of lines as parallel, perpendicular, or neither.

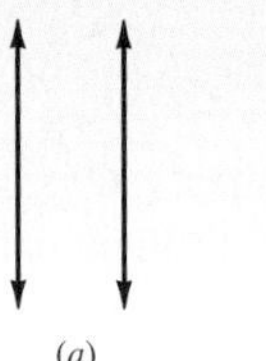
(*a*)

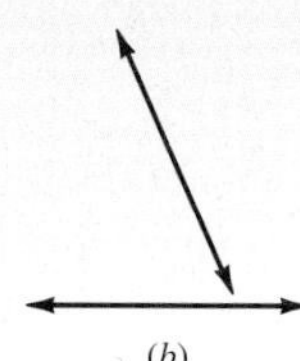
(*b*)

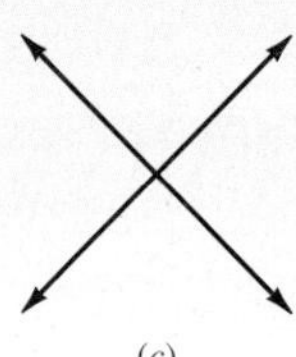
(*c*)

We call the angle formed by two perpendicular lines or line segments a *right angle.* We designate a right angle by forming a small square.

We can refer to a specific angle by naming three points. The middle point is the vertex of the angle.

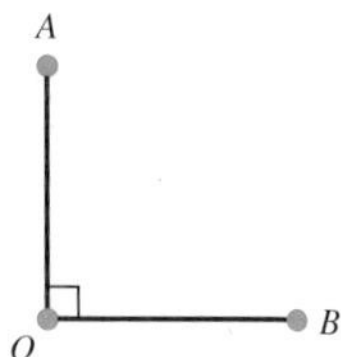

Example 4 Naming an Angle

< Objective 3 >

Name the indicated angle.

The vertex of the angle is O, and the angle begins at C and ends at B, so we would name the angle $\angle COB$.

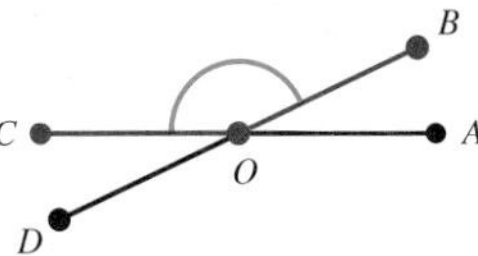

NOTE

We could also call this angle $\angle BOC$.

Check Yourself 4

Name the indicated angle.

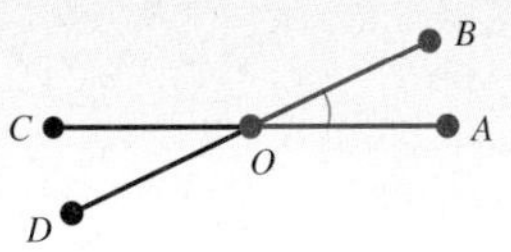

When there is no possibility of confusion, we may refer to an angle by its vertex or by using a symbol that appears in the angle. In the figure, the angle shown may be named $\angle JKL$ or $\angle LKJ$ (as noted earlier), or by simply writing $\angle K$ or $\angle x$.

One way to measure an angle is to use a unit called a *degree*. There are 360 degrees (we write this as 360°) in a complete circle. You can see from the picture on the left that there are four right angles in a circle. If we divide 360° by 4, we find that each right angle measures 90°. Here are some other angles with their measurements.

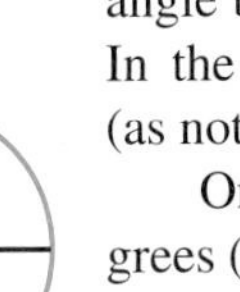

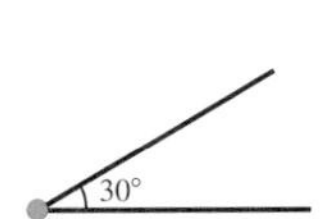

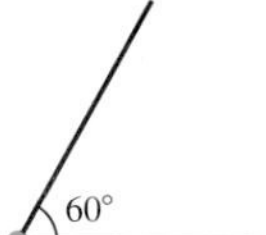

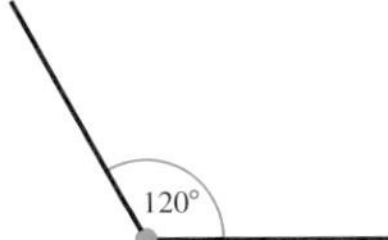

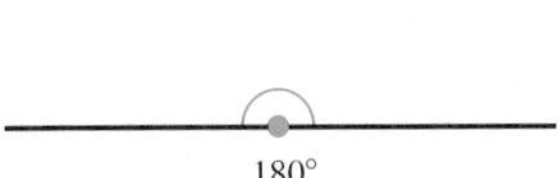

An *acute angle* measures between 0° and 90°. An *obtuse angle* measures between 90° and 180°. A *straight angle* measures 180°.

Example 5 Labeling Types of Angles

Label each angle as acute, obtuse, right, or straight.

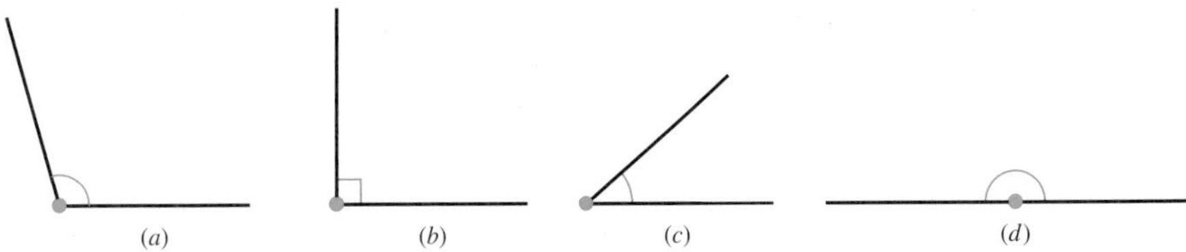

Angle (a) is obtuse (the angle is more than 90°). Angle (b) is a right angle (designated by the small square). Angle (c) is an acute angle (it is less than 90°), and angle (d) is a straight angle.

Check Yourself 5

Label each angle as an acute, obtuse, right, or straight angle.

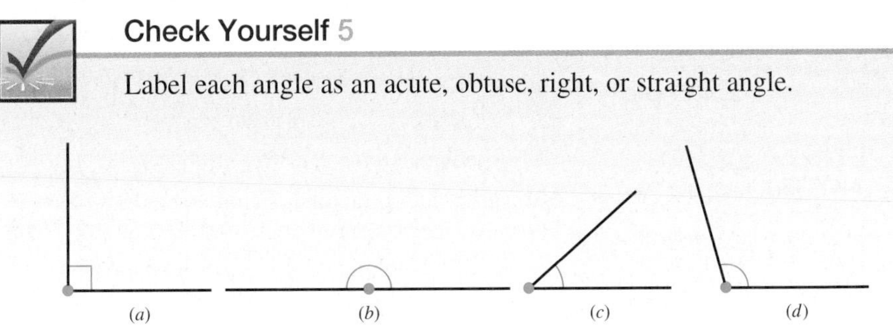

NOTE

Your protractor may show the degree measures in both directions.

We can use a tool called a *protractor* to measure an angle.

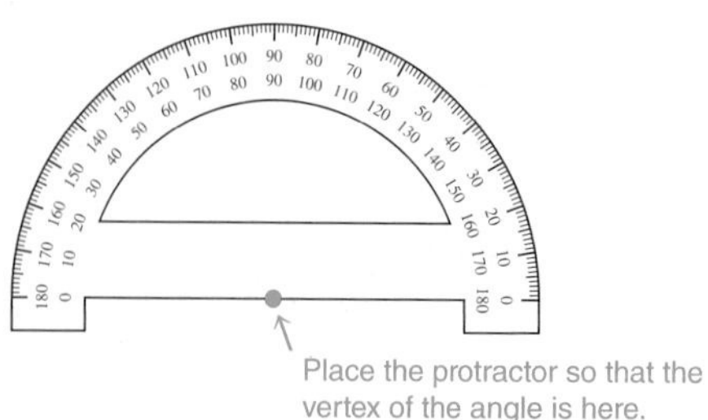

We read the protractor by placing one line segment of the angle at 0°. We then read the number that the other line segment passes through. This number represents the degree measurement of the angle. The point at the center of the protractor, the endpoint of the two line segments, is the vertex of the angle.

Example 6 Measuring an Angle

< Objective 4 >

Use a protractor to measure each angle.

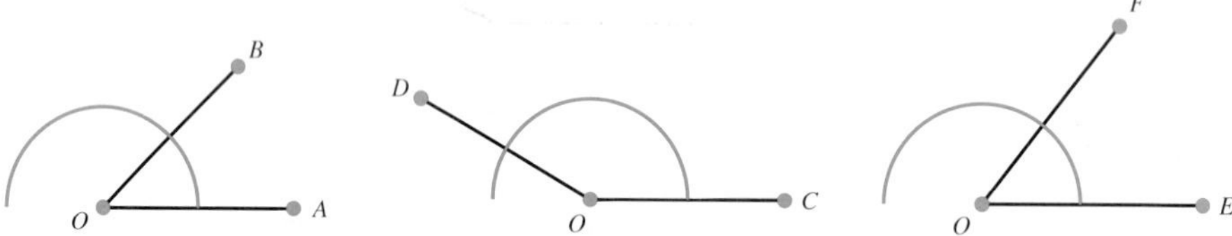

$\angle AOB$ measures 45° and $\angle COD$ is 150°. The measure of $\angle EOF$ is between 50° and 55°. We estimate it to be 52°.

Check Yourself 6

Use a protractor to measure each angle.

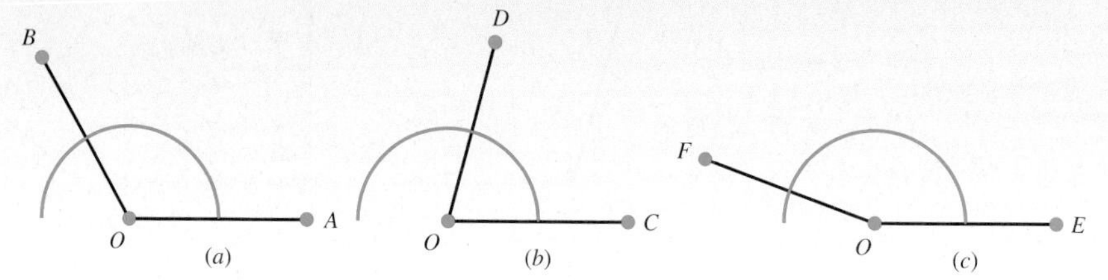

If we wish to refer to the degree measure of $\angle ABC$, we write $m\angle ABC$.

Example 7 Measuring an Angle

NOTE

$m\angle AOB = 20°$ is read "the measure of angle *AOB* is 20 degrees."

Find $m\angle AOB$.

Using a protractor, we find $m\angle AOB = 20°$.

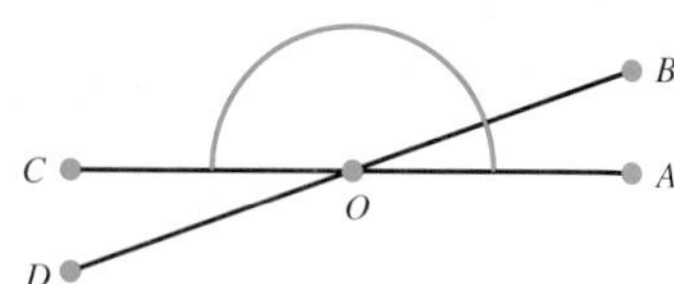

Check Yourself 7

Find $m\angle AOC$.

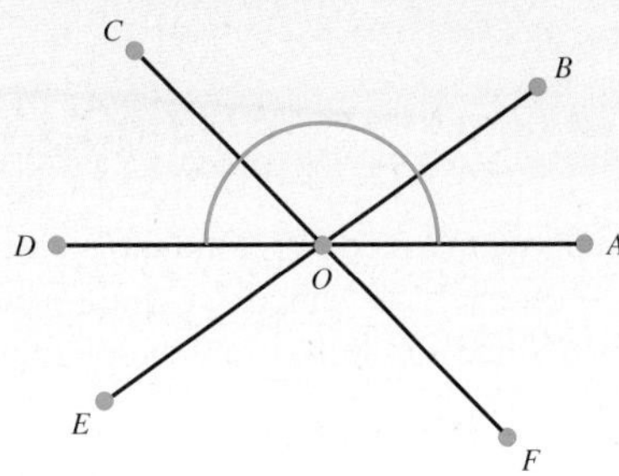

If the sum of the measures of two angles is 90°, the two angles are said to be **complementary.** In the figure at the right, $\angle x$ and $\angle y$ are complementary angles.

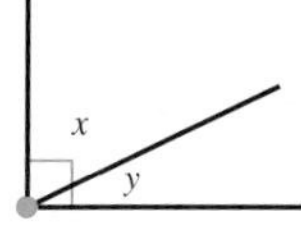

If the sum of the measures of two angles is 180°, the two angles are said to be **supplementary.** In the figure below, $\angle a$ and $\angle b$ are supplementary angles.

Example 8 Finding the Measure of an Angle

< Objective 5 >

In each case, find the measure of angle x.

(a)

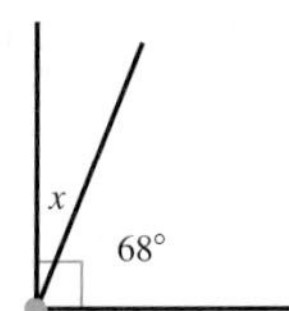

(b)

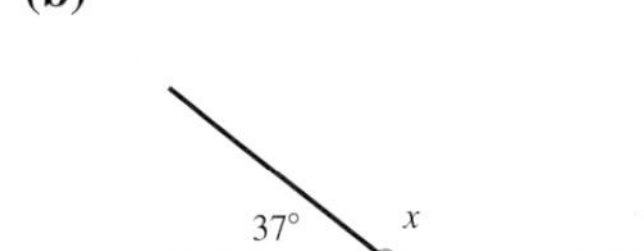

(a) Since $m\angle x + 68° = 90°$ (complementary angles), we must have $m\angle x = 90° - 68° = 22°$

(b) Since $m\angle x + 37° = 180°$ (supplementary angles), we must have $m\angle x = 180° - 37° = 143°$

Check Yourself 8

In each case, find the measure of angle x.

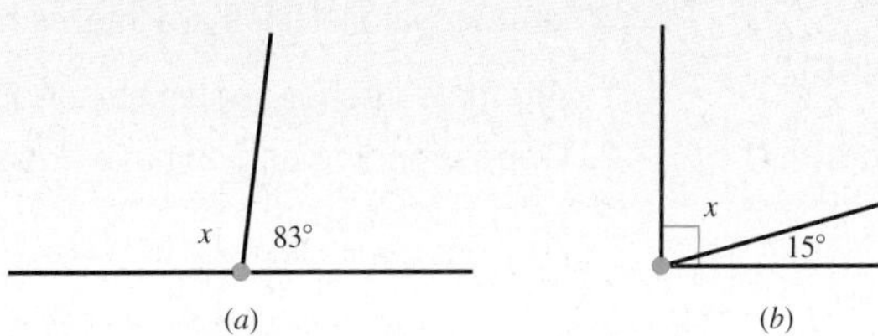

Suppose that two lines intersect as shown in the figure, forming four angles.

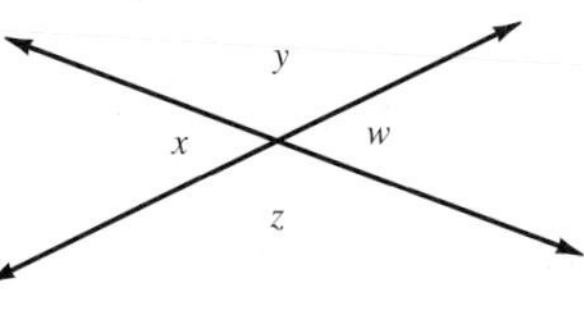

We say that $\angle x$ and $\angle w$ are **vertical angles.** Likewise, $\angle y$ and $\angle z$ are vertical angles. Vertical angles have a special property.

Property

Vertical Angles Vertical angles have equal measure.

Example 9 Finding the Measures of Angles

Suppose $m\angle w = 59°$. Find the measures of $\angle x$, $\angle y$, and $\angle z$.

$m\angle x = 59°$ since $\angle x$ and $\angle w$ are vertical angles.

Since $\angle x$ and $\angle y$ are supplementary,

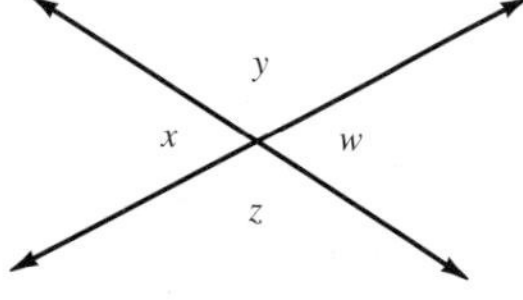

$$\begin{aligned} m\angle y &= 180° - m\angle x \\ &= 180° - 59° \\ &= 121° \end{aligned}$$

$\angle y$ and $\angle z$ are vertical angles, so $m\angle z = m\angle y$.

Check Yourself 9

Find the measures of $\angle x$, $\angle y$, and $\angle z$, if $m\angle w = 32°$.

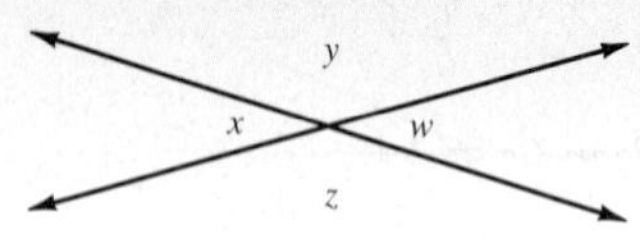

Suppose now that two parallel lines are intersected by a third line p as in the figure at the right.

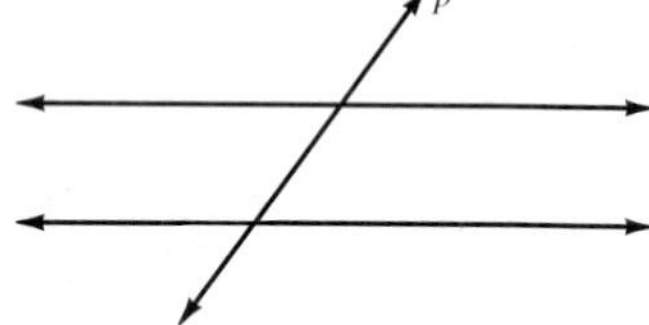

The line p is called a **transversal.** Several angles are created in this situation.

In the figure below, the two indicated angles, $\angle x$ and $\angle y$, are called **alternate interior angles.**

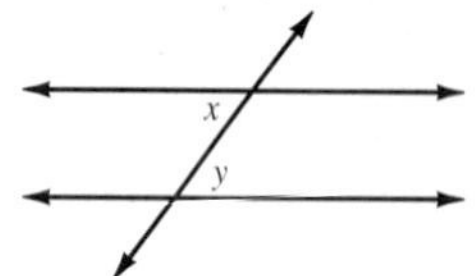

In this figure, $\angle a$ and $\angle b$ are called **corresponding angles.**

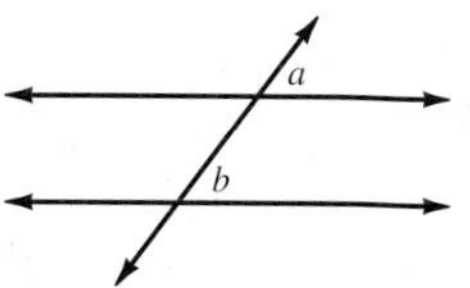

Property

Parallel Lines and a Transversal

When two parallel lines are intersected by a transversal,

1. Alternate interior angles have equal measure.
2. Corresponding angles have equal measure.

Example 10 Finding the Measures of Angles

Given that $m\angle x = 125°$, find the measures of $\angle a$, $\angle b$, and $\angle c$.

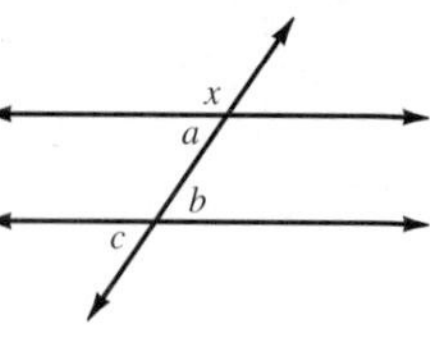

$m\angle x + m\angle a = 180°$ Supplementary angles

So $m\angle a = 180° - 125° = 55°$.

$m\angle a = m\angle b$ Alternate interior angles

So $m\angle b = 55°$.

$m\angle a = m\angle c$ Corresponding angles

So $m\angle c = 55°$.

Check Yourself 10

Given that $m\angle x = 67°$, find $m\angle y$.

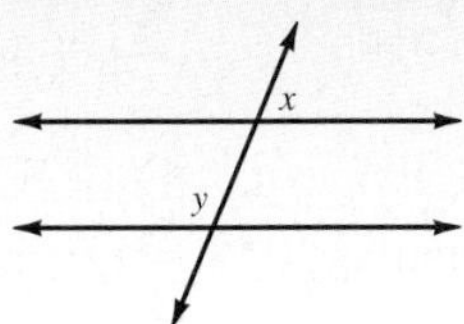

Check Yourself ANSWERS

1. **(a)** Line segment; **(b)** line segment; **(c)** line **2.** Points M and N; line $\overleftrightarrow{MN}$; and line segment $\overline{MN}$ **3.** **(a)** Parallel; **(b)** neither; **(c)** perpendicular **4.** $\angle BOA$ or $\angle AOB$ **5.** **(a)** Right; **(b)** straight; **(c)** acute; **(d)** obtuse **6.** **(a)** 120°; **(b)** 75°; **(c)** 160° **7.** 135° **8.** **(a)** 97°; **(b)** 75° **9.** $m\angle x = 32°$; $m\angle y = 148°$; $m\angle z = 148°$ **10.** 113°

Reading Your Text

These fill-in-the-blank exercises will help you understand some of the key vocabulary used in this section. The answers to these exercises are in the Answers Appendix at the back of the text.

(a) *Geo* means __________, just as it does in the words *geography* and *geology*.

(b) If two lines intersect such that four right angles are formed, we say that the two lines are __________.

(c) An __________ angle measures between 90° and 180°.

(d) If the sum of the measures of two angles is 90°, the two angles are said to be __________.

Skills Calculator/Computer Career Applications Above and Beyond

8.1 exercises

< Objective 1 >

1. Draw line segment AB.

A B

2. Draw line EF.

E F

3. Draw line AC.

A C

4. Draw line segment BC.

B C

Identify each object as a line or line segment.

5.

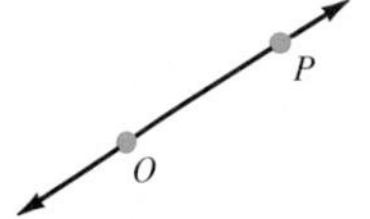

6.

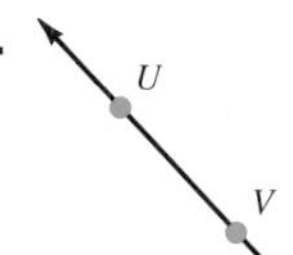

7.

8.

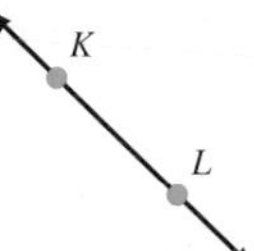

9. 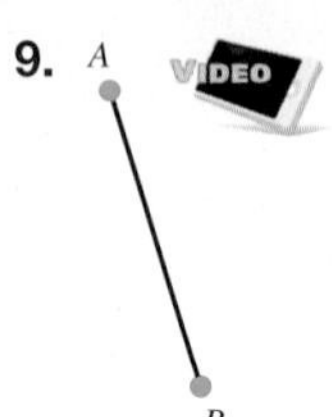

10. W, X

11. 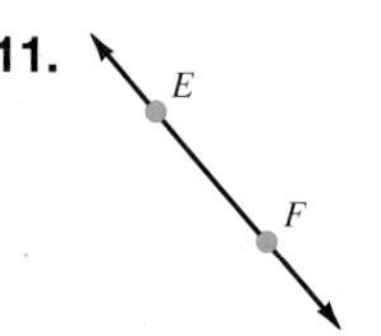

12. G, H

< Objective 2 >

13. Are the two lines parallel, perpendicular, or neither?

14. Are the two lines parallel, perpendicular, or neither?

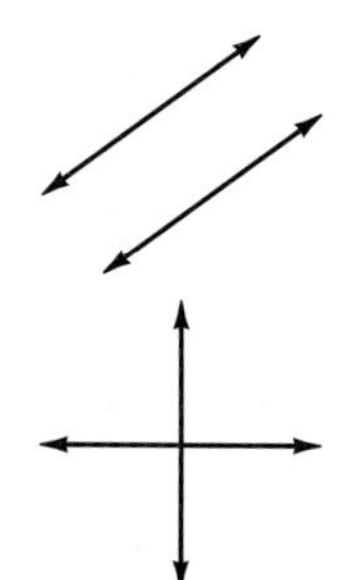

< Objective 3 >

Give an appropriate name for each indicated angle.

15. 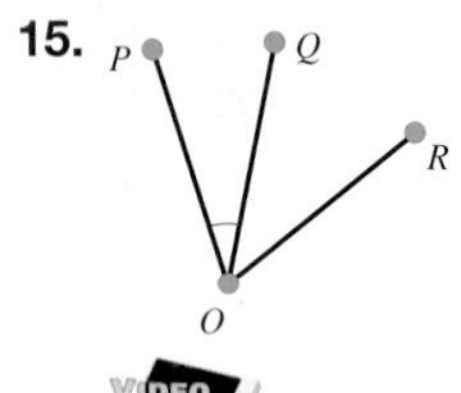

16. 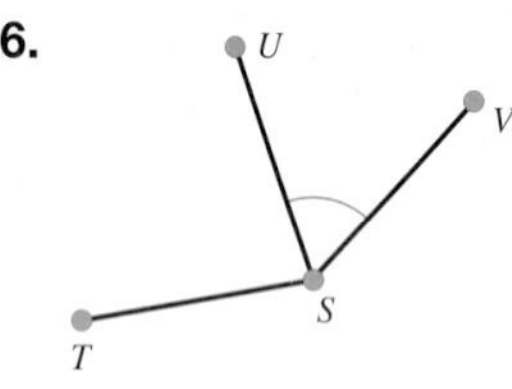

17. 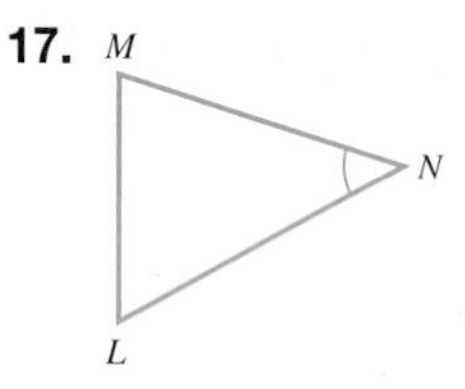

18. 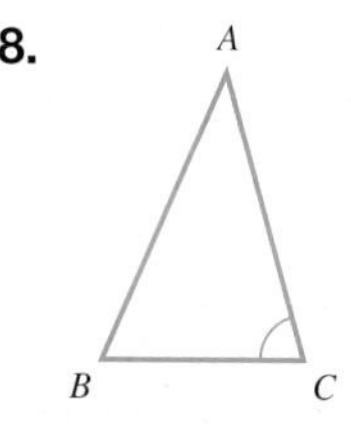

19. F, G, E, H

20. 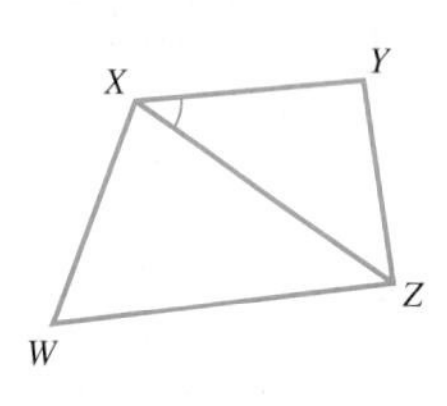

21. 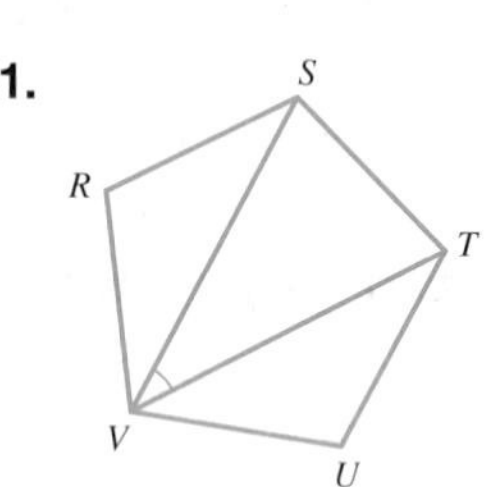

22. 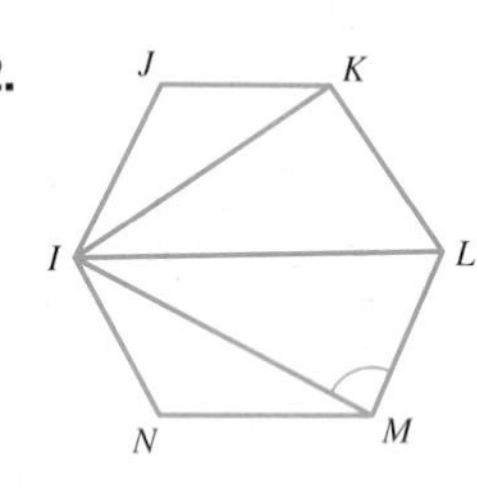

< Objective 4 >

Measure each angle with a protractor. Identify the angle as acute, right, obtuse, or straight.

23. 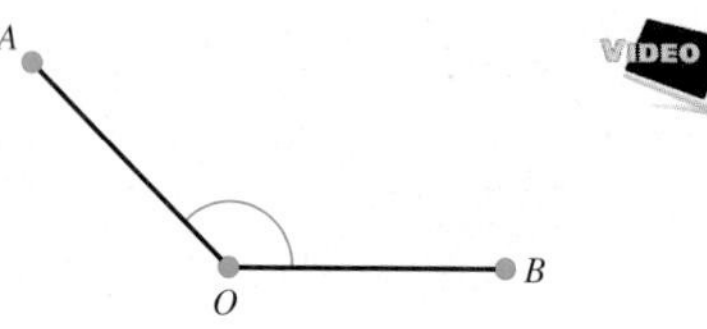

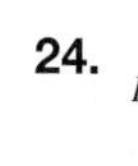

24. 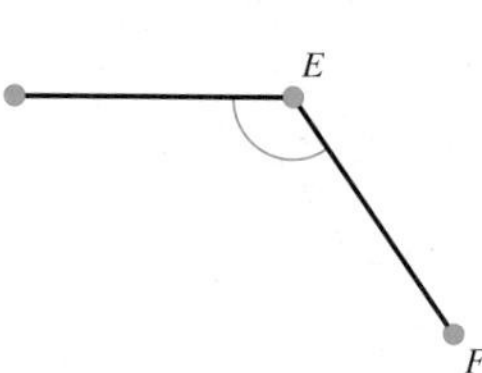

25. 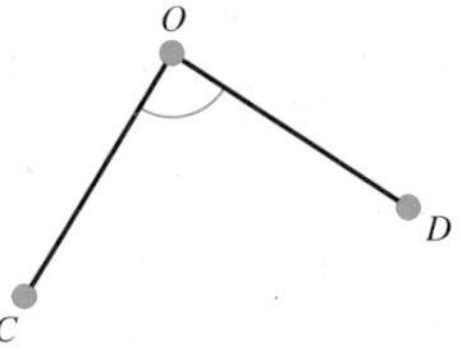

26. 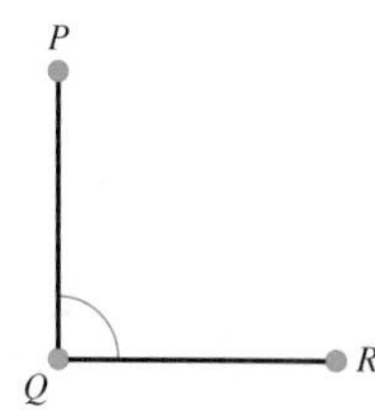

27. 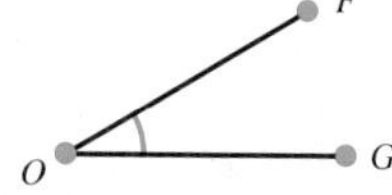

28. 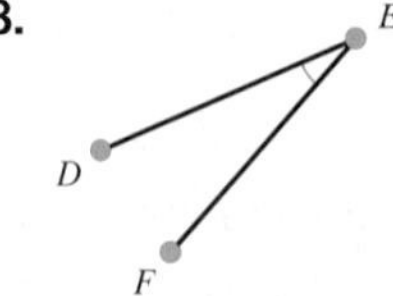

< Objective 5 >

Suppose that $m\angle x = 29°$.

29. Find the complement of $\angle x$. 

30. Find the supplement of $\angle x$.

Suppose that $m\angle y = 53°$.

31. Find the supplement of $\angle y$.

32. Find the complement of $\angle y$.

33. Find $m\angle x$.

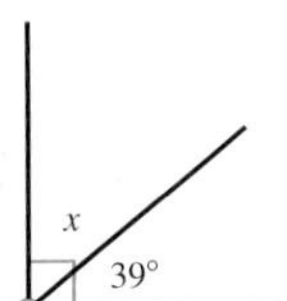

34. Find $m\angle y$.

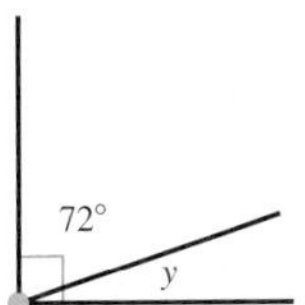

35. Find $m\angle x$ and $m\angle y$.

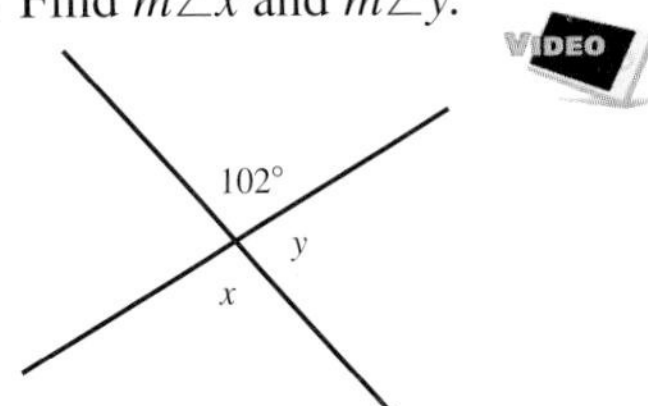

36. Find $m\angle a$ and $m\angle b$.

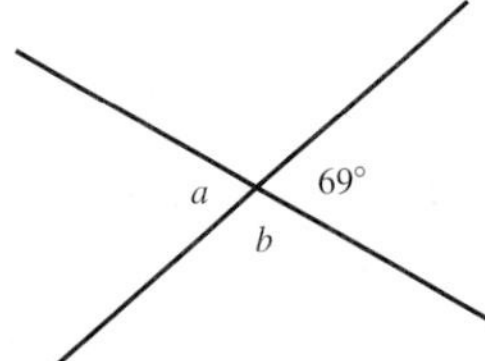

37. Find $m\angle w$.

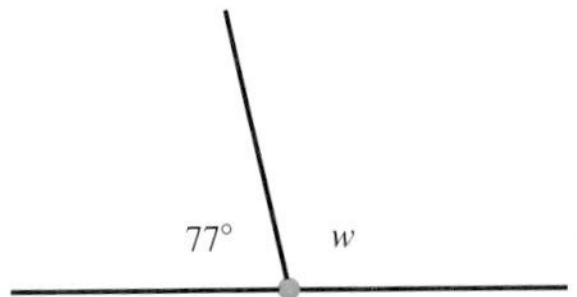

38. Find $m\angle z$.

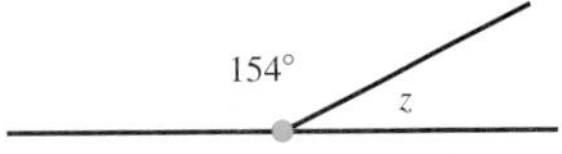

Find $m\angle a$, $m\angle b$, and $m\angle c$.

39.

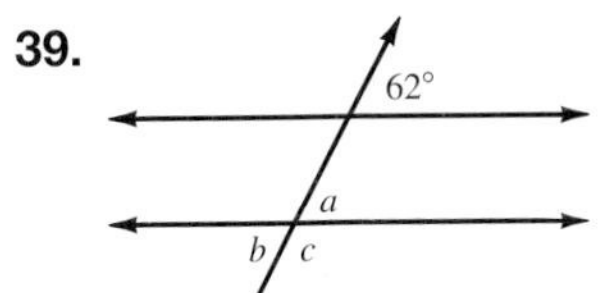

40.

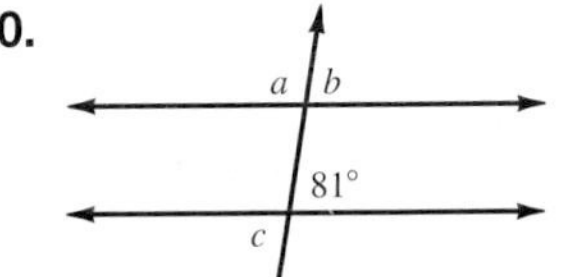

41.

42.

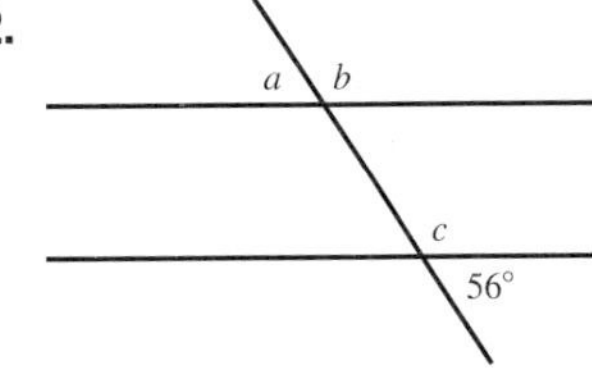

Find the measure of, and sketch, each angle.

43. $\angle A$ represents $\frac{1}{6}$ of a complete circle.

44. $\angle B$ represents $\frac{1}{3}$ of a complete circle.

45. $\angle C$ represents $\frac{7}{12}$ of a complete circle.

46. $\angle D$ represents $\frac{11}{12}$ of a complete circle.

Label each statement as **true** *or* **false.**

47. There are exactly two different line segments that can be drawn through two points.

48. There are exactly two different lines that can be drawn through two points.

49. Two opposite sides of a square are parallel line segments.

50. Two adjacent sides of a square are perpendicular line segments.

51. $\angle ABC$ always has the same measure as $\angle CAB$.

52. Two acute angles have the same measure.

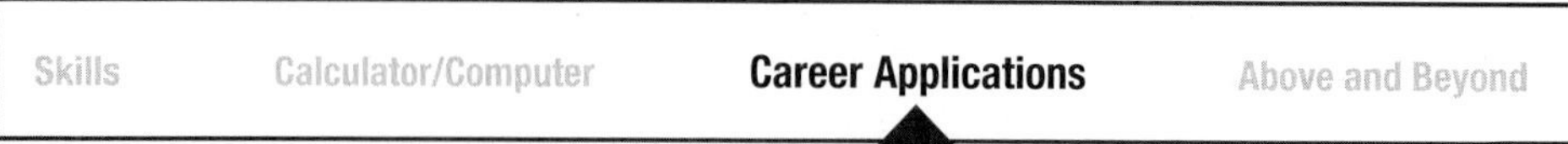

53. **Electronics** This is a picture of an analog voltmeter. The needle rotates clockwise as the voltage increases. The total angular distance covered by the needle as it travels from 0 volts (V) to 10 V is 100°.

   (a) Assuming that the angle is proportional to the voltage, what angular distance do you estimate the needle would travel from the initial 0-V position if 5 V are measured?

   (b) If the angular distance traveled from the original 0-V location is 85°, estimate the voltage.

54. **Automotive Technology** In a 4-stroke engine, the crankshaft turns two complete turns for each cycle of a cylinder. How many degrees does the crankshaft turn for one cycle?

Answers

1. A —— B 3. ↔ A C 5. Line 7. Line segment 9. Line segment 11. Line 13. Parallel 15. $\angle POQ$ or $\angle QOP$ 17. $\angle MNL$ or $\angle LNM$ 19. $\angle FEG$ or $\angle GEF$ 21. $\angle SVT$ or $\angle TVS$ 23. 135°; obtuse 25. 90°; right 27. 30°; acute 29. 61° 31. 127° 33. 51° 35. $m\angle x = 102°$; $m\angle y = 78°$ 37. 103° 39. $m\angle a = 62°$; $m\angle b = 62°$; $m\angle c = 118°$ 41. $m\angle a = 48°$; $m\angle b = 48°$; $m\angle c = 132°$ 43. 60°; 45. 210°; 47. False 49. True 51. False 53. (a) 50°; (b) 8.5 V

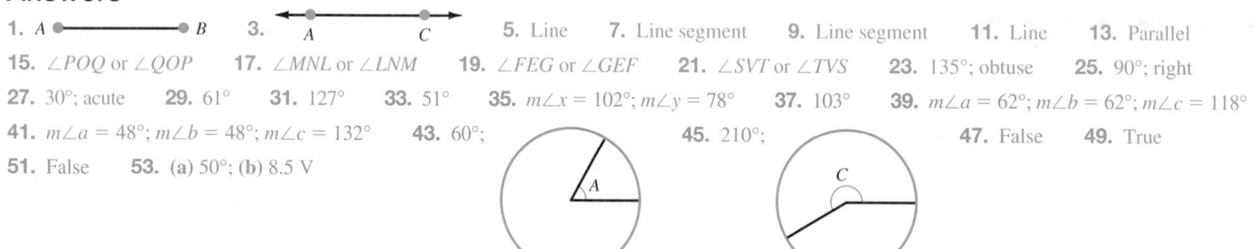

8.2 Perimeter and Area

< 8.2 Objectives >

1 > Identify polygons
2 > Find the perimeter of a polygon
3 > Find the area of a polygon
4 > Convert units of area

You have been working with many of the important concepts in geometry throughout this text. As early as Section 1.2, you were finding the perimeters of squares, rectangles, and other figures. By Section 1.5, you were even finding the areas of many of these objects.

In this section, we review these geometry concepts, add some new ones, and tie them together as a single unit. We begin by identifying some of the figures that we work with.

Definition

Polygon A polygon is a simple closed figure with three or more sides in which each side is a line segment.

By this, we mean that there are no openings, that nonadjacent sides do not intersect, and that there are no curves.

Example 1 **Identifying Polygons**

< Objective 1 >

Determine whether or not each object is a polygon. Explain your answer.

(a)  **(b)**  **(c)**  **(d)**

(a) This is a polygon because there are four straight sides and the figure is closed.

(b) This is not a polygon because the geometric figure is not closed.

(c) This is not a polygon because the sides intersect (it is not simple); it is two polygons (triangles) that meet at a point.

(d) This is a polygon. The figure has six sides and it is closed and simple.

Check Yourself 1

Determine which figures are polygons. Explain your answers.

(a)  **(b)**  **(c)**  **(d)** 

You probably know the names of the more common polygons.

Definition

Types of Polygons

NOTE

Other polygon names, based on the number of sides, include
Pentagon: Five sides
Hexagon: Six sides
Octagon: Eight sides
Stop signs are typically octagons.

Triangle: Any polygon with exactly three sides is a triangle.

Quadrilateral: Any polygon with exactly four sides is a quadrilateral.

Parallelogram: A quadrilateral in which opposite sides are parallel is a parallelogram.

Equivalently, a parallelogram is a quadrilateral in which the opposite sides have the same length.

Rectangle: A quadrilateral in which the interior angles are right angles is a rectangle.

Square: A rectangle in which all four sides have the same length is a square.

A square is an example of a **regular polygon.** A regular polygon is one in which all of the sides have the same length, and all of the interior angles have the same measure.

Example 2 Identifying Polygons

NOTE

We take a more in-depth look at triangles in Section 8.4.

Name the polygon and determine if it is regular. If it is not on the list, give the number of sides.

(a)  **(b)**  **(c)** 

(a) This is a triangle because it has three sides.
The sides are all the same length, so it is regular.
When a triangle is regular, we call it equilateral.

(b) This is a parallelogram, but it is not regular.

(c) This six-sided polygon (commonly called a hexagon) is regular.

Check Yourself 2

Name the polygon and determine if it is regular. If it is not on the list, give the number of sides.

(a)  **(b)**  **(c)** 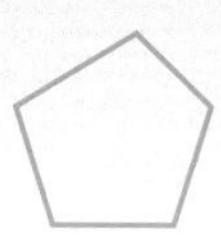

There are two important properties that we associate with polygons: perimeter and area. We look at the perimeter of a polygon first.

Definition

The Perimeter of a Polygon

The perimeter of a polygon is the distance around the polygon.
It is found by taking the sum of the lengths of its sides.

Example 3 Finding the Perimeter of a Polygon

< Objective 2 >

Find the perimeter of each figure.

(a)

8 in.

14 in.

The opposite sides of a rectangle have the same length, so its perimeter is

$8 + 14 + 8 + 14 = 44$ in.

(b)

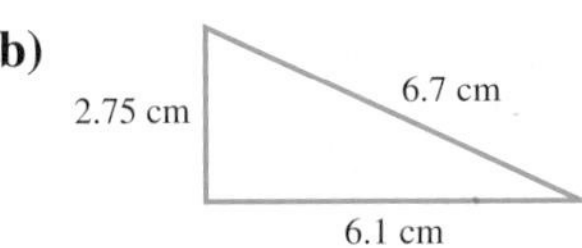

The perimeter of this triangle is $2.75 + 6.7 + 6.1 = 15.55$ cm.

Check Yourself 3

Find the perimeter of each polygon.

(a)

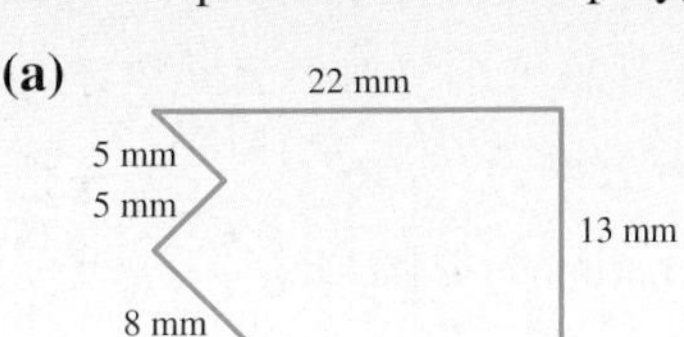

(b) Each side of the square measures $\frac{4}{5}$ in.

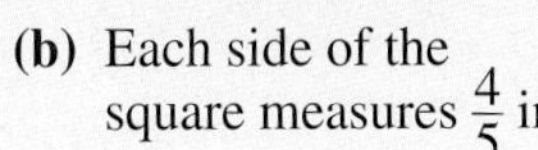

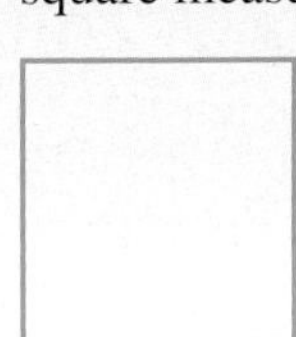

Property

Perimeter Formulas

Square

Each side has length s.

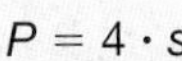

$P = 4 \cdot s$

The perimeter of a square is 4 times the length of a side.

Rectangle

If we label the sides as L and W (length and width),

$P = 2 \cdot L + 2 \cdot W = 2(L + W)$

The perimeter of a rectangle is the sum of twice the length and twice the width. Equivalently, the perimeter is twice the sum of the length and width.

Adjacent sides have length b and h (base and height).

$P = 2 \cdot b + 2 \cdot h = 2(b + h)$

Note: All four ways of writing the rectangle formula give the same result.

Regular Polygon

Each of the n sides has length s.

$P = n \cdot s$

Example 4 Finding the Perimeter of a Polygon

Find the perimeter of each polygon.

(a) Each side measures 2 cm.

(b)

32 in.

60 in.

(a) There are six sides, so the perimeter is

$$P = 6 \cdot s$$
$$= 6(2)$$
$$= 12$$

The perimeter is 12 cm.

(b) We use the rectangle formula with a 60-in. length and 32-in. width.

$$P = 2 \cdot L + 2 \cdot W$$
$$= 2(60) + 2(32)$$
$$= 120 + 64$$
$$= 184$$

Its perimeter is 184 in.

Check Yourself 4

Find the perimeter of each regular polygon.

(a) 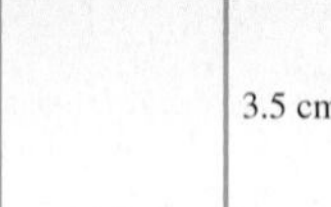

(b) 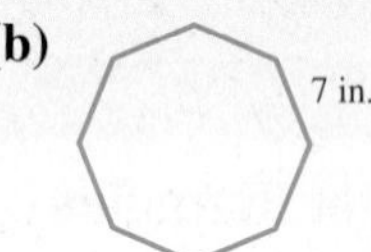

Another important property of a polygon is its **area.** By area we mean the amount of space that a two-dimensional object takes up, or how much it takes to fill it.

While perimeter is measured in one dimension because it is a length, area is a two-dimensional measure (space filled). If our units are inches, for example, then we measure area in the number of one-inch squares that it takes to "cover" the interior space of the figure. Example 5 illustrates this idea.

Example 5 Computing Area

< Objective 3 >

Find the area of the rectangle shown.

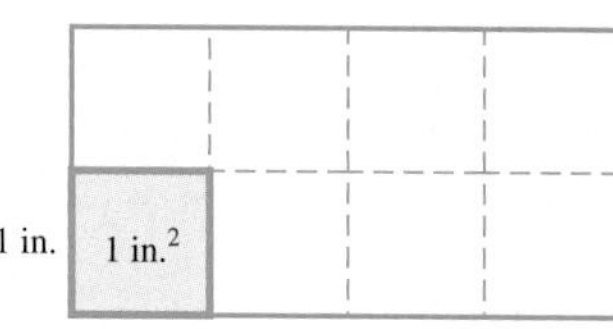

RECALL

We also use exponent notation: 8 in.².

Because the rectangle holds eight one-inch squares, its area is 8 sq in.

Check Yourself 5

Find the area of the square shown.

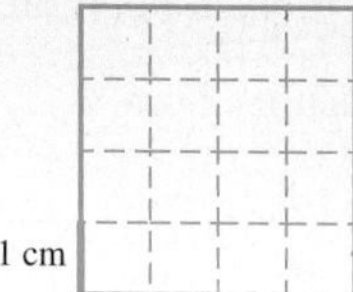

As you learned in Section 1.5, we can find the area of a rectangle by taking the product of the lengths of its base and height. In Example 5, we had two rows of 4 1-inch squares, so we had

$A = 2 \text{ in.} \times 4 \text{ in.} = 8 \text{ in.}^2$

Property

The Area of a Rectangle

If the dimensions of the rectangle are given as length L and width W,

$A = L \cdot W$

The area of a rectangle is the product of the length and the width.

The area of a rectangle with base b and height h is

$A = b \cdot h$

Because the base and height of a square are the same length, we describe its area differently.

Property

The Area of a Square

The area of a square with side s is

$A = s^2$

The area of a square is the square of the length of a side.

Example 6 Computing Area

Find the area of each figure.

(a) 5.0 mm, 13.1 mm

(b) 71 in.

(a) We find the product of the dimensions.

$$A = L \cdot W$$
$$= (13.1 \text{ mm})(5.0 \text{ mm})$$
$$= 65.5 \text{ mm}^2$$

(b) According to the formula for the area of a square, we square the length of a side.

$$A = s^2$$
$$= (71 \text{ in.})^2$$
$$= 5{,}041 \text{ in.}^2$$

Check Yourself 6

Find the area of each figure.

(a)

(b)

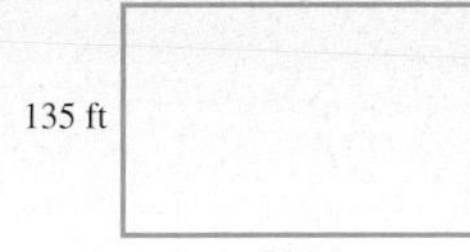

NOTE

Both rectangles and squares are parallelograms.

Two other important figures are parallelograms and triangles. The figure shown at the right is a **parallelogram.** Opposite sides are parallel and have the same length.

Because opposite sides are parallel and have the same length, if we include the height, and "cut off" a triangle from one side, it fits perfectly onto the other side, giving us a rectangle.

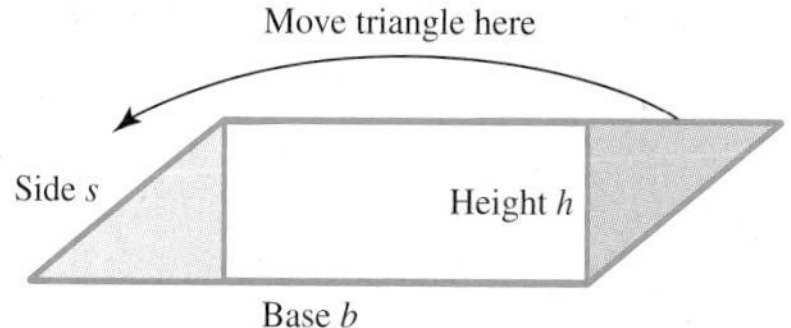

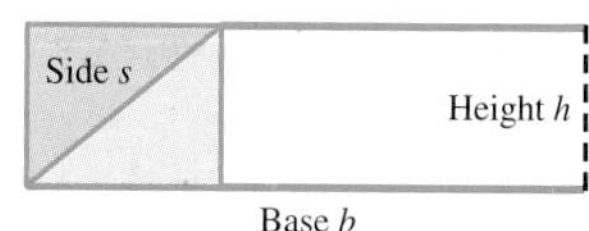

You should see that even though we use the length of a side to compute the perimeter of a parallelogram, we use the height to find its area.

Property

The Perimeter and Area of a Parallelogram

The **perimeter** of a parallelogram can be found by adding the lengths of the sides.

$P = 2 \cdot b + 2 \cdot s = 2(b + s)$

The **area** of a parallelogram can be found by taking the product of the lengths of its base and its height.

$A = b \cdot h$

Example 7 Parallelograms

Find the perimeter and area of the parallelogram shown.

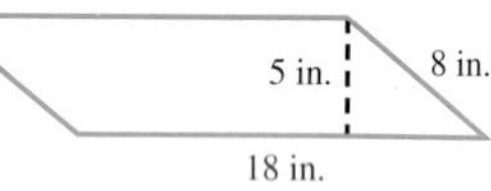

The parallelogram's base is 18 in., its side measures 8 in., and its height is 5 in.

Perimeter

$P = 2 \cdot b + 2 \cdot s$

$= 2(18 \text{ in.}) + 2(8 \text{ in.})$

$= 36 \text{ in.} + 16 \text{ in.}$

$= 52 \text{ in.}$

Area

$A = b \cdot h$

$= (18 \text{ in.})(5 \text{ in.})$

$= 90 \text{ in.}^2$

Check Yourself 7

Find the perimeter and area of the parallelogram shown.

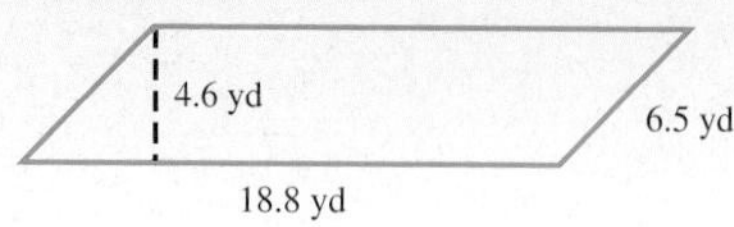

Triangles are three-sided polygons. Every triangle can be formed by halving some parallelogram along a diagonal.

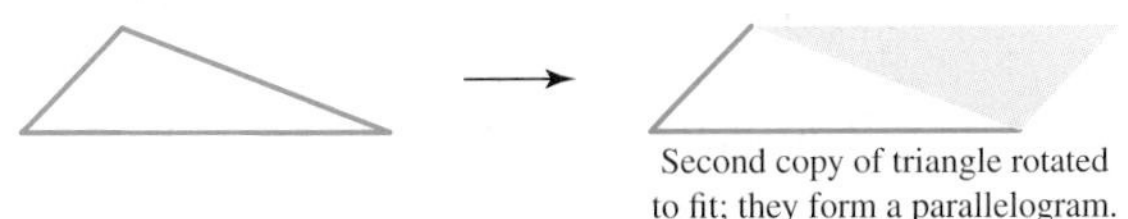

Second copy of triangle rotated to fit; they form a parallelogram.

We can think of a triangle as half a parallelogram. This gives us a formula for the area of a triangle.

Property

The Area of a Triangle

The **area** of a triangle can be found by taking half the product of the length of its base and its height.

$$A = \frac{1}{2} \cdot b \cdot h$$

Example 8 Triangles

NOTES

The base and height of a **right triangle** are the sides that form the right angle.

We look at triangles more closely in Section 8.4.

Find the perimeter and area of the triangle shown.

To find the perimeter of the triangle, we add the lengths of the sides.

$$P = 11.1 \text{ mm} + 17.2 \text{ mm} + 26.5 \text{ mm}$$
$$= 54.8 \text{ mm}$$

We use the triangle's base, 17.2 mm, and its height, 7.3 mm, to find its area.

$$A = \frac{1}{2} b \cdot h$$
$$= \frac{1}{2}(17.2 \text{ mm})(7.3 \text{ mm})$$
$$= 62.78 \text{ mm}^2$$

Check Yourself 8

Find the perimeter and area of the triangle shown.

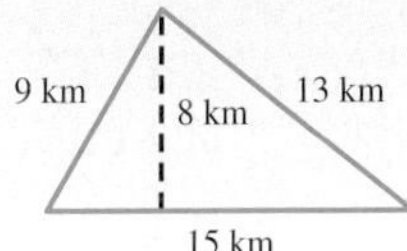

Moving on from triangles, we look at a quadrilateral called a **trapezoid.** A trapezoid is a four-sided polygon with exactly one pair of parallel sides.

Several such figures are shown. The first is a general trapezoid. The second is an *isosceles trapezoid* because the nonparallel sides are the same length. The third is a *right trapezoid* because one of its sides is perpendicular to the parallel sides.

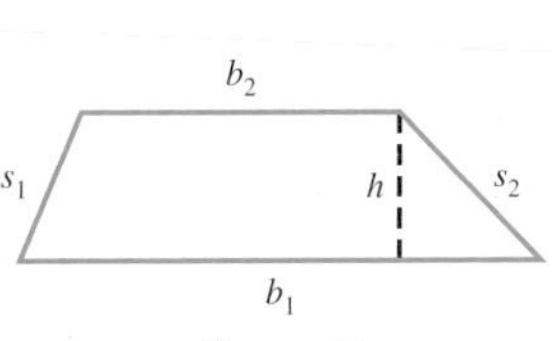

Trapezoid

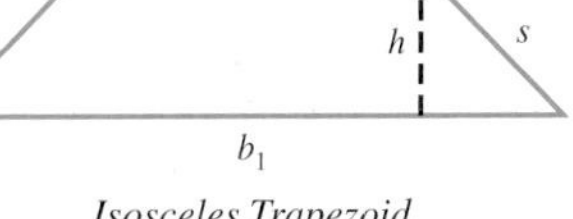

Isosceles Trapezoid

Right Trapezoid

We find a trapezoid's perimeter in the usual manner, by adding the lengths of all four sides.

To find the area of a trapezoid, we treat the trapezoid as a rectangle with the same height as the trapezoid, but with a base determined as the *average* of the lengths of the two parallel sides of the trapezoid. You will learn to justify this formula in a future math class.

Property

The Area of a Trapezoid

The area of a trapezoid is given by

$$A = \frac{1}{2}h(b_1 + b_2)$$

Example 9 **Trapezoids**

Find the perimeter and area of each trapezoid.

(a)

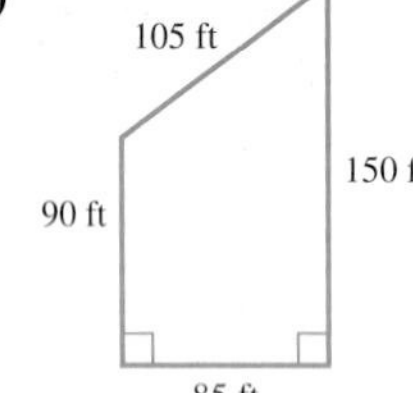

(b)

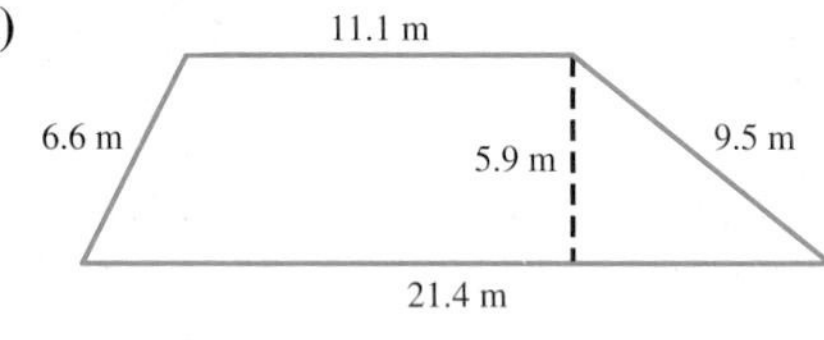

(a) The perimeter is simply the sum of the sides.

$P = 90 \text{ ft} + 105 \text{ ft} + 150 \text{ ft} + 85 \text{ ft} = 430 \text{ ft}$

To find the area, we identify the "height" as the 85-ft side.

RECALL

Follow the order of operations. Add inside the parentheses before multiplying.

$$A = \frac{1}{2}h(b_1 + b_2)$$
$$= \frac{1}{2}(85 \text{ ft})(90 \text{ ft} + 150 \text{ ft})$$
$$= 10{,}200 \text{ ft}^2$$

(b) $P = 6.6 \text{ m} + 11.1 \text{ m} + 9.5 \text{ m} + 21.4 \text{ m} = 48.6 \text{ m}$

$$A = \frac{1}{2}h(b_1 + b_2)$$
$$= \frac{1}{2}(5.9 \text{ m})(21.4 \text{ m} + 11.1 \text{ m})$$
$$= 95.875 \text{ m}^2$$

Check Yourself 9

Find the perimeter and area of each trapezoid (round to the nearest tenth, if necessary).

(a)

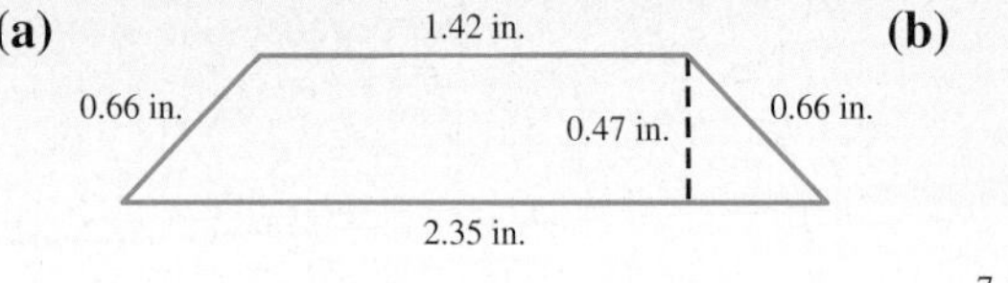

(b)

10 cm
13 cm
7 cm
8 cm

You learned to convert measurements in Chapter 7. We need another step, when converting square units. Consider the square yards shown.

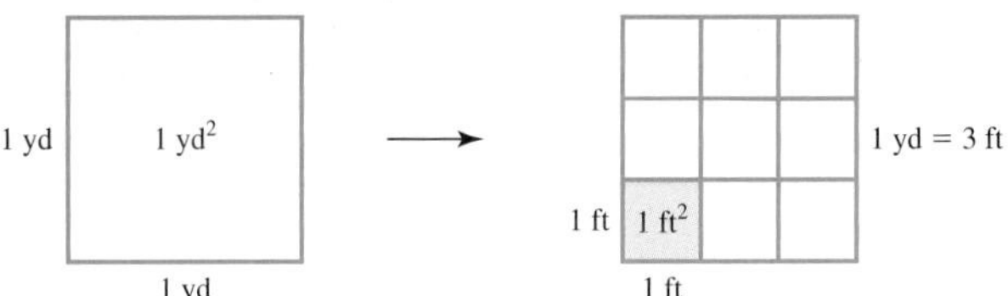

NOTE

$3^2 = 9$
$12^2 = 144$

Because 1 yd = 3 ft, we can see that 1 yd^2 = 9 ft^2. Similarly, 1 ft^2 = 144 in.^{2} More generally, we need to square conversion factors when converting units of area (square units).

Property

Conversion Factors

U.S. System of Measurement	Metric System of Measurement	Conversions
1 ft^2 = 144 in.^{2}	1 cm^2 = 100 mm^2	1 in.^{2} ≈ $(2.54)^2$ cm^2
1 yd^2 = 9 ft^2	1 m^2 = 10,000 cm^2	≈ 6.45 cm^2
1 mi^2 = 27,878,400 ft^2	= 1,000,000 mm^2	1 cm^2 ≈ $(0.39)^2$ in.^{2}
= 3,097,600 yd^2	1 km^2 = 1,000,000 m^2	≈ 0.15 in.^{2}
= 640 acres	1 hectare (ha) = $(100)^2$ m^2	1 mi^2 ≈ $(1.61)^2$ km^2
1 acre = 43,560 ft^2	= 10,000 m^2	≈ 2.59 km^2
= 4,840 yd^2		1 km^2 ≈ $(0.62)^2$ mi^2
		≈ 0.38 mi^2

Example 10 Converting Area Measurements

< Objective 4 >

(a) A room measures 12 ft by 15 ft. How many square yards of carpeting are needed to cover the floor?

Begin by finding the area of the floor.

$$A = L \cdot W$$
$$= (12 \text{ ft})(15 \text{ ft})$$
$$= 180 \text{ ft}^2$$

We then convert to square yards by using the appropriate conversion factor.

$$180 \cancel{\text{ft}^2} \times \frac{1 \text{ yd}^2}{9 \cancel{\text{ft}^2}} = \frac{180}{9} \text{ yd}^2$$
$$= 20 \text{ yd}^2$$

(b) A rectangular field is 220 yd long and 110 yd wide. Find its area, in acres.

Begin by finding the area in square yards.

$$A = L \cdot W$$
$$= (220 \text{ yd})(110 \text{ yd})$$
$$= 24{,}200 \text{ yd}^2$$

Then, we use the conversion factor to convert this to acres.

$$24{,}200 \cancel{\text{yd}^2} \times \frac{1 \text{ acre}}{4{,}840 \cancel{\text{yd}^2}} = \frac{24{,}200}{4{,}840} \text{ acre}$$
$$= 5 \text{ acres}$$

Check Yourself 10

(a) A hallway is 27 ft long and 4 ft wide. How many square yards of linoleum are needed to cover the hallway?

(b) A proposed site for an elementary school is 200 yd by 198 yd. Find its area in acres (round to the nearest tenth acre).

Check Yourself ANSWERS

1. **(a)** Yes; **(b)** yes; **(c)** no; **(d)** no **2.** **(a)** Square, regular; **(b)** triangle, not regular; **(c)** 5 sides (pentagon), not regular **3.** **(a)** 69 mm; **(b)** $3\frac{1}{5}$ in. or $\frac{16}{5}$ in. **4.** **(a)** 14 cm; **(b)** 56 in. **5.** 16 cm^2 **6.** **(a)** 169 m^2; **(b)** 30,375 ft^2 **7.** $P = 50.6$ yd; $A = 86.48$ yd^2 **8.** $P = 37$ km; $A = 60$ km^2 **9.** **(a)** $P = 5.1$ in.; $A = 0.9$ $in.^2$; **(b)** $P = 38$ cm; $A = 80$ cm^2 **10.** **(a)** 12 yd^2; **(b)** 8.2 acres

Reading Your Text

These fill-in-the-blank exercises will help you understand some of the key vocabulary used in this section. The answers to these exercises are in the Answers Appendix at the back of the text.

(a) A ________ is a polygon with exactly three sides.

(b) A polygon is ________ if all of the sides have the same length and all of the interior angles have the same measure.

(c) You can find the perimeter of a polygon by taking the ________ of the lengths of the sides of the polygon.

(d) One square yard is equal to ________ square feet.

8.2 exercises

Skills Calculator/Computer Career Applications Above and Beyond

< Objective 1 >

Determine whether each object is a polygon or not. If an object is a polygon, determine its type and whether or not it is regular.

1. 

2. 

3. 

4. 

5. 

6. 

7. 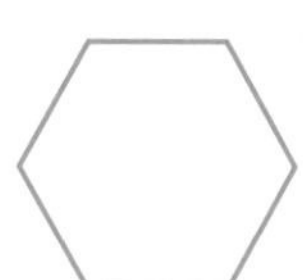

8. 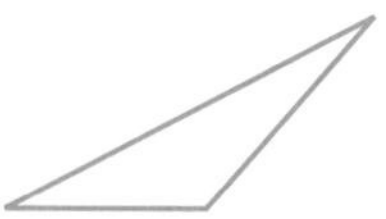

9. 

10. 

11. 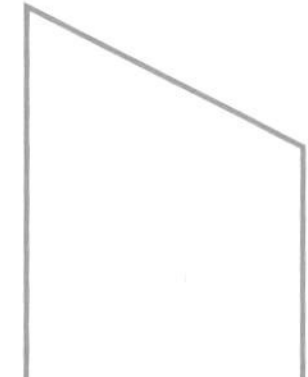

12. 

13. 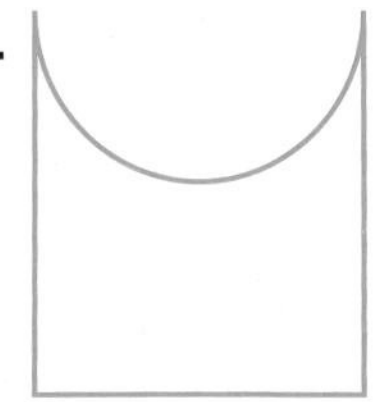

14. 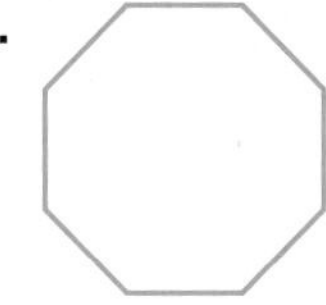

15. 

16. 

< Objective 2 >

Find the perimeter of each figure.

17. 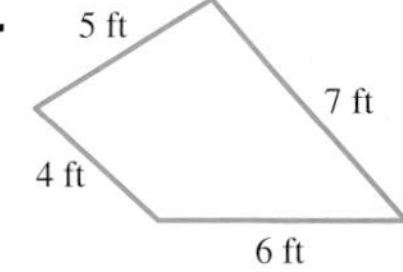

18. 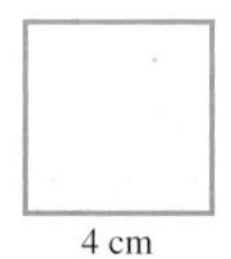

19. 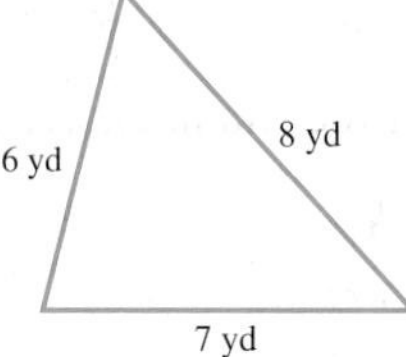

20. 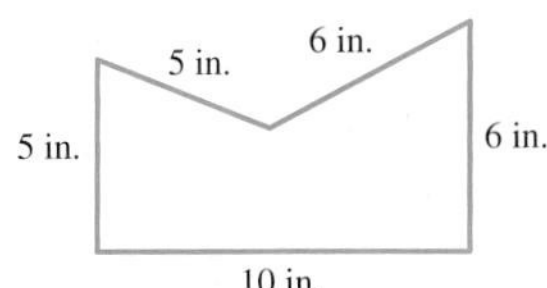

21. 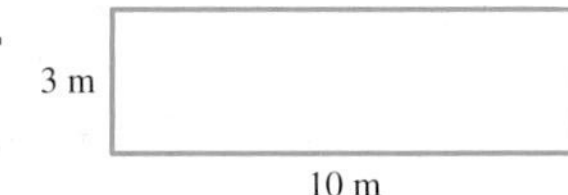

22. 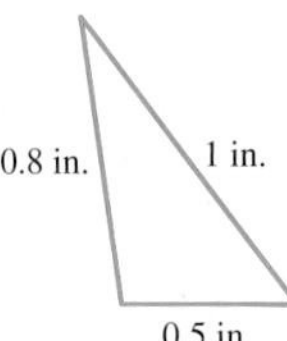

23. 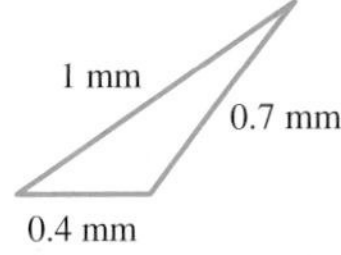

24. 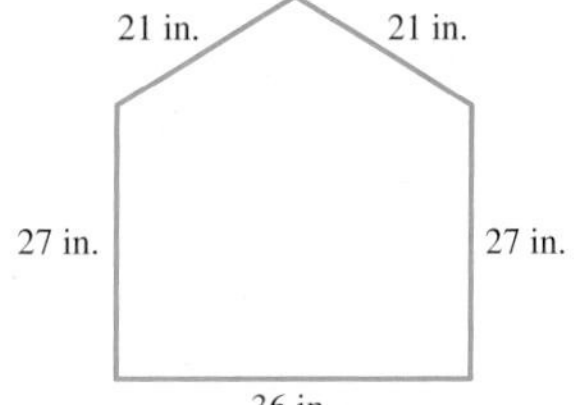

25. 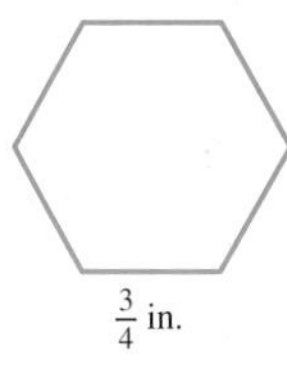

26. 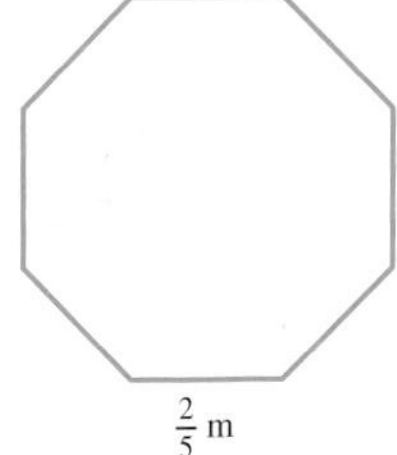

27. 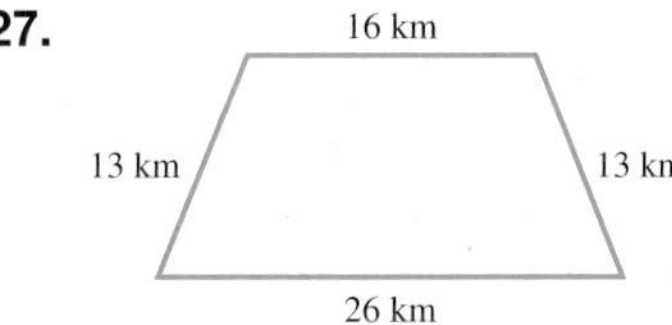

28. 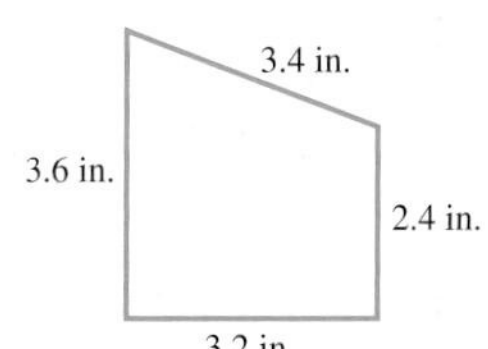

< Objective 3 >

Find the area of each figure.

29.

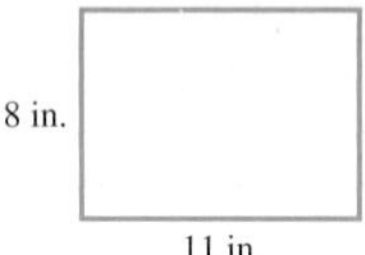

30.

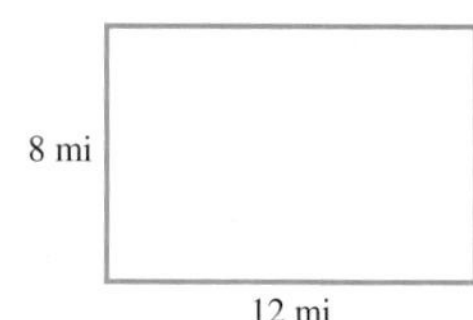

31.

32.

33.

34.

35.

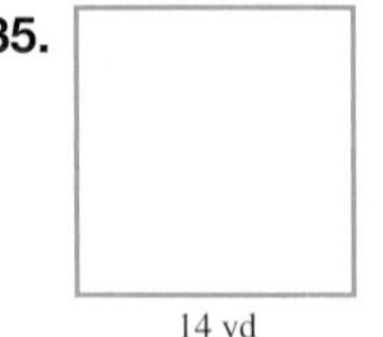

36.

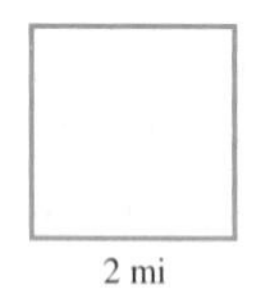

37.

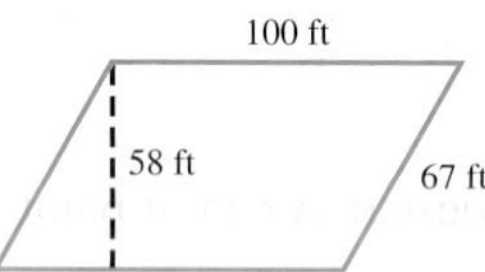

38.

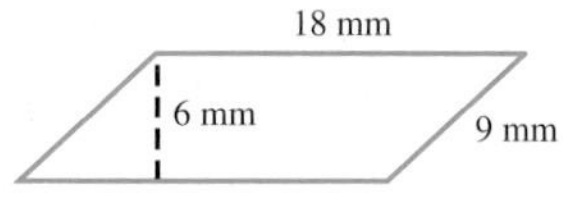

39.

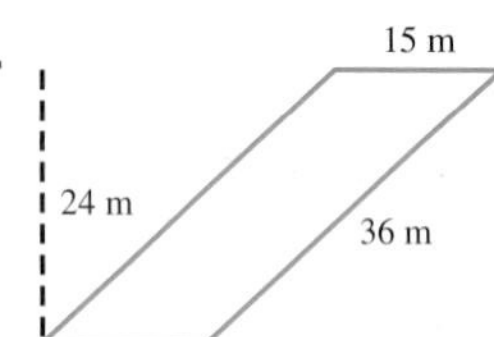

40.

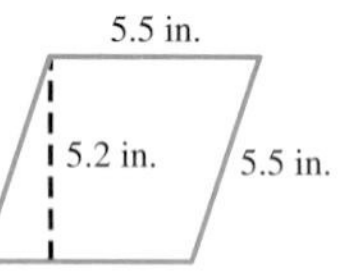

41.

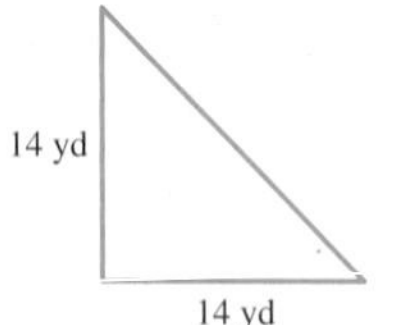

42.

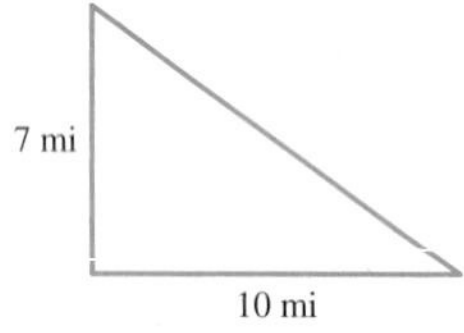

43.

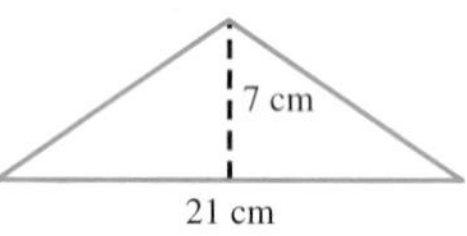

44.

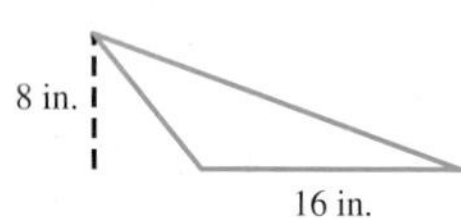

45.

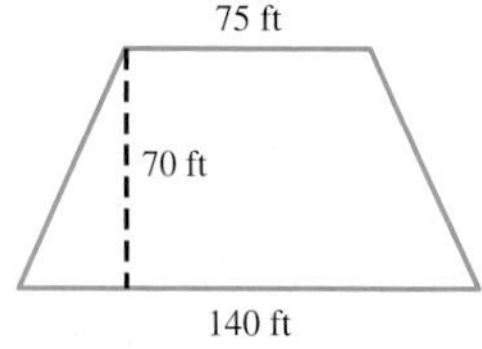

46.

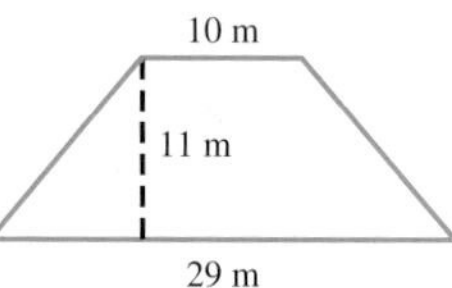

47.

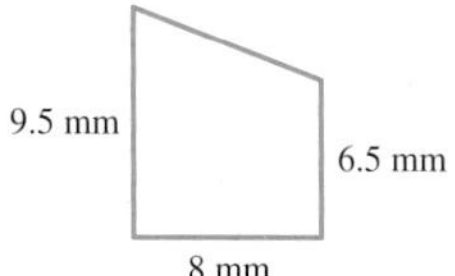

48.

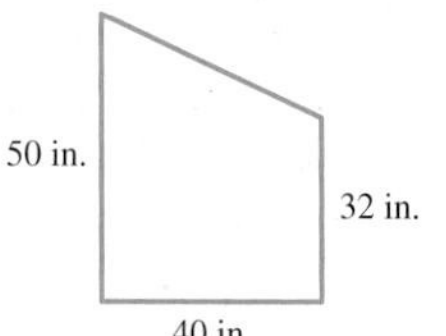

< Objective 4 >

Convert each measurement.

49. 15 ft^2 to in.^2

50. 1,050 in.^2 to ft^2

51. 360 ft^2 to yd^2

52. 20 yd^2 to in.^2

53. 8 km^2 to m^2

54. 10,000,000 m^2 to km^2

55. 30 m^2 to cm^2 **56.** 4,200 mm^2 to cm^2 **57.** 16 in.^{2} to cm^2

58. 6 km^2 to mi^2

Convert each acre and hectare measurement (round your results to one decimal place).

59. 18,000 acres to mi^2

60. 4,200,000 ft^2 to acres

61. 12,000,000 m^2 to ha

62. 1,500 ha to km^2

63. 2,500 ha to acres

64. 2,500 acres to ha

Solve each application.

65. CRAFTS A Tetra-Kite uses 12 triangular pieces of plastic for its surface. Each triangle has a 12-in. base and a 12-in. height. How much material (square inches) is needed for such a kite?

66. CONSTRUCTION How much does it cost to carpet a 12-ft by 18-ft recreation room with $15 per square yard carpeting?

Hint: You cannot purchase fractions of a square yard of carpeting.

67. CONSTRUCTION A triangular hole is to be punched into a piece of sheet metal. The base of the triangle is 1.6 in. and the height is 0.85 in. What is the area of the hole to be punched?

68. CONSTRUCTION A can of paint covers 600 ft^2. How many cans of paint are needed to paint a 12-ft by 14-ft room if it needs **two coats** and has 8-ft high ceilings (paint the four walls, but not the ceiling or floor)?

69. CRAFTS A vegetable garden measures 180 ft by 265 ft.

(a) How many acres make up the garden?

(b) How many feet of fencing does it take to enclose the garden?

70. BUSINESS AND FINANCE A shopping center sits on a rectangular lot with dimensions 550 yd by 440 yd. Find its size, in acres.

71. CONSTRUCTION A square lot measures 160 yd per side.

(a) What is its area, to the nearest tenth acre?

(b) What is its area, to the nearest tenth hectare?

72. CONSTRUCTION A lot measures 6.3 km by 9.1 km.

(a) What is its area, to the nearest hectare?

(b) What is its area, to the nearest tenth acre?

Determine whether each statement is **true** *or* **false.**

73. A polygon can have as few as two sides.

74. All regular triangles are equilateral triangles.

75. All regular quadrilaterals are squares.

76. Circles are a type of polygon.

Fill in each blank with **always, sometimes,** *or* **never.**

77. Two copies of the same triangle can __________ be fit together to form a parallelogram.

78. One of the sides of a trapezoid __________ gives the height of the trapezoid.

79. **Manufacturing Technology** A piece of rectangular stock is to be coated on the top side with sealer; the top of the piece measures 2.3 in. by 4.8 in.

(a) Calculate the area to be covered (to the nearest square inch).

(b) One can of sealer covers 500 ft^2. How many (whole) parts does one can of sealer cover?

80. **Mechanical Engineering** The allowable compressive stress of steel is 2,900 pounds per square inch (psi). Can a rectangular piece of stock measuring $\frac{3}{4}$ in. by $\frac{1}{2}$ in. handle a force of 1,100 lb?

81. A piece of 3-in.-wide strapping material is to be cut into parallelograms, each with a 4.75-in. base (see figure). What is the area of each piece?

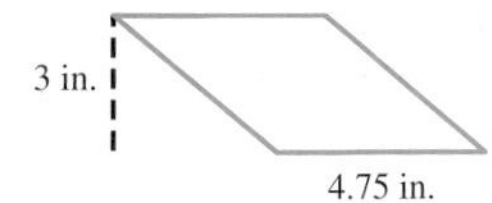

82. **Information Technology** A printed circuit board (PCB) measures 45 mm by 67 mm.

(a) What is the area of the board?

(b) Report the area of the PCB in square centimeters.

83. **Automotive Technology** A rectangular tire patch is 4 cm by 7 cm. Find its area.

84. **Automotive Technology** The air filter in a car measures 6 in. by 9 in.

(a) Find the area of the filter.

(b) To wrap the filter in an aluminum frame, find the perimeter of the filter.

85. **Manufacturing Technology** A 4-in. by 14-in. steel plate weighs 0.048 pounds per square inch. What is the weight of the plate?

86. **Construction** How many feet of baseboard does it take to go around an 11-ft 4-in. by 13-ft 8-in. room?

Skills Calculator/Computer Career Applications **Above and Beyond**

87. What is the effect on the area of a rectangle if one dimension is doubled? Construct some examples to demonstrate your answer.

88. What is the effect on the area of a rectangle if both dimensions are doubled? Construct some examples to demonstrate your answer.

89. What is the effect on the area of a triangle if the base is doubled and the height is halved? Construct some examples to demonstrate your answer.

90. What is the effect on the area of a triangle if both the base and height are doubled? Construct some examples to demonstrate your answer.

Answers

1. Yes; quadrilateral; not regular **3.** No **5.** Yes; rectangle; not regular **7.** Yes; hexagon; regular **9.** No **11.** Yes; trapezoid; not regular **13.** No **15.** No **17.** 22 ft **19.** 21 yd **21.** 26 m **23.** 2.1 mm **25.** $4\frac{1}{2}$ in. **27.** 68 km **29.** 88 in.^{2} **31.** 84 ft^2 **33.** 36 mm^2 **35.** 196 yd^2 **37.** 5,800 ft^2 **39.** 360 m^2 **41.** 98 yd^2 **43.** 73.5 cm^2 **45.** 7,525 ft^2 **47.** 64 mm^2 **49.** 2,160 in.^{2} **51.** 40 yd^2 **53.** 8,000,000 m^2 **55.** 300,000 cm^2 **57.** 103.2 cm^2 **59.** 28.1 mi^2 **61.** 1,200 ha **63.** 6,080 acres **65.** 864 in.^{2} **67.** 0.68 in.^{2} **69.** (a) 1.1 acres; (b) 890 ft **71.** (a) 5.3 acres; (b) 2.1 ha **73.** False **75.** True **77.** always **79.** (a) 11 in.^{2}; (b) 6,545 parts **81.** 14.25 in.^{2} **83.** 28 cm^2 **85.** 2.688 lb **87.** Above and Beyond **89.** Above and Beyond

Activity 22 ::

Know the Angles

Work with a small group of students to complete this activity.

1. Draw a triangle. Use a protractor to carefully measure the interior angles. Find the sum of these three angles.

   $\angle A =$ $\angle B =$ $\angle C =$ Sum =

   Sketch a second triangle. Measure the interior angles and find their sum.

   $\angle A =$ $\angle B =$ $\angle C =$ Sum =

   What do you notice about the sum of the three angles in a triangle?

   Make a conjecture about the sum of the three interior angles of any triangle. Test your conjecture on another triangle.

2. A **quadrilateral** is a four-sided polygon. Draw any quadrilateral and measure the four interior angles with a protractor. Record these and find their sum.

   $\angle A =$ $\angle B =$ $\angle C =$ $\angle D =$ Sum =

   Make a conjecture concerning the sum of the interior angles of *any* quadrilateral and test your conjecture on another quadrilateral.

3. A **pentagon** is a five-sided polygon. Draw any pentagon and measure the five interior angles with a protractor. Record these and find their sum.

   $\angle A =$ $\angle B =$ $\angle C =$ $\angle D =$ $\angle E =$ Sum =

   Make a conjecture concerning the sum of the interior angles of *any* pentagon and test your conjecture on another pentagon.

4. A **hexagon** is a six-sided polygon. Draw any hexagon and measure the six interior angles with a protractor. Record these and find their sum.

   $\angle A =$ $\angle B =$ $\angle C =$ $\angle D =$ $\angle E =$ $\angle F =$ Sum =

   Make a conjecture concerning the sum of the interior angles of *any* hexagon and test your conjecture on another hexagon.

5. Now try to generalize. Suppose we have a polygon with k sides. Give a formula for the sum of the interior angles. Sum =

8.3 Circles and Composite Figures

< 8.3 Objectives >

1 > Find the circumference of a circle

2 > Find the area of a circle

3 > Find the perimeter of a composite figure

4 > Find the area of a composite figure

We have already seen how to find the perimeters and areas of several figures. In this section we apply these concepts to circles and composite figures. The distance around the outside of a circle is closely related to perimeter. We call the perimeter of a circle its **circumference.**

Definition

Circumference of a Circle

The *circumference* of a circle is the distance around that circle.

NOTE

The circumference formula comes from the ratio

$\frac{C}{D} = \pi$

We begin by defining some terms. In the circle, d represents the **diameter.** This is the distance across the circle through its center (labeled with the letter O, for **origin**). The **radius** r is the distance from the center to a point on the circle. The diameter is always twice the radius.

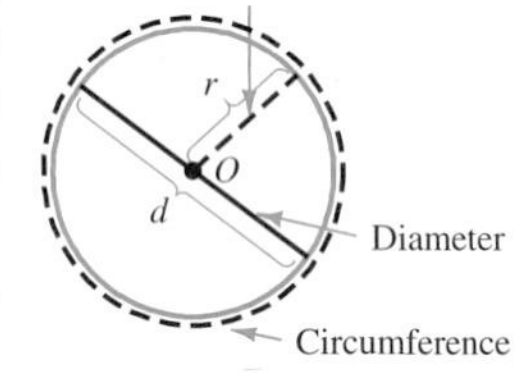

It was discovered long ago that the ratio of the circumference of a circle to its diameter always stays the same.

We use the Greek letter π (pi) to name this special ratio. Pi is approximately 3.14, rounded to two decimal places. This ratio gives us a formula to find the circumference of a circle.

Property

The Circumference of a Circle

$C = \pi d$

Example 1 Finding the Circumference of a Circle

< Objective 1 >

NOTE

Because 3.14 is an approximation for pi, we can only say that the circumference is approximately 14.1 ft. The symbol ≈ means "approximately."

A circle has a diameter of 4.5 ft. Find its circumference, using 3.14 for π. If your calculator has a $\boxed{\pi}$ key, use that key instead of a shorter decimal approximation for π.

By the formula,

$$
\begin{aligned}
C &= \pi d \\
&\approx 3.14 \times 4.5 \text{ ft} \\
&\approx 14.1 \text{ ft} \quad \text{(rounded to one decimal place)}
\end{aligned}
$$

4.5 ft

Check Yourself 1

A circle has a diameter of $3\frac{1}{2}$ in. Find its circumference

Because $d = 2r$ (the diameter is twice the radius) and $C = \pi d$, we have a second way of writing the formula for the circumference of a circle, $C = \pi(2r)$, or $C = 2\pi r$.

Property

The Circumference of a Circle $C = 2\pi r$

Example 2 **Finding the Circumference of a Circle**

NOTE

If you want to approximate π, you do not need to worry about running out of decimal places. The value for pi has been calculated to over 10 trillion decimal places on a computer (the printout would be over 30 million pages long).

A circle has a radius of 8 in. Find its circumference, using 3.14 for π.

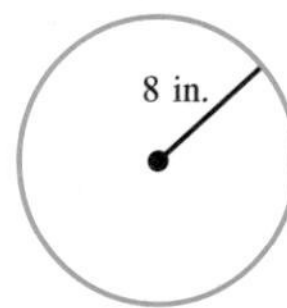

From the formula,

$$C = 2\pi r$$
$$\approx 2 \times 3.14 \times 8 \text{ in.}$$
$$\approx 50.2 \text{ in.} \quad \text{(rounded to one decimal place)}$$

Check Yourself 2

Find the circumference of a circle with a radius of 2.5 in.

The number pi (π), which we used to find circumference, is also used in finding the area of a circle.

Property

The Area of a Circle $A = \pi r^2$

This is read, "Area equals pi r squared." You can multiply the radius by itself and then by pi.

Example 3 **Finding the Area of a Circle**

< Objective 2 >

A circle has a 7-in. radius. What is its area?

Use the area formula with 3.14 for π and $r = 7$ in.

$$A \approx 3.14 \times (7 \text{ in.})^2$$
$$\approx 153.86 \text{ in.}^2$$

Again, the area is an approximation because we use 3.14, an approximation for π.

Check Yourself 3

Find the area of a circle whose diameter is 4.8 cm. Remember that the formula refers to the radius. Use 3.14 for π and round your result to the nearest tenth of a square centimeter.

Many objects have odd shapes—that is, they are not the basic shapes we have studied so far. In Chapter 1, we called these "oddly shaped figures." The proper name for these types of objects is **composite geometric figures** or just **composite figures.** More exactly, a composite figure is formed by joining simple geometric figures together.

Definition

Composite Geometric Figures A *composite figure* is a geometric figure that is formed by adjoining two or more basic geometric figures.

To help you to understand what we mean, consider the floor plan in the next example.

Example 4 Perimeter of a Composite Figure

< Objective 3 >

The floor plan shown is a living room adjoining a foyer. Find the length of molding (perimeter) needed for the room.

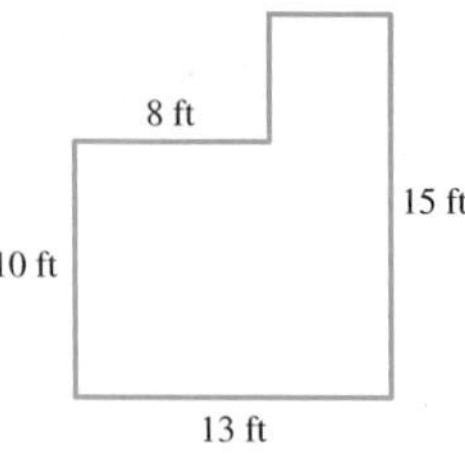

NOTE

We are simplifying this by omitting gaps such as doors.

First, we compute the missing lengths. For example, the top wall of the foyer is

13 ft − 8 ft = 5 ft

Similarly, the left wall of the foyer area is

15 ft − 10 ft = 5 ft

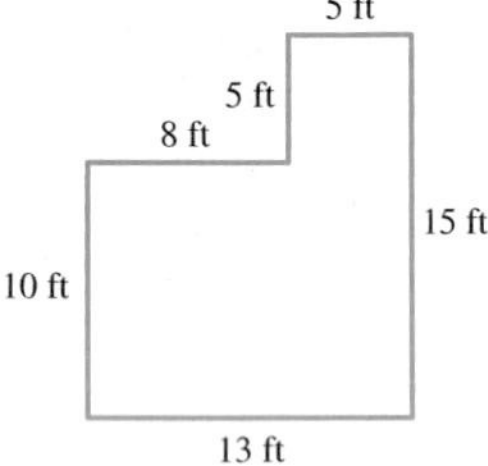

As usual, the perimeter (or the amount of molding needed) is the sum of the lengths of the sides.

$P = 13 + 15 + 5 + 5 + 8 + 10 = 56$

The room needs 56 ft of molding.

Check Yourself 4

Find the perimeter of the figure shown.

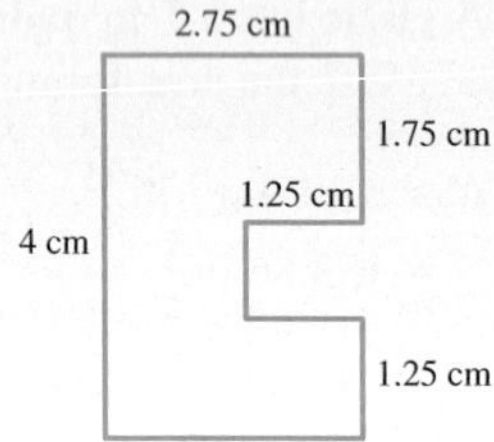

To find the area of a composite figure, we usually consider the more basic shapes separately. That is, we find the area of each of the simple figures and combine them to produce the area of the larger composite figure. We illustrate this in Example 5.

Example 5 Area of a Composite Figure

< Objective 4 >

Find the amount of carpeting (area) needed for the living room and foyer shown in Example 4.

First, we subdivide the room into more basic shapes—in this case, we have a square and a rectangle. Since we already found the missing dimensions of the room in Example 4, we include them in the sketch.

NOTE

We could have divided the room differently, but the result is the same.

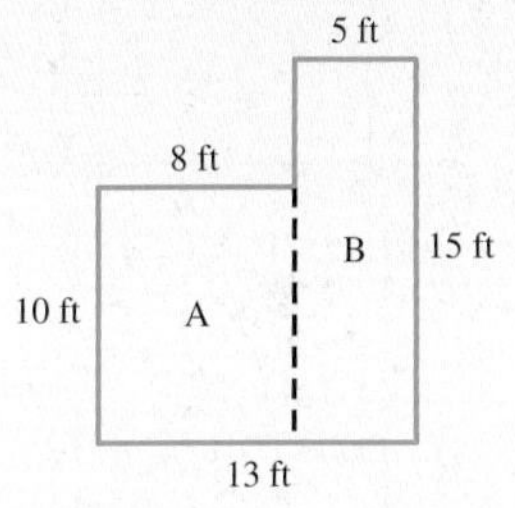

$A_A = 8 \times 10$
$= 80 \text{ ft}^2$
$A_B = 15 \times 5$
$= 75 \text{ ft}^2$
$A_A + A_B = 80 + 75$
$= 155 \text{ ft}^2$

To compute the area, we consider the foyer (a square) and the living room (rectangle) separately. Compute the area of each part, and then add the areas together to get a total area.

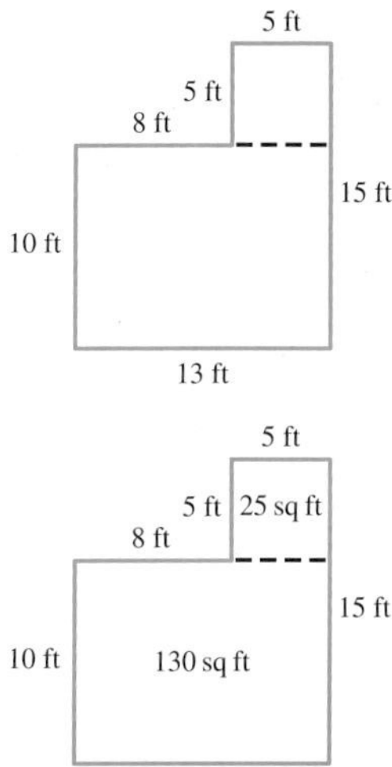

$A_{\text{Foyer}} = 5 \text{ ft} \times 5 \text{ ft} = 25 \text{ ft}^2$

$A_{\text{LvgRm}} = 10 \text{ ft} \times 13 \text{ ft} = 130 \text{ ft}^2$

$A = A_{\text{Foyer}} + A_{\text{LvgRm}} = 25 \text{ ft}^2 + 130 \text{ ft}^2 = 155 \text{ ft}^2$

As a general rule, carpeting is sold by the square yard. There are 9 ft^2 to the square yard.

$\frac{155}{9} = 17\frac{2}{9}$

Because we are sold whole square yards, we must purchase 18 yd^2 of carpet.

Check Yourself 5

Find the area of the figure shown.

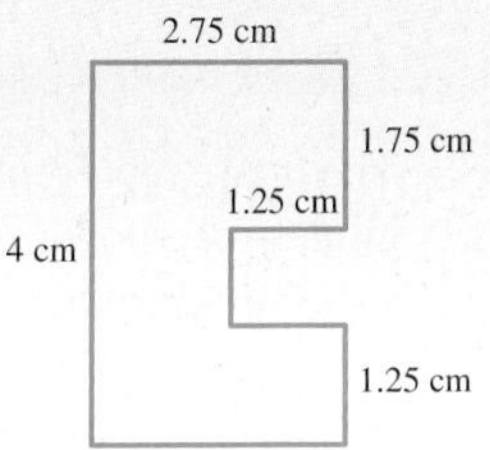

Often, composite figures involve parts of circles coupled with polygons. Consider the track in Example 6.

Example 6 Composite Figures

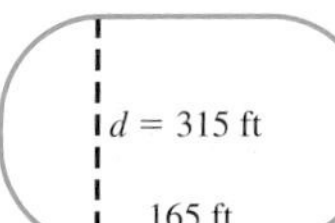

A high school track is shown. Compute the distance around the track and the area contained by it.

There are two 165-ft lengths and two semicircles (making one full circle). To compute the distance around the track, we add the two lengths to the circumference of the circle.

$C = \pi d \approx 3.14 \times 315 \approx 989 \text{ ft}$

$989 + 2 \times 165 = 1{,}319$

The track is approximately 1,319 ft long (about one-quarter mile).

To compute the area, we compute the areas of the interior rectangle and the circle separately, and then add the total. Remember to compute the radius of the circle for the area ($d = 2r$).

$r = \frac{315}{2} = 157.5$

$A_{\text{Circle}} = \pi r^2 \approx 3.14(157.5)^2 = 77{,}891.625$

$A_{\text{Rect}} = LW = 165 \times 315 = 51{,}975$

$A = A_{\text{Circle}} + A_{\text{Rect}} = 77{,}891.625 + 51{,}975 = 129{,}866.625$

The area is approximately 129,867 ft^2.

Check Yourself 6

(a) We wish to build a wrought-iron gate frame according to the sketch. How many feet of material do we need? Use 3.14 for π and round to the nearest foot.
Hint: We are looking for the distance around the object.

(b) Find the perimeter and area of the figure shown. Use 3.14 for π and round to the nearest tenth yard.

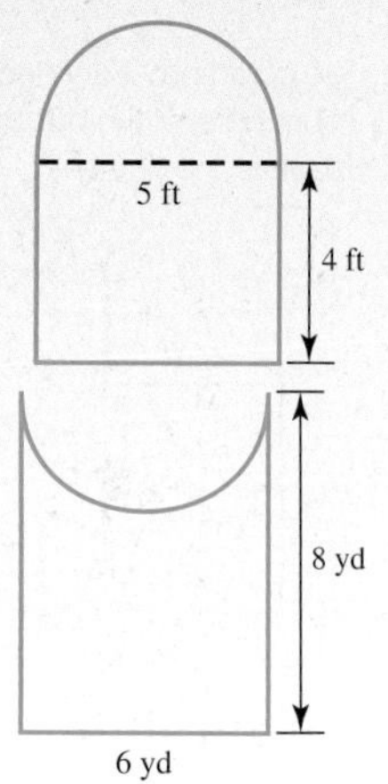

In manufacturing applications, it is frequently necessary to find the difference between two areas (or volumes) in order to calculate the area (or volume) of a production piece.

Example 7 A Manufacturing Technology Application

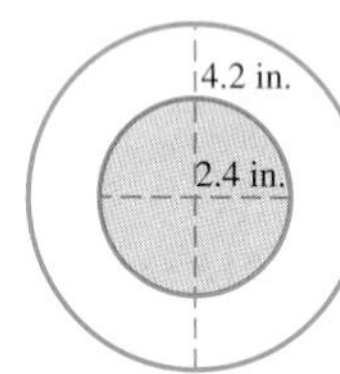

The figure represents a cross section of an O-ring. The diameters of the inner and outer circles are shown. Find the area of the O-ring shown (the white part in the reverse-color image of two circles). Report your result to the nearest hundredth in.^{2}.

The area of the larger circle, which includes the O-ring, can be computed using its radius. When we halve the diameter, we get a radius of 2.1 in.

$A_{\text{outer}} = \pi r^2 = \pi(2.1)^2 \approx 13.85 \text{ in.}^2$

Similarly, we compute the area of the inner circle, which has a radius of 1.2 in.

$A_{\text{inner}} = \pi r^2 = \pi(1.2)^2 \approx 4.52 \text{ in.}^2$

The area of the O-ring is the difference between these two areas.

$A_{\text{O-ring}} = A_{\text{outer}} - A_{\text{inner}} \approx 13.85 - 4.52 = 9.33 \text{ in.}^2$

Check Yourself 7

Find the outer and inner circumferences of the O-ring pictured in Example 7. Use 3.14 for π and round to the nearest hundredth in.

Check Yourself ANSWERS

1. 11 in. **2.** 15.7 in. **3.** 18.1 cm^2 **4.** 16 cm **5.** 9.75 cm^2 **6. (a)** 21 ft; **(b)** $P \approx 31.4$ yd; $A \approx 33.9$ yd^2 **7.** Outer: 13.19 in.; inner: 7.54 in.

Reading Your Text

These fill-in-the-blank exercises will help you understand some of the key vocabulary used in this section. The answers to these exercises are in the Answers Appendix at the back of the text.

(a) We call the perimeter of a circle its ________.

(b) The ________ is the distance from the center to a point on the circle.

(c) Joining basic figures into a single shape forms a ________ geometric figure.

(d) You can find the ________ of a composite figure by taking the sum of the areas of each of its basic-figure components.

< Objective 1 >

Find the circumference of each figure. Use 3.14 for π and round your answer to one decimal place.

1. 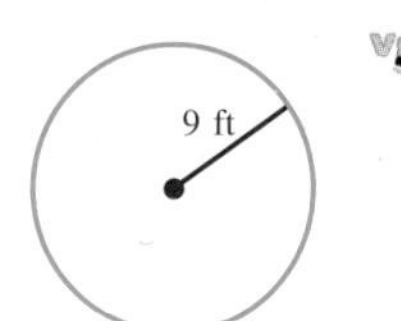

2. 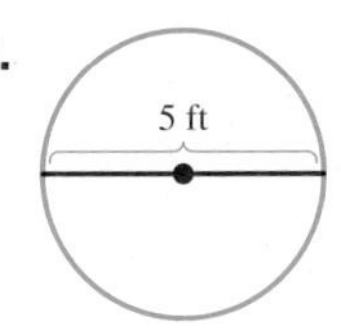

3. 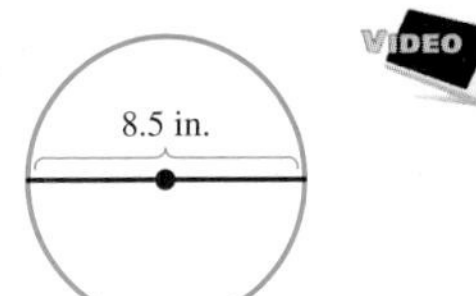

4. 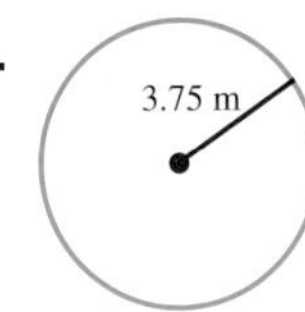

5. 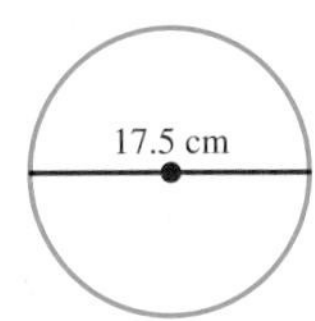

6. 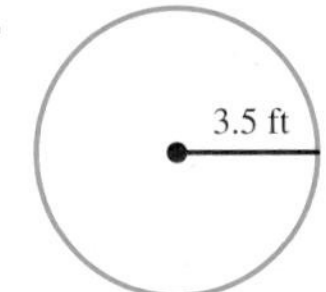

7. 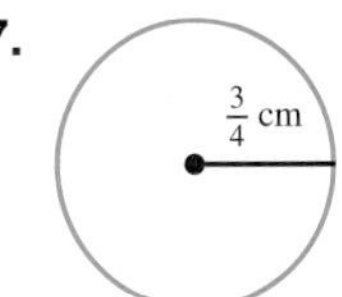

8. 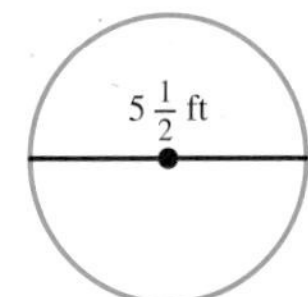

< Objective 2 >

Find the area of each figure. Use 3.14 for π and round your answer to one decimal place.

9. 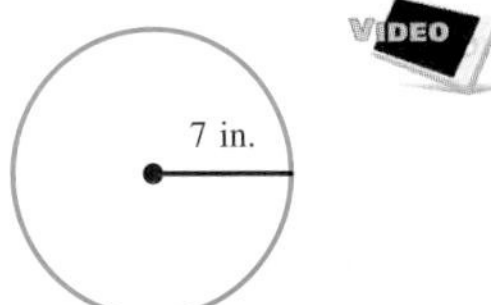

10. 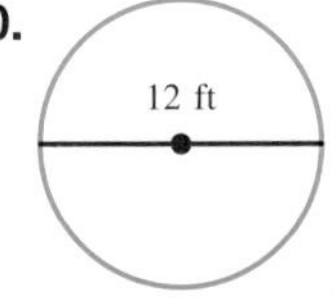

11. 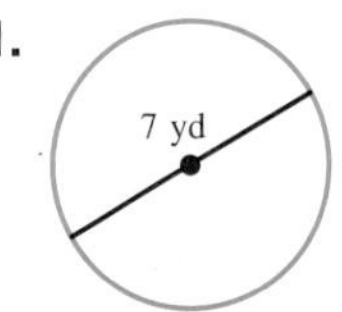

12. 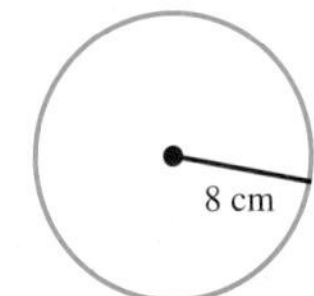

13. 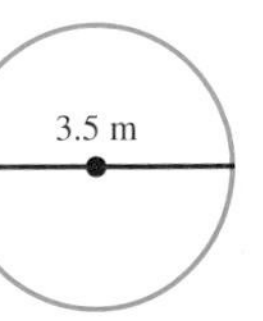

14. 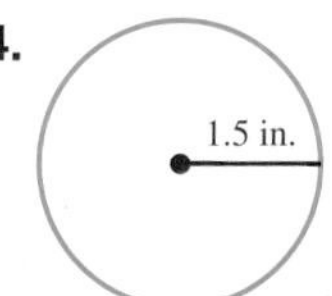

15. 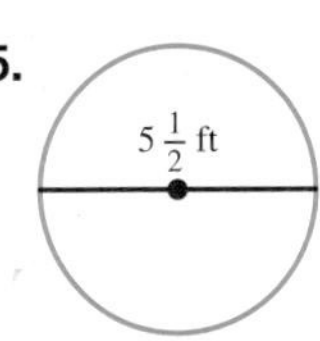

16. 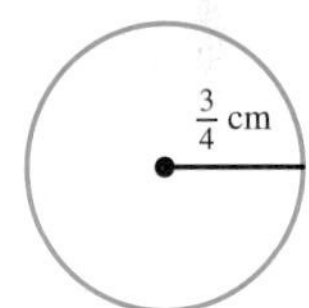

Solve each application.

17. **Science and Medicine** A path runs around a circular lake with a diameter of 1,000 yd. Robert jogs around the lake three times for his morning run. How far has he run?

18. **Crafts** A circular rug is 6 ft in diameter. Binding for the edge costs $4.50 per yard. What does it cost to bind the rug?

19. **Business and Finance** A circular lawn has a radius of 28 ft. You have a bag of fertilizer that covers 2,500 ft^2 of lawn. Do you have enough?

20. **Crafts** A circular coffee table has a diameter of 5 ft. What does it cost to have the top refinished if the company charges $8 per square foot for the refinishing?

21. **Construction** A circular terrace has a radius of 6 ft. If it costs $4.50 per square foot to pave the terrace with brick, what does it cost to pave the whole terrace?

22. **Construction** A house addition is in the shape of a semicircle (a half-circle) with a radius of 9 ft. What is its area?

< Objective 3 >

Find the perimeter of each figure.

23. 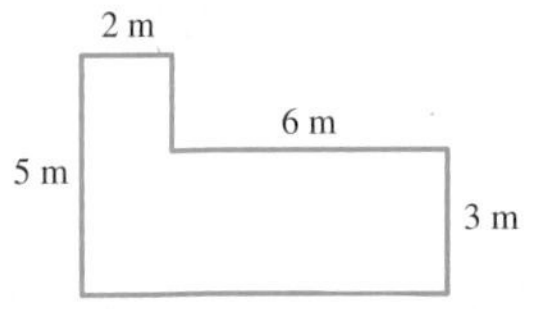

24. 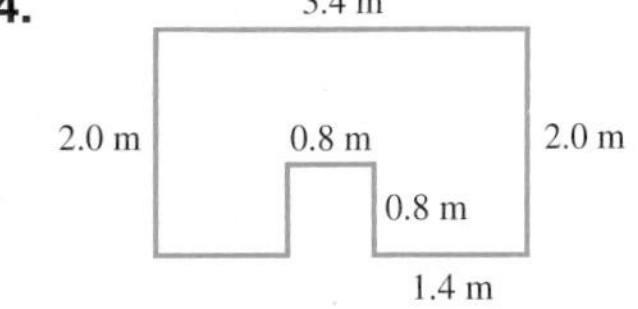

25. 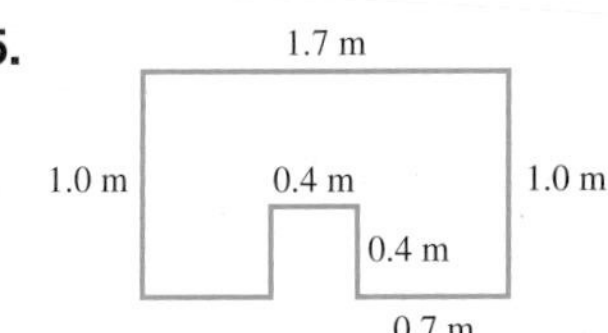

26. 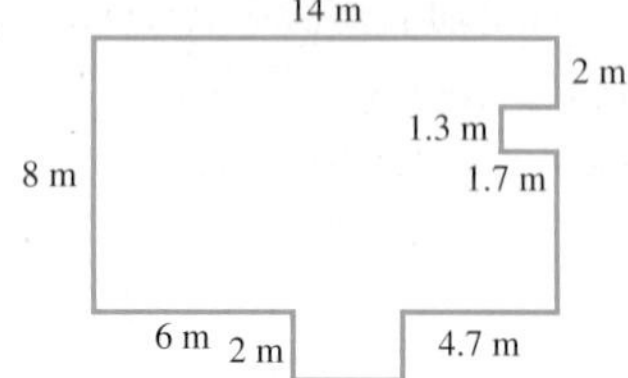

27. 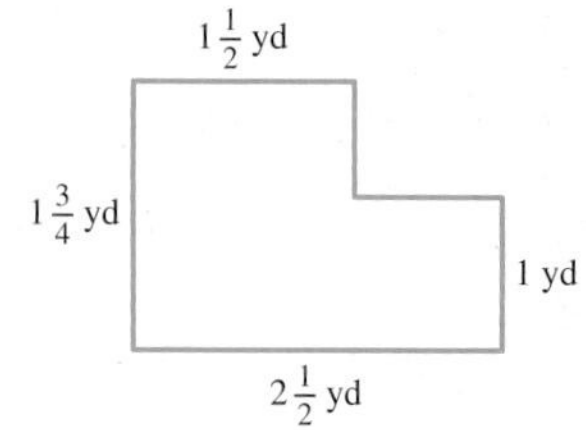

28. 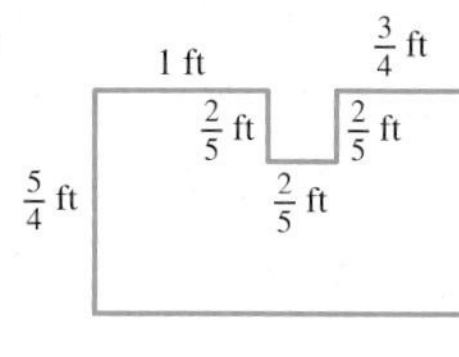

Find the perimeter or circumference of each figure. Use 3.14 for π and round your results to one decimal place.

29. 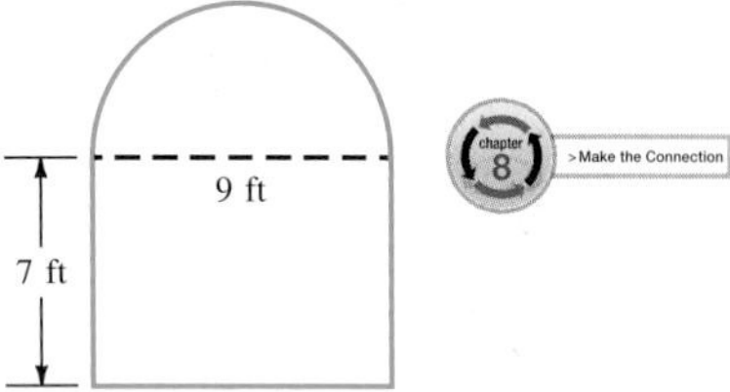

30. 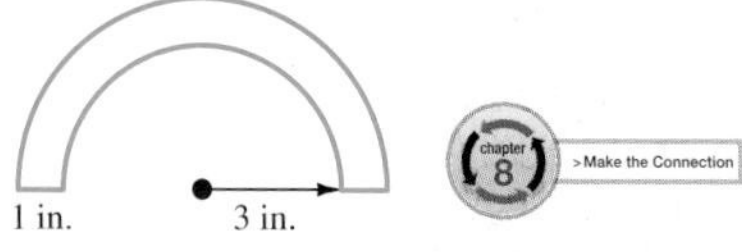

31. 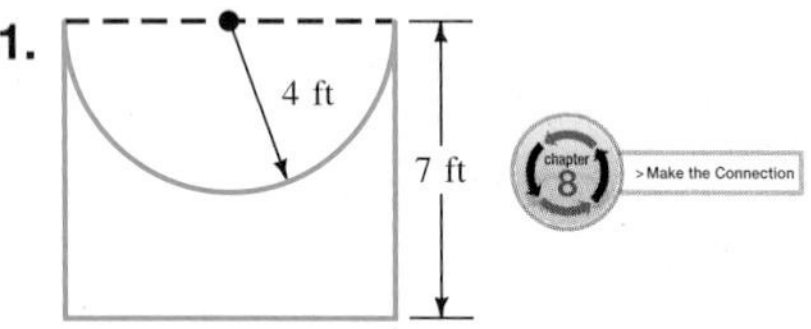

32. 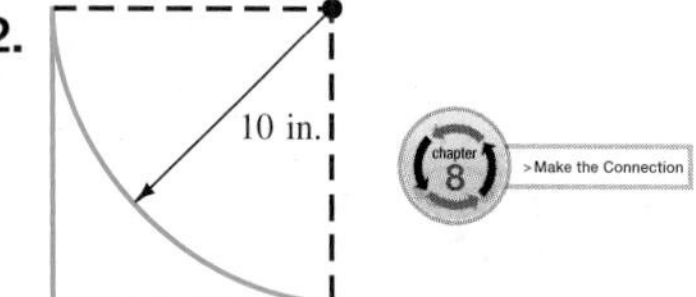

< Objective 4 >

Find the area of each figure.

33. 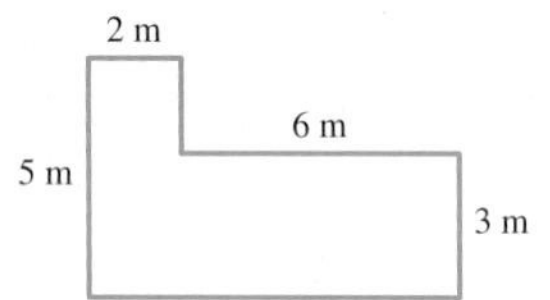

34. 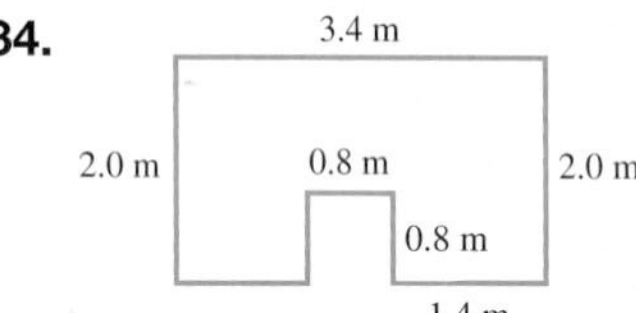

35. 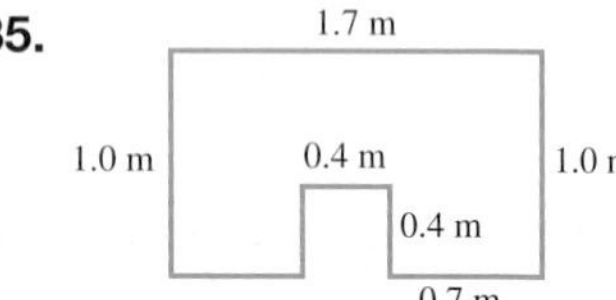

36. 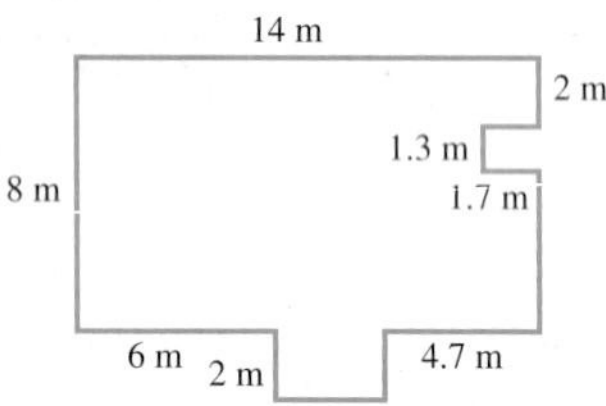

37. 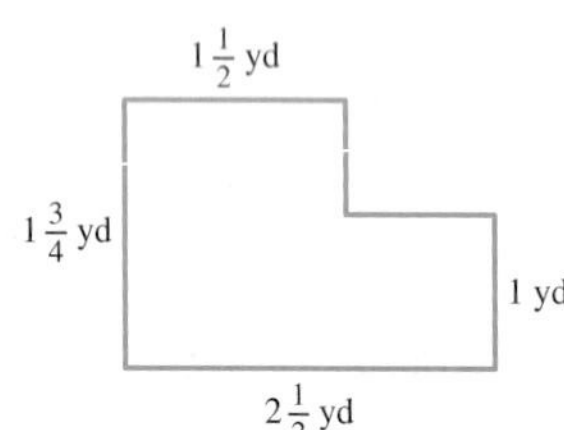

38. 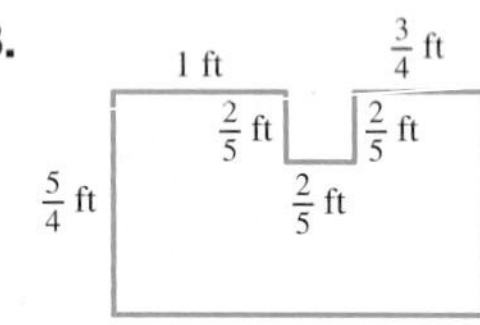

Find the area of the shaded part in each figure. Use 3.14 for π and round your answers to one decimal place, when appropriate.

39. 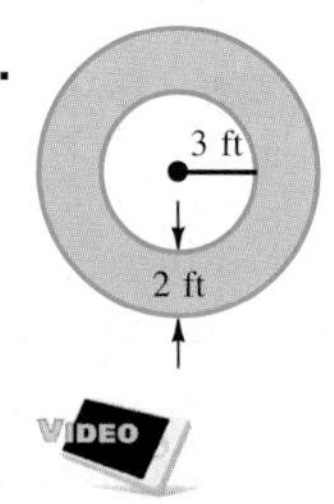

40. 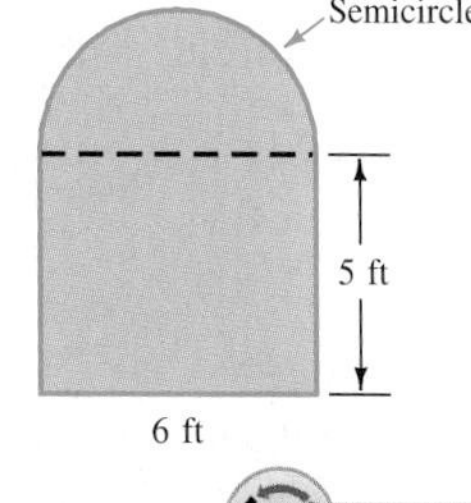

41. 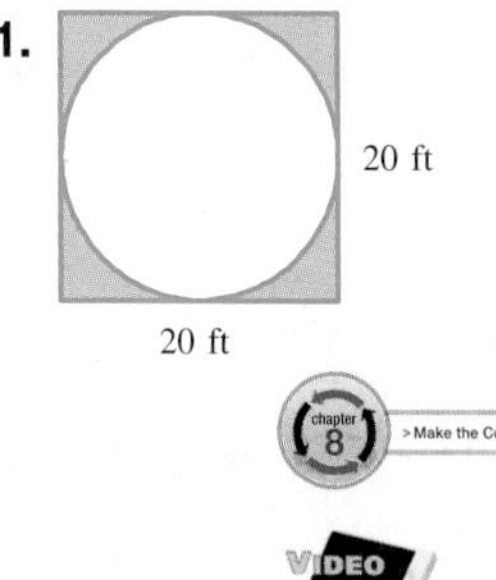

42. 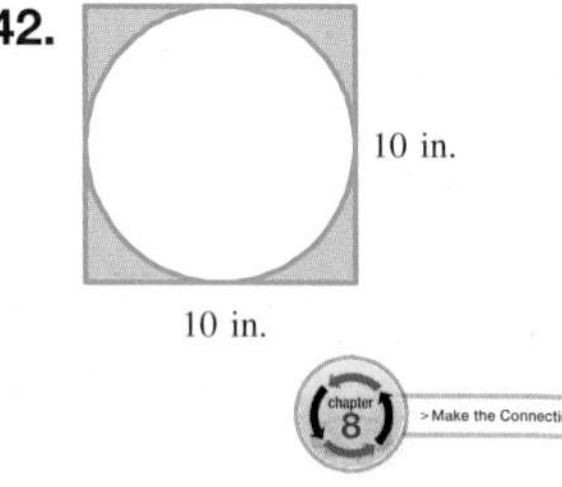

We use 3.14, or even $\frac{22}{7}$, as approximations for π when calculating by hand; we use technology when we need more precise answers. Since π is an *irrational number*, we cannot write it exactly as a decimal or fraction. Technology makes it easier to use π rounded to any decimal place required in an application. Calculators and computers do not have an exact value for π either, but they carry many more decimal places in their computations—often as many as 16 or more.

On many graphing calculators, π is found as a *2nd function* above the caret key [^]. In this case, we first press [2nd] to access π. We express the keystrokes necessary to enter π as [2nd] [π].

If the radius of a circle is 6 in., we use a calculator to approximate its area, πr^2 with $r = 6$.

[2nd] [π] [×] 6 [x^2] [ENTER]

```
π*6²
        113.0973355
```

We see that its area is approximately 113.097 in.^{2}.

Use the π key on your calculator to approximate the circumference and area of each circle. Round your results to the nearest thousandth.

43. 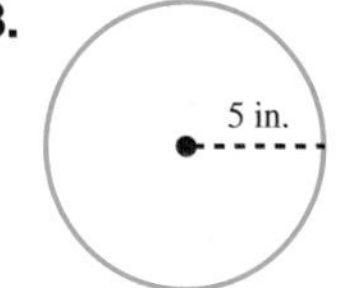

44. 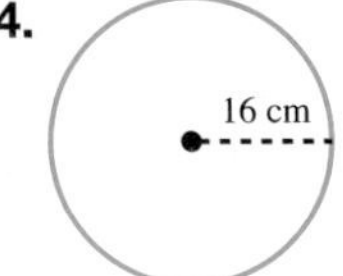

45. 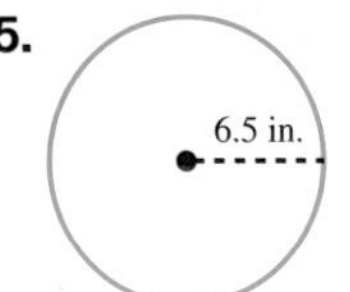

46. 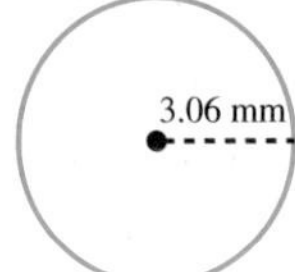

47. 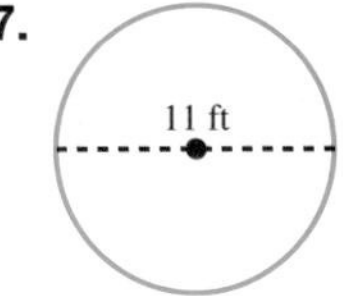

48. 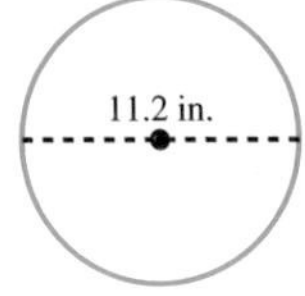

49. **ALLIED HEALTH** An ultrasound technician measures the biparietal (head) diameter of a 14-week-old fetus as 2.5 cm. Determine the head circumference of the fetus.

50. **ALLIED HEALTH** An ultrasound technician measures the biparietal (head) diameter of a 28-week-old fetus as 7.3 cm. Determine the head circumference of the fetus.

51. **ALLIED HEALTH** During a high-risk pregnancy, amniotic fluid levels are monitored. To measure fluid levels, the ultrasound technician looks for roughly circular areas of fluid on the ultrasound image. One such fluid pocket had a diameter of 1.7 cm. Determine the area (cross-sectional area) of this circular fluid pocket.

52. **ALLIED HEALTH** During a high-risk pregnancy, amniotic fluid levels are monitored. To measure fluid levels, the ultrasound technician will look for roughly circular areas of fluid on the ultrasound image. One such fluid pocket had a diameter of 0.9 cm. Determine the area (cross-sectional area) of this circular fluid pocket.

53. **MANUFACTURING TECHNOLOGY** The cross section of a shaft key takes the shape of a quarter-circle. Find the perimeter and area of this cross section of the shaft key.

6 mm

54. **MANUFACTURING TECHNOLOGY** The surface of this piece needs to be coated with a nonstick coating. A container of the coating can cover 28 ft^2. How many parts can be coated with one can? chapter 8 > Make the Connection

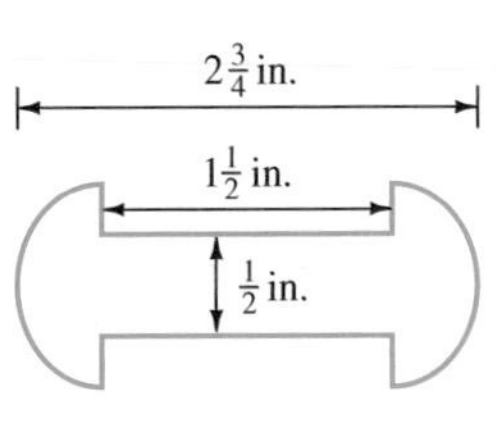

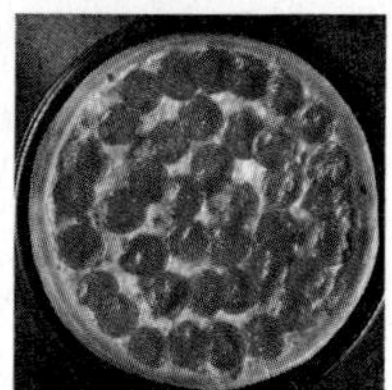

55. Papa Doc's delivers pizza. The 8-in.-diameter pizza is \$8.99, and the price of a 16-in.-diameter pizza is \$17.98. Write a plan to determine which is the better buy.

56. The distance from Philadelphia to Sea Isle City is 100 mi. A car was driven this distance using tires with a radius of 14 in. How many revolutions of each tire occurred on the trip?

57. Find the area and the circumference (or perimeter) of each object.

(a) A penny; **(b)** a nickel; **(c)** a dime; **(d)** a quarter; **(e)** a half-dollar; **(f)** a dollar coin; **(g)** a dollar bill; and **(h)** one face of the pyramid on the back of a \$1 bill.

58. How would you determine the cross-sectional area of a Douglas fir tree (at, say, 3 ft above the ground), without cutting it down? Use your method to solve the following problem:

If the circumference of a Douglas fir is 6 ft 3 in., measured at a height of 3 ft above the ground, compute the cross-sectional area of the tree at that height.

Answers

1. 56.5 ft **3.** 26.7 in. **5.** 55 cm **7.** 4.7 cm **9.** 153.9 in.^{2} **11.** 38.5 yd^2 **13.** 9.6 m^2 **15.** 23.7 ft^2 **17.** 9,420 yd **19.** Yes **21.** \$508.68 **23.** 26 m **25.** 6.2 m **27.** $8\frac{1}{2}$ yd **29.** 37.1 ft **31.** 34.6 ft **33.** 28 m^2 **35.** 1.54 m^2 **37.** $3\frac{5}{8}$ yd^2 **39.** 50.2 ft^2 **41.** 86 ft^2 **43.** $C \approx 31.416$ in.; $A \approx 78.54$ in.^{2} **45.** $C \approx 40.841$ in.; $A \approx 132.732$ in.^{2} **47.** $C \approx 34.558$ ft; $A \approx 95.033$ ft.^{2} **49.** 7.9 cm **51.** 2.3 cm^2 **53.** Perimeter: 21.42 mm; area: 28.27 mm^2 **55.** Above and Beyond **57.** Above and Beyond

Exploring Circles

In this activity, we ask the question, How does multiplying the radius of a circle by a certain factor change the circumference? And how does this same action change the area?

Complete lines two and three for Circle 1. Use 3.14 for π and round to one decimal place.

	Radius	Circumference	Factor	New Radius	New Circumference
Circle 1	10 ft	62.8 ft	2	20 ft	125.6 ft
	10 ft		3		
	10 ft		$\frac{1}{2}$		
Circle 2					

Construct a second circle by choosing a new radius and multiplying factors. Use this circle to complete Circle 2 in the table.

How does the new circumference compare to the original circumference in each case? Try to describe your observations with a general statement.

Now check the effect on the area of each circle. Use 3.14 for π and complete the table.

	Radius	Area	Factor	New Radius	New Area
Circle 1	10 ft		2		
	10 ft		3		
	10 ft		$\frac{1}{2}$		
Circle 2					

How does the new area compare to the original area in each case? Try to describe your observations with a general statement.

Challenge: If you want to double the area of a circle, what should you multiply the radius by?

Hint: You will probably have to solve this by trial and error. You will not find an exact answer, but you can determine an answer accurate to the nearest thousandth.

8.4 Triangles

< 8.4 Objectives >

1 > Classify triangles by angles
2 > Classify triangles by sides
3 > Find the measures of the angles of a triangle
4 > Identify similar triangles
5 > Find the lengths of the sides of a triangle

Now that you know something about angles, it is interesting to look again at triangles. Why is this shape called a triangle?

Literally, *triangle* means "three angles."

The same classifications we used for angles can be used for triangles. If a triangle has a right angle, we call it a *right triangle*.

If it has three acute angles, it is called an *acute triangle*.

If it has an obtuse angle, it is called an *obtuse triangle*.

Example 1 Identifying an Acute Triangle

< Objective 1 >

Identify any acute triangles.

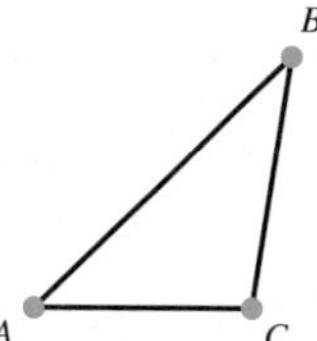

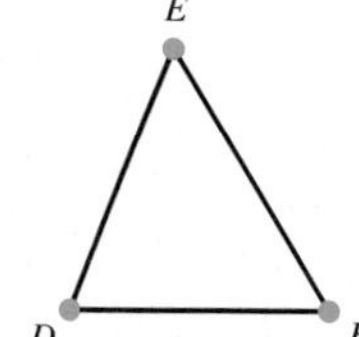

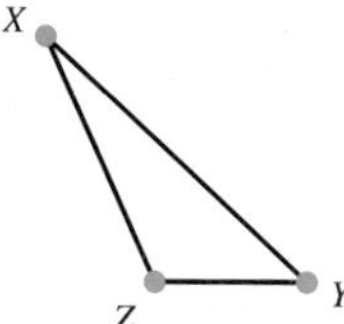

Only $\triangle DEF$ is an acute triangle. Both $\triangle ABC$ and $\triangle XYZ$ have one obtuse angle.

Check Yourself 1

Identify any obtuse triangles.

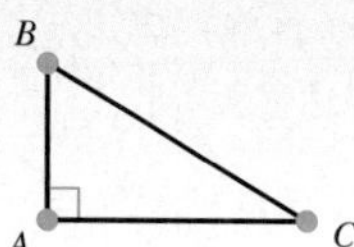

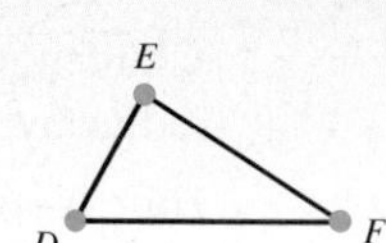

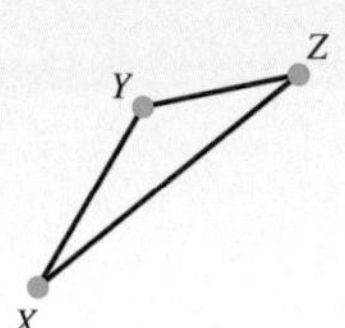

NOTES

All three angles of an equilateral triangle have the same measure.

Exactly two angles of an isosceles triangle have equal measure.

The angles with equal measure are opposite the sides with equal length.

All three angles of a scalene triangle have different measures.

We can also classify triangles based on how many sides have the same length.

A triangle is called an *equilateral triangle* if all three sides have the same length.

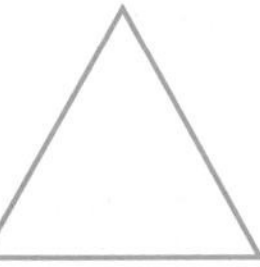

A triangle is called an *isosceles triangle* if exactly two sides have the same length.

A triangle is called a *scalene triangle* if no two sides have the same length.

Example 2 Labeling Types of Triangles

< Objective 2 >

NOTE

Each of these triangles can be classified in different ways. $\triangle XYZ$ is a right triangle, but it is also scalene.

Of these triangles, which are equilateral? Isosceles? Scalene?

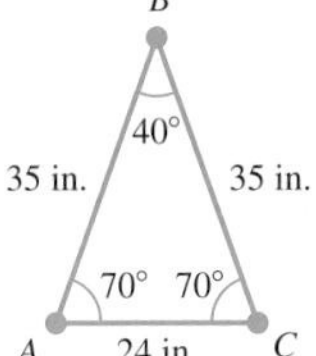

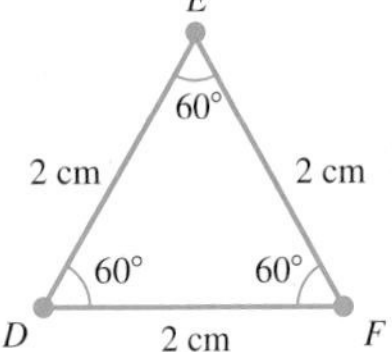

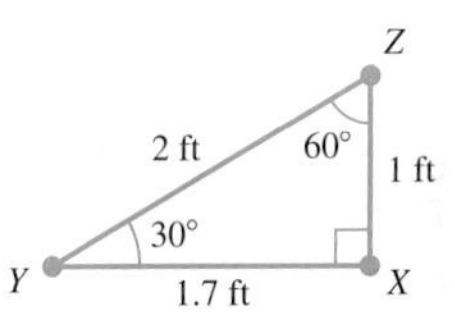

$\triangle ABC$ is an isosceles triangle because two of the sides are the same length. And $\triangle DEF$ is an equilateral triangle because all three sides are the same length. $\triangle XYZ$ is a scalene triangle because all three sides have different lengths.

Check Yourself 2

Label each triangle as equilateral, isosceles, or scalene.

(a)

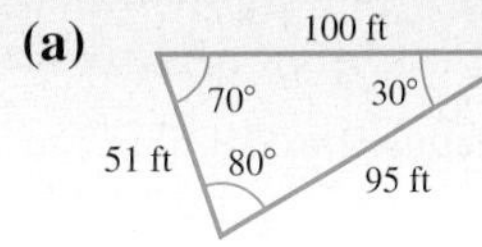

(b)

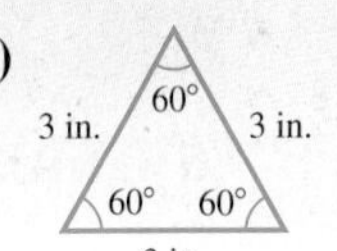

(c)

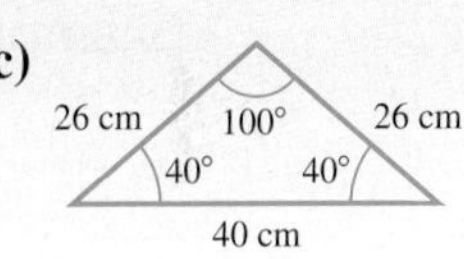

Go back and look at the sum of the angles inside each of the triangles in Example 2. You will note that they always add to 180°. No matter how we draw a triangle, the sum of the three angles inside the triangle is *always* 180°.

Here is an experiment that might convince you that this is always the case.

1. Using a straight edge, draw any triangle on a sheet of paper.

2. Use scissors to cut out the triangle.
3. Cut the three vertices off of the triangle.

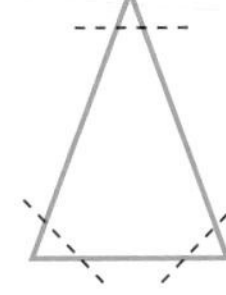

4. Lay the three vertices (with the points of the triangle touching) together. They always form a straight angle, which we saw in Section 8.1 measures 180°. ✓

Property

Angles of a Triangle

For any triangle *ABC*,

$m\angle A + m\angle B + m\angle C = 180°$

We say that the sum of the (interior) angles of every triangle is 180°.

Example 3 Finding an Angle Measure

< Objective 3 >

Find the measure of the third angle in this triangle.

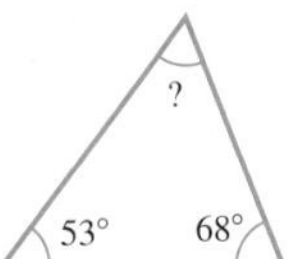

We need the three measurements to add up to 180°, so we add the two given measurements ($53° + 68° = 121°$). Then we subtract that from 180° ($180° - 121° = 59°$). This gives us the measure of the third angle, 59°.

Check Yourself 3

Find the measure of $\angle ABC$.

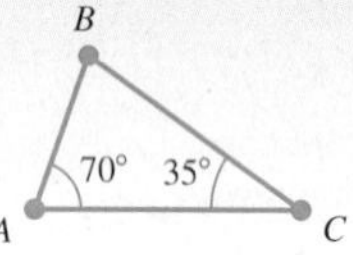

Definition

Similar Triangles

If the measurements of the three angles in two different triangles are the same, we say the two triangles are **similar triangles.**

Similar triangles have exactly the same shape, but their sizes may differ. In this case, they are "scale" versions of each other.

Example 4 Identifying Similar Triangles

< Objective 4 >

NOTE

This can be written

$\triangle ABC \sim \triangle XYZ$

Which two triangles are similar?

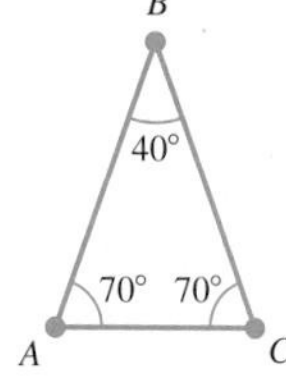

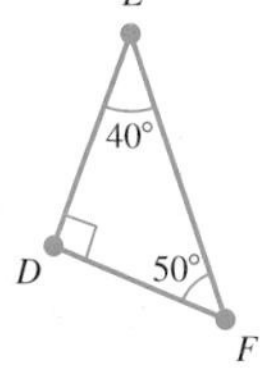

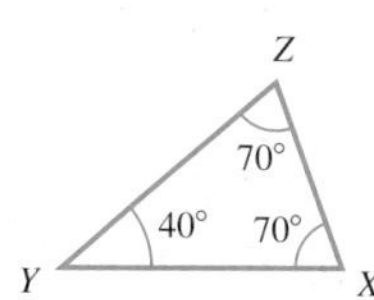

Although they are of different size, $\triangle ABC$ and $\triangle XYZ$ are similar because they have the same angle measurements.

Check Yourself 4

Find the two triangles that are similar.

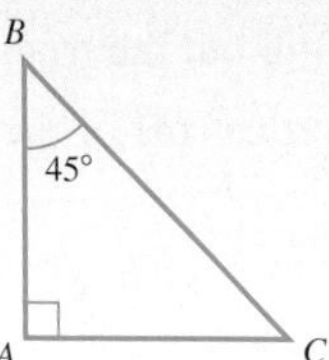

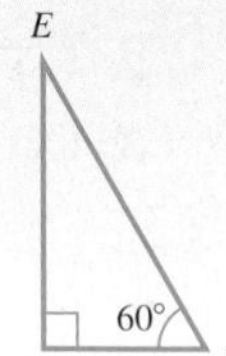

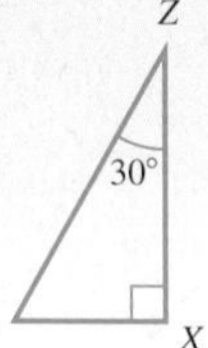

Finally, we return to an idea that we first saw in Chapter 5.

Property

Similar Triangles If two triangles are similar, their corresponding sides have the same ratio.

This property of similar triangles is often used to find the heights of tall objects as illustrated in Example 5.

Example 5 Finding the Height of a Tree

< Objective 5 >

If a 180-cm man casts a shadow that is 60 cm long, how tall is a tree that casts a shadow that is 9 m long?

Because of the angle of the sun, the man and his shadow form a similar triangle to the tree and its shadow. Therefore, we can use ratios to find the height of the tree.

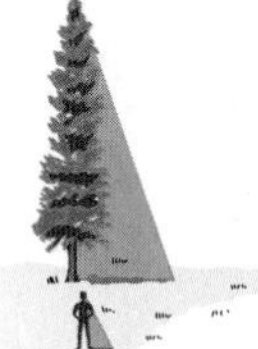

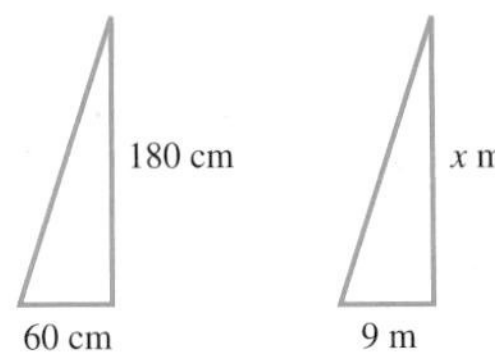

$$\frac{180 \text{ cm}}{60 \text{ cm}} = \frac{x \text{ m}}{9 \text{ m}}$$

$$x \cdot 60 = 180 \cdot 9$$

$$x = \frac{180 \cdot 9}{60}$$

$$x = 27$$

The tree is 27 m tall.

Check Yourself 5

If a 160-cm man casts a 120-cm shadow, how tall is a building if its shadow is 60 m?

Check Yourself ANSWERS

1. $\triangle XYZ$ **2.** **(a)** Scalene; **(b)** equilateral; **(c)** isosceles **3.** 75° **4.** $\triangle DEF$ and $\triangle XZY$ **5.** 80 m

Reading Your Text

These fill-in-the-blank exercises will help you understand some of the key vocabulary used in this section. The answers to these exercises are in the Answers Appendix at the back of the text.

(a) A triangle is called an ______________ triangle if all three sides are the same length.

(b) A triangle is called an ______________ triangle if exactly two sides are the same length.

(c) If the measurements of the three angles in two different triangles are the same, we say that the two triangles are ______________ triangles.

(d) If two triangles are similar, their ______________ sides have the same ratio.

8.4 exercises

Skills | Calculator/Computer | Career Applications | Above and Beyond

< Objective 1 >

Label each triangle as acute or obtuse.

1. 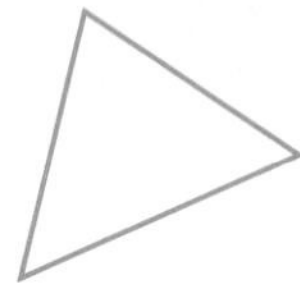

2. 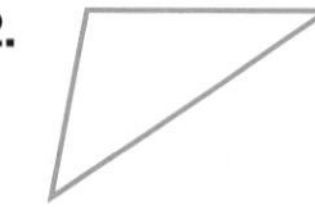

3. 

4. 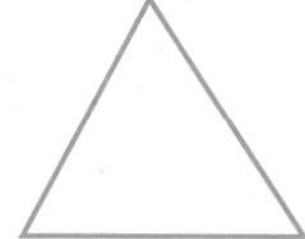

< Objective 2 >

Label each triangle as equilateral, isosceles, or scalene.

5. 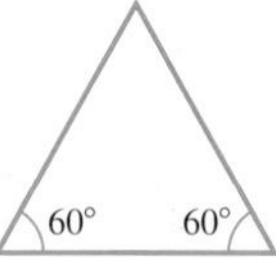

6. 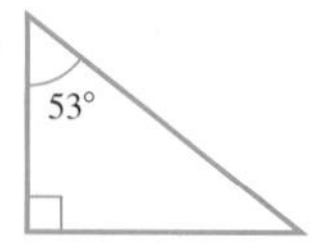

7. 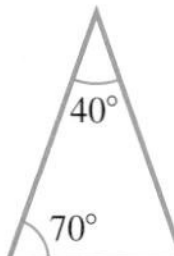

8. 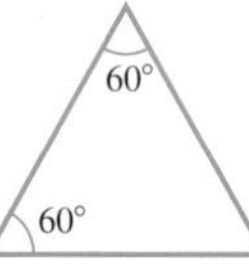

9. 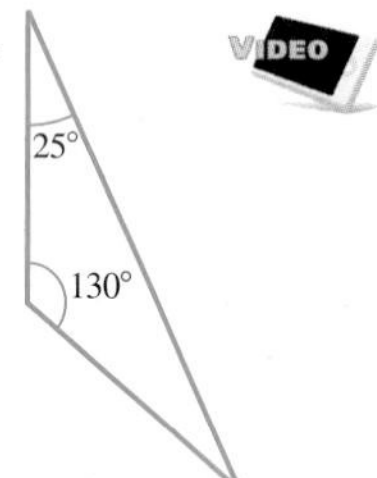

10. 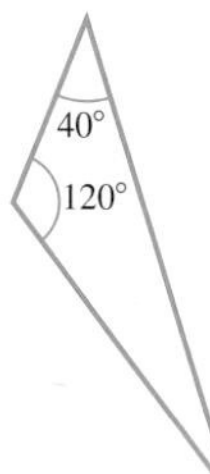

< Objective 3 >

Find the missing angle and then label the triangle as equilateral, isosceles, or scalene.

11. 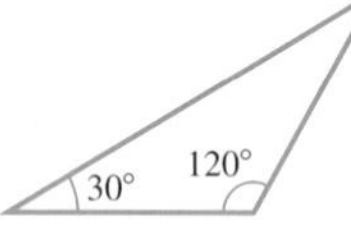

12. 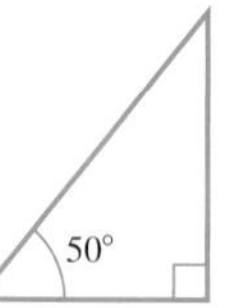

13. 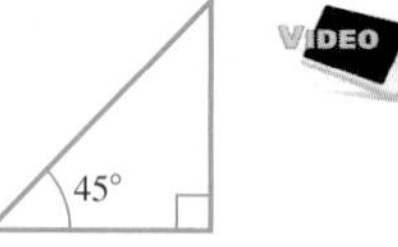

14. 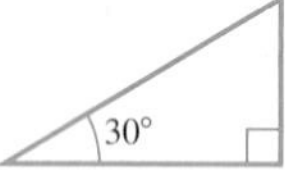

15. 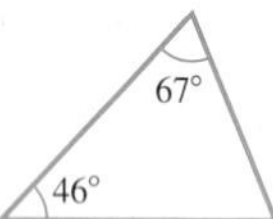

16. 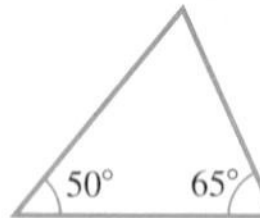

For each triangle shown, find the indicated angle(s).

17. Find $m\angle C$.

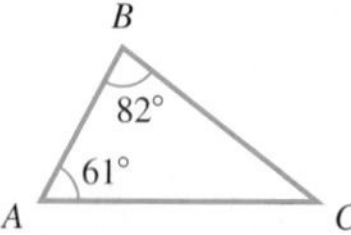

18. Find $m\angle B$.

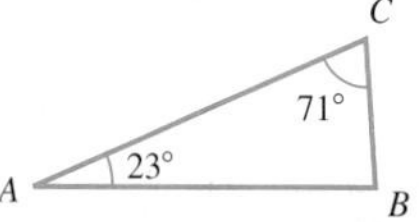

19. Find $m\angle A$.

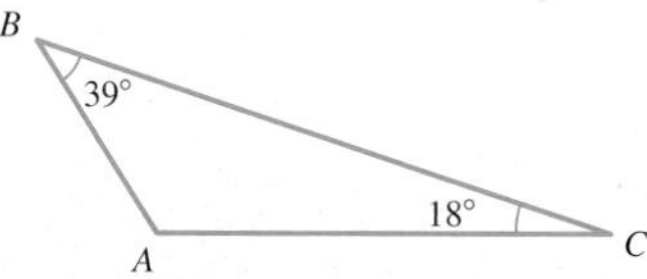

20. Find $m\angle B$.

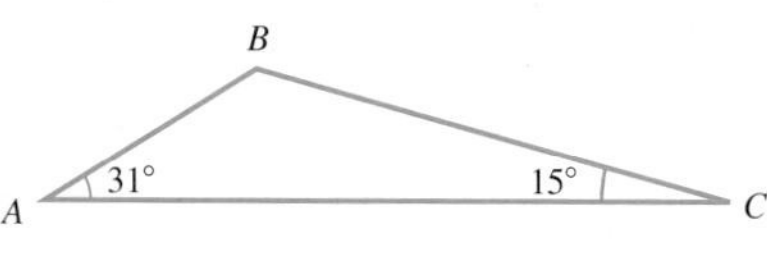

21. Find $m\angle B$.

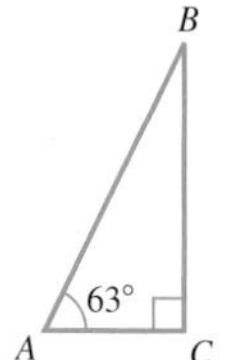

22. Find $m\angle A$.

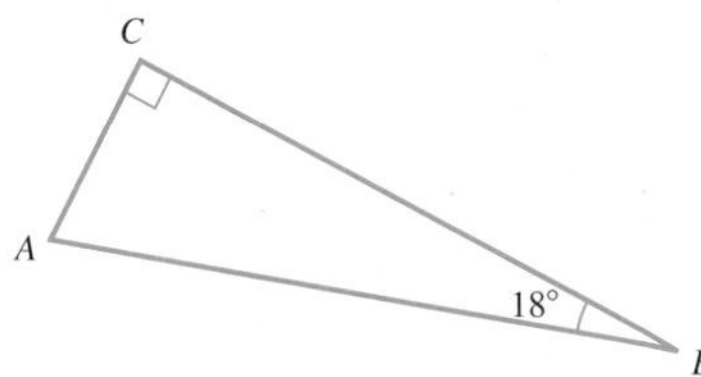

23. Find $m\angle A$ and $m\angle C$.

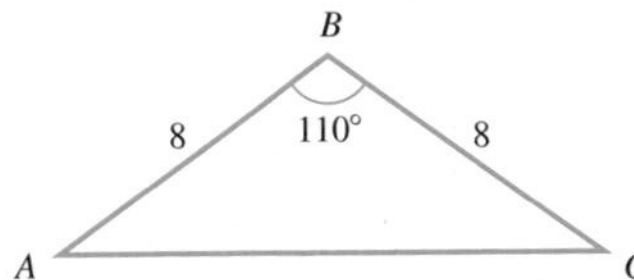

24. Find $m\angle D$ and $m\angle F$.

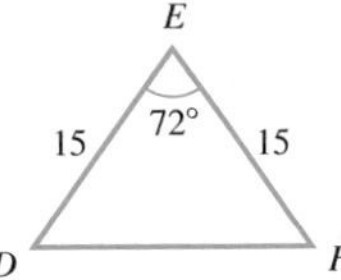

< Objective 4 >

25. Which two triangles are similar?

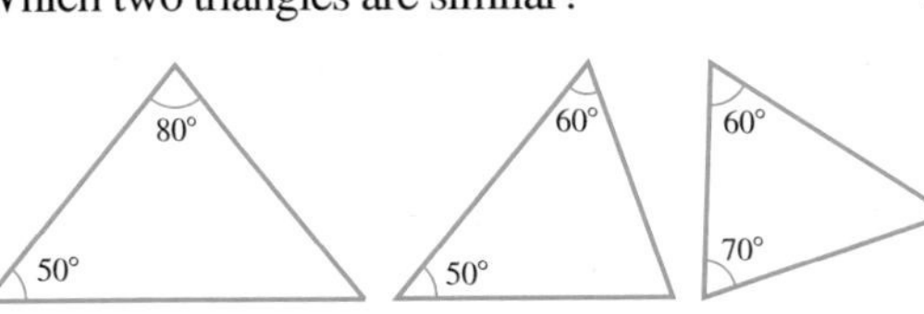

26. Which two triangles are similar?

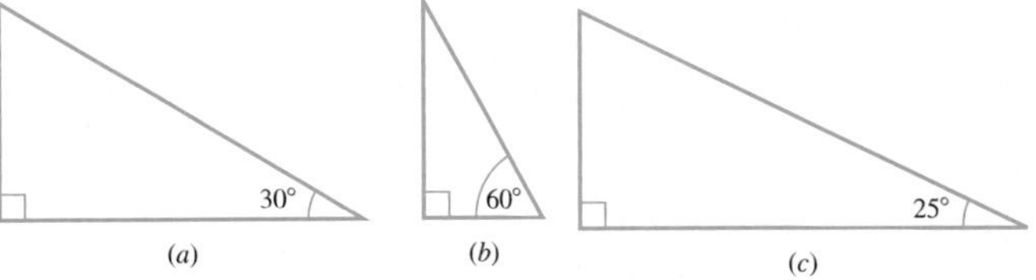

< Objective 5 >

Each pair of triangles is similar. Find the indicated side.

27. Find v. VIDEO

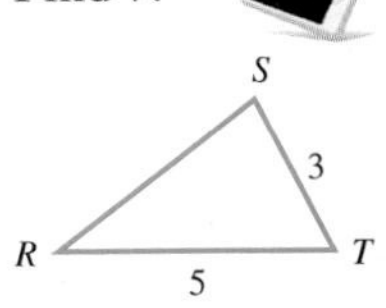

28. Find f.

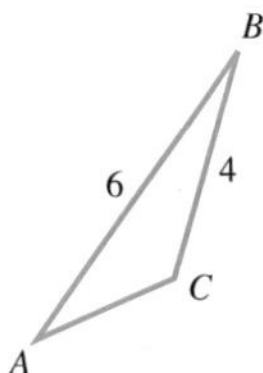

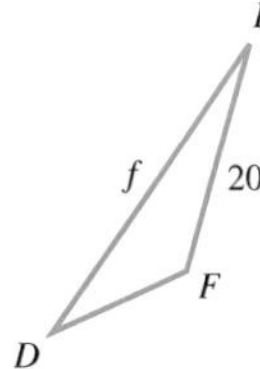

29. Find g.

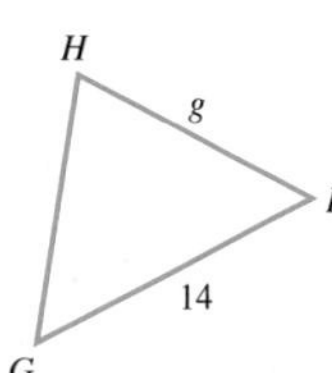

30. Find m.

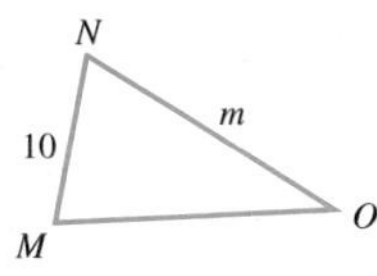

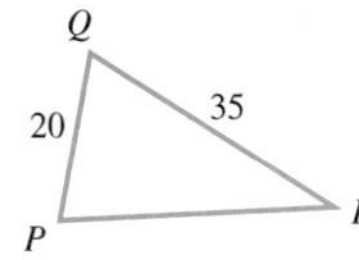

31. Find t.

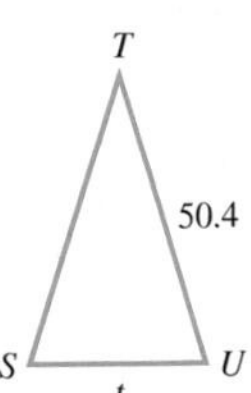

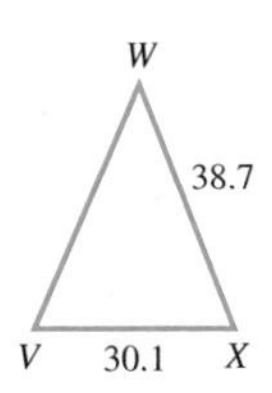

32. Find e.

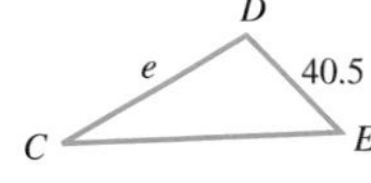

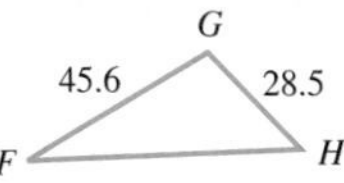

Identify a pair of similar triangles in each figure and find the length of the indicated side. Round your results to the nearest hundredth.

33. Find $\overline{KL}$.

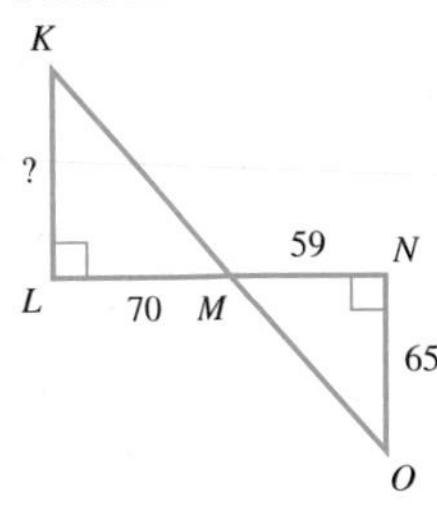

34. Find $\overline{PQ}$.

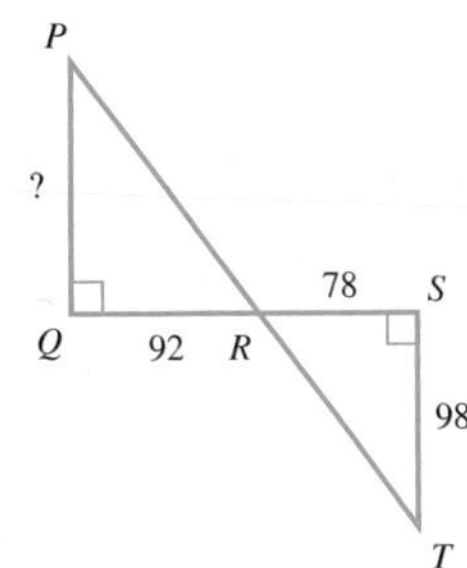

35. Find $\overline{VX}$.

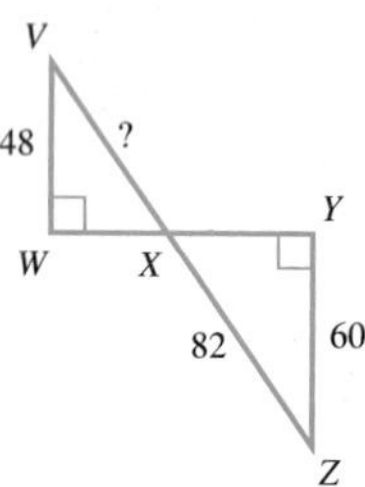

36. Find $\overline{AC}$.

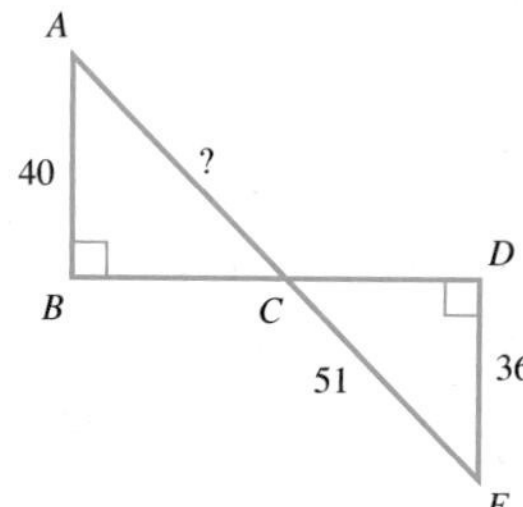

Find the indicated side. If necessary, round to the nearest tenth of a unit.

37. Find $\overline{DE}$.

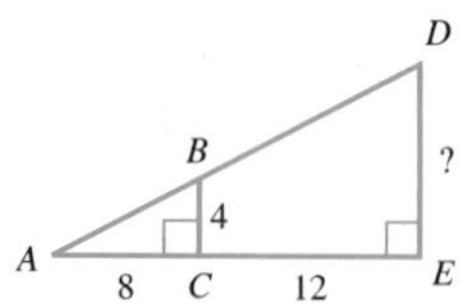

38. Find $\overline{IJ}$.

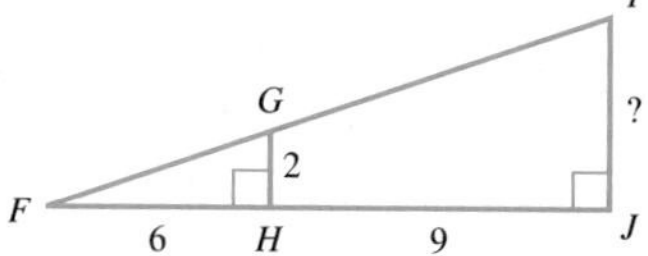

39. Find $\overline{KL}$.

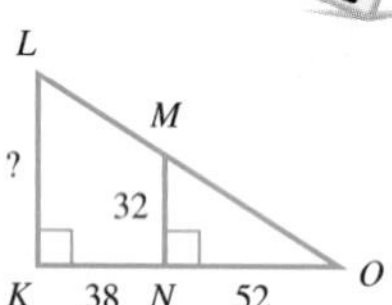

40. Find $\overline{PQ}$.

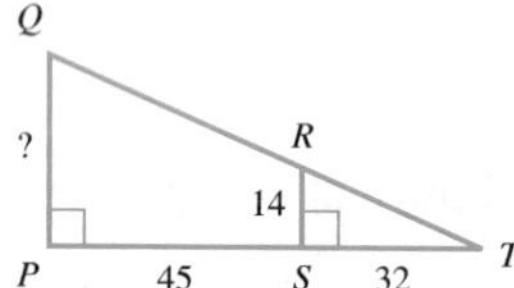

41. Given: $m\angle BCA = m\angle DEA$. Find $\overline{DE}$.

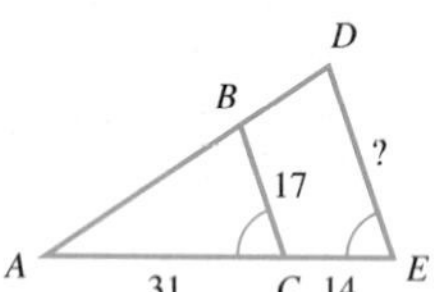

42. Given: $m\angle GHF = m\angle IJF$. Find $\overline{IJ}$.

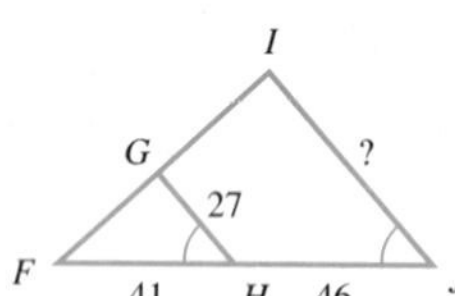

43. **Geometry** A lamppost casts a 4-ft shadow. At the same time, a yardstick casts a 9-in. shadow. How tall is the pole?

44. **Geometry** A tree casts a 5-m shadow. At the same time, a meter stick casts a 0.4-m shadow. How tall is the tree?

Skills	Calculator/Computer	**Career Applications**	Above and Beyond

45. **Manufacturing Technology** In a common truss, the slope triangle tells you how high the truss is compared to the run. Find the height of this common truss. *Hint:* The run is half the span.

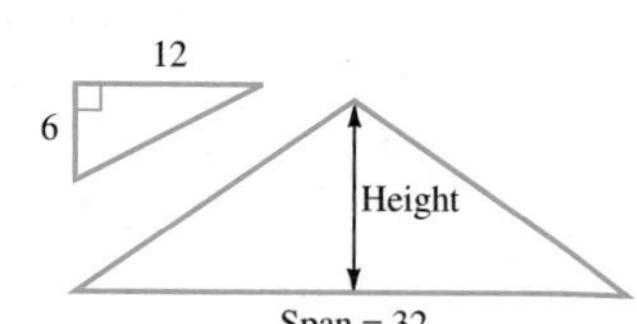

46. **Manufacturing Technology** A chipping hammer is shaped like a wedge with a tip angle of 25°. Find the angle at the base of the isosceles triangle.

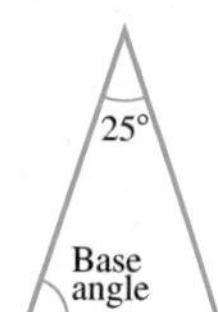

47. **Manufacturing Technology** The slope triangle on a truss tells you how high the truss is compared to the run.

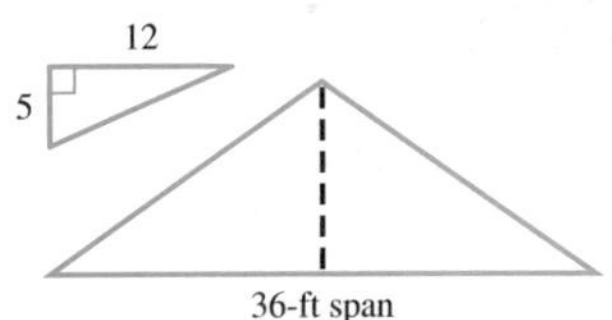

What is the height of this truss, given the $\frac{5}{12}$-slope triangle? Remember that the run is half the span.

Skills | Calculator/Computer | Career Applications | **Above and Beyond**

48. Use the ideas of similar triangles to determine the height of a pole or tree on your campus. Work with one or two partners.

One side of each triangle has been extended, forming what is called an exterior angle. *In each case, find the measure of the indicated exterior angle.*

49.

VIDEO

50.

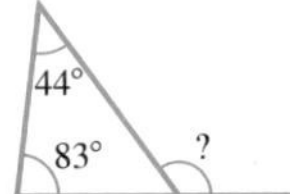

51.

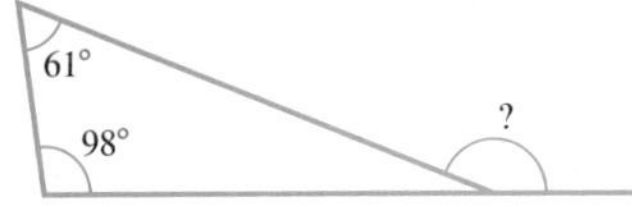

52. What do you observe from exercises 49 to 51? Write a general conjecture about an exterior angle of a triangle.

53. Write an argument to show that an equilateral triangle cannot have a right angle.

54. Argue that, given an equilateral triangle, the measure of each angle must be 60°.

55. Argue that a triangle cannot have more than one obtuse angle.

56. Is it possible to have an isosceles right triangle? If such a triangle exists, what can be said about the angles? Defend your statements.

57. Create an argument to support the statement:

If $\triangle ABC$ is a right triangle, with $m\angle C = 90°$, then $\angle A$ and $\angle B$ must be acute and complementary.

Answers

1. Acute **3.** Acute **5.** Equilateral **7.** Isosceles **9.** Isosceles **11.** 30°; isosceles **13.** 45°; isosceles **15.** 67°; isosceles **17.** 37° **19.** 123° **21.** 27° **23.** $m\angle A = 35°$; $m\angle C = 35°$ **25.** (b) and (c) **27.** 20 **29.** 12 **31.** 39.2 **33.** 77.12 **35.** 65.6 **37.** 10 **39.** 55.4 **41.** 24.7 **43.** 16 ft **45.** 8 ft **47.** 7.5 ft **49.** 143° **51.** 159° **53.** Above and Beyond **55.** Above and Beyond **57.** Above and Beyond

Activity 24 ::

Composite Geometric Figures

When first introduced to geometry, you worked with fairly straightforward figures such as squares, circles, and triangles. Most real-world objects cannot be described by such simple geometric figures. Consider the chair that you are using right now. The chair is probably made up of several shapes put together.

Composite geometric figures are figures formed by combining two or more simple geometric figures. One example of a composite geometric figure is the Norman window. Norman windows are windows constructed by combining a rectangle with a half-circle (see the figure). Given such a figure, there are several questions we could ask. We might ask an area question to find the amount of glass used, a perimeter question to find the amount of frame, or a combination question to determine the amount of wall space necessary to accommodate such a window.

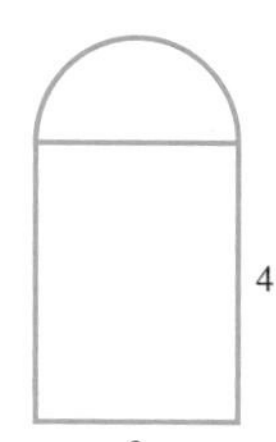

Assume all measurements are in feet and round answers to two decimal places when necessary.

1. Find the area of the rectangular piece of glass in the Norman window pictured.
2. Find the area of the half-circle piece of glass.
3. Find the total area of the glass.
4. If the glass costs \$3 per square foot for a rectangular piece and \$4.75 per square foot for the circular piece, find the total cost of the glass.
5. Find the outer perimeter of the figure.
6. Find the length of framework needed for the Norman window (do not forget to include the strip of frame separating the rectangle and half-circle).
7. Find the cost of the framework if it costs \$1.25 per foot for straight pieces and \$3.25 per foot for curved pieces.
8. Use your answers to exercises 4 and 7 to determine the total cost of the Norman window pictured.
9. Find the dimensions of the smallest rectangle that would completely contain the Norman window.
10. Find the area of the "leftover" created by cutting the Norman window from the rectangle found in exercise 9.

8.5 Square Roots and the Pythagorean Theorem

< 8.5 Objectives >

1 > Find the square root of a perfect square
2 > Identify the hypotenuse of a right triangle
3 > Identify a Pythagorean triple
4 > Use the Pythagorean theorem
5 > Approximate the square root of a number

Some numbers can be written as the product of two identical factors, for example,

$9 = 3 \times 3$

The factor is called a **square root** of the number. The symbol $\sqrt{\ }$ (called a **radical sign**) is used to indicate a square root. Thus, $\sqrt{9} = 3$ because $3 \times 3 = 9$.

Example 1 Finding the Square Root

< Objective 1 >

NOTE

To use the $\sqrt{\ }$ key with a scientific calculator, first enter the 49 and then press the key. With a graphing calculator, press the radical key first and then enter the 49 and a closing parenthesis.

Find the square root of 49 and of 16.

(a) $\sqrt{49} = 7$ Because $7 \times 7 = 49$.

(b) $\sqrt{16} = 4$ Because $4 \times 4 = 16$.

Check Yourself 1

Find the indicated square root.

(a) $\sqrt{121}$ **(b)** $\sqrt{36}$

The most frequently used theorem in geometry is undoubtedly the Pythagorean theorem. In this section we use that theorem. You will also learn a little about the history of the theorem. It is a theorem that applies only to right triangles.

The side opposite the right angle of a right triangle is called the **hypotenuse.** The other two sides are called **legs.** Note that the legs are perpendicular to each other.

Example 2 Identifying the Hypotenuse

< Objective 2 >

In the right triangle, the side labeled c is the hypotenuse.

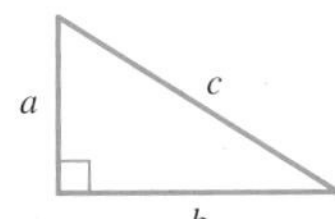

Check Yourself 2

Which side represents the hypotenuse of the right triangle?

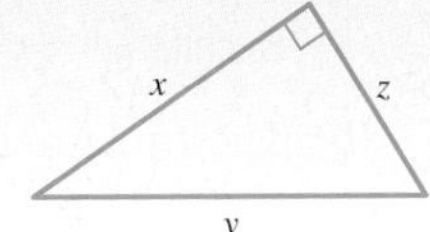

NOTE

Sometimes, Pythagorean triples are called *perfect triples.*

The numbers 3, 4, and 5 have a special relationship. Together they are called a **Pythagorean triple,** which means that when you square all three numbers, the sum of the smaller squares equals the squared value of the largest number.

Example 3 Identifying Pythagorean Triples

< Objective 3 >

Show that each set of numbers is a Pythagorean triple.

(a) 3, 4, and 5

$3^2 = 9 \qquad 4^2 = 16 \qquad 5^2 = 25$

and $9 + 16 = 25$, so $3^2 + 4^2 = 5^2$.

(b) 7, 24, and 25

$7^2 = 49 \qquad 24^2 = 576 \qquad 25^2 = 625$

and $49 + 576 = 625$, so $7^2 + 24^2 = 25^2$.

Check Yourself 3

Show that each set of numbers is a Pythagorean triple.

(a) 5, 12, and 13 **(b)** 6, 8, and 10

The triples above, and many more, were known to the Babylonians more than 4,000 years ago. We have found stone tablets with dozens of Pythagorean triples carved into them. The basis of the Pythagorean theorem was understood long before the time of Pythagoras (ca. 540 B.C.). The Babylonians not only understood perfect triples but also knew how triples related to right triangles.

Property

The Pythagorean Theorem

The sum of the squares of the length of the legs of a right triangle equals the square of the length of its hypotenuse.

As a formula, if a right triangle has legs with lengths a and b, and its hypotenuse has length c, then

$a^2 + b^2 = c^2$

This means that the lengths of the sides of a right triangle always form a Pythagorean triple.

Example 4 Finding the Length of a Side of a Right Triangle

< Objective 4 >

Find the missing length for each right triangle.

(a)

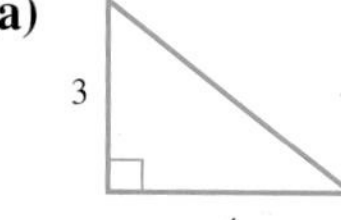

(b)

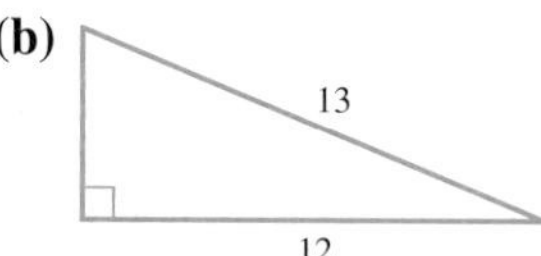

(a) A perfect triple is formed if the hypotenuse is 5 units long, creating the triple 3, 4, 5. Recall that $3^2 + 4^2 = 9 + 16 = 25 = 5^2$.

(b) The triple must be 5, 12, 13, which makes the missing length 5 units. Here, $5^2 + 12^2 = 25 + 144 = 169 + 13^2$.

Check Yourself 4

Find the length of the unlabeled side for each right triangle.

(a)

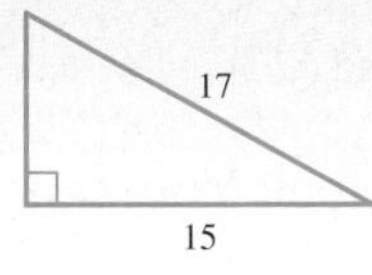

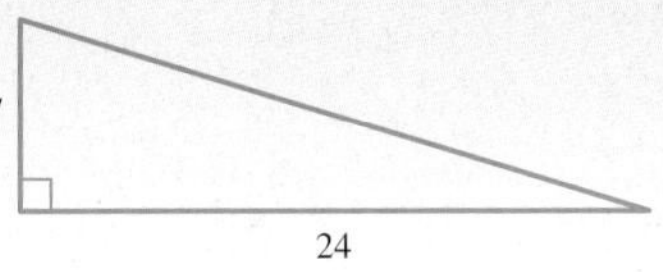

Example 5 Using the Pythagorean Theorem

NOTE

The triangle has sides 6, 8, and 10.

If the lengths of the legs of a right triangle are 6 and 8, find the length of the hypotenuse.

$c^2 = a^2 + b^2$ The value of the hypotenuse is found from the Pythagorean theorem with $a = 6$ and $b = 8$.

$c^2 = (6)^2 + (8)^2 = 36 + 64 = 100$

$c^2 = \sqrt{100} = 10$ The length of the hypotenuse is 10 (because $10^2 = 100$).

Check Yourself 5

Find the hypotenuse of a right triangle whose legs measure 9 and 12.

In some right triangles, the lengths of the hypotenuse and one side are given and we are asked to find the length of the missing side.

Example 6 Using the Pythagorean Theorem

Find the missing length.

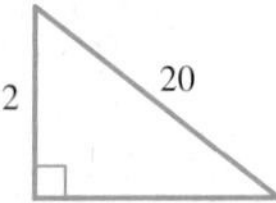

$a^2 + b^2 = c^2$ Use the Pythagorean theorem with $a = 12$ and $c = 20$.

$(12)^2 + b^2 = (20)^2$

$144 + b^2 = 400$

$b^2 = 400 - 144 = 256$

$b = \sqrt{256} = 16$ The missing side is 16.

Check Yourself 6

Find the missing length for a right triangle with one leg measuring 8 cm and the hypotenuse measuring 10 cm.

Not every square root is a whole number. In fact, there are only 10 whole-number square roots for the numbers from 1 to 100. They are the square roots of 1, 4, 9, 16, 25, 36, 49, 64, 81, and 100. However, we can approximate square roots that are not whole numbers. For example, we know that the square root of 12 is not a whole number. We also know that its value must lie somewhere between the square root of 9 $(\sqrt{9} = 3)$ and the square root of 16 $(\sqrt{16} = 4)$. That is, $\sqrt{12}$ is between 3 and 4.

Example 7 Approximating Square Roots

< Objective 5 >

Approximate $\sqrt{29}$.

$\sqrt{25} = 5$ and $\sqrt{36} = 6$, so $\sqrt{29}$ must be between 5 and 6.

Check Yourself 7

$\sqrt{19}$ is between which pair of numbers?

(a) 4 and 5 **(b)** 5 and 6 **(c)** 6 and 7

Check Yourself ANSWERS

1. (a) 11; **(b)** 6 **2.** Side y **3. (a)** $5^2 + 12^2 = 25 + 144 = 169$, $13^2 = 169$, so $5^2 + 12^2 = 13^2$; **(b)** $6^2 + 8^2 = 36 + 64 = 100$, $10^2 = 100$ so $6^2 + 8^2 = 10^2$ **4. (a)** 8; **(b)** 25 **5.** 15 **6.** 6 cm **7. (a)** 4 and 5

Reading Your Text

These fill-in-the-blank exercises will help you understand some of the key vocabulary used in this section. The answers to these exercises are in the Answers Appendix at the back of the text.

(a) The symbol $\sqrt{}$ (called a ______________ sign) is used to indicate a square root.

(b) The side opposite the right angle of a right triangle is called the ______________.

(c) The ______________ theorem says that the square of the hypotenuse of a right triangle is equal to the sum of the squares of the other two sides.

(d) Not every square root is a ______________ number.

8.5 exercises

Skills | Calculator/Computer | Career Applications | Above and Beyond

< Objective 1 >

Find each square root.

1. $\sqrt{64}$  **2.** $\sqrt{121}$ **3.** $\sqrt{169}$ 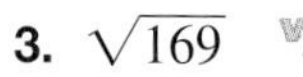 **4.** $\sqrt{196}$

< Objective 2 >

Identify the hypotenuse of each triangle.

5.

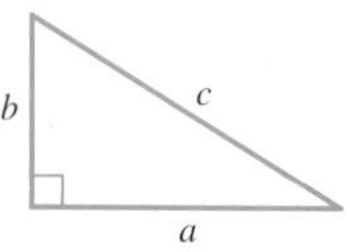

6.

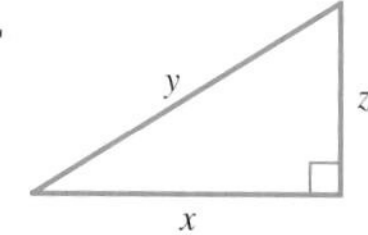

< Objective 3 >

Identify the Pythagorean triples.

7. 3, 4, 5 **8.** 4, 5, 6 **9.** 7, 12, 13 **10.** 5, 12, 13

11. 8, 15, 17 **12.** 9, 12, 15

< Objective 4 >

Find the missing length for each right triangle.

13.

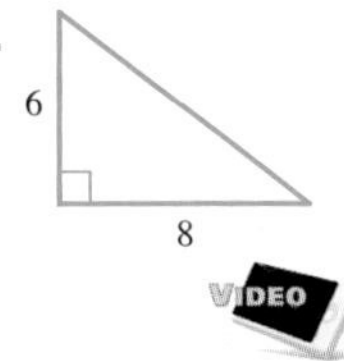

14.

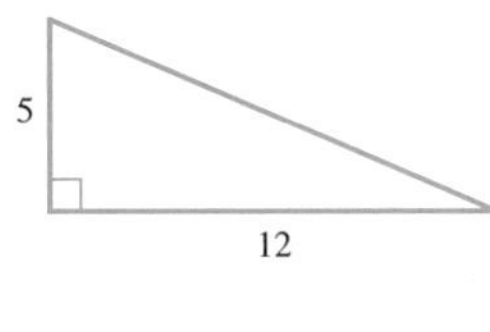

15.

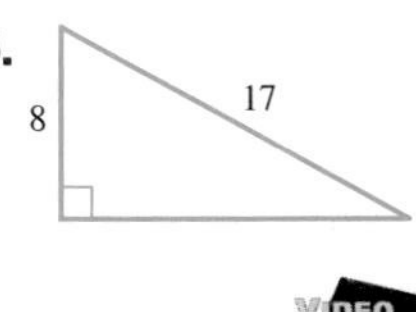

16.

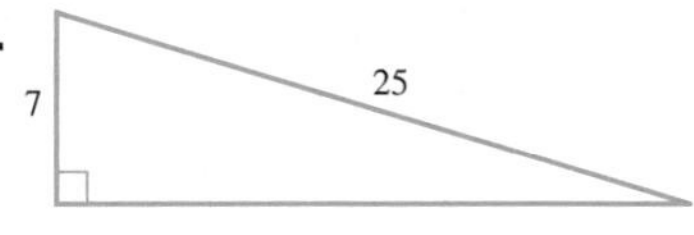

< Objective 5 >

Select the correct approximation for each square root.

17. Is $\sqrt{23}$ between **(a)** 3 and 4, **(b)** 4 and 5, or **(c)** 5 and 6?

18. Is $\sqrt{15}$ between **(a)** 1 and 2, **(b)** 2 and 3, or **(c)** 3 and 4?

19. Is $\sqrt{44}$ between **(a)** 6 and 7, **(b)** 7 and 8, or **(c)** 8 and 9?

20. Is $\sqrt{31}$ between **(a)** 3 and 4, **(b)** 4 and 5, or **(c)** 5 and 6?

Find the perimeter of each triangle.
Hint: First find the missing side.

21.

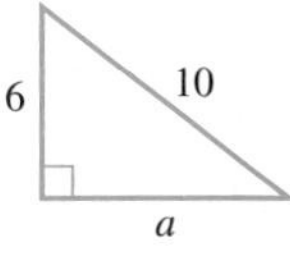

22.

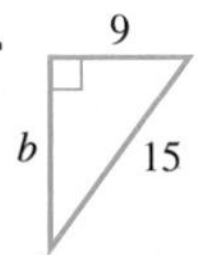

23.

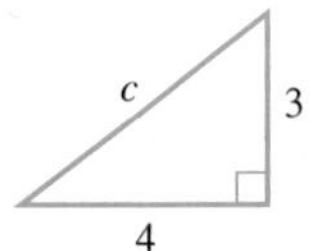

24.

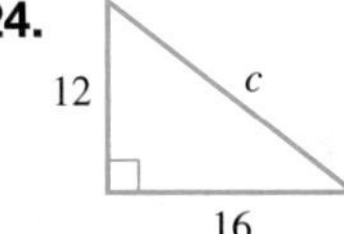

25. Find the altitude, h, of the isosceles triangle shown.
Hint: The altitude bisects the base.

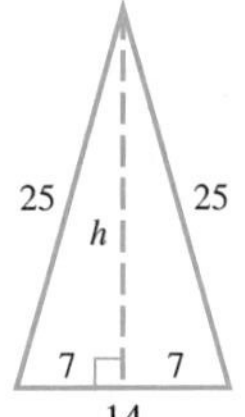

26. Find the altitude of the isosceles triangle shown.
Hint: The altitude bisects the base.

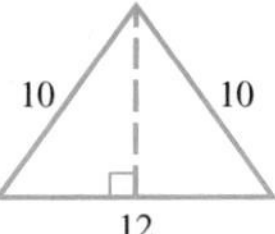

Find the length of the diagonal of each rectangle.

27.

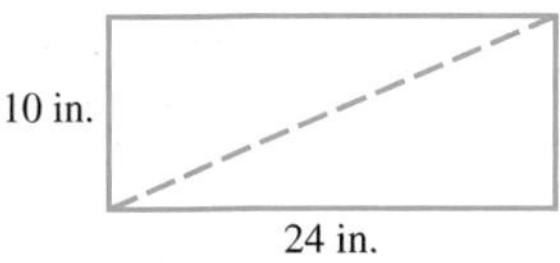

28.

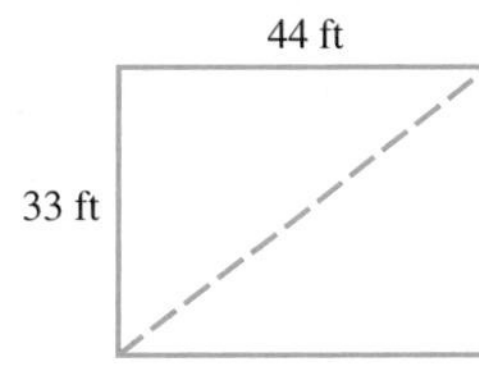

Determine whether each statement is ***true*** *or* ***false.***

29. For any triangle with sides a, b, and c, it is true that $a^2 + b^2 = c^2$.

30. If we know the lengths of two of the sides of a right triangle, we can always find the length of the third side.

In each statement, fill in the blank with **always, sometimes,** *or* **never.**

31. The hypotenuse is __________ the longest side of a right triangle.

32. The square root of a whole number is ________ a whole number.

Skills | **Calculator/Computer** | Career Applications | Above and Beyond

We use the *square root* key to find a square root with a calculator. How you use this key depends on whether you have a scientific or graphing calculator.

Scientific Calculator

With scientific calculators, you enter the number first, and then tell the calculator to find its square root using the square root key.

For instance, if we press

256 [√]

the display reads 16.

This tells us that $16^2 = 256$ or $\sqrt{256} = 16$.

Graphing Calculator

With graphing calculators, we enter the square root symbol first, then the number and contain it with a closing parenthesis. Usually, we then need to enter the computation in the calculator.

The example above looks like

[√] 256 [ )] [ENTER]

On many graphing calculators, the square root utility is found as the *2nd function* above the [x^2] key. In this case, we first press [2nd] to access the square root operation.

√(256)
16

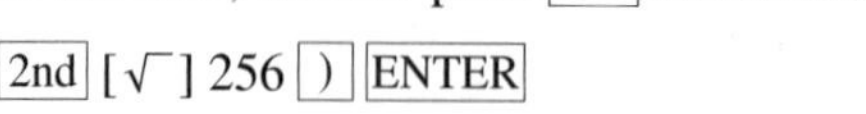

In any case, the display reads 16.

Use a calculator to find the square root of each number.

33. 64 **34.** 144 **35.** 289 **36.** 1,024

37. 1,849 **38.** 784 **39.** 8,649 **40.** 5,329

Use a calculator to approximate each square root. Round to the nearest tenth.

41. $\sqrt{23}$ **42.** $\sqrt{31}$ **43.** $\sqrt{51}$ **44.** $\sqrt{42}$ **45.** $\sqrt{134}$ **46.** $\sqrt{251}$

47. A 24-ft-high castle wall is surrounded by a moat 7 ft across. Will a 26-ft ladder, placed at the edge of the moat, be long enough to reach the top of the wall?

48. A baseball diamond is the shape of a square with 90-ft sides. Find the distance from home plate to second base. Round to the nearest tenth.

Skills | Calculator/Computer | **Career Applications** | Above and Beyond

49. **ELECTRONICS** The image represents a small portion of a printed circuit board that is in the layout stage. The conductive trace being plotted between the solder pads is comprised of a vertical and a horizontal trace that meet at a right angle. The horizontal component is 0.86 in., and the vertical component is 0.92 in. If the trace could be run from point to point diagonally, its distance could be determined by finding $d = \sqrt{(0.86)^2 + (0.92)^2}$. How long would it be?

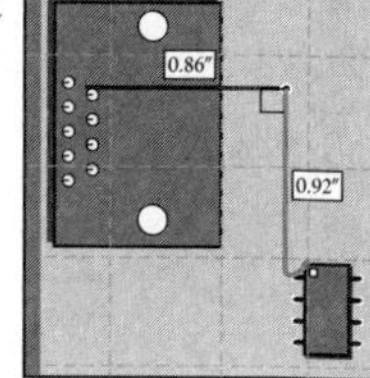

50. **MANUFACTURING TECHNOLOGY** Find the height of this truss.

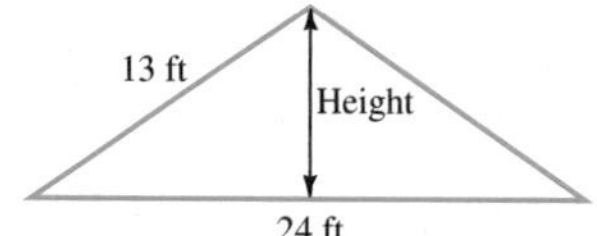

Answers

1. 8 **3.** 13 **5.** c **7.** Yes **9.** No **11.** Yes **13.** 10 **15.** 15 **17.** (b) **19.** (a) **21.** 24 **23.** 12 **25.** 24 **27.** 26 in. **29.** False **31.** always **33.** 8 **35.** 17 **37.** 43 **39.** 93 **41.** 4.8 **43.** 7.1 **45.** 11.6 **47.** Yes **49.** 1.26 in.

8.6 Solid Geometry

< 8.6 Objectives >

1 > Find the total surface area and volume of a rectangular solid

2 > Find the total surface area and volume of a cylinder

3 > Find the lateral surface area of a solid

4 > Find the surface area and volume of a sphere

5 > Convert between cubic units

We are now ready to look at objects that appear in the real world—those that sit in three-dimensional space. **Solids** can usually be thought of as a set of two-dimensional objects joined to enclose space.

Definition

Solids A solid is a three-dimensional figure. It has length, width, and height.

There are two properties of solids that we consider first—total surface area and volume. The **total surface area,** or **TSA,** of a solid is a two-dimensional measure of the area of each of the surfaces of the solid. For example, the amount of wrapping paper that it would take to cover a box is a measure of its surface area.

Volume is a three-dimensional measure that refers to the amount of space the solid takes up. This measures how much it would take to completely fill the box.

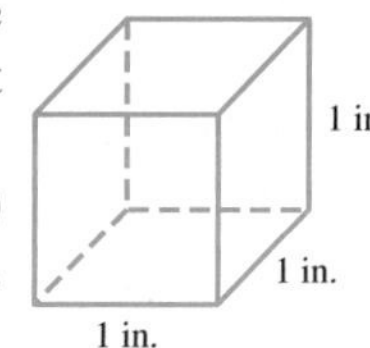

RECALL

We looked at volume when studying multiplication in Section 1.5.

Volume is measured in **cubic units,** such as cubic inches (in.^{3}) and cubic meters (m^3). A cubic unit is exactly what it sounds like, a cube measuring exactly one unit on every side.

When finding the volume of a figure, we want to know how many cubic units are contained in that figure.

When finding the surface area, we are looking for the number of square units that would completely cover the surface or outside of the figure.

We start with a straightforward example, a **rectangular solid** (such as a box, crate, or most rooms). A rectangular solid consists of three pairs of rectangles positioned to enclose space.

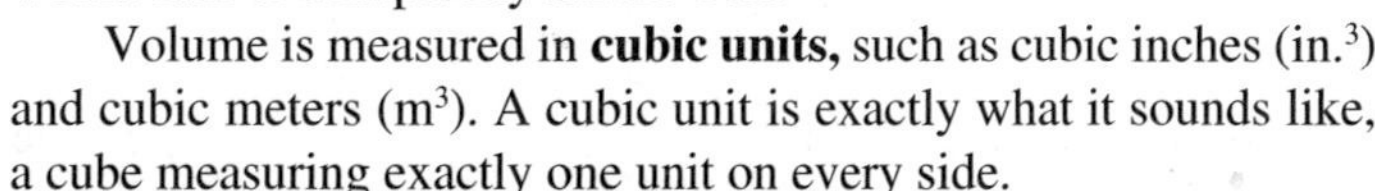

We compute the total surface area by taking the area of each of the six sides (or faces) and adding them together. With a rectangular solid, there are three pairs of sides.

We compute the volume by taking the product of its three dimensions. This gives the number of cubic units that it takes to fill the figure.

Property

Surface Area and Volume of Rectangular Solids

For a rectangular solid with length L, width W, and height H,

$$\text{TSA} = 2LW + 2LH + 2WH = 2(LW + LH + WH)$$

$$V = LWH$$

Note: These dimensions are sometimes called base, depth, and height.

Example 1 Finding Surface Area and Volume

< Objective 1 >

Find the total surface area and volume of a 2-in. by 3-in. by 5-in. rectangular solid.

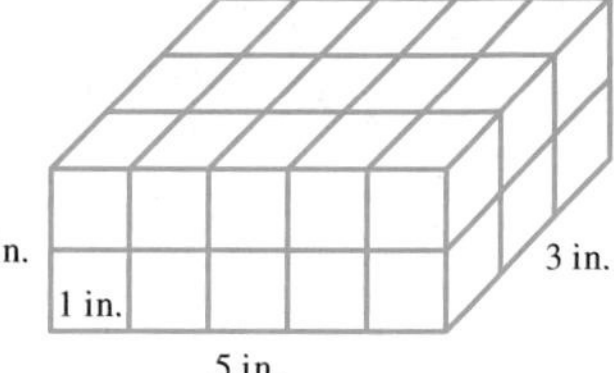

Surface Area

- The front and back faces are 2-in. by 5-in. rectangles.
- The left and right sides are both 2-in. by 3-in. rectangles.
- The top and bottom are 3-in. by 5-in. rectangles.

$$\begin{aligned} \text{TSA} &= 2(2 \text{ in.} \times 5 \text{ in.}) + 2(2 \text{ in.} \times 3 \text{ in.}) + 2(3 \text{ in.} \times 5 \text{ in.}) \\ &= 2(10 \text{ in.}^2) + 2(6 \text{ in.}^2) + 2(15 \text{ in.}^2) \\ &= 20 \text{ in.}^2 + 12 \text{ in.}^2 + 30 \text{ in.}^2 \\ &= 62 \text{ in.}^2 \end{aligned}$$

Volume

The volume is the product of its three dimensions.

$$\begin{aligned} V &= LWH \\ &= (3 \text{ in.})(5 \text{ in.})(2 \text{ in.}) \\ &= 30 \text{ in.}^3 \end{aligned}$$

Check Yourself 1

Find the total surface area and volume of the figure.

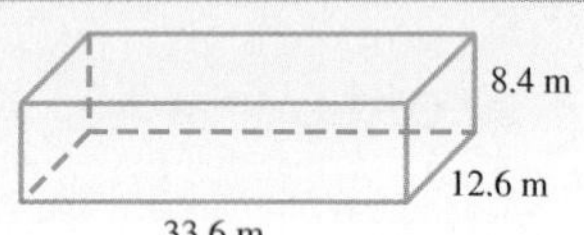

A **cube** is a special type of rectangular solid; it is the three-dimensional equivalent of a square. A cube is a rectangular solid whose length, width, and height are the same.

Because there is only one unique dimension, *side s*, we can simplify the total surface area and volume formulas in the case of a cube.

Property

Formulas for Cubes

$\text{TSA} = 6 \cdot s^2$ There are six square faces, each with sides s and area s^2.

$V = s^3$ In the case of a cube, $LWH = s \cdot s \cdot s = s^3$.

Example 2 Cubes

Find the total surface area and volume of the cube.

We use the total surface area and volume formulas with $s = 5$ in.

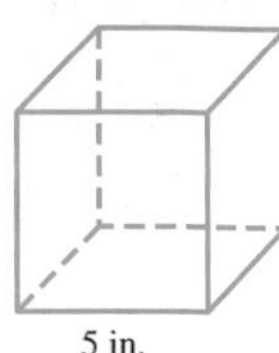

$$\begin{aligned} \text{TSA} &= 6s^2 \\ &= 6(5 \text{ in.})^2 \\ &= 6 \cdot (25 \text{ in.}^2) \quad \text{Follow the order of operations.} \\ &= 150 \text{ in.}^2 \end{aligned}$$

The total surface area of a 5-in. cube is 150 in.^{2}.

$V = s^3$

$= (5 \text{ in.})^3$ $\quad s = 5$ in.

$= 125 \text{ in.}^3$ $\quad 5^3 = 125$

The volume of a 5-in. cube is 125 in.^{3}.

Check Yourself 2

Find the total surface area and volume of the cube.

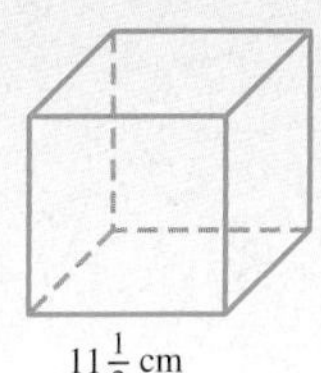

The next solid to consider is the **cylinder**. A cylinder, such as a soup can, is made up of two circles wrapped by a rectangle, as shown.

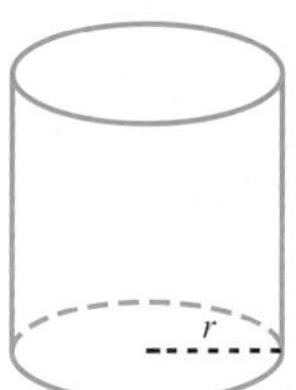

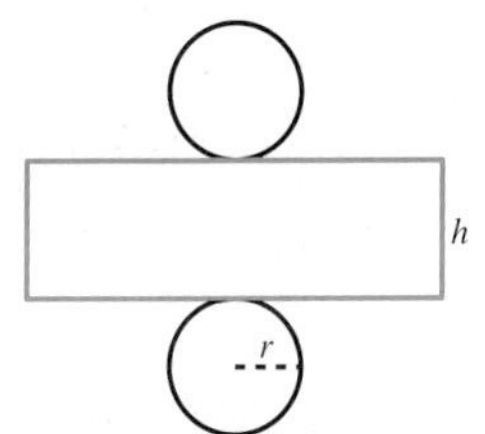

Because the cylinder is a rectangle and two circles, we find its total surface area by adding together the area of the rectangle and the two circles. Note that because the rectangle "wraps around" the circles, its width is the circumference of the circle, $2\pi r$.

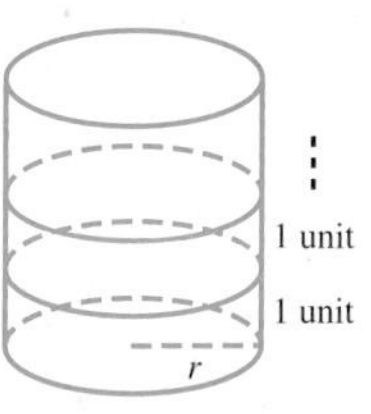

To find its volume, we begin by taking a single unit of height because it has the same volume as the area of the circle, but in cubic units. In fact, each unit of height equals one more "circle" of volume.

This leads to the conclusion that the volume can be found by multiplying the area of one of the circles by the height of the cylinder.

Property

Surface Area and Volume of Cylinders

For a cylinder with radius r and height h,

$\text{TSA} = 2\pi r^2 + 2\pi rh$ $\quad$ Recall: $\pi \approx 3.14$.

$V = \pi r^2 h$

Example 3 Cylinders

< Objective 2 >

Find the total surface area and volume of the cylinder.

We use the formulas with $r = 4.2$ cm and $h = 8.4$ cm.

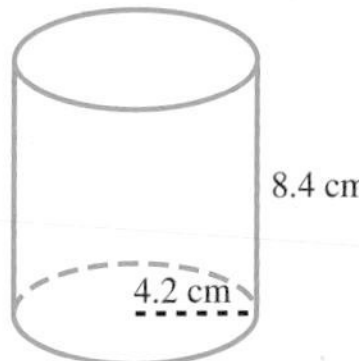

$\text{TSA} = 2\pi r^2 + 2\pi rh$

$\approx 2(3.14)(4.2 \text{ cm})^2 + 2(3.14)(4.2 \text{ cm})(8.4 \text{ cm})$

$= 110.7792 \text{ cm}^2 + 221.5584 \text{ cm}^2$

$= 332.3376 \text{ cm}^2 \approx 332.34 \text{ cm}^2$

$V = \pi r^2 h$

$\approx (3.14)(4.2 \text{ cm})^2 (8.4 \text{ cm})$ Follow the order of operations.

$= (3.14)(17.64 \text{ cm}^2) (8.4 \text{ cm})$ $4.2^2 = 17.64$

$= 465.27264 \text{ cm}^3 \approx 465.27 \text{ cm}^3$

Check Yourself 3

Find the total surface area and volume for the cylinder.

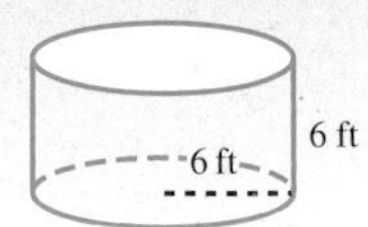

Before we continue, you may have noticed that rather than simply saying *surface area* or *area*, we are calling this measure the *total surface area*. This is because there is another surface area measurement called the **lateral surface area (LSA).**

The lateral surface area is a measure of the surface area without the top or bottom faces. To picture this, consider a soup can. The total surface area would be used to measure the amount of sheet metal needed to make the can. Lateral surface area would be used to measure the amount of labeling needed to go around the can.

Property

Lateral Surface Area

The lateral surface area of a rectangular solid is given by

$\text{LSA} = 2LH + 2WH$

The lateral surface area of a cylinder is

$\text{LSA} = 2\pi rh$

Example 4 Lateral Surface Area

< Objective 3 >

Compute the lateral surface area of each figure.

(a)

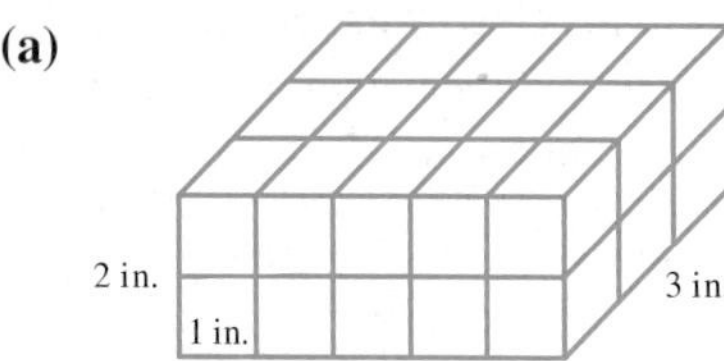

(b)

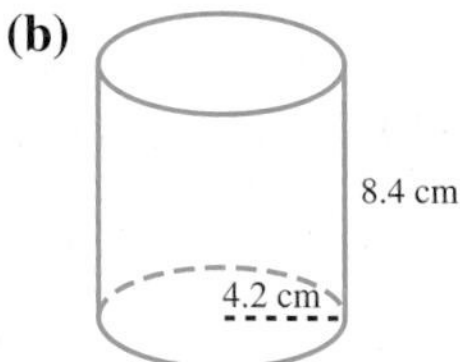

(a) The lateral surface area is the front and back as well as the two sides, but not the top or bottom.

$\text{LSA} = 2(2 \text{ in.})(5 \text{ in.}) + 2(2 \text{ in.})(3 \text{ in.})$

$= 20 \text{ in.}^2 + 12 \text{ in.}^2$

$= 32 \text{ in.}^2$

(b) The lateral surface area is the area of the rectangle (the base of which is equal to the circumference of the circles).

$\text{LSA} = 2\pi rh$

$\approx 2(3.14)(4.2 \text{ cm})(8.4 \text{ cm})$

$\approx 221.56 \text{ cm}^2$

Check Yourself 4

Compute the lateral surface area of each figure.

(a)

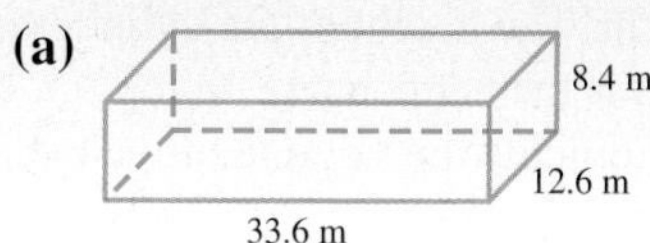

(b)

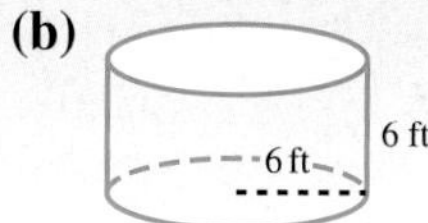

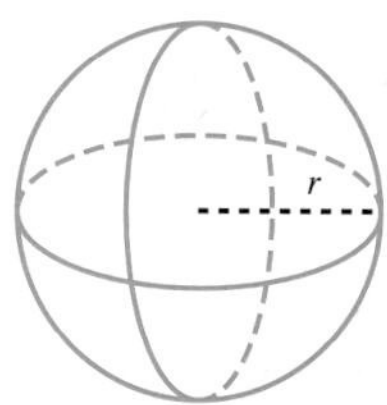

Finally, the three-dimensional analog of a circle is a **sphere.** A ball is a good example of a sphere. Just as a circle is the set of points that are the same distance from some fixed center point in two dimensions, a sphere is the set of points equidistant from some fixed center point in three dimensions.

We can compute the surface area and volume of a sphere using formulas derived from calculus. Note that a sphere doesn't have a top and bottom, so there is no distinction between total surface area and lateral surface area—there is just the surface area.

Property

Surface Area and Volume of Spheres

The surface area and volume of a sphere with radius r are given by,

$$A = 4\pi r^2$$

$$V = \frac{4}{3}\pi r^3$$

Example 5 Spheres

< Objective 4 >

RECALL

We use 3.14 to approximate π.

Find the surface area and volume of the sphere.

We use the formulas with $r = 5$ m.

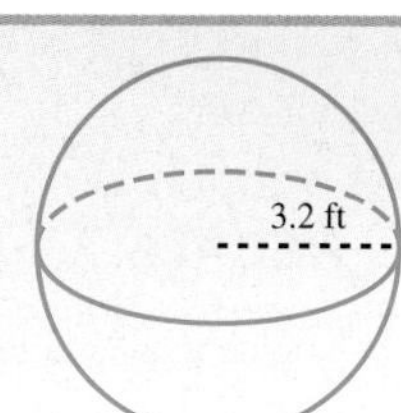

$$A = 4\pi r^2 \approx 4(3.14)(5\text{ m})^2$$
$$= 314\text{ m}^2$$

$$V = \frac{4}{3}\pi r^3 \approx \frac{4}{3}(3.14)(5\text{ m})^3$$
$$\approx 523.33\text{ m}^3$$

Check Yourself 5

Find the surface area and volume of the sphere.

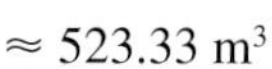

Many real-world solids are composed of two or more basic solid figures. We treat *composite solids* in the same way as we did their two-dimensional counterparts. We work with each simpler part separately before combining them, as needed.

Example 6 An Agricultural Technology Application

A grain silo can be constructed by attaching a dome (a hemisphere or half sphere) to the top of a cylinder. If the diameter of such a grain silo is 20 m and its height is 55 m, including the hemisphere on top, find its volume.

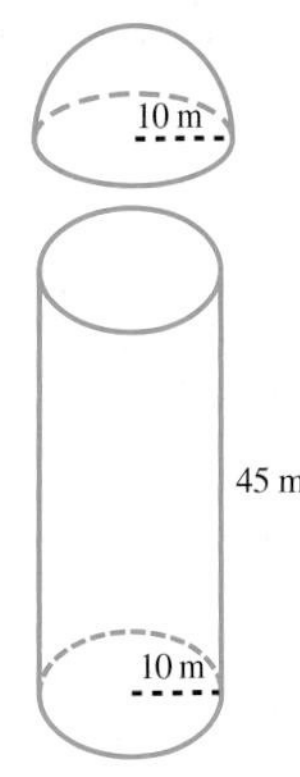

We consider the cylinder and hemisphere parts separately. It is always a good idea to make a rough sketch to help see the problem.

The radius of the cylinder and hemisphere is half the diameter or 10 m. The height of the cylinder is 45 m because 10 m of the total height is the hemisphere on top.

Compute the volume of the cylinder and sphere separately.

$$\begin{aligned} V_{\text{cyl}} &= \pi r^2 h \\ &\approx (3.14)(10 \text{ m})^2(45 \text{ m}) \\ &= 14{,}130 \text{ m}^3 \end{aligned}$$

$$\begin{aligned} V_{\text{Sphere}} &= \frac{4}{3}\pi r^3 \\ &\approx \frac{4}{3}(3.14)(10 \text{ m})^3 \\ &= 4{,}186\frac{2}{3} \text{ m}^3 \end{aligned}$$

Because the silo only contains half a sphere, we halve this last volume.

$$\begin{aligned} V_{\text{Hemi}} &= \frac{1}{2} \times 4{,}186\frac{2}{3} \text{ m}^3 \\ &= 2{,}093\frac{1}{3} \text{ m}^3 \end{aligned}$$

Finally, we add the two measurements.

$$\begin{aligned} V &= V_{\text{Cyl}} + V_{\text{Hemi}} \\ &\approx 14{,}130 \text{ m}^3 + 2{,}093\frac{1}{3} \text{ m}^2 \\ &= 16{,}223\frac{1}{3} \text{ m}^3 \end{aligned}$$

The silo holds about $16{,}223 \text{ m}^3$.

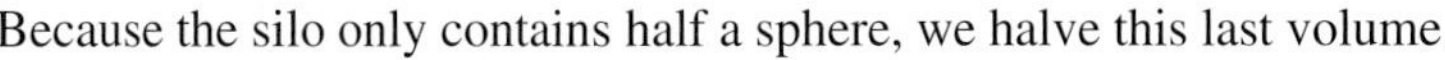

Check Yourself 6

A propane tank is constructed by attaching hemispheres to both ends of a cylinder. If the total length of a 250-gal tank is 92 in. and its diameter is 30 in., find the tank's volume. Use 3.14 for π and round your result to the nearest in.^3.

RECALL

Cubing means to take to the third power.

Recall from Section 8.2 that we needed to square conversion factors when converting area measurements or square units. For instance, while 1 yd = 3 ft, we have $1 \text{ yd}^2 = 9 \text{ ft}^2$.

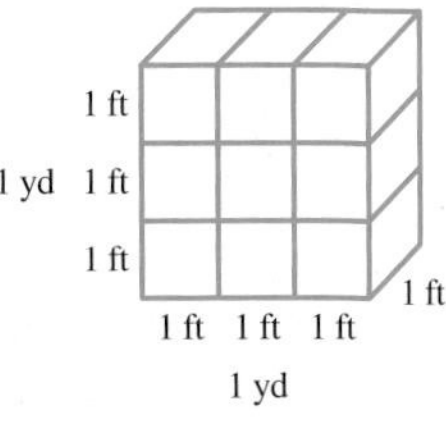

This relationship extends to cubic units, as well. However, when converting between cubic units, we need to cube the conversion factors. Consider the cubic yard. It should be clear that if we take a square yard that is only 1 ft deep, we have 9 ft^3.

This is not a cubic yard, yet, since it is only 1 ft deep. We need two more of these layers to make a cubic yard.

Because there are three layers of these 1-ft slices, we have

$$1 \text{ yd}^3 = 3 \times 9 \text{ ft}^3 = 27 \text{ ft}^3$$

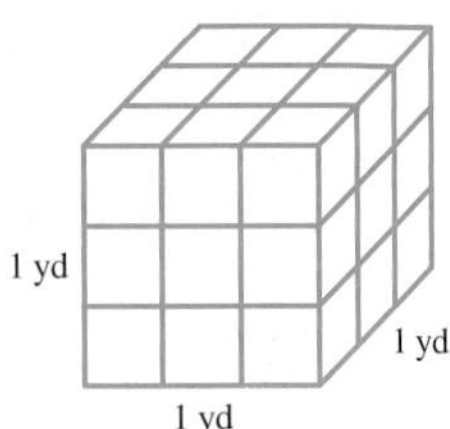

Property

Converting Cubic Units

U.S. System of Measurement	Metric System of Measurement	Conversions
$1 \text{ ft}^3 = (12)^3 \text{ in.}^3$	$1 \text{ cm}^3 = (10)^3 \text{ mm}^3$	$1 \text{ in.}^3 \approx (2.54)^3 \text{ cm}^3$
$= 1{,}728 \text{ in.}^3$	$= 1{,}000 \text{ mm}^3$	$\approx 16.387 \text{ cm}^3$
$1 \text{ yd}^3 = 27 \text{ ft}^3$	$1 \text{ m}^3 = (100)^3 \text{ cm}^3$	$1 \text{ cm}^3 \approx (0.39)^3 \text{ in.}^3$
	$= 1{,}000{,}000 \text{ cm}^3$	$\approx 0.059 \text{ in.}^3$
		$1 \text{ ft}^3 \approx 0.027 \text{ m}^3$
		$1 \text{ m}^3 \approx 34.966 \text{ ft}^3$

Example 7 Converting Cubic Units

< Objective 5 >

A professor's office measures 9-ft by 11-ft and has 8-ft-high ceilings. Find its volume, in cubic yards.

Begin by finding the volume of the room, in cubic feet.

$$V = (9 \text{ ft})(11 \text{ ft})(8 \text{ ft})$$
$$= 792 \text{ ft}^3$$

Since $1 \text{ yd}^3 = 27 \text{ ft}^3$, we have the unit ratio

$$1 = \frac{1 \text{ yd}^3}{27 \text{ ft}^3}$$

RECALL

We use the ratio with yd³ on top and ft³ in the bottom so that ft³ "cancel" and we are left with yd³.

We use the ratio with yd^3 in the numerator and ft^3 in the denominator so that we can simplify the expression to cubic yards.

$$792 \cancel{\text{ft}^3} \times \frac{1 \text{ yd}^3}{27 \cancel{\text{ft}^3}} = \frac{792}{27} \text{ yd}^3$$
$$= \frac{88}{3} \text{ yd}^3$$
$$= 29\frac{1}{3} \text{ yd}^3$$

Check Yourself 7

A soup can (cylinder) has a $7\frac{1}{2}$-cm diameter and a height of 11 cm. Find the volume of the can, in cubic inches. Use $\pi \approx 3.14$ and round your result to two decimal places.

Check Yourself ANSWERS

1. TSA $= 1{,}622.88 \text{ m}^2$; $V = 3{,}556.224 \text{ m}^3$ **2.** TSA $= 793\frac{1}{2} \text{ cm}^2$; $V = 1{,}520\frac{7}{8} \text{ cm}^3$
3. TSA $\approx 452.16 \text{ ft}^2$; $V \approx 678.24 \text{ ft}^3$ **4.** **(a)** 776.16 m^2; **(b)** 226.08 ft^2
5. $A \approx 128.61 \text{ ft}^2$; $V \approx 137.19 \text{ ft}^3$ **6.** $57{,}933 \text{ in.}^3$ **7.** 28.81 in.^3 or 28.66 in.^3

Reading Your Text

These fill-in-the-blank exercises will help you understand some of the key vocabulary used in this section. The answers to these exercises are in the Answers Appendix at the back of the text.

(a) A solid is a __________ dimensional figure.

(b) Volume is measured in __________ units.

(c) A __________ is the three-dimensional analog of a circle.

(d) When converting between cubic units, we need to __________ the conversion factors.

< Objectives 1–3 >

*Find the **(a)** lateral surface area; **(b)** total surface area; and **(c)** volume of each figure. Use $\pi \approx 3.14$ and round your results to two decimal places, when appropriate.*

1. 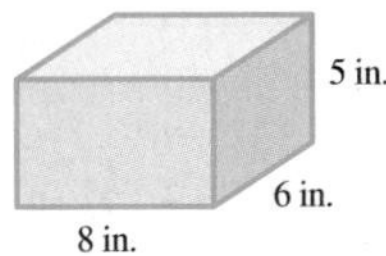

2. 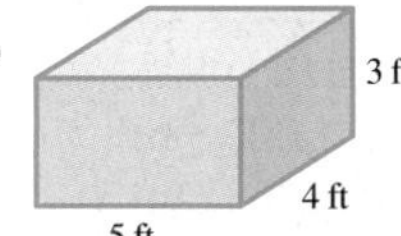

3. 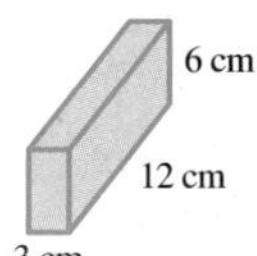

4. 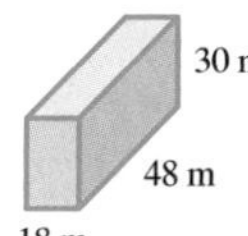

5. 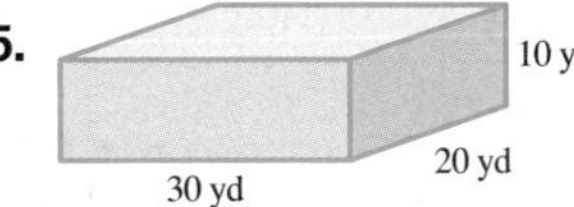

6. 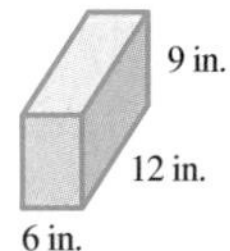

7. 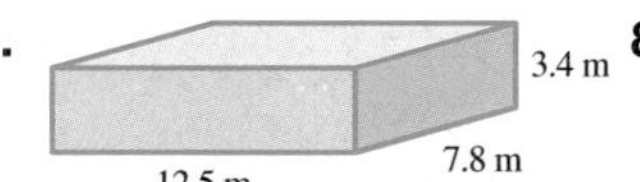

8. 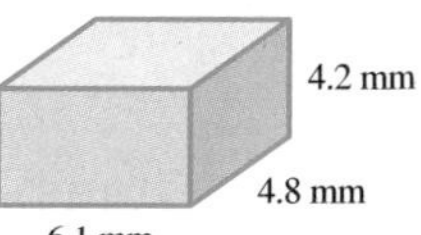

9. 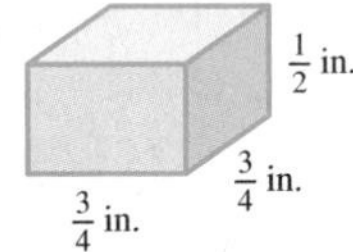

10. 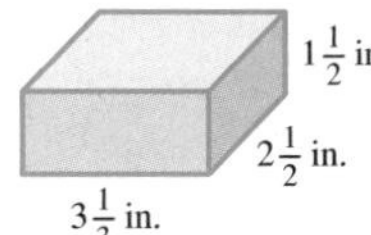

11. 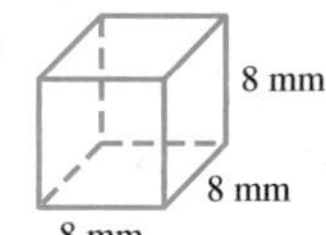

12. 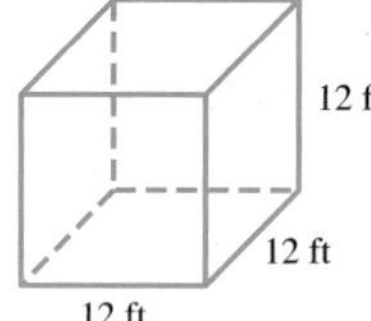

13. 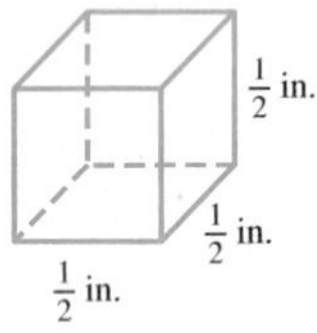

14. 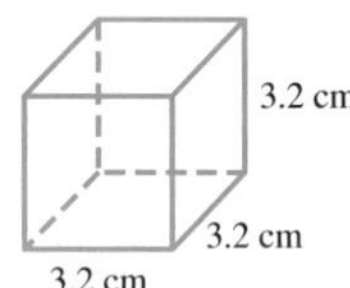

15. 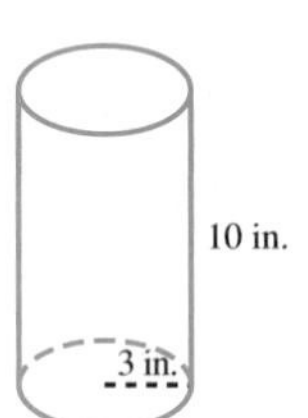

16. 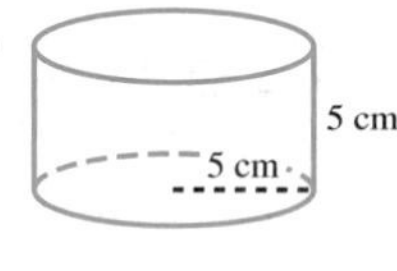

17. 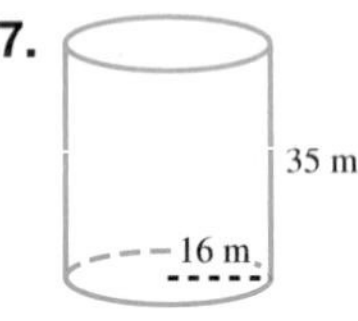

18. 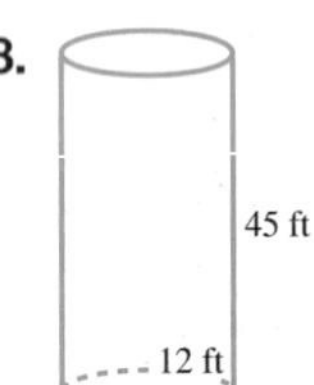

19. 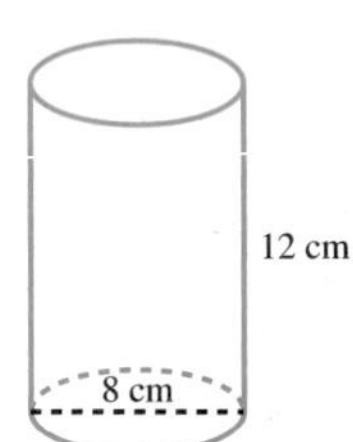

20. 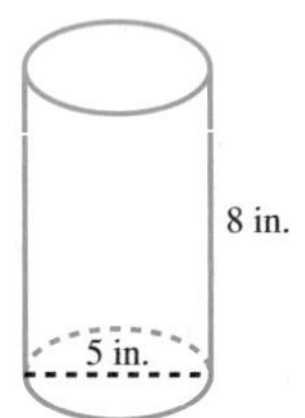

< Objective 4 >

*Find the **(a)** total surface area, and **(b)** volume of each figure. Use $\pi \approx 3.14$ and round your results to two decimal places, when appropriate.*

21. 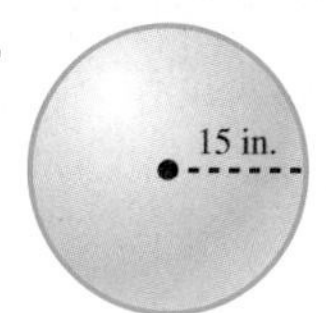

22. 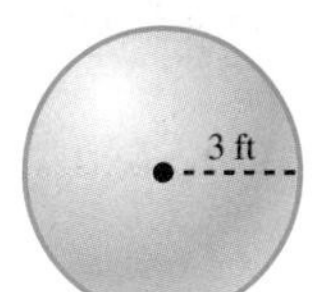

23. 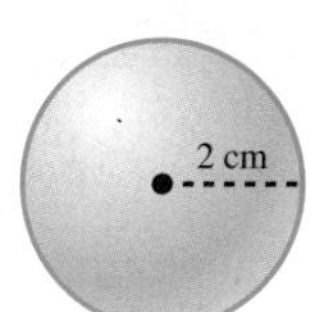

24. 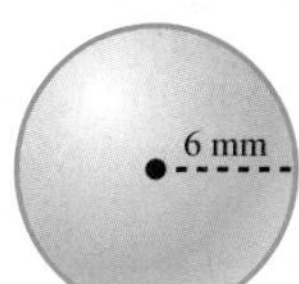

25. 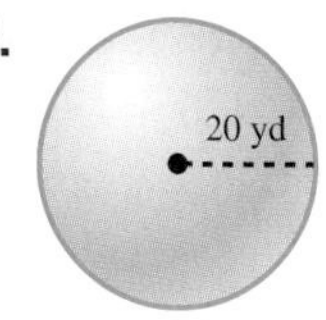

26. 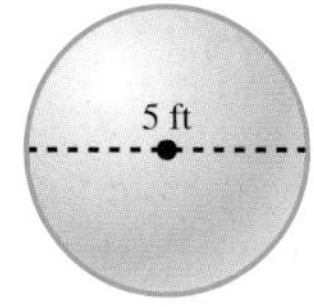

27. 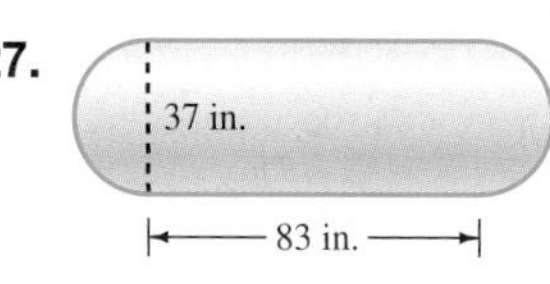

28. 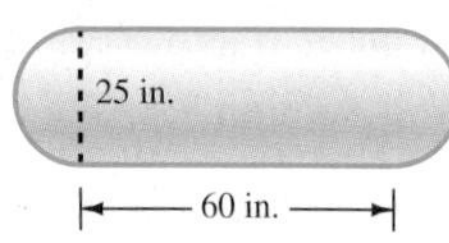

< Objective 5 >

Convert each measurement. Round your results to two decimal places when necessary.

29. 25 ft^3 to $in.^3$

30. 150 yd^3 to ft^3

31. 16 yd^3 to ft^3

32. 16 ft^3 to $in.^3$

33. 6 yd^3 to $in.^3$

34. 15 yd^3 to $in.^3$

35. 25 cm^3 to mm^3

36. 80 cm^3 to mm^3

37. 12 m^3 to cm^3

38. 12 m^3 to mm^3

39. 18 $in.^3$ to cm^3

40. 18 cm^3 to $in.^3$

41. 42 m^3 to ft^3

42. 42 ft^3 to m^3

43. 10 m^3 to $in.^3$

44. 10 ft^3 to cm^3

Complete each application.

45. **Geometry** A room measures 12 ft by 16 ft with 8-ft-high ceilings.

(a) Find the volume of the room.

(b) A reasonable estimation of the size of a furnace is 3 BTUs per cubic foot of air to be heated. How many BTUs are needed to heat this room?

46. **Science and Medicine** A veterinarian needs to medicate a tank full of fish. The fish are housed in a rectangular tank that is 12 in. wide, 30 in. long, and 18 in. high. Find the volume of the tank.

47. **Geometry** A silo with a 20-ft diameter is 60 ft tall (shaped like a cylinder, topped with a hemisphere).

(a) How many cubic feet of silage will it hold (to the nearest whole cubic foot)?

(b) What is the surface area of the silo (including the hemisphere, but not the bottom)?

48. **Construction** A propane tank is designed as a right circular cylinder with hemispheres on both ends. The radius of the tank is 2 ft and the total length (including hemispheres) is 16 ft. Find the volume of the tank (to the nearest whole cubic foot).

Determine whether each statement is **true** *or* **false.**

49. 1 yd^3 is larger than 1 m^3.

50. The total surface area of a cylinder includes the circles on both ends.

Complete each statement with **always, sometimes,** *or* **never.**

51. The number of cubic units of volume of a solid is __________ larger than the number of square units of total surface area of the solid.

52. The total surface area of a solid is __________ less than the lateral surface area of the solid.

8.6 exercises

Skills | **Calculator/Computer** | Career Applications | Above and Beyond

You learned to use the π key on your calculator to compute the circumference and area of a circle in the exercises that followed Section 8.3. We use the π key in the same way when working with solids.

Use a calculator to approximate the **(a)** *total surface area and* **(b)** *volume of each figure. Round your results to three decimal places, when appropriate.*

53.

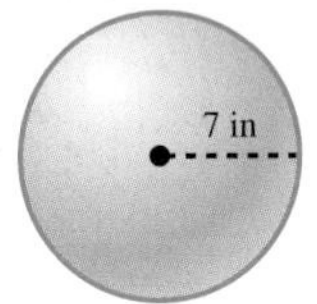

54.

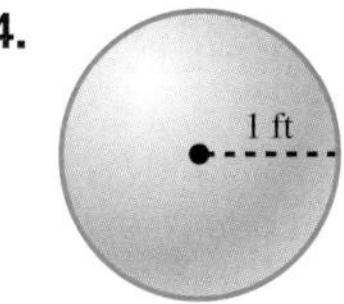

55.

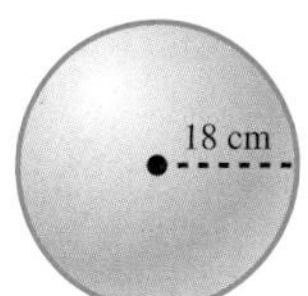

56.

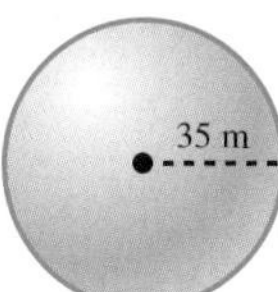

57.

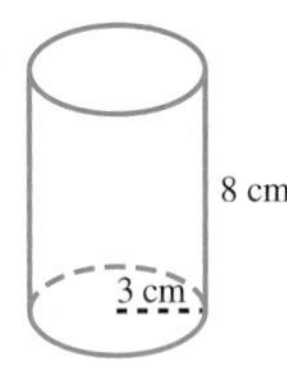

58.

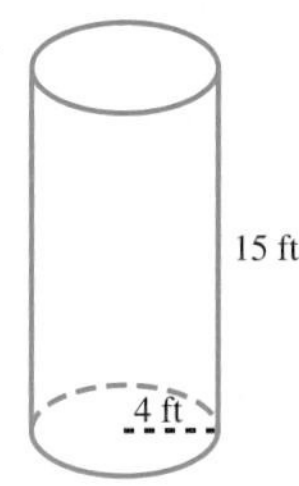

Skills | Calculator/Computer | **Career Applications** | Above and Beyond

59. **Manufacturing Technology** A $\frac{1}{8}$-in. drill bit is ground from a $2\frac{1}{2}$-in. long cylinder-shaped piece of steel with a $\frac{1}{8}$-in. diameter. Find the volume of the steel stock (round your result to two decimal places).

60. **Manufacturing Technology** A $\frac{5}{32}$-in. drill bit is ground from a 3-in. long cylinder-shaped piece of steel with a $\frac{5}{32}$-in. diameter. Find the volume of the steel stock.

61. **Manufacturing Technology** A piece of metal weighs 0.092 lb/in.^{3}. Find the weight of a 5-in. by 12-in. by 3-in. piece.

62. **Manufacturing Technology** A piece of metal weighs 0.092 lb/in.^{3}. Find the weight of a 4-in. by 18-in. by 6-in. piece.

63. **Health Sciences** A veterinarian needs to medicate a 12-in. by 30-in. by 18-in. tank full of fish. If 1 gal is equivalent to 231 in.^{3}, how many gallons does the tank hold?

64. **Health Sciences** A pet store owner needs to treat a tank full of tropical fish. The fish are kept in a cylindrical tank with a diameter of 18 in. and a height of 60 in. The dosage depends on the volume of water in the tank. Determine the volume of the tank, to the nearest whole gallon, if 1 gal is 231 in.^{3}.

Answers

1. **(a)** 156 in.^{2}; **(b)** 236 in.^{2}; **(c)** 240 in.^{3} **3.** **(a)** 180 cm^2; **(b)** 252 cm^2; **(c)** 216 cm^3 **5.** **(a)** 1,000 yd^2; **(b)** 2,200 yd^2; **(c)** 6,000 yd^3 **7.** **(a)** 248.04 m^2; **(b)** 333.04 m^2; **(c)** 331.5 m^3 **9.** **(a)** $1\frac{1}{2}$ in.^{2}; **(b)** $2\frac{5}{8}$ in.^{2}; **(c)** $\frac{9}{32}$ in.^{3} **11.** **(a)** 256 mm^2; **(b)** 384 mm^2; **(c)** 512 mm^3 **13.** **(a)** 1 in.^{2}; **(b)** $1\frac{1}{2}$ in.^{2}; **(c)** $\frac{1}{8}$ in.^{3} **15.** **(a)** 188.4 in.^{2}; **(b)** 244.92 in.^{2}; **(c)** 282.6 in.^{3} **17.** **(a)** 3,516.8 m^2; **(b)** 5,124.48 m^2; **(c)** 28,134.4 m^3 **19.** **(a)** 301.44 cm^2 **(b)** 401.92 cm^2; **(c)** 602.88 cm^3 **21** **(a)** 2,826 in.^{2}; **(b)** 14,130 in.^{3} **23.** **(a)** 50.24 cm^2; **(b)** 33.49 cm^3 **25.** **(a)** 5,024 yd^2; **(b)** 33,493.33 yd^3 **27.** **(a)** 13,941.6 in.^{2}; **(b)** 115,705.6 in.^{3} **29.** 43,200 in.^{3} **31.** 432 ft^3 **33.** 279,936 in.^{3} **35.** 25,000 mm^3 **37.** 12,000,000 cm^3 **39.** 294.97 cm^3 **41.** 1,468.57 ft^3 **43.** 604,212.48 in.^{3} **45.** **(a)** 1,536 ft^3; **(b)** 4,608 BTUs **47.** **(a)** 17,793 ft^3; **(b)** 3,768 ft^2 **49.** False **51.** sometimes **53.** **(a)** 615.752 in.^{2}; **(b)** 1,436.755 in.^{3} **55.** **(a)** 4,071.504 cm^2; **(b)** 24,429.024 cm^3 **57.** **(a)** 207.345 cm^2; **(b)** 226.195 cm^3 **59.** 0.03 in.^{3} **61.** 16.56 lb **63.** 28 gal

Definition/Procedure	Example	Reference
Lines and Angles		**Section 8.1**
Line A series of points that goes straight forever. **Line segment** A piece of a line that has two endpoints.	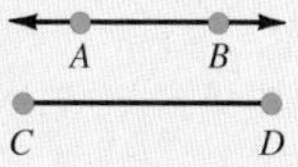 A B C D	p. 428
Angle A figure consisting of two line segments that share a common endpoint.	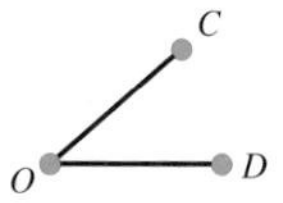 C O D	p. 429
Perpendicular lines Lines are *perpendicular* if they intersect to form four equal angles. **Parallel lines** Lines are *parallel* if they never intersect.	These lines are perpendicular. 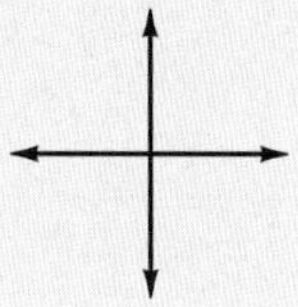	p. 429
Acute angles have a measure less than 90°. **Obtuse angles** have a measure between 90° and 180°. **Right angles** have a measure of 90°.	$\angle CEF$ is obtuse. 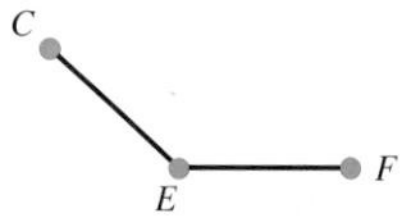 C E F	p. 430
Straight angles have a measure of 180°.	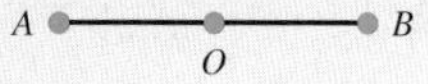 A B O	p. 431
Complementary angles Two angles are complementary if the sum of their measures is 90°.	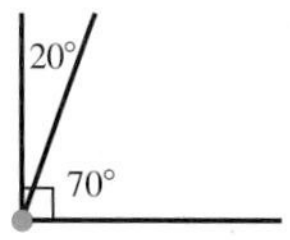 20° 70°	p. 433
Supplementary angles Two angles are supplementary if the sum of their measures is 180°.	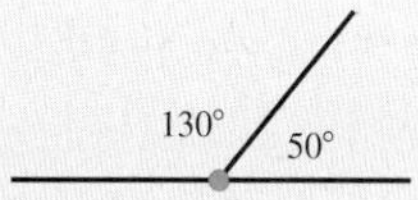 130° 50°	p. 433
Vertical angles When two lines intersect, two pairs of vertical angles are formed. Vertical angles have equal measures.	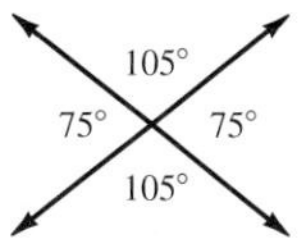 105° 75° 75° 105°	p. 434
Parallel lines and a transversal When two parallel lines are intersected by a transversal, **1.** Alternate interior angles have equal measure. **2.** Corresponding angles have equal measure.	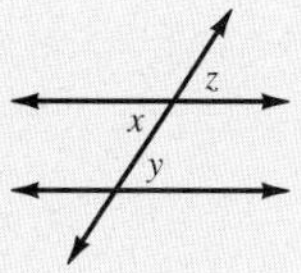 z x y $m\angle x = m\angle y$ $m\angle y = m\angle z$	p. 434

Continued

Definition/Procedure	Example	Reference
Perimeter and Area		Section 8.2
The **perimeter** of a figure is the distance around the figure. It can be found by taking the sum of its sides. The **area** of a figure is the amount of space it fills. **Rectangle** $P = 2L + 2W$ $A = L \cdot W$ **Square** $P = 4s$ $A = s^2$ **Parallelogram** $P = 2b + 2s$ $A = b \cdot h$ **Triangle** $A = \frac{1}{2}b \cdot h$ **Trapezoid** $A = \frac{1}{2} \cdot h \cdot (b_1 + b_2)$	4 in. 7 in. $P = 2L + 2W$ $= 2(4\text{ in.}) + 2(7\text{ in.})$ $= 8\text{ in.} + 14\text{ in.}$ $= 22\text{ in.}$ $A = L \times W$ $= (4\text{ in.}) \times (7\text{ in.})$ $= 28\text{ in.}^2$ 12 cm 15 cm 9 cm $P = (9\text{ cm}) + (12\text{ cm}) + (15\text{ cm})$ $= 36\text{ cm}$ $A = \frac{1}{2}b \cdot h$ $= \frac{1}{2}\underbrace{(9\text{ cm})}_{\text{base}} \cdot \underbrace{(12\text{ cm})}_{\text{height}}$ $= 54\text{ cm}^2$	p. 440
To convert units of area, use the square of the one-dimensional conversion factor. *U.S. Customary System* $1\text{ ft}^2 = 144\text{ in.}^2$ $1\text{ yd}^2 = 9\text{ ft}^2$ $1\text{ mi}^2 = 640\text{ acres}$ $1\text{ acre} = 43{,}560\text{ ft}^2$ *Metric System* $1\text{ cm}^2 = 100\text{ mm}^2$ $1\text{ m}^2 = 10{,}000\text{ cm}^2$ $1\text{ hectare (ha)} = 10{,}000\text{ m}^2$ *U.S. Customary-Metric* $1\text{ in.}^2 = (2.54)^2\text{ cm}^2$ $\approx 6.45\text{ cm}^2$ $1\text{ cm}^2 \approx 0.15\text{ in.}^2$ $1\text{ mi}^2 \approx 2.59\text{ km}^2$ $1\text{ km}^2 \approx 0.38\text{ mi}^2$	$15\text{ ft}^2 = 15(144\text{ in.}^2)$ $= 2{,}160\text{ in.}^2$ $40\text{ cm}^2 = 40\,\cancel{\text{cm}^2} \times \frac{1\text{ m}^2}{10{,}000\,\cancel{\text{cm}^2}}$ $= \frac{40}{10{,}000}\text{ m}^2$ $= 0.004\text{ m}^2$	p. 447
Circles and Composite Figures		Section 8.3
The **circumference** of a circle is the distance around that circle. The **radius** is the distance from the center to a point on the circle. The **diameter** is twice the radius. The circumference is found using the formula $C = \pi d = 2\pi r$ where π is approximately 3.14. The area of a circle is found using the formula $A = \pi r^2$	If $r = 4.5$ cm, then $C = 2\pi r$ $\approx (2)(3.14)(4.5)$ ≈ 28.3 cm (rounded) If $r = 4.5$ cm, then $A = \pi r^2$ $\approx (3.14)(4.5)^2$ $\approx 63.6\text{ cm}^2$ (rounded)	p. 454

Continued

Definition/Procedure	Example	Reference
Triangles		Section 8.4
A triangle is **acute** if all three angles are less than 90°. A triangle is **right** if it has a right angle. A triangle is **obtuse** if it has an obtuse angle. A triangle is **equilateral** if all three sides have the same length. A triangle is **isosceles** if exactly two sides have the same length. A triangle is **scalene** if all three sides have different lengths.	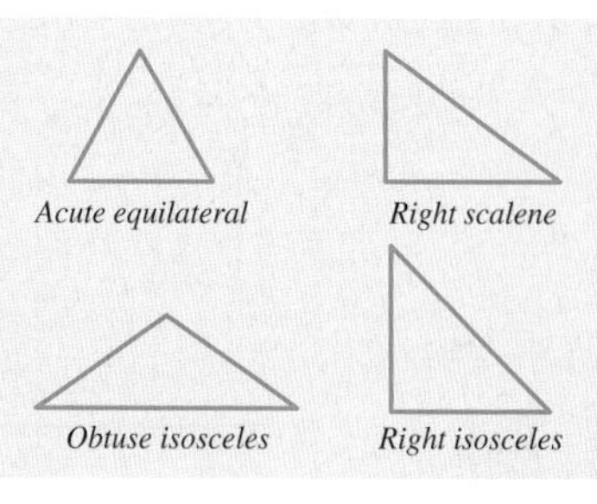 	*p.* 464
Similar triangles Two triangles are *similar* if the measures of the three angles in the two different triangles are the same. Corresponding sides of similar triangles are proportional.	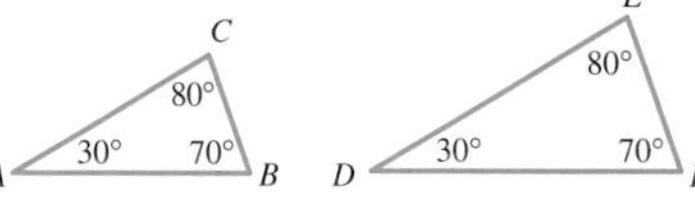 $\triangle ACB$ and $\triangle DEF$ are similar triangles.	*p.* 446
Square Roots and the Pythagorean Theorem		Section 8.5
The square root of a number is a value that, when squared, gives us that number. The length of the three sides of a right triangle form a Pythagorean triple. 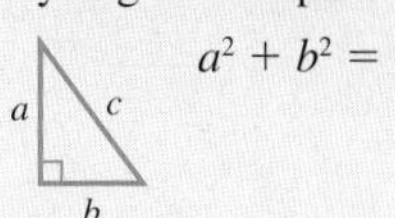 $a^2 + b^2 = c^2$	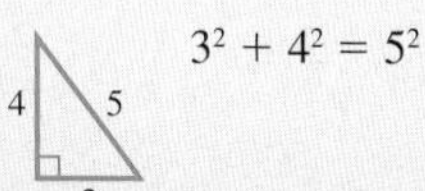 $3^2 + 4^2 = 5^2$	*p.* 474
Solid Geometry		Section 8.6
Solid A three-dimensional figure. **Total Surface Area (TSA)** The area of the surface of a solid. **Volume (V)** The amount of three-dimensional space filled by a solid. **Lateral Surface Area (LSA)** The surface area of a solid without the top or bottom faces. **Rectangular solids, cubes, cylinders,** and **spheres** are all examples of solid figures. **Rectangular Solid** $\text{TSA} = 2LW + 2LH + 2WH$ $\text{LSA} = 2LH + 2WH$ $V = LWH$ **Cube** $\text{TSA} = 6s^2$ $\text{LSA} = 4s^2$ $V = s^3$ **Cylinder** $\text{TSA} = 2\pi r^2 + 2\pi rh$ $\text{LSA} = 2\pi rh$ $V = \pi r^2 h$ **Sphere** $A = 4\pi r^2$ $V = \frac{4}{3}\pi r^3$ To convert units of volume, use the cube of the one-dimensional conversion factor.	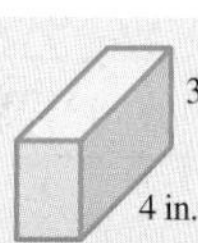 $\text{TSA} = 2(2)(4) + 2(3)(4) + 2(2)(3)$ $= 16 + 24 + 12$ $= 52 \text{ in.}^2$ $\text{LSA} = 2(2)(4) + 2(3)(4)$ $= 16 + 24$ $= 40 \text{ in.}^2$ $V = (2)(3)(4)$ $= 24 \text{ in.}^3$ 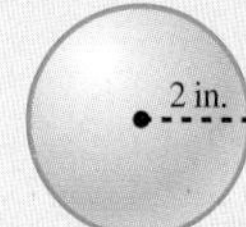 $A \approx 4(3.14)(2)^2$ $= 50.24 \text{ in.}^2$ $V \approx \frac{4}{3}(3.14)(2)^3$ $\approx 33.49 \text{ in.}^3$ $1 \text{ ft}^3 = (12)^3 \text{ in.}^3$ $= 1{,}728 \text{ in.}^3$	*p.* 479

summary exercises :: chapter 8

This summary exercise set will help ensure that you have mastered each of the objectives of this chapter. The exercises are grouped by section. You should reread the material associated with any exercises that you find difficult. The answers to the odd-numbered exercises are in the Answers Appendix at the back of the text.

8.1 *Name the angle; label it as acute, obtuse, right, or straight; then estimate its measure with a protractor.*

1. 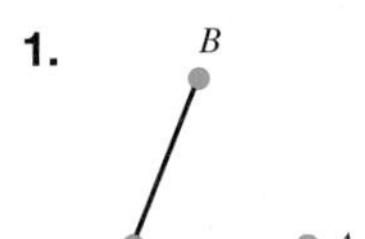

2. 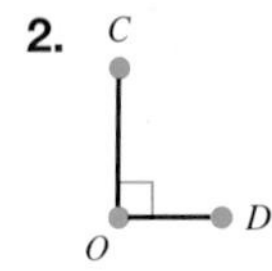

3. 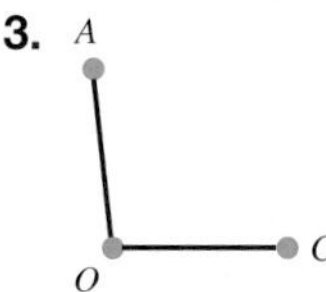

4. 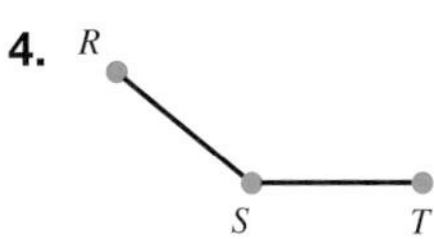

5. 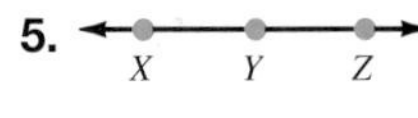

6. 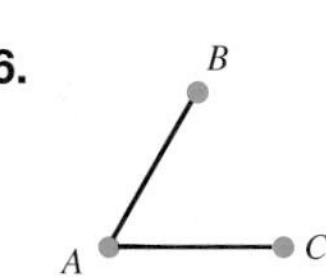

Give the measure of each angle in degrees.

7. $\angle A$ represents $\frac{3}{8}$ of a complete circle.

8. $\angle B$ represents $\frac{7}{10}$ of a complete circle.

9. If $m\angle x = 43°$, find the complement of $\angle x$.

10. If $m\angle y = 82°$, find the supplement of $\angle y$.

Find $m\angle x$.

11. 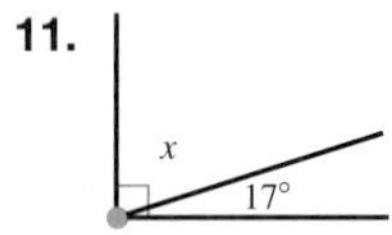

12. 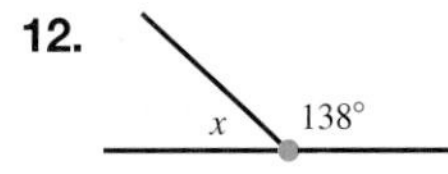

13. 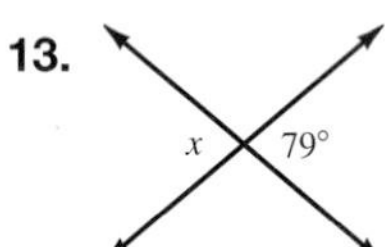

14. 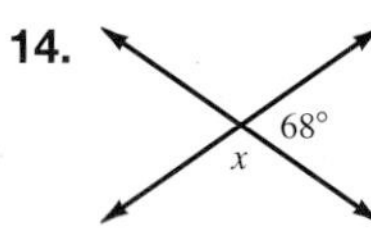

15. 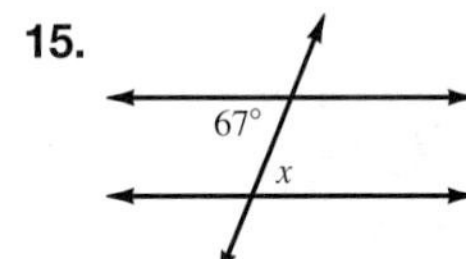

16. 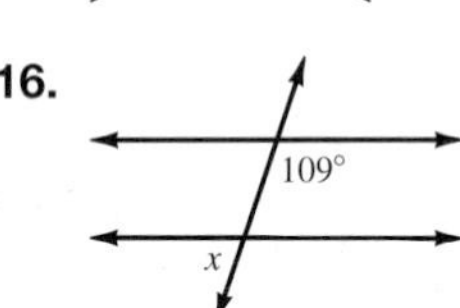

8.2 *Find the area of each figure.*

17. 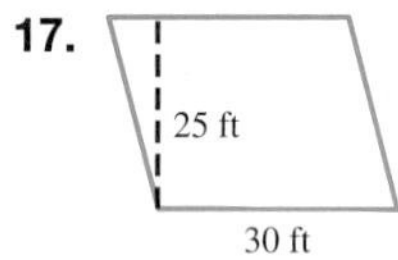

18. 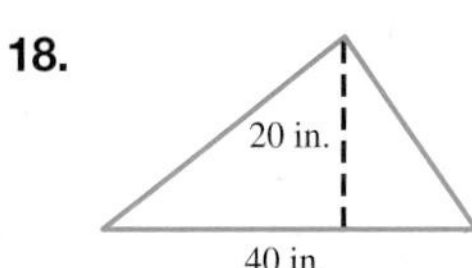

Find the perimeter and area of each figure.

19. 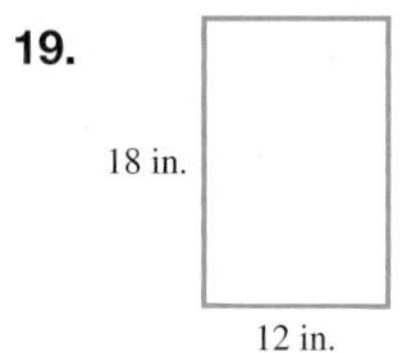

20. 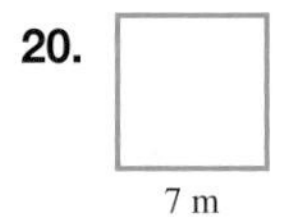

21. 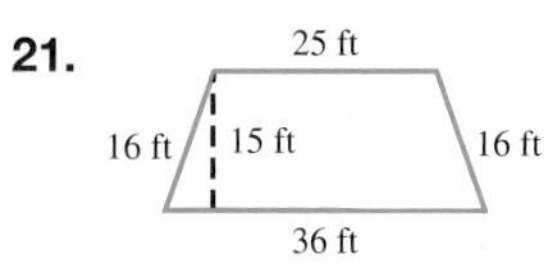

22. 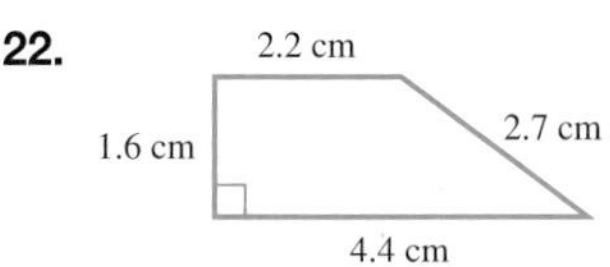

Solve each application.

23. CRAFTS How many square feet of vinyl floor covering will be needed to cover the floor of a room that is 10 ft by 18 ft? How many square yards will be needed?
Hint: How many square feet are in a square yard?

24. CONSTRUCTION A rectangular roof for a house addition measures 15 ft by 30 ft. A roofer will charge \$175 per "square" (100 ft^2). Find the cost of the roofing for the addition.

8.3 *Find the circumference and area of each figure. Use 3.14 for π and round to the nearest tenth.*

25.

12 in.

26.

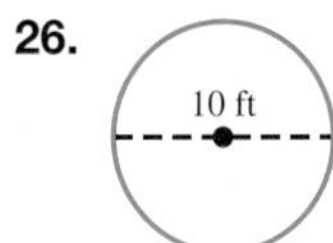

Find the perimeter and area of each figure. Use 3.14 for π and round to the nearest tenth.

27.

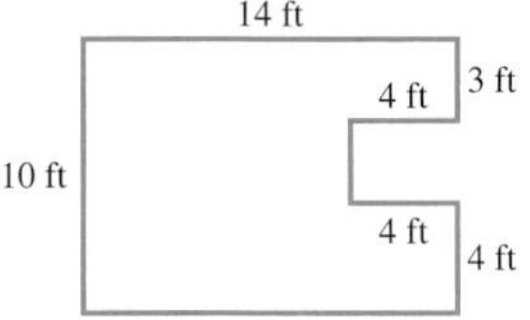

28.

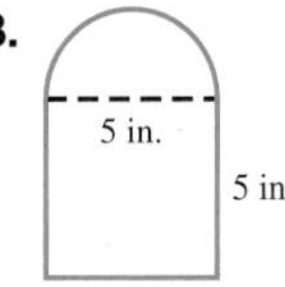

8.4 *Find the missing angle and then label the triangle as equilateral, isosceles, or scalene.*

29.

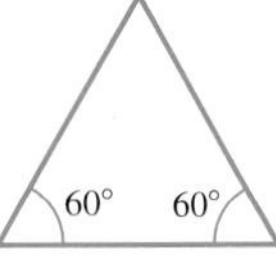

30.

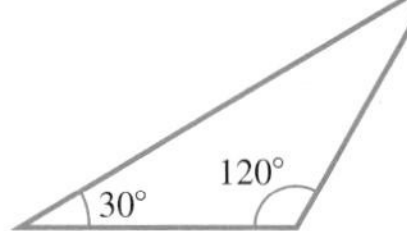

31.

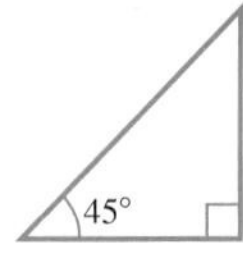

32.

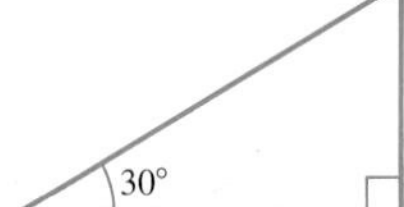

Find the indicated side. Round results to the nearest tenth.

33. Given: $\triangle MNO$ is similar to $\triangle PQR$. Find the length of $\overline{QR}$.

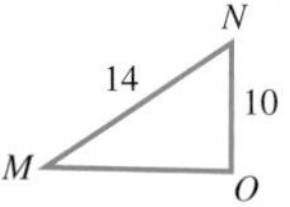

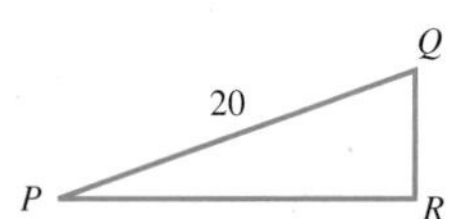

34. Given: $\triangle SUT$ is similar to $\triangle VWX$. Find the length of $\overline{WX}$.

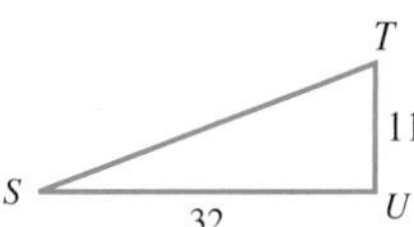

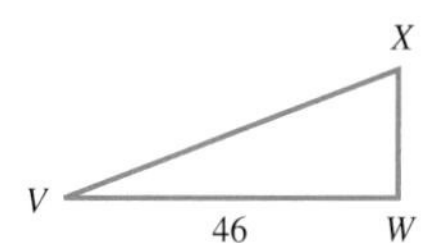

35. GEOMETRY A tree casts a shadow that is 11.2 m long at the same time that a 4.0-m pole casts a shadow that is 1.4 m long. How tall is the tree?

36. GEOMETRY A 5 ft 4 in. woman casts a 2 ft 6 in. shadow. How tall is a Japanese maple tree if its shadow is 16 ft 9 in. (to the nearest inch)?

8.5 *Find each square root. Where necessary, use a calculator and round to the nearest hundredth.*

37. $\sqrt{324}$ **38.** $\sqrt{784}$ **39.** $\sqrt{189}$ **40.** $\sqrt{91}$

Find the length of the unknown side.

41.

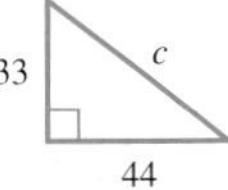

42.

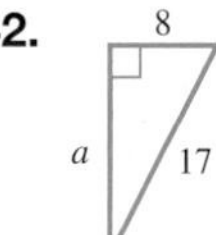

8.6 *Find the **(a)** lateral surface area, **(b)** total surface area, and **(c)** volume of each figure. Use $\pi \approx 3.14$ and round your results to two decimal places, when appropriate.*

43.
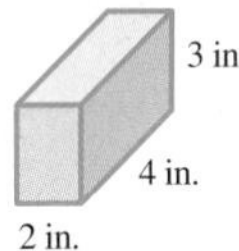

44.
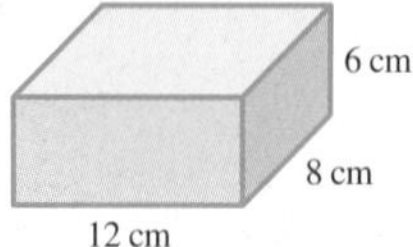

45.
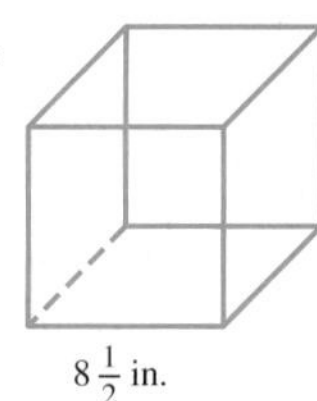

46.
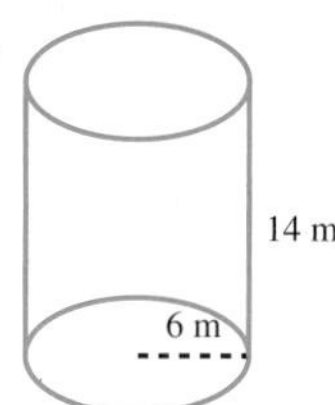

*Find the **(a)** surface area and **(b)** volume of each figure. Use $\pi \approx 3.14$ and round your results to two decimal places, when appropriate.*

47.
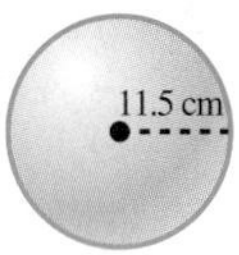

48.
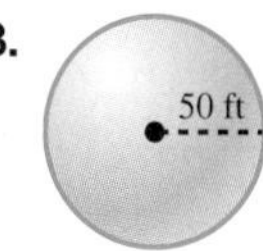

Convert each measurement. Round your results to the nearest hundredth, when appropriate.

49. 12 ft^3 to $in.^3$

50. 16 m^3 to cm^3

51. 30 $in.^3$ to cm^3

52. 30 cm^3 to $in.^3$

chapter test 8

CHAPTER 8

Use this chapter test to assess your progress and to review for your next exam. Allow yourself about an hour to take this test. The answers to these exercises are in the Answers Appendix at the back of the text.

1. Label each pair of lines as parallel, perpendicular, or neither.

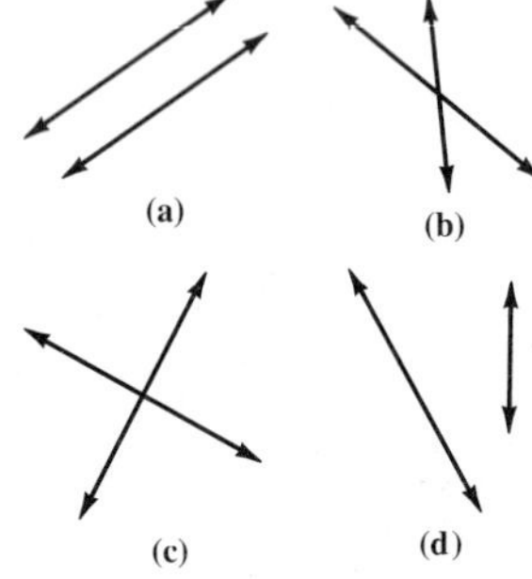

2. Label each of the angles as acute, obtuse, right, or straight.

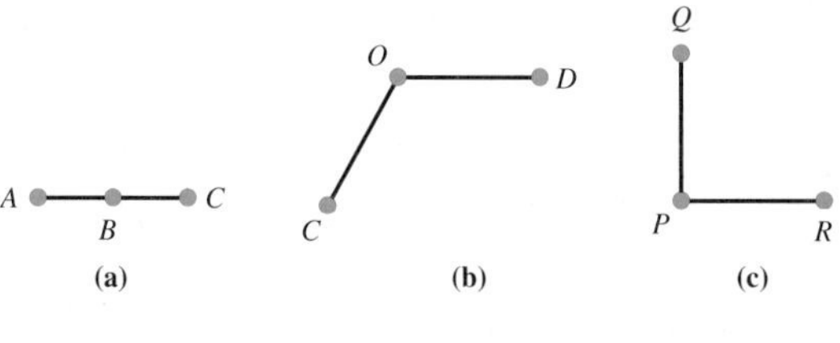

Label each triangle as acute, obtuse, or right.

3.
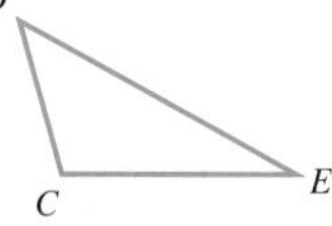

4.
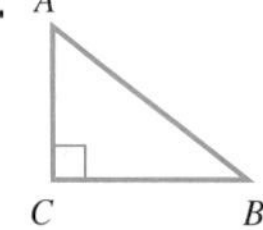

Use a protractor to measure each angle.

5.

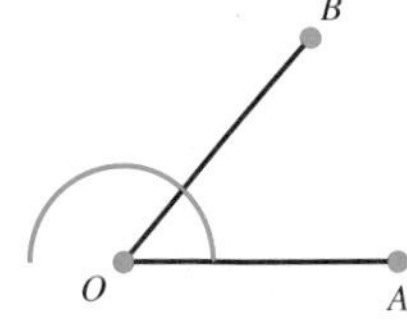

6.

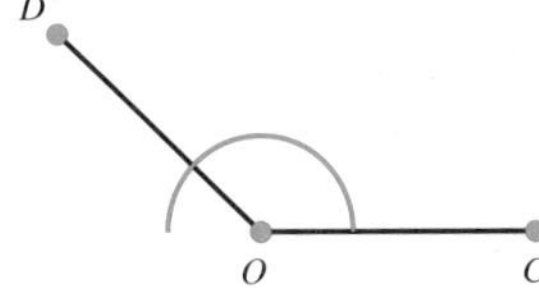

Find $m\angle x$.

7.

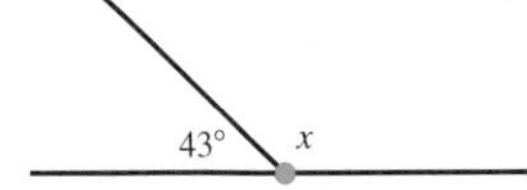

8.

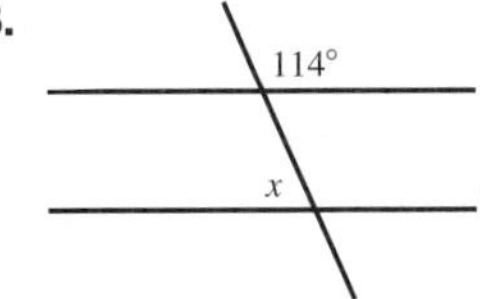

9. Find $m\angle A$.

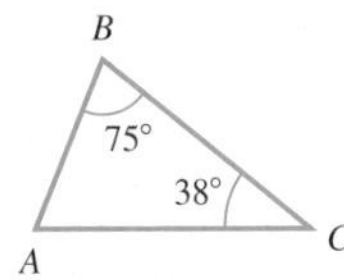

10. Find the square root of 441.

Convert each measurement. Round your result to two decimal places when appropriate.

11. 20 in.^{3} to cm^3
Hint: 1 in. = 2.54 cm

12. 8 yd^3 to ft^3

Find the perimeter or circumference of each figure. Use 3.14 *for* π *and round to the nearest tenth when appropriate.*

13. 9 mm; 21 mm

14. 4 in.; 1.75 in.; 2 in.

15.

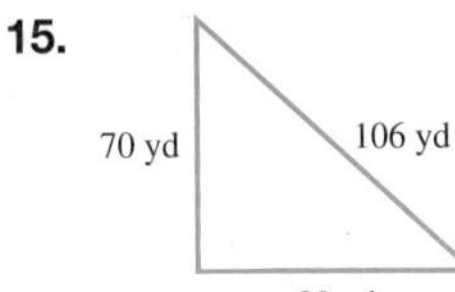

16. 9 m; 30 m; 20 m; 12 m

17.

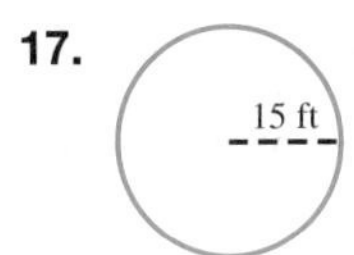

18.

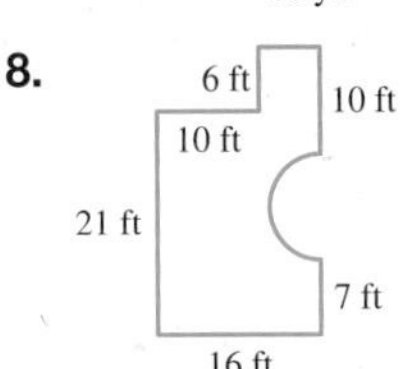

Find the area of each figure. Use 3.14 *for* π *and round to the nearest tenth when appropriate.*

19.

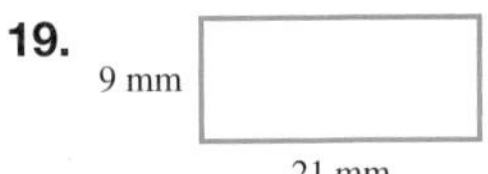

20. 4 in.; 1.75 in.; 2 in.

21.

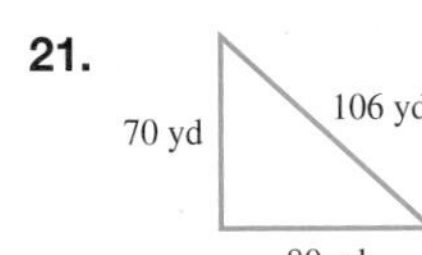

22.

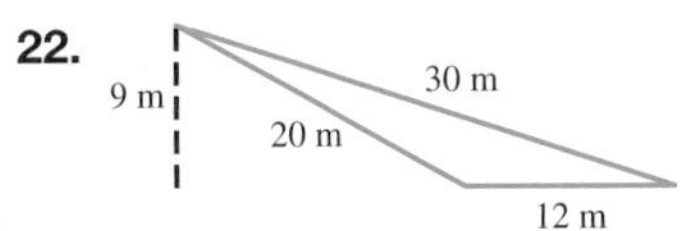

23.

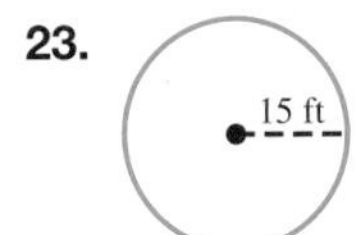

24.

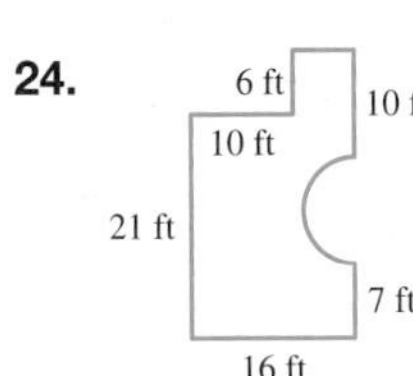

Find the **(a)** *lateral surface area,* **(b)** *total surface area, and* **(c)** *volume of each figure. Use* $\pi \approx 3.14$ *and round your results to two decimal places when necessary.*

25.

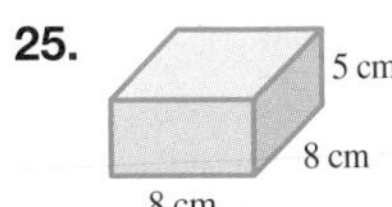

26.

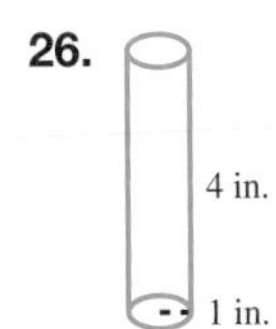

27. Find the **(a)** surface area and **(b)** volume of the sphere. Use $\pi \approx 3.14$ and round your results to two decimal places when necessary.

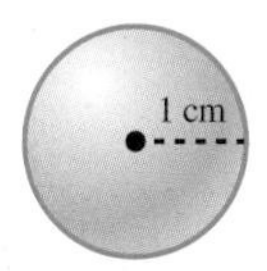

28. The legs of a right triangle are 39 m and 52 m in length. Find the length of the hypotenuse.

29. Given that $\triangle DEF$ is similar to $\triangle GHI$, find the length of $\overline{GH}$. Round to the nearest tenth.

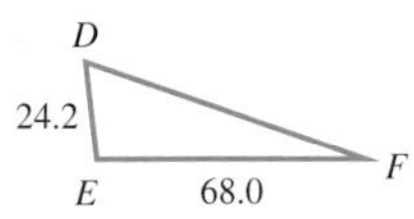

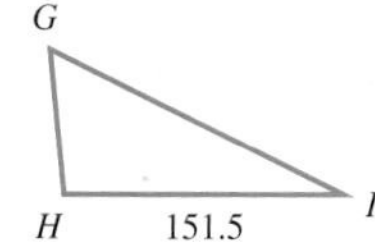

30. **Mechanical Engineering** A reasonable estimation of the size of a furnace is 3 BTUs per cubic foot of air to be heated. How many BTUs are needed to heat an office that measures 9-ft by 11-ft with 8-ft ceilings.

cumulative review chapters 1–8

Use this exercise set to review concepts from earlier chapters. While it is not a comprehensive exam, it will help you identify any material that you need to review before moving on to the next chapter. In addition to the answers, you will find section references for these exercises in the Answers Appendix in the back of the text.

1.1

1. Give the place value of 6 in the number 4,865,201.

1.7

2. Evaluate $82 - 2 \times 3^2$.

2.1

3. Find the prime factorization for 630.

4. Find the greatest common factor (GCF) of 20, 24, and 32.

5. Find the least common multiple (LCM) of 20 and 24.

2.5

6. Multiply $\frac{2}{3} \times 1\frac{4}{5} \times \frac{5}{8}$.

2.6

7. Divide $2\frac{5}{8} \div \frac{7}{12}$.

8. **Business and Finance** Your living room measures $5\frac{2}{3}$ yd by $4\frac{1}{4}$ yd. If carpeting costs $24 per square yard, what does it cost to carpet the room?

9. **Science and Medicine** If you drive 270 mi in $4\frac{1}{2}$ hr, what is your average speed?

3.3

10. Add $\frac{3}{5} + \frac{1}{6} + \frac{2}{3}$

3.4

11. Subtract $7\frac{3}{8} - 3\frac{5}{6}$

12. **Statistics** Adam's goal is to run 20 mi per week. So far he has run $3\frac{1}{2}$ mi, $4\frac{2}{3}$ mi, and $5\frac{1}{4}$ mi. How much more does he need to run to reach his goal?

4.1

Find the indicated place values.

13. 3 in 17.2396

14. 5 in 8.0915

4.3

15. Find the perimeter and area of the figure.

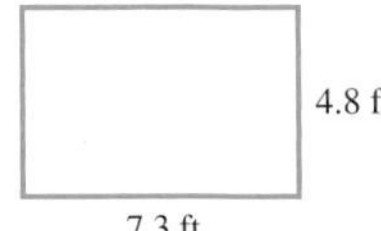

4.5

16. Write 0.125 as a simplified fraction.

17. Write the decimal equivalent of $\frac{5}{16}$.

5.2

18. Simplify $\frac{81{,}000 \text{ dollars}}{4 \text{ years}}$

5.4

19. Solve $\frac{35}{14} = \frac{10}{w}$

20. **Geometry** If 1 gal of paint covers 250 ft^2, how many square feet does $4\frac{1}{2}$ gal cover?

6.1

21. Write 8.5% as a decimal.

22. Write 37.5% as a fraction.

6.2

23. Write $\frac{27}{40}$ as a percent.

6.3

24. Find 16% of 320.

25. 35% of what number is 525?

6.4

26. **Statistics** The number of students at a certain high school dropped by 6% since last year. There are now 1,269 students. How many were there last year?

7.1–7.4

Complete each statement.

27. 3 mi = ______________ yd

28. 250 mg = ______________ g

29. 5.8 km = ______________ m

8.1

30. Use a protractor to find the measure of the given angle.

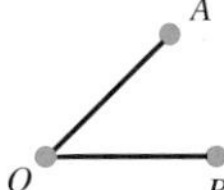

8.2–8.3

31. Find the circumference of a circle whose radius is 7.9 ft. Use 3.14 for π and round the result to the nearest tenth of a foot.

Find the area of each figure. Use 3.14 *for* π.

32.

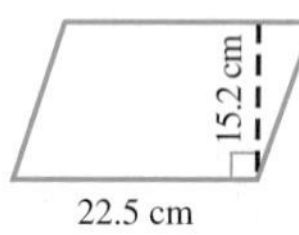

33.

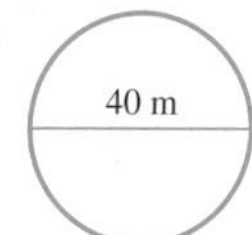

34.

8.4

35. Find the missing angle and identify the triangle as equilateral, isosceles, or scalene.

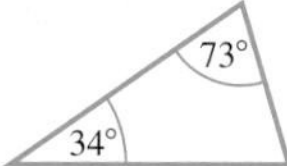

8.5

36. The given triangles are similar. Find x.

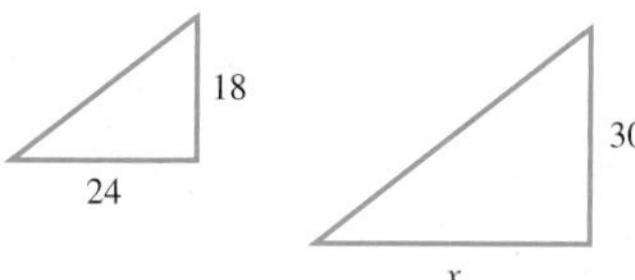

37. If the legs of a right triangle have lengths 8 ft and 15 ft, find the length of the hypotenuse.

8.6

38. Find the **(a)** lateral surface area, **(b)** total surface area, and **(c)** volume of the cube.

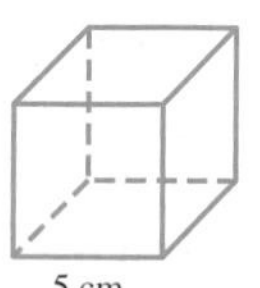

39. Find the **(a)** surface area and **(b)** volume of the sphere. Use $\pi \approx 3.14$ and round your results to two decimal places when necessary.

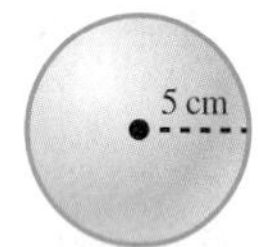

40. GEOMETRY A fish tank in the shape of a rectangular solid is 48 in. wide, 24 in. deep, and 30 in. high. Determine the volume of the tank, to the nearest gallon, if 1 gal is 231 in.^{3}.

Data Analysis and Statistics

INTRODUCTION

Thanks to our many technological advances, we are now "awash" in data. In fact, we have gathered so much data about so many topics that we cannot even analyze it all. To add to the difficulty, more data are gathered every time anyone uses the Internet.

Every year we develop new techniques and advance new technologies to help us analyze our data, but the pace at which we accumulate data far outstrips our ability to understand the data we gather.

In this chapter, we look at some of the ways we describe data sets to help us understand them. One difficulty we encounter is the *outlier*. That is, a data point that does not fit with the rest of a data set. You will have the opportunity to look at outliers in data sets when you complete Activity 26.

CHAPTER 9 OUTLINE

9 prerequisite check

CHAPTER 9

This Prerequisite Check highlights the skills you will need in order to be successful in this chapter. The answers to these exercises are in the Answers Appendix at the back of the text.

Simplify each fraction.

1. $\frac{20}{25}$

2. $\frac{45,000}{60,000}$

Evaluate each expression.

3. $\frac{3 + 5 + 7 + 9}{4}$

4. $\frac{6 + 11}{2}$

5. $3(8 - 5)^2$

6. $2 + 3 \times 5^2$

7. $\frac{1}{4} \times 360$

8. $\frac{5}{6} \times 360$

Write each fraction as a percent.

9. $\frac{5}{8}$

10. $\frac{1}{6}$

Write each list of numbers in ascending order (from smallest to largest).

11. 7, 1, 2, 11, 5, 0, 10, 13

12. 21, 50, 123, 81, 12, 7, 55, 56

Use a ruler (with U.S. Customary units) to find the length of the line segment.

13. 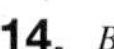

Use a protractor to find the measure of the angle.

14. 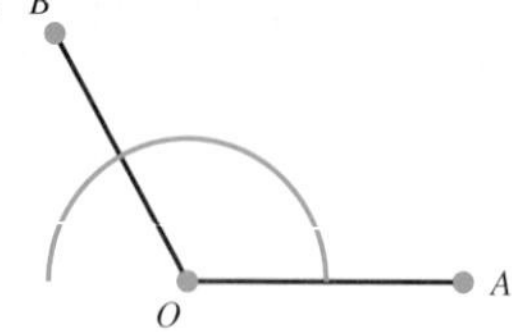

9.1 Mean, Median, and Mode

< 9.1 Objectives >

1 > Calculate the mean of a data set

2 > Find the median of a data set

3 > Compare the mean and median of a data set

4 > Find the mode of a data set

When looking at a set of objects, it is common to ask what the typical member of the set looks like. For instance, given a set of marbles, we might reasonably ask which color we would most likely choose if we picked up one marble at random. Given a set of numbers, we might ask what number is most representative of the numbers in the set.

These are both instances where we are looking for an *average* of a set. However, we would most likely use different methods to find the average member of a set of marbles and a set of numbers. We call an average a *measure of center* and use it as one way to distinguish between sets of objects.

The first measure of center or average we look at is the most common. It is called the **mean** and is used to find an average of a set of numbers. Most often, when people talk about an average, they are looking at the mean of a set of numbers.

Step by Step

Finding a Mean

Step 1 Add all the numbers in the set.

Step 2 Divide that sum by the number of items in the set.

Example 1 Finding a Mean

< Objective 1 >

RECALL

On a calculator, you need to use parentheses to group the numerator.

```
(12+19+15+14)/4
              15
```

Find the mean of the set of numbers 12, 19, 15, and 14.

Step 1 Add all the numbers.

$12 + 19 + 15 + 14 = 60$

Step 2 Divide that sum by the number of items.

$60 \div 4 = 15$ There are four items in this group.

The mean of this set is 15.

You should see that we grouped the numbers together to form their sum before dividing.

$$\frac{12 + 19 + 15 + 14}{4} = \frac{60}{4} = 15$$

Check Yourself 1

Find the mean of the set of numbers 17, 24, 19, and 20.

Next, we apply the concept of mean to a word problem.

Example 2 Finding a Mean

The ticket prices (in dollars) for the nine concerts held at the Civic Arena this school year were

33, 31, 30, 59, 32, 35, 32, 36, 56

What was the mean price for these tickets?

NOTE

We round to the nearest hundredth because we are dealing with money.

Step 1 Add all the numbers.

$33 + 31 + 30 + 59 + 32 + 35 + 32 + 36 + 56 = 344$

Step 2 Divide by 9.

$344 \div 9 = 38.22$ Divide by 9 because there are 9 ticket prices.

The mean ticket price was $38.22.

Check Yourself 2

The costs (in dollars) of the six textbooks that Aaron needs for the fall quarter are

75, 69, 125, 156, 104, 80

Find the mean cost of these books.

The mean is the most common way of measuring the center or average of a set of numbers. A problem arises if the set has an extreme value. In such a case, the mean may **not be typical** of the data set.

Example 3 Computing a Mean with an Extreme Value

Recently, a university looked at the annual average salaries of students who graduated with a B.S. degree in geology 10 years earlier. They were able to make the claim that the average student earned over $7 million.

To verify this claim, we look at the graduates.

B.S. Geology Recipients

Occupation	Salary
Science teacher	$47,500
Graduate student	$14,500
Graduate student	$16,500
Oil prospector	$96,000
Pro basketball player	$35,000,000

Step 1 $47{,}500 + 14{,}500 + 16{,}500 + 96{,}000 + 35{,}000{,}000 = 35{,}174{,}500$

Step 2 $35{,}174{,}500 \div 5 = 7{,}034{,}900$

The mean annual salary was $7,034,900.

Check Yourself 3

A realtor has a list of six homes for sale. Their prices are $269,000, $249,900, $225,000, $290,000, $254,900, and $2,450,000.

Find the mean price of the houses listed.

NOTE

The median is often used to describe salaries, home prices, and people's ages.

In Example 3, the mean annual salary for the degree recipients is over \$7 million. However, the typical student majoring in geology would not expect to earn \$7 million per year after receiving a degree.

In situations such as these, it is better to use a different measure of the average or center of the set of numbers called the *median*.

The **median** of a set of numbers is the value in the middle when the numbers are sorted in (ascending) order. This way, half the numbers are above the median and half are below the median.

Step by Step

Finding a Median

Step 1 Sort the numbers in ascending order (lowest value to highest value).

Case 1 There is an odd number of data points.

Step 2 Select the middle data value; this is the median.

Case 2 There is an even number of data points.

Step 2 Select the two middle data values.

Step 3 Compute the mean of these two numbers; this is the median.

NOTE

To find the mean of two middle values, add them together and divide by 2.

We demonstrate finding a median for both cases in Example 4.

Example 4 Finding the Median

< Objective 2 >

Find the median of each set of numbers.

(a) 35, 18, 27, 38, 19, 63, 22

Step 1 Rewrite the numbers in order from smallest to largest.

18, 19, 22, 27, 35, 38, 63

Step 2 There is an odd number of data points (7), so this is an example of the first case and we simply select the middle value.

NOTE

Three numbers are less than 27 and three are more than 27.

18, 19, 22, 27, 35, 38, 63

↑ Middle value (27)

The median is 27.

(b) 29, 88, 73, 81, 62, 37

Step 1 Rewrite the numbers in order from smallest to largest.

29, 37, 62, 73, 81, 88

Step 2 There is an even number of data points (6), so this is an example of the second case. We select the two middle values.

29, 37, 62, 73, 81, 88

Middle values (62, 73)

Step 3 Find the mean of the pair of middle values.

$$\frac{62 + 73}{2} = \frac{135}{2}$$

$$= 67\frac{1}{2}$$

The median is $67\frac{1}{2}$.

Check Yourself 4

Find the median of each set of numbers.

(a) 8, 6, 19, 4, 21, 5, 27 **(b)** 43, 29, 13, 38, 29, 53

In Example 5, we find the median of the data set that we first saw in Example 3. You should look at how the mean compares to the median for the salaries in question.

Example 5 Finding a Median

> CAUTION

You must sort the numbers before finding the median!

(a) Find the median salary for the B.S. recipients given in Example 3.

Step 1 Sort the data.

$14,500
$16,500
$47,500
$96,000
$35,000,000

Because there is an odd number of data points (5), this is an example of the first case.

Step 2 Select the middle value.

$14,500, $16,500, $47,500, $96,000, $35,000,000

↑ Middle value

The median salary for these degree recipients is $47,500. This is a much more reasonable expectation for someone who graduates with this degree.

(b) If there were a sixth graduate from the program (as shown below), find the median.

B.S. Geology Recipients

Occupation	Salary
Graduate student	$14,500
Graduate student	$16,500
Science teacher	$47,500
Researcher	$87,500
Oil prospector	$96,000
Pro basketball player	$35,000,000

Step 1 The data set is sorted already.

There is now an even number of data points (6), so this is an example of the second case.

Step 2 Select the two middle data points.

$14,500, $16,500, $47,500, $87,500, $96,000, $35,000,000

Middle values

$47,500 and $87,500 are the two points in the middle.

Step 3 Compute the mean of these two middle values.

$$\frac{47{,}500 + 87{,}500}{2} = \frac{135{,}000}{2} = 67{,}500$$

The median income for this set is $67,500.

Check Yourself 5

In each case, find the median of the data set given.

(a) The total wheat production (in millions of bushels) in Kansas between the years 2006 and 2010 is given.

Wheat Supply (Kansas)

	2006	2007	2008	2009	2010
Production	1,808	2,051	2,499	2,218	2,207

Source: Kansas Agricultural Statistics Service, Kansas Department of Agriculture

(b) A realtor has a list of six homes for sale. Their prices are \$269,000, \$249,900, \$225,000, \$290,000, \$254,900, and \$2,450,000.

Note: You found the mean of this set in Check Yourself 3. How does the mean compare to the median? Which is a better measure of the average home price on the realtor's list?

Now let us look more explicitly at comparing means and medians.

Example 6 Comparing the Mean and the Median

< Objective 3 >

We give the hourly wages of seven employees of a local chip manufacturing plant.

15, 18, 16, 20, 47, 20, 18

(a) Find the mean hourly wage.

Step 1 Add the numbers in the set.

$15 + 18 + 16 + 20 + 47 + 20 + 18 = 154$

Step 2 Divide that sum by the number of items in the set.

$154 \div 7 = 22$

The mean wage is \$22 an hour.

(b) Find the median wage for the seven workers.

Step 1 Rewrite the numbers in order from smallest to largest.

15, 16, 18, 18, 20, 20, 47

Step 2 There are 7 employees (odd), so we select the middle value.

15, 16, 18, 18, 20, 20, 47

↑ Middle value (18)

The median wage is \$18 per hour.

(c) Compare the mean and median of this data set. Which better describes the typical hourly wage paid at this plant?

The mean hourly wage is \$22 and the median is \$18. The median is a better description of the typical wage because only one employee earns more than \$22. The average worker should not expect to earn \$22 or more at this plant because most of the employees earn closer to \$18 per hour.

Check Yourself 6

Jessica compiled her company's phone bills for a year.

86, 85, 90, 74, 242, 112, 98, 95, 86, 102, 104, 95

(a) Find the mean amount of the phone bills.
(b) Find the median amount of the phone bills.
(c) Compare the mean and median of this data set. Which better describes her company's typical monthly phone bill?

Mean and median are both useful ways of determining a typical member of many sets of numbers. In general, we prefer to use the mean because the techniques of inferential statistics are more effective when applied to the mean than with the median. However, as we stated earlier, if there are extreme values, the median may be more typical of a data set, and thus, the median may be a better measure of center.

Both the mean and median as measures of the typical member of a data set work with sets of numbers. When our data is not made up of numbers, then we cannot use either method to describe the *typical member*.

For instance, we can not add together a set of colors and divide by the size of the set. Nor can we get anything useful by sorting a set of phone numbers from smallest to largest.

As an example, if we surveyed respondents as to the color they prefer for new automobiles, we might get this data set.

NOTE

When trying to find an average for sets of numbers that are not quantitative, such as zip codes or phone numbers, we use the mode rather than the mean or median.

This makes more sense than adding up a set of phone numbers and dividing.

red	blue
blue	red
blue	black
purple	black
orange	blue

We cannot compute a mean for this set (adding these ten colors, then dividing by 10 does not give anything meaningful). Similarly, there is no natural order in which to sort the data, so we are unable to find a middle value.

Another measure used as an average is the **mode.** The mode is always used when the data are not numbers.

Definition

Mode The *mode* of a set of data is the item or number that appears most frequently.

Example 7 Finding a Mode

< Objective 4 >

For the data set given above, the mode is blue (which appears four times).

Check Yourself 7

A researcher studying the amphibian population in Yellowstone National Park recorded her sightings during the course of her visit. Find the mode of her data set.

Columbia spotted frog	boreal toad
Columbia spotted frog	blotched tiger salamander
chorus frog	boreal toad
boreal toad	chorus frog
Columbia spotted frog	chorus frog
Columbia spotted frog	chorus frog
chorus frog	Columbia spotted frog
blotched tiger salamander	Columbia spotted frog
Columbia spotted frog	Columbia spotted frog
chorus frog	blotched tiger salamander

Adapted from information provided by the National Park Service; U.S. Department of the Interior

We can also find the mode of a set of numbers.

Example 8 Finding a Mode

Find the mode of the set of numbers.

22, 24, 24, 24, 24, 27, 28, 32, 32

The mode, 24, is the number that appears most frequently.

Check Yourself 8

Find the mode of the set of numbers.

7, 7, 7, 9, 11, 13, 13, 15, 15, 15, 15, 21

If a data set has two values that appear the most, we say it is *bimodal*. If there are three or more values that appear the most, or no value that appears more than once, the data set has no mode.

Example 9 Finding Modes

Find the mode of each set.

(a) 4, 6, 1, 5, 5, 4, 2

Both 4 and 5 appear twice. The other members of the set (1, 2, and 6) appear only once. Therefore, 4 and 5 are both modes and we say the set is bimodal.

(b) The computers in a lab, by manufacturer.

Apple, HP, HP, Toshiba, Dell, Apple, Toshiba, Dell

Each computer manufacturer appears twice. In this case, we say the data set does not have a mode.

Check Yourself 9

Find the mode of each set.

(a) A set of eye colors for a group of people. blue, brown, brown, hazel, green, hazel, blue, brown, hazel
(b) 17, 43, 6, 22, 23, 19, 65, 51

To summarize our work so far, we have three methods for determining the typical member of a data set, or measuring its center. These are the mean, median, and mode.

Mean and median are used only for sets of numbers, whereas mode can be used for any set.

The mean is the more common, and desirable, method for determining an average, but median is more resistant to extreme values.

Example 10 A Health Sciences Application

A pregnant patient tested positive for gestational diabetes during the last 3 months of her pregnancy. Blood glucose levels (in milligrams per 100 milliliters) were gathered and recorded on a regular basis. The results are tabulated here.

78	104	103	101	78	120	103
97	75	128	90	98	80	128
119	106	83	99	101	108	127
118	96	125	78	123	124	92

Compute the mean.

To do this, we must first find the sum of the numbers. In this case, a calculator (or computer) proves very useful, since there are 28 numbers to add. The sum of these is 2,882. Dividing by 28 produces 102.928. . . . To the nearest whole number, the mean glucose level is 103.

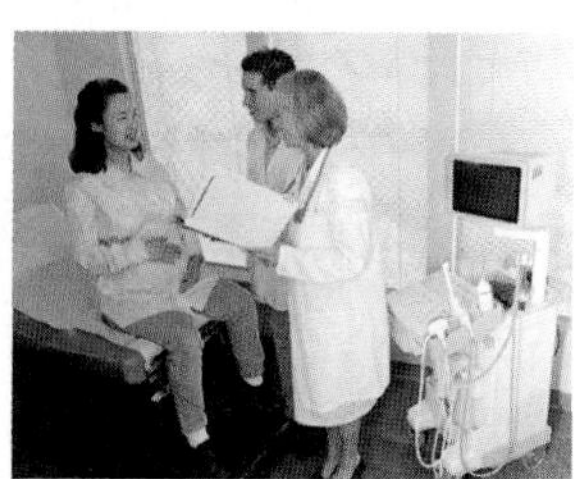

Check Yourself 10

For the data given in Example 10, find the median.

Check Yourself ANSWERS

1. 20 **2.** $101.50 **3.** $623,133.33 **4.** **(a)** 8; **(b)** $33\frac{1}{2}$ **5.** **(a)** 2,207 million bushels; **(b)** $261,950 **6.** **(a)** $105.75; **(b)** $95; **(c)** The median is a better measure of her company's typical phone bill because 10 of 12 bills were below the mean. **7.** Columbia spotted frog **8.** 15 **9.** **(a)** Brown and hazel (bimodal); **(b)** no mode **10.** 102

Reading Your Text

These fill-in-the-blank exercises will help you understand some of the key vocabulary used in this section. The answers to these exercises are in the Answers Appendix at the back of the text.

(a) To find the ____________ of a set of numbers, add the numbers in the set and divide by the number of items in the set.

(b) The ____________ is the middle value when an odd number of numbers are arranged in order.

(c) The ____________ of a set of data is the item or number that appears most frequently.

(d) A set with two different modes is called ____________.

9.1 exercises

Skills | Calculator/Computer | Career Applications | Above and Beyond

< Objective 1 >

Find the mean of each set of numbers.

1. 6, 9, 10, 8, 12 
2. 13, 15, 17, 17, 18
3. 13, 15, 17, 19, 24, 25
4. 41, 43, 56, 67, 69, 72
5. 12, 14, 15, 16, 16, 16, 17, 22, 25, 27
6. 21, 25, 27, 32, 36, 37, 43, 43, 44, 51
7. 5, 8, 9, 11, 12 
8. 7, 18, 11, 7, 12
9. 9, 8, 11, 14, 9
10. 21, 23, 25, 27, 22, 20

< Objective 2 >

Find the median of each set of numbers.

11. 2, 3, 5, 6, 10 
12. 12, 13, 15, 17, 18
13. 23, 31, 24, 36, 27, 38
14. 25, 15, 4, 36, 1, 9
15. 46, 13, 47, 25, 68, 51
16. 26, 71, 69, 33, 71, 25, 75

< Objective 3 >

17. 45, 60, 70, 38, 54, 64, 70

(a) Calculate the mean of the set of numbers (to the nearest hundredth).

(b) Find the median of the set of numbers.

(c) Is the mean or median a better indicator of the average number in the set?

18. 140, 125, 128, 150, 810, 112, 144, 153

(a) Calculate the mean of the set of numbers.

(b) Find the median of the set of numbers.

(c) Is the mean or median a better indicator of the average number in the set?

19. **Business and Finance** A real estate agent lists eight houses for sale at the prices shown.

\$209,000	\$224,900	\$249,900	\$215,000
\$289,900	\$265,000	\$274,900	\$749,900

(a) Calculate the mean home price of the agent's listings.

(b) Find the median home price of the agent's listings.

(c) Is the mean or median a better indicator of the typical home price listed?

20. **Statistics** A group of students takes an exam. Their test scores are shown.

82, 76, 90, 94, 88, 64, 72, 92, 88, 76, 80

(a) Calculate the mean test score.

(b) Find the median test score.

(c) Is the mean or median a better indicator of the typical test score?

21. **Science and Medicine** A physician sees 12 female patients in one day. As part of her examination, she measures the height of each patient. Their heights are listed below, in inches.

66, 60, 62, 64, 64, 58, 63, 68, 65, 64, 63, 65

(a) Calculate the mean height of her female patients.

(b) Find the median height of her female patients.

(c) Is the mean or median a better indicator of the typical height of her female patients?

22. **Information Technology** An IT firm forms a group to solve server issues encountered by their customer base. The group is selected from throughout the firm's ranks.

The annual salary of each group member is shown below.

\$68,000	\$46,000	\$52,000
\$85,400	\$89,500	\$105,000
\$72,000	\$68,000	\$740,000

(a) Calculate the mean salary of the group members.

(b) Find the median salary of the group members.

(c) Is the mean or median a better indicator of the typical group member's salary?

< Objective 4 >

Find the mode of each set of numbers.

23. 17, 13, 16, 18, 17 

24. 41, 43, 56, 67, 69, 72

25. 21, 44, 25, 27, 32, 36, 37, 44

26. 9, 8, 10, 9, 9, 10, 8

27. 12, 13, 7, 14, 4, 11, 9

28. 8, 2, 3, 3, 4, 9, 9, 3

29. 18, 36, 36, 23, 38, 28, 18, 41

30. 125, 96, 134, 125, 125, 96, 105, 105, 96, 101

31. **Science and Medicine** A nurse records the eye color of the first eight patients she sees one morning. Which color is the mode?

hazel, green, brown, brown, blue, green, hazel, green

32. **Statistics** The weather in Philadelphia over the last 7 days was as follows:

rain, sunny, cloudy, rain, sunny, rain, rain

What type of weather was the mode?

Solve each application.

33. **Statistics** High temperatures of 86°, 91°, 92°, 103°, and 98°F were recorded for the first 5 days of July. What was the mean high temperature? VIDEO

34. **Statistics** A sales associate drove 238, 159, 87, 163, and 198 miles on a 5-day trip. What was the mean number of miles driven per day?

35. **Statistics** Highway mileage ratings for seven new cars are 43, 29, 51, 36, 33, 42, and 32 miles per gallon (mi/gal). What is the mean rating?

36. **Statistics** The enrollments in four schools are 278, 153, 215, and 198 students. What is the mean enrollment?

37. **Statistics** To get an A in history, you must have a mean of 90 on five tests. Your scores thus far are 83, 93, 88, and 91. How many points must you have on the final test to receive an A?
Hint: First find the total number of points you need to get an A. VIDEO

38. **Statistics** To pass biology, you must have a mean of 70 on six quizzes. So far your scores have been 65, 78, 72, 66, and 71. How many points must you have on the final quiz to pass biology?

39. **Statistics** Louis had scores of 87, 82, 93, 89, and 84 on five tests. Tamika had scores of 92, 83, 89, 94, and 87 on the same five tests. Who had the higher mean score? By how much?

40. **Statistics** The Wong family had heating bills of \$105, \$110, \$90, and \$67 in the first 4 months of 2010. The bills for the same months of 2011 were \$110, \$95, \$75, and \$76. In which year was the mean monthly bill higher? By how much?

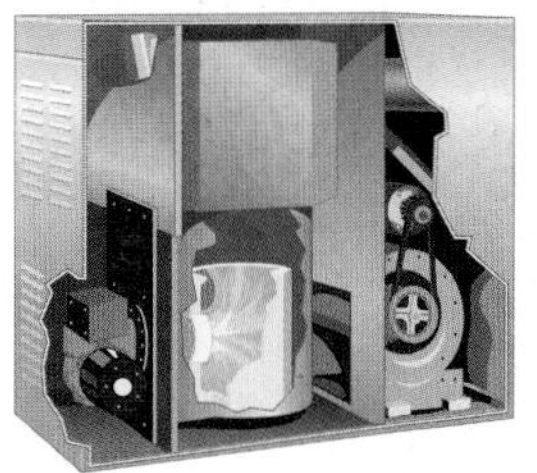

Science and Medicine *Monthly energy use, in kilowatt-hours (kWh), by appliance type for four typical U.S. families is shown in the table.*

	Wong Family	McCarthy Family	Abramowitz Family	Gregg Family
Electric range	97	115	80	96
Electric heat	1,200	1,086	1,103	975
Water heater	407	386	368	423
Refrigerator	127	154	98	121
Lights	75	99	108	94
Air conditioner	123	117	96	120
TV	39	45	21	47

41. What is the mean number of kilowatt-hours used each month by the four families for heating their homes?

42. What is the mean number of kilowatt-hours used each month by the four families for hot water?

43. What is the mean number of kilowatt-hours used per appliance by the McCarthy family?

44. What is the mean number of kilowatt-hours used per appliance by the Gregg family?

Determine whether each statement is **true** *or* **false.**

45. The mode can only be found for sets of numbers.

46. The mean can be larger than most of the data.

In each statement, fill in the blank with **always, sometimes,** *or* **never.**

47. The mean and the median are __________ the same number.

48. To find the median, you must ________ put the numbers in order.

Skills	**Calculator/Computer**	Career Applications	Above and Beyond

Many calculators have built-in statistical functions to compute the mean, median, and mode as well as to perform more advanced analyses of data sets.

Because these features vary widely between calculator models, you should consult the instruction manual or your instructor if you choose to use these capabilities (you can usually download an instruction manual from your calculator manufacturer's website).

You can use your calculator to compute the mean of a data set without using its built-in statistical functions. In order to do this, you need to remember to treat the fraction bar as a grouping symbol. That is, you need to tell the calculator to add the numbers in the data set before performing division. Because this "violates" the order of operations, you must use parentheses to perform the addition before the division.

To find the mean of the set

2,253 3,451 2,157 4,126 967

we add the numbers and contain them in parentheses and then divide by 5.

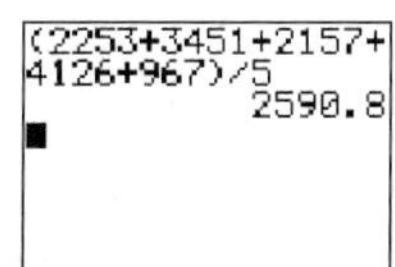

[( ] 2253 [+] 3451 [+] 2157 [+] 4126 [+] 967 [ )] [÷] 5 [ENTER]

The display should read 2590.8.

Use a calculator to find the mean of each set of numbers.

49. 48, 50, 51, 52, 49, 50

50. 20, 18, 17, 24, 22, 19

51. 346, 351, 353, 347, 341, 382, 373, 363

52. 1,560, 1,540, 1,570, 1,555, 1,565, 1,545, 1,557

53. 16,430, 15,487, 17,982, 11,290, 21,908, 16,545

54. 311,431, 286,356, 356,090, 292,007, 301,857, 299,005

55. 18, 21, 20, 22

56. 356, 371, 366, 373, 359, 363

57. 1,898, 1,913, 1,875, 1,937

58. 15,865, 16,270, 16,090, 15,904

59. **Business and Finance** Revenue for the leading apparel companies in the United States in 2011 is given in the table.

Company	Revenue (in millions)
Vf	$7,703
Polo Ralph Lauren	5,660
Phillips-Van Heusen	4,637
Itochu International	4,587
Levi Strauss	4,411
Marubeni America	3,668
The Jones Group	3,643
Coach	3,608
Brown Shoes	2,504
Liz Clairborne	2,500

Source: Manta Media Inc.

What is the mean revenue taken in by these companies?

60. **Business and Finance**

Unemployment in the United States (in thousands)

Year	Employed	Unemployed
2001	136,933	6,801
2002	136,485	8,378
2003	137,736	8,774
2004	139,252	8,149
2005	141,730	7,591
2006	144,427	7,001
2007	146,047	7,078
2008	145,362	8,924
2009	139,877	14,265
2010	139,064	14,825

Source: U.S. Department of Labor, Bureau of Labor Statistics

Find the mean number of employed and unemployed people per year from 2001 to 2010. Round to the nearest thousand.

Business and Finance *The number of large work stoppages (strikes and lockouts) in the United States and the resulting number of work days idle (in thousands) are shown in the table.*

Year	Stoppages	Days Idle
2003	14	4,091
2004	17	3,344
2005	22	1,736
2006	20	2,688
2007	21	1,265
2008	15	1,954
2009	5	124
2010	11	302

Source: U.S. Department of Labor, Bureau of Labor Statistics

61. Find the mean number of large work stoppages per year from 2003 to 2010.

62. Find the mean number of work days idle from 2003 to 2010.

Skills | Calculator/Computer | **Career Applications** | Above and Beyond

63. **Information Technology** Response times in milliseconds (ms) from your computer to a local router using ping are given by the table.

2.2	2.5
2.3	2.4
1.9	2.2
2.0	2.4
2.5	2.5

(a) Compute the mean.

(b) Compute the median.

(c) Compute the mode.

64. **Construction** The thicknesses (in millimeters) of several parts are as follows:

30.9, 30.7, 29, 30.6, 29.3, 31.2, 29.3

(a) Calculate the mean. Round to the nearest hundredth.

(b) Find the median.

(c) Find the mode.

65. **Automotive** Early in 2011, twenty gas stations from around the United States were surveyed to determine the price of regular gasoline.

3.14	2.93	3.16	3.19
2.99	3.36	3.25	3.02
3.48	3.31	3.04	3.15
3.33	3.55	3.22	3.24
3.18	3.19	3.38	3.33

(a) Calculate the mean gas price.

(b) Calculate the median gas price.

(c) Should you use the mean or median to describe the typical gas price?

66. **Welding** Throughout the day, welds are randomly chosen and tested for strength. The results are shown in the table. [Tensile strength is expressed in pounds per square inch (psi or lb/in.^{2}).]

2,314	2,318	2,307	2,291
2,289	2,301	2,320	2,318
2,322	2,297	2,314	2,296
2,309	2,311	2,304	2,321

(a) Calculate the mean tensile strength.

(b) Calculate the median tensile strength.

(c) Should you use the mean or median to describe the typical strength of a weld in this data set?

Skills | Calculator/Computer | Career Applications | **Above and Beyond**

67. **Business and Finance** Fred compiled records of his utility bills for 12 months: $153, $151, $143, $137, $132, $129, $134, $141, $158, $155, $149, and $158.

(a) Find the mean of Fred's monthly utility bills.

(b) Find the median of Fred's monthly utility bills.

(c) Is the mean or median a more useful representative of Fred's monthly utility bills? Write a brief paragraph justifying your response.

68. **Business and Finance** These scores were recorded on a 200-point final examination: 193, 185, 163, 186, 192, 135, 158, 174, 188, 172, 168, 183, 195, 165, 183.

   (a) Find the mean final examination score.

   (b) Find the median final examination score.

   (c) Is the mean or median a more useful representative of the final examination scores? Write a brief paragraph justifying your response.

69. List the advantages and disadvantages of the mean, median, and mode.

70. In a certain math class, you take four tests and the final, which counts as two tests. Your grade is the average of the six tests. At the end of the course, you compute both the mean and the median.

   (a) You want to convince the professor to use the mean to compute your average. Write a note to your professor explaining why this is a better choice. Choose numbers that make a convincing argument.

   (b) You want to convince the professor to use the median to compute your average. Write a note to your professor explaining why this is a better choice. Choose numbers that make a convincing argument.

71. Create a set of five numbers such that the mean is equal to the median.

72. Create a set of five numbers such that the mean is greater than the median.

73. Create a set of five numbers such that the mean is less than the median.

74. Write a paragraph describing the conditions necessary for the mean of a data set to be greater than the median, less than the median, and equal to the median. How do you think the mode would compare to the mean and median in each of these situations?

Answers

1. 9 **3.** $18\frac{5}{6}$ **5.** 18 **7.** 9 **9.** 10.2 **11.** 5 **13.** 29 **15.** $46\frac{1}{2}$ **17.** (a) 57.29; (b) 60; (c) mean **19.** (a) $309,812.50; (b) $257,450; (c) median **21.** (a) 63.5; (b) 64; (c) mean **23.** 17 **25.** 44 **27.** No mode **29.** Bimodal: 18 and 36 **31.** Green **33.** 94°F **35.** 38 mi/gal **37.** 95 points **39.** Tamika, by two points **41.** 1,091 kWh **43.** 286 kWh **45.** False **47.** sometimes **49.** 50 **51.** 357 **53.** 16,607 **55.** 20.25 **57.** 1,905.75 **59.** $4,292,100,000 **61.** $15\frac{5}{8}$ stoppages **63.** (a) 2.29 ms; (b) 2.35 ms; (c) 2.5 ms **65.** (a) $3.22; (b) $3.21; (c) mean **67.** (a) $145; (b) $146; (c) Above and Beyond **69.** Above and Beyond **71.** Answers will vary. **73.** Answers will vary.

Activity 25 ::

Car Color Preferences

While we tend to use the mean and median to describe the center of a data set, people who work in marketing and manufacturing use the mode at least as often. In many such applications, the mode of a data set is the natural way to describe the center.

Out-of-Class Component

You should find a safe spot to observe cars. This could be an intersection, street, or even a parking lot. Record the color of the first 10 cars you see. Use broad color categories (such as *blue*), rather than more specific categories (such as *light blue* and *dark blue*). Make a second list of the next 25 cars you see.

In-Class Component

You should have two data sets: one list of 10 colors and one list of 25 colors.

1. Find the mode of each of the two data sets.
2. **(a)** Do the modes differ?

   (b) If so, which do you feel is more accurate, and why?
3. Create a data set of 35 colors by combining your two lists. Find the mode of this data set.
4. The method you used to gather your data is called **convenience sampling.** Briefly describe why the method might take on that name.
5. **(a)** Describe two benefits and two weaknesses of convenience sampling as a method of gathering data.

   (b) Statisticians generally avoid convenience sampling, believing its weaknesses outweigh its strengths. Briefly describe how you might create a sample of car colors that more accurately mirrors the preferences of the population as a whole.

9.2 Describing Data Sets

< 9.2 Objectives >

1 > Compute the quartiles of a data set

2 > Give the five-number summary of a data set

3 > Construct a box-and-whisker plot

4 > Interpret a box-and-whisker plot

In Section 9.1, you learned to describe a data set using its measures of center: mean, median, and mode.

This is usually not enough to provide a clear picture of a data set.

Example 1 Describing Data with Measures of Center

< Objective 1 >

Find the mean and median of each set of numbers.

(a) 1, 2, 4, 3, 5, 1, 1, 3

Recall from Section 9.1 that the mean is computed by adding the numbers together and dividing by the number of elements in the set.

$$\text{Mean} = \frac{1 + 2 + 4 + 3 + 5 + 1 + 1 + 3}{8} = \frac{20}{8} = 2.5$$

To compute the median, we list the numbers in increasing order and find the midpoint.

1, 1, 1, **2, 3,** 3, 4, 5

RECALL

There are always two values in the "middle" when there is an even number of terms in the list.

Because we have two elements in the "middle," the median is the mean of the pair of middle numbers.

$$\text{Median} = \frac{2 + 3}{2} = \frac{5}{2} = 2.5$$

(b) 2, 2, 2, **2, 3,** 3, 3, 3

$$\text{Mean} = \frac{2 + 2 + 2 + 2 + 3 + 3 + 3 + 3}{8} = \frac{20}{8} = 2.5$$

$$\text{Median} = \frac{2 + 3}{2} = \frac{5}{2} = 2.5$$

Check Yourself 1

Find the median and mode of each set of numbers.

(a) 1, 2, 2, 7, 7, 7, 7, 10 **(b)** 24, 7, 7, 10, 6, 156, 7

In Example 1, both sets had the same mean and the same median, yet they are very different sets of numbers. Clearly, neither the mean nor the median alone is sufficient to distinguish a list of numbers.

NOTE

In fact, the median is the second quartile Q_2.

To assist us, we describe a data set using several numbers. One set of numbers we use is the **quartiles.** Just as the median divides a data set into halves, the quartiles divide the set into quarters (four equal parts). We use the notation Q_1, Q_2, and Q_3 to represent each of the three quartiles.

Step by Step

Finding Quartiles

NOTE

Graphing calculators and spreadsheets do not always agree on the quartiles. We chose the method used by spreadsheets such as Excel.

Step 1 Rewrite the data set in ascending order.

Step 2 Find the median.

Case 1 There is an odd number of data points.

Step 3 Make a list of the median and those numbers to the *left* of the median.

Step 4 Find the median of the list created in step 3. This is the first quartile Q_1.

Step 5 Repeat steps 3 and 4 with the median and those numbers to the *right* of the median. This is the third quartile Q_3.

Case 2 There is an even number of data points.

Step 3 Make a list of only those numbers to the *left* of the median.

Step 4 Find the median of the list created in step 3. This is the first quartile Q_1.

Step 5 Repeat steps 3 and 4 with only those numbers to the *right* of the median. This is the third quartile Q_3.

Example 2 Finding Quartiles

Find the first and third quartiles, Q_1 and Q_3, of each data set.

NOTE

We use a similar process to divide a set using deciles (tenths) and percentiles (hundredths).

(a) 1, 2, 4, 3, 5, 1, 1, 3

We sorted this list and found the median in Example 1.

1, 1, 1, 2, 3, 3, 4, 5

Median = 2.5

Since the data set has an even number of elements, this is an example of the second case.

To find the first quartile, we construct a list of only those numbers to the left of the median.

1, 1, 1, 2 are all to the left of the median. The median of this list is 1, which is the first quartile.

$Q_1 = 1$

3, 3, 4, 5 are all to the right of the median. The median of this list is 3.5, which is the third quartile.

$Q_3 = 3.5$

(b) 2, 2, 2, 2, 3, 3, 3, 3

We found the median to be 2.5 in Example 1.

This data set also has an even number of elements, so it too is an example of the second case.

2, 2, 2, 2 are all to the left of the median. The median of this set gives the first quartile

$Q_1 = 2$

Similarly, the third quartile is 3.

(c) 8, 11, 6, 9, 13, 10, 12, 9, 13

Sort the list: 6, 8, 9, 9, 10, 11, 12, 13, 13.

The median is the middle element, 10.

Since the data set has an odd number of elements, this is an example of the first case. To find the first quartile, we construct a list of the median and those numbers to the left of the median.

6, 8, 9, 9, 10

The median of this list is 9, so $Q_1 = 9$.

We find the third quartile in a similar manner: we list the median and those numbers to the right of the median.

10, 11, 12, 13, 13

The third quartile is the median of this list, 12.

Check Yourself 2

Find the first and third quartiles of each set of numbers.

(a) 1, 2, 2, 7, 7, 7, 7, 10 **(b)** 24, 7, 7, 10, 6, 156, 7

NOTE

We refer to the smallest and largest members of a set as the Min and Max, respectively.

The final pair of numbers we use is the smallest (minimum) and largest (maximum) elements of the data set.

We are now ready to give a **five-number summary** of a data set.

Definition

Five-Number Summary

The five-number summary associated with a data set is given by

Min, Q_1, Median, Q_3, Max

The five-number summary serves to distinguish between sets of data where a single number is not sufficient. We now find some five-number summaries.

Example 3 Finding Five-Number Summaries

< Objective 2 >

Find the five-number summary of each data set.

(a) 1, 2, 4, 3, 5, 1, 1, 3

The smallest number is 1, so 1 is the Min. The largest number is 5, so 5 is the Max. Using our results from Examples 1 and 2, we have

1, 1, 2.5, 3.5, 5

(b) 2, 2, 2, 2, 3, 3, 3, 3

The five-number summary for this set is simply 2, 2, 2.5, 3, 3.

(c) 8, 11, 6, 9, 13, 10, 12, 9, 13

The minimum is 6 and the maximum is 13. Using our result from Example 2 gives us the five-number summary.

6, 9, 10, 12, 13

Check Yourself 3

Give the five-number summary for each set of numbers.

(a) 1, 2, 2, 7, 7, 7, 7, 10 **(b)** 24, 7, 7, 10, 6, 156, 7

We now see that our two data sets from Example 1 are distinct. We can use a picture to display these differences based on the five-number summary. The graph we create is called a **box-and-whisker plot.**

Step by Step

Constructing Box-and-Whisker Plots

Step 1 Find the five-number summary of the given set of data.

Step 2 Construct a horizontal number line from the Min to the Max values in the summary.

Step 3 Mark off each of the numbers in the summary on the number line to scale (use small vertical lines).

Step 4 Draw a box from Q_1 to Q_3 (so the number line is in the middle of the box).

Example 4 Constructing Box-and-Whisker Plots

< Objective 3 >

Construct box-and-whisker plots for each data set.

(a) 1, 2, 4, 3, 5, 1, 1, 3

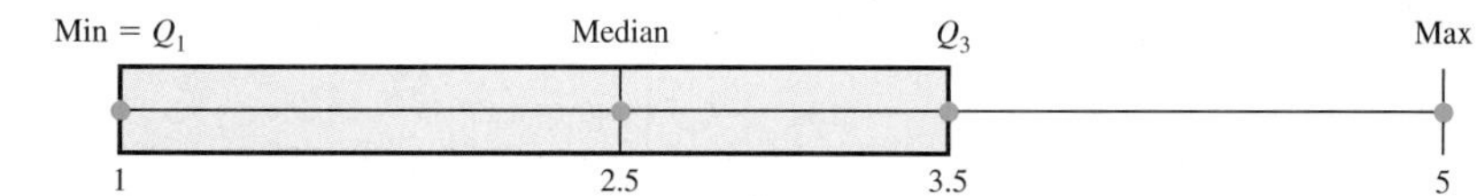

> **NOTE**
>
> It is also possible to construct a vertical box-and-whisker plot.

(b) 2, 2, 2, 2, 3, 3, 3, 3

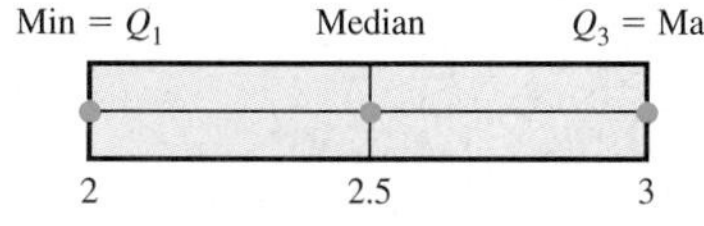

(c) 8, 11, 6, 9, 13, 10, 12, 9, 13

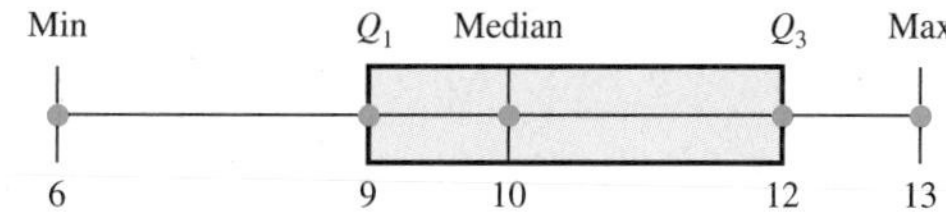

> **NOTE**
>
> Many graphing calculators compute five-number summaries and even produce box-and-whisker plots.

Check Yourself 4

Construct a box-and-whisker plot for each set of numbers.

(a) 1, 2, 2, 7, 7, 7, 7, 10 **(b)** 24, 7, 7, 10, 6, 56, 7

Our box-and-whisker plots for each set of data look different, which reflects differences between the sets.

A common application of box-and-whisker plots involves stock prices.

Example 5 Using Box-and-Whisker Plots

< Objective 4 >

We tracked two stocks on the New York Stock Exchange (NYSE) over a two-week period: Microsoft Corp (MSFT) and United Natural Foods, Inc. (UNFI). The closing price of each stock is given in the table.

Stock	Mon.	Tues.	Wed.	Thu.	Fri.	Mon.	Tues.	Wed.	Thu.	Fri.
MSFT	26.60	26.96	26.00	26.08	26.71	26.56	26.74	26.07	25.54	25.30
UNFI	35.10	34.90	34.51	35.18	35.96	35.73	36.10	35.45	35.06	34.99

Source: Yahoo! Finance

Construct box-and-whisker plots to compare the two stocks.

The five-number summaries are given for each stock.

MSFT: 25.30, 26.00, 26.32, 26.71, 26.96

UNFI: 34.51, 34.99, 35.14, 35.73, 36.10

We use these summaries to construct box-and-whisker plots.

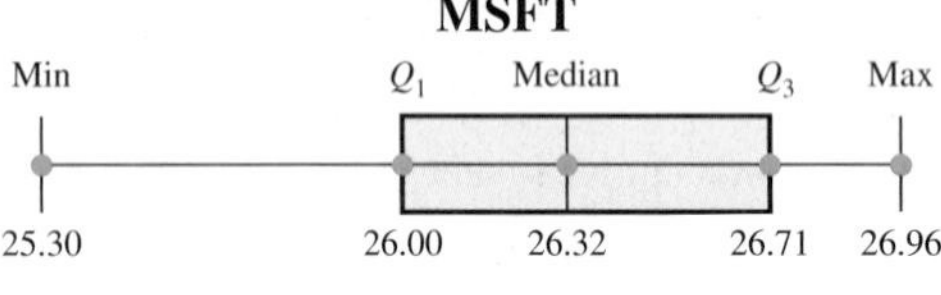

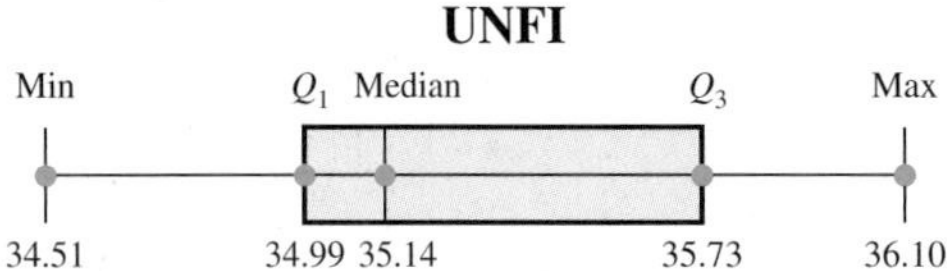

NOTE

We used the same scale for both plots.

Check Yourself 5

We tracked the closing stock prices over a different 2-week period. Construct box-and-whisker plots to compare the results. Interpret your findings.

Stock	Mon.	Tues.	Wed.	Thu.	Fri.
MSFT	26.23	27.16	26.69	26.73	27.15
UNFI	43.82	45.17	44.53	44.62	44.52
MSFT	27.53	27.70	26.96	27.34	27.02
UNFI	43.98	43.41	42.37	42.48	41.75

Check Yourself ANSWERS

1. **(a)** Median 7, mode 7; **(b)** median 7, mode 7
2. **(a)** $Q_1 = 2$, $Q_3 = 7$; **(b)** $Q_1 = 7$, $Q_3 = 17$
3. **(a)** 1, 2, 7, 7, 10; **(b)** 6, 7, 7, 17, 156
4. **(a)**

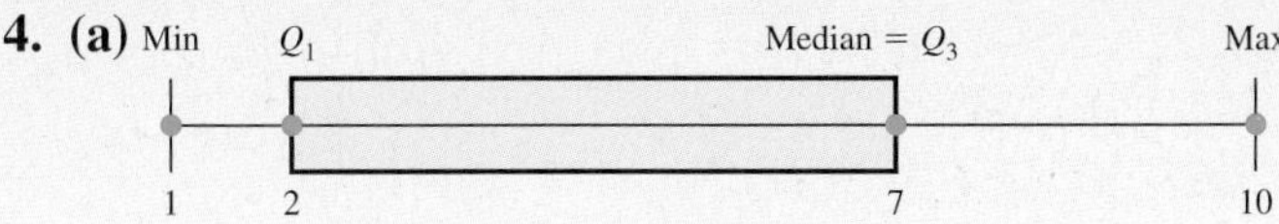

(b)

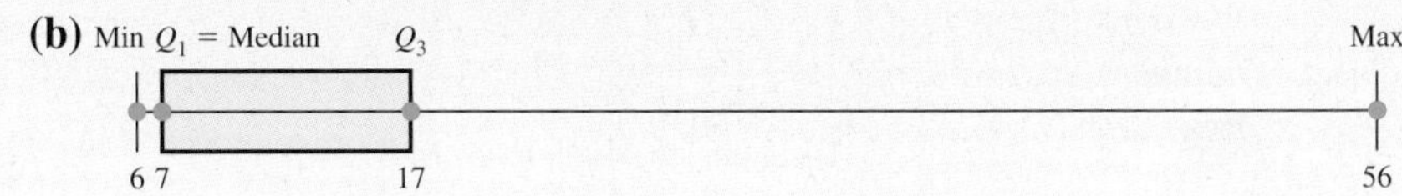

5.

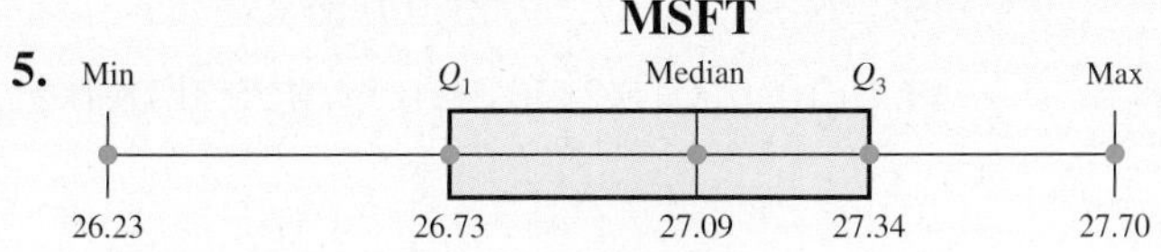

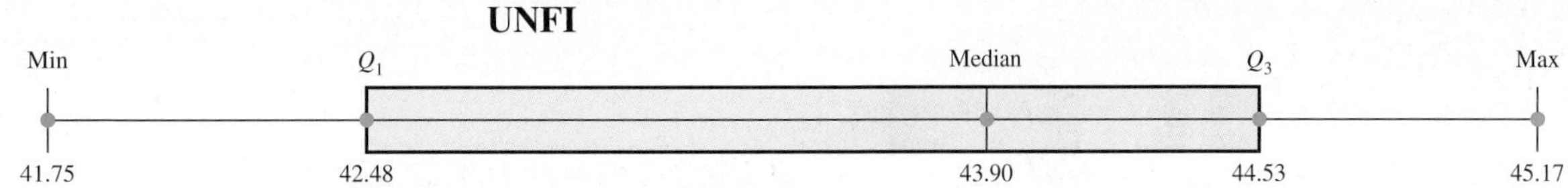

The box-and-whisker plot for UNFI is much wider than MSFT's plot. This indicates a wider range of stock prices. We say that the UNFI stock is more *volatile* than the MSFT stock.

Reading Your Text

These fill-in-the-blank exercises will help you understand some of the key vocabulary used in this section. The answers to these exercises are in the Answers Appendix at the back of the text.

(a) The ______________ divide a data set into quarters (four equal parts).

(b) Just as a single number (the median) separates the data into two groups, it takes ______________ numbers to separate data into four groups.

(c) The five-number ______________ associated with a data set is given by Min, Q_1, Median, Q_3, Max.

(d) A graph based on the five-number summary is called a ___________ plot.

Skills Calculator/Computer Career Applications Above and Beyond

9.2 exercises

< Objective 1 >

Find the median of each set of numbers.

1. 2, 8, 5, 6, 9, 7, 4, 4, 5, 4, 3 VIDEO
2. 7, 7, 5, 4, 1, 9, 8, 8, 8, 5, 2
3. 11, 12, 16, 14, 14, 14, 8, 12, 10, 18
4. 26, 30, 38, 67, 59, 21, 17, 85, 22, 22
5. 326, 245, 123, 222, 245, 300, 350, 602, 256
6. 0.10, 0.25, 0.24, 0.24, 0.30, 0.20, 0.18, 0.21, 0.28, 0.26

Find the first and third quartiles, Q_1 and Q_3, of each set of numbers.

7. 2, 8, 5, 6, 9, 7, 4, 4, 5, 4, 3 

8. 7, 7, 5, 4, 1, 9, 8, 8, 8, 5, 2

9. 11, 12, 16, 14, 14, 14, 8, 12, 10, 18

10. 26, 30, 38, 67, 59, 21, 17, 85, 22, 22

11. 326, 245, 123, 222, 245, 300, 350, 602, 256

12. 0.10, 0.25, 0.24, 0.24, 0.30, 0.20, 0.18, 0.21, 0.28, 0.26

Find the Min and Max of each set of numbers.

13. 2, 8, 5, 6, 9, 7, 4, 4, 5, 4, 3 

14. 7, 7, 5, 4, 1, 9, 8, 8, 8, 5, 2

15. 11, 12, 16, 14, 14, 14, 8, 12, 10, 18

16. 26, 30, 38, 67, 59, 21, 17, 85, 22, 22

17. 326, 245, 123, 222, 245, 300, 350, 602, 256

18. 0.10, 0.25, 0.24, 0.24, 0.30, 0.20, 0.18, 0.21, 0.28, 0.26

< Objective 2 >

Give the five-number summary of each set of numbers.

19. 2, 8, 5, 6, 9, 7, 4, 4, 5, 4, 3 

20. 7, 7, 5, 4, 1, 9, 8, 8, 8, 5, 2

21. 11, 12, 16, 14, 14, 14, 8, 12, 10, 18

22. 26, 30, 38, 67, 59, 21, 17, 85, 22, 22

23. 326, 245, 123, 222, 245, 300, 350, 602, 256

24. 0.10, 0.25, 0.24, 0.24, 0.30, 0.20, 0.18, 0.21, 0.28, 0.26

< Objective 3 >

Construct a box-and-whisker plot for each set of numbers.

25. 2, 8, 5, 6, 9, 7, 4, 4, 5, 4, 3

26. 7, 7, 5, 4, 1, 9, 8, 8, 8, 5, 2

27. 11, 12, 16, 14, 14, 14, 8, 12, 10, 18

28. 26, 30, 38, 67, 59, 21, 17, 85, 22, 22

29. 326, 245, 123, 222, 245, 300, 350, 602, 256

30. 0.10, 0.25, 0.24, 0.24, 0.30, 0.20, 0.18, 0.21, 0.28, 0.26

< Objective 4 >

This table gives the maximum depth and total bottom time for 25 recreational scuba dives. Use this table to complete exercises 31 and 32.

Diving Data

Dive no.	Max Depth (ft)	Bottom Time (s)
1	24	55
2	24	10
3	26	22
4	30	35
5	45	31
6	58	45
7	109	25
8	40	35
9	42	30
10	42	26
11	48	29
12	64	31
13	50	32
14	72	24
15	42	35
16	55	33
17	64	24
18	71	32
19	63	27
20	45	30
21	43	30
22	59	29
23	59	26
24	51	20
25	78	23

31. **Science and Medicine**

(a) Give the five-number summary of the depth data.

(b) Construct a box-and-whisker plot for the depth data.

(c) Describe the depth data based on the box-and-whisker plot.

32. **Science and Medicine**

(a) Give the five-number summary of the bottom time data.

(b) Construct a box-and-whisker plot for the bottom time data.

(c) Describe the bottom time data based on the box-and-whisker plot.

33. **Business and Finance** Closing prices for Adobe Systems stock are shown during a 2-week period in 2010. Construct and interpret a box-and-whisker plot for these data.

ADBE (Adobe Systems)										
Date	10/18	10/19	10/20	10/21	10/22	10/25	10/26	10/27	10/28	10/29
Price	28.06	27.58	28.21	27.70	28.21	28.20	28.14	28.17	28.10	28.15

Source: Yahoo! Finance

34. **Business and Finance** The closing prices for Kellogg Company stock are shown during the same 2-week period in 2010 as the Adobe Systems data in exercise 33.

K (Kellogg Co.)										
Date	10/18	10/19	10/20	10/21	10/22	10/25	10/26	10/27	10/28	10/29
Price	48.14	47.95	48.06	47.80	47.61	47.63	47.33	47.16	47.45	48.29

Source: Yahoo! Finance

(a) Construct a box-and-whisker plot for the Kellogg Company closing stock price data.

(b) Describe any distinctive features shown by the plot.

(c) Compare your plot with the one constructed in exercise 33.

35. **Statistics** This table gives the mean temperature (in degrees Fahrenheit) for the month of July over a 20-year period in Roanoke, VA.

Year	1991	1992	1993	1994	1995	1996	1997	1998	1999	2000
Temp	77.9° F	76.9°	80.2°	77.5°	76.9°	74.4°	76.1°	77.5°	79.1°	73.3°
Year	2001	2002	2003	2004	2005	2006	2007	2008	2009	2010
Temp	74.3°	78.0°	75.1°	76.5°	77.8°	77.5°	75.7°	76.1°	73.3°	79.3°

Source: NOAA; NCDC

(a) Construct a box-and-whisker plot based on the data.

(b) Discuss any significant features of the plot.

36. **Statistics** This table gives the mean temperature (in degrees Fahrenheit) for the month of January over a 20-year period in Dickinson, ND.

Year	1991	1992	1993	1994	1995	1996	1997	1998	1999	2000
Temp	11.6° F	27.7°	9.6°	6.1°	16.5°	6.2°	8.5°	17.1°	14.7°	19.3°
Year	2001	2002	2003	2004	2005	2006	2007	2008	2009	2010
Temp	23.0°	20.4°	15.9°	9.3°	10.4°	29.2°	21.2°	16.8°	13.3°	12.4°

Source: NOAA; NCDC

(a) Construct a box-and-whisker plot based on the data.

(b) Discuss any significant features of the plot.

(c) Compare the box-and-whisker plot constructed in exercise 35 with the one constructed in this exercise.

Determine whether each statement is **true** *or* **false.**

37. The quartiles divide the data set into three equal parts.

38. The median is equal to the second quartile.

In each statement, fill in the blank with **always, sometimes,** *or* **never.**

39. Before finding the quartiles, we must ____________ sort the data.

40. The third quartile is ____________ smaller than the median.

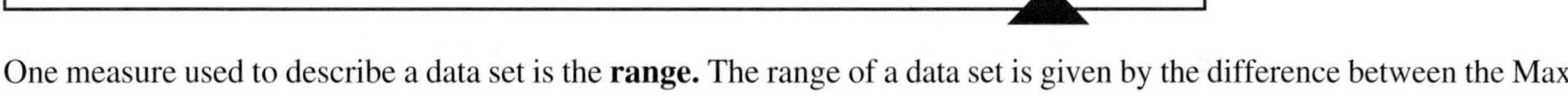

One measure used to describe a data set is the **range.** The range of a data set is given by the difference between the Max and the Min of the set. The range describes the variability of the data (that is, how much do the numbers vary).

Range = Max − Min

Find the range of each set of numbers.

41. 2, 8, 5, 6, 9, 7, 4, 4, 5, 4, 3 

42. 7, 7, 5, 4, 1, 9, 8, 8, 8, 5, 2

43. 11, 12, 16, 14, 14, 14, 8, 12, 10, 18

44. 26, 30, 38, 67, 59, 21, 17, 85, 22, 22

45. 326, 245, 123, 222, 245, 300, 350, 602, 256

46. 0.10, 0.25, 0.24, 0.24, 0.30, 0.20, 0.18, 0.21, 0.28, 0.26

Another measure that we use is the **interquartile range** (IQR). The IQR is given by the difference between the third quartile and the first quartile. The IQR measures how large an interval is needed to contain the middle 50% of the data. It is used to measure variability and to assist in determining if there are any *outliers* in the data.

$\text{IQR} = Q_3 - Q_1$ 

Find the IQR of each set of numbers.

47. 2, 8, 5, 6, 9, 7, 4, 4, 5, 4, 3

48. 7, 7, 5, 4, 1, 9, 8, 8, 8, 5, 2

49. 11, 12, 16, 14, 14, 14, 8, 12, 10, 18

50. 26, 30, 38, 67, 59, 21, 17, 85, 22, 22

51. 326, 245, 123, 222, 245, 300, 350, 602, 256

52. 0.10, 0.25, 0.24, 0.24, 0.30, 0.20, 0.18, 0.21, 0.28, 0.26

One characteristic we look for when describing and analyzing a data set is the presence of **outliers.** An outlier of a data set is a number that is "far away" from most of the other numbers in the set. We use the IQR to determine whether a data set has any outliers.

A number from a set is an outlier if it is more than 1.5 IQRs from either the first quartile or the third quartile. That is, we compute "boundaries" for outliers.

Lower boundary: $Q_1 - 1.5 \times \text{IQR}$
Upper boundary: $Q_3 + 1.5 \times \text{IQR}$

Any number in the data set that is less than the lower boundary or greater than the upper boundary is considered an outlier.

Find any outliers in each set of numbers.

53. 2, 8, 5, 6, 9, 7, 4, 4, 5, 4, 3

54. 7, 7, 5, 4, 1, 9, 8, 8, 8, 5, 2

55. 11, 12, 16, 14, 14, 14, 8, 12, 10, 18

56. 26, 30, 38, 67, 59, 21, 17, 85, 22, 22

57. 326, 245, 123, 222, 245, 300, 350, 602, 256

58. 0.10, 0.25, 0.24, 0.24, 0.30, 0.20, 0.18, 0.21, 0.28, 0.26

Some outliers are so far from the rest of the data that we call them **extreme outliers.** An extreme outlier is more than 3 IQRs from one of the quartiles. Outliers that are not extreme outliers are called **mild outliers.**

To determine the extreme outliers, again we set boundaries.

Lower boundary: $Q_1 - 3 \times \text{IQR}$
Upper boundary: $Q_3 + 3 \times \text{IQR}$

Classify any outliers of each data set as extreme or mild.

59. 2, 8, 5, 6, 9, 7, 4, 4, 5, 4, 3

60. 7, 7, 5, 19, 4, 1, 9, 8, 8, 8, 5, 2

61. 11, 12, 16, 14, 14, 14, 8, 12, 10, 18

62. 26, 30, 38, 67, 59, 21, 17, 85, 22, 22

63. 326, 245, 123, 222, 245, 300, 350, 602, 256

64. 0.10, 0.25, 0.24, 0.24, 0.92, 0.20, 0.18, 0.21, 0.28, 0.26

65. Research Microsoft Corporation's (MSFT) closing stock prices for the most recent 2-week period (see Example 5). Give the five-number summary of these data, construct a box-and-whisker plot for the data, and interpret your display.

66. Research United Natural Foods, Inc (UNFI) closing stock prices for the most recent 2-week period (see Example 5). Give the five-number summary of these data, construct a box-and-whisker plot for the data, and interpret your display.

Answers

1. 5 **3.** 13 **5.** 256 **7.** 4; 6.5 **9.** 11; 14 **11.** 245; 326 **13.** 2; 9 **15.** 8; 18 **17.** 123; 602 **19.** 2, 4, 5, 6.5, 9
21. 8, 11, 13, 14, 18 **23.** 123, 245, 256, 326, 602

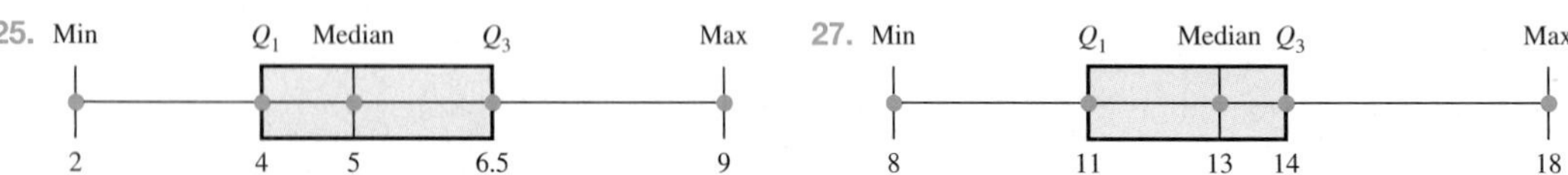

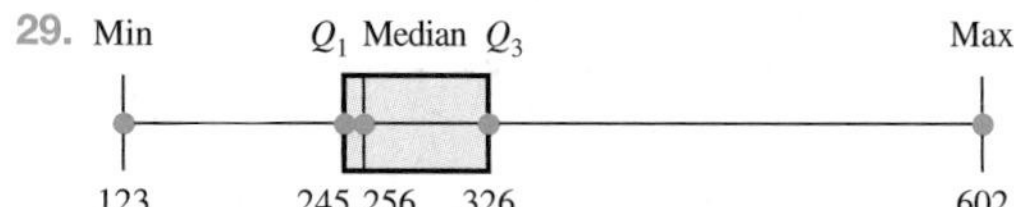

31. (a) 24, 42, 50, 63, 109 (b)

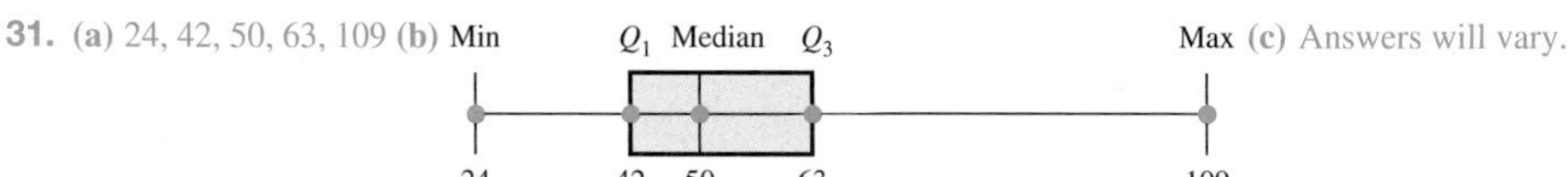

(c) Answers will vary.

(b) Answers will vary. **37.** False **39.** always **41.** 7 **43.** 10 **45.** 479 **47.** 2.5 **49.** 3 **51.** 81 **53.** None
55. None **57.** 123, 602 **59.** None **61.** None **63.** Mild: 123; extreme: 602 **65.** Above and Beyond

Outliers in Scientific Data

Marine scientists have placed numerous buoys in the oceans. These buoys are equipped to measure various properties of the water and to relay these measurements to satellites at regular intervals. The data are then collected and analyzed.

The table shows midday temperature data (in degrees Fahrenheit) collected off the coast of Cape Charles, VA, over a 12-day period.

Date	7/1	7/2	7/3	7/4	7/5	7/6	7/7	7/8	7/9	7/10	7/11	7/12
Temp.	72.0°	73.2°	71.8°	72.6°	73.3°	74.1°	74.8°	76.0°	57.7°	77.5°	77.1°	78.2°

Source: NODC

1. Find the mean temperature for the period shown.
2. Give the five-number summary for these data.
3. Compute the interquartile range of these data.
4. Identify any outliers in the data set.
5. Compute the mean of the data set without the outlier.
6. Give the five-number summary for the data set formed by removing the outlier from the table.

   Investigation showed that the buoy responsible for the given data was nonterminally damaged sometime between 7/8 and 7/10. It is believed that a small personal watercraft struck the buoy about when it sent the July 9 data.

7. Describe how you might use this information to analyze salinity data measured by the same buoy.

9.3 Tables and Bar Graphs

< 9.3 Objectives >

1 > Read and interpret tables

2 > Read and interpret bar graphs

3 > Create a bar graph

NOTE

Rows read left to right.
Columns read top to bottom.

A **table** is a display of information in rows or columns. Tables can be used anywhere we need to summarize information.

Here, we have a table describing land area and world population. Each entry in the table is called a **cell.** We use this table in Examples 1 and 2.

Continent or Region	Land Area (1,000 mi²)	Percent of Total Land	Population 1900 (millions)	Population 1950 (millions)	Population 2000 (millions)
North America	9,400	16.2	106	221	305
South America	6,900	11.9	38	111	515
Europe	3,800	6.6	400	392	510
Asia (including Russia)	17,400	30.1	932	1,591	4,028
Africa	11,700	20.2	118	229	889
Oceania (including Australia)	3,300	5.7	6	12	32
Antarctica	5,400	9.3	Uninhabited	—	—
World total	57,900		1,600	2,556	6,279

Source: Bureau of the Census, U.S. Dept. of Commerce

Example 1 **Reading a Table**

< Objective 1 >

Use the land area and world population table to answer each question.

(a) What was the population of Africa in 1950?

Looking at the cell that is in the row labeled Africa and the column labeled 1950, we find the population listed as 229. Because the population figures are given in millions, the population was 229,000,000.

(b) What is the land area of Asia in square miles?

The cell in the row Asia and column labeled land area says 17,400. Because the column is labeled "1,000 mi²," the land area is 17,400 thousand square miles, or 17,400,000 mi².

Check Yourself 1

Use the land area and world population table to answer each question.

(a) What was the population of South America in 1900?
(b) What is the land area of Africa as a percent of Earth's total land area?

We can frequently use a table to find answers to questions that are not directly answered as part of the table.

Example 2 Interpreting a Table

Use the world population and land area table to answer each question.

(a) To the nearest tenth of a percent, what percent of the world's population was in North America in the year 2000?

305,000,000 of the world's 6,279,000,000 people lived in North America.

$$\frac{305{,}000{,}000}{6{,}279{,}000{,}000} \approx 0.04857 \approx 4.9\%$$

RECALL

Compute North America's portion of Earth's total land area similarly.

$$\frac{9{,}400}{57{,}900} \approx 0.1623 \approx 16\%$$

Although North America has more than 16% of Earth's land area, it had less than 5% of the world's population.

(b) What percent of Earth's habitable land is in Asia?

First, we must decide what is meant by "habitable land." We will assume anything outside of Antarctica is habitable. To find the amount of habitable land, we take the total of 57,900,000 and subtract Antarctica's 5,400,000. This leaves total habitable land of 52,500,000 mi^2.

$$\frac{17{,}400{,}000}{52{,}500{,}000} \approx 0.3314 \approx 33.1\%$$

(c) What was the mean population for the six populated regions in 1900?

Although we could add the six numbers, you should see that they have already been totaled. Using that total, we find the average.

$$\frac{1{,}600{,}000{,}000}{6} \approx 267{,}000{,}000$$

Check Yourself 2

Use the world population and land area table to answer each question.

(a) To the nearest percent, what was the increase in the population of Africa between 1950 and 2000?
(b) Did world population increase by a greater percent between 1900 and 1950 or between 1950 and 2000?

NOTE

Spreadsheets are valuable tools for working with tables and their associated graphs.

If a table has only one or two columns of numeric information, it is often easier to interpret it in picture form. A **graph** is a diagram that represents the connection between two or more things. **Bar graphs** are probably the most common type of graph used to display data.

Example 3 Reading a Bar Graph

< Objective 2 >

This bar graph represents the response to a Gallup poll that asked people what their favorite spectator sport was. In the graph, the information at the bottom describes the sport, and the information along the side describes the percentage of people surveyed. The height of the bar indicates the percentage of people who favor that particular sport.

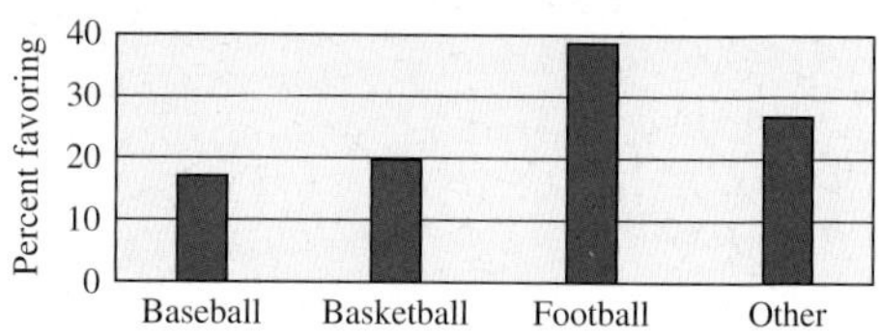

(a) Find the percentage of people for whom football is their favorite spectator sport.

We frequently have to estimate our answer when reading a bar graph. In this case, 38% is a good estimate.

(b) Find the percentage of people for whom baseball is their favorite spectator sport.

Again, we can only estimate our answer. It appears to be approximately 17% of the people responding who favor baseball.

Check Yourself 3

This bar graph represents the number of students who majored in each of five areas at Experimental Community College.

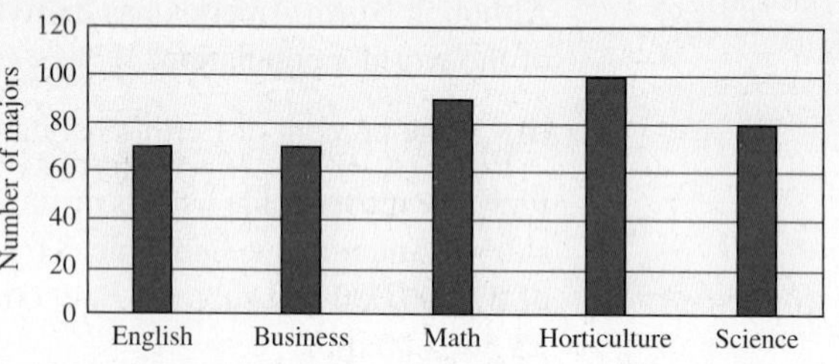

(a) How many mathematics majors were there?
(b) How many English majors were there?

Some bar graphs display additional information by using different colors or shading for different bars. With such graphs it is important to read the legend. The **legend** is the key that describes what each color or shade of the bar represents.

Example 4 Reading the Legend of a Graph

This bar graph represents the average student age at ECC.

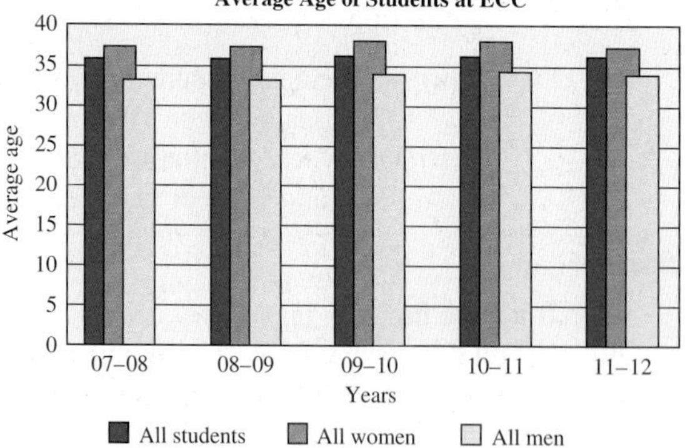

(a) What was the average age of female students in 2011–2012?

The legend tells us that the ages of all women are represented as the medium blue color. Looking at the height of the medium blue column for the year 2011–2012, we see the average age was about 37.

(b) Who tends to be older, male students or female students?

The medium blue bar is higher than the light blue bar in every year. Female students tend to be older than male students at ECC.

Check Yourself 4

Use the graph in Example 4 to answer each question.

(a) Did the average age of female students increase or decrease between 2010–2011 and 2011–2012?
(b) What was the average age of male students in 2009–2010?

Now that we can read bar graphs, it is time to create one.

Example 5 Creating a Bar Graph

< Objective 3 >

This table represents the 2010 population of the six most populated urban areas in the world. Population figures represent the city and all of its suburbs. Create a bar graph from the information in the table.

Population of the World's Largest Urban Areas

City	2010 Population
Tokyo, Japan	37,700,000
Mexico City, Mexico	23,600,000
New York City, USA	23,300,000
Seoul, South Korea	22,700,000
Mumbai, India	21,900,000
São Paulo, Brazil	20,800,000

Source: United Nations

We let the vertical axis, the vertical line to the left of the graph, represent population and place the six urban areas along the horizontal axis. To create a graph, we must decide on the scale for the vertical axis.

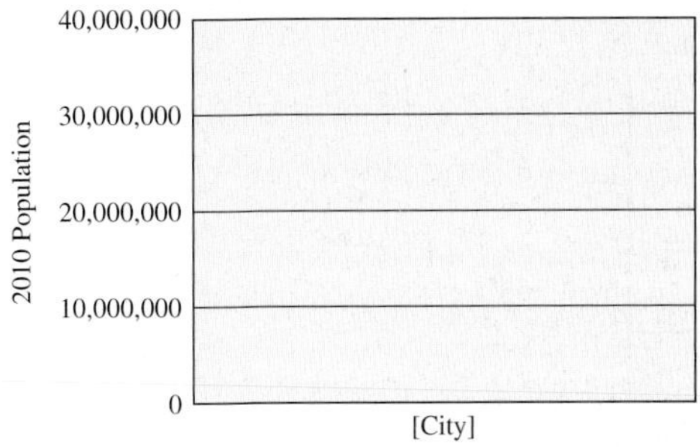

1. Pick a number that is slightly larger than the biggest number we are to graph. 40,000,000 is slightly larger than 37,700,000.
2. Decide how long the axis will be. It is best if this length easily divides into the number of step 1. To accomplish this division, we choose 4 units.
3. Scale the axis by dividing it with hashmarks. Label each hashmark with the appropriate number. In this graph, each unit represents 10,000,000 people (the 40,000,000 divided by the 4 units results in 10,000,000 people per unit).

Now, the height of each bar is determined by using the scale created for the axis. Remembering that we have 10,000,000 people per unit, we divide each population by 10,000,000. The result is the height of each bar. The height for Mexico City is

2.36 units. Remember, all we can read from a bar graph is a rough approximation of the actual number.

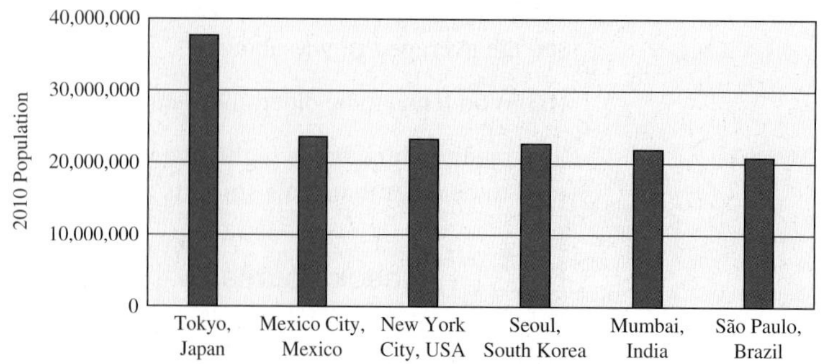

Check Yourself 5

This table represents the 2010 population of the six most populated cities in the United States. Each population is the population within the city limits, which is why the New York population is so different from that in the table in Example 5. Create a bar graph from the information in the table.

Population of the Largest Cities in the United States

City	2010 Population
New York City, NY	8,175,000
Los Angeles, CA	3,793,000
Chicago, IL	2,696,000
Houston, TX	2,099,000
Philadelphia, PA	1,526,000
Phoenix, AZ	1,446,000

Source: U.S. Census Bureau

Check Yourself ANSWERS

1. **(a)** 38,000,000; **(b)** 20.2% **2.** **(a)** 288%; **(b)** 1950–2000 (146% vs. 60%)
3. **(a)** 90; **(b)** 70 **4.** **(a)** It decreased; **(b)** 34
5.

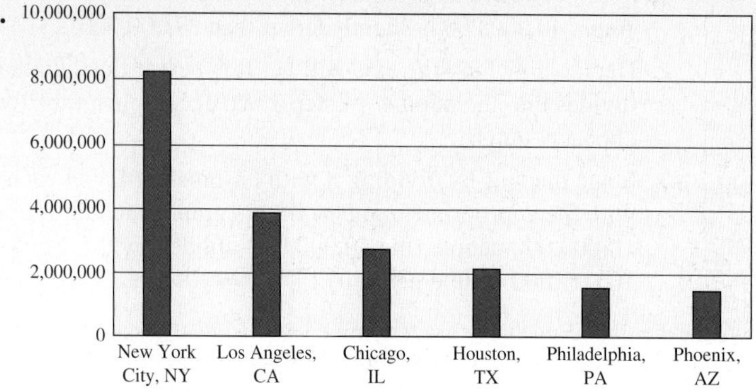

Reading Your Text

These fill-in-the-blank exercises will help you understand some of the key vocabulary used in this section. The answers to these exercises are in the Answers Appendix at the back of the text.

(a) A table is a display of information in ______________ or columns.

(b) Each entry in a table is called a ______________.

(c) A ______________ is a diagram that represents the connection between two or more things.

(d) A ______________ is the key that describes what each color or shade of a bar in a bar graph represents.

Skills Calculator/Computer Career Applications Above and Beyond

9.3 exercises

< Objective 1 >

Use the world population and land area table reproduced here for exercises 1 to 10. Round answers to the nearest tenth or tenth of a percent.

Continent or Region	Land Area (1,000 mi^2)	Percent of Total Land	Population 1900 (millions)	Population 1950 (millions)	Population 2000 (millions)
North America	9,400	16.2	106	221	305
South America	6,900	11.9	38	111	515
Europe	3,800	6.6	400	392	510
Asia (including Russia)	17,400	30.1	932	1,591	4,028
Africa	11,700	20.2	118	229	889
Oceania (including Australia)	3,300	5.7	6	12	32
Antarctica	5,400	9.3	Uninhabited	—	—
World total	57,900		1,600	2,556	6,279

Source: Bureau of the Census, U.S. Dept. of Commerce

1. **(a)** What was the population in North America in 1950?
   (b) What is the land area of North America as a percent of Earth's total land area?
2. **(a)** What was the population of Europe in 2000?
   (b) What is the land area of Europe?
3. **(a)** What was the percent increase in population in Asia from 1900 to 1950?
   (b) What was the percent increase in population in Asia from 1950 to 2000?
   (c) What was the population per square mile in Asia in 1950?
   (d) What was the population per square mile in Asia in 2000?
4. Compare the population per square mile in Asia to the population per square mile in North America for the year 2000.
5. What was the average population per continent for the five habitable continents besides Asia in 1950?

6. What was the percent increase in the population for all five inhabited continents besides Asia from 1950 to 2000?

7. (a) What percent of Earth's inhabitable land is in North America?
   (b) What percent of the world population in the year 2000 was in North America?

8. What was the percent increase in the population in South America from 1900 to 2000?

9. (a) What was the number of people per square mile for the entire world in 1950?
   (b) What was the number of people per square mile for the entire world in 2000?
   (c) What was the percent increase in the number of people per square mile for the entire world from 1950 to 2000?

10. (a) What was the mean population of the six continents or land masses that were habitable in 2000?
   (b) What was the mean population in 1950?
   (c) What was the percent increase in the mean population from 1950 to 2000?

Use the table to complete exercises 11 to 14.

Gasoline Retail Prices, U.S. City Average, 1996–2010

Year	(cost per gallon, including taxes) Regular	Premium	All
1996	$1.199	$1.381	$1.245
1997	1.199	1.380	1.244
1998	1.030	1.214	1.072
1999	1.136	1.320	1.176
2000	1.484	1.663	1.523
2001	1.420	1.602	1.460
2002	1.345	1.530	1.386
2003	1.561	1.748	1.603
2004	1.852	2.042	1.895
2005	2.270	2.468	2.314
2006	2.572	2.779	2.618
2007	2.796	3.008	2.843
2008	3.246	3.485	3.299
2009	2.353	2.590	2.406
2010	2.782	3.022	2.835

Source: U.S. Dept. of Energy

11. (a) What was the mean cost of a gallon of regular gas in 2000?
   (b) What was the mean cost of a gallon of premium gas in 2005?

12. (a) What was the decrease in the price of a gallon of regular gas from 2007 to 2009?
   (b) What was the percent decrease in price of all types of gas from 2008 to 2009? Round your result to the nearest tenth of a percent.

13. (a) What was the increase in the price of a gallon of regular gas from 1996 to 2010?
   (b) What was the increase in the price of a gallon of premium gas from 1996 to 2010?

14. (a) What was the percent increase in the cost of a gallon of regular gas from 1996 to 2010 (to the nearest whole percent)?
   (b) What was the average annual increase in the cost of a gallon of regular gas from 1996 to 2010 (to the nearest tenth of a cent)?

< Objective 2 >

Use the graph, showing the total U.S. motor vehicle production for the years 2001 to 2010, to complete exercises 15 to 18.

U.S. Motor Vechicle Production (in millions)

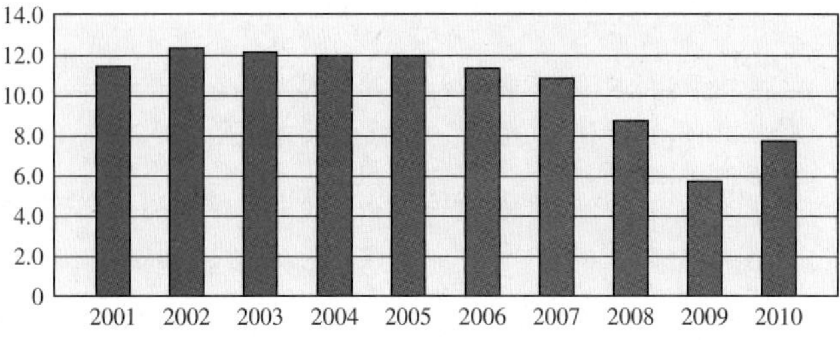

Source: Ward's Auto

15. What was the production in 2008?

16. In what year did the greatest production occur?

17. Find the median number of vehicles produced in the 10 years.

18. In what year was the production decline the greatest?

Use the bar graph, showing the attendance at a circus for 7 days in August, to complete exercises 19 to 22.

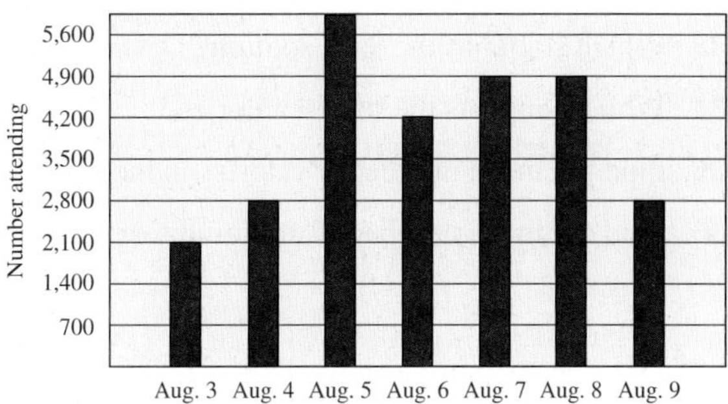

19. Find the attendance on August 4.

20. Which day had the greatest attendance?

21. Which day had the lowest attendance?

22. Find the median attendance over the 7 days.

Use the bar graph to complete exercises 23 to 26.

Small and midsize wagons made up between 5% and 10% of all vehicles sold in the United States between 2001 and 2010.

Wagons Sold (U.S.) (in thousands)

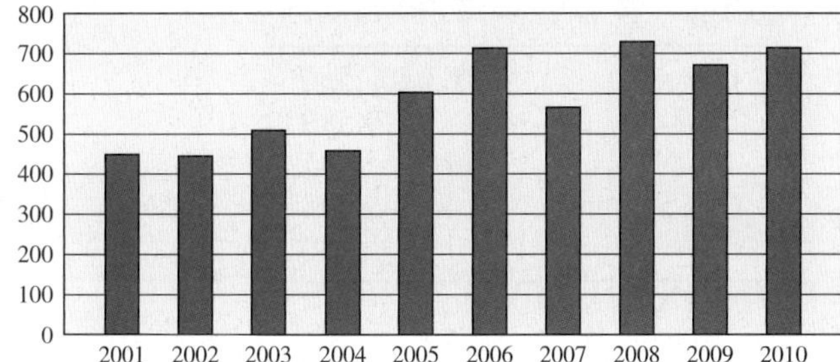

Source: Bureau of Transportation Statistics

23. How many wagons were sold in 2008?

24. In what year did the greatest sales occur?

25. What was the percent increase in sales from 2001 to 2010?

26. In what year did the greatest decrease in sales occur?

Consider the nutritional facts given for Campbell's cream of mushroom soup. Assume you consumed 1 cup of soup.

Nutrition Facts	Amount / serving	%DV*	Amount / serving	%DV*
Serv. Size 1/2 cup (120mL) condensed soup	**Total Fat** 7g	11%	**Total Carb.** 9g	3%
Servings about 2.5	Sat. Fat 2.5g	13%	Fiber 1g	4%
Calories 110	**Cholest.** Less than 5mg	1%	Sugars 1g	
Fat Cal. 60	**Sodium** 870mg	36%	**Protein** 2g	
*Percent Daily Values (DV) are based on a 2,000 calorie diet.	Vitamin A 0% • Vitamin C 0% • Calcium 2% • Iron 2%			

Satisfaction guaranteed. For questions or comments, please call 1-800-257-8443. Please have code and date information on can end available. For recipes, information & more, visit Campbell's Community at www.campbellsoup.com 1261-56

27. How many calories did you consume?

28. What percent of the daily value of saturated fat did you consume?

29. What percent of fiber did you get?

30. How many grams of sodium did you get?

Use the table to complete exercises 31 to 36.

Soup	Calories	Fat	Total Protein	Sodium
Cream of mushroom	110	7 g	2 g	870 mg
Cream of chicken	130	8 g	11 g	890 mg
Split pea	180	3.5 g	10 g	860 mg
Tomato	100	0 g	2 g	760 mg

31. Which soup has the least fat?

32. Which soup has the most sodium?

33. Which soup has the least sodium?

34. Which soup has the fewest calories?

35. Find the mean number of calories in the soups.

36. Find the mean number of milligrams of sodium in the soups.

< Objective 3 >

37. The table shows the U.S. population by age in 2010. Construct a bar graph to display this information.

Age	Population
Under 20	83,267,556
20–39	82,829,589
40–59	115,562,485
60–79	45,849,148
80+	11,236,760

Source: U.S. Census Bureau

38. Median annual earnings for full-time employees by education levels are shown in the table. Create a bar graph to display this information.

Education	Earnings
High School Diploma	$34,197
Associate's Degree	$44,086
Bachelor's Degree	$57,026
Master's Degree	$69,958
Doctorate Degree	$88,867

Source: U.S. Census Bureau

39. The table shows the average number of rainy days in San Juan, Puerto Rico, by season. Construct a bar graph to display this information.

Season	Rainy Days
Winter	42
Spring	44
Summer	56
Fall	55

Source: University of North Carolina—Chapel Hill

40. Median starting salaries for selected entry-level positions are given in the table. Construct a bar graph to display this information.

Position	Earnings
Actuary	$58,271
Biomedical Engineer	$49,741
LAN Support	$49,682
Contracts Administrator	$49,874
Telecommunications Technician	$45,082

Source: Salary.com

Skills | Calculator/Computer | **Career Applications** | Above and Beyond

41. Mechanical Engineering

American Wire Gauge	Wire Diameter (in.)	American Wire Gauge	Wire Diameter (in.)
2	0.2576	12	0.0808
3	0.2294	14	0.0640
4	0.2043	16	0.0508
6	0.1620	18	0.0403
8	0.1285	20	0.0319
10	0.1019		

(a) What is the diameter of a 12-gauge wire?

(b) What is the difference in diameter between 14-gauge and 10-gauge wire?

42. Automotive

Temperature Protection (°F)	Required Percent of Ethylene Glycol
15	22%
10	26%
5	29%
0	34%
−10	39%
−20	43%
−30	48%
−40	53%

(a) What percent ethylene glycol is required to provide protection down to −20°F?

(b) What is the temperature protection provided by a mixture of 29% ethylene glycol?

(c) If the percent of ethylene glycol is increased from 29% to 39%, how much does it change the temperature protection?

43. Welding

Metal	Density (g/cm³)	Melting Point (°C)
Iron	7.87	1,538
Aluminum	2.699	660.4
Copper	8.93	1,084.9
Tin	5.765	231.9
Titanium	4.507	1,668

(a) What is the density of titanium?

(b) What is the difference in melting points between copper and iron?

(c) Which metal has the highest melting point?

44. Health Sciences In 2010, overall health care costs came to approximately \$2.3 trillion dollars in the United States. The percentage of those costs and dollar amounts spent (in billions of dollars) for selected services are shown in the table.

Service	Percent	Cost (in billions)
Hospital Care	31%	\$713
Physician Care	20%	\$460
Retail Drugs	10%	\$230
Nursing Home Care	5%	\$115

Source: Kaiser Family Foundation

(a) What percent of overall health care costs was spent on retail drugs?

(b) How much did physician care cost?

(c) How much more was spent on hospital care than nursing home care?

45. Automotive Create a bar graph to display the data in the mileage table.

Model	Mileage
Taurus	27
Malibu	29
Stratus	26
Camry	32

46. Welding Create a bar graph to display the melting points of the metals listed.

Metal	Density (g/cm³)	Melting Point (°C)
Iron	7.87	1,538
Aluminum	2.699	660.4
Copper	8.93	1,084.9
Tin	5.765	231.9
Titanium	4.507	1,668

AUTOMOTIVE TECHNOLOGY *Bar graphs may be positioned horizontally rather than vertically. Consider the bar graph showing the top five auto manufacturers, based on March 2012 sales figures.*

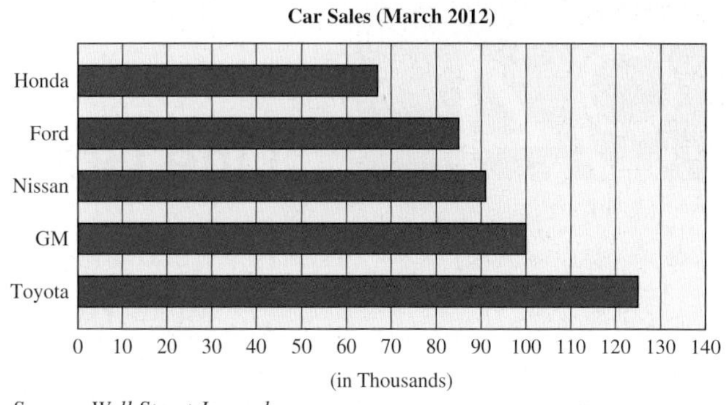

Source: Wall Street Journal

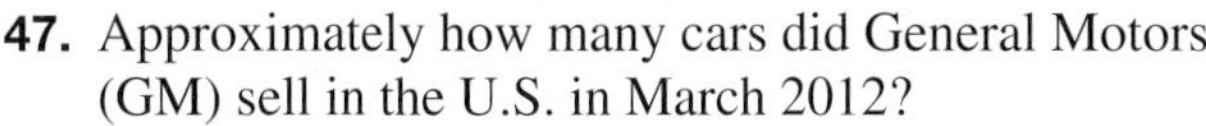

47. Approximately how many cars did General Motors (GM) sell in the U.S. in March 2012?

48. Approximately how many cars did Toyota sell in the U.S. in March 2012?

49. Approximately how many more cars did Toyota sell than GM?

50. Approximately how many more cars did Ford sell than Honda?

BUSINESS AND FINANCE *A professional association recorded the income and expenses from their annual conference over a five-year period.*

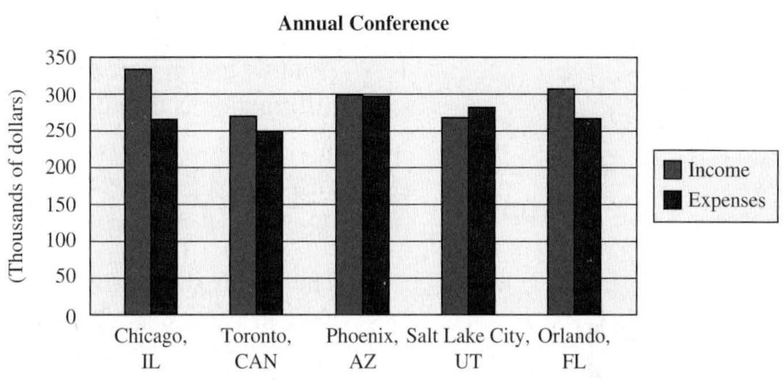

Source: AMATYC

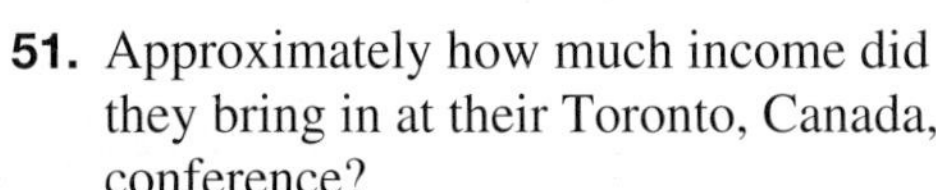

51. Approximately how much income did they bring in at their Toronto, Canada, conference?

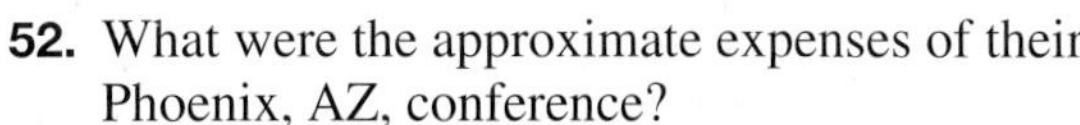

52. What were the approximate expenses of their Phoenix, AZ, conference?

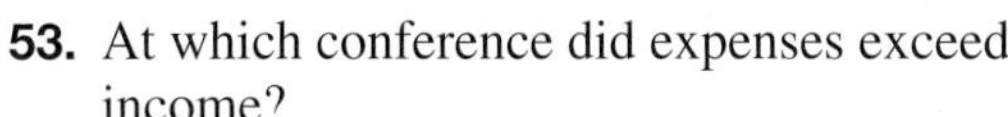

53. At which conference did expenses exceed income?

54. By approximately how much did income exceed expenses at their Chicago, IL, conference?

55. Compare current gas prices in your area to those in the table given for exercises 11 to 14. Describe the differences.

56. Research national gas prices and compare them to the prices in the table given for exercises 11 to 14.

Answers

1. (a) 221,000,000; (b) 16.2% **3.** (a) 70.7%; (b) 153.2%; (c) 91.4 people; (d) 231.5 people **5.** 193,000,000 people per continent **7.** (a) 17.9%; (b) 4.9% **9.** (a) 44.1; (b) 108.4; (c) 145.8% **11.** (a) \$1.484; (b) \$2.468 **13.** (a) \$1.583; (b) \$1.641 **15.** 8,700,000 vehicles **17.** 11.4 million vehicles **19.** 2,800 people **21.** August 3 **23.** 729,000 wagons **25.** 57.8% **27.** 220 **29.** 8% **31.** Tomato **33.** Tomato **35.** 130 cal

37.

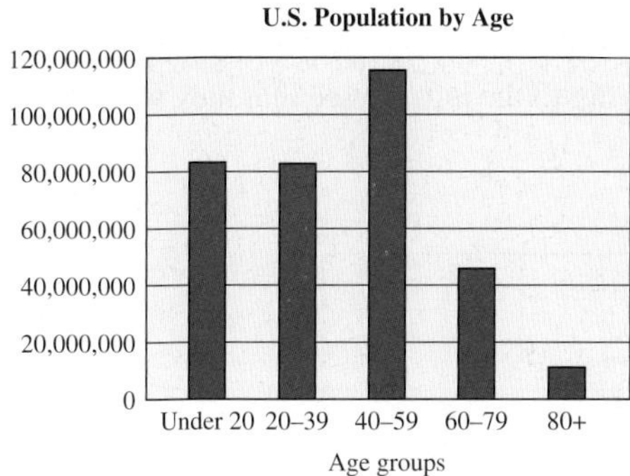

39.

Rainy Days, Puerto Rico

60 50 40 30 20 10 0

Winter Spring Summer Fall

41. (a) 0.0808 in.; (b) 0.0379 in. **43.** (a) 4.507 g/cm³; (b) 453.1°C; (c) Titanium **45.**

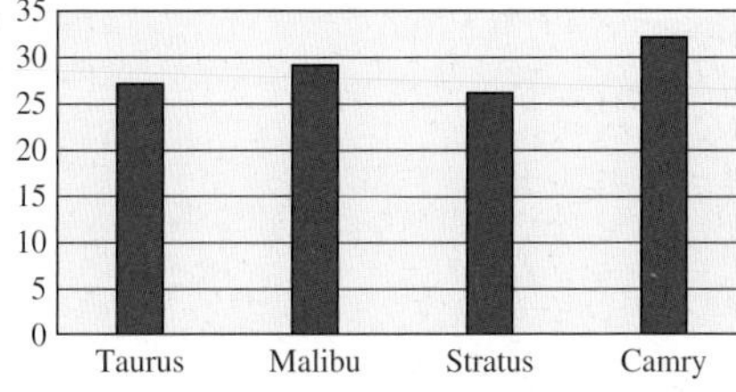

47. 100,000 cars **49.** 25,000 more cars **51.** \$270,000 **53.** Salt Lake City, UT **55.** Above and Beyond

9.4

Line Graphs and Pie Charts

< 9.4 Objectives >

1 > Read a line graph

2 > Make a prediction from a line graph

3 > Construct a line graph

4 > Read and interpret a pie chart

5 > Create a pie chart

We have already seen that data can be represented graphically with a bar graph. Another useful type of graph is called a **line graph.** In a line graph, one of the types of information is usually related to time (clock time, day, month, or year).

Example 1 Reading a Line Graph

< Objective 1 >

The graph shows the number of regular season games that the Pittsburgh Steelers won in each season over the 10-year period, 2001–2010. The years are displayed on the bottom (*horizontal axis*) and the number of victories is given along the side (*vertical axis*).

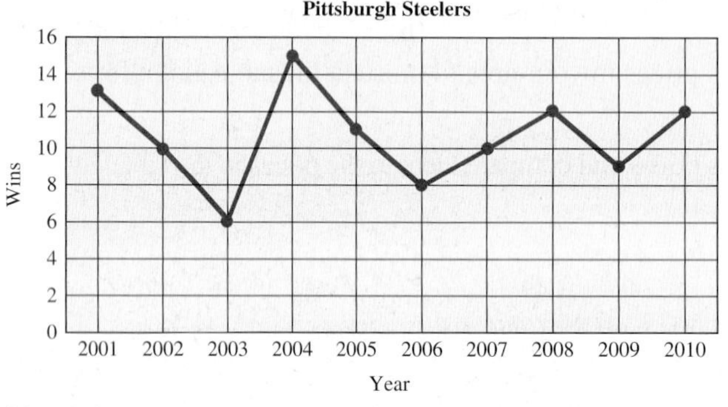

(a) How many games did the Steelers win in 2007?

We look across the bottom and find 2007. Then, we move straight up until we see the point indicated by the graph. Following to the left, we see that they won 10 games in 2007.

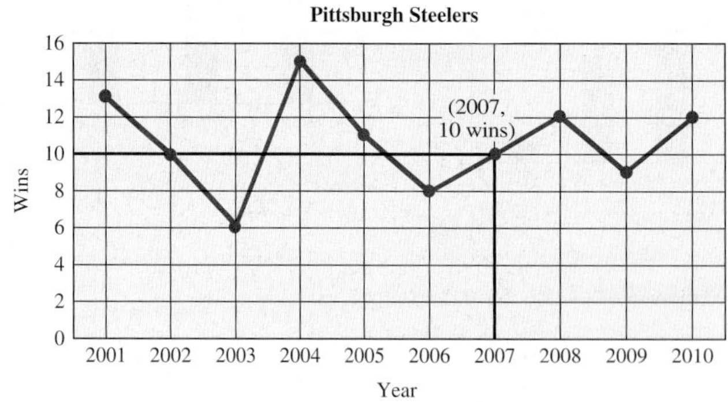

(b) Find the mean number of games won by the Steelers over this 10-year period.

For each dot on the line, we look to the left side to see the number of victories the dot represents. We then find the mean of these numbers.

$$\text{Mean} = \frac{13 + 10 + 6 + 15 + 11 + 8 + 10 + 12 + 9 + 12}{10}$$
$$= \frac{106}{10} = 10.6$$

They averaged over 10 wins a season between 2001 and 2010.

Check Yourself 1

The graph indicates the high temperatures in Baltimore, MD, for a week in September.

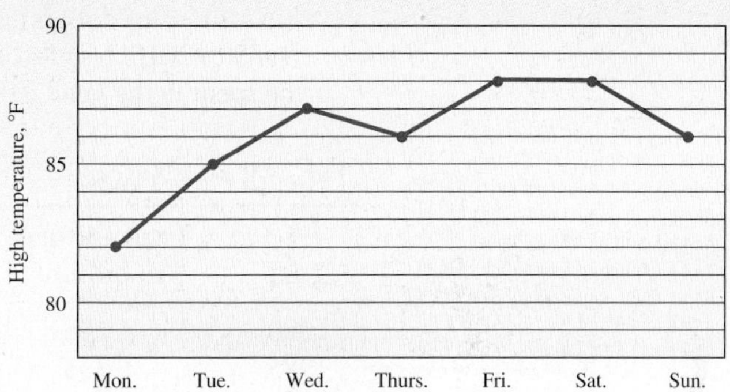

(a) What was the high temperature on Friday?
(b) Find the mean high temperature for that week.

It is often tempting, and sometimes useful, to use a line graph to predict a future value. Using an earlier trend to predict a future value is called **extrapolation.** This is something that statisticians warn us not to rely on, but it is done anyway. The key is not to predict very far from the data.

Example 2 Making a Prediction

< Objective 2 >

Use the line graph and table to predict the number of Social Security beneficiaries in the year 2015.

Social Security Beneficiaries

Year	Beneficiaries (in millions)
1980	35.5
1985	37.0
1990	39.8
1995	43.4
2000	45.4
2005	48.4
2010	54.0
2015	?

Source: Social Security Administration

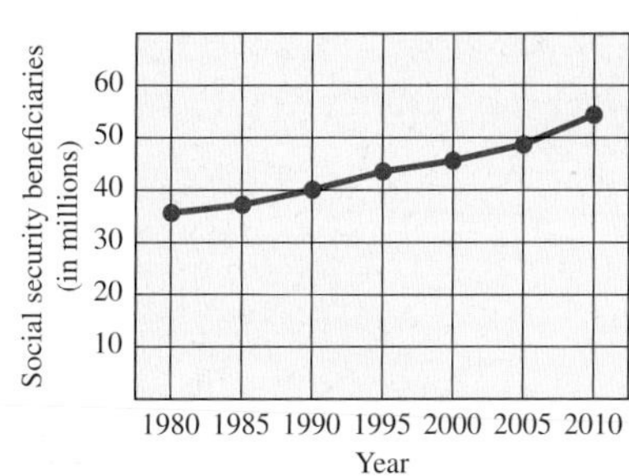

NOTE

You will learn better methods for estimating the answer to this question in future math classes.

The number of beneficiaries grew from 48.4 million to 54.0 million between 2005 and 2010. It is not unreasonable to expect at least this much growth between 2010 and 2015.

$54.0 - 48.4 = 5.6$

Therefore, we would expect at least another 5.6 million beneficiaries by 2015.

$54.0 + 5.6 = 59.6$

This gives us 59.6 million beneficiaries in 2015.

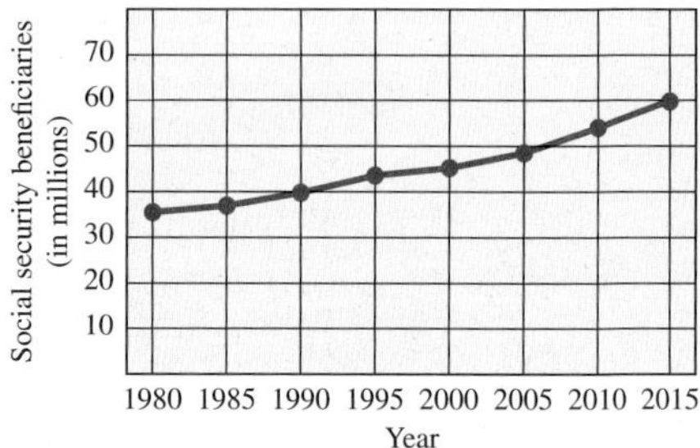

Check Yourself 2

The graph and table show the amount spent on health care (in billions of dollars) in the United States every 5 years from 1980 to 2010. Use this information to predict the amount that will be spent in the year 2015.

Health Care Costs

Year	Expenditures (in billions)
1980	256
1985	431
1990	724
1995	1,028
2000	1,377
2005	2,029
2010	2,594
2015	?

Source: Centers for Medicare and Medicaid Services

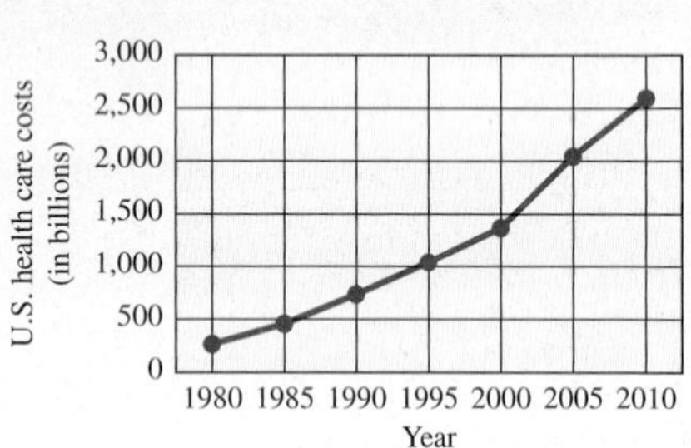

We construct line graphs in much the same way that we made bar graphs in Section 9.3. Instead of the horizontal axis giving categories, though, we list the time units.

We still find the height of each data point, but instead of bars, we plot points and then connect the dots. Consider Example 3.

Example 3 Constructing a Line Graph

< Objective 3 >

The table gives the average height of healthy girls, by age. Construct a line graph to display this information.

We begin by scaling the vertical axis. We need the vertical axis to include the smallest and largest numbers, so we choose to have it range from 35 to 60 in. We then include *grid lines* to make it easier to see where points are.

Girl's Heights (in inches)

Age	Height
2	35.5
5	44.3
8	51.5
11	59.6

Source: Centers for Disease Control and Prevention

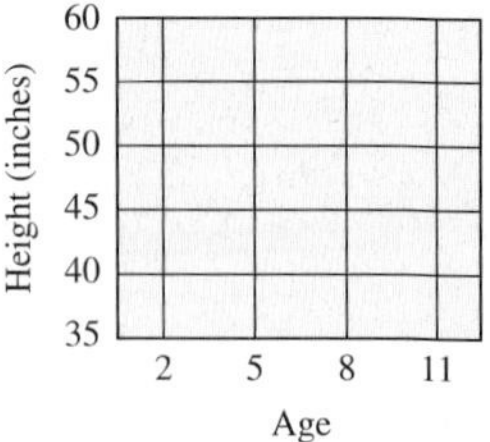

Now we plot points. Because the heights fall between the grid lines, we do our best to estimate where they belong. In general, if we need to be more precise, we use software to accomplish this task.

Finally, we connect the dots.

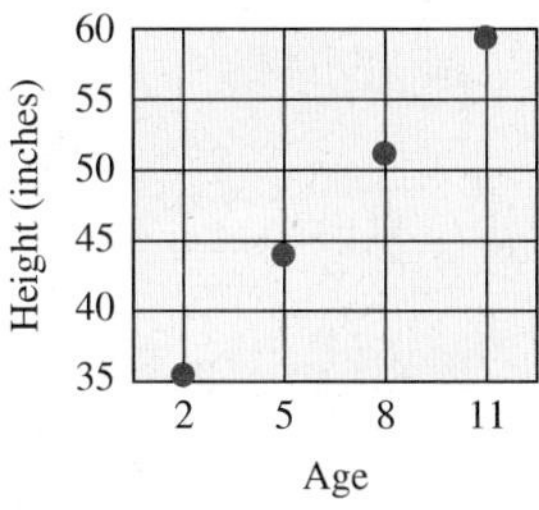

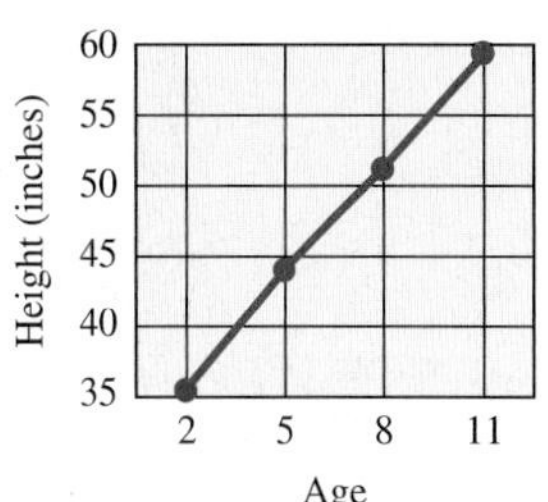

Check Yourself 3

The table gives the average height of healthy boys, by age. Construct a line graph to display this information.

Boy's Heights (in inches)

Age	Height
3	38.8
6	46.9
9	54.4
12	60.9

Source: Centers for Disease Control and Prevention

When a graph represents how some unit is divided, we usually use a *pie chart.*

As you might expect, a **pie chart** is a circle. Wedges (or sectors) are drawn in the circle to show how much of the whole each part makes up.

Example 4 Reading a Pie Chart

< Objective 4 >

This pie chart represents the results of a survey that asked students how they most often get to school.

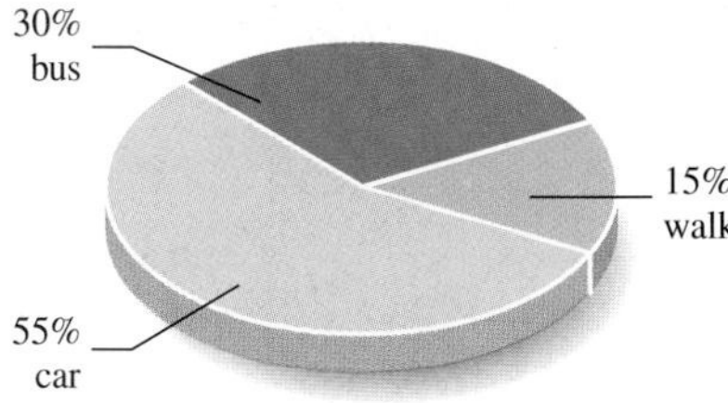

(a) What percent of the students walk to school?

We see that 15% walk to school.

(b) What percent of the students do not arrive by car?

Because 55% arrive by car, 100% − 55%, or 45%, do not.

Check Yourself 4

This pie chart represents the results of a survey that asked students whether they bought lunch, brought it, or skipped lunch altogether.

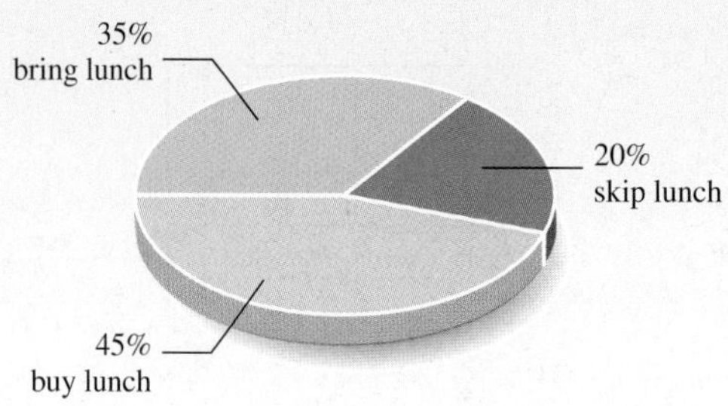

(a) What percent of the students skipped lunch?
(b) What percent of the students did not buy lunch?

If we know what the whole pie represents, we can also find out more about each wedge or category. Example 5 illustrates this point.

Example 5 Interpreting a Pie Chart

This pie chart shows how Sarah spent her $12,000 college scholarship.

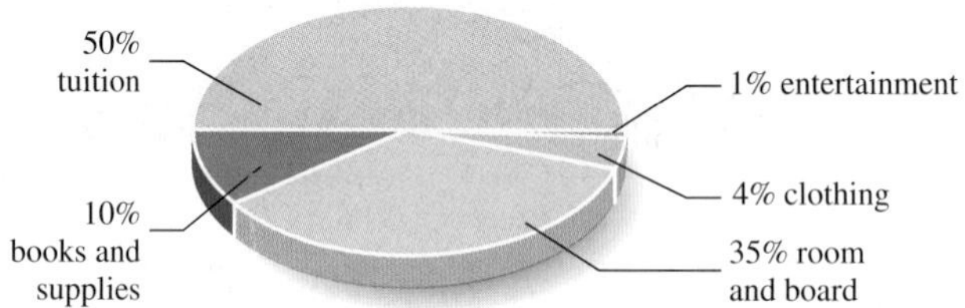

(a) How much did she spend on tuition?

50% of her $12,000 scholarship, or $6,000.

(b) How much did she spend on clothing and entertainment?

Together, 5% of the money was spent on clothing and entertainment, and $0.05 \times 12{,}000 = 600$. Therefore, she spent $600 on clothing and entertainment.

Check Yourself 5

This pie chart shows how Rebecca spends an average 24-hr school day.

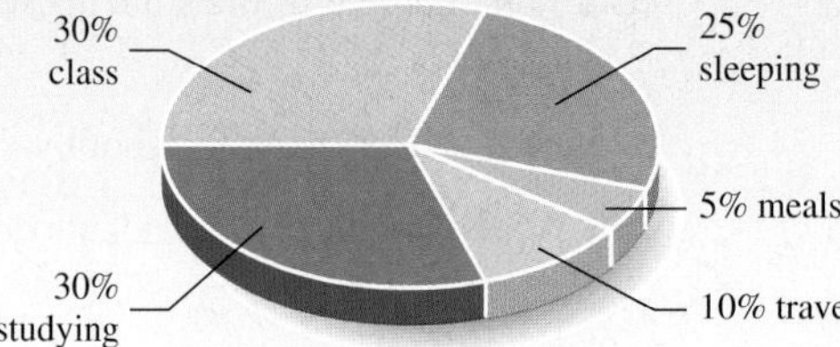

(a) How many hours does she spend sleeping each day?
(b) How many hours does she spend altogether studying and in class?

If we are creating a pie chart, how do we know how much of the circle to use for each piece? Making this decision requires a scale for the circle. A standard scale has been established for all circles. As we saw in Chapter 8, each circle has 360°. That means that $\frac{1}{4}$ of the circle has $\frac{1}{4}$ of 360°, which is 90°.

With a protractor, we can create our own pie chart.

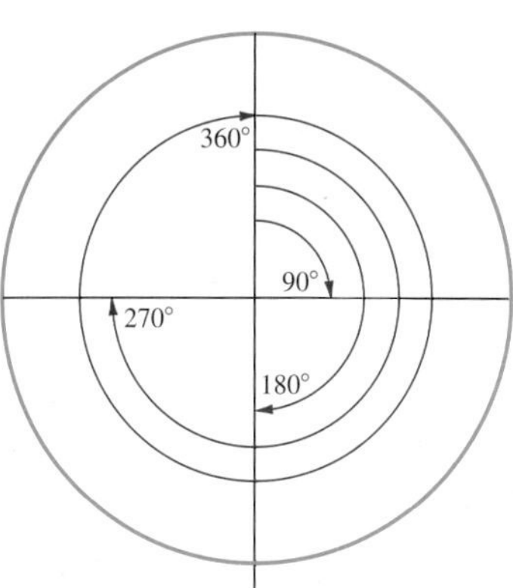

Example 6 Creating a Pie Chart

< Objective 5 >

This table gives the source of automobiles purchased in the United States in one year. Create a pie chart to display this information.

Source of Automobiles Purchased

Country of Origin	Number	% of Total
United States	6,500,000	80
Japan	800,000	10
Germany	400,000	5
All others	400,000	5

Source: American Automotive Manufacturers' Association

To find the size of the slice for each country, we take the given percent of 360°. We create another table column to represent the degrees needed.

Source of Automobiles Purchased

Country of Origin	Number	% of Total	Degrees
United States	6,500,000	80	288
Japan	800,000	10	36
Germany	400,000	5	18
All others	400,000	5	18

Source: American Automotive Manufacturers' Association

NOTE

Because 80% of the cars purchased were manufactured in the United States, we take 80% of 360° to find the size of the wedge.

$0.80 \times 360 = 288$

So, we need a 288° wedge.

We compute the number of degrees needed for the other three wedges in a similar manner.

Using a protractor, we start with Japan and mark a section that is 36°.

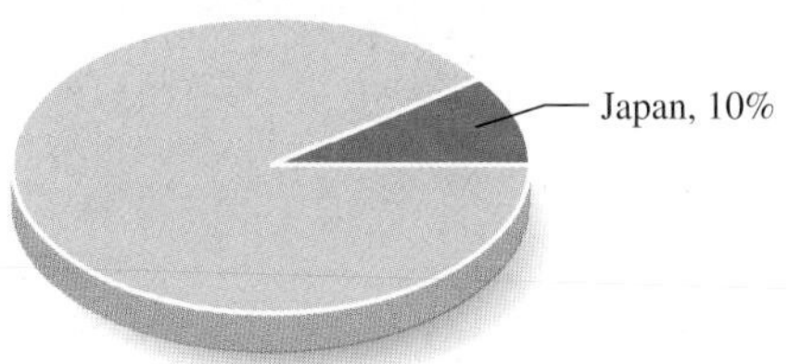

Again, using the protractor, we mark the 18° section for Germany and the 18° section for the other countries.

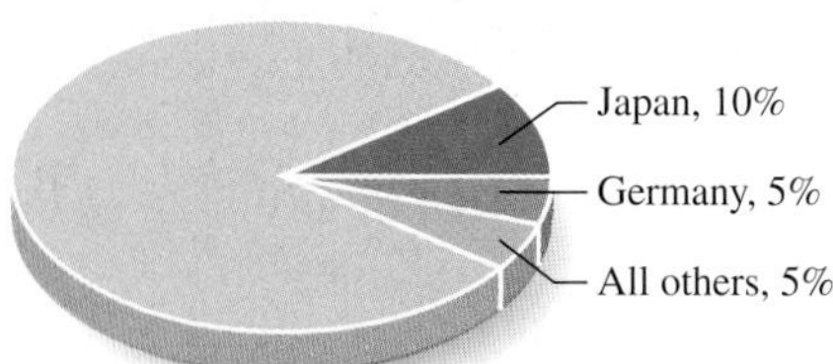

There is no need to measure the remainder of the pie. What is left is the 288° section for U.S.-made cars. Note that we saved the largest section for last. It is much easier to mark the smaller sections and leave the largest for last.

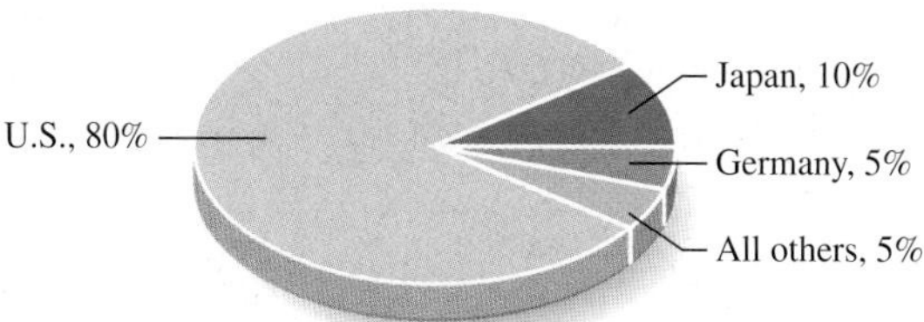

Check Yourself 6

Create a pie chart for the table, showing TV ownership for all U.S. homes.

TV Ownership

Number of TVs	% of U.S. Homes
0	2
1	22
2	34
3 or more	42

Source: Nielsen Media Research

Check Yourself ANSWERS

1. **(a)** 88°F; **(b)** 86°F **2.** \$3,159 billion (or \$3.159 trillion) is a reasonable prediction.

3. Boys Height

(Height in inches): 35, 40, 45, 50, 55, 60, 65; Age: 3, 6, 9, 12

4. **(a)** 20%; **(b)** 55% **5.** **(a)** 6 hr; **(b)** 14.4 hr

6.

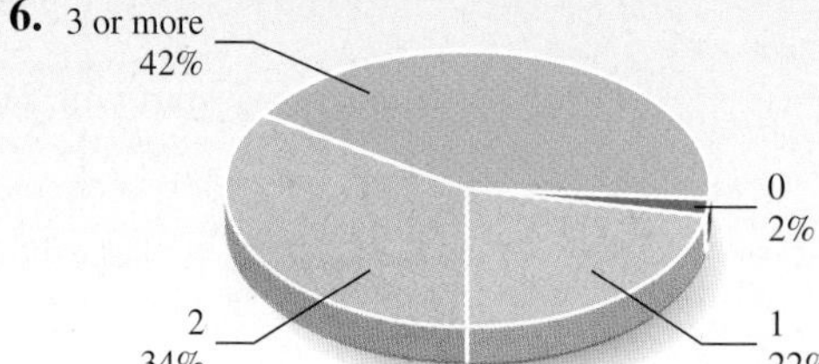

Reading Your Text 

These fill-in-the-blank exercises will help you understand some of the key vocabulary used in this section. The answers to these exercises are in the Answers Appendix at the back of the text.

(a) We often use a line graph to predict a ____________ value.

(b) Using an earlier trend to predict a future value is called ____________.

(c) In a ____________ chart, wedges are drawn in a circle to show how much of the whole each part makes up.

(d) Each ____________ has 360°.

Skills Calculator/Computer Career Applications Above and Beyond

9.4 exercises

< Objective 1 >

Statistics *A family tracked their electricity costs over a five-year period. Their annual cost, rounded to the nearest ten dollars, is displayed in the graph. Use this graph to complete exercises 1 to 4.*

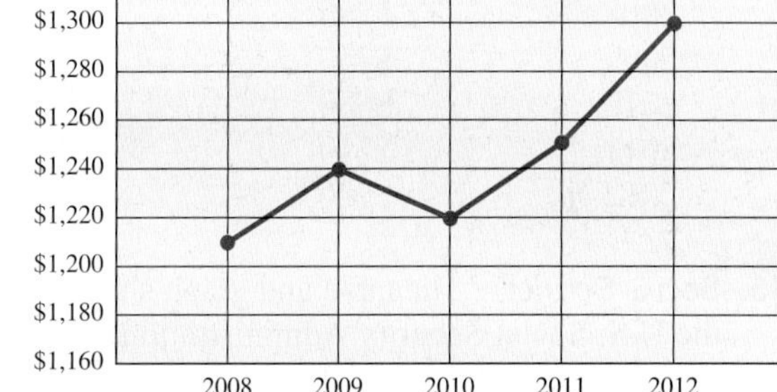

1. What was their electricity cost in 2010?
2. What was their mean annual cost of electricity during this period?
3. What was the decrease in their annual cost of electricity from 2009 to 2010?
4. In which year(s) did they experience the greatest increase in cost?

Statistics *A family tracked their natural gas costs over a 5-year period. Their annual cost, rounded to the nearest ten dollars, is displayed in the graph. Use this graph to complete exercises 5 to 8.*

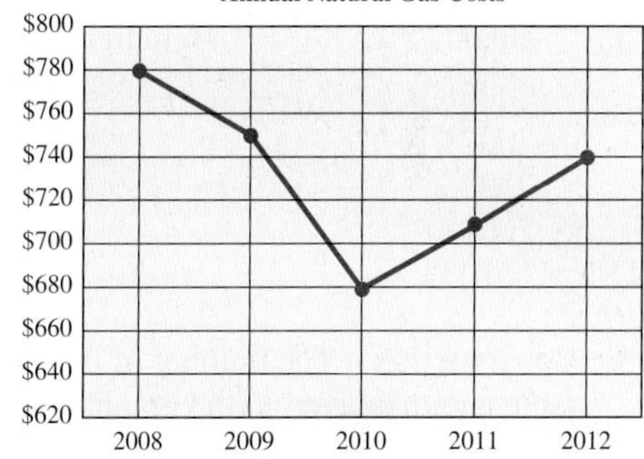

5. What was their natural gas cost in 2010?
6. What was their mean annual cost of natural gas during this period?
7. What was the decrease in their annual cost of natural gas from 2009 to 2010?
8. In which year(s) did they experience the greatest increase in cost?

Social Science *Use the graph, showing the number of robberies in a city during the last 6 months of a year, to complete exercises 9 to 12.*

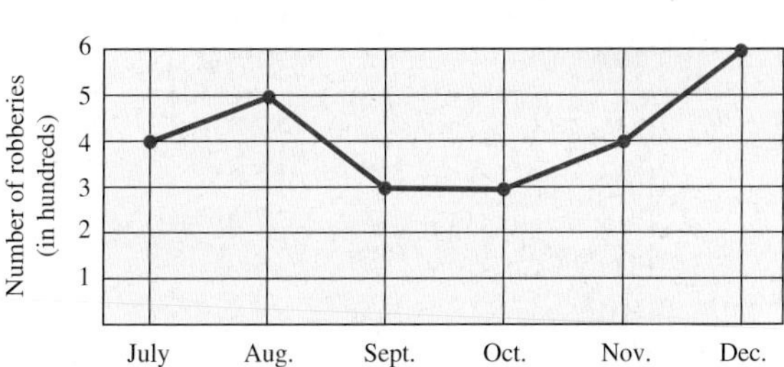

9. In which month did the greatest number of robberies occur? VIDEO
10. How many robberies occurred in November?
11. Find the decrease in the number of robberies between August and September. VIDEO
12. What was the mean number of robberies per month over these 6 months?

< Objective 2 >

13. STATISTICS Use the table and graph to predict the 2013 electicity cost for this family.

Annual Electricity Costs

Year	Cost
2008	$1,210
2009	$1,240
2010	$1,220
2011	$1,250
2012	$1,300

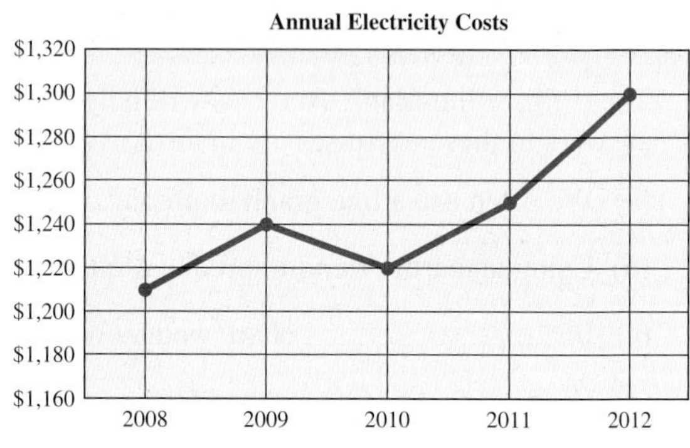

14. STATISTICS Use the table and graph to predict the 2013 natural gas cost for this family.

Annual Natural Gas Costs

Year	Cost
2008	$780
2009	$750
2010	$680
2011	$710
2012	$740

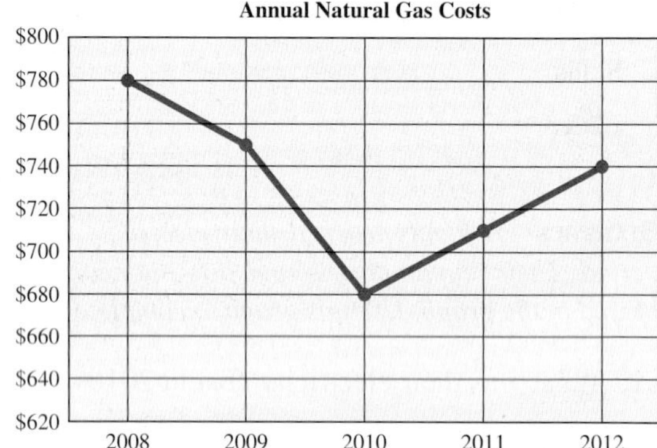

15. SOCIAL SCIENCE The table and graph show the number of beneficiaries of Disability Insurance (administered by the U.S. Social Security Administration). Use this information to predict the number of beneficiaries in 2011.

Disability Insurance Beneficiaries (in millions)

Year	Beneficiaries
2006	8.6
2007	8.9
2008	9.3
2009	9.7
2010	10.2

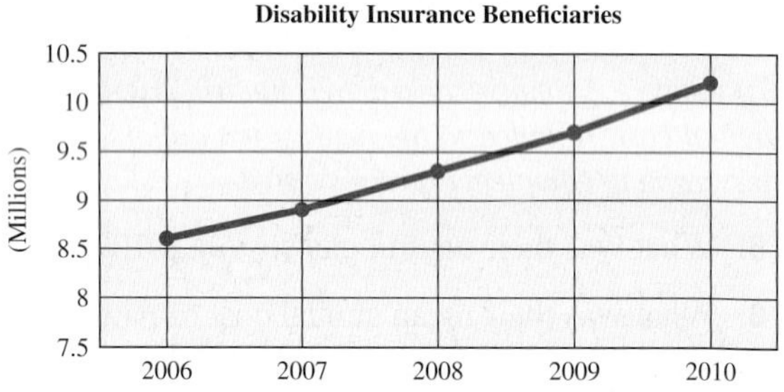

16. CONSTRUCTION The number of privately owned housing starts of one-unit structures each year are shown on the table and displayed in the graph. Use this information to predict the number of such starts in 2011.

Housing Starts (single unit) (in millions)

Year	Number
2006	17.7
2007	12.4
2008	7.4
2009	5.3
2010	5.7

Source: U.S. Census Bureau

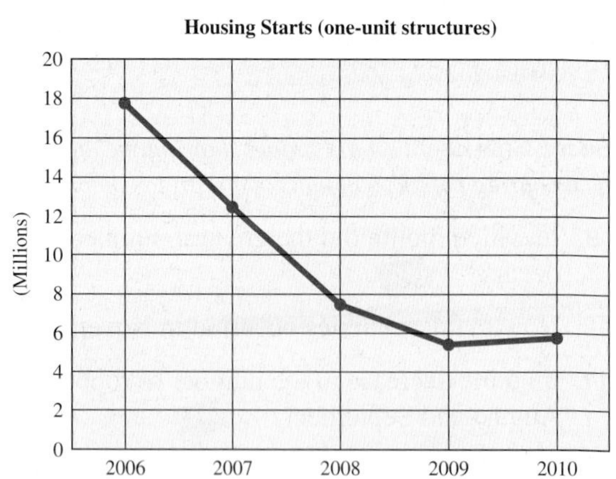

< Objective 3 >

17. Science and Medicine The table gives the average height of healthy girls, by age. Construct a line graph to display this information.

Girl's Heights (in inches)	
Age	**Height**
3	38.4
6	46.0
9	54.4
12	61.4
Source: Centers for Disease Control and Prevention	

18. Science and Medicine The table gives the average height of healthy boys, by age. Construct a line graph to display this information.

Boy's Heights (in inches)	
Age	**Height**
2	35.9
5	44.5
8	52.2
11	58.5
Source: Centers for Disease Control and Prevention	

19. Technology The table gives the highway fuel efficiency (in miles per gallon) of a particular automobile based on its speed (in miles per hour). Construct a line graph to display this information with the car's speed on the horizontal axis and the fuel efficiency on the vertical axis.

Fuel Efficiency vs. Speed	
Speed (mi/hr)	**Fuel Efficiency (mi/gal)**
40	28.2
45	29.6
50	30.0
55	30.0
60	29.1
65	27.6
70	24.9
75	23.1
Source: U.S. Department of Energy	

20. **Technology** The table gives the highway fuel efficiency (in miles per gallon) of a particular light truck based on its speed (in miles per hour). Construct a line graph to display this information with the truck's speed on the horizontal axis and the fuel efficiency on the vertical axis.

Fuel Efficiency vs. Speed

Speed (mi/hr)	Fuel Efficiency (mi/gal)
40	21.0
45	21.6
50	22.0
55	22.2
60	22.3
65	22.0
70	20.5
75	18.0

Source: U.S. Department of Energy

21. **Business and Finance** A college publishes its salary scale for one category of professor, based on the number of years of service. Construct a line graph to display this information and predict the salary of a professor with 25 years of experience.

Salary (in thousands of dollars)

Years	Salary
0	$41.9
5	47.4
10	53.6
15	60.7
20	68.7

Source: College of Southern Nevada

22. **Statistics** The table shows the population of Martin County, NC, over time. Construct a line graph to display this information and predict the 2020 population.

Population (in thousands)

Year	Population
1970	24.7
1980	25.9
1990	25.1
2000	25.6
2010	24.5

Source: U.S. Census Bureau

< Objective 4 >

Technology *A local utility company broke down annual electricity usage for the typical household in its area. The results are summarized in the pie chart.*

Electricity Use

Lighting 11%
Hot Water 12%
Food Storage 8%
Cooling 12%
Other 26%
Heating 31%

Source: Portland General Electric

23. What percentage of a typical household's electricity usage goes toward heating?

24. What percentage of a typical household's electricity usage goes toward food storage?

25. If a household used 11,600 kilowatt-hours (kWh) of electricity in a year, how many kWh were used for heating (round your result to the nearest kWh)?

26. If a household used 11,600 kilowatt-hours (kWh) of electricity in a year, how many kWh were used for lighting (round your result to the nearest kWh)?

Business and Finance *The owner of a small side business broke down last year's expenses. The results are summarized in the pie chart. Use this information to complete exercises 27 to 30.*

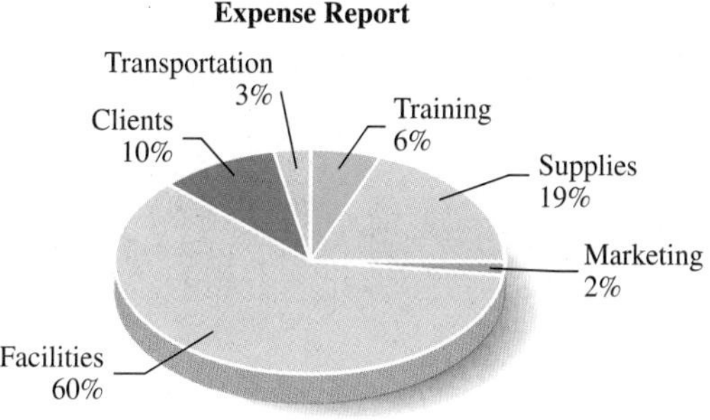

27. What percentage of the year's expenses went toward supplies?

28. What percentage of the year's expenses went toward facilities?

29. If the year's expenses totaled $35,000, how much was spent on supplies?

30. If the year's expenses totaled $35,000, how much was spent on training?

< Objective 5 >

31. **Business and Finance** A local company budgets $600,000 for the coming year, broken down as shown in the table. Complete the table and construct a pie chart to display this information.

Category	Dollars	Percent	Degrees
Production	$270,000	45%	162°
Research	90,000		
Taxes	60,000		
Operations	120,000		
Miscellaneous	60,000		

32. **Science and Medicine** A researcher studying the amphibian population in Yellowstone National Park recorded the number of sightings during the course of her visit. Complete the table and construct a pie chart to display this information.

Amphibian	Count	Percent	Degrees
Columbia Spotted Frog	120	40%	144°
Boreal Toad	45		
Chorus Frog	90		
Blotched Tiger Salamander	45		

Source: U.S. Department of the Interior

33. **Statistics** Of the 100 U.S. Senators in the 112th Congress, 29 served in the military (including reserve and guard service). Construct a pie chart showing their service, by branch.

Military Service

Branch	Count
Army	14
Navy	7
Air Force	5
Marine Corps	3
No Service	71

34. **Business and Finance** 1,000 people who bought a particular ice cream maker responded to a request to review their purchase. Their responses are shown in the table. Construct a pie chart displaying this information.

Ice Cream Maker Reviews

Review	Count
5-Star	640
4-Star	180
3-Star	80
2-Star	40
1-Star	60

Skills | Calculator/Computer | **Career Applications** | Above and Beyond

35. **Health Sciences** The number of recorded polio cases worldwide has dropped sharply since coordinated efforts to eradicate the disease began in 1988, as shown in the table. Construct a line graph to display this information and use it to estimate the number of polio cases that will be reported in 2015.

Recorded Polio Cases (in thousands)

Year	Cases
1990	23.5
1995	7.0
2000	3.0
2005	2.0
2010	1.4

Source: The World Health Organization

36. **Agricultural Technology** The table shows the average urban price for a pound of navel oranges in March of each year listed (March is generally considered the height of navel orange season). Construct a line graph to display this information and predict the price in 2011.

Navel Oranges

Year	Cost
2006	\$0.888
2007	\$1.301
2008	\$0.898
2009	\$0.889
2010	\$0.858

Source: U.S. Bureau of Labor Statistics

37. **AUTOMOTIVE TECHNOLOGY** The top five light vehicle manufacturers, by March 2012 U.S. sales (in thousands), are shown in the table. Construct a pie chart to show the share of all such light vehicle sales each manufacturer received.

Light Vehicle Sales (in thousands)

Manufacturer	Sales
General Motors	231
Ford	223
Toyota	203
Chrysler	163
Nissan	136
Other	447

Source: Wall Street Journal

38. **AUTOMOTIVE TECHNOLOGY** The top five car manufacturers, by March 2012 U.S. sales (in thousands), are shown in the table. Construct a pie chart to show the share of all such car sales each manufacturer received.

Car Sales (in thousands)

Manufacturer	Sales
Toyota	125
General Motors	100
Nissan	91
Ford	85
Honda	67
Other	294

Source: Wall Street Journal

Skills | Calculator/Computer | Career Applications | **Above and Beyond**

BUSINESS AND FINANCE *A professional association recorded the income and expenses from their annual conference over a 5-year period. Use the line graph of this information to complete exercises 39 to 44.*

39. Which conference brought in the most income?

40. Which conference brought in the least income?

41. Which conference produced the largest difference between income and expenses?

42. At which conference did expenses exceed income?

43. At which conference were income and expenses approximately equal?

44. Approximate the difference between their income and expenses in Orlando, FL.

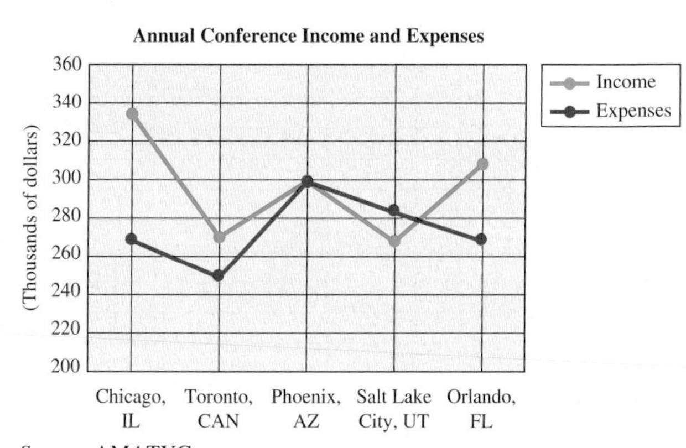

Source: AMATYC

STATISTICS *One of the things you learned in this section is to make predictions based on data. Statisticians have many techniques they use to accomplish this, most of which are more complex than what we did in this section.*

Most predictions are one of two types: ***interpolation*** *and* ***extrapolation.*** *The difference between these two types of predictions is relatively straightforward.*

As its name indicates, predictions within a data set are called interpolation. For instance, if you know a baby's weight when she is 2 months, 5 months, and 8 months old, then predicting the baby's weight at 6 months is an example of interpolation.

Conversely, extrapolation considers predictions outside a given data set. In the case above, predicting the baby's weight at 11 months is extrapolation. Extrapolation is often less accurate than interpolation and care must be taken to determine if making predictions outside of a data set is warranted.

Consider the average high temperature (in degrees Fahrenheit) for Lincoln, NE, for a series of months.

High Temperature: Lincoln, NE

Month	March	April	June	July
Temperature	50.0°F	64.9°F	84.8°F	91.1°F

Source: Weatherbase

45. Predict the average high temperature in May by taking the mean of the high temperatures for April and June. This is an example of interpolation.

46. Predict the average high temperature in August based on the high temperatures for June and July. This is an example of extrapolation.

47. Continue to extrapolate to predict the high temperature in December.

48. Does your answer to exercise 47 seem reasonable? Justify your answer.

A kitten weighs 4 pounds at 3 months of age. At 8 months it weights 9 lb and at 10 months it weighs 11 lb.

49. Predict the kitten's weight at 12 months. Does this seem reasonable?

50. Predict the kitten's weight at 5 years old (60 months). Is this reasonable?

Answers

1. $1,220 **3.** $20 **5.** $680 **7.** $70 **9.** December **11.** 200 **13.** $1,350 **15.** 10,700,000 beneficiaries

17.

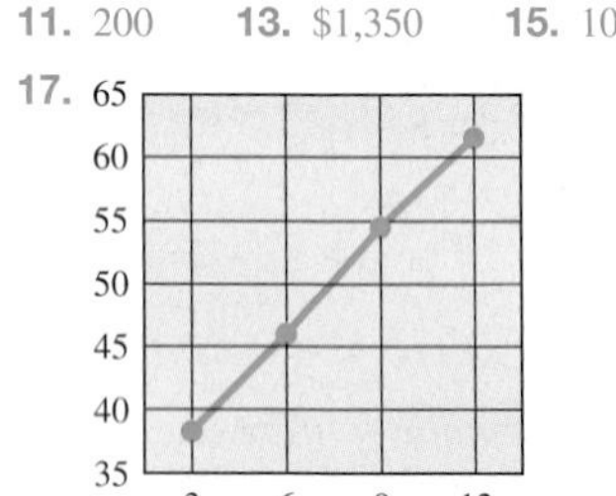

19.

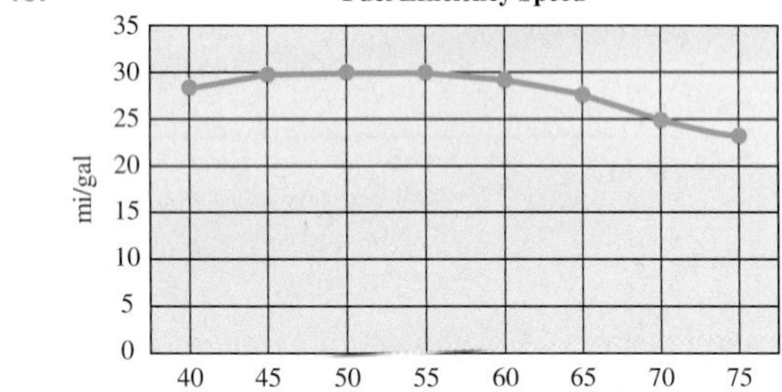

21. $76,700

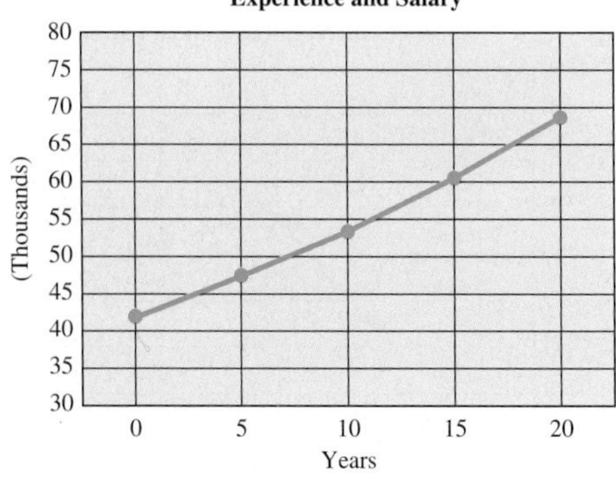

23. 31% **25.** 3,596 kWh **27.** 19% **29.** $6,650

31.

Category	Dollars	Percent	Degrees
Production	$270,000	45%	162°
Research	90,000	15%	54°
Taxes	60,000	10%	36°
Operations	120,000	20%	72°
Miscellaneous	60,000	10%	36°

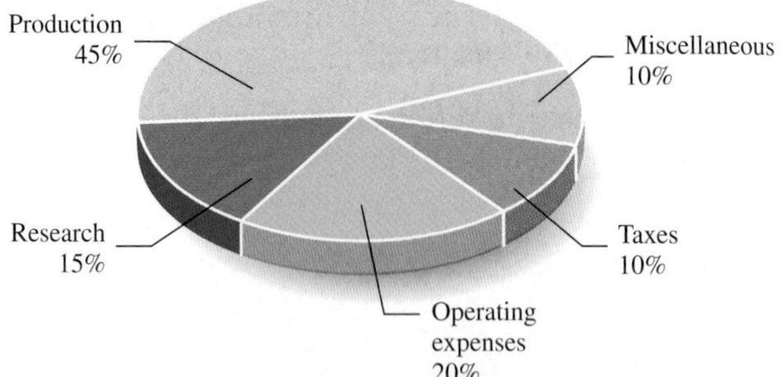

33.

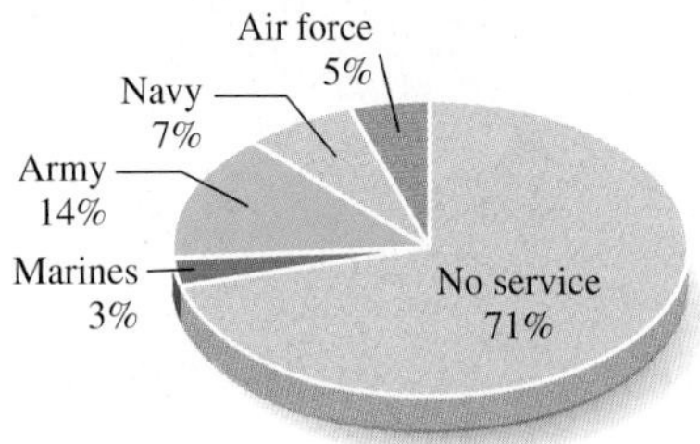

35. 800 cases

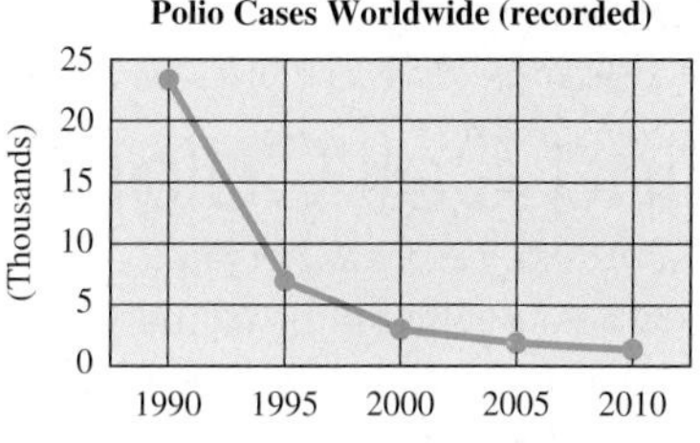

37.

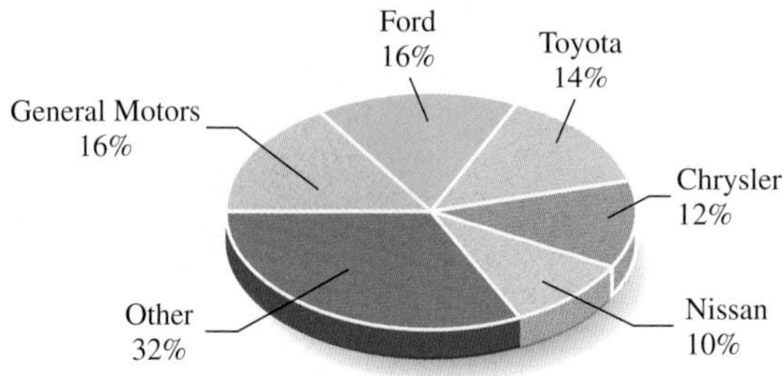

39. Chicago, IL 41. Chicago, IL 43. Phoenix, AZ 45. 74.85°F 47. 122.6°F 49. 13 lb; yes

Activity 27 ::

Graphing Car Color Data

Consider the data you gathered in Activity 25. If you worked in the automobile industry, information concerning customer color preferences would be important. To form conclusions, you need to present your data to other people (who are probably not statisticians). Bar graphs and pie charts are useful ways of presenting such data.

1. Create a bar graph using the set of 35 car colors compiled in Activity 25.
2. Create a pie chart using the set of 35 car colors compiled in Activity 25.
3. **(a)** Briefly describe the view of your data given by the two graphs.
   (b) In which graph is the mode more easily distinguished?
   (c) In which is it easier to get a sense of the "whole" data set?
4. Compare your graphs to those of a classmate. Briefly describe how yours differ from your classmate's graphs (consider the differences in data sets and in presentation).
5. Write a short letter to the manager of an auto body and paint shop, recommending levels of inventory for different color paints. Include either the bar graph or the pie chart in your letter.

Definition/Procedure	Example	Reference
Mean, Median, and Mode		**Section 9.1**
Computing a Mean		
Step 1 Add all the numbers in the set. **Step 2** Divide that sum by the number of items in the set.	Given the numbers 4, 8, 17, 23 $4 + 8 + 17 + 23 = 52$ Mean $= \frac{52}{4} = 13$	*p.* 501
Finding a Median		
Step 1 Sort the numbers in ascending order (lowest value to highest value). **Case 1** There is an odd number of data points. **Step 2** Select the middle data value; this is the median. **Case 2** There is an even number of data points. **Step 2** Select the two middle data values. **Step 3** Compute the mean of these two numbers; this is the median.	Given the 5 numbers 6, 4, 10, 7, 5 Rewrite the list in ascending order: $4, 5, \underbrace{6,}_{\text{Median}} 7, 10$ The middle value is the median: 6. Given the 6 numbers 9, 2, 5, 13, 7, 3 Rewrite the list in ascending order: $2, 3, \underbrace{5, 7,}_{\text{Middle values}} 9, 13$ Take the mean of the middle values: $\frac{5+7}{2} = \frac{12}{2} = 6$ The median is 6.	*p.* 503
Finding a Mode		
The *mode* is the number that occurs most frequently in a set of numbers.	Given the numbers 2, 3, 3, 3, 5, 5, 7, 7, 9, 11 3 is the mode.	*p.* 506
Describing Data Sets		**Section 9.2**
Finding Quartiles		
Step 1 Rewrite the data set in ascending order. **Step 2** Find the median. **Case 1** There is an odd number of data points. **Step 3** Make a list of the median and those numbers to the *left* of the median. **Step 4** Find the median of the list created in step 3. This is the first quartile Q_1. **Step 5** Repeat steps 3 and 4 with the median and those numbers to the *right* of the median. This is the third quartile Q_3. **Case 2** There is an even number of data points. **Step 3** Make a list of only those numbers to the *left* of the median. **Step 4** Find the median of the list created in step 3. This is the first quartile Q_1. **Step 5** Repeat steps 3 and 4 with only those numbers to the *right* of the median. This is the third quartile Q_3.	Consider 2, 3, 5, 7, 9, 13. The median is 6. There is an even number of data points, so we only look at the numbers to the left of 6 to find the first quartile. The median of this set, 2, 3, and 5, is 3. $Q_1 = 3$ Similarly, $Q_3 = 9$.	*p.* 517

Continued

Definition/Procedure	Example	Reference
Finding a Five-Number Summary		
The five-number summary is given by the list min, Q_1, median, Q_3, max	The five-number summary of the preceding list is 2, 3, 6, 9, 13	p. 518
Box-and-Whisker Plots		
Mark off the five-number summary on a number line from min to max and draw a rectangle between the quartiles.	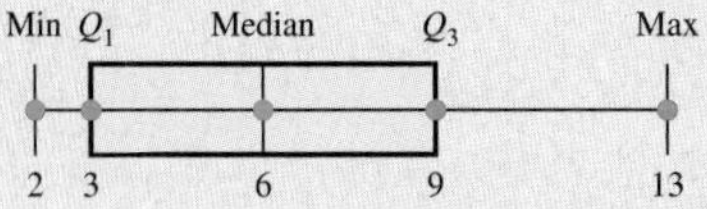 Min Q_1 Median Q_3 Max; 2 3 6 9 13	p. 519
Tables and Bar Graphs		Section 9.3
A **table** is a display of information in parallel columns or rows.		p. 528

U.S. Dept. of Transportation

Year	Cars Registered
1970	89,243,557
1975	106,705,934
1980	121,600,843
1985	131,664,029
1990	143,549,627
1995	136,066,045
2000	133,621,245
2005	136,568,083
2010	193,979,654

A **graph** is a diagram that relates two different pieces of information. One of the most common graphs is the **bar graph.** — p. 529

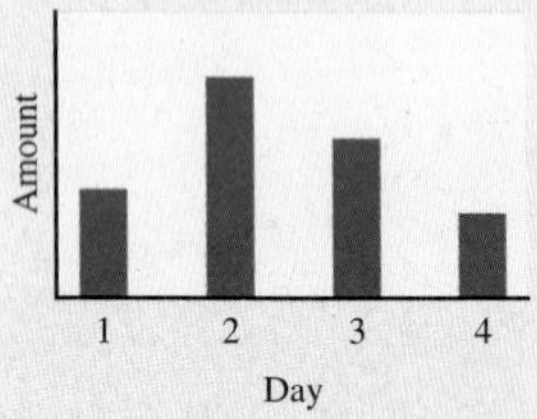

Amount
1 2 3 4
Day

Line Graphs and Pie Charts — Section 9.4

Line graph In line graphs, one of the axes is usually related to time. — p. 540

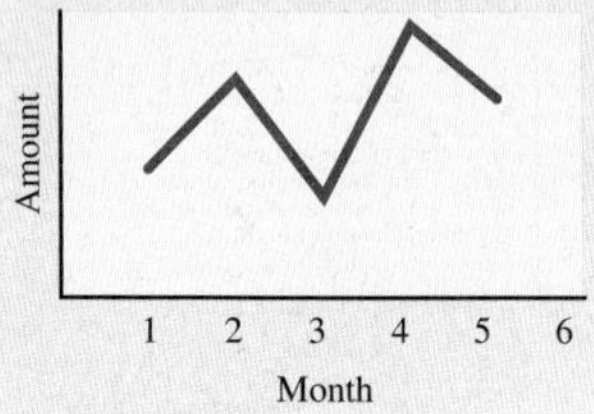

Amount
1 2 3 4 5 6
Month

Definition/Procedure	Example	Reference
Pie chart Pie charts are graphs that show the component parts of a whole. 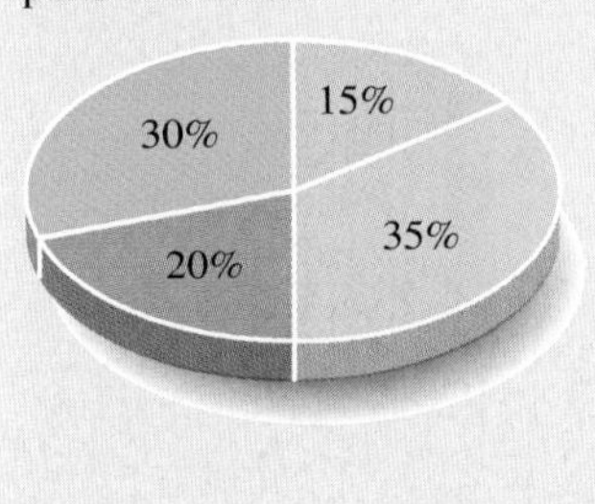	Each percent is shown as the percent of a 360° circle. 30% of 360° = $0.30 \times 360° = 108°$ 20% of 360° = $0.20 \times 360° = 72°$ 15% of 360° = $0.15 \times 360° = 54°$ 35% of 360° = $0.35 \times 360° = 126°$	p. 543

summary exercises :: chapter 9

This summary exercise set will help ensure that you have mastered each of the objectives of this chapter. The exercises are grouped by section. You should reread the material associated with any exercises that you find difficult. The answers to the odd-numbered exercises are in the Answers Appendix at the back of the text.

9.1 *Find the mean of each set.*

1. 8, 6, 7, 4, 5

2. 12, 14, 17, 19, 13

3. 117, 121, 122, 118, 115, 125, 123, 119

4. 134, 126, 128, 129, 133, 125, 122, 127

5. Elmer had test scores of 89, 71, 93, and 87 on his four math tests. What was his mean score?

6. The costs (in dollars) of the seven textbooks that Jacob needs for the spring semester are 77, 66, 55, 49, 85, 80, and 78. Find the mean cost of these books.

Find the median and mode of each set.

7. 16, 20, 20, 19, 18

8. 8, 9, 9, 11, 11, 8, 7, 11, 12, 14, 10

9. 26, 31, 28, 35, 27, 28, 31, 30, 28, 30

10. 15, 18, 21, 23, 17, 19, 30, 35, 15, 32

11. Anita's first four test scores in her mathematics class were 88, 91, 86, and 93. What score must she get on her next test to have a mean of 90?

12. The sales of a small company over 3 days were \$2,400, \$2,800, and \$3,300. How much do sales need to be in the fourth day to achieve a mean of \$3,000?

9.2 *Give the five-number summary of each set of numbers.*

13. 30, 32, 21, 35, 28, 28, 24, 23, 26, 30

14. 93, 79, 84, 62, 66, 94, 90, 87, 74, 76, 77, 72, 68, 62, 74, 85, 98, 69, 97, 78, 71

15. Construct a box-and-whisker plot for the set of numbers

30, 32, 21, 35, 28, 28, 24, 23, 26, 30

16. The scores on the first examination for an algebra class are

93, 79, 84, 62, 66, 94, 90, 87, 74, 76, 77, 72, 68, 62, 74, 85, 98, 69, 97, 78, 71

Construct a box-and-whisker plot for the examination grades and describe the results.

9.3 *Use the table to complete exercises 17 to 24. Round percentages to the nearest tenth of a percent, when necessary.*

World Motor Vehicle Production, 1960–2010

			(in thousands)			
Year	**United States**	**Canada**	**Europe**	**Japan**	**Other**	**World Total**
2010	7,763	2,068	17,493	9,629	40,676	77,629
2005	11,947	2,688	19,285	10,800	21,762	66,482
2000	12,778	2,966	15,176	10,145	15,978	57,043
1995	11,985	2,408	17,045	10,196	8,349	49,983
1990	9,783	1,928	18,866	13,487	4,496	48,560
1985	11,653	1,933	16,113	12,271	2,939	44,909
1980	8,010	1,324	15,496	11,043	2,692	38,565
1970	8,284	1,160	13,049	5,289	1,637	29,419
1960	7,905	398	6,837	482	866	16,488

Source: American Automobile Manufacturers Association; International Organization of Automobile Manufacturers

17. What was the motor vehicle production in Japan in 1960? 2010?

18. What was the motor vehicle production in countries outside the United States in 1960? 2010?

19. What was the percent decrease in motor vehicle production in the United States from 1960 to 2010?

20. What was the percent increase in motor vehicle production in countries outside the United States from 1960 to 2010?

21. What percent of world motor vehicle production occurred in Japan in 2010?

22. What percent of world motor vehicle production occurred in the United States in 2010?

23. What percent of world motor vehicle production occurred outside the United States and Japan in 2010?

24. Between 1960 and 2010, did the production of motor vehicles increase by a greater percent in Canada or Europe?

The number of students attending Berndt Community College in selected years is displayed in the bar graph. Use the graph to complete exercises 25 and 26.

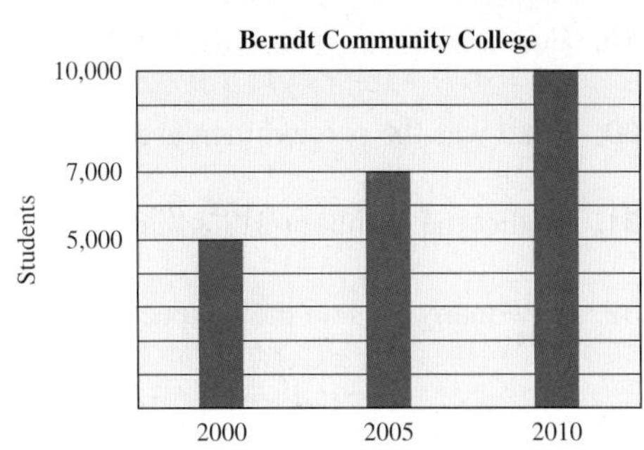

25. How many more students were enrolled in 2010 than in 2000?

26. What was the percent increase from 2000 to 2005?

27. The five least violent states in the United States and the per resident cost of violence in each state, rounded to the nearest $10 per resident are shown in the table. Display this information in a bar graph.

Least Violent States

State	Cost per Resident (to the nearest $10)
Maine	$600
Vermont	$720
New Hampshire	$720
Minnesota	$880
Utah	$850

Source: Reuters; U.S. Census Bureau Cost include indirect costs

28. The five most violent states in the United States and the per resident cost of violence in each state, rounded to the nearest $100 per resident are shown in the table. Display this information in a bar graph.

Most Violent States

State	Cost per Resident (to the nearest $100)
Louisiana	$2,200
Tennessee	$1,800
Nevada	$2,000
Florida	$1,800
Arizona	$1,700

Source: Reuters; U.S. Census Bureau Cost include indirect costs

9.4 *A large chain store tracked the number of PCs (in thousands) it sold over several years and used a line graph to display that information. Use the graph to complete exercises 29 to 32.*

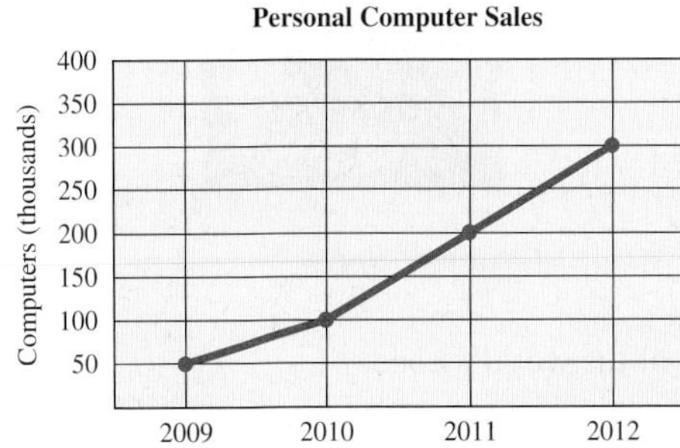

29. How many more PCs were sold in 2012 than in 2009?

30. What was the percent increase in sales from 2009 to 2012?

31. Predict the number of PCs that they will sell in 2013.

32. How many PCs do they expect to sell in 2015?

A homeowner used an online service to track the change in his home's value over a 6-month period and recorded this information in the table shown (in thousands of dollars). Use the table to complete exercises 33 to 36.

Home Value (in thousands)

Month	Value
Nov	\$259
Dec	\$259
Jan	\$262
Feb	\$267
Mar	\$279
Apr	\$286

Source: Zillow.com

33. Construct a line graph to display this information.

34. By what percent did the home's value increase from March to April (round your result to the nearest tenth of a percent)?

35. By what percent did the home's value increase over the 6-month period (round your result to the nearest tenth of a percent)?

36. Predict the home's value in May.

Business & Finance The pie chart shows the top cocoa producing nations in 2011. Use this chart to complete exercises 37 and 38.

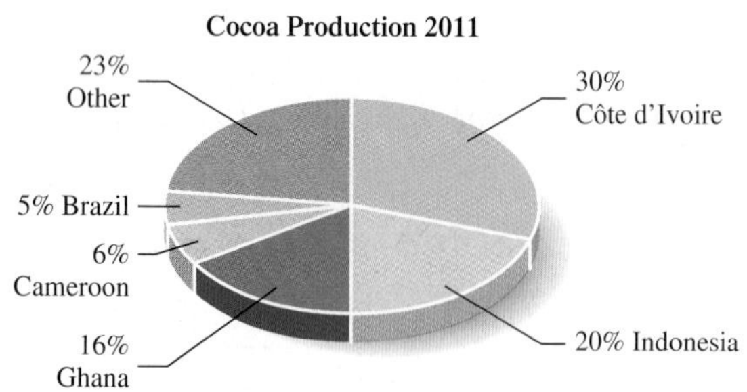

Source: U.N. Food & Agricultural Organization

37. Which country was the largest cocoa bean producer in 2011? What percent of the world's cocoa beans were grown by this country?

38. A global total of 8,980 million pounds of cocoa beans were grown in 2011. How many pounds did Cameroon produce?

39. **Statistics** A professor compiled the end-of-term grades from her algebra course. Complete the table and construct a pie chart to display this information.

Grade	Count	Percent	Degrees
A	7	17.5%	63°
B	12		
C	13		
D	5		
F	3		

40. **Statistics** Approximately 850,000 newspapers are bought each day in a particular state. The sales are broken down by individual paper (in thousands). Complete the table and construct a pie chart to display this information.

Newspaper	Sales	Percent	Degrees
Daily	255	30%	108°
Tribune	170		
Star	127.5		
Mercury	42.5		
Other	255		

Source: Audit Bureau of Circulations

CHAPTER 9

chapter test 9

Use this chapter test to assess your progress and to review for your next exam. Allow yourself about an hour to take this test. The answers to these exercises are in the Answers Appendix at the back of the text.

1. Find the mean of the numbers 12, 19, 15, 20, 11, and 13.

2. Find the median of the numbers 8, 9, 15, 3, 1.

3. Find the median of the numbers 12, 18, 9, 10, 16, 6.

4. Find the mode of the numbers 6, 2, 3, 6, 2, 9, 2, 6, 6.

5. **Science and Medicine** Give the five-number summary of the average monthly temperatures in Akron, OH, in 2011.

Average Monthly Temperatures: Akron, OH; 2011

Jan	Feb	Mar	Apr	May	Jun	Jul	Aug	Sep	Oct	Nov	Dec
23°F	29°F	37°F	50°F	62°F	69°F	76°F	71°F	64°F	52°F	46°F	36°F

Source: The University of Dayton

6. **Construction** A bus carried 234 passengers on the first day of a newly scheduled route. The next 4 days there were 197, 172, 203, and 214 passengers. What was the mean number of riders per day?

7. **Statistics** To earn an A in biology, you must have a mean of 90 on four tests. Your scores thus far are 87, 89, and 91. Find the minimum number of points needed on the final exam to earn an A.

8. These hair colors are from the students in a school's Scuba club. What color is the mode?

   brown, black, red, blonde, brown, brown, blue, gray

Agricultural *The table gives the cocoa bean production in a recent growing season, in millions of pounds, along with the value of that season's crop, in millions of dollars. Use this table to complete exercises 9 to 12.*

Cocoa Bean Production and Value

Nation	Production (millions of pounds)	Share of World's Total	Value (millions of dollars)
Côte d'Ivoire	2,706	34.74%	$3,287
Ghana	1,608	20.64%	1,951
Indonesia	1,078	13.84%	1,309
Cameroon	462	5.93%	561
Nigeria	462	5.93%	561
Brazil	363	4.66%	441
Ecuador	286	3.67%	347
Malaysia	70	0.90%	86
Other	755	9.69%	917

Source: International Cocoa Organization; IndexMundi

9. What was the world's total production of cocoa beans that growing season?

10. What was the total value of the world's cocoa production that season?

11. In a subsequent year, Côte d'Ivoire's production fell to 2,688 million pounds. Find the percent decrease this represents (round to the nearest hundredth of a percent).

12. In that same year, Indonesia's production increased to 1,760 million pounds. What percent increase does this represent (round to the nearest percent)?

Science and Medicine *The total number of severe Atlantic hurricanes (Categories 4 and 5) are given for each 5-year period. Use this table to complete exercises 13 to 16.*

Severe Atlantic Hurricanes

1981–85	1986–90	1991–95	1996–2000	2001–05	2006–10
4	5	5	11	14	11

Source: National Weather Service

13. Construct a bar graph to display this information.

14. How many severe Atlantic hurricanes occurred between 2006 and 2010?

15. What was the percent increase in the number of severe Atlantic hurricanes between the periods 1996–2000 and 2001–2005 (to the nearest percent)?

16. Which period saw the largest increase over the period that came before it?

Business and Finance *The graph shows ticket sales for the last 6 months of the year. Use the line graph to complete exercises 17 and 18.*

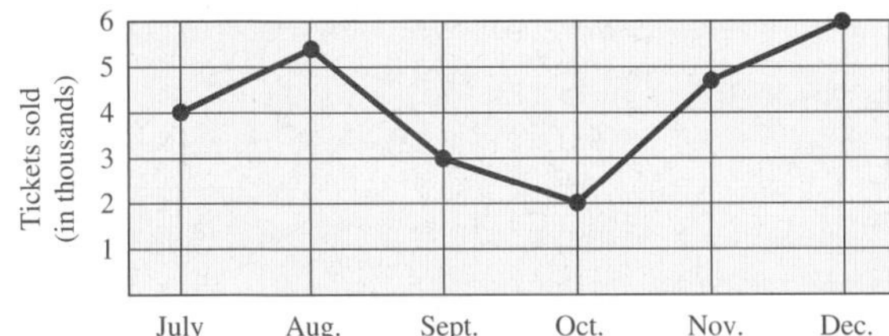

17. What month had the greatest number of ticket sales?

18. Between what two months did the greatest decrease in ticket sales occur?

Business and Finance *A manufacturing company tracked the relationship between the number of workers absent during a shift and the number of defective products coming off their production line. Use this table to complete exercises 19 and 20.*

Worker Absences	Defective Products
0	9
1	10
2	12
3	16
4	18

19. Construct a line graph to display this information.

20. Predict the number of defective products that would come off the line if five workers were absent.

Business and Finance *The pie chart represents the way a new company ships its goods.*

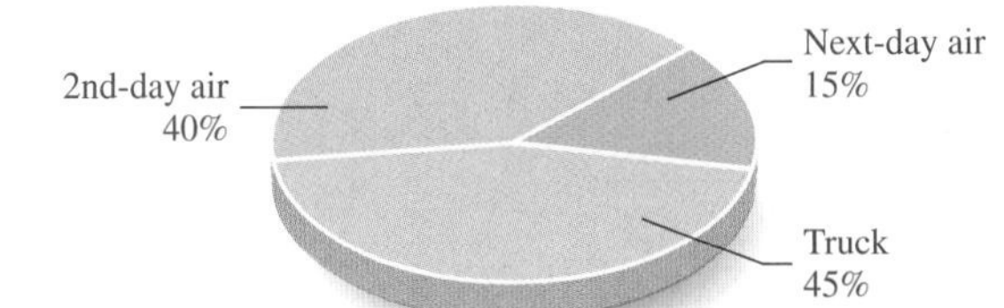

21. What percentage was shipped by truck?

22. What percentage was shipped by truck or second-day air?

23. If the company shipped 1,200 items in one month, how many did they ship by next-day air?

24. If the company shipped 1,200 items in one month, how many did they ship by truck?

25. Use your answer to exercise 24 to determine the total cost of shipping by truck that month if the average cost of a ground shipment is $52.50.

cumulative review chapters 1–9

Use this exercise set to review concepts from earlier chapters. While it is not a comprehensive exam, it will help you identify any material that you need to review before moving on to the next chapter. In addition to the answers, you will find section references for these exercises in the Answers Appendix in the back of the text.

1.1

1. What is the place value of 6 in the numeral 126,489?

Perform the indicated operation.

1.2

2. $\begin{array}{r} 5{,}306 \\ 389 \\ +\ 26{,}583 \\ \hline \end{array}$

1.3

3. $\begin{array}{r} 74{,}983 \\ -\ 35{,}695 \\ \hline \end{array}$

1.5

4. 86×305

1.6

5. $27\overline{)8{,}322}$

1.3

6. $86{,}135 - 37{,}547$

4.3

7. 2.45×30.7

2.5

8. $\frac{4}{7} \times \frac{28}{24}$

2.7

9. $\frac{11}{15} \div \frac{121}{90}$

3.4

10. $3\frac{2}{3} + 5\frac{5}{6} - 2\frac{5}{12}$

5.4

Solve for the unknown.

11. $\frac{4}{7} = \frac{8}{x}$

12. $\frac{3}{5} = \frac{x}{15}$

6.1

13. Write 18% as a decimal and fraction.

6.2

14. Write $\frac{17}{40}$ as a decimal and percent.

7.1

Perform the indicated operations.

15. $\begin{array}{r} 7\text{ lb}\ \ 9\text{ oz} \\ +\ 3\text{ lb}\ 12\text{ oz} \\ \hline \end{array}$

16. $\begin{array}{r} 4\text{ min}\ 10\text{ s} \\ -\ 2\text{ min}\ 35\text{ s} \\ \hline \end{array}$

Complete each statement.

7.2

17. 8 km = ________ m

18. 3,000 mg = ________ g

19. 500 cm = ________ m

20. 25 cL = ________ mL

9.4

21. **Statistics** According to the line graph, between what two years was the increase in benefits the greatest?

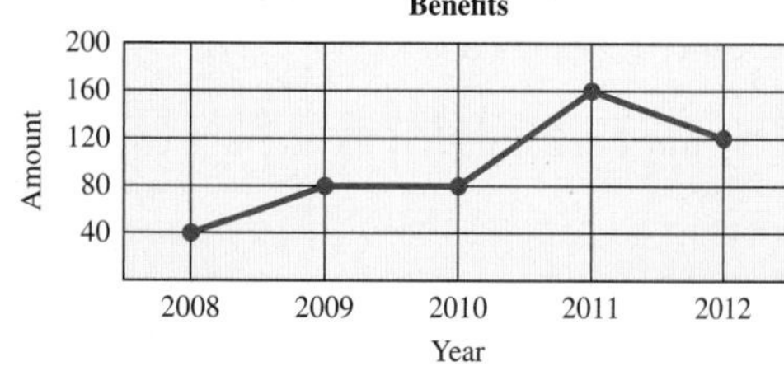

9.3

22. **BUSINESS AND FINANCE** Construct a bar graph to represent the data.

Type of Stock	Number of Stocks
Industrial capital goods	110
Industrial consumer's goods	184
Public utilities	60
Railroads	15
Banks	25
Property liability insurance	16

9.1

23. Calculate the mean, median, and mode for the data.

11, 9, 3, 6, 7, 9, 8, 11, 12, 13, 11, 11, 4, 8, 12

5.2

24. If a boat uses 14 gal of gas to go 102 mi, how many gallons would be needed to go 510 mi?

Express each as a simplified rate.

25. $\frac{1{,}760 \text{ ft}}{20 \text{ s}}$

26. $\frac{133 \text{ pitches}}{7 \text{ innings}}$

8.2

27. **GEOMETRY** The floor of a room that is 12 ft by 18 ft is to be carpeted. If the price of the carpet is \$17 per square yard, what will the carpet cost?

2.6

28. If you drive 152 miles in $3\frac{1}{6}$ hours, what is your average speed?

8.2

29. **GEOMETRY** A rectangle has length $8\frac{3}{5}$ cm and width $5\frac{7}{10}$ cm. Find its perimeter.

30. **GEOMETRY** The sides of a square each measure $13\frac{5}{6}$ ft. Find the perimeter of the square.

6.3

31. What is $9\frac{1}{2}\%$ of 1,400?

32. 15 is what percent of 7,500?

33. 111 is 60% of what number?

2.5

34. Find $\frac{2}{3}$ of $6\frac{1}{2}$.

6.4

35. **SOCIAL SCIENCE** The number of students attending a small college increased 6% since last year. This year there are 2,968 students. How many students attended last year?

The Real Number System

CHAPTER 10 OUTLINE

INTRODUCTION

In this chapter, we expand our numbers to include those that are less than zero. We call them negative numbers and place them to the left of zero on a number line.

Of course, you are already familiar with negative numbers. We often need negative numbers to describe very cold temperatures. We use negative numbers when doing accounting or working with money, when describing performance in many sports, and even when describing the behavior of elements and molecules in chemistry.

While we develop the skills to perform computations with negative numbers, we will also learn to model applications. This leads us into an introduction to algebra in Chapter 11.

In Activity 28, you will have the opportunity to explore and work with negative and positive numbers as you gather weather data where you live.

10 prerequisite check

This Prerequisite Check highlights the skills you will need in order to be successful in this chapter. The answers to these exercises are in the Answers Appendix at the back of the text.

Evaluate each expression.

1. $12 + 9$

2. $147 - 68$

3. 23×16

4. $\frac{122}{8}$

5. $3 \times 8^2 - 5$

6. $56 - 3 \times 2^3$

Name each property.

7. $12 + 5 = 5 + 12$

8. $3(3 + 1) = 3 \times 3 + 3 \times 1$

9. $(9 \times 7) \times 2 = 9 \times (7 \times 2)$

10. $\frac{14 + 23}{3} = \frac{14}{3} + \frac{23}{3}$

Use the circle shown to complete exercises 11–12. Use 3.14 for π and round your answers to two decimal places.

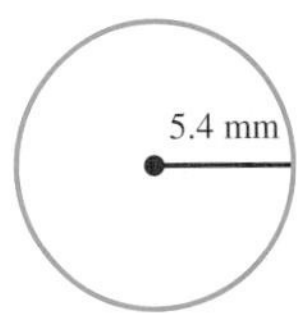

11. Find the circumference of the circle.

12. Find the area of the circle.

10.1

Real Numbers and Order

< 10.1 Objectives >

1 > Represent an integer on a number line

2 > Order a set of real numbers

3 > Identify extreme values

4 > Evaluate absolute value expressions

The numbers used to count things—1, 2, 3, 4, 5, and so on—are called the **natural** (or **counting**) **numbers.** The **whole numbers** consist of the natural numbers and zero—0, 1, 2, 3, 4, 5, and so on. They can be represented on a number line like the one shown. Zero (0) is considered the origin.

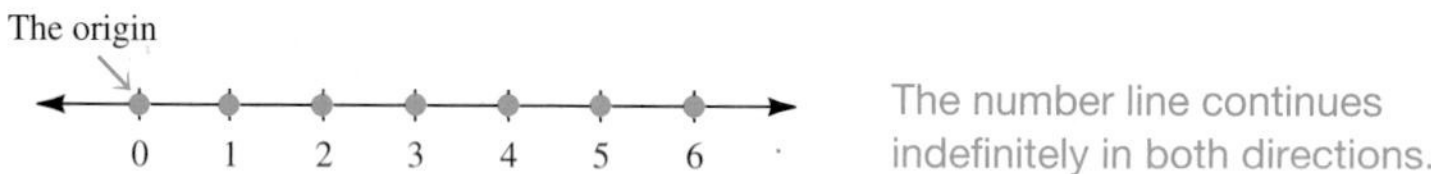

The number line continues indefinitely in both directions.

When numbers are used to represent physical quantities (such as altitude, temperature, and amount of money), it is often necessary to distinguish between *positive* and *negative* quantities. It is convenient to represent these quantities with plus (+) or minus (−) signs. For instance,

The Empire State building is 1,250 feet tall (+1,250).

The altitude at Badwater in Death Valley is 282 ft *below* sea level (−282).

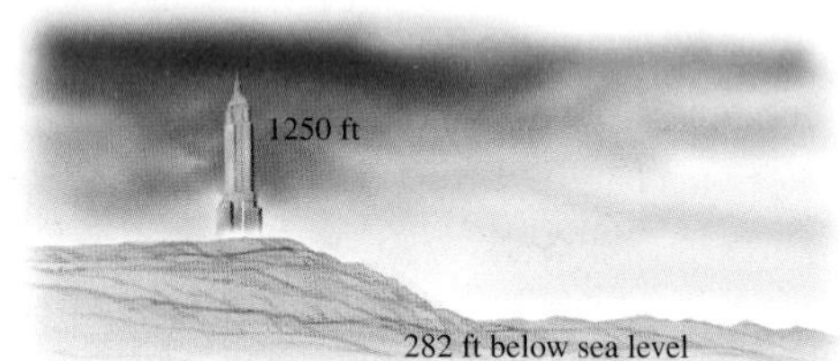

The temperature in Chicago might be 10° *below* zero (−10°).

An account could show a *gain* of \$100 (+100), or a *loss* of \$100 (−100).

These numbers suggest the need to extend the whole numbers to include both positive numbers (such as +100) and negative numbers (such as −282).

To represent the negative numbers, we extend the number line to the *left* of zero and name equally spaced points.

Numbers used to name points to the right of zero are positive numbers. They are written with a positive (+) sign or with no sign at all.

+6 and 9 are positive numbers

Numbers used to name points to the left of zero are negative numbers. They are always written with a negative (−) sign.

−3 and −20 are negative numbers

Read "negative 3."

Positive and negative numbers considered together (along with zero) are **real numbers.**

Zero is neither positive nor negative.

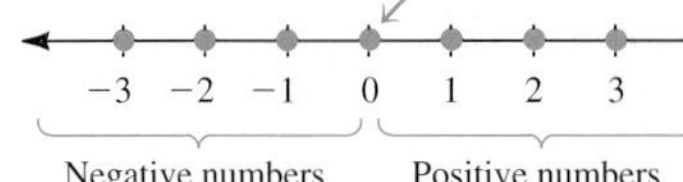

Here is a number line extended to include both positive and negative numbers.

The numbers used to name the points shown on the number line are called the **integers.** The integers consist of the natural numbers, their negatives, and the number 0. We can write

$\ldots, -3, -2, -1, 0, 1, 2, 3, \ldots$

A set of three dots is called an *ellipsis* and indicates that a pattern continues.

Example 1 Representing Integers on the Number Line

< Objective 1 >

Represent the integers on the number line shown.

$-3, -12, 8, 15, -7$

−12 −7 −3 8 15

−15 −10 −5 0 5 10 15

Check Yourself 1

Represent the integers on the given number line.

$-1, -9, 4, -11, 7, 20$

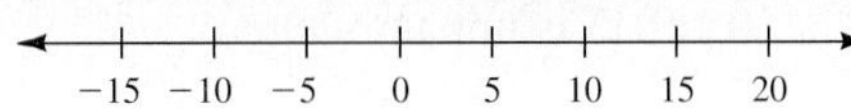

The set of numbers on the number line is *ordered.* The numbers get smaller as you move to the left on the number line and larger as you move to the right. When a set of numbers is written from smallest to largest, the numbers are said to be in *ascending order.*

Example 2 Ordering Real Numbers

< Objective 2 >

Place each set of numbers in ascending order.

(a) $9, -5, -8, 3, 7$

From smallest to largest, the numbers are

$-8, -5, 3, 7, 9$ Note that this is the order in which the numbers appear on a number line.

(b) $3, -2, 18, -20, -13$

From smallest to largest, the numbers are

$-20, -13, -2, 3, 18$

Check Yourself 2

Place each set of numbers in ascending order.

(a) $12, -13, 15, 2, -8, -3$ **(b)** $3, 6, -9, -3, 8$

RECALL

In Section 9.2, we called these values the Min and Max.

The least and greatest numbers in a set are called the **extreme values.** The least element is called the **minimum,** and the greatest element is called the **maximum.**

Example 3 Identifying Extreme Values

< Objective 3 >

Determine the minimum and maximum values of each set of numbers.

(a) 9, −5, −8, 3, 7

From our previous ordering of these numbers, we see that −8, the least element, is the minimum, and 9, the greatest element, is the maximum.

(b) 3, −2, 18, −20, −13

−20 is the minimum and 18 is the maximum.

Check Yourself 3

Determine the minimum and maximum values of each set of numbers.

(a) 12, −13, 15, 2, −8, −3 **(b)** 3, 6, −9, −3, 8

Integers are not the only kind of real numbers. Decimals and fractions are also real numbers.

Example 4 Identifying Integers

Which real numbers are integers?

(a) 145 is an integer.

(b) −28 is an integer.

(c) 0.35 is not an integer.

(d) $-\frac{2}{3}$ is not an integer.

Check Yourself 4

Which real numbers are integers?

−23 1,054 −0.23 0 −500 $-\frac{4}{5}$

RECALL

A common fraction can be written as the ratio of two integers $\frac{a}{b}$ in which $b \neq 0$.

We used 3.14 to approximate π in many calculations in Section 8.3.

The set of real numbers corresponds to the points on a number line. That is, *every* point on a number line is associated with a real number and *every* real number can be represented by a point on a number line.

We divide the real numbers into two categories. You have worked primarily with one of the categories, the **rational numbers.** Every number that can be written as a common fraction is a rational number. This includes all repeating or finite decimals, all integers, and all common fractions.

Numbers that are not rational are called **irrational numbers.** You have already encountered some of the irrational numbers. For example, we worked with π (pi) in Chapter 8. Pi is given by the ratio of the circumference of a circle and its diameter. There is no nice fraction or decimal we can use to represent π. We can approximate π in calculations, but we do not have an exact decimal representation.

NOTE

1 in.

$\sqrt{2}$ in.

1 in.

There are many other irrational numbers, some of which are useful enough to "name," just as we named π. For instance, the square root of any whole number that is not a perfect square names an irrational number. We can use the Pythagorean theorem to find these numbers in nature.

There are even irrational numbers that we have not named. One example requires you to recognize the pattern that emerges in the following decimal.

0.101101110111110 . . .

You will learn much more about this rich set of numbers when you enroll in future math courses.

RECALL

We call zero (0) the origin.

An important idea for our work in this chapter is the **absolute value** of a number. This represents the distance of the point named by the number from the origin on a number line.

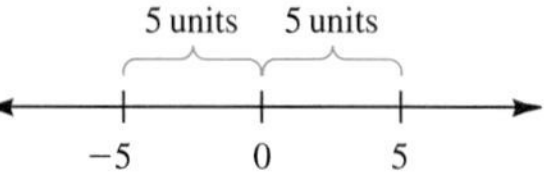

The absolute value of 5 is 5. The absolute value of −5 is also 5.

In symbols we write

$|5| = 5$ and $|-5| = 5$

Read "the absolute value of 5." Read "the absolute value of negative 5."

NOTE

Distance and length are always given as positive numbers.

The absolute value of a number does *not* depend on whether the number is to the right or to the left of the origin, only its *distance* from the origin.

Example 5 Evaluating Absolute Value Expressions

< Objective 4 >

NOTE

$-|-7|$ is the opposite of $|-7|$. In (b), we see that $|-7| = 7$, so its opposite $-|-7| = -7$.

This is different from $-(-7)$, which is the opposite of -7, which is 7.

(a) $|7| = 7$

(b) $|-7| = 7$

(c) $-|-7| = -7$

(d) $|-10| + |10| = 10 + 10 = 20$

(e) $|8 - 3| = |5| = 5$ Absolute value bars serve as another set of grouping symbols, so we do the operation inside first.

(f) $|8| - |3| = 8 - 3 = 5$

Here, evaluate the absolute values and then subtract.

Check Yourself 5

Evaluate.

(a) $|8|$ **(b)** $|-8|$ **(c)** $-|-8|$

(d) $|-9| + |4|$ **(e)** $|9 - 4|$ **(f)** $|9| - |4|$

The language of real numbers can be applied to many situations. Example 6 gives you an idea of the variety of problems that are best modeled with real numbers.

Example 6 Real Numbers in the Real World

Represent each quantity with an integer.

RECALL

You should include units in your answer to an application.

(a) A sick infant's temperature drops by 4 degrees Fahrenheit (°F) after being given acetaminophen.

After taking the medication, the infant's temperature went down, so we need a negative number.

−4°F

(b) A project is $1,500 over budget.

The cost of the project went up, so we use a positive number.

+$1,500 (or simply $1,500)

Check Yourself 6

Represent each quantity with an integer.

(a) After an hour, it is noticed that an intravenous solution (IV) is running 30 mL ahead of schedule.

(b) The bridge is on schedule to be completed 7 months early.

Check Yourself ANSWERS

1. 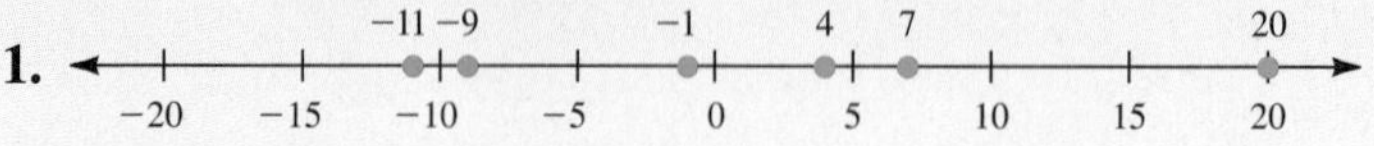

2. **(a)** $-13, -8, -3, 2, 12, 15$; **(b)** $-9, -3, 3, 6, 8$
3. **(a)** Minimum is -13; maximum is 15; **(b)** minimum is -9; maximum is 8
4. -23, 1,054, 0, and -500 5. **(a)** 8; **(b)** 8; **(c)** -8; **(d)** 13; **(e)** 5; **(f)** 5
6. **(a)** $+30$ mL; **(b)** -7 months

Reading Your Text

These fill-in-the-blank exercises will help you understand some of the key vocabulary used in this section. The answers to these exercises are in the Answers Appendix at the back of the text.

(a) The whole numbers consist of the natural numbers and __________.

(b) __________ numbers are used to describe below-zero temperatures.

(c) When a set of numbers is written from smallest to largest, the numbers are said to be in __________ order.

(d) The __________ of a number is given by its distance from the origin on the number line.

10.1 exercises

Skills | Calculator/Computer | Career Applications | Above and Beyond

< Objective 1 >

Represent each set of numbers on the given number line.

1. $5, -15, 18, -8, 3$

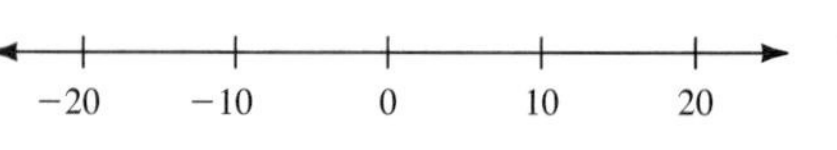

2. $-18, 4, -5, 13, 9$

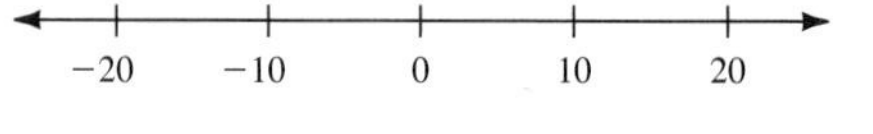

3. $-\frac{3}{4}, 2\frac{1}{2}, -\frac{1}{2}, -3\frac{1}{4}, \frac{2}{3}$

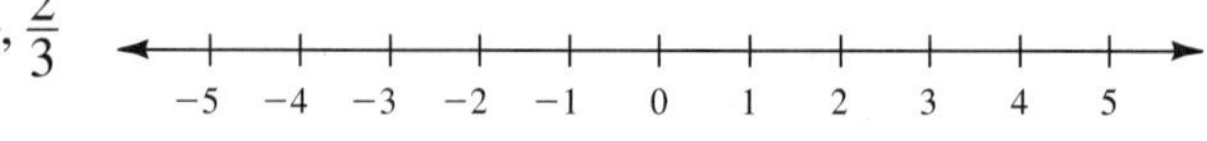

4. $3.2, -1.4, 0.7, -0.9, -2.1$

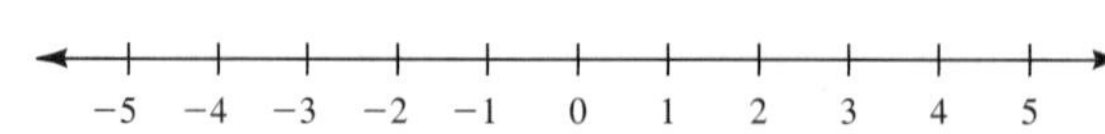

Which numbers are integers?

5. $5, -\frac{2}{9}, 175, -234, -0.64$

6. $-45, 0.35, \frac{3}{5}, 700, -26$

< Objective 2 >

Place each set in ascending order.

7. $3, -5, 2, 0, -7, -1, 8$

8. $-2, 7, 1, -8, 6, -1, 0$

9. $9, -2, -11, 4, -6, 1, 5$

10. $23, -18, -5, -11, -15, 14, 20$

11. $-\frac{1}{2}, \frac{3}{4}, -\frac{5}{6}, \frac{2}{3}, -\frac{1}{3}$

12. $\frac{3}{7}, -\frac{6}{7}, \frac{1}{7}, -\frac{1}{2}, \frac{2}{7}$

13. $-6.1, -5.9, 6.1, 5.9, -6.0$

14. $3.5, -5.3, -3.5, 5.3, 4$

< Objective 3 >

Determine the minimum and maximum values of each set.

15. $5, -6, 0, 10, -3, 15, 1, 8$

16. $9, -1, 3, 11, -4, 2, 5, -2$

17. $21, -15, 0, 7, -9, 16, -3, 11$

18. $-22, 0, 22, -31, 18, -5, 3$

19. $3, 0, \frac{1}{2}, -\frac{2}{3}, 5, \frac{3}{4}, -\frac{1}{6}$

20. $-3, 2, \frac{7}{12}, -\frac{3}{4}, \frac{5}{6}, -\frac{10}{3}, \frac{5}{2}$

21. $-3.3, 4\frac{1}{2}, -3, -2.8, 4.3, 4.8$

22. $-11, 4\frac{1}{2}, \frac{15}{4}, -10.9, -11.1, 0$

< Objective 4 >

Evaluate.

23. $|17|$

24. $|28|$

25. $|-10|$

26. $|-7|$

27. $-|3|$

28. $-|5|$

29. $-|-8|$

30. $-|-13|$

31. $|-2| + |3|$

32. $|4| + |-3|$

33. $|-9| + |9|$

34. $|11| + |-11|$

35. $|4| - |-4|$

36. $|5| - |-5|$

37. $|15| - |8|$

38. $|11| - |3|$

39. $|15 - 8|$

40. $|11 - 3|$

41. $|-9| + |2|$

42. $|-7| + |4|$

43. $|-8| - |-7|$

44. $|-9| - |-4|$

Fill in each blank with $>$, $<$, or $=$ to make a true statement.

45. -9 ___ -6

46. $|-9|$ ___ $|-6|$

47. $|-9|$ ___ 6

48. 9 ___ $|-6|$

Place absolute value bars in the proper location on the left side of each equation in order to make it true.

49. $6 + (-2) = 4$

50. $8 + (-3) = 5$

51. $6 + (-2) = 8$ 

52. $8 + (-3) = 11$

Represent each quantity with a real number.

53. An altitude of 400 ft above sea level

54. An altitude of 80 ft below sea level

55. A loss of $200

56. A profit of $400

57. A decrease in population of 25,000

58. An increase in population of 12,500

59. **Business and Finance** The withdrawal of $50 from a checking account. 

60. **Business and Finance** The deposit of $200 into a savings account.

61. **Science and Medicine** A temperature decrease of 10°F in 1 hr. 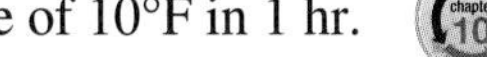

62. **Social Science** An increase of 25,000 in a city's population.

63. **Business and Finance** An increase of 75 points in the Dow-Jones average.

64. **Statistics** An eight-game losing streak by the local baseball team.

65. **Business and Finance** A country exported $90,000,000 more than it imported, creating a positive trade balance.

66. **Business and Finance** A stock lost 8.5% of its value.

Determine whether each statement is **true** *or* **false.**

67. All whole numbers are integers.

68. All integers are real numbers.

69. All integers are whole numbers.

70. All real numbers are integers.

Skills | Calculator/Computer | **Career Applications** | Above and Beyond

Represent each quantity with a real number.

71. **Agricultural Technology** The erosion of 5 cm of topsoil from an Iowa cornfield. 

72. **Agricultural Technology** The formation of 2.5 cm of new topsoil on the African savanna.

73. **Construction Technology** The elevations, in inches, of several points on a jobsite are

$-18, 27, -84, 37, 59, -13, 4, 92, 49, 66, -45$

Arrange the elevations in ascending order.

74. **Electronics** Several 12-volt (V) batteries were tested using a voltmeter. The voltage values were entered into a table indicating their value in reference to 12 V. Determine the maximum and minimum voltage measurements taken.

Battery	Variance from 12 V (in V)
Cell 1	+1
Cell 2	0
Cell 3	−1
Cell 4	−3
Cell 5	+2

75. **Electrical Engineering** Several resistors were tested using an ohmmeter. The resistance values were entered into a table indicating their value in reference to 10,000 ohms (10 kΩ). List the resistors in ascending order according to their measured resistance.

Battery	Variance from 10,000 Ω (in Ω)
Resistor 1	+175
Resistor 2	−60
Resistor 3	−188
Resistor 4	+10
Resistor 5	+218
Resistor 6	−65
Resistor 7	−302

76. **Electrical Engineering** Which of the resistors in exercise 75 had a measured value furthest from 10,000 Ω?

77. **(a)** Plot −3 and 4 on the given number line.

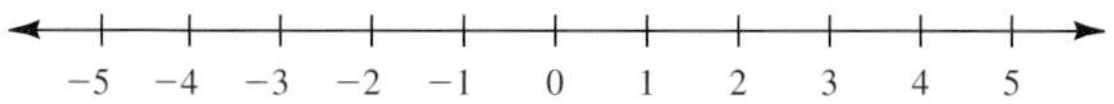

(b) How far apart are −3 and 4 on the number line?

78. **(a)** Plot $-\frac{1}{2}$ and $4\frac{1}{4}$ on the given number line.

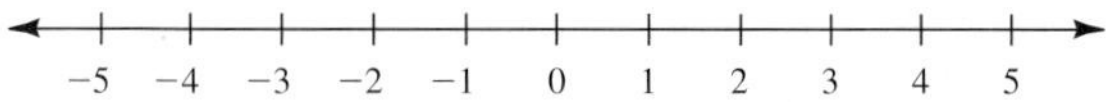

(b) How far apart are $-\frac{1}{2}$ and $4\frac{1}{4}$ on the number line?

79. **(a)** Plot $-3\frac{1}{2}$ on the given number line.

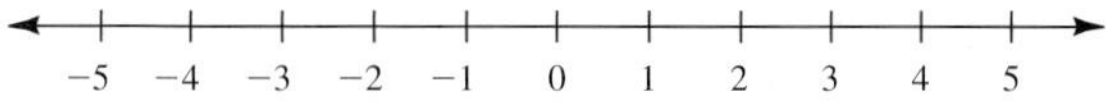

(b) Is $-3\frac{1}{2}$ less than or equal to −3?

(c) How far apart are $-3\frac{1}{2}$ and −3 on the number line?

80. You learned that mixed numbers like $3\frac{1}{2}$ mean $3 + \frac{1}{2}$ and $3\frac{1}{2}$ is one half unit to the right of 3 on a number line.

(a) How might you describe a negative mixed number such as $-3\frac{1}{2}$?

(b) How far apart is a negative mixed number from its integer part?

(c) Is a negative mixed number to the left or right of its integer part on a number line?

Answers

1.

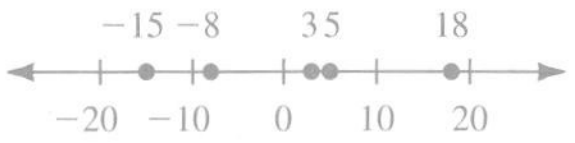

3.

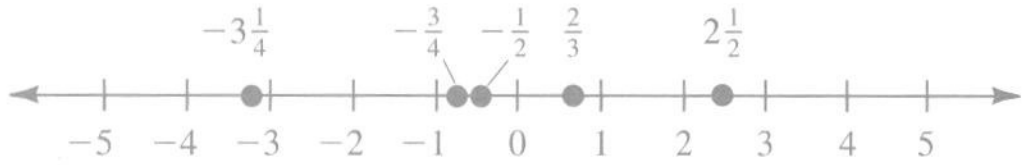

5. 5, 175, −234 **7.** −7, −5, −1, 0, 2, 3, 8 **9.** −11, −6, −2, 1, 4, 5, 9 **11.** $-\frac{5}{6}, -\frac{1}{2}, -\frac{1}{3}, \frac{2}{3}, \frac{3}{4}$ **13.** −6.1, −6.0, −5.9, 5.9, 6.1 **15.** Min: −6; max: 15 **17.** Min: −15; max: 21 **19.** Min: $-\frac{2}{3}$; max: 5 **21.** Min: −3.3; max: 4.8 **23.** 17 **25.** 10 **27.** −3 **29.** −8 **31.** 5 **33.** 18 **35.** 0 **37.** 7 **39.** 7 **41.** 11 **43.** 1 **45.** < **47.** > **49.** $6 + (-2) = 4$ or $|6 + (-2)| = 4$ **51.** $6 + |(-2)| = 8$ or $|6| + |(-2)| = 8$ **53.** 400 ft or +400 ft **55.** −\$200 **57.** −25,000 people **59.** −\$50 **61.** −10°F **63.** +75 points **65.** +\$90,000,000 **67.** True **69.** False **71.** −5 cm **73.** −84 in., −45 in., −18 in., −13 in., 4 in., 27 in., 37 in., 49 in., 59 in., 66 in., 92 in. **75.** 9,698 Ω, 9,812 Ω, 9,935 Ω, 9,940 Ω, 10,010 Ω, 10,175 Ω, 10,218 Ω

77. **(a)**

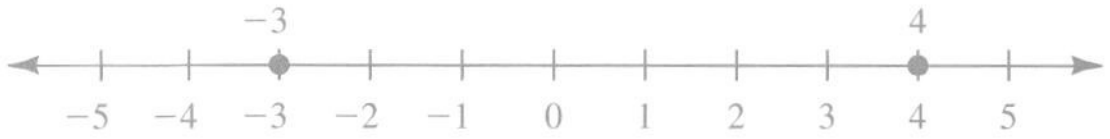

(b) 7 units

79. **(a)**

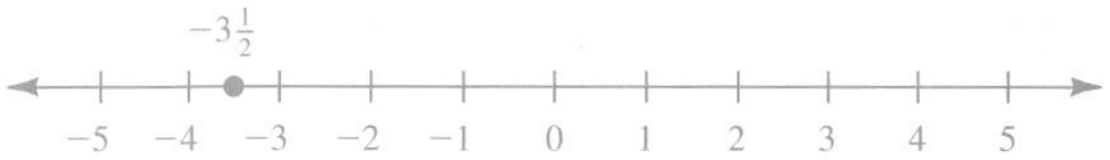

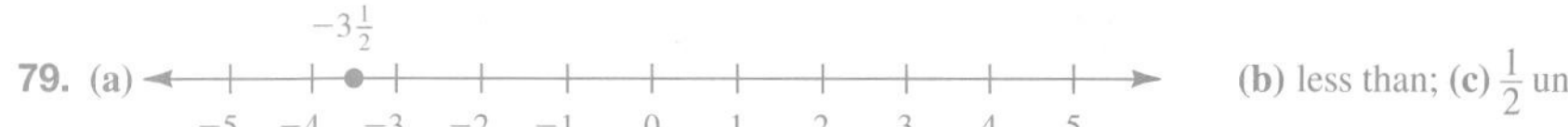

(b) less than; **(c)** $\frac{1}{2}$ unit

10.2 Adding Real Numbers

< 10.2 Objectives >

1 > Add two numbers with the same sign

2 > Add two numbers with opposite signs

In Section 10.1, we introduced negative numbers. Now we examine the four arithmetic operations (addition, subtraction, multiplication, and division) and see how those operations are performed when real numbers are involved. We start by considering addition.

An application may help. As before, we represent a gain of money as a positive number and a loss as a negative number.

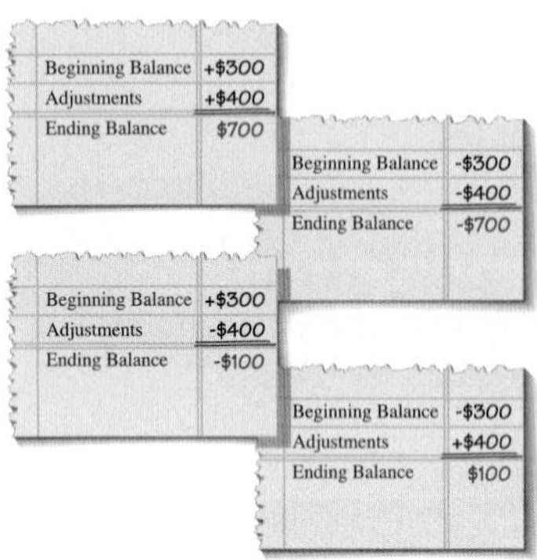

If you gain \$300 and then gain \$400, the result is a gain of \$700.

$300 + 400 = 700$

If you lose \$300 and then lose \$400, the result is a loss of \$700.

$-300 + (-400) = -700$

If you gain \$300 and then lose \$400, the result is a loss of \$100.

$300 + (-400) = -100$

If you lose \$300 and then gain \$400, the result is a gain of \$100.

$-300 + 400 = 100$

We can use a number line to illustrate adding real numbers. Starting at the origin, we move to the *right* for positive numbers and to the *left* for negative numbers.

Example 1 Adding Real Numbers on a Number Line

< Objective 1 >

(a) Add $3 + 2$.

Start at the origin and move 3 units to the right. Then, move 2 units more to the right to find the sum.

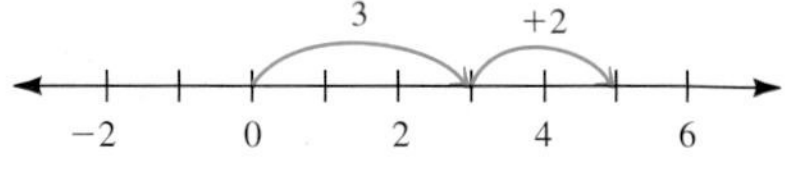

$3 + 2 = 5$

(b) Add $\frac{4}{3} + \frac{2}{3}$.

Start at the origin and move $\frac{4}{3}$ units to the right. Then move $\frac{2}{3}$ more to the right to find the sum. So we have

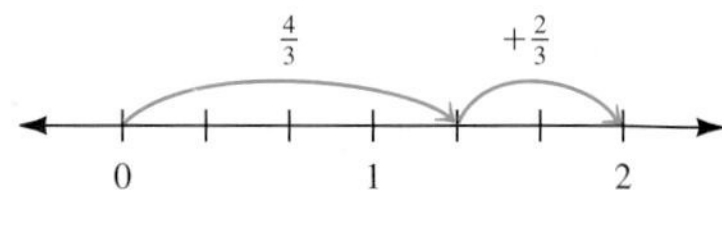

$$\frac{4}{3} + \frac{2}{3} = \frac{6}{3} = 2$$

Check Yourself 1

Add.

(a) $5 + 6$ **(b)** $\frac{5}{4} + \frac{7}{4}$

A number line also helps you visualize the sum of two negative numbers. Remember to move left for negative numbers.

Example 2 Adding Numbers with the Same Sign

(a) Add $-3 + (-4)$.

Start at the origin and move 3 units to the left. Then move 4 more units to the left to find the sum. From the graph we see that the sum is

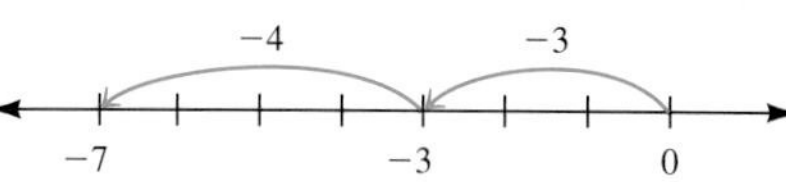

$$-3 + (-4) = -7$$

(b) Add $-\frac{3}{2} + \left(-\frac{1}{2}\right)$.

As before, we start at the origin. From that point move $\frac{3}{2}$ units left. Then move another $\frac{1}{2}$ unit left to find the sum. In this case

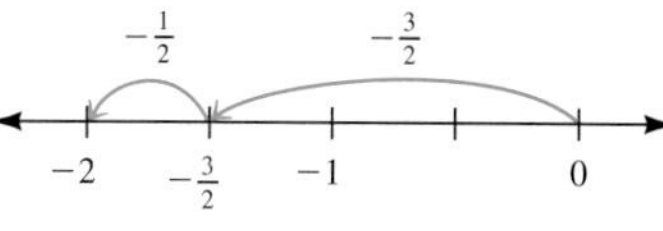

$$-\frac{3}{2} + \left(-\frac{1}{2}\right) = -2$$

Check Yourself 2

Add.

(a) $-4 + (-5)$ **(b)** $-3 + (-7)$

(c) $-5 + (-15)$ **(d)** $-\frac{5}{2} + \left(-\frac{3}{2}\right)$

You should notice some helpful patterns in Examples 1 and 2. These patterns allow you to do the work mentally without having to use a number line.

Property

Adding Numbers with the Same Sign

If two numbers have the same sign, add their absolute values. Give the sum the sign of the original numbers.

Example 3 Adding Real Numbers

NOTE

The sum of two positive numbers is positive; the sum of two negative numbers is negative.

(a) $-8 + (-5) = -13$ Add the absolute values ($8 + 5 = 13$) and give the sum the sign ($-$) of the original numbers.

(b) $[-3 + (-4)] + (-6)$ Add inside the brackets as your first step.

$= -7 + (-6) = -13$

Check Yourself 3

Add mentally.

(a) $7 + 9$ **(b)** $-7 + (-9)$

(c) $-5.8 + (-3.2)$ **(d)** $[-5 + (-2)] + (-3)$

We can also use a number line to illustrate adding numbers with *different* signs.

Example 4 Adding Numbers with Opposite Signs

< Objective 2 >

(a) Add $3 + (-6)$.

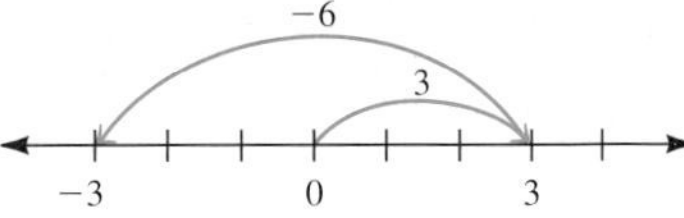

First move 3 units to the right of the origin. Then move 6 units to the left.

$3 + (-6) = -3$

(b) Add $-4 + 7$.

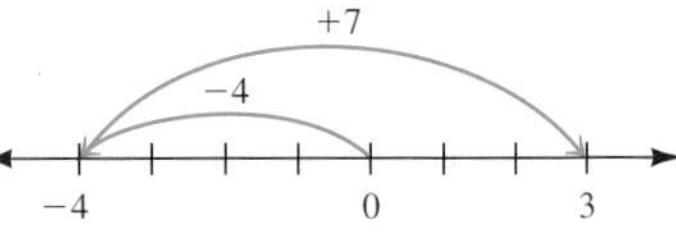

This time move 4 units to the left of the origin as the first step. Then move 7 units to the right.

$-4 + 7 = 3$

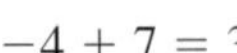

Check Yourself 4

Add.

(a) $7 + (-5)$ **(b)** $4 + (-8)$
(c) $-4 + 9$ **(d)** $-7 + 3$

You have no doubt noticed that, in adding a positive number and a negative number, sometimes the sum is positive and sometimes it is negative. The result depends on which of the numbers has the larger absolute value. This leads us to the second part of our addition rule.

Property

Adding Numbers with Different Signs

If two numbers have different signs, subtract their absolute values, the smaller from the larger. Give the result the sign of the number with the larger absolute value.

Example 5 Adding Real Numbers

Add.

(a) $7 + (-19) = -12$

Because the two numbers have different signs, subtract their absolute values ($19 - 7 = 12$). The sum has the sign ($-$) of the number with the larger absolute value, -19.

(b) $-13 + 7 = -6$

Subtract the absolute values ($13 - 7 = 6$). The sum has the sign ($-$) of the number with the larger absolute value, -13.

(c) $-4.5 + 8.2 = 3.7$

Subtract the absolute values ($8.2 - 4.5 = 3.7$). The sum has the sign ($+$) of the number with the larger absolute value, 8.2.

RECALL

Real numbers can be fractions and decimals as well as integers.

Check Yourself 5

Add mentally.

(a) $5 + (-14)$ **(b)** $-7 + (-8)$ **(c)** $-8 + 15$

(d) $7 + (-8)$ **(e)** $-\frac{2}{3} + \left(-\frac{7}{3}\right)$ **(f)** $5.3 + (-2.3)$

There are two properties of addition that we should mention before concluding this section. First, the sum of any number and 0 is always that number.

Property

Additive Identity Property

For any number *a*,

$a + 0 = 0 + a = a$

Example 6 **Adding Zero**

Add.

(a) $9 + 0 = 9$

(b) $0 + (-8) = -8$

(c) $-25 + 0 = -25$

NOTE

Zero is called the additive identity.

Check Yourself 6

Add.

(a) $8 + 0$ **(b)** $0 + (-7)$ **(c)** $-36 + 0$

NOTES

The opposite of a number is also called its additive inverse.

3 and −3 are opposites.

We need one further definition to state our second property. Every number has an *opposite.* It corresponds to a point that is the same distance from the origin as the given number, but in the opposite direction.

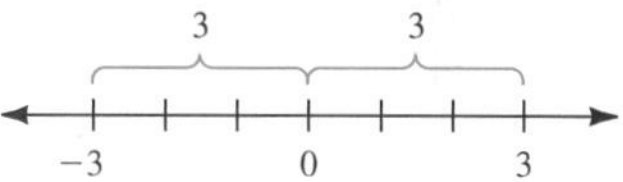

The opposite of 9 is −9.

The opposite of −15 is 15.

Our next property states that the sum of any number and its opposite is 0.

Property

Additive Inverse Property

For any number a, there exists a number $-a$ such that

$a + (-a) = -a + a = 0$

The sum of any number and its opposite, or additive inverse, is 0.

Example 7 **Adding Inverses**

Add.

(a) $9 + (-9) = 0$

(b) $-15 + 15 = 0$

(c) $-2.3 + 2.3 = 0$

(d) $\frac{4}{5} + \left(-\frac{4}{5}\right) = 0$

NOTE

All properties of addition from Section 1.2 apply when negative numbers are involved.

Check Yourself 7

Add.

(a) $-17 + 17$ **(b)** $12 + (-12)$

(c) $\frac{1}{3} + \left(-\frac{1}{3}\right)$ **(d)** $-1.6 + 1.6$

We can now use the associative and commutative properties of addition, first introduced in Section 1.2, to find a sum when more than two numbers are involved. Example 8 illustrates these properties.

Example 8 Adding Real Numbers

NOTE

We use the commutative property to reverse the order of addition for −3 and 5. We then group −5 and 5. Do you see why?

$$\begin{aligned} -5 + (-3) + 5 &= -5 + 5 + (-3) \\ &= [-5 + 5] + (-3) \\ &= 0 + (-3) \\ &= -3 \end{aligned}$$

Check Yourself 8

Add.

(a) $-4 + 5 + (-3)$ **(b)** $-8 + 4 + 8$

Real numbers appear in many situations.

Example 9 An Application of Real Numbers

A vendor earned profits of −$86.75, $111.50, and $123 one weekend (Friday through Sunday). What was the vendor's total weekend profit?

We add the vendor's daily profits to get the weekend profit.

$-86.75 + 111.50 + 123 = 147.75$

So the vendor earned a weekend profit of $147.75.

Check Yourself 9

A softball team scored 3 runs in one inning, gave up 4 runs in another inning, scored 2 runs after that, and gave up 3 runs in the final inning. How far was the team ahead at the end of the game?

Check Yourself ANSWERS

1. **(a)** 11; **(b)** 3 **2.** **(a)** −9; **(b)** −10; **(c)** −20; **(d)** −4 **3.** **(a)** 16; **(b)** −16; **(c)** −9; **(d)** −10 **4.** **(a)** 2; **(b)** −4; **(c)** 5; **(d)** −4 **5.** **(a)** −9; **(b)** −15; **(c)** 7; **(d)** −1; **(e)** −3; **(f)** 3 **6.** **(a)** 8; **(b)** −7; **(c)** −36 **7.** **(a)** 0; **(b)** 0; **(c)** 0; **(d)** 0 **8.** **(a)** −2; **(b)** 4 **9.** −2 runs (they lost by 2)

Reading Your Text

These fill-in-the-blank exercises will help you understand some of the key vocabulary used in this section. The answers to these exercises are in the Answers Appendix at the back of the text.

(a) The sum of two negative numbers is always ________.

(b) Adding a ________ number can be illustrated on a number line by moving to the left.

(c) When adding numbers with different signs, the result has the same sign as the number with the larger ________ value.

(d) The sum of any number and its opposite, or additive inverse, is ________.

Skills | Calculator/Computer | Career Applications | Above and Beyond

10.2 exercises

< Objectives 1 and 2 >

Evaluate each expression.

1. $3 + 6$
2. $5 + 9$
3. $11 + 5$
4. $8 + 7$
5. $-2 + (-3)$
6. $-1 + (-9)$
7. $9 + (-3)$
8. $10 + (-4)$
9. $8 + (-14)$
10. $7 + (-11)$
11. $-4 + 17$
12. $-9 + 12$
13. $-15 + 8$
14. $-23 + 6$
15. $-13 + (-24)$
16. $-87 + (-23)$
17. $-13 + 24$
18. $-87 + 23$
19. $36 + (-45)$
20. $45 + (-36)$
21. $-458 + (-179)$
22. $-912 + (-312)$
23. $432 + (-243)$
24. $861 + (-902)$
25. $-689 + 471$
26. $-333 + 851$
27. $-732 + 1{,}104$
28. $732 + (-1{,}104)$
29. $2,417 + (-7{,}332)$
30. $-4{,}387 + 5{,}008$
31. $-1{,}056 + (-4{,}879)$
32. $-32{,}678 + (-81{,}092)$
33. $\frac{3}{4} + \frac{5}{4}$
34. $\frac{1}{2} + \frac{4}{5}$
35. $-\frac{3}{5} + \left(-\frac{7}{5}\right)$
36. $-\frac{1}{8} + \left(-\frac{3}{8}\right)$
37. $-\frac{2}{3} + \left(-\frac{1}{4}\right)$
38. $-\frac{3}{4} + \left(-\frac{5}{12}\right)$
39. $-2\frac{2}{3} + \left(-1\frac{1}{2}\right)$
40. $-6\frac{1}{2} + \left(-\frac{3}{5}\right)$
41. $\frac{3}{4} + \left(-\frac{1}{4}\right)$
42. $-\frac{7}{12} + \frac{1}{3}$
43. $-\frac{2}{5} + \frac{13}{20}$
44. $\frac{2}{3} + \left(-\frac{5}{6}\right)$
45. $5\frac{1}{3} + \left(-4\frac{4}{5}\right)$
46. $-17\frac{3}{4} + 21\frac{1}{3}$
47. $-1.6 + (-2.3)$
48. $-3.5 + (-2.6)$
49. $-3.6 + 7.6$
50. $13.4 + (-11.4)$
51. $-9 + 0$
52. $0 + (-15)$
53. $14 + (-14)$
54. $-5 + 5$
55. $-9 + (-17) + 9$
56. $15 + (-3) + (-15)$
57. $2 + 5 + (-11) + 4$
58. $7 + (-9) + (-5) + 6$
59. $1 + (-2) + 3 + (-4)$
60. $(-9) + 0 + (-2) + 12$
61. $\frac{5}{3} + \left(-\frac{4}{3}\right) + \frac{5}{3}$
62. $-\frac{6}{5} + \left(-\frac{13}{5}\right) + \frac{4}{5}$
63. $-\frac{3}{2} + \left(-\frac{7}{4}\right) + \frac{3}{4}$
64. $\frac{2}{3} + \left(-\frac{5}{6}\right) + \left(-\frac{1}{2}\right)$
65. $2.8 + (-5.5) + (-2.9)$
66. $-5.4 + (-2.1) + (-3.5)$
67. $|3 + (-4)|$
68. $|-11 + 9|$
69. $|-17 + 8|$
70. $|-27 + 14|$
71. $|-5 + (-6)|$
72. $|-17 + (-14)|$
73. $|-3 + 2 + (-4)|$
74. $|-2 + 7 + (-5)|$
75. $|2 + (-3)| + |-3 + 2|$
76. $|8 + (-10)| + |-12 + 14|$

Evaluate and round each result to the nearest tenth.

77. $-4.1967 + 5.2943 + (-3.1698)$
78. $5.3297 + 4.1897 + (-3.2869)$
79. $-7.19863 + 4.8629 + 3.2689 + (-5.7936)$
80. $3.6829 + 4.5687 + 7.28967 + (-5.1623)$

81. **Statistics** Beach Channel High School's football team scored one field goal (3 points) and gave up a touchdown (7 points) in the first quarter. In the third quarter, the team scored another field goal and gave up a safety (2 points). The team gave up a final field goal in the fourth quarter. By how much did the team lose?

82. **Business and Finance** Jean deposited a check for \$625, wrote two for \$68.74 and \$29.95, and used her debit card to pay for a purchase of \$57.65. What is her new account balance?

83. **Science and Medicine** The temperature dropped by 23°F from a high of 8°F. What was the low temperature?

84. **Science and Medicine** The overnight low temperature was listed as −14°C. The temperature rose 19°C by noon. What was the noontime temperature?

chapter 10 > Make the Connection

Label each statement as **true** *or* **false.**

85. $-10 + 6 = 6 + (-10)$

86. $5 + (-9) = -9 + 5$

87. $|-3| + |2| = |-3 + 2|$

88. $|-8| + |3| = |-8 + 3|$

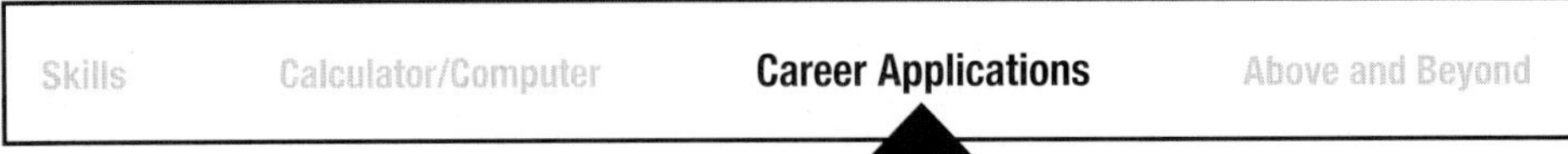

89. **Information Technology** Amir, a network administrator, has a budget of \$50,000 at the beginning of April. He enters his expenditures and receipts for the month: \$1,000 for travel, \$9,550 for technology, \$542 for miscellaneous expenses, \$443 received from returns, \$123 for supplies, and \$150 for subscriptions. How much money does Amir have left in his budget? Make sure to write an integer expression that represents the change in the budget.

90. **Information Technology** Fred has been hired to redesign a database that is having performance issues. He finds that one table in the database called CUSTOMERS has field sizes of FIRST NAME and LAST NAME to be 100 bytes each. He knows from experience that first names average around 30 bytes and last names average around 45 bytes. A character is a byte. Fred knows that wasting space on a very large database can cause performance issues. Write an integer expression that represents the change in the field sizes. By how much does Fred need to modify the field sizes?

91. **Mechanical Engineering** A pneumatic actuator is operated by a pressurized air reservoir. At the beginning of the operator's shift, the pressure in the reservoir was 126 lb/in.^{2}. At the end of each hour, the operator records the change in pressure of the reservoir. The values (in pounds per square inch) recorded for this shift were a drop of 12, a drop of 7, a rise of 32, a drop of 17, a drop of 15, a rise of 31, a drop of 4, and a drop of 14. What is the pressure in the tank at the end of the shift?

92. **Mechanical Engineering** A diesel engine for an industrial shredder has an 18-qt oil capacity. When the maintenance technician checked the oil, it was 7 qt low. Later that day, she added 4 qt to the engine. What was the oil level after the 4 qt were added?

Electrical Engineering Dry cells or batteries have a positive and negative terminal. When correctly connected in series (positive to negative), the voltage of each cell can be added together. If a cell is connected and its terminals are reversed, the current will flow in the opposite direction.

For example, if three 3-V cells are connected in series and one cell is inserted backwards, the resulting voltage is 3 V.

$3\text{ V} + 3\text{ V} + (-3)\text{ V} = 3\text{ V}$

The voltages are added together because the cells are in series, but you must pay attention to the current flow.

Now complete exercises 93 and 94.

93. Assume you have a 24-V cell and a 12-V cell with their negative terminals connected. What would the resulting voltage be if measured from the positive terminals?

+ 24 V − − 12 V +

94. If a 24-V cell, an 18-V cell, and 12-V cell are supposed to be connected in series and the 18-V cell is accidentally reversed, what would the total voltage be? 18 V

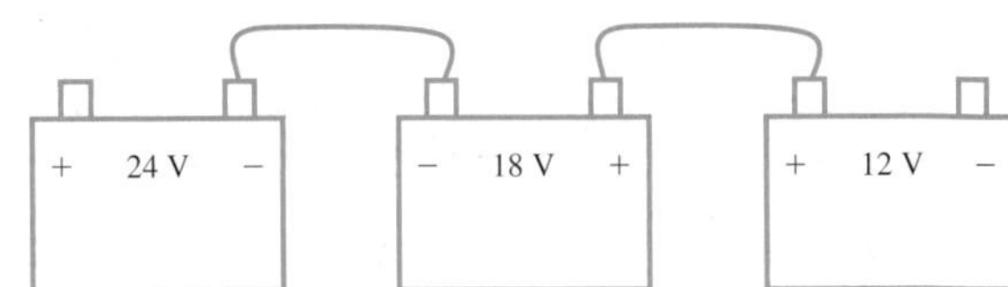

Place absolute value bars in the proper location on the left side of each equation in order to make it true.

95. $-3 + 7 = 10$

96. $-5 + 9 = 14$

97. $-6 + 7 + (-4) = 3$

98. $-10 + 15 + (-9) = 4$

99. **(a)** Evaluate $14 + (-8)$.

(b) Evaluate $14 - 8$.

(c) What do you notice about your answers to parts (a) and (b)?

(d) Is this always true? Try some combinations and make a conjecture.

(e) What do you think happens when we evaluate $14 - (-8)$?

100. **(a)** Evaluate $8 + (-14)$.

(b) Evaluate $8 - 14$.

(c) What do you notice about your answers to parts (a) and (b)?

(d) Is this always true? Try some combinations and make a conjecture.

(e) What do you think happens when we evaluate $8 - (-14)$?

Answers

1. 9 **3.** 16 **5.** -5 **7.** 6 **9.** -6 **11.** 13 **13.** -7 **15.** -37 **17.** 11 **19.** -9 **21.** -637 **23.** 189 **25.** -218 **27.** 372 **29.** $-4{,}915$ **31.** $-5{,}935$ **33.** 2 **35.** -2 **37.** $-\frac{11}{12}$ **39.** $-4\frac{1}{6}$ **41.** $\frac{1}{2}$ **43.** $\frac{1}{4}$ **45.** $\frac{8}{15}$ **47.** -3.9 **49.** 4 **51.** -9 **53.** 0 **55.** -17 **57.** 0 **59.** -2 **61.** 2 **63.** $-\frac{5}{2}$ **65.** -5.6 **67.** 1 **69.** 9 **71.** 11 **73.** 5 **75.** 2 **77.** -2.1 **79.** -4.9 **81.** 6 points **83.** $-15°$F **85.** True **87.** False **89.** \$39,078 **91.** 120 lb/in.^{2} **93.** 12 V **95.** $|-3| + 7 = 10$ **97.** $|-6 + 7 + (-4)| = 3$ **99.** **(a)** 6; **(b)** 6; **(c)** They are the same. **(d)** Above and Beyond; **(e)** Above and Beyond

Activity 28 ::

Hometown Weather

The local weather provides us with many interesting applications of real numbers and data gathering.

1. Collect the daily high and low temperatures in your locale for the previous week.
2. Compute the mean high and mean low temperatures for the week.
3. List each day's high and low temperatures as their distance from the mean for the week. For example, if the weekly mean high temperature was 65°F, and Tuesday's high temperature was 62°F, then list it as −3°F.
4. Consider the differences from the weekly mean high temperature listed in exercise 3. What is the mean of this set of numbers?
5. Consider the differences from the weekly mean low temperature listed in exercise 3. What is the mean of this set of numbers?
6. Explain your answers to exercises 4 and 5.
7. Find the average annual high temperature for your locale.
8. Use real numbers to describe the difference between each of the high and low temperatures for the previous week and the annual averages.
9. Construct a line graph of the high temperatures for the previous week.
10. Add a second line graph to the graph constructed in exercise 9 for the low temperatures.
11. Add horizontal lines to your graph, one for the annual average high temperature and one for the annual average low temperature.
12. Describe the relation between the answers to exercise 8 and the graph.

10.3 Subtracting Real Numbers

< 10.3 Objectives >

1 > Find the difference between two real numbers

2 > Add and subtract mixed numbers

To begin our discussion of subtraction when real numbers are involved, we look back at a problem using natural numbers. Of course, we know that

$$8 - 5 = 3$$

From our work in adding real numbers in Section 10.2, we know that it is also true that

$$8 + (-5) = 3$$

Comparing these equations, we see that they have the same result. This leads us to an important pattern. Any subtraction problem can be written as an addition problem. Subtracting 5 is the same as adding the opposite of 5, or -5. We write this fact as

$$8 - 5 = 8 + (-5) = 3$$

This leads us to a rule for subtracting real numbers.

Step by Step

Subtracting Real Numbers

Step 1 Rewrite the subtraction problem as an addition problem.

a. Change the subtraction symbol ($-$) to an addition symbol ($+$)

b. Replace the number being subtracted with its opposite

Step 2 Add the resulting real numbers as before.

$a - b = a + (-b)$

Example 1 illustrates this process.

Example 1 Subtracting Real Numbers

< Objective 1 >

NOTE

We rewrite each subtraction problem as an addition problem and then add as we did in Section 10.2.

Change the subtraction symbol ($-$) to an addition symbol ($+$).

(a) $15 - 7 = 15 + (-7)$

$= 8$

Replace 7 with its opposite, -7.

(b) $9 - 12 = 9 + (-12) = -3$

(c) $-6 - 7 = -6 + (-7) = -13$

(d) $-\frac{3}{5} - \frac{7}{5} = -\frac{3}{5} + \left(-\frac{7}{5}\right) = -\frac{10}{5} = -2$

(e) $2.1 - 3.4 = 2.1 + (-3.4) = -1.3$

(f) Subtract 5 from -2.

We write the statement as $-2 - 5$ and proceed as before.

$$-2 - 5 = -2 + (-5) = -7$$

Check Yourself 1

Subtract.

(a) $18 - 7$ **(b)** $5 - 13$ **(c)** $-7 - 9$

(d) $-\frac{5}{6} - \frac{7}{6}$ **(e)** $-2 - 7$ **(f)** $5.6 - 7.8$

We use the subtraction rule in the same way when the number being subtracted is negative. Change the subtraction to addition and replace the negative number being subtracted with its opposite, which is positive.

Example 2 Subtracting Real Numbers

Subtract.

Change the subtraction to addition.

(a) $5 - (-2) = 5 + (+2) = 5 + 2 = 7$

Replace −2 with its opposite, +2 or 2.

(b) $7 - (-8) = 7 + (+8) = 7 + 8 = 15$

(c) $-9 - (-5) = -9 + 5 = -4$

(d) $-12.7 - (-3.7) = -12.7 + 3.7 = -9$

(e) $-\frac{3}{4} - \left(-\frac{7}{4}\right) = -\frac{3}{4} + \left(+\frac{7}{4}\right) = \frac{4}{4} = 1$

(f) Subtract −4 from −5. We write

$-5 - (-4) = -5 + 4 = -1$

Check Yourself 2

Subtract.

(a) $8 - (-2)$ **(b)** $3 - (-10)$ **(c)** $-7 - (-2)$

(d) $-9.8 - (-5.8)$ **(e)** $7 - (-7)$

We are now ready to describe negative mixed numbers, such as $-2\frac{1}{2}$.

Recall that we define a positive mixed number as the sum of a whole number part and a proper fraction part. For instance,

$5\frac{1}{4} = 5 + \frac{1}{4}$

This, of course, agrees with the number line approach, in which $5\frac{1}{4}$ is $\frac{1}{4}$ of a unit to the right of 5 on a number line.

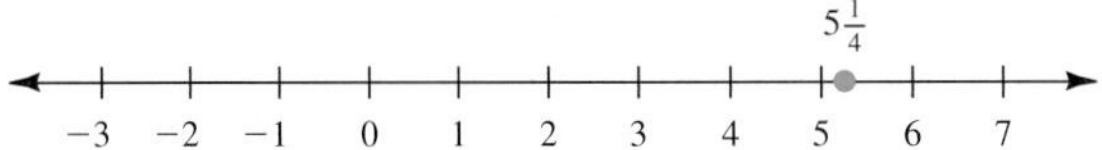

Looking again at a number line, we can also locate $-2\frac{1}{2}$, which is $\frac{1}{2}$ of a unit to the left of −2 on a number line.

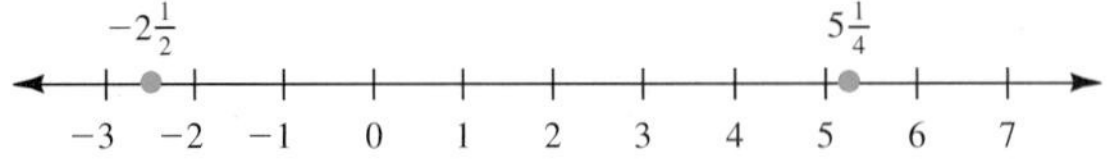

This suggests that $-2\frac{1}{2} = -2 - \frac{1}{2}$ or $-2 + \left(-\frac{1}{2}\right)$.

We can add and subtract mixed numbers in several different ways. In Chapters 2 and 3, you learned to rewrite mixed numbers as improper fractions and then to add or subtract them as you add fractions. You also learned to add or subtract the whole number parts and to add or subtract the fraction parts. When doing so, we might need to borrow or simplify the result.

We need to pay special attention when working with negative mixed numbers. Based on the discussion above, we know that

$$2\frac{1}{2} = 2 + \frac{1}{2} \quad \text{and} \quad -2\frac{1}{2} = -2 - \frac{1}{2} = -2 + \left(-\frac{1}{2}\right)$$

are opposites.

In Example 3, we add and subtract the whole number parts and the fraction parts separately. This example is a review of Chapter 3 material integrated with what you have learned about positive and negative numbers.

Example 3 Adding and Subtracting Mixed Numbers

< Objective 2 >

Evaluate each expression.

(a) $5\frac{1}{4} + 3\frac{1}{2}$

We add the whole number parts and the mixed number parts separately. Then, we combine them and simplify, if necessary.

RECALL

To add fractions with different denominators, rewrite them as equivalent fractions by using their *least common denominator*.

$$5\frac{1}{4} + 3\frac{1}{2} = (5 + 3) + \left(\frac{1}{4} + \frac{1}{2}\right)$$

$$= 8 + \left(\frac{1}{4} + \frac{2}{4}\right) \qquad \text{The LCD of } \tfrac{1}{4} \text{ and } \tfrac{1}{2} \text{ is 4: } \tfrac{1}{2} = \tfrac{2}{4}.$$

$$= 8 + \frac{3}{4} = 8\frac{3}{4}$$

(b) $5\frac{1}{4} - 3\frac{1}{2}$

In order to subtract $\frac{1}{2}$ from $\frac{1}{4}$, we need to *borrow*.

$$5\frac{1}{4} = 5 + \frac{1}{4}$$

$$= 4 + 1 + \frac{1}{4}$$

$$= 4 + \frac{4}{4} + \frac{1}{4}$$

$$= 4 + \frac{5}{4} = 4\frac{5}{4}$$

We continue as before.

NOTE

We could write

$4 - 3$ as $4 + (-3)$ and

$\frac{5}{4} - \frac{2}{4}$ as $\frac{5}{4} + \left(-\frac{2}{4}\right)$.

$$5\frac{1}{4} - 3\frac{1}{2} = 4\frac{5}{4} - 3\frac{2}{4} \qquad \tfrac{1}{2} = \tfrac{2}{4}$$

$$= 4\frac{5}{4} + \left(-3\frac{2}{4}\right) \qquad \text{Rewrite the subtraction problem as an addition problem.}$$

$$= 4 + \frac{5}{4} + (-3) + \left(-\frac{2}{4}\right)$$

$$= (4 - 3) + \left(\frac{5}{4} - \frac{2}{4}\right) \qquad \text{Subtract the whole number and fraction parts separately.}$$

$$= 1 + \frac{3}{4} = 1\frac{3}{4}$$

(c) $3\frac{1}{2} - 5\frac{1}{4}$

We rewrite the subtraction problem as an addition problem.

$$3\frac{1}{2} - 5\frac{1}{4} = 3\frac{1}{2} + \left(-5\frac{1}{4}\right)$$

$$= 3 + \frac{1}{2} + (-5) + \left(-\frac{1}{4}\right) \qquad -5\frac{1}{4} = -5 + \left(-\frac{1}{4}\right)$$

$$= 3 + (-5) + \frac{1}{2} + \left(-\frac{1}{4}\right)$$

$$= -2 + \frac{1}{4}$$

$$= -1\frac{3}{4}$$

NOTE

$-2 + \frac{1}{4}$ is $\frac{1}{4}$ to the *right* of -2 on the number line.

Check Yourself 3

Evaluate each expression.

(a) $6\frac{1}{3} + 4\frac{3}{4}$ **(b)** $6\frac{1}{3} - 4\frac{3}{4}$ **(c)** $4\frac{3}{4} - 6\frac{1}{3}$

When the first mixed number is negative, we proceed similarly, as illustrated in Example 4.

Example 4 Adding and Subtracting Mixed Numbers

Evaluate each expression.

(a) $-5\frac{1}{4} + 3\frac{1}{2}$

We work with the whole number parts and the fraction parts separately, again.

$$-5\frac{1}{4} + 3\frac{1}{2} = -5 + \left(-\frac{1}{4}\right) + 3 + \frac{1}{2}$$

$$= -5 + 3 + \left(-\frac{1}{4}\right) + \frac{1}{2}$$

$$= -2 + \frac{1}{4} = -1\frac{3}{4}$$

(b) $-5\frac{1}{4} + \left(-3\frac{1}{2}\right)$

$$-5\frac{1}{4} + \left(-3\frac{1}{2}\right) = -5 + \left(-\frac{1}{4}\right) + (-3) + \left(-\frac{1}{2}\right)$$

$$= -5 + (-3) + \left(-\frac{1}{4}\right) + \left(-\frac{1}{2}\right)$$

$$= -8 + \left(-\frac{3}{4}\right) = -8\frac{3}{4}$$

(c) $-5\frac{1}{4} - \left(-3\frac{1}{2}\right)$

We add the opposite of $-3\frac{1}{2}$, which is $3\frac{1}{2}$.

$$-5\frac{1}{4} - \left(-3\frac{1}{2}\right) = -5\frac{1}{4} + 3\frac{1}{2}$$

$$= -5 + \left(-\frac{1}{4}\right) + 3 + \frac{1}{2}$$

$$= -5 + 3 + \left(-\frac{1}{4}\right) + \frac{1}{2}$$

$$= -2 + \frac{1}{4} = -1\frac{3}{4}$$

Check Yourself 4

Evaluate each expression.

(a) $6\frac{1}{3} + \left(-4\frac{3}{4}\right)$ **(b)** $-6\frac{1}{3} - \left(-4\frac{3}{4}\right)$

(c) $-6\frac{1}{3} + \left(-4\frac{3}{4}\right)$ **(d)** $6\frac{1}{3} - \left(-4\frac{3}{4}\right)$

Example 5 An Application of Subtraction

From a seaside cliff 1,700 ft above sea level in the Cayman Islands, Nicole looks south to where the Cayman Trench is located. The Cayman Trench is the deepest part of the Caribbean, reaching a depth of 24,576 ft.

How far above the bottom of the Cayman Trench is Nicole standing?

To gauge the distance, we subtract, treating the depth of the trench as a negative number because the trench is below sea level.

$$1{,}700 - (-24{,}576) = 1{,}700 + (+24{,}576)$$
$$= 26{,}276$$

Nicole is standing 26,276 feet above the bottom of the Cayman Trench.

Check Yourself 5

The high temperature one year for a mountain town was 88°F. During the winter, the temperature dipped as low as -14°F. What was the temperature range experienced by the town that year?

Check Yourself ANSWERS

1. (a) 11; **(b)** -8; **(c)** -16; **(d)** -2; **(e)** -9; **(f)** -2.2 **2. (a)** 10; **(b)** 13; **(c)** -5; **(d)** -4; **(e)** 14
3. (a) $11\frac{1}{12}$; **(b)** $1\frac{7}{12}$; **(c)** $-1\frac{7}{12}$ **4. (a)** $1\frac{7}{12}$; **(b)** $-1\frac{7}{12}$; **(c)** $-11\frac{1}{12}$; **(d)** $11\frac{1}{12}$ **5.** 102°F

Reading Your Text

These fill-in-the-blank exercises will help you understand some of the key vocabulary used in this section. The answers to these exercises are in the Answers Appendix at the back of the text.

(a) Any subtraction problem can be written as an __________ problem.

(b) We define the __________ of real numbers by $a - b = a + (-b)$.

(c) To subtract real numbers, change the operation to addition and replace the second number with its __________.

(d) The opposite of a negative number is a __________ number.

10.3 exercises

Skills | Calculator/Computer | Career Applications | Above and Beyond

< Objectives 1 and 2 >

Evaluate each expression.

1. $21 - 13$ **2.** $36 - 22$ **3.** $82 - 45$ **4.** $103 - 56$

5. $8 - 10$ VIDEO **6.** $14 - 19$ **7.** $24 - 45$ **8.** $136 - 352$

9. $-5 - 3$ **10.** $-15 - 8$ **11.** $-9 - 14$ VIDEO **12.** $-8 - 12$

13. $3 - (-4)$ VIDEO **14.** $6 - (-8)$ **15.** $5 - (-11)$ **16.** $7 - (-5)$

17. $7 - (-12)$ **18.** $3 - (-10)$ **19.** $-36 - (-24)$ **20.** $-28 - (-11)$

21. $-19 - (-27)$ VIDEO **22.** $-11 - (-16)$ **23.** $-11 - (-11)$ **24.** $-15 - (-15)$

25. $0 - (-8)$ **26.** $0 - (-11)$ **27.** $\frac{15}{7} - \frac{8}{7}$ **28.** $\frac{17}{8} - \frac{9}{8}$

29. $\frac{7}{6} - \frac{19}{6}$ **30.** $\frac{5}{9} - \frac{32}{9}$ **31.** $-\frac{2}{5} - \frac{7}{10}$ VIDEO **32.** $-\frac{5}{9} - \frac{7}{18}$

33. $\frac{3}{4} - \left(-\frac{3}{2}\right)$ VIDEO **34.** $\frac{5}{6} - \left(-\frac{7}{6}\right)$ **35.** $\frac{6}{7} - \left(-\frac{5}{14}\right)$ **36.** $\frac{11}{16} - \left(-\frac{7}{8}\right)$

37. $-\frac{3}{4} - \left(-\frac{11}{4}\right)$ **38.** $-\frac{1}{2} - \left(-\frac{5}{8}\right)$ **39.** $-\frac{2}{3} - \left(-\frac{3}{4}\right)$ **40.** $\frac{3}{10} - \frac{2}{3}$

41. $2\frac{3}{8} - 5\frac{3}{4}$ **42.** $11\frac{3}{4} - 6\frac{4}{5}$ **43.** $6\frac{1}{2} - \left(-5\frac{4}{5}\right)$ **44.** $3\frac{1}{6} - \left(-4\frac{2}{3}\right)$

45. $-2\frac{5}{6} - 3\frac{1}{2}$ **46.** $-4\frac{7}{10} - 3\frac{1}{3}$ **47.** $-8\frac{2}{3} - \left(-1\frac{1}{5}\right)$ **48.** $-3\frac{1}{2} - \left(-4\frac{1}{4}\right)$

49. $-2\frac{3}{8} + 5\frac{3}{4}$ **50.** $11\frac{3}{4} + \left(-6\frac{4}{5}\right)$ **51.** $6\frac{1}{2} + \left(-5\frac{4}{5}\right)$ **52.** $3\frac{1}{6} + \left(-4\frac{2}{3}\right)$

53. $-2\frac{5}{6} + \left(-3\frac{1}{2}\right)$ **54.** $-4\frac{7}{10} + \left(-3\frac{1}{3}\right)$ **55.** $8\frac{2}{3} + \left(-1\frac{1}{5}\right)$ **56.** $3\frac{1}{2} + \left(-4\frac{1}{4}\right)$

57. $7.9 - 5.4$ **58.** $11.7 - 4.5$ **59.** $7.8 - 11.6$ **60.** $14.3 - 25.5$

61. $-3.4 - 4.7$ **62.** $-8.1 - 7.6$ **63.** $8.3 - (-5.7)$ **64.** $6.5 - (-4.3)$

65. $8.9 - (-11.7)$ **66.** $14.5 - (-24.6)$ **67.** $-12.7 - (-5.7)$ **68.** $-5.6 - (-2.6)$

69. $-6.9 - (-10.1)$ **70.** $-3.4 - (-7.6)$

71. **Statistics** On April 7, 2012, the high temperature in the United States was recorded as 91°F in Gila Bend, AZ, and both Harlingen and Laredo, TX. The low temperature for the day was −18°F, recorded in Deering, AK (at −1°F, Stanley, ID, recorded the lowest temperature in the contiguous United States). What was the temperature range in the United States on that day?

72. **Statistics** What was the temperature range in the contiguous 48 states on April 7, 2012 (see exercise 71)?

73. **Statistics** The lowest temperature ever recorded in the state of Oregon was −54°F (in Seneca, on February 10, 1933). The state's record high temperature occurred in Pendleton on August 10, 1898, when it reached 119°F. What is the historical temperature range in the state of Oregon? chapter 10 > Make the Connection

74. **Business and Finance** A government agency is operating despite a \$2.3 million budget deficit. Congress authorizes a \$3.5 million allocation in order for the agency to meet an additional \$1.9 million in payroll costs. How much is in the agency's budget after these transactions?

Fill in each blank with **always, sometimes,** *or* **never.**

75. The difference between two negative numbers is ______________ negative.

76. A positive number subtracted from a negative number is ______________ negative.

77. The difference between two positive numbers is ______________ negative.

78. A negative number subtracted from a positive number is ______________ negative.

Skills | **Calculator/Computer** | Career Applications | Above and Beyond

As you would expect, you can use a calculator to check your arithmetic even when working with negative numbers. Calculators work with negative numbers in one of two ways.

With scientific calculators, you enter the number and then use the [+/−] key to change its sign to negative.

With graphing calculators, you first enter the sign using the [(−)] key and then enter the number that follows.

In both cases, the key for negative numbers is *different* than the subtraction key [−].

We show how to evaluate $132 + 547 + (-234) - 112 - (-327)$ using either type of calculator. You do not need to enter the parentheses. They are included in the written expression in order to avoid careless errors with signs.

Scientific Calculator

132 [+] 547 [+] 234 [+/−] [−] 112 [−] 327 [+/−] [=]

Graphing Calculator

132 [+] 547 [+] [(−)] 234 [−] 112 [−] [(−)] 327 [ENTER]

In either case, the display reads 660.

Use a calculator to evaluate each expression.

79. $8 + 4 - (-3) - 2$

80. $-27 - 43 - (-29) + 13$

81. $145 - (-547) + (-92) - 234$

82. $10{,}945 - (-2{,}347) + (-7{,}687) + 41$

Skills | Calculator/Computer | **Career Applications** | Above and Beyond

Construction The elevation of the reference point used to set up a job is 362 in. Find the difference in elevation at the given points (use a negative sign for elevations below the reference point).

83. 311 in.

84. 491 in.

Manufacturing Technology At the beginning of the week, there were 2,489 lb of steel in inventory. Report the weekly change in steel inventory for the given end-of-week inventories.

85. 2,581 lb

86. 2,111 lb

87. **Electrical Engineering** A certain electric motor spins at 5,400 rotations per minute (rpm) when unloaded. When a load is applied, the motor spins at 4,250 rpm. What is the change in rpm after loading?

88. **Electrical Engineering** A cooling fan used to help dissipate heat from an electronic device has three modes of operation: off, low speed, and high speed. Low speed moves air at 34 cubic feet per minute (ft^3/min). High speed moves air at 52 ft^3/min. What is the difference in the volumes of air moved by the low and high speeds?

Answers

1. 8 **3.** 37 **5.** −2 **7.** −21 **9.** −8 **11.** −23 **13.** 7 **15.** 16 **17.** 19 **19.** −12 **21.** 8 **23.** 0 **25.** 8 **27.** 1 **29.** −2 **31.** $-\frac{11}{10}$ or $-1\frac{1}{10}$ **33.** $\frac{9}{4}$ or $2\frac{1}{4}$ **35.** $\frac{17}{14}$ or $1\frac{3}{14}$ **37.** 2 **39.** $\frac{1}{12}$ **41.** $-3\frac{3}{8}$ **43.** $12\frac{3}{10}$ **45.** $-6\frac{1}{3}$ **47.** $-7\frac{7}{15}$ **49.** $3\frac{3}{8}$ **51.** $\frac{7}{10}$ **53.** $-6\frac{1}{3}$ **55.** $7\frac{7}{15}$ **57.** 2.5 **59.** −3.8 **61.** −8.1 **63.** 14 **65.** 20.6 **67.** −7 **69.** 3.2 **71.** 109°F **73.** 173°F **75.** sometimes **77.** sometimes **79.** 13 **81.** 366 **83.** −51 in. **85.** +92 lb **87.** −1,150 rpm

Activity 29 ::

Plus/Minus Ratings in Hockey

The plus/minus statistic in professional hockey provides us with one application of the arithmetic of signed numbers. A player is awarded one point $(+1)$ if the player is on the ice when the player's team scores an even-strength or shorthanded goal (regardless of who actually scores the goal). A player is awarded -1 point if the opposing team scores such a goal while the player is on the ice.

Boston Bruin center Patrice Bergeron finished the 2011–2012 season as the overall leader in the plus/minus category with $+36$.

1. Midway through the 2011–2012 season, Patrice Bergeron entered a game against the Ottawa Senators with a plus/minus rating of $+20$. In the game, which Boston won 4–3, Bergeron was on the ice for one of his team's qualifying goals as well as all three qualifying goals by the Senators. What was Bergeron's plus/minus rating for the game?

2. After the game, what was Bergeron's plus/minus rating for the season?

3. In the same game, Ottawa defenseman Filip Kuba was on the ice for two of Ottawa's qualifying goals, as well as for one such goal by Boston. What was Kuba's plus/minus rating for the game?

4. If Kuba entered the game with a plus/minus rating of $+11$, what was his rating after the game?

5. Washington Capital forward Alex Ovechkin finished the 2011–2012 season as one of the league's top five goal scorers. However, his plus/minus rating for the season was -8.
   (a) Ovechkin entered a week in which he played three games with a plus/minus rating of –5. The Capitals lost the first game to the Florida Panthers 5–4. Ovechkin was on the ice for one qualifying Washington goal and three such Florida goals. Washington won the week's second game against the Ottawa Senators, 5–3. Ovechkin was on the ice for one qualifying goal by each team. In the week's third game, in which Washington beat the Toronto Maple Leafs 4–2, Ovechkin was on the ice for one qualifying Washington goal and both qualifying Toronto goals. What was Ovechkin's plus/minus rating for the week?
   (b) What was Ovechkin's plus/minus rating for the season after the three games that week?

Source: ESPN

10.4 Multiplying Real Numbers

< 10.4 Objectives >

1 > Find the product of real numbers

2 > Find the reciprocal of a real number

3 > Evaluate expressions involving real numbers

When you first considered multiplication, you thought of it as repeated addition. As you might expect, how we add real numbers leads us to rules for multiplying them. For example,

$$3 \cdot 4 = \underbrace{4 + 4 + 4}_{} = 12$$

We interpret multiplication as repeated addition to find the product, 12.

NOTE

3(−4) is different than the subtraction problem 3 − 4.

Now, consider the product $3(-4)$.

$$3(-4) = (-4) + (-4) + (-4) = -12$$

Because multiplication is *commutative,* we know that the order of the factors does not matter. Therefore,

$$-4 \cdot 3 = 3 \cdot (-4) = -12$$

Looking at these products suggests the first portion of our rule for multiplying real numbers. The product of a positive number and a negative number is negative.

Property

Multiplying Numbers with Different Signs The product of two numbers with different signs is negative.

To use this rule when multiplying two numbers with different signs, multiply their absolute values and attach a negative sign.

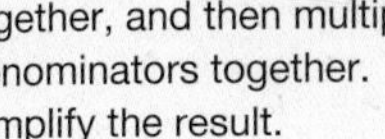

Example 1 Multiplying Real Numbers

< Objective 1 >

Multiply.

(a) $5(-6) = -30$

The product is negative.

NOTE

Multiply numerators together, and then multiply denominators together. Simplify the result.

(b) $-10(10) = -100$

(c) $8(-12) = -96$

(d) $-\frac{3}{4}\left(\frac{2}{5}\right) = -\frac{3}{10}$

Check Yourself 1

Multiply.

(a) $-7(5)$ **(b)** $-12(9)$ **(c)** $-15(8)$ **(d)** $\left(-\frac{5}{7}\right)\left(\frac{4}{5}\right)$

The product of two negative numbers is harder to visualize. This pattern may help you see how we can determine the sign of the product.

NOTES

$-1(-2)$ is the opposite of -2.

We present a more detailed explanation at the end of the section.

This number is decreasing by 1.

$$3(-2) = -6$$
$$2(-2) = -4$$
$$1(-2) = -2$$
$$0(-2) = 0$$
$$-1(-2) = 2$$

Do you see that the product is *increasing* by 2 each time?

What should the product $-2(-2)$ be? Continuing the pattern shown, we see that

$-2(-2) = 4$

This suggests that the product of two negative numbers is positive, which is the case.

Property

Multiplying Numbers with the Same Sign

The product of two numbers with the same sign is positive.

Example 2 **Multiplying Real Numbers**

Multiply.

(a) $9 \cdot 7 = 63$ The product of two positive numbers (same sign, +) is positive.

(b) $-8(-5) = 40$ The product of two negative numbers (same sign, −) is positive.

(c) $-\frac{1}{2}\left(-\frac{1}{3}\right) = \frac{1}{6}$

Check Yourself 2

Multiply.

(a) $10 \cdot 12$ **(b)** $-8(-9)$ **(c)** $-\frac{2}{3}\left(-\frac{6}{7}\right)$

Two numbers, 0 and 1, have special properties in multiplication.

Property

Multiplicative Identity Property

The product of 1 and any number is that number. The number 1 is called the **multiplicative identity.** In symbols,

$a \cdot 1 = 1 \cdot a = a$

Property

Multiplicative Property of Zero

The product of 0 and any number is 0. In symbols,

$a \cdot 0 = 0 \cdot a = 0$

Example 3 Multiplying Real Numbers

Find each product.

(a) $1(-7) = -7$

(b) $15(1) = 15$

(c) $-7(0) = 0$

(d) $0 \cdot 12 = 0$

(e) $-\frac{4}{5}(0) = 0$

Check Yourself 3

Multiply.

(a) $-10(1)$ **(b)** $0(-17)$ **(c)** $\frac{5}{7}(1)$ **(d)** $0\left(\frac{3}{4}\right)$

We include one more property of multiplication in our list.

Property

Multiplicative Inverse Property

For any nonzero number a, there is a number $\frac{1}{a}$ such that

$a \cdot \frac{1}{a} = 1$ $\quad$ $\frac{1}{a}$ is called the multiplicative inverse, or the reciprocal, of a. The product of any nonzero number and its reciprocal is 1.

Example 4 Finding a Reciprocal

< Objective 2 >

RECALL

The product of two negative numbers is a positive number.

(a) The reciprocal of 3 is $\frac{1}{3}$ because $3 \cdot \frac{1}{3} = 1$.

(b) The reciprocal of -5 is $\frac{1}{-5}$ or $-\frac{1}{5}$ because $-5\left(-\frac{1}{5}\right) = 1$.

(c) The reciprocal of $\frac{2}{3}$ is $\frac{1}{\frac{2}{3}}$ or $\frac{3}{2}$ because $\frac{2}{3} \cdot \frac{3}{2} = 1$.

Check Yourself 4

Find the multiplicative inverse (or reciprocal) of each number.

(a) 6 **(b)** -4 **(c)** $\frac{1}{4}$ **(d)** $-\frac{3}{5}$

In addition to the properties just mentioned, we can extend the commutative and associative properties for multiplication to real numbers. Example 5 is an application of the associative property of multiplication.

Example 5 Multiplying Real Numbers

< Objective 3 >

Find the product.

$-3(2)(-7)$

Applying the associative property, we can group the first two factors to write

$[(-3)(2)](-7)$ Evaluate first.

$= (-6)(-7)$

$= 42$

NOTE

This "grouping" can be done mentally.

Check Yourself 5

Find the product.

$-5(-8)(-2)$

In Section 10.5, we take a closer look at how the order of operations comes into play when there are negative numbers. To do that, we need to learn some basic skills involving negative numbers.

The first such skill involves quantities with multiple negative signs. Our approach takes into account two ways of looking at positive and negative numbers.

First, a negative sign indicates the opposite of the number which follows. For instance, we have already said that the opposite of 5 is -5, whereas the opposite of -5 is 5. This last instance can be translated as $-(-5) = 5$.

Second, any number must correlate to some point on the number line. That is, any nonzero number is either positive or negative. No matter how many negative signs a quantity has, you can always simplify it so that it is represented by a negative or a positive number (one negative sign or none).

Example 6 Simplifying Real Numbers

Simplify the expression $-(-(-(-4)))$.

The opposite of -4 is 4, so $-(-4) = 4$.

The opposite of 4 is -4, so $-(-(-4)) = -4$. The opposite of this last number, -4, is 4, so

$-(-(-(-4))) = 4$

NOTE

You should see a pattern emerge. An even number of negative signs gives a positive number, whereas an odd number of negative signs produces a negative number.

Check Yourself 6

Simplify the expression $-(-(-(-(-(-(-12))))))$.

We should also learn to evaluate expressions that contain both an exponent and a negative sign. Example 7 provides us with the opportunity to do just that.

Example 7 Evaluating Expressions

Evaluate each expression.

(a) $(-5)^2$

This means $(-5) \cdot (-5)$, which is 25.

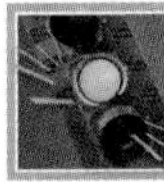

> CAUTION

Many students make careless errors when evaluating these types of expressions. Remember that $(-5)^2 \neq -5^2$.

(b) -5^2

This is not the same as the example in part (a). One way to look at this is to say that -5^2 is the *opposite* of $5^2 = 25$ so that

$-5^2 = -25$

Alternatively, we can look at -5 as shorthand notation for $-1 \cdot 5$. In which case, the order of operations requires that we compute the exponent prior to performing the multiplication. Therefore,

$-5^2 = -(5^2) = -(25) = -25$

(c) $(-5)^3$

This means $(-5) \cdot (-5) \cdot (-5)$.

From our earlier work in Example 5 of this section, we know we can use the associative property.

RECALL

In Example 6, we saw that multiplying an odd number of negative signs yields a negative number.

$$\begin{aligned}(-5)^3 &= (-5) \cdot (-5) \cdot (-5)\\ &= [(-5) \cdot (-5)] \cdot (-5)\\ &= 25 \cdot (-5)\\ &= -125\end{aligned}$$

Check Yourself 7

Evaluate each expression.

(a) -7^3 **(b)** $(-7)^3$ **(c)** $(-3)^4$

(d) -3^4 **(e)** $\left(-\frac{2}{3}\right)^2$ **(f)** $-\frac{2^2}{3}$

Of course, there are many applications that involve both positive and negative numbers.

Example 8 An Application of Real Numbers

The manager responsible for worker productivity at the TarCo manufacturing plant conducts a study on the amount of time workers do not engage in productive work. The manager finds that the average worker begins working 10 min after the shift begins, leaves for lunch 5 min early, and returns 10 min late. The manager also finds that the average employee works 15 min after the shift is over. Finally, the manager finds that employees spend an average of 22 min engaged in their 15-min coffee break. If the plant employs 230 people, how much productivity does the plant lose each day?

We begin by computing the productivity lost in a day by the average worker.

$-10 + (-5) + (-10) + 15 + (15 - 22) = -17$

Because the plant employs 230 people, the total lost productivity can be found by multiplication.

$230(-17) = -3{,}910$

The plant loses a total of 3,910 min of productivity per day (or about 65.2 hr).

Check Yourself 8

A math professor grades an exam as follows. Each correct answer is worth 4 points, each question left blank is worth 0 points, and each incorrect answer is worth -2 points. If a student answered 21 questions correctly, leaves 1 question blank, and answers 3 questions incorrectly, what score did the student earn on the exam?

Property

The Product of Two Negative Numbers

This is a detailed explanation of why the product of two negative numbers is positive.

From our earlier work, we know that the sum of a number and its opposite is 0:

$5 + (-5) = 0$

Multiply both sides of the equation by -3:

$(-3)[5 + (-5)] = (-3)(0)$

Because the product of 0 and any number is 0, on the right we have 0.

$(-3)[5 + (-5)] = 0$

We use the distributive property on the left.

$(-3)(5) + (-3)(-5) = 0$

We know that $(-3)(5) = -15$, so the equation becomes

$-15 + (-3)(-5) = 0$

We now have a statement of the form

$-15 + \square = 0$

in which $\square$ is the value of $(-3)(-5)$. We also know that $\square$ is the number that must be added to -15 to get 0, so $\square$ is the opposite of -15, or 15. This means that

$(-3)(-5) = 15$ The product is positive!

Regardless of which numbers we use in this argument the resulting product of two negative numbers will always be positive.

Check Yourself ANSWERS

1. **(a)** -35; **(b)** -108; **(c)** -120; **(d)** $-\frac{4}{7}$ **2.** **(a)** 120; **(b)** 72; **(c)** $\frac{4}{7}$
3. **(a)** -10; **(b)** 0; **(c)** $\frac{5}{7}$; **(d)** 0 **4.** **(a)** $\frac{1}{6}$; **(b)** $-\frac{1}{4}$; **(c)** 4; **(d)** $-\frac{5}{3}$ **5.** -80
6. -12 **7.** **(a)** -343; **(b)** -343; **(c)** 81; **(d)** -81; **(e)** $\frac{4}{9}$; **(f)** $-\frac{4}{3}$ **8.** 78

Reading Your Text

These fill-in-the-blank exercises will help you understand some of the key vocabulary used in this section. The answers to these exercises are in the Answers Appendix at the back of the text.

(a) The product of two numbers with different signs is ______________.

(b) The product of two negative numbers is ______________.

(c) The number 1 is called the multiplicative ______________.

(d) $\frac{1}{a}$ is called the multiplicative inverse or ______________ of a.

Skills | Calculator/Computer | Career Applications | Above and Beyond

10.4 exercises

< Objective 1 >

Evaluate each expression.

1. $4 \cdot 10$

2. $3 \cdot 14$

3. $5(-12)$ VIDEO

4. $10(-2)$

5. $8(-10)$

6. $13(-7)$

7. $-8(9)$

8. $-12(3)$

9. $-11(12)$

10. $-17(5)$

11. $-8(-7)$ VIDEO

12. $-9(-8)$

13. $-5(-12)$

14. $-7(-3)$

< Objective 2 >

15. $1(-18)$

16. $-3(1)$

17. $\frac{3}{4} \cdot \frac{4}{3}$

18. $-\frac{5}{3}\left(-\frac{3}{5}\right)$

19. $-5\left(-\frac{1}{5}\right)$ VIDEO

20. $7 \cdot \frac{1}{7}$

21. $5\left(-\frac{1}{5}\right)$

22. $-7\left(\frac{1}{7}\right)$

23. $-5(0)$

24. $0\left(-\frac{2}{3}\right)$

< Objective 3 >

25. $-5(3)(-2)$

26. $-4(2)(-3)$

27. $8(-3)(7)$

28. $13(-2)(6)$

29. $-3(-5)(-2)$

30. $-6(4)(-3)$

31. $-5(-4)(2)$

32. $-5(2)(-6)$

33. $-9(-12)(0)$

34. $-13(0)(-7)$

35. $-(-3)$

36. $-(-9)$

37. $-(-(-1))$

38. $-(-(-11))$

39. $-(-(-(-(-123))))$

40. $-(-(-(-(-(-80)))))$

41. -6^3

42. $(-6)^3$

43. -6^2

44. $(-6)^2$

45. $(-8)^2$ VIDEO

46. -8^2 VIDEO

47. -8^3

48. $(-8)^3$

49. -10^2

50. $(-10)^2$

51. $4\left(-\frac{3}{2}\right)$

52. $9\left(-\frac{2}{3}\right)$

53. $-\frac{1}{4}(8)$

54. $-\frac{3}{2}(4)$ VIDEO

55. $-9\left(-\frac{2}{3}\right)$

56. $-6\left(-\frac{3}{2}\right)$

57. $\frac{4}{5}\left(-\frac{3}{8}\right)$

58. $-\frac{2}{3}\left(-\frac{6}{7}\right)$

59. $-\frac{3}{4}\left(-\frac{10}{21}\right)$

60. $-\frac{1}{5}\left(\frac{3}{4}\right)$

61. $-\frac{1}{3}\left(\frac{6}{5}\right)(-10)$

62. $-\frac{1}{2}\left(\frac{4}{3}\right)(-6)$

63. $\left(-2\frac{5}{6}\right)\left(3\frac{1}{2}\right)$

64. $\left(-4\frac{7}{10}\right)\left(3\frac{1}{3}\right)$

65. $\left(-8\frac{2}{3}\right)\left(-1\frac{1}{5}\right)$

66. $\left(-3\frac{1}{2}\right)\left(-4\frac{1}{4}\right)$

67. $3.25(-4)$

68. $5.4(-5)$

69. $-1.25(-12)$

70. $-1.5(-20)$

STATISTICS A professor grades an exam by awarding 5 points for each correct answer and subtracting 2 points for each incorrect answer. Points are neither added nor subtracted for answers left blank. What is the exam score of each student?

71. A student answers 14 questions correctly and 4 incorrectly while leaving 2 questions blank.

72. A student answers 16 questions correctly and 3 incorrectly while leaving 1 question blank.

73. **Social Science** A gambler lost \$45 per hour at a slot machine over a 4-hr period. How much money did the gambler lose?

74. **Social Science** A poker player loses \$325 per hour at a poker table over a 3-hr period. After a break, she wins \$145 per hour for 2 hr. How did she do, overall?

Determine whether each statement is **true** *or* **false.**

75. The square of a negative number is negative.

76. The opposite of the square of a number is negative.

77. The opposite of the opposite of a negative number is negative.

78. The cube of a negative number is negative.

Skills | **Calculator/Computer** | Career Applications | Above and Beyond

79. Use a calculator to complete the table.

$4 \cdot 3$	12
$4 \cdot 2$	8
$4 \cdot 1$	
$4 \cdot 0$	
$4(-1)$	
$4(-2)$	
$4(-3)$	
$4(-4)$	

80. Use a calculator to complete the table.

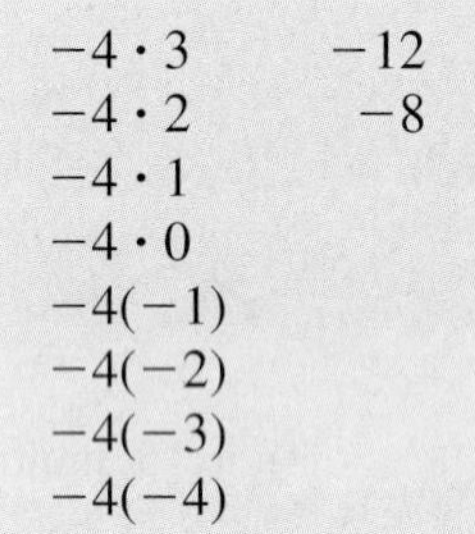

$-4 \cdot 3$	-12
$-4 \cdot 2$	-8
$-4 \cdot 1$	
$-4 \cdot 0$	
$-4(-1)$	
$-4(-2)$	
$-4(-3)$	
$-4(-4)$	

Skills | Calculator/Computer | **Career Applications** | Above and Beyond

81. **Manufacturing Technology** Companies occasionally sell products at a loss to draw in customers or to reward good customers. The theory is that customers buy other products along with the discounted product and the net result is a profit.

Beguhn Industries sells five different products. For each unit of product sold, the company makes or loses money. Product A, makes \$18; product B, loses \$4; product C, makes \$11; product D, makes \$38; and product E, loses \$15. During the previous month, Beguhn Industries sold 127 units of product A, 273 units of product B, 201 units of product C, 377 units of product D, and 43 units of product E.

Calculate the profit or loss for the month.

82. **Mechanical Engineering** The bending moment created by a center support on a steel beam is approximated by the formula $-\frac{1}{4}PL^3$, in which P is the load on each side of the center support and L is the length of the beam on each side of the center support (assuming a symmetrical beam and load).

If the total length of the beam is 24 ft (12 ft on each side of the center) and the total load is 4,124 lb (2,062 lb on each side of the center), what is the bending moment (in lb-ft³) at the center support?

Answers

1. 40 **3.** −60 **5.** −80 **7.** −72 **9.** −132 **11.** 56 **13.** 60 **15.** −18 **17.** 1 **19.** 1 **21.** −1 **23.** 0 **25.** 30 **27.** −168 **29.** −30 **31.** 40 **33.** 0 **35.** 3 **37.** −1 **39.** −123 **41.** −216 **43.** −36 **45.** 64 **47.** −512 **49.** −100 **51.** −6 **53.** −2 **55.** 6 **57.** $-\frac{3}{10}$ **59.** $\frac{5}{14}$ **61.** 4 **63.** $-9\frac{11}{12}$ **65.** $10\frac{2}{5}$ **67.** −13 **69.** 15 **71.** 62 **73.** \$180 **75.** False **77.** True **79.**

$4 \cdot 3$	12
$4 \cdot 2$	8
$4 \cdot 1$	4
$4 \cdot 0$	0
$4(-1)$	−4
$4(-2)$	−8
$4(-3)$	−12
$4(-4)$	−16

81. +\$17,086

10.5 Dividing Real Numbers and the Order of Operations

< 10.5 Objectives >

1 > Find the quotient of two real numbers

2 > Recognize that division by zero is undefined

3 > Use the order of operations to evaluate expressions with real numbers

We know that multiplication and division are related operations. We can use this fact, and our work from Section 10.4, to determine rules for the division of real numbers. Every division problem can be stated as an equivalent multiplication problem. For instance,

$\frac{15}{5} = 3$ because $15 = 5 \cdot 3$

$\frac{-24}{6} = -4$ because $-24 = (6)(-4)$

$\frac{-30}{-5} = 6$ because $-30 = (-5)(6)$

These examples illustrate that because the two operations are related, the rules of signs for multiplication are also true for division.

Property

Dividing Real Numbers	The quotient of two numbers with different signs is negative. The quotient of two numbers with the same sign is positive.

To divide two real numbers, divide their absolute values. Then attach the proper sign according to the division rule.

Example 1 — Dividing Real Numbers

< Objective 1 >

Divide.

(a) $\frac{28}{7} = 28 \div 7 = 4$ ⟵ Positive (numerator: Positive; denominator: Positive)

(b) $\frac{-36}{-4} = -36 \div (-4) = 9$ ⟵ Positive (numerator: Negative; denominator: Negative)

(c) $\frac{-42}{7} = -42 \div 7 = -6$ ⟵ Negative (numerator: Negative; denominator: Positive)

(d) $\frac{75}{-3} = 75 \div (-3) = -25$ ⟵ Negative (numerator: Positive; denominator: Negative)

(e) $\dfrac{15.2}{-3.8} = -4$ ← Negative

(Positive → 15.2; Negative → −3.8)

Check Yourself 1

Divide.

(a) $\dfrac{-55}{11}$ (b) $\dfrac{80}{20}$ (c) $\dfrac{-48}{-8}$ (d) $\dfrac{144}{-12}$ (e) $\dfrac{-13.5}{-2.7}$

You should be very careful when 0 is involved in a division problem. Remember that 0 divided by any nonzero number is just 0.

$\dfrac{0}{-7} = 0$ because $0 = (-7)(0)$

However, if zero is the *divisor,* we have a special problem. Consider

$\dfrac{9}{0} = ?$

This means that $9 = 0 \cdot ?$.

Can 0 times a number ever be 9? No, so there is no solution.

Because $\dfrac{9}{0}$ cannot be replaced by any number, we agree that *division by* 0 *is not allowed.*

Property

Division by Zero Division by 0 is undefined.

Example 2 Division and Zero

< Objective 2 >

Divide, if possible.

(a) $\dfrac{7}{0}$ is undefined.

(b) $\dfrac{0}{5} = 0$

(c) $\dfrac{-9}{0}$ is undefined.

(d) $\dfrac{0}{-8} = 0$

NOTE

The expression $\frac{0}{0}$ is called an **indeterminate form.** You will learn more about this in later mathematics classes.

Check Yourself 2

Divide if possible.

(a) $\dfrac{0}{3}$ (b) $\dfrac{5}{0}$ (c) $\dfrac{-7}{0}$ (d) $\dfrac{0}{-9}$

In Section 10.4, we began to develop the skills to evaluate more complex expressions involving negative numbers. We extend that work here.

To begin, remember that every number must correlate to some point on a number line. That is, any nonzero number is either positive or negative.

With fractions, this fact gives us

$$-\frac{3}{5} = \frac{-3}{5} = \frac{3}{-5}$$

All three quantities represent the same point on the number line

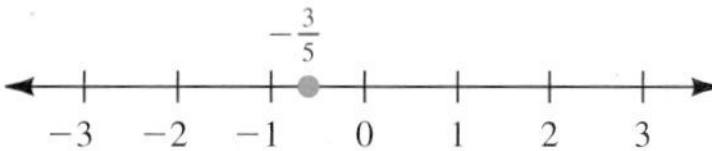

In this text, we generally choose to write negative fractions with the minus sign outside the fraction, such as $-\frac{3}{5}$.

In Section 10.4, we used the facts about multiple negative signs to help us evaluate integers. Here, we evaluate fractions with multiple negative signs.

Example 3 Simplifying Fractions with Negative Signs

Simplify each expression.

(a) $-\frac{-3}{4}$

This is the opposite of $\frac{-3}{4}$, which is $\frac{3}{4}$, a positive number.

(b) $-\frac{-3}{-4}$

The fraction part represents a negative number divided by another negative number, which is positive.

$$\frac{-3}{-4} = \frac{3}{4}$$

Therefore,

$$-\frac{-3}{-4} = -\left(\frac{-3}{-4}\right) = -\frac{3}{4}$$

Check Yourself 3

Simplify each expression.

(a) $-\frac{7}{-10}$ **(b)** $-\frac{-2}{-3}$

When symbols of grouping, or more than one operator, are involved in an expression, we must always remember to follow the rules for the order of operations.

Property

Order of Operations

We presented the order of operations in Section 1.7.

"Please Excuse My Dear Aunt Sally."

1. Perform all operations inside grouping symbols. Grouping symbols include parentheses, brackets, absolute value signs, fraction bars, and radicals.
2. Apply all exponents.
3. Perform all multiplication and division operations, from left to right.
4. Perform all addition and subtraction operations, from left to right.

Example 4 Order of Operations

< Objective 3 >

RECALL

In the expression $(-5)^2$, the base is -5 and the exponent applied to this base gives $(-5)(-5) = 25$.

In the expression -5^2, the base is 5, so the exponent is not applied to the negative sign. This gives $-5^2 = -5 \cdot 5 = -25$.

Evaluate each expression.

(a) $7(-9 + 12)$ Evaluate inside the parentheses first.
$= 7(3) = 21$

(b) $-8(-7) - 40$ Multiply first, then subtract.
$= 56 - 40$
$= 16$

(c) $(-5)^2 - 3$ Evaluate the power first.
$= (-5)(-5) - 3$ Note that $(-5)^2 = (-5)(-5) = 25$
$= 25 - 3$
$= 22$

(d) $-5^2 - 3$ Note that $-5^2 = -25$. The power applies *only* to the 5.
$= -25 - 3$
$= -28$

Check Yourself 4

Evaluate each expression.

(a) $8(-9 + 7)$ **(b)** $-3(-5) + 7$
(c) $(-4)^2 - (-4)$ **(d)** $-4^2 - (-4)$

Because the fraction bar also serves as a grouping symbol, all operations in the numerator or denominator should each be done first, as illustrated in Example 5.

Example 5 Order of Operations

Evaluate each expression.

(a) $\dfrac{-6(-7)}{3} = \dfrac{42}{3} = 14$ Multiply in the numerator and then divide.

(b) $\dfrac{3 + (-12)}{3} = \dfrac{-9}{3} = -3$ Add in the numerator and then divide.

(c) $\dfrac{-4 + 2(-6)}{-6 - 2} = \dfrac{-4 + (-12)}{-6 - 2}$ Multiply in the numerator. Then add in the numerator and subtract in the denominator.

$= \dfrac{-16}{-8} = 2$ Divide as the last step.

Check Yourself 5

Evaluate each expression.

(a) $\dfrac{-4 + (-8)}{6}$ **(b)** $\dfrac{3 - 2(-6)}{-5}$ **(c)** $\dfrac{-2(-4) - (-6)(-5)}{(-4)(11)}$

Many students have difficulty applying the distributive property when negative numbers are involved. Just remember that the sign of a number "travels" with that number.

Example 6 Applying the Distributive Property with Negative Numbers

RECALL

We usually enclose negative numbers in parentheses in the middle of an expression to avoid careless errors.

RECALL

We use brackets rather than nesting parentheses to avoid careless errors.

Use the distributive property to evaluate each expression.

(a) $-7(3 + 6) = -7 \cdot 3 + (-7) \cdot 6$ Apply the distributive property.

$= -21 + (-42)$ Multiply first and then add.

$= -63$

(b) $-3(5 - 6) = -3[5 + (-6)]$ First, change the subtraction to addition.

$= -3 \cdot 5 + (-3)(-6)$ Distribute the −3.

$= -15 + 18$ Multiply first and then add.

$= 3$

(c) $5(-2 - 6) = 5[-2 + (-6)]$

$= 5 \cdot (-2) + 5 \cdot (-6)$

$= -10 + (-30)$

$= -40$ The sum of two negative numbers is negative.

Check Yourself 6

Use the distributive property to evaluate each expression.

(a) $-2(-3 + 5)$ **(b)** $4(-3 + 6)$ **(c)** $-7(-3 - 8)$

Combining the elements from Examples 3 to 6 requires that we carefully apply the order of operations. You must remain vigilant with any negative signs.

Example 7 Evaluating Expressions

RECALL

Fraction bars are grouping symbols. Evaluate the numerator and denominator separately.

NOTE

In future courses you will find that we rarely change an improper fraction to a mixed number.

Use the order of operations to evaluate each expression.

(a) $4 + 2 \cdot (5 - 7)^2$

$= 4 + 2 \cdot (-2)^2$ Evaluate inside parentheses first.

$= 4 + 2 \cdot 4$ Apply the exponent.

$= 4 + 8$ Multiply.

$= 12$ Add.

(b) $\frac{3 - (-2)^3}{-7 + 3}$

$= \frac{3 - (-8)}{-4}$

$= \frac{3 + 8}{-4}$

$= \frac{11}{-4}$

$= -\frac{11}{4}$

Check Yourself 7

Use the order of operations to evaluate each expression.

(a) $-35 - (3 - 7)^3$ **(b)** $\frac{2 - 3 \cdot (1 - 5)^2}{(-3)^3 + (-2)^4}$

Check Yourself ANSWERS

1. **(a)** -5; **(b)** 4; **(c)** 6; **(d)** -12; **(e)** 5 **2.** **(a)** 0; **(b)** undefined; **(c)** undefined; **(d)** 0

3. **(a)** $\frac{7}{10}$; **(b)** $-\frac{2}{3}$ **4.** **(a)** -16; **(b)** 22; **(c)** 20; **(d)** -12 **5.** **(a)** -2; **(b)** -3; **(c)** $\frac{1}{2}$

6. **(a)** -4; **(b)** 12; **(c)** 77 **7.** **(a)** 29; **(b)** $\frac{46}{11}$

Reading Your Text

These fill-in-the-blank exercises will help you understand some of the key vocabulary used in this section. The answers to these exercises are in the Answers Appendix at the back of the text.

(a) The quotient of two numbers with different signs is ______________.

(b) The quotient of two negative numbers is ______________.

(c) ______________ by 0 is undefined.

(d) Every nonzero number is either ______________ or negative.

10.5 exercises

Skills Calculator/Computer Career Applications Above and Beyond

< Objectives 1–3 >

Evaluate each expression.

1. $\frac{70}{14}$ **2.** $\frac{48}{6}$ **3.** $\frac{-20}{-4}$ VIDEO

4. $\frac{-75}{-3}$ **5.** $\frac{-24}{8}$ **6.** $\frac{56}{-7}$

7. $\frac{50}{-5}$ **8.** $\frac{-52}{4}$ **9.** $\frac{0}{-8}$ VIDEO

10. $\frac{-9}{-1}$ **11.** $\frac{-17}{1}$ **12.** $\frac{18}{0}$ VIDEO

13. $\frac{-27}{-1}$ **14.** $\frac{0}{8}$ **15.** $\frac{-10}{0}$

16. $\frac{-32}{1}$ **17.** $\frac{-8}{32}$ **18.** $\frac{-6}{-30}$

19. $\frac{24}{-16}$ **20.** $\frac{-25}{10}$ **21.** $\frac{-28}{-42}$

22. $\frac{-125}{-75}$ **23.** $-\frac{-13}{52}$ **24.** $-\frac{-52}{-13}$

25. $-\frac{-12}{-15}$ **26.** $-\frac{-91}{-7}$ **27.** $5(7 - 2)$

28. $7(8 - 5)$ **29.** $2(5 - 8)$ **30.** $6(14 - 16)$

31. $(-3)(9 - 7)$ **32.** $(-6)(12 - 9)$ **33.** $(-3)(-2 - 5)$ VIDEO

34. $(-2)(-7 - 3)$ **35.** $\frac{-6(-3)}{2}$ **36.** $\frac{-9(5)}{-3}$

37. $\frac{24}{-4 - 8}$ VIDEO **38.** $\frac{36}{-7 + 3}$ **39.** $\frac{55 - 19}{-12 - 6}$

40. $\frac{-11 - 7}{-14 + 8}$ **41.** $\frac{7 - 5}{2 - 2}$ **42.** $\frac{-11 - (-3)}{-4 - (-4)}$

43. $\frac{-9(-6) - 10}{18 - (-4)}$ **44.** $\frac{4 - 2(-6)}{-14 - (-6)}$

Compute, as indicated.

45. $(-2)(-7) + (2)(-3)$

46. $(-3)(-6) + (4)(-2)$

47. $(-7)(3) - (-2)(-8)$

48. $(-5)(2) - (3)(-4)$

49. $\dfrac{3(6) - (-4)(8)}{6 - (-4)}$

50. $\dfrac{5(-2) - (-4)(-5)}{-4 - 2}$

51. $\dfrac{2(-5) + 4(6 - 8)}{3(-4 + 2)}$

52. $\dfrac{-3(-5) - 3(5 - 8)}{4(-8 + 6)}$

53. $(-7)^2 - 17$

54. $(-6)^2 - 20$

55. $-7^2 - 17$

56. $-6^2 - 20$

57. $(-4)^2 - (-2)(-5)$

58. $-4^2 - (-2)(-5)$ VIDEO

59. $(-6)^2 - (-3)^2$

60. $-6^2 - 3^2$

61. $(-8)^2 - 8^2$

62. $-11^2 - (-11)^2$

63. $5 + 3(4 - 6)^2$

64. $8 - 2(3 - 6)^3$

65. $-20 \div 2 + 10 \cdot 2$

66. $-6 \cdot 3 \div 2 - 9$

67. $\dfrac{4 - (-3)^2}{-7(4) + 3}$

68. $\dfrac{4 - (-2)^3}{5^2 - (-2)^2}$

69. $-60 \div (-3)(4) - 2^3 + 4$

70. $\dfrac{16 \div (-2)(-4)}{2^3 + 4}$

71. $\dfrac{7 - (-1)}{3^2 - 9 \div (-3)}$

72. $\dfrac{4 - (-3) + 1}{2 - 2 \div (-2)}$

73. **Business and Finance** Michelle deposits \$1,000 in her checking account each month. She writes a check for \$100 for car insurance and \$200 for her car payment. She also makes a \$55 payment on her student loan. How much money is left for her to use each week (assume there are 4 weeks in the month)?

74. **Business and Finance** An advertising agency lost \$42,000 in revenue last year when a client left for another agency. What was the agency's monthly loss in revenue?

75. **Science and Medicine** At noon, the temperature was 70°F. It dropped at a constant rate until 5 P.M., when it was 58°F. What was the hourly change in temperature? chapter 10 >Make the Connection VIDEO

76. **Science and Medicine** A chemist has 84 oz of a solution, which she pours into test tubes. If the chemist pours $\frac{2}{3}$ oz in each test tube, how many can she fill?

Fill in each blank with **always, sometimes,** *or* **never.**

77. The sum of a positive number and the square of a negative number is ____________ negative.

78. The product of three negative numbers is ____________ negative.

79. The sum of a negative number and the product of negative numbers is ____________ negative.

80. A negative number subtracted from the square of a negative number is ____________ negative.

Skills	**Calculator/Computer**	Career Applications	Above and Beyond

In Section 10.3, we saw how to use calculators to evaluate expressions when there are negative numbers. We build on that here.

Multiplying and dividing negative numbers is no different than adding and subtracting them.

To evaluate $457(-734)$, enter the numbers along with the multiplication key.

Scientific Calculator

457 [×] 734 [+/−] [=]

Graphing Calculator

457 [×] [(−)] 734 [ENTER]

In either case, the answer is −335,438.

Similarly, $\dfrac{457}{-734}$ is entered with the division key.

Scientific Calculator

457 [÷] 734 [+/–] [=]

Graphing Calculator

457 [÷] [(–)] 734 [ENTER]

To four decimal places, we get $\frac{457}{-734} \approx -0.6226$.

We have to be more careful when evaluating exponential expressions if there is a negative sign. You must determine whether the base is positive or negative (does the exponent apply to the negative sign, as well?).

If the base is negative, then the whole base needs to be contained by parentheses. To apply the exponent, scientific calculators use the [y^x] key. Graphing calculators use the caret key [^] to indicate an exponent follows.

To evaluate $(-3)^6$, we enter

Scientific Calculator

[(] 3 [+/–] [)] [y^x] 6 [=]

Graphing Calculator

[(] [(–)] 3 [)] [^] 6 [ENTER]

to get 729.

To evaluate -3^6, we enter

Scientific Calculator

3 [+/–] [y^x] 6 [=]

Graphing Calculator

[(–)] 3 [^] 6 [ENTER]

to get –729.

Use your calculator to evaluate each expression. Round your answers to two decimal places.

81. $25(-21)$

82. $15(-45)$

83. $-34(-28)$

84. $-71(-19)$

85. $345 \div (-25)$

86. $128 \div (-28)$

87. $-564 \div 36$

88. $-232 \div 52$

89. $-28 \div (-14)$

90. $-456 \div (-124)$

91. $(-4)^5$

92. $(-5)^4$

Skills | Calculator/Computer | **Career Applications** | Above and Beyond

93. **Manufacturing Technology** Peer's Pipe Fitters started the month of July with 1,789 gal of liquified petroleum gas (LP) in their tank. After 21 working days, there were 676 gal left in the tank. What was the average amount of LP consumed each day?

94. **Business and Finance** Three friends bought equal shares in an investment for a total of $21,000. They sold it later for $17,232. How much did each person profit?

Answers

1. 5 **3.** 5 **5.** −3 **7.** −10 **9.** 0 **11.** −17 **13.** 27 **15.** Undefined **17.** $-\frac{1}{4}$ **19.** $-\frac{3}{2}$ **21.** $\frac{2}{3}$ **23.** $\frac{1}{4}$ **25.** $-\frac{4}{5}$ **27.** 25 **29.** −6 **31.** −6 **33.** 21 **35.** 9 **37.** −2 **39.** −2 **41.** Undefined **43.** 2 **45.** 8 **47.** −37 **49.** 5 **51.** 3 **53.** 32 **55.** −66 **57.** 6 **59.** 27 **61.** 0 **63.** 17 **65.** 10 **67.** $\frac{1}{5}$ **69.** 76 **71.** $\frac{2}{3}$ **73.** $161.25 **75.** −2.4°F **77.** never **79.** sometimes **81.** −525 **83.** 952 **85.** −13.8 **87.** −15.67 **89.** 2 **91.** −1,024 **93.** $53 \frac{\text{gal}}{\text{day}}$

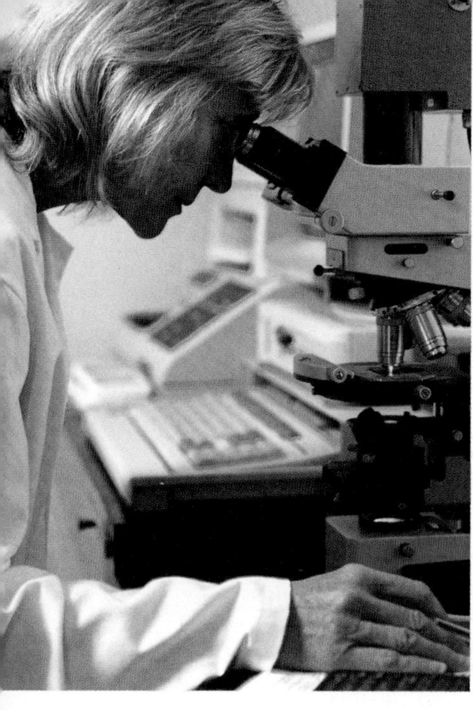

Activity 30 ::

Building Molecules

Every atom has an associated **valence number.** The valence of an atom is the number of electrons in its outermost electron-energy level.

From the valence number, we can determine the number of electrons that an atom must exchange in a *covalent bond* to form a stable molecule. We refer to this number as the **covalent bonding number.**

If an atom's valence is less than 4, then its covalent bonding number is the same as its valence number. In this case, the atom needs to give up electrons to another atom in order to form a covalent bond. The number of electrons that such an atom must give up is equal to its covalent bonding number.

Note that it is common, in chemistry, to include a "+" symbol before the covalent bonding number of an atom when it is a positive number.

If an atom's valence is greater than 4, then its covalent bonding number is equal to its valence minus 8 (this will always be zero or a negative number). These atoms gain electrons when forming covalent bonds. Specifically, they must gain the same number of electrons as the absolute value of their covalent bonding number in order to form a stable molecule.

An atom whose valence is exactly 4 can either gain or give up 4 electrons when forming a stable molecule. These atoms have a covalent bonding number of either $+4$ or -4, as the situation requires.

A stable molecule is formed when the sum of all the covalent bonding numbers of the atoms in the molecule equals 0.

The possible covalent bonding numbers for atoms are

$-4, -3, -2, -1, 0, 1, 2, 3, 4$

1. Find the covalent bonding number for each of the atoms given in the following table.

Atom	Valence Number	Covalent Bonding Number
Boron	3	
Calcium	2	
Carbon	4	
Chlorine	7	
Hydrogen	1	
Nitrogen	5	
Phosphorus	5	
Sulfur	6	

2. If the covalent bonding numbers of the atoms in a molecule add up to -2, and only one more atom is to be added, what covalent bonding number should it have? What valence number should it have?

3. A molecule contains 2 boron atoms. How many sulfur atoms would need to be added to make a stable molecule?

4. When combining hydrogen and chlorine, what is the fewest number of each atom that can be used to create a stable molecule?

5. When combining hydrogen and nitrogen, what is the fewest number of each atom that can be used to create a stable molecule?

6. When combining phosphorus and calcium, what is the fewest number of each atom that can be used to create a stable molecule?

7. When combining carbon and boron, what is the fewest number of each atom that can be used to create a stable molecule?

8. When combining carbon and nitrogen, what is the fewest number of each atom that can be used to create a stable molecule?

Definition/Procedure	Example	Reference
Real Numbers and Order		**Section 10.1**
Positive numbers Numbers used to name points to the right of 0 on a number line. **Negative numbers** Numbers used to name points to the left of 0 on a number line. **Real numbers** The set containing all of the numbers corresponding to points on a number line.	Negative numbers · Positive numbers −3 −2 −1 0 1 2 3 Zero is neither positive nor negative.	p. 571
Integers The set consisting of the natural numbers, their opposites, and 0.	The integers are $\{\ldots, -3, -2, -1, 0, 1, 2, 3, \ldots\}$	
Absolute value The distance on a number line between the point named by a number and 0. The absolute value of a number is always positive or 0.	The absolute value of a number a is written $\|a\|$. $\|7\| = 7 \qquad \|-8\| = 8$	p. 573
Adding Real Numbers *To Add Real Numbers*		**Section 10.2**
1. If two numbers have the same sign, add their absolute values. Give the sum the sign of the original numbers. **2.** If two numbers have different signs, subtract the smaller absolute value from the larger. Give the result the sign of the number with the larger absolute value.	$5 + 8 = 13$ $-3 + (-7) = -10$ $5 + (-3) = 2$ $7 + (-9) = -2$	p. 579
Opposites Two numbers are opposites if the points name the same distance from 0 on a number line, but in opposite directions. The opposite of a positive number is negative. The opposite of a negative number is positive. 0 is its own opposite.	5 units · 5 units −5 0 5 The opposite of 5 is −5. 3 units · 3 units −3 0 3 The opposite of −3 is 3.	p. 580
Subtracting Real Numbers *To Subtract Real Numbers*		**Section 10.3**
To subtract real numbers, add the first number and the opposite of the number being subtracted.	$4 - (-2) = 4 + 2 = 6$ Replace −2 with its opposite, 2.	p. 587
Multiplying Real Numbers *To Multiply Real Numbers*		**Section 10.4**
To multiply real numbers, multiply the absolute values of the numbers. Then use the multiplication rules to determine the sign of the product. **1.** If the numbers have different signs, the product is negative. **2.** If the numbers have the same sign, the product is positive.	$5 \cdot 7 = 35$ $(-4)(-6) = 24$ $(8)(-7) = -56$	p. 595

Continued

Definition/Procedure	Example	Reference
Dividing Real Numbers and the Order of Operations		Section 10.5
To Divide Real Numbers		
To divide real numbers, divide the absolute values of the numbers. Then use the division rules to determine the sign of the quotient. **1.** If the numbers have the same sign, the quotient is positive. **2.** If the numbers have different signs, the quotient is negative.	$\frac{-8}{-2} = 4$ $27 \div (-3) = -9$ $\frac{-16}{8} = -2$	*p.* 603
Always follow the proper **order of operations** when evaluating an expression. **1.** Perform all operations inside grouping symbols. **2.** Apply any exponents. **3.** Perform all multiplication and division operations, from left to right. **4.** Perform all addition and subtraction operations, from left to right.	$4 + 2(5 - 7)^2$ $= 4 + 2(-2)^2$ **P**lease $= 4 + 2 \cdot (4)$ **E**xuse $= 4 + 8$ **M**y **D**ear $= 12$ **A**unt **S**ally	*p.* 605

summary exercises :: chapter 10

This summary exercise set will help ensure that you have mastered each of the objectives of this chapter. The exercises are grouped by section. You should reread the material associated with any exercises that you find difficult. The answers to the odd-numbered exercises are in the Answers Appendix at the back of the text.

10.1 *Represent each set of numbers on the number line.*

1. $6, -18, -3, 2, 15, -9$

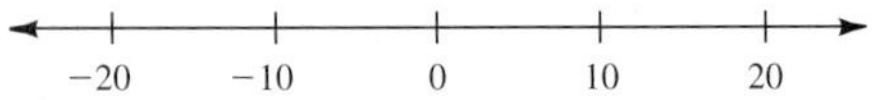

2. $4.7, -\frac{3}{4}, \pi, 3\frac{1}{2}, -3, -4.25$

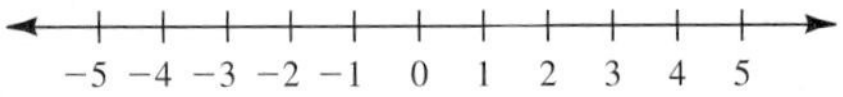

Place each set in ascending order.

3. $4, -3, 6, -7, 0, 1, -2$

4. $-\frac{1}{2}, -\frac{2}{3}, \frac{3}{5}, -\frac{4}{5}, \frac{5}{6}, \frac{7}{10}$

Determine the maximum and minimum values of each set of numbers.

5. $4, -2, 5, 1, -6, 3, -4$

6. $-4, 2, 5, -9, 8, 1, -6$

Evaluate.

7. $|9|$

8. $|-9|$

9. $-|9|$

10. $-|-9|$

11. $|12 - 8|$

12. $|8| - |12|$

13. $-|8 - 12|$

14. $|-8| - |-12|$

10.2 *Add.*

15. $-3 + (-8)$

16. $10 + (-4)$

17. $6 + (-6)$

18. $-16 + (-16)$

19. $-18 + 0$

20. $\frac{3}{8} + \left(-\frac{11}{8}\right)$

21. $5.7 + (-9.7)$

22. $-18 + 7 + (-3)$

10.3 *Subtract.*

23. $8 - 13$

24. $-7 - 10$

25. $10 - (-7)$

26. $-5 - (-1)$

27. $-9 - (-9)$

28. $0 - (-2)$

29. $-\frac{5}{4} - \left(-\frac{17}{4}\right)$

30. $7.9 - (-8.1)$

Evaluate each expression.

31. $|4 - 8|$

32. $|4| - |8|$

33. $|-4 - 8|$

34. $|-4| - |-8|$

35. $-6 - (-2) + 3$

36. $-5 - (5 - 8)$

37. Subtract -7 from -8.

38. Subtract -9 from the sum of 6 and -2.

39. **STATISTICS** On February 13, 2012, the high temperature in the United States was recorded as 84°F in Kahului, Hawaii (at 73°F, Fort Myers, Florida, recorded the highest temperature in the contiguous United States). The low temperature for the day was -16°F, recorded in Embarrass, Minnesota. What was the temperature range in the United States on that day?

40. **STATISTICS** What was the temperature range in the contiguous 48 states on February 13, 2012 (see exercise 39)?

10.4 *Multiply.*

41. $10(-7)$

42. $-8(-5)$

43. $-3(-15)$

44. $1(-15)$

45. $0(-8)$

46. $\frac{2}{3}\left(-\frac{3}{2}\right)$

47. $-4\left(\frac{3}{8}\right)$

48. $-\frac{5}{4}(-1)$

49. $-8(-2)(5)$

50. $-4(-3)(2)$

51. $\frac{2}{5}(-10)\left(-\frac{5}{2}\right)$

52. $\frac{4}{3}(-6)\left(-\frac{3}{4}\right)$

10.5 *Divide.*

53. $\frac{80}{16}$

54. $\frac{-63}{7}$

55. $\frac{-81}{-9}$

56. $\frac{0}{-5}$

57. $\frac{32}{-8}$

58. $\frac{-7}{0}$

Evaluate each expression.

59. $\frac{-8 + 6}{-8 - (-10)}$

60. $\frac{2(-3) - 1}{5 - (-2)}$

61. $\frac{(-5)^2 - (-2)^2}{-5 - (-2)}$

Evaluate each expression.

62. $2(-4 + 3)$

63. $2(-3) - (-5)(-3)$

64. $(2 - 8)(2 + 8)$

chapter test 10

CHAPTER 10

Use this chapter test to assess your progress and to review for your next exam. Allow yourself about an hour to take this test. The answers to these exercises are in the Answers Appendix at the back of the text.

Represent each integer on the number line.

1. 5, −12, 4, −7, 18, −17

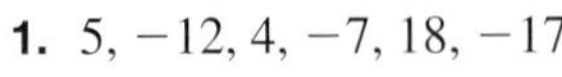

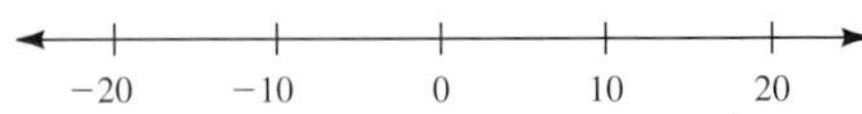

2. Place the numbers in ascending order: 4, −3, −6, 5, 0, $\frac{3}{4}$, $\frac{1}{2}$, 2, −2

3. Determine the maximum and minimum of the data set.

3, 2, −5, 6, 1, −2

4. **STATISTICS** On January 2, 2011, the high temperature in the United States was recorded as 83°F in Kona, Hawaii (at 82°F, both Melbourne and Fort Pierce, Florida, recorded the highest temperature in the contiguous United States). The low temperature for the day was −27°F, recorded in Barter Island, Alaska (at −26°F, Gunnison, Colorado, recorded the lowest temperature in the contiguous United States).
(a) What was the temperature range in the United States on that day?
(b) What was the temperature range in the contiguous 48 states on January 2, 2011?

Evaluate each expression.

5. $|7|$

6. $|-7|$

7. $-8 + (-5)$

8. $6 + (-9)$

9. $9 - 15$

10. $-9 - 15$

11. $(-8)(5)$

12. $(-9)(-7)$

13. $\frac{75}{-3}$

14. $\frac{-27}{-9}$

15. $|18 - 7|$

16. $|18| - |-7|$

17. $-9 + (-12)$

18. $5 - (-4)$

19. $-7 - (-7)$

20. $(4.5)(-6)$

21. $(-2)(-3)(-4)$

22. $\frac{-45}{9}$

23. $\frac{9}{0}$

24. $-8 - (-3 + 7)^2$

25. $\frac{5 + (-9) - 6}{(-3)^2 + (-2)^3}$

cumulative review chapters 1–10

Use this exercise set to review concepts from earlier chapters. While it is not a comprehensive exam, it will help you identify any material that you need to review before moving on to the next chapter. In addition to the answers, you will find section references for these exercises in the Answers Appendix in the back of the text.

1.1

1. What is the place value of 9 in the numeral 4,593,657?

Evaluate each expression.

1.2

2. $\begin{array}{r} 7{,}623 \\ 3{,}006 \\ +\ 131{,}602 \\ \hline \end{array}$

1.3

3. $\begin{array}{r} 125{,}678 \\ -\ 96{,}105 \\ \hline \end{array}$

1.5

4. 105×509

1.6

5. $56\overline{)22{,}540}$

4.2

6. $103.456 - 89.769$

4.3

7. 30.45×60.34

2.5

8. $\frac{8}{11} \times \frac{33}{56}$

2.7

9. $\frac{7}{15} \div \frac{21}{45}$

3.4

10. $6\frac{1}{9} - 3\frac{4}{27} + 5\frac{7}{18}$

5.4

11. Solve for the unknown: $\frac{6}{3} = \frac{8}{x}$.

6.1

12. Write 58% as a decimal and fraction.

6.2

13. Write $\frac{12}{25}$ as a decimal and percent.

7.1

Simplify.

14. 7 ft 22 in.

15. 8 lb 20 oz

Compute, as indicated.

16. 5 ft 8 in. + 6 ft 10 in.

17. 5 lb 8 oz − 2 lb 10 oz

18. $3 \times (3 \text{ hr } 30 \text{ min})$

19. $\frac{10 \text{ min } 45 \text{ s}}{5}$

Solve each application.

20. **Crafts** A plan for a bookcase requires three pieces of lumber 2 ft 8 in. long and two pieces 3 ft 4 in. long. What is the total length of material that is needed?

5.2

21. **Business and Finance** You can buy three bottles of dishwashing liquid, each containing 1 pt 6 fl oz, on sale for $2.40. For the same price you can buy a large container holding 2 qt. Which is the better buy?

4.3

22. **Geometry** A rectangle has length 8.6 cm and width 5.7 cm. Find the area of the rectangle.

8.2

23. **Geometry** The sides of a square each measure $8\frac{1}{2}$ in. Find the area of the square.

8.3

24. **Geometry** Find the circumference of a circle whose diameter is 8.2 ft. Use 3.14 for π, and round the result to the nearest tenth.

6.3

25. 116 is 145% of what number?

6.4

26. **Business and Finance** The sales tax on an item costing $136 is $10.20. What is the sales tax rate?

Complete each statement.

7.3

27. 17 g = ________ kg

7.2

28. 82 cm = ________ mm

8.1

29. Use a protractor to find the measure of the given angle.

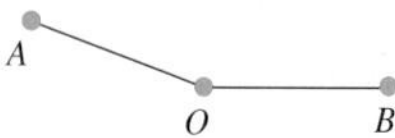

9.3

30. **Social Science** According to the bar graph, how many more students were enrolled in the university in 2010 than in 2000?

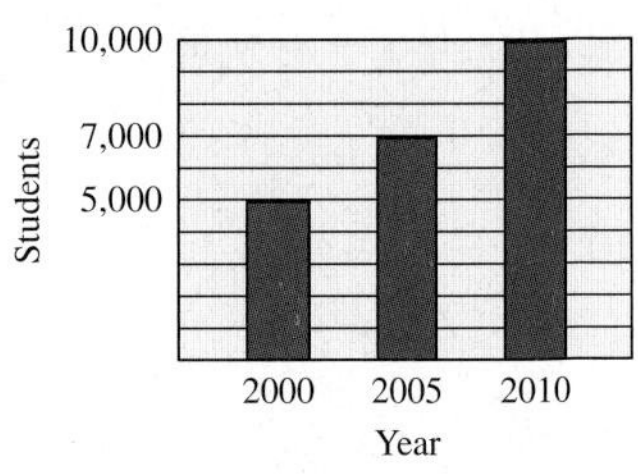

9.1

31. Calculate the mean, median, and mode for the data.

15, 16, 18, 13, 17, 19, 17, 21

Evaluate each expression.

10.2

32. $-9 + 13$

10.3

33. $17 - (-3)$

34. $-9 - 23$

35. $|-8| - |-23|$

36. $|-8 - 23|$

10.4

37. $(-9)(-12)$

10.5

38. $\frac{-36}{-9}$

39. $(-7)^2 - (16 \div 2)$

40. $45 \div 5 \times 2^3$

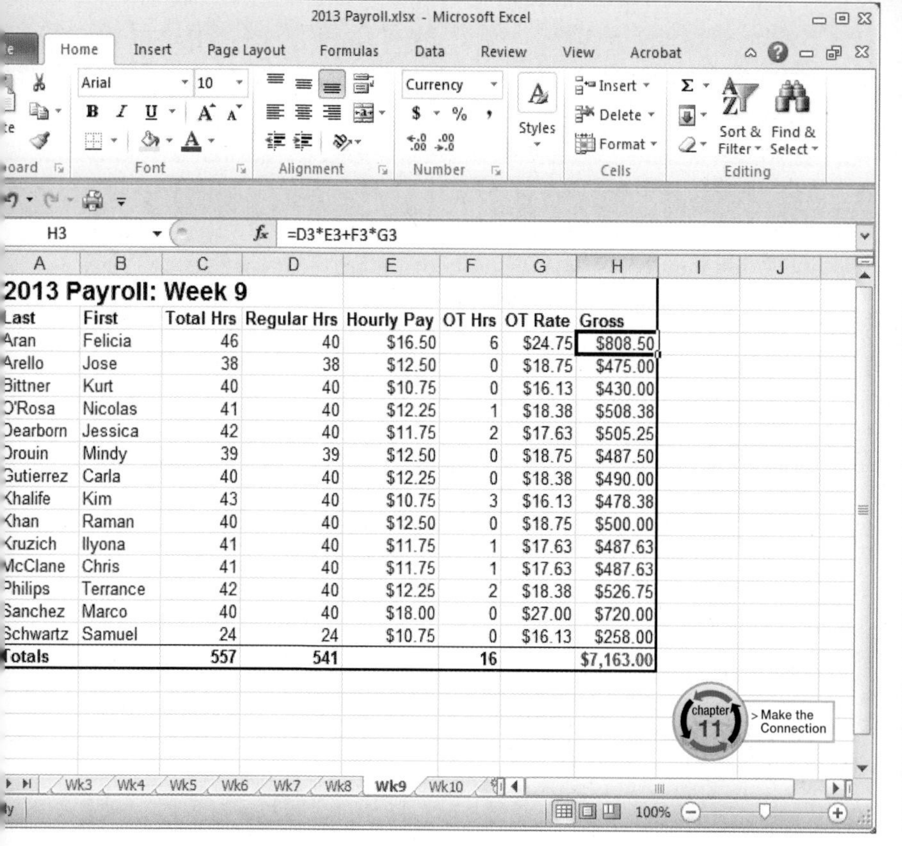

2013 Payroll.xlsx - Microsoft Excel

H3 =D3*E3+F3*G3

2013 Payroll: Week 9

Last	First	Total Hrs	Regular Hrs	Hourly Pay	OT Hrs	OT Rate	Gross
Aran	Felicia	46	40	$16.50	6	$24.75	$808.50
Arello	Jose	38	38	$12.50	0	$18.75	$475.00
Bittner	Kurt	40	40	$10.75	0	$16.13	$430.00
D'Rosa	Nicolas	41	40	$12.25	1	$18.38	$508.38
Dearborn	Jessica	42	40	$11.75	2	$17.63	$505.25
Drouin	Mindy	39	39	$12.50	0	$18.75	$487.50
Gutierrez	Carla	40	40	$12.25	0	$18.38	$490.00
Khalife	Kim	43	40	$10.75	3	$16.13	$478.38
Khan	Raman	40	40	$12.50	0	$18.75	$500.00
Kruzich	Ilyona	41	40	$11.75	1	$17.63	$487.63
McClane	Chris	41	40	$11.75	1	$17.63	$487.63
Philips	Terrance	42	40	$12.25	2	$18.38	$526.75
Sanchez	Marco	40	40	$18.00	0	$27.00	$720.00
Schwartz	Samuel	24	24	$10.75	0	$16.13	$258.00
Totals		557	541		16		$7,163.00

CHAPTER

11

An Introduction to Algebra

INTRODUCTION

Small businesses often use spreadsheet software to keep track of things. This is especially true of the many home-based businesses that have cropped up since computers became so common.

Spreadsheets, such as Microsoft Excel and OpenOffice Calc, became popular because they are easy to use. Most people can learn to use spreadsheet software by taking a single course at a local community college, or even by working through a tutorial on their own. While larger firms need more complex software, requiring extensive training, small businesses find that spreadsheets (and perhaps database software, such as Access and Base) can help them with most of their bookkeeping needs.

While many people have come to rely on spreadsheets, those who can realize their full power have a strong background in algebra. This is because spreadsheets can be thought of as multidimensional algebra machines. Consider a typical payroll spreadsheet.

While it may seem that each "cell" is simply typed in, the truth is that only the name, total hours, and hourly rate are manually entered on each line. The other fields, such as the gross pay, are determined by formulas. You can see such a formula, which refers to other cells in the spreadsheet, in the formula line. This allows someone to copy the formula to every employee, without having to retype it each time.

CHAPTER 11 OUTLINE

11 prerequisite check

This Prerequisite Check highlights the skills you will need in order to be successful in this chapter. The answers to these exercises are in the Answers Appendix at the back of the text.

Write each phrase as an arithmetic expression and solve.

1. 8 less than 10

2. The sum of 3 and the product of 5 and 6

Find the reciprocal of each number.

3. -12

4. $4\frac{5}{8}$

Evaluate each expression.

5. $\left(-\frac{3}{2}\right) \times \left(-\frac{2}{3}\right)$

6. $4\left(\frac{1}{4}\right)$

7. $\frac{2}{2}$

8. $5 + 2 \times 3^2$

9. -8^2

10. $(-8)^2$

11. **Business and Finance** An $8\frac{1}{2}$-acre plot of land is on sale for \$120,000. What is the price per acre?

12. **Business and Finance** A grocery store adds a 30% markup to the wholesale price of goods to determine their retail price. What is the retail price of a snack bar if its wholesale price is \$1.19?

11.1

From Arithmetic to Algebra

< 11.1 Objectives >

1 > Use the symbols and language of algebra

2 > Identify algebraic expressions

3 > Use algebra to model an application

In arithmetic, you learned to calculate with numbers using addition, subtraction, multiplication, and division.

In algebra, we still use numbers and the same four operations. However, we also use letters to represent numbers. Letters such as x, y, L, and W are called **variables** when they represent numerical values.

Here we see two rectangles whose lengths and widths are labeled with numbers.

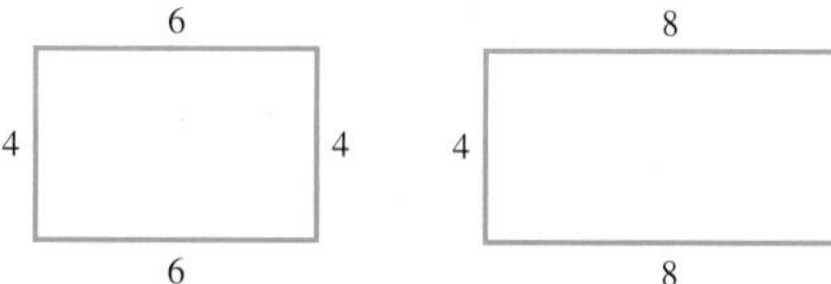

If we want to represent the length and width of *any* rectangle, we can use the variables L for length and W for width.

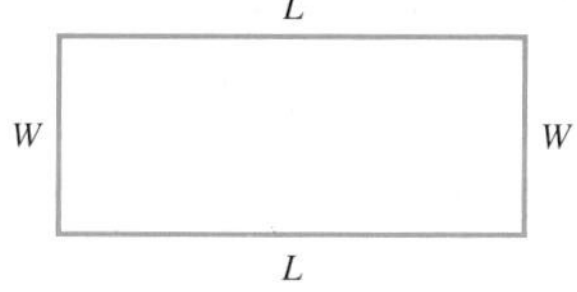

RECALL

In arithmetic
+ denotes addition
− denotes subtraction
× denotes multiplication
÷ denotes division

You are familiar with the four symbols (+, −, ×, ÷) used to indicate the fundamental operations of arithmetic.

To see how these operations are indicated in algebra, we begin by looking at addition.

Definition

Addition — $x + y$ means the *sum* of x and y, or x *plus* y.

Example 1 Writing Expressions That Indicate Addition

< Objective 1 >

(a) The *sum* of a and 3 is written as $a + 3$.

(b) L *plus* W is written as $L + W$.

(c) 5 *more than* m is written as $m + 5$.

(d) x *increased by* 7 is written as $x + 7$.

Check Yourself 1

Write each phrase symbolically.

(a) The sum of y and 4 **(b)** a plus b

(c) 3 more than x **(d)** n increased by 6

Now look at how subtraction is indicated in algebra.

Definition

Subtraction

$x - y$ means the *difference* of x and y, or x *minus* y. Subtracting y is the same as adding its opposite, so

$x - y = x + (-y)$

$x - y$ is not the same as $y - x$.

Example 2 Writing Expressions That Indicate Subtraction

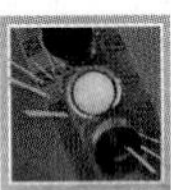

> CAUTION

"x minus y," "the difference of x and y," "x decreased by y," and "x take away y" are all written in the same order as the instructions are given, $x - y$.

However, we reverse the order when writing "x less than y" and "x subtracted from y." These two phrases are translated as $y - x$.

(a) r *minus* s is written as $r - s$.

(b) The *difference* of m and 5 is written as $m - 5$.

(c) x *decreased by* 8 is written as $x - 8$.

(d) 4 *less than* a is written as $a - 4$.

(e) x *subtracted from* 5 is written as $5 - x$.

(f) 7 *take away* y is written as $7 - y$.

Check Yourself 2

Write each phrase symbolically.

(a) w minus z **(b)** The difference of a and 7
(c) y decreased by 3 **(d)** 5 less than b
(e) b subtracted from 8 **(f)** 4 take away x

You have seen that the operations of addition and subtraction are written exactly the same way in algebra as in arithmetic. This is not true for multiplication because the symbol × looks like the letter x, so we use other symbols to show multiplication to avoid confusion. Here are some ways to write multiplication.

Definition

Multiplication

NOTE

x and y are called the **factors** of the product xy.

A centered dot	$x \cdot y$	All these indicate the *product* of x and y, or x *times* y.
Writing the letters next to each other or separated only by parentheses	xy $x(y)$ $(x)(y)$	

Example 3 Writing Expressions That Indicate Multiplication

NOTE

You can place letters next to each other or numbers and letters next to each other to show multiplication. But you *cannot* place numbers side by side to show multiplication: 37 means the number thirty-seven, not 3 times 7.

(a) The product of 5 and a is written as $5 \cdot a$, $(5)(a)$, or $5a$. The last expression, $5a$, is the shortest and the most common way of writing the product.

(b) 3 times 7 can be written as $3 \cdot 7$ or $(3)(7)$.

(c) Twice z is written as $2z$.

(d) The product of 2, s, and t is written as $2st$.

(e) 4 more than the product of 6 and x is written as $6x + 4$.

Check Yourself 3

Write each phrase symbolically.

(a) m times n
(b) The product of h and b
(c) The product of 8 and 9
(d) The product of 5, w, and y
(e) 3 more than the product of 8 and a

Before moving on to division, look at how we combine the symbols we have learned so far.

Definition

Expression An **expression** is a meaningful collection of numbers, variables, and operations.

Example 4 Identifying Expressions

< Objective 2 >

NOTES

Not every collection of symbols is an expression.

(a) $2m + 3$ is an expression. It means that we multiply 2 and m, then add 3.

(b) $x + \cdot + 3$ is not an expression. The three operations in a row have no meaning.

(c) $y = 2x - 1$ is not an expression, it is an *equation*. The equal sign is not an operation sign.

(d) $3a + 5b - 4c$ is an expression.

Check Yourself 4

Identify the expressions.

(a) $7 - \cdot x$
(b) $6 + y = 9$
(c) $a + b - c$
(d) $3x - 5yz$

To write more complicated expressions in algebra, we need some "punctuation marks." Parentheses ( ) mean that an expression is to be thought of as a single quantity. Brackets [ ] are used in exactly the same way as parentheses in algebra. Example 5 shows expressions with grouping symbols.

Example 5 Expressions with More Than One Operation

NOTES

This can be read as "3 times the quantity *a* plus *b*."

No parentheses are needed in part (b) since the 3 multiplies *only a*.

(a) 3 times the sum of a and b is written as

$3(a + b)$

The sum of a and b is a single quantity, so it is enclosed in parentheses.

(b) The sum of 3 times a and b is written as $3a + b$.

(c) 2 times the difference of m and n is written as $2(m - n)$.

(d) The product of s plus t and s minus t is written as $(s + t)(s - t)$.

(e) The product of b and 3 less than b is written as $b(b - 3)$.

Check Yourself 5

Write each phrase symbolically.

(a) Twice the sum of p and q
(b) The sum of twice p and q
(c) The product of a and the quantity $b - c$
(d) The product of x plus 2 and x minus 2
(e) The product of x and 4 more than x

NOTE

In algebra the fraction form is usually used.

Now we look at division. In arithmetic, you see the division sign ÷, the long division symbol $\overline{)\ }$, and fraction notation. For example, to indicate the quotient when 9 is divided by 3, you could write

$9 \div 3$ or $3\overline{)9}$ or $\frac{9}{3}$

Definition

Division $\frac{x}{y}$ means x *divided* by y or the *quotient* of x and y.

Example 6 Writing Expressions That Indicate Division

RECALL

The fraction bar is a grouping symbol.

(a) m divided by 3 is written as $\frac{m}{3}$.

(b) The quotient of a plus b, divided by 5 is written as $\frac{a+b}{5}$.

(c) The quantity p plus q divided by the quantity p minus q is written as $\frac{p+q}{p-q}$.

Check Yourself 6

Write each phrase symbolically.

(a) r divided by s

(b) The quotient when x minus y is divided by 7

(c) The quantity a minus 2 divided by the quantity a plus 2

We can use many different letters to represent variables. In Example 6, the letters m, a, b, p, and q represented different variables. We often choose a letter that reminds us of what it represents, for example, L for *length* or W for *width*. These variables may be uppercase or lowercase letters, although lowercase is used more often.

Example 7 Writing Geometric Expressions

(a) *Length* times *width* is written $L \cdot W$.

(b) One-half of *altitude* times *base* is written $\frac{1}{2} \cdot a \cdot b$.

(c) *Length* times *width* times *height* is written $L \cdot W \cdot H$.

(d) Pi (π) times *diameter* is written πd.

Check Yourself 7

Write each geometric expression symbolically.

(a) 2 times *length* plus two times *width*

(b) 2 times pi (π) times *radius*

Algebra can be used to model a variety of applications, such as the one shown in Example 8.

Example 8 Modeling Applications with Algebra

< Objective 3 >

Carla earns $10.25 per hour in her job. Write an expression that describes her weekly gross pay in terms of the number of hours she works.

NOTE

We are asked to describe her pay given that her hours may vary.

We represent the number of hours she works in a week by the variable h. Carla's pay is figured by taking the product of her hourly wage and the number of hours she works.

So, the expression

$$10.25h$$

describes Carla's weekly gross pay.

NOTE

The words "twice" and "doubled" indicate multiplication by 2.

Check Yourself 8

The specifications for an engine cylinder call for the stroke length to be two more than twice the diameter of the cylinder. Write an expression for the stroke length of a cylinder based on its diameter.

We close this section by listing many of the common words used to indicate arithmetic operations.

Words Indicating Operations

The operations listed are usually indicated by the words shown.

Addition (+)	Plus, and, more than, increased by, sum
Subtraction (−)	Minus, from, less than, decreased by, difference, take away
Multiplication (·)	Times, of, by, product
Division (÷)	Divided, into, per, quotient

Check Yourself ANSWERS

1. **(a)** $y + 4$; **(b)** $a + b$; **(c)** $x + 3$; **(d)** $n + 6$ **2.** **(a)** $w - z$; **(b)** $a - 7$; **(c)** $y - 3$; **(d)** $b - 5$; **(e)** $8 - b$; **(f)** $4 - x$ **3.** **(a)** mn; **(b)** hb; **(c)** $8 \cdot 9$ or $(8)(9)$; **(d)** $5wy$; **(e)** $8a + 3$ **4.** **(a)** not an expression; **(b)** not an expression; **(c)** expression; **(d)** expression **5.** **(a)** $2(p + q)$; **(b)** $2p + q$; **(c)** $a(b - c)$; **(d)** $(x + 2)(x - 2)$; **(e)** $x(x + 4)$ **6.** **(a)** $\frac{r}{s}$; **(b)** $\frac{x - y}{7}$; **(c)** $\frac{a - 2}{a + 2}$ **7.** **(a)** $2L + 2W$; **(b)** $2\pi r$ **8.** $2d + 2$

Reading Your Text

These fill-in-the-blank exercises will help you understand some of the key vocabulary used in this section. The answers to these exercises are in the Answers Appendix at the back of the text.

(a) In algebra, we use letters, called ________, to represent unknown quantities.

(b) $x + y$ means the ________ of x and y.

(c) $x \cdot y$, $(x)(y)$, and xy are all ways of indicating ________ in algebra.

(d) An ________ is a meaningful collection of numbers, variables, and symbols of operation.

11.1 exercises

Skills | Calculator/Computer | Career Applications | Above and Beyond

< Objective 1 >

Write each phrase symbolically.

1. The sum of c and d VIDEO

2. a plus 7

3. w plus z

4. The sum of m and n

5. x increased by 2

6. 3 more than b

7. 10 more than y VIDEO

8. m increased by 4

9. a minus b

10. 5 less than s

11. b decreased by 7

12. r minus 3

13. 6 less than r VIDEO

14. x decreased by 3

15. w times z

16. The product of 3 and c

17. The product of 5 and t VIDEO

18. 8 times a

19. The product of 8, m, and n

20. The product of 7, r, and s

21. The product of 3 and the quantity p plus q VIDEO

22. The product of 5 and the sum of a and b

23. Twice the sum of x and y VIDEO

24. 3 times the sum of m and n

25. The sum of twice x and y

26. The sum of 3 times m and n

27. Twice the difference of x and y

28. 3 times the difference of c and d

29. The quantity a plus b times the quantity a minus b

30. The product of x plus y and x minus y

31. The product of m and 3 less than m VIDEO

32. The product of a and 7 more than a

33. x divided by 5 VIDEO

34. The quotient when b is divided by 8

35. The quotient of a plus b, and 7

36. The difference x minus y, divided by 9

37. The difference of p and q, divided by 4 VIDEO

38. The sum of a and 5, divided by 9

39. The sum of a and 3, divided by the difference of a and 3

40. The difference of m and n, divided by the sum of m and n

Use x as the variable to write each phrase symbolically.

41. 5 more than a number

42. A number increased by 8

43. 7 less than a number

44. A number decreased by 10

45. 9 times a number

46. Twice a number

47. 6 more than 3 times a number

48. 5 times a number, decreased by 10

49. Twice the sum of a number and 5

50. 3 times the difference of a number and 4 VIDEO

51. The product of 2 more than a number and 2 less than that same number

52. The product of 5 less than a number and 5 more than that same number

53. The quotient of a number and 7

54. A number divided by 3

55. The sum of a number and 5, divided by 8

56. The quotient when 7 less than a number is divided by 3

57. 6 more than a number divided by 6 less than that same number VIDEO

58. The quotient when 3 less than a number is divided by 3 more than that same number

Write each geometric expression symbolically.

59. Four times the length of a side s

60. $\frac{4}{3}$ times π times the cube of the radius r VIDEO

61. π times the radius r squared times the height h

62. Twice the length L plus twice the width W

63. One-half the product of the height h and the sum of two unequal sides b_1 and b_2

64. Six times the length of a side s squared

< Objective 2 >

Identify the expressions.

65. $2(x + 5)$ 

66. $4 + (x - 3)$

67. $4 + \div m$

68. $6 + a = 7$

69. $2b = 6$

70. $x(y + 3)$

71. $2a + 5b$

72. $4x + \cdot 7$

< Objective 3 >

73. **Number Problem** Two numbers have a sum of 35. If one number is x, express the other number in terms of x.

74. **Science and Medicine** It is estimated that the Earth is losing 4,000 species of plants and animals every year. If S represents the number of species living last year, how many species are on the Earth this year?

75. **Business and Finance** The simple interest earned when a principal P is invested at a rate r for a time t is calculated by multiplying the principal by the rate by the time. Write an expression for the interest earned.

76. **Science and Medicine** The kinetic energy of a particle of mass m is found by taking one-half of the product of the mass and the square of the velocity v. Write an expression for the kinetic energy of a particle.

77. **Business and Finance** Four hundred tickets were sold for a school play. The tickets were of two types: general admission and student. There were x general admission tickets sold. Write an expression for the number of student tickets sold.

78. **Business and Finance** Nate has $375 in his bank account. He wrote a check for x dollars for a concert ticket. Write an expression that represents the remaining money in his account.

Match each phrase with the proper expression.

79. 8 decreased by x — **(a)** $x - 8$

80. 8 less than x — **(b)** $8 - x$

81. The difference between 8 and x

82. 8 from x

Determine whether each statement is **true** *or* **false.**

83. The phrase "7 more than x" indicates addition.

84. A product is the result of dividing two numbers.

Complete each statement with **always, sometimes,** *or* **never.**

85. An expression is _________ an equation.

86. A number written in front of a variable _________ indicates multiplication.

Skills	Calculator/Computer	**Career Applications**	Above and Beyond

87. **Allied Health** The standard dosage given to a patient is equal to the product of the desired dose D and the available quantity Q divided by the available dose H. Write the standard dosage calculation formula.

88. **Information Technology** Mindy is the manager of the help desk at a large cable company. She notices that, on average, her staff can handle 50 calls per hour. Last week, during a thunderstorm, the call volume increased from 65 to 150 calls per hour.

To determine the average number of customers in the system, she needs to take the quotient of the average rate of customer arrivals (the call volume) a and the average rate at which customers are served h minus the average rate of customer arrivals a. Write a formula for the average number of customers in the system.

89. **CONSTRUCTION TECHNOLOGY** K Jones Manufacturing produces hex bolts and carriage bolts. It sold 284 more hex bolts than carriage bolts last month. Write a formula that describes the number of carriage bolts it sold last month. Let H be the number of hex bolts sold last month.

90. **ELECTRICAL ENGINEERING** Electrical power P is the product of voltage V and current I. Write an expression for the electrical power.

Skills	Calculator/Computer	Career Applications	**Above and Beyond**

91. Rewrite each algebraic expression using English phrases. Exchange papers with another student to edit your writing. Be sure the meaning in English is the same as in algebra. These expressions are not complete sentences, so your English does not have to be in complete sentences. Here is an example.

Algebra: $2(x - 1)$

English: We could write "double 1 less than a number." Or we might write "a number diminished by 1 and then multiplied by 2."

(a) $n + 3$ **(b)** $\frac{x + 2}{5}$ **(c)** $3(5 + a)$ **(d)** $3 - 4n$ **(e)** $\frac{x + 6}{x - 1}$

92. Use the Internet to find the origins of the symbols $+$, $-$, $\times$, and $\div$. Summarize your findings.

Answers

1. $c + d$ **3.** $w + z$ **5.** $x + 2$ **7.** $y + 10$ **9.** $a - b$ **11.** $b - 7$ **13.** $r - 6$ **15.** wz **17.** $5t$ **19.** $8mn$ **21.** $3(p + q)$ **23.** $2(x + y)$ **25.** $2x + y$ **27.** $2(x - y)$ **29.** $(a + b)(a - b)$ **31.** $m(m - 3)$ **33.** $\frac{x}{5}$ **35.** $\frac{a + b}{7}$ **37.** $\frac{p - q}{4}$ **39.** $\frac{a + 3}{a - 3}$ **41.** $x + 5$ **43.** $x - 7$ **45.** $9x$ **47.** $3x + 6$ **49.** $2(x + 5)$ **51.** $(x + 2)(x - 2)$ **53.** $\frac{x}{7}$ **55.** $\frac{x + 5}{8}$ **57.** $\frac{x + 6}{x - 6}$ **59.** $4s$ **61.** $\pi r^2 h$ **63.** $\frac{1}{2}h(b_1 + b_2)$ **65.** Expression **67.** Not an expression **69.** Not an expression **71.** Expression **73.** $35 - x$ **75.** Prt **77.** $400 - x$ **79.** (b) **81.** (b) **83.** True **85.** never **87.** $\frac{DQ}{H}$ **89.** $H - 284$ **91.** Above and Beyond

11.2 Evaluating Algebraic Expressions

< 11.2 Objectives >

1 > Evaluate an algebraic expression

2 > Use the order of operations to evaluate an expression

3 > Use a calculator to evaluate an expression

4 > Use an algebraic expression to solve an application

Algebra gives us a powerful tool to solve problems. To use algebra, we need to learn to *evaluate an algebraic expression.* This means that we have values for the variables in an expression and use them to compute, as indicated by the expression.

Step by Step

Evaluating an Algebraic Expression

Step 1 Replace each variable with its given number value.

Step 2 Compute, following the rules for the order of operations.

Example 1 Evaluating Algebraic Expressions

< Objective 1 >

NOTE

We use parentheses when we make the initial substitution. This helps us avoid some careless errors.

Let $a = 5$ and $b = 7$.

(a) To evaluate $a + b$, we replace a with 5 and b with 7.

$a + b = (5) + (7) = 12$

(b) To evaluate $3ab$, we again replace a with 5 and b with 7.

$3ab = 3(5)(7) = 105$

Check Yourself 1

If $x = 6$ and $y = 7$, evaluate.

(a) $y - x$ **(b)** $5xy$

We follow the rules for the order of operations when evaluating an expression.

Example 2 Evaluating Algebraic Expressions

< Objective 2 >

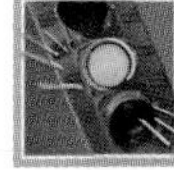

>CAUTION

The expression in part (b) is different than

$(3c)^2 = [3(4)]^2$
$= 12^2 = 144$

Evaluate each expression if $a = 2$, $b = 3$, $c = 4$, and $d = 5$.

(a) $5a + 7b = 5(2) + 7(3)$ Multiply first.
$= 10 + 21 = 31$ Then add.

(b) $3c^2 = 3(4)^2$ Apply the exponent.
$= 3 \cdot 16 = 48$ Then multiply.

(c) $7(c + d) = 7[(4) + (5)]$ Add inside the parentheses.
$= 7 \cdot 9 = 63$

(d) $5a^4 - 2d^2 = 5(2)^4 - 2(5)^2$ Apply the exponents.
$= 5 \cdot 16 - 2 \cdot 25$ Multiply.
$= 80 - 50 = 30$ Subtract.

Check Yourself 2

If $x = 3$, $y = 2$, $z = 4$, and $w = 5$, evaluate each expression.

(a) $4x^2 + 2$ **(b)** $5(z + w)$ **(c)** $7(z^2 - y^2)$

We follow the same steps when one of the variables is replaced by a fraction.

Example 3 Evaluating Expressions

RECALL

The rules for the order of operations require that we multiply first and then add.

Evaluate $5a + 4b$ if $a = -2$ and $b = \frac{3}{4}$.

Replace a with -2 and b with $\frac{3}{4}$.

$$5a + 4b = 5(-2) + 4\left(\frac{3}{4}\right)$$
$$= -10 + 3$$
$$= -7$$

Check Yourself 3

Evaluate $3x + 5y$ if $x = -2$ and $y = -\frac{4}{5}$.

When an algebraic expression contains parentheses or other grouping symbols, remember to follow the proper order of operations.

Example 4 Evaluating Expressions

NOTE

We can interchange parentheses and brackets as convenient.

Evaluate each expression if $x = 3$, $y = -4$, and $z = 1$.

(a) $3x(y + 2z)$

Replace x with 3, y with -4, and z with 1 and evaluate properly.

$$3x(y + 2z) = 3(3)[(-4) + 2(1)]$$
$$= 3(3)(-4 + 2)$$
$$= 3(3)(-2)$$
$$= -18$$

(b) $(1 - 2x)(z - y)$

$$(1 - 2x)(z - y) = [1 - 2(3)][(1) - (-4)]$$
$$= (-5)(5)$$
$$= -25$$

Check Yourself 4

Evaluate each expression if $a = 3$, $b = 0$, and $c = -6$.

(a) $3c(c - a)$ **(b)** $\frac{ab}{c - a}$

We follow the same rules no matter how many variables are in the expression.

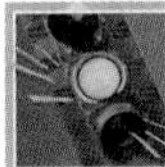

Example 5 Evaluating Expressions

Evaluate each expression if $a = -4$, $b = 2$, $c = -5$, and $d = 6$.

(a) $7a - 4c = 7(-4) - 4(-5)$ ← This becomes $-(-20)$, or $+20$.

$= -28 + 20$

$= -8$

(b) $7c^2 = 7(-5)^2 = 7 \cdot 25$ ← Apply the exponent first, and then multiply by 7.

$= 175$

> CAUTION

When a squared variable is replaced by a negative number, square the negative.

$(-5)^2 = (-5)(-5) = 25$

The exponent applies to -5!

$-5^2 = -(5 \cdot 5) = -25$

The exponent applies only to 5!

(c) $b^2 - 4ac = (2)^2 - 4(-4)(-5)$

$= 4 - 4(-4)(-5)$

$= 4 - 80$

$= -76$

(d) $b(a + d) = (2)[(-4) + (6)]$ ← Add inside the parentheses first.

$= 2(2)$

$= 4$

Check Yourself 5

Evaluate if $p = -4$, $q = 3$, and $r = -2$.

(a) $5p - 3r$ **(b)** $2p^2 + q$ **(c)** $p(q + r)$
(d) $-q^2$ **(e)** $(-q)^2$

When an expression includes division, indicated by a fraction bar, evaluate the numerator and denominator separately. Divide as the final step.

Example 6 Evaluating Algebraic Expressions

If $p = 2$, $q = 3$, and $r = 4$, evaluate:

RECALL

The fraction bar, like parentheses, is a grouping symbol. Work first in the numerator and denominator before dividing.

(a) $\frac{8p}{r}$

Replace p with 2 and r with 4.

$\frac{8p}{r} = \frac{8(2)}{(4)} = \frac{16}{4} = 4$ Divide as the last step.

(b) $\frac{7q + r}{p + q} = \frac{7(3) + (4)}{(2) + (3)}$ Evaluate the top and bottom separately.

$= \frac{21 + 4}{2 + 3} = \frac{25}{5} = 5$

Check Yourself 6

Evaluate each expression if $c = 5$, $d = 8$, and $e = 3$.

(a) $\frac{6c}{e}$ **(b)** $\frac{4d + e}{c}$ **(c)** $\frac{10d - e}{d + e}$

We follow the same steps, even when working with more complicated expressions.

Example 7 Evaluating Expressions

Evaluate each expression if $x = 4$, $y = -5$, $z = 2$, and $w = -3$.

(a) $\dfrac{z - 2y}{x} = \dfrac{(2) - 2(-5)}{(4)} = \dfrac{2 + 10}{4}$

$= \dfrac{12}{4} = 3$

(b) $\dfrac{3x - w}{2x + w} = \dfrac{3(4) - (-3)}{2(4) + (-3)} = \dfrac{12 + 3}{8 + (-3)}$

$= \dfrac{15}{5} = 3$

Check Yourself 7

Evaluate each expression if $m = -6$, $n = 4$, and $p = -3$.

(a) $\dfrac{m + 3n}{p}$ **(b)** $\dfrac{4m + n}{m + 4n}$

We can use a calculator or computer to evaluate an expression. Calculators are programmed to correctly follow the order of operations.

	Algebraic Notation	Calculator Notation
Addition	$6 + 2$	6 [+] 2
Subtraction	$4 - 8$	4 [−] 8
Multiplication	$(3)(-5)$	3 [×] [(−)] 5 or 3 [×] 5 [+/−]
Division	$\dfrac{8}{6}$	8 [÷] 6
Exponential	3^4	3 [^] 4 or 3 [y^x] 4
	$(-3)^4$	[(] [(−)] 3 [)] [^] 4 or [(] 3 [+/−] [)] [y^x] 4

Calculators are especially helpful when working with decimals, as we see in Example 8.

Example 8 Evaluating Expressions

< Objective 3 >

Evaluate each expression if $A = 2.3$, $B = 8.4$, and $C = 4.5$. Round your answer to the nearest tenth.

(a) $A + B(-C)$

Letting A, B, and C take on the given values, we have

2.3 [+] 8.4 [×] [(−)] 4.5 [ENTER] −35.5

(b) $-B + (-A)C^2$

Substituting the given values, we have

[(−)] 8.4 [+] [(−)] 2.3 [×] 4.5 [^] 2 [ENTER] −54.975

Rounding to the nearest tenth gives us −55.0.

NOTE

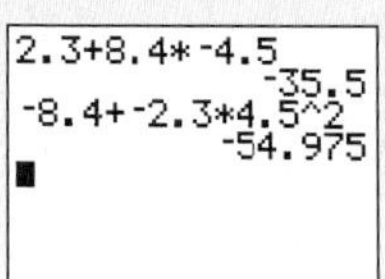

Check Yourself 8

Evaluate each expression when $A = -2$, $B = 3$, and $C = 5$.

(a) $A + B(-C)$ (b) $C + BA^3$ (c) $\frac{4(B - C)}{2A}$

Calculators are also helpful when evaluating complicated expressions. If there is a fraction bar, you must remember to use parentheses to group the numerator or denominator properly.

Example 9 Using a Calculator to Evaluate an Expression

Use a calculator to evaluate each expression.

(a) $\frac{4x + y}{z}$ if $x = 2$, $y = 1$, and $z = 3$

> **RECALL**
>
> Graphing calculators usually use an [ENTER] key instead of an [=] key.

Begin by writing the expression with the values substituted for the variables.

$$\frac{4x + y}{z} = \frac{4(2) + (1)}{(3)}$$

Then, enter the numerical expression into a calculator.

[(] 4 [×] 2 [+] 1 [)] [÷] 3 [ENTER] Remember to enclose the entire numerator in parentheses.

The display should read 3.

```
(4*2+1)/3
                3
(7*2-6)/(3*-2-2)
               -1
```

(b) $\frac{7x - y}{3z - x}$ if $x = 2$, $y = 6$, and $z = -2$

Again, we begin by substituting.

$$\frac{7x - y}{3z - x} = \frac{7(2) - (6)}{3(-2) - 2}$$

Then, we enter the expression into a calculator.

[(] 7 [×] 2 [−] 6 [)] [÷] [(] 3 [×] [(−)] 2 [−] 2 [)] [ENTER]

The display should read -1.

Check Yourself 9

Use a calculator to evaluate each expression if $x = 2$, $y = -6$, and $z = 5$.

(a) $\frac{2x + y}{z}$ (b) $\frac{4y - 2z}{3x}$

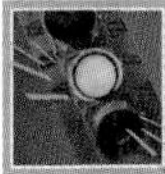

> CAUTION

A calculator follows the correct order of operations when evaluating an expression. If we omit the parentheses in Example 9(b) and enter

7 [×] 2 [−] 6 [÷] 3 [×] [(−)] 2 [−] 2 [ENTER]

the calculator will interpret our input as $7 \cdot 2 - \frac{6}{3} \cdot (-2) - 2$, which is not what we wanted.

Whether working with a calculator or pencil and paper, you must remember to take care both with signs and with the order of operations.

As you might expect, we use algebraic expressions in many applications. We conclude this section with two such applications.

Example 10 An Application of Algebra

< Objective 4 >

A car is advertised for rent at a cost of \$59 per day plus 20 cents per mile. The total cost can be found by evaluating the expression

$59d + 0.20m$

in which d represents the number of days and m the number of miles. Find the total cost for a 3-day rental if 250 miles are driven.

$59(3) + 0.20(250)$

$= 177 + 50$

$= 227$

The total cost is \$227.

Check Yourself 10

The cost to hold a wedding reception at a certain cultural arts center is \$195 per hour plus \$27.50 per guest. The total cost can be found by evaluating the expression

$195h + 27.50g$

in which h represents the number of hours and g the number of guests. Find the total cost for a 4-hour reception with 220 guests.

Example 11 Applying Algebra

The lens formula (in the study of *optics*) states that the *focal length* (the distance between a *lens* and the *focal point*) of a lens is given by the formula

$$\frac{d_o d_i}{d_o + d_i}$$

in which d_o is the distance of an object from a thin lens and d_i is the distance of the object's image from the lens. Find the focal length of a lens if an object 24 in. from a lens produces an image 1 in. from the lens.

We substitute as before.

$$\frac{d_o d_i}{d_o + d_i} = \frac{(24)(1)}{(24) + (1)}$$

$$= \frac{24}{25}$$

So the focal length is $\frac{24}{25}$ in. (or 0.96 in.).

Check Yourself 11

In an electric circuit with electromotive force of E volts and resistance R ohms, the rate of change in the current with respect to resistance is given by

$-\frac{E}{R^2}$ amperes per ohm

Find the rate of change in the current with respect to resistance if $E = 100$ volts and $R = 12$ ohms.

Check Yourself ANSWERS

1. **(a)** 1; **(b)** 210 **2.** **(a)** 38; **(b)** 45; **(c)** 84 **3.** -10 **4.** **(a)** 162; **(b)** 0
5. **(a)** -14; **(b)** 35; **(c)** -4; **(d)** -9; **(e)** 9 **6.** **(a)** 10; **(b)** 7; **(c)** 7 **7.** **(a)** -2; **(b)** -2
8. **(a)** -17; **(b)** -19; **(c)** 2 **9.** **(a)** -0.4; **(b)** -5.67 **10.** \$6,830
11. $-\frac{25}{36}$ amperes per ohm (approximately -0.69)

Reading Your Text

These fill-in-the-blank exercises will help you understand some of the key vocabulary used in this section. The answers to these exercises are in the Answers Appendix at the back of the text.

(a) Finding the value of an expression is called ________ the expression.

(b) If a squared variable is replaced by a negative number, the result is ________.

(c) Always follow the order of ________ when evaluating an algebraic expression.

(d) A fraction bar is a ________ symbol.

Skills Calculator/Computer Career Applications Above and Beyond

11.2 exercises

< Objectives 1 and 2 >

Evaluate each expression if $a = -2$, $b = 5$, $c = -4$, and $d = 6$.

1. $3c - 2b$ **2.** $4c - 2b$ **3.** $8b + 2c$ **4.** $7a - 2c$

5. $-b^2 + b$ **6.** $(-b)^2 + b$ **7.** $3a^2$ **8.** $6c^2$

9. $c^2 - 2d$ **10.** $3a^2 + 4c$ **11.** $2a^2 + 3b^2$ **12.** $4b^2 - 2c^2$

13. $2(a + b)$ **14.** $5(b - c)$ **15.** $4(2a - d)$ **16.** $6(3c - d)$

17. $a(b + 3c)$ **18.** $c(3a - d)$ **19.** $\frac{6d}{c}$ **20.** $\frac{8b}{5c}$

21. $\frac{3d + 2c}{b}$ **22.** $\frac{2b + 3d}{2a}$ **23.** $\frac{2b - 3a}{c + 2d}$ **24.** $\frac{3d - 2b}{5a + d}$

25. $d^2 - b^2$ **26.** $c^2 - a^2$ **27.** $(d - b)^2$ **28.** $(c - a)^2$

29. $(d - b)(d + b)$ **30.** $(c - a)(c + a)$ **31.** $d^3 - b^3$ **32.** $c^3 + a^3$

33. $(d - b)^3$ **34.** $(c + a)^3$ **35.** $(d - b)(d^2 + db + b^2)$ **36.** $(c + a)(c^2 - ac + a^2)$

37. $b^2 + a^2$ **38.** $d^2 - a^2$ **39.** $(b + a)^2$ **40.** $(d - a)^2$

41. $a^2 + 2ad + d^2$ **42.** $b^2 - 2bc + c^2$

Evaluate each expression if $x = -2$, $y = -3$, and $z = 4$.

43. $x^2 - 2y^2 + z^2$ **44.** $4yz + 6xy$ **45.** $2xy - (x^2 - 2yz)$ **46.** $3yz - 6xyz + x^2y^2$

47. $2y(z^2 - 2xy) + yz^2$ **48.** $-z - (-2x - yz)$

Evaluate each expression if $x = -3, y = 5,$ *and* $z = \frac{2}{3}$.

49. $x^2 - y$

50. $\frac{y - x}{z}$

51. $z - y^2$

52. $z - \frac{z + x}{y - x}$

< Objective 4 >

53. **Geometry** The formula for the area of a triangle is given by $A = \frac{1}{2}bh$. Find the area of a triangle if $b = 4$ cm and $h = 8$ cm.

54. **Geometry** The perimeter of a rectangle with length L and width W is given by the formula $P = 2L + 2W$. Find the perimeter of a rectangle if its length is 10 in. and its width is 5 in.

55. **Business and Finance** The simple interest I on a principal P dollars at interest rate r for time t is given by $I = Prt$. Find the simple interest earned on a principal of \$6,000 at 3% for 2 years. VIDEO
Hint: 3% = 0.03.

56. **Business and Finance** Use the simple interest formula in exercise 55 to find the interest earned on a principal of \$12,500 at 4.5% for 3 years.

57. **Business and Finance** Use the formula $P = \frac{I}{r \cdot t}$ to find the principal invested if the total interest earned was \$150 and the interest rate was 4% for 2 years.

58. **Business and Finance** Use the formula $r = \frac{I}{P \cdot t}$ to find the interest rate if \$5,000 earns \$1,500 interest in 6 years.

59. **Science and Medicine** A formula that relates Celsius and Fahrenheit temperatures is $F = \frac{9}{5}C + 32$. If the low temperature was $-10°$C one day, what was the Fahrenheit equivalent?

60. **Geometry** The area of a circle with radius r is $A = \pi r^2$. Use $\pi \approx 3.14$ to approximate the area of a circle if the radius is 3 ft.

61. **Business and Finance** A local telephone company offers a long-distance telephone plan that charges \$5.25 per month and \$0.08 per minute of calling time. The expression $0.08t + 5.25$ represents the monthly long-distance bill for a customer who makes t minutes (min) of long-distance calling on this plan. Find the monthly bill for a customer who makes 173 min of long-distance calls on this plan.

62. **Science and Medicine** The speed of a model car as it slows down is given by $v = 20 - 4t$, where v is the speed in meters per second (m/s) and t is the time in seconds (s) during which the car has slowed. Find the speed of the car 1.5 s after it has begun to slow.

In each problem, decide if the given numbers make the statement **true** *or* **false.**

63. $x - 7 = 2y + 5; x = 22, y = 5$

64. $3(x - y) = 6; x = 5, y = -3$

65. $2(x + y) = 2x + y; x = -4, y = -2$

66. $x^2 - y^2 = x - y; x = 4, y = -3$

Determine whether each statement is **true** *or* **false.**

67. When evaluating an expression that has a fraction bar, dividing the numerator by the denominator is the first step.

68. The value of w^2 is always nonnegative.

Complete each statement with **always, sometimes,** *or* **never.**

69. When n is replaced with a number, the value of $-n^2$ is ________ positive.

70. When x is replaced with a number, the value of $-5x$ is ________ negative.

< Objective 3 >

Use your calculator to evaluate each expression if $x = -2.34$, $y = -3.14$, and $z = 4.12$. Round your results to the nearest tenth.

71. $x + yz$ **72.** $y - 2z$ **73.** $x^2 - z^2$ **74.** $x^2 + y^2$

75. $\frac{xy}{z - x}$ **76.** $\frac{y^2}{zy}$ **77.** $\frac{2x + y}{2x + z}$ **78.** $\frac{x^2y^2}{xz}$

Use a calculator to evaluate the expression $x^2 - 4x^3 + 3x$ for each value.

79. $x = 3$ **80.** $x = 12$ **81.** $x = 27$ **82.** $x = 48$

Skills | Calculator/Computer | **Career Applications** | Above and Beyond

83. **Allied Health** The concentration, in micrograms per milliliter (μg/mL), of an antihistamine in a patient's bloodstream can be approximated using the formula $-2t^2 + 13t + 1$, in which t is the number of hours since the drug was administered. Approximate the concentration of the antihistamine 1 hour after it has been administered. VIDEO

84. **Allied Health** Use the formula given in exercise 83 to approximate the concentration of the antihistamine 3 hours after it has been administered.

85. **Electrical Engineering** Evaluate $\frac{rT}{5,252}$ for $r = 1,180$ and $T = 3$ (round to the nearest thousandth).

86. **Mechanical Engineering** The kinetic energy (in joules) of a particle is given by $\frac{1}{2}mv^2$. Find the kinetic energy of a particle if its mass is 60 kg and its velocity is 6 m/s.

Skills | Calculator/Computer | Career Applications | **Above and Beyond**

87. Write an English interpretation for each algebraic expression.

(a) $(2x^2 - y)^3$ (b) $3n - \frac{n - 1}{2}$ (c) $(2n + 3)(n - 4)$

88. Is $a^n + b^n = (a + b)^n$? Try a few numbers and decide if you think this is true for all numbers, true for some numbers, or never true. Write an explanation of your findings and give examples.

Answers

1. −22 **3.** 32 **5.** −20 **7.** 12 **9.** 4 **11.** 83 **13.** 6 **15.** −40 **17.** 14 **19.** −9 **21.** 2 **23.** 2 **25.** 11 **27.** 1 **29.** 11 **31.** 91 **33.** 1 **35.** 91 **37.** 29 **39.** 9 **41.** 16 **43.** 2 **45.** −16 **47.** −72 **49.** 4 **51.** $-\frac{73}{3}$ **53.** 16 cm^2 **55.** \$360 **57.** \$1,875 **59.** 14°F **61.** \$19.09 **63.** True **65.** False **67.** False **69.** never **71.** −15.3 **73.** −11.5 **75.** 1.1 **77.** 14 **79.** −90 **81.** −77,922 **83.** 12 μg/mL **85.** 0.674 **87.** Above and Beyond

Activity 31 ::

Evaluating Net Pay

Many people are paid based on the number of hours they work. However, while a person may earn a fixed number of dollars per hour, the person's actual paycheck differs from this straightforward multiplication.

The **gross pay** of an hourly employee is determined by multiplying the number of hours worked by the amount paid per hour. More generally, gross pay is the amount earned before any money is deducted. A person's **net pay** is the amount the person actually receives, after all deductions.

1. Ilyona earns \$12.50 per hour working at her local library. Find her gross pay if she works a 35-hour week.
2. The federal government deducts 6% of her gross pay for taxes and an additional 7% for FICA. The state also deducts 5% of her gross for state taxes. How much do the federal and state governments deduct from her pay?
3. Ilyona contributes \$25 each week to her benefits package, and \$8 each week is paid as city employees' union dues. Find Ilyona's net weekly pay.
4. Find her yearly gross and net earnings. Assume she is paid for 52 weeks.

Obviously, it is not efficient for a large company to compute steps 1 to 4 manually, one at a time, for each employee. By creating and using formulas, the process can be made more efficient.

5. Create an expression using r for hourly pay and t for the number of hours worked that describes a person's gross pay.
6. Create an expression that describes a person's net pay. Assume the deductions stated in part 2 apply.
7. Ilyona's supervisor is paid \$15.75 per hour and works 40 hours per week. Use the expression found in exercise 6 to find the supervisor's net pay.

11.3 Simplifying Algebraic Expressions

< 11.3 Objectives >

1 > Identify terms and coefficients

2 > Identify like terms

3 > Combine like terms

To find the perimeter of a rectangle, we add 2 times the length and 2 times the width. In algebra, we write this as

RECALL

The perimeter of a figure is the distance around that figure.

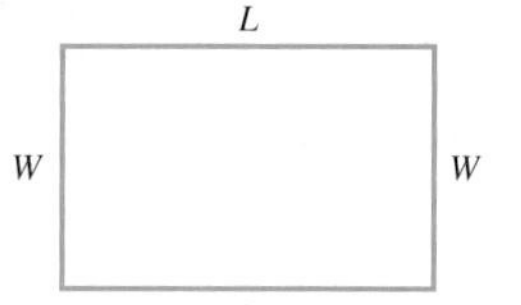

Perimeter $= 2L + 2W$

We call $2L + 2W$ an **algebraic expression,** or more simply an **expression.** Recall that an expression is a mathematical idea written in symbols. It is a meaningful collection of letters, numbers, and operation signs.

Some expressions are

$5x^2$

$3a + 2b$

$4x^3 - 2y + 1$

$3(x^2 + y^2)$

Addition and subtraction signs break expressions into smaller parts called *terms.*

Definition

Term — A **term** is an expression that can be written as a number, or the product of a number and one or more variables and their exponents.

In an expression, each sign (+ or −) is a part of the term that follows the sign.

Example 1 — Identifying Terms

< Objective 1 >

RECALL

Each term "owns" the sign that precedes it.

(a) $5x^2$ has one term.

(b) $\underbrace{3a}_{\text{Term}} \underbrace{+ 2b}_{\text{Term}}$ has two terms: $3a$ and $2b$.

(c) $\underbrace{4x^3}_{\text{Term}} \underbrace{- 2y}_{\text{Term}} \underbrace{+ 1}_{\text{Term}}$ has three terms: $4x^3$, $-2y$, and 1.

Check Yourself 1

List the terms of each expression.

(a) $2b^4$ **(b)** $5m + 3n$ **(c)** $2s^2 - 3t - 6$

A term may have any number of factors. For instance, $5xy$ is a term. Its factors are 5, x, and y. The number factor of a term is called the **numerical coefficient.** So for the term $5xy$, the numerical coefficient is 5.

Example 2 Identifying the Numerical Coefficient

NOTE

We usually simply say the *coefficient* when referring to the numerical coefficient.

(a) $4a$ has the numerical coefficient 4.

(b) $6a^3b^4c^2$ has the numerical coefficient 6.

(c) $-7m^2n^3$ has the coefficient -7.

(d) Because $1 \cdot x = x$, the coefficient of x is understood to be 1.

Check Yourself 2

Give the numerical coefficient of each term.

(a) $8a^2b$ **(b)** $-5m^3n^4$ **(c)** y

If terms contain exactly the *same letters* (or variables) raised to the *same powers,* they are called **like terms.**

Example 3 Identifying Like Terms

< Objective 2 >

(a) Each pair represents like terms.

$6a$ and $7a$

$5b^2$ and b^2

$10x^2y^3z$ and $-6x^2y^3z$

$-3m^2$ and m^2

Each pair of terms has the same variables, with each variable raised to the same power—the coefficients do not need to be the same.

(b) These are *not* like terms.

Different variables

$6a$ and $7b$

Different exponents

$5b^2$ and b^3

Different exponents

$3x^2y$ and $4xy^2$

Check Yourself 3

List the like terms.

$5a^2b$ ab^2 a^2b $-3a^2$ $4ab$ $3b^2$ $-7a^2b$

Like terms of an expression can always be combined into a single term.

$$\underbrace{x + x}_{2x} + \underbrace{x + x + x + x + x}_{5x} = \underbrace{x + x + x + x + x + x + x}_{7x}$$

RECALL

Here we use the distributive property from Section 1.5.

Rather than having to write out all those x's, try

$2x + 5x = (2 + 5)x = 7x$

In the same way,

$9b + 6b = (9 + 6)b = 15b$

and

$10a - 4a = (10 - 4)a = 6a$

This leads us to a procedure for *combining like terms.*

Step by Step

Combining Like Terms

To combine like terms:

Step 1 Add or subtract the numerical coefficients.

Step 2 Attach the common variables.

Example 4 Combining Like Terms

< Objective 3 >

RECALL

When any factor is multiplied by 0, the product is 0.

Combine like terms:

(a) $8m + 5m = (8 + 5)m = 13m$

(b) $5pq^3 - 4pq^3 = 1pq^3 = pq^3$ Multiplication by 1 is understood.

(c) $7a^3b^2 - 7a^3b^2 = 0a^3b^2 = 0$

Check Yourself 4

Combine like terms.

(a) $6b + 8b$ **(b)** $12x^2 - 3x^2$

(c) $8xy^3 - 7xy^3$ **(d)** $9a^2b^4 - 9a^2b^4$

The idea is the same for expressions involving more than two terms.

Example 5 Combining Like Terms

NOTES

The distributive property can be used over any number of like terms.

With practice you will do these steps mentally instead of writing them out.

Combine like terms.

(a) $4xy - xy + 2xy$

$= (4 - 1 + 2)xy$

$= 5xy$

Only like terms can be combined.

(b) $8x - 2x + 5y$

$= 6x + 5y$

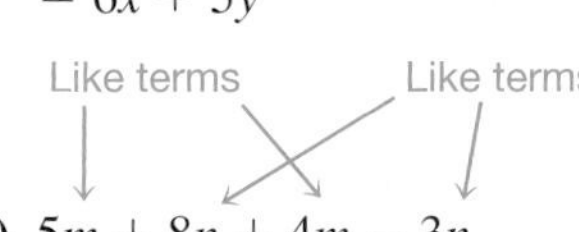

(c) $5m + 8n + 4m - 3n$ We use the associative and commutative properties.

$= (5m + 4m) + (8n - 3n)$

$= 9m + 5n$

(d) $4x^2 + 2x - 3x^2 + x$

$= (4x^2 - 3x^2) + (2x + x)$

$= x^2 + 3x$

Check Yourself 5

Combine like terms.

(a) $4m^2 - 3m^2 + 8m^2$ **(b)** $9ab + 3a - 5ab$

(c) $4p + 7q + 5p - 3q$

As these examples illustrate, combining like terms often means changing the grouping and the order in which the terms are written. Again, all this is possible because of the properties of addition that we introduced in Section 1.2.

You may not realize it, but adding and subtracting algebraic expressions occurs all the time in the world. You have probably combined like terms successfully many times before ever taking this course.

Example 6 An Application of Algebra

In anticipation of a holiday rush, a produce market receives six cases of apples and four cases of oranges from their supplier. The market already had two cases of apples and two cases of oranges in stock. How many cases of each does the market have after the delivery?

We add

6 apples + 4 oranges + 2 apples + 2 oranges

and combine like terms using the *commutative* property.

(6 apples + 2 apples) + (4 oranges + 2 oranges) = 8 apples + 6 oranges

Therefore, the market begins the day with eight cases of apples and six cases of oranges.

If we let a represent the number of cases of apples and r represent the number of cases of oranges, then we write

$$6a + 4r + 2a + 2r = (6a + 2a) + (4r + 2r)$$
$$= 8a + 6r$$

NOTE

We cannot add apples and oranges.

Check Yourself 6

An electronics store has 8 single packages and 20 two-packs of flash drives in stock. They receive a shipment of 48 single packs and 24 two-packs. Algebraically represent the number and type of packages of flash drives that the store has after the shipment arrives.

Check Yourself ANSWERS

1. (a) $2b^4$; **(b)** $5m, 3n$; **(c)** $2s^2, -3t, -6$ **2. (a)** 8; **(b)** -5; **(c)** 1
3. The like terms are $5a^2b$, a^2b, and $-7a^2b$ **4. (a)** $14b$; **(b)** $9x^2$; **(c)** xy^3; **(d)** 0
5. (a) $9m^2$; **(b)** $4ab + 3a$; **(c)** $9p + 4q$ **6.** $56x + 44y$

Reading Your Text

These fill-in-the-blank exercises will help you understand some of the key vocabulary used in this section. The answers to these exercises are in the Answers Appendix at the back of the text.

(a) The product of a number and a variable is called a ____________.

(b) The ____________ factor of a term is called the numerical coefficient.

(c) Terms that contain exactly the same variables raised to the same powers are called ____________ terms.

(d) The ____________ property enables us to combine like terms into a single term.

< Objective 1 >

List the terms of each expression.

1. $5a + 2$ **2.** $7a - 4b$ **3.** $4x^3$

4. $3x^2$ **5.** $3x^2 + 3x - 7$ **6.** $2a^3 - a^2 + a$

< Objective 2 >

List the like terms in each group of terms.

7. $5ab, 3b, 3a, 4ab$ **8.** $9m^2, 8mn, 5m^2, 7m$

9. $4xy^2, 2x^2y, 5x^2, -3x^2y, 5y, 6x^2y$  **10.** $8a^2b, 4a^2, 3ab^2, -5a^2b, 3ab, 5a^2b$

< Objective 3 >

Combine the like terms.

11. $3m + 7m$  **12.** $6a^2 + 8a^2$ **13.** $7b^3 + 10b^3$

14. $7rs + 13rs$ **15.** $21xyz + 7xyz$ **16.** $4mn^2 + 15mn^2$

17. $9z^2 - 3z^2$ **18.** $7m - 6m$ **19.** $5a^3 - 5a^3$

20. $13xy - 9xy$ **21.** $19n^2 - 18n^2$ 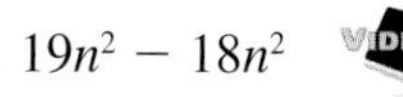 **22.** $7cd - 7cd$

23. $21p^2q - 6p^2q$ **24.** $17r^3s^2 - 8r^3s^2$ **25.** $10x^2 - 7x^2 + 3x^2$

26. $13uv + 5uv - 12uv$ **27.** $9a - 7a + 4b$ 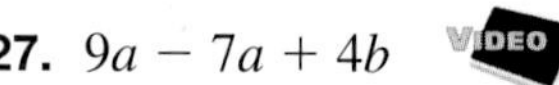 **28.** $5m^2 - 3m + 6m^2$

29. $7x + 5y - 4x - 4y$ **30.** $6a^2 + 11a + 7a^2 - 9a$ **31.** $4a + 7b + 3 - 2a + 3b - 2$

32. $5p^2 + 2p + 8 + 4p^2 + 5p - 6$ **33.** $\frac{2}{3}m + 3 + \frac{4}{3}m$ **34.** $\frac{1}{5}a - 2 + \frac{4}{5}a$

35. $\frac{13}{5}x + 2 - \frac{3}{5}x + 5$ **36.** $\frac{17}{12}y + 7 + \frac{7}{12}y - 3$ **37.** $2.3a + 7 + 4.7a + 3$

38. $5.8m + 4 - 2.8m + 11$

Perform the indicated operations.

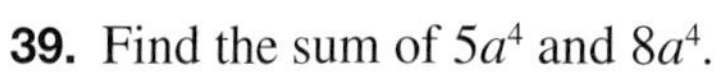

39. Find the sum of $5a^4$ and $8a^4$.  **40.** What is the sum of $9p^2$ and $12p^2$? 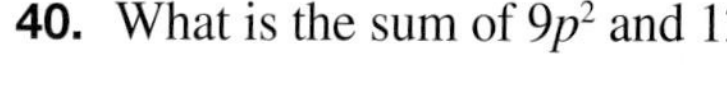

41. Subtract $12a^3$ from $15a^3$. **42.** Subtract $5m^3$ from $18m^3$.

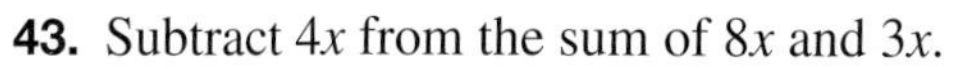

43. Subtract $4x$ from the sum of $8x$ and $3x$. **44.** Subtract $8ab$ from the sum of $7ab$ and $5ab$.

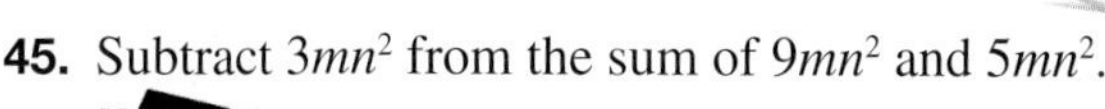

45. Subtract $3mn^2$ from the sum of $9mn^2$ and $5mn^2$. **46.** Subtract $4x^2y$ from the sum of $6x^2y$ and $12x^2y$.

Use the distributive property to remove the parentheses in each expression. Then, simplify the expression by combining like terms.

47. $2(3x + 2) + 4$ **48.** $3(4z + 5) - 9$ **49.** $5(6a - 2) + 12a$ 

50. $7(4w - 3) - 25w$ **51.** $4s + 2(s + 4) + 4$ **52.** $5p + 4(p + 3) - 8$

Evaluate each expression if $a = 2$, $b = 3$, and $c = 5$. Be sure to combine like terms, when possible, as the first step.

53. $7a^2 + 3a$ **54.** $11b^2 - 9b$ **55.** $3c^2 + 5c^2$ **56.** $9b^3 - 5b^3$

57. $5b + 3a - 2b$ **58.** $7c - 2b + 3c$ **59.** $5ac^2 - 2ac^2$ **60.** $5a^3b - 2a^3b$

61. **Geometry** A rectangle has sides that measure $8x + 9$ and $6x - 7$. Find the simplified expression that represents its perimeter.

62. **Geometry** A triangle has sides measuring $3x + 7$, $4x - 9$, and $5x + 6$. Find the simplified expression that represents its perimeter.

63. **Construction** A wooden beam is $(3y^2 + 3y - 2)$ meters (m) long. If a piece $(y^2 - 8)$ m is cut, find an expression that represents the length of the remaining piece of beam.

64. **Construction** A steel girder is $(9y^2 + 6y - 4)$ m long. Two pieces are cut from the girder. One has length $(3y^2 + 2y - 1)$ m and the other has length $(4y^2 + 3y - 2)$ m. Find the length of the remaining piece.

65. **Geometry** Find an expression for the perimeter of the given triangle.

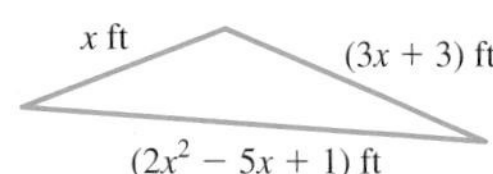

66. **Geometry** Find an expression for the perimeter of the given rectangle.

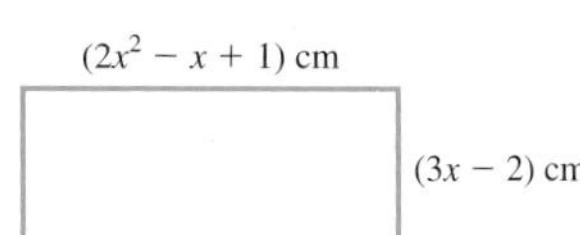

67. **Geometry** Find an expression for the perimeter of the given figure.

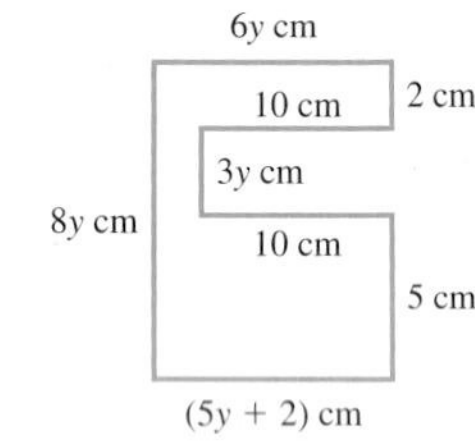

68. **Geometry** Find the perimeter of the accompanying figure.

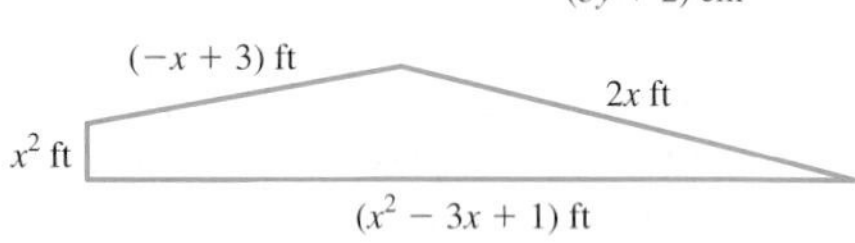

69. **Business and Finance** The cost of producing x units of an item is $150 + 25x$. The revenue from selling x units is $90x - x^2$. The profit is given by the revenue minus the cost. Find the simplified expression that represents the profit. chapter 11 > Make the Connection

70. **Business and Finance** The revenue from selling y units is $3y^2 - 2y + 5$ and the cost of producing y units is $y^2 + y - 3$. Find the simplified expression that represents the profit. chapter 11 > Make the Connection

Determine whether each statement is **true** *or* **false.**

71. For two terms to be *like terms,* the numerical coefficients must match.

72. The key property that allows like terms to be combined is the distributive property.

Complete each statement with **always, sometimes,** *or* **never.**

73. Like terms can ____________ be combined.

74. When adding two expressions, the terms can ____________ be rearranged.

Skills | **Calculator/Computer** | Career Applications | Above and Beyond

Use a calculator to evaluate each expression for the given values of the variables. Round your results to the nearest tenth.

75. $7x^2 - 5y^3$; $x = 7.1695$, $y = 3.128$

76. $2x^2 + 3y + 5x$; $x = 3.61$, $y = 7.91$

77. $(4x^2y)(2xy^2) - 5x^3y$; $x = 1.29$, $y = 2.56$

78. $3x^3y - 4xy + 2x^2y^2$; $x = 3.26$, $y = 1.68$

79. **Allied Health** A person's body mass index (BMI) can be calculated using their height h, in inches, and their weight w, in pounds, with the formula

$$\frac{703w}{h^2}$$

Compute the BMI of a 69-in., 190-lb man (to the nearest tenth).

80. **Allied Health** A person's body mass index (BMI) can be calculated using their height h, in centimeters, and their weight w, in kilograms, with the formula

$$\frac{10{,}000w}{h^2}$$

Compute the BMI of a 160-cm, 70-kg woman (to the nearest tenth).

81. **Mechanical Engineering** A primary beam can support a load of $54p$. A second beam is added that can support a load of $32p$. What is the total load that the two beams can support? VIDEO

82. **Mechanical Engineering** Two objects are spinning on the same axis. The moment of inertia of the first object is $\frac{6^3}{12}b$. The moment of inertia of the second object is given by $\frac{30^3}{36}b$. The total moment of inertia is given by the sum of the moments of inertia of the two objects. Write a simplified expression for the total moment of inertia for the two objects described.

Skills Calculator/Computer Career Applications **Above and Beyond**

83. A toy store begins the day with four Frisbees and eight basketballs in stock. During the morning shift, two Frisbees and one basketball are sold. In the afternoon, a shipment containing six Frisbees arrived. The afternoon shift sells three Frisbees and two basketballs.

Algebraically represent the number of Frisbees and of basketballs that are left at the end of the day (use f to represent the number of Frisbees and b to represent the number of basketballs).

84. Determine the number of pounds of each type of coffee that a retailer has at the end of the day, given the following information.

A retailer begins the day with 24 lb of Kona coffee, 17 lb of Italian roast, and 12 lb of Sumatran roast.

The retailer sells 8 lb of the Kona variety, 11 lb of the Italian, and 7 lb of the Sumatran. A delivery brings 4 lb of Kona and 16 lb of Sumatran coffees.

Express your answer algebraically, using K, I, and S to represent the number of pounds of Kona, Italian, and Sumatran coffees, respectively.

85. Write a paragraph explaining the difference between n^2 and $2n$.

86. Complete the explanation: "x^3 and $3x$ are not the same because . . . ."

87. Complete the statement: "$x + 2$ and $2x$ are different because . . . ."

88. Write an English phrase for each algebraic expression.

(a) $2x^3 + 5x$ **(b)** $(2x + 5)^3$ **(c)** $6(n + 4)^2$

89. Work with another student to complete this exercise. Place >, <, or = in the blank in these statements.

1^2 _____ 2^1

2^3 _____ 3^2

3^4 _____ 4^3

4^5 _____ 5^4

What happens as the table of numbers is extended? Try more examples.

What sign seems to occur the most in your table: >, <, or =?

Write an algebraic statement for the pattern of signs in this table. Do you think this is a pattern that continues? Add more lines to the table and extend the pattern to the general case by writing the pattern in algebraic notation. Write a short paragraph stating your conjecture.

90. Work with other students on this exercise.

Part 1: Evaluate the three expressions $\frac{n^2 - 1}{2}$, n, and $\frac{n^2 + 1}{2}$, using odd values of n: 1, 3, 5, 7, etc. Make a chart like the following one and complete it.

n	$a = \frac{n^2 - 1}{2}$	$b = n$	$c = \frac{n^2 + 1}{2}$	a^2	b^2	c^2
1						
3						
5						
7						
9						
11						
13						
15						

Part 2: The numbers, a, b, and c that you get in each row have a surprising relationship to each other. Complete the last three columns and work together to discover this relationship. You may want to find out more about the history of this famous number pattern.

Answers

1. $5a, 2$ **3.** $4x^3$ **5.** $3x^2, 3x, -7$ **7.** $5ab, 4ab$ **9.** $2x^2y, -3x^2y, 6x^2y$ **11.** $10m$ **13.** $17b^3$ **15.** $28xyz$ **17.** $6z^2$ **19.** 0 **21.** n^2 **23.** $15p^2q$ **25.** $6x^2$ **27.** $2a + 4b$ **29.** $3x + y$ **31.** $2a + 10b + 1$ **33.** $2m + 3$ **35.** $2x + 7$ **37.** $7a + 10$ **39.** $13a^4$ **41.** $3a^3$ **43.** $7x$ **45.** $11mn^2$ **47.** $6x + 8$ **49.** $42a - 10$ **51.** $6s + 12$ **53.** 34 **55.** 200 **57.** 15 **59.** 150 **61.** $28x + 4$ **63.** $(2y^2 + 3y + 6)$ m **65.** $(2x^2 - x + 4)$ ft **67.** $(22y + 29)$ cm **69.** $-x^2 + 65x - 150$ **71.** False **73.** always **75.** 206.8 **77.** 260.6 **79.** 28.1 **81.** $86p$ **83.** $5f + 5b$ **85.** Above and Beyond **87.** Above and Beyond **89.** Above and Beyond

Writing Equations

In Section 11.1, you learned to translate phrases to algebraic expressions. In most applications, you need more than an expression; you need an equation.

1. Write an algebraic equation for the statement "Three more than a number is 9."
2. Write an algebraic equation describing "an employee's gross pay is the hourly pay times the number of hours worked."
3. Use the equation in exercise 2 to determine the gross pay of someone who works 40 hours, earning \$9.75 per hour.
4. Create an equation that determines the net pay if the employee in exercise 2 pays a total of 16% of the gross pay to the federal and state governments.
5. Determine the net pay for the employee in exercise 3.
6. Write a paragraph describing some reasons why, or situations in which, forming an equation to compute net pay might be useful.
7. **(a)** Describe another situation in which constructing an equation would be useful.
   (b) Construct an equation for the situation described in part (a).

11.4 Solving Equations with the Addition Property

< 11.4 Objectives >

1 > Determine whether a number is a solution of an equation

2 > Use the addition property to solve equations

3 > Use equations to solve applications

In the remainder of this chapter we work with one of the most important tools of mathematics—the equation. The ability to recognize and solve various types of equations is probably the most useful algebraic skill you will learn. To start, we define an *equation*.

Definition

Equation An **equation** is a mathematical statement that two expressions are equal.

Some examples are $3 + 4 = 7$, $x + 3 = 5$, $P = 2L + 2W$. As you can see, an equal sign (=) separates the two expressions. These expressions are often called the *left side* and the *right side* of the equation.

$$\underbrace{x + 3}_{\text{Left side}} \underset{\text{Equals}}{=} \underset{\text{Right side}}{5}$$

NOTE

An equation such as

$x + 3 = 5$

is called a **conditional equation** because it can be either true or false depending on the value of the variable.

An equation may be either true or false. For instance, $3 + 4 = 7$ is true because both sides name the same number. What about an equation such as $x + 3 = 5$ that has a variable on one side? Any number can replace x in the equation. However, only one number makes this equation a true statement.

$$\text{If} \quad x = \begin{cases} 1 & 1 + 3 = 5 \text{ is false} \\ 2 & 2 + 3 = 5 \text{ is true} \\ 3 & 3 + 3 = 5 \text{ is false} \end{cases}$$

The number 2 is called a *solution* (or *root*) of the equation $x + 3 = 5$ because substituting 2 for x gives a true statement. 2 is the only solution to this equation.

Definition

Solution A **solution** to an equation is any value for the variable that makes the equation a true statement.

Example 1 Verifying a Solution

< Objective 1 >

(a) Is 3 a solution of the equation $2x + 4 = 10$?

To find out, replace x with 3 and evaluate $2x + 4$ on the left.

Left side		Right side
$2 \cdot (3) + 4$	$\stackrel{?}{=}$	10
$6 + 4$	$\stackrel{?}{=}$	10
10	=	10

Since $10 = 10$ is a true statement, 3 is a solution of the equation.

NOTE

Until the left side equals the right side, we place a question mark over the equal sign.

(b) Is $\frac{5}{3}$ a solution of the equation $3x - \frac{2}{3} = 2x + 1$?

To find out, replace x with $\frac{5}{3}$ and evaluate each side separately.

Left side		Right side
$3 \cdot \left(\frac{5}{3}\right) - \frac{2}{3}$	$\stackrel{?}{=}$	$2 \cdot \left(\frac{5}{3}\right) + 1$
$\frac{15}{3} - \frac{2}{3}$	$\stackrel{?}{=}$	$\frac{10}{3} + \frac{3}{3}$
$\frac{13}{3}$	$=$	$\frac{13}{3}$

RECALL

Always apply the rules for the order of operations. Multiply first; then add or subtract.

Because the two sides name the same number, we have a true statement, and $\frac{5}{3}$ is a solution.

Check Yourself 1

For the equation $2x - 1 = x + 5$,

(a) Is 6 a solution? **(b)** Is $\frac{8}{3}$ a solution?

You may be wondering whether an equation can have more than one solution. It certainly can. For instance,

$$x^2 = 9$$

has two solutions. They are 3 and -3 because

$$(3)^2 = 9 \qquad \text{and} \qquad (-3)^2 = 9$$

In this chapter, we generally work with *linear equations.* These are equations that can be put into the form

$$ax + b = 0$$

in which the variable is x, a and b are numbers, and a is not equal to 0. In a linear equation, the variable can only appear to the first power. No other power (x^2, x^3, etc.) can appear. Linear equations are also called **first-degree equations.** The *degree* of an equation in one variable is the highest power to which the variable is raised. In the equation $5x^4 - 9x^2 + 7x - 2 = 0$, the highest power to which the x is raised is four. Therefore, it is a fourth-degree equation.

Property

Solutions for Linear Equations

Linear equations in one variable that can be written in the form

$$ax + b = 0 \qquad a \neq 0$$

have exactly one solution.

Example 2 Identifying Expressions and Equations

Label each statement as an expression, a linear equation, or a nonlinear equation. Recall that an equation is a statement in which an equal sign separates two expressions.

(a) $4x + 5$ is an expression.
(b) $2x + 8 = 0$ is a linear equation.
(c) $3x^2 - 9 = 0$ is a nonlinear equation.
(d) $5x = 15$ is a linear equation.
(e) $\frac{3}{x} + 2 = 0$ is a nonlinear equation.

NOTE

There can be no variable in the denominator of a linear equation.

Check Yourself 2

Label each as an expression, a linear equation, or a nonlinear equation.

(a) $2x^2 = 8$ **(b)** $2x - 3 = 0$ **(c)** $5x - 10$

(d) $2x + 1 = 7$ **(e)** $5 - \frac{6}{x} = 2x$

You can find the solution to an equation such as $x + 3 = 8$ by guessing the answer to the question "What plus 3 is 8?" Here the answer to the question is 5, which is also the solution to the equation. But for more complicated equations we need something more than guesswork. A better method is to transform the given equation to an *equivalent equation* whose solution can be found by inspection.

Definition

Equivalent Equations Equations that have exactly the same solutions are called **equivalent equations.**

NOTE

In some cases we write the equation in the form

$\square = x$

The number is the solution when the variable is isolated on either the left or the right.

These are all equivalent equations.

$2x + 3 = 5 \qquad 2x = 2 \qquad \text{and} \qquad x = 1$

They all have the same solution, 1. We say that a linear equation is *solved* when it is transformed to an equivalent equation of the form

$$x = \square$$

The variable is alone on one side. — The other side is some number, the solution.

The addition property of equality is the first property you need to transform an equation to an equivalent form.

Property

The Addition Property of Equality

If $a = b$

then $a + c = b + c$

In words, adding the same quantity to both sides of an equation gives an equivalent equation.

An equation is a statement that the two sides are equal. Adding the same quantity to both sides does not change the equality or "balance."

In Example 3 we apply this idea to solve an equation.

Example 3 Using the Addition Property to Solve an Equation

< Objective 2 >

Solve.

$x - 3 = 9$

Remember that our goal is to isolate x on one side of the equation. Because 3 is being subtracted from x, we can add 3 to remove it. We use the addition property to add 3 to both sides of the equation.

$$\begin{aligned} x - 3 &= 9 \\ +3 &\quad +3 \\ x &= 12 \end{aligned}$$

Adding 3 "undoes" the subtraction and leaves x alone on the left.

NOTE

To check, replace x with 12 in the original equation.

$x - 3 = 9$

$(12) - 3 \stackrel{?}{=} 9$

$9 = 9$ True

Because we have a true statement, 12 is the solution.

Because 12 is the solution of the equivalent equation $x = 12$, it is the solution to our original equation.

Check Yourself 3

Solve and check.

$x - 5 = 4$

The addition property also allows us to add a negative number to both sides of an equation. This is really the same as subtracting the same quantity from both sides.

Example 4 Solving an Equation

Solve.

$x + 2 = \frac{11}{2}$

In this case, 2 is *added* to x on the left. We can use the addition property to subtract 2 from both sides. This "undoes" the addition and leaves the variable x alone on one side of the equation.

NOTE

Because subtraction is defined in terms of addition, we can add *or* subtract the same quantity from both sides of the equation.

$$\begin{array}{rcl} x + 2 &=& \frac{11}{2} \\ -2 && -\frac{4}{2} \\ \hline x &=& \frac{7}{2} \end{array}$$

We subtract 2 from each side.

$-\frac{4}{2} = -2$

The solution is $\frac{7}{2}$. To check, replace x with $\frac{7}{2}$.

$\left(\frac{7}{2}\right) + 2 = \frac{11}{2}$ True

Check Yourself 4

Solve and check.

$x + 6 = \frac{11}{3}$

What if the equation has variable terms on both sides? You can use the addition property to add or subtract a term involving the variable to get the desired result.

Example 5 Solving an Equation

Solve.

$5x = 4x + 7$

We start by subtracting $4x$ from both sides of the equation. Do you see why? Remember that an equation is solved when we have an equivalent equation of the form $x = \square$.

NOTE

Subtracting $4x$ is the same as adding $-4x$.

$$\begin{array}{rcl} 5x &=& 4x + 7 \\ -4x && -4x \\ \hline x &=& 7 \end{array}$$

Subtracting $4x$ from both sides *removes* $4x$ from the right.

To check: Since 7 is a solution to the equivalent equation $x = 7$, it should be a solution of the original equation. To find out, replace x with 7.

$5 \cdot (7) \stackrel{?}{=} 4 \cdot (7) + 7$

$35 \stackrel{?}{=} 28 + 7$

$35 = 35$ True

Check Yourself 5

Solve and check.

$7x = 6x + 3$

Recall that addition can be set up either in a vertical format such as

$$\begin{array}{r} 256 \\ +192 \\ \hline 448 \end{array}$$

or in a horizontal format

$256 + 192 = 448$

When we use the addition property to solve an equation, the same choices are available. In our examples to this point we used the vertical format. In Example 6 we use the horizontal format. In the remainder of this text, we assume that you are familiar with both formats.

Example 6 Solving an Equation

Solve.

$7x - 8 = 6x$

We want all variables on *one* side of the equation. If we choose the left, we subtract $6x$ from both sides of the equation. This removes $6x$ from the right side.

$$7x - 8 - 6x = 6x - 6x$$
$$x - 8 = 0$$

We want the variable alone, so we add 8 to both sides. This isolates x on the left.

$$x - 8 + 8 = 0 + 8$$
$$x \qquad = 8$$

We leave it to you to check that 8 is the solution.

Check Yourself 6

Solve and check.

$9x + 3 = 8x$

Often an equation has more than one variable term *and* more than one number, as in Example 7.

Example 7 Solving an Equation

Solve.

$5x - 7 = 4x + 3$

We would like the variable terms on the left, so we start by subtracting $4x$ to remove that term from the right side of the equation.

$$\begin{array}{rcr} 5x - 7 = & & 4x + 3 \\ -4x & & -4x \\ \hline x - 7 = & & 3 \end{array}$$

NOTE

You could just as easily have added 7 to both sides and *then* subtracted $4x$. The result would be the same. In fact, some students prefer to combine the two steps.

To isolate the variable, we add 7 to both sides to undo the subtraction on the left.

$$\begin{array}{rcl} x - 7 &=& 3 \\ +7 & & +7 \\ \hline x &=& 10 \end{array}$$

The solution is 10. To check, replace x with 10 in the original equation.

$$5 \cdot (10) - 7 \stackrel{?}{=} 4 \cdot (10) + 3$$

$$43 = 43 \quad \text{True}$$

RECALL

By *simplify,* we mean to combine all like terms.

Check Yourself 7

Solve and check.

(a) $4x - 5 = 3x + 2$ **(b)** $6x + 2 = 5x - 4$

When solving an equation, you should always simplify each side as much as possible before using the addition property.

Example 8 Simplifying an Equation

Solve $5 + 8x - 2 = 2x - 3 + 5x$.

Like terms Like terms

$$5 + 8x - 2 = 2x - 3 + 5x$$

Notice that like terms appear on both sides of the equation. We start by combining the numbers on the left (5 and -2). Then we combine the like terms ($2x$ and $5x$) on the right.

$$3 + 8x = 7x - 3$$

Now we can apply the addition property, as before.

$$\begin{array}{rcl} 3 + 8x &=& 7x - 3 \\ -7x &=& -7x \\ \hline 3 + x &=& -3 \\ -3 & & -3 \\ \hline x &=& -6 \end{array}$$

Subtract $7x$.

Subtract 3 to isolate x.

The solution is -6. To check, always return to the original equation. That catches any possible errors in simplifying. Replacing x with -6 gives

$$5 + 8(-6) - 2 \stackrel{?}{=} 2(-6) - 3 + 5(-6)$$

$$5 - 48 - 2 \stackrel{?}{=} -12 - 3 - 30$$

$$-45 = -45 \quad \text{True}$$

Check Yourself 8

Solve and check.

(a) $3 + 6x + 4 = 8x - 3 - 3x$ **(b)** $5x + 21 + 3x = 20 + 7x - 2$

We may have to apply some other properties when solving equations. In Example 9, we use the distributive property to clear an equation of parentheses.

Example 9 Using the Distributive Property to Solve Equations

RECALL

$2(3x + 4) = 2(3x) + 2(4)$
$= 6x + 8$

Solve.

$2(3x + 4) = 5x - 6$

Applying the distributive property on the left gives

$6x + 8 = 5x - 6$

We proceed as before.

$$
\begin{array}{rcl}
6x + 8 &=& 5x - 6 \\
-5x \quad &=& -5x \qquad \text{Subtract } 5x. \\
\hline
x + 8 &=& -6 \\
-8 &=& -8 \qquad \text{Subtract } 8. \\
\hline
x &=& -14
\end{array}
$$

The solution is -14. We leave it to you to check this result.

Remember: Always return to the original equation to check.

Check Yourself 9

Solve and check each equation.

(a) $4(5x - 2) = 19x + 4$ **(b)** $3(5x + 1) = 2(7x - 3) - 4$

Given an expression such as

$-2(x - 5)$

we use the distributive property to create the equivalent expression

$-2x + 10$

Distributing negative numbers is shown in Example 10.

Example 10 Distributing a Negative Number

Solve each equation.

(a)
$$
\begin{array}{rcll}
-2(x - 5) &=& -3x + 2 & \\
-2x + 10 &=& -3x + 2 & \text{Distribute the } -2. \\
+3x \quad && +3x & \text{Add } 3x. \\
\hline
x + 10 &=& 2 & \\
-10 &=& -10 & \text{Subtract 10.} \\
\hline
x &=& -8 & \text{The solution is } -8.
\end{array}
$$

(b)
$$
\begin{array}{rcll}
-3(3x + 5) &=& -5(2x - 2) & \\
-9x - 15 &=& -5(2x - 2) & \text{Distribute the } -3. \\
-9x - 15 &=& -10x + 10 & \text{Distribute the } -5. \\
+10x \quad && +10x & \text{Add } 10x. \\
\hline
x - 15 &=& 10 & \\
+15 &=& +15 & \text{Add 15.} \\
\hline
x &=& 25 & \text{The solution is 25.}
\end{array}
$$

Check:

$$
\begin{array}{rcll}
-3[3(25) + 5] &\stackrel{?}{=}& -5[2(25) - 2] & \\
-3(75 + 5) &\stackrel{?}{=}& -5(50 - 2) & \text{Follow the order of operations.} \\
-3(80) &\stackrel{?}{=}& -5(48) & \\
-240 &=& -240 & \text{True}
\end{array}
$$

RECALL

Return to the original equation to check your solution.

Check Yourself 10

Solve each equation.

(a) $-2(x - 3) = -x + 5$ **(b)** $-4(2x - 1) = -3(3x + 2)$

The main reason for learning how to set up and solve algebraic equations is so that we can use them to solve word problems. In fact, algebraic equations were *invented* to make solving applications much easier. The first word problems that we know about are over 4,000 years old. They were literally "written in stone," on Babylonian tablets, about 500 years before the first algebraic equation makes its appearance.

Before algebra, people solved word problems primarily by **substitution,** which is a method of finding unknown values by using trial and error in a logical way. Example 11 shows how to solve a word problem by using substitution.

Example 11 Solving a Word Problem

< Objective 3 >

NOTE

Consecutive integers are integers that follow each other, such as 8 and 9.

The sum of two consecutive integers is 37. Find the two integers.

We take a guess-and-check approach to this problem.

When we try 20 and 21, we get 41 as the sum, which is too large.

Trying smaller numbers, say 15 and 16, gives a sum of 31, which is too small.

We continue to guess until we come to 18 and 19. They are consecutive and their sum is 37. The answer is 18 and 19.

Check Yourself 11

The sum of two consecutive integers is 91. Find the two integers.

Most word problems are not quite so easy to solve. For more complicated word problems, we use a five-step procedure. Using this step-by-step approach helps to organize our work. Organization is a key to solving word problems.

Step by Step

Solving Word Problems

Step 1 Read the problem carefully. Then reread it to decide what you are asked to find.

Step 2 Choose a letter to represent one of the unknowns in the problem. Then represent all other unknowns of the problem with expressions that use the same variable.

Step 3 Translate the problem to the language of algebra to form an equation.

Step 4 Solve the equation.

Step 5 Answer the question and include units in your answer, when appropriate. Check your solution by returning to the original problem.

The third step is usually the hardest. We must translate words to the language of algebra. Before we look at a complete example, the next table may help you review that translation step.

RECALL

We discussed these translations in Section 11.1. You might find it helpful to review that section before going on.

Translating Words to Algebra

Words	Algebra
The sum of x and y	$x + y$
3 plus a	$3 + a$ or $a + 3$
5 more than m	$m + 5$
b increased by 7	$b + 7$
The difference of x and y	$x - y$
4 less than a	$a - 4$
s decreased by 8	$s - 8$
The product of x and y	$x \cdot y$ or xy
5 times a	$5 \cdot a$ or $5a$
Twice m	$2m$
The quotient of x and y	$\frac{x}{y}$
a divided by 6	$\frac{a}{6}$
One-half of b	$\frac{b}{2}$ or $\frac{1}{2}b$

Now let's look at some typical examples of translating phrases to algebra.

Example 12 Translating Statements

Translate each English expression to an algebraic expression.

(a) The sum of a and 2 times b

$a + 2b$

Sum 2 times b

(b) 5 times m, increased by 1

$5m + 1$

5 times m Increased by 1

(c) 5 less than 3 times x

$3x - 5$

3 times x 5 less than

(d) The product of x and y, divided by 3

$\frac{xy}{3}$

Check Yourself 12

Write each expression symbolically.

(a) 2 more than twice x
(b) 4 less than 5 times n
(c) The product of twice a and b
(d) The sum of s and t, divided by 5

Now we work through a complete example. Although this problem can be solved by guessing, we present it to help you practice the five-step approach.

Example 13 Solving an Application

The sum of a number and 5 is 17. What is the number?

Step 1 *Read carefully.* We must find the unknown number.

Step 2 *Choose letters or variables.* Let x represent the unknown number. There are no other unknowns.

Step 3 *Translate.*

$$x + 5 = 17$$

(The sum of → +; is → =)

Step 4 *Solve.*

$$x + 5 = 17 \quad \text{Subtract 5.}$$
$$x + 5 - 5 = 17 - 5$$
$$x = 12$$

Step 5 *Answer.* The number is 12.

Check. Is the sum of 12 and 5 equal to 17? Yes, $12 + 5 = 17$.

> CAUTION

Always return to the *original problem* to check your result and *not* to the equation in step 3. This helps prevent possible errors!

Check Yourself 13

The sum of a number and 8 is 35. What is the number?

Of course, there are many applications requiring the addition property to solve an equation. Consider the consumer application in Example 14.

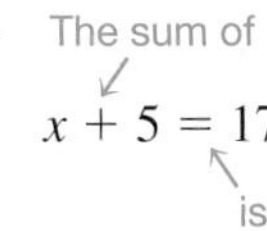

Example 14 A Consumer Application

An appliance store is having a sale on washers and dryers. They are charging \$999 for a washer and dryer combination. If the washer sells for \$649, how much is someone paying for the dryer as part of the combination?

Step 1 *Read carefully.* We are asked to find the cost of a dryer in this application.

Step 2 *Choose letters or variables.* Let d represent the cost of a dryer as part of the washer-dryer combination. This is the only unknown quantity in the problem.

Step 3 *Translate.*

$$\underbrace{d + 649} = 999$$

The washer costs \$649.
Together, they cost \$999.

Step 4 *Solve.*

$$d + 649 = 999$$
$$d + 649 - 649 = 999 - 649 \quad \text{Subtract 649 to isolate the variable.}$$
$$d = 350$$

Step 5 *Answer.* The dryer costs \$350 as part of this combination.

Check. A \$649 washer and a \$350 dryer cost a total of \$649 + \$350 = \$999.

RECALL

Always answer an application with a full sentence.

Check Yourself 14

Of 18,540 votes cast in the school board election, 11,320 went to Carla. How many votes did her opponent Marco receive? Who won the election?

Check Yourself ANSWERS

1. **(a)** 6 is a solution; **(b)** $\frac{8}{3}$ is not a solution. **2.** **(a)** nonlinear equation; **(b)** linear equation; **(c)** expression; **(d)** linear equation; **(e)** nonlinear equation **3.** 9 **4.** $-\frac{7}{3}$ **5.** 3 **6.** -3 **7.** **(a)** 7; **(b)** -6 **8.** **(a)** -10; **(b)** -3 **9.** **(a)** 12; **(b)** -13 **10.** **(a)** 1; **(b)** -10 **11.** 45 and 46 **12.** **(a)** $2x + 2$; **(b)** $5n - 4$; **(c)** $2ab$; **(d)** $\frac{s+t}{5}$
13. The equation is $x + 8 = 35$. The number is 27.
14. Marco received 7,220 votes; Carla won the election.

Reading Your Text

These fill-in-the-blank exercises will help you understand some of the key vocabulary used in this section. The answers to these exercises are in the Answers Appendix at the back of the text.

(a) An ________ is a mathematical statement that two expressions are equal.

(b) A solution to an equation is a value for the variable that makes the equation a ________ statement.

(c) Equivalent equations have the same ________.

(d) The answer to an application should always be given using a full ________.

11.4 exercises

Skills | Calculator/Computer | Career Applications | Above and Beyond

< Objective 1 >

Is the number shown in parentheses a solution to the given equation?

1. $x + 4 = 9$ (5)

2. $x + 2 = 11$ (8)

3. $x - 15 = 6$ (-21) 

4. $x - 11 = 5$ (16)

5. $5 - x = 2$ (4)

6. $10 - x = 7$ (3)

7. $4 - x = 6$ (-2)

8. $5 - x = 6$ (-3)

9. $3x + 4 = 13$ (8) 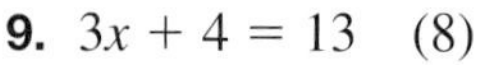 

10. $5x + 6 = 31$ (5)

11. $4x - 5 = 7$ (2)

12. $2x - 5 = 1$ (3)

13. $5 - 2x = 7$ (-1)

14. $4 - 5x = 9$ (-2)

15. $4x - 5 = 2x + 3$ (4)

16. $5x + 4 = 2x + 10$ (4)

17. $x + 3 + 2x = 5 + x + 8$ (5)

18. $5x - 3 + 2x = 3 + x - 12$ (-2)

19. $\frac{3}{4}x = 18$ (20)

20. $\frac{3}{5}x = 24$ (40)

21. $\frac{3}{5}x + 5 = 11$ (10)

22. $\frac{2}{3}x + 8 = -12$ (-6)

Label each as an expression or an equation.

23. $2x + 1 = 9$

24. $7x + 14$

25. $2x - 8$

26. $5x - 3 = 12$

27. $7x + 2x + 8 - 3$

28. $x + 5 = 13$

29. $2x - 8 = 3$

30. $12x - 5x + 2 + 5$

< Objective 2 >

Solve and check each equation.

31. $x + 9 = 11$ 

32. $x - 4 = 6$

33. $x - 8 = 3$

34. $x + 11 = 15$

35. $x - 8 = -10$

36. $x + 5 = 2$

37. $x + 4 = -3$

38. $x - 5 = -4$

39. $11 = x + 5$

40. $x + 7 = 0$

41. $4x = 3x + 4$

42. $7x = 6x - 8$

43. $11x = 10x - 10$

44. $9x = 8x + 5$

45. $6x + 3 = 5x$

46. $12x - 6 = 11x$

47. $8x - 4 = 7x$

48. $9x - 7 = 8x$

49. $2x + 3 = x + 5$

50. $3x - 2 = 2x + 1$

51. $5x - 7 = 4x - 3$

52. $8x + 5 = 7x - 2$

53. $7x - 2 = 6x + 4$

54. $10x - 3 = 9x - 6$

55. $3 + 6x + 2 = 3x + 11 + 2x$

56. $6x - 3 + 2x = 7x + 8$

57. $4x + 7 + 3x = 5x + 13 + x$

58. $5x + 9 + 4x = 9 + 8x - 7$

59. $3x - 5 + 2x - 7 + x = 5x + 2$

60. $5x + 8 + 3x - x + 5 = 6x - 3$

61. $4(3x + 4) = 11x - 2$

62. $2(5x - 3) = 9x + 7$

63. $3(7x + 2) = 5(4x + 1) + 17$

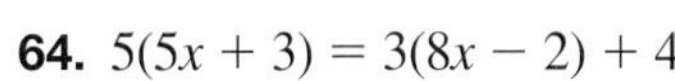

64. $5(5x + 3) = 3(8x - 2) + 4$

65. $\frac{5}{4}x - 1 = \frac{1}{4}x + 7$

66. $\frac{7}{5}x + 3 = \frac{2}{5}x - 8$

67. $\frac{9}{2}x - \frac{3}{4} = \frac{7}{2}x + \frac{5}{4}$

68. $\frac{11}{3}x + \frac{1}{6} = \frac{8}{3}x + \frac{19}{6}$

Translate each English statement to an algebraic equation. Use x as the variable in each case.

69. 3 more than a number is 7.

70. 5 less than a number is 12.

71. 7 less than 3 times a number is twice that same number.

72. 4 more than 5 times a number is 6 times that same number.

73. 2 times the sum of a number and 5 is 18 more than that same number. 

74. 3 times the sum of a number and 7 is 4 times that same number.

< Objective 3 >

Solve each word problem. Show the equation you use for the solution.

75. **Number Problem** The sum of a number and 7 is 33. What is the number?

76. **Number Problem** The sum of a number and 15 is 22. What is the number?

77. **Number Problem** The sum of a number and -15 is 7. What is the number? 

78. **Number Problem** The sum of a number and -8 is 17. What is the number? 

79. **Social Science** In an election, the winning candidate has 1,840 votes. If the total number of votes cast was 3,260, how many votes did the losing candidate receive?

80. **Business and Finance** Mike and Stefanie work at the same company and make a total of $6,760 per month. If Stefanie makes $3,400 per month, how much does Mike earn every month?

81. **Business and Finance** A washer-dryer combination costs \$650. If the washer costs \$360, what does the dryer cost?

82. **Technology** You have \$2,350 saved for the purchase of a new computer system that costs \$3,675. How much more must you save?

83. **Crafts** Jeremiah had found 50 bones for a Halloween costume. In order to complete his 62-bone costume, how many more does he need?
Let b be the number of bones he needs, and use the equation $b + 50 = 62$ to solve the problem.

84. **Business and Finance** Four hundred tickets to the opening of an art exhibit were sold. General admission tickets cost \$5.50, whereas students were only required to pay \$4.50 for tickets. If total ticket sales were \$1,950, how many of each type of ticket were sold?
Let x be the number of general admission tickets sold so that $400 - x$ is the number of student tickets sold. Use the equation $5.5x + 4.5(400 - x) = 1{,}950$ to solve the problem.

85. **Business and Finance** A shop pays \$2.25 for each copy of a magazine and sells the magazines for \$3.25 each. If the fixed costs associated with the sale of these magazines are \$50 per month, how many must the shop sell in order to realize \$175 in profit from the magazines?
Let m be the number of magazines the shop must sell, and use the equation $3.25m - 2.25m - 50 = 175$ to solve the problem.

86. **Number Problem** The sum of a number and 15 is 22. Find the number.
Let x be the number and solve the equation $x + 15 = 22$ to find the number.

87. Which equation is equivalent to $8x + 5 = 9x - 4$?
(a) $17x = -9$ (b) $x = -9$ (c) $8x + 9 = 9x$ (d) $9 = 17x$

88. Which equation is equivalent to $5x - 7 = 4x - 12$? VIDEO
(a) $9x = 19$ (b) $9x - 7 = -12$ (c) $x = -18$ (d) $x - 7 = -12$

89. Which equation is equivalent to $12x - 6 = 8x + 14$?
(a) $4x - 6 = 14$ (b) $x = 20$ (c) $20x = 20$ (d) $4x = 8$

90. Which equation is equivalent to $7x + 5 = 12x - 10$?
(a) $5x = -15$ (b) $7x - 5 = 12x$ (c) $-5 = 5x$ (d) $7x + 15 = 12x$

Determine whether each statement is **true** *or* **false.**

91. Every linear equation with one variable has exactly one solution.

92. Isolating the variable on the right side of an equation results in a negative solution.

93. If we add the same number to both sides of an equation, we always obtain an equivalent equation.

94. The equations $x^2 = 9$ and $x = 3$ are equivalent equations.

Complete each statement with **always, sometimes,** *or* **never.**

95. An equation __________ has one solution.

96. If a first-degree equation has a variable term on both sides, we _______ use the addition property to solve the equation.

Skills	Calculator/Computer	**Career Applications**	Above and Beyond

97. **Construction Technology** K-Jones Manufacturing produces hex bolts and carriage bolts.
(a) They sold 284 more cases of hex bolts than carriage bolts last month. Write an equation for the number of cases of carriage bolts sold last month given the number of cases of hex bolts sold h.
(b) If they sold 2,680 carriage bolts, how many cases of hex bolts did they sell?

98. **Engineering Technology** The specifications for an engine cylinder of a particular ship calls for the stroke length to be two more than twice the diameter of the cylinder.

(a) Write an equation for the required stroke length given a cylinder's diameter d.

(b) Find the stroke length specified for a cylinder with a 52 in. diameter.

99. **Electronics Technology** Berndt Electronics earns a marginal profit of \$560 on the sale of each server. If other costs amount to \$4,500, will they earn a profit of at least \$5,000 on the sale of 15 servers? VIDEO

100. **Electronics Technology** How much profit does Berndt Electronics earn on the sale of 15 servers (see exercise 99)? How many do they need to sell in order to reach a \$5,000 profit?

Skills | Calculator/Computer | Career Applications | **Above and Beyond**

101. An algebraic equation is a complete sentence. It has a subject, a verb, and a predicate. For example, $x + 2 = 5$ can be written in English as "Two more than a number is five" or "A number added to two is five." Write an English version of each equation. Be sure to write complete sentences and that the sentences express the same idea as the equations. Exchange sentences with another student and see if your interpretations of each other's sentences result in the same equation.

(a) $2x - 5 = x + 1$ **(b)** $2(x + 2) = 14$ **(c)** $n + 5 = \frac{n}{2} - 6$ **(d)** $7 - 3a = 5 + a$

102. Complete the explanation in your own words: "The difference between $3(x - 1) + 4 - 2x$ and $3(x - 1) + 4 = 2x$ is . . . ."

103. "I make \$2.50 an hour more in my new job." If $x =$ the amount I used to make per hour and $y =$ the amount I now make, which of the equations say the same thing as the previous statement? Explain your choices by translating the equation to English and comparing with the original statement.

(a) $x + y = 2.50$ **(b)** $x - y = 2.50$ **(c)** $x + 2.50 = y$

(d) $2.50 + y = x$ **(e)** $y - x = 2.50$ **(f)** $2.50 - x = y$

104. "The river rose 4 feet above flood stage last night." If $a =$ the river's height at flood stage and $b =$ the river's height now (the morning after), which of the equations say the same thing as the previous statement? Explain your choices by translating the equations to English and comparing the meaning with the original statement.

(a) $a + b = 4$ **(b)** $b - 4 = a$ **(c)** $a - 4 = b$

(d) $a + 4 = b$ **(e)** $b + 4 = a$ **(f)** $b - a = 4$

105. **(a)** Do you think this is a linear equation in one variable?

$3(2x + 4) = 6(x + 2)$

(b) What happens when you use the properties of this section to solve the equation?

(c) Pick *any* number to substitute for x in this equation. Now try a different number to substitute for x in the equation. Try yet another number to substitute for x in the equation. Summarize your findings.

(d) Can this equation be called *linear in one variable?* Refer to the definition as you explain your answer.

106. **(a)** Do you think this is a linear equation in one variable?

$4(3x - 5) = 2(6x - 8) - 3$

(b) What happens when you use the properties of this section to solve the equation?

(c) Do you think it is possible to find a solution for this equation?

(d) Can this equation be called *linear in one variable?* Refer to the definition as you explain your answer.

107. "Surprising Results!" Work with other students to try this experiment. Each person should do the six steps mentally, not telling anyone else what his or her calculations are.

(a) Think of a number. **(b)** Add 7. **(c)** Multiply by 3. **(d)** Add 3 more than the original number.

(e) Divide by 4. **(f)** Subtract the original number.

What number do you end up with? Compare your answer with everyone else's. Does everyone have the same answer? Make sure that everyone followed the directions accurately. How do you explain the results? Algebra makes the explanation clear. Work together to do the problem again, using a variable for the number. Make up another series of computations that give "surprising results."

Answers

1. Yes **3.** No **5.** No **7.** Yes **9.** No **11.** No **13.** Yes **15.** Yes **17.** Yes **19.** No **21.** Yes **23.** Equation **25.** Expression **27.** Expression **29.** Equation **31.** 2 **33.** 11 **35.** -2 **37.** -7 **39.** 6 **41.** 4 **43.** -10 **45.** -3 **47.** 4 **49.** 2 **51.** 4 **53.** 6 **55.** 6 **57.** 6 **59.** 14 **61.** -18 **63.** 16 **65.** 8 **67.** 2 **69.** $x + 3 = 7$ **71.** $3x - 7 = 2x$ **73.** $2(x + 5) = x + 18$ **75.** 26; $x + 7 = 33$ **77.** 22; $x - 15 = 7$ **79.** 1,420; $1{,}840 + x = 3{,}260$ **81.** \$290; $x + 360 = 650$ **83.** 12 **85.** 225 magazines **87.** (c) **89.** (a) **91.** True **93.** True **95.** sometimes **97.** **(a)** $c = h - 284$; **(b)** 2,964 cases **99.** No **101.** Above and Beyond **103.** Above and Beyond **105.** Above and Beyond **107.** Above and Beyond

Activity 33 ::

Graphing Solutions

You have now solved many equations using algebra. Often, it is convenient to present a picture of the solutions to an equation instead of giving a set of numbers. One method to present a picture uses the familiar number line.

1. Plot the points $\{-2, 0.5, 3\}$ on a number line.

2. Solve the equation $x + 5 = 8$.

3. Plot the solution to the equation in exercise 2 on a number line.

We often use a number line to present sets of numbers. For example, the set of numbers greater than 2 is written algebraically as $\{x \mid x > 2\}$, and is shown on a number line as

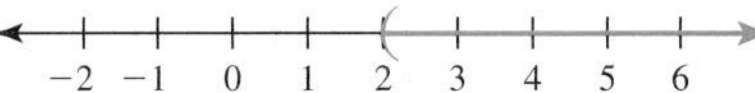

whereas the set of numbers less than or equal to 3 is written as $\{x \mid x \leq 3\}$ and shown on a number line as

NOTE

We read $\{x \mid x > 2\}$ as "the set of every value of x for which x is greater than 2."

4. Graph the set of numbers $\{x \mid x < 3\}$ on a number line.

5. Graph the set of numbers $\{x \mid x \geq -1\}$ on a number line.

6. Solve the inequality $x + 2 > 0$.

7. Graph every solution to exercise 6 on a number line.

11.5 Solving Equations with the Multiplication Property

< 11.5 Objectives >

1 > Use the multiplication property to solve an equation

2 > Combine like terms before solving an equation

3 > Use algebra to solve percent problems

NOTE

Subtracting 6 from both sides yields $6x - 6 = 12$

In this section, we look at a different type of equation. What if we wanted to solve the following equation?

$6x = 18$

The addition property that you just learned does not help us in this situation. We need a second property for solving such an equation.

Property

The Multiplication Property of Equality

If $a = b$ then $ac = bc$ when $c \neq 0$

In words, multiplying both sides of an equation by the same nonzero number gives an equivalent equation.

We work through some examples using this rule.

Example 1 **Using the Multiplication Property to Solve Equations**

< Objective 1 >

RECALL

Multiplying both sides by $\frac{1}{6}$ is equivalent to dividing both sides by 6.

Solve

$6x = 18$

Here the variable x is multiplied by 6. So we apply the multiplication property and multiply both sides by $\frac{1}{6}$. Keep in mind that we want an equation of the form

$$x = \square$$

$$6x = 18$$

$$\frac{1}{6}(6x) = \frac{1}{6}(18)$$

$$\frac{1}{6} \cdot \frac{6x}{1} = \frac{1}{6} \cdot \frac{18}{1}$$

$$\frac{\cancel{6}x}{\cancel{6}} = \frac{\cancel{18}^{3}}{\cancel{6}_{1}} \qquad 18 \div 6 = 3$$

$$x = 3$$

NOTE

$\frac{1}{6}(6x) = \frac{\cancel{6}x}{\cancel{6}} = x$

We then have x alone on the left, which is what we want.

The solution is 3. To check, replace x with 3:

$6(3) \stackrel{?}{=} 18$

$18 = 18$ True

Check Yourself 1

Solve and check.

$8x = 32$

In Example 1 we solved the equation by multiplying both sides by the reciprocal of the coefficient of the variable. Example 2 illustrates a slightly different approach to solving an equation by using the multiplication property.

Example 2 **Solving Equations**

NOTE

Because division is defined in terms of multiplication, we can divide both sides of an equation by the same nonzero number.

Solve.

$5x = -35$

The variable x is multiplied by 5. We *divide* both sides by 5 to "undo" that multiplication.

$\frac{5x}{5} = \frac{-35}{5}$

$x = -7$ The right side simplifies to -7. Be careful with the rules for signs.

The solution is -7.

We leave it to you to check the solution.

Check Yourself 2

Solve and check.

$7x = -42$

Example 3 **Solving Equations**

Solve.

$-9x = 54$

In this case, x is multiplied by -9, so we divide both sides by -9 to isolate x on the left.

RECALL

Dividing by -9 and multiplying by $-\frac{1}{9}$ produce the same result—they are the same operation.

$\frac{-9x}{-9} = \frac{54}{-9}$

$x = -6$

The solution is -6. Check.

$(-9)(-6) \stackrel{?}{=} 54$

$54 = 54$ True

Check Yourself 3

Solve and check.

$-10x = -60$

Example 4 illustrates the multiplication property when there are fractions in an equation.

Example 4 **Solving Equations**

(a) Solve $\frac{x}{3} = 6$.

Here x is *divided* by 3. We use multiplication to isolate x.

RECALL

$\frac{x}{3} = \frac{1}{3}x$

$3\left(\frac{x}{3}\right) = 3 \cdot (6)$ This leaves x alone on the left because

$x = 18$ $3\left(\frac{x}{3}\right) = \frac{3}{1} \cdot \frac{x}{3} = \frac{x}{1} = x$

The solution is 18.

Check.

$\frac{(18)}{3} \stackrel{?}{=} 6$

$6 = 6$ True

(b) Solve $\frac{x}{5} = -9$.

$5\left(\frac{x}{5}\right) = 5(-9)$ Because x is divided by 5, we multiply both sides by 5.

$x = -45$

The solution is -45. To check, we replace x with -45.

$\frac{(-45)}{5} \stackrel{?}{=} -9$

$-9 = -9$ True

The solution is verified.

Check Yourself 4

Solve and check.

(a) $\frac{x}{7} = 3$ **(b)** $\frac{x}{4} = -8$

When the variable is multiplied by a fraction that has a numerator other than 1, there are two approaches to finding the solution.

Example 5 Using Reciprocals to Solve Equations

Solve.

$\frac{3}{5}x = 9$

One approach is to multiply by 5 as the first step.

$5\left(\frac{3}{5}x\right) = 5 \cdot (9)$

$3x = 45$

Now we divide by 3.

$\frac{3x}{3} = \frac{45}{3}$

$x = 15$

To check the solution 15, substitute 15 for x.

$\frac{3}{5} \cdot (15) \stackrel{?}{=} 9$

$9 = 9$ True

NOTE

We multiply by $\frac{5}{3}$ because it is the reciprocal of $\frac{3}{5}$, and the product of a number and its reciprocal is 1.

$\left(\frac{5}{3}\right)\left(\frac{3}{5}\right) = 1$

A second approach combines the multiplication and division steps and is generally a bit more efficient. We multiply by $\frac{5}{3}$.

$\frac{5}{3}\left(\frac{3}{5}x\right) = \frac{5}{3} \cdot (9)$

$x = \frac{5}{\cancel{3}_1} \cdot \frac{\cancel{9}^3}{1} = 15$

So $x = 15$, as before.

Check Yourself 5

Solve and check.

$$\frac{2}{3}x = 18$$

You may sometimes have to simplify an equation before applying the methods of this section. Example 6 illustrates this procedure.

Example 6 Combining Like Terms and Solving Equations

< Objective 2 >

Solve and check.

$$3x + 5x = 40$$

Using the distributive property, we combine the like terms on the left to write

$$8x = 40$$

We now proceed as before.

$$\frac{8x}{8} = \frac{40}{8} \quad \text{Divide by 8.}$$

$$x = 5$$

The solution is 5. To check, we return to the original equation. Substituting 5 for x yields

$$3 \cdot (5) + 5 \cdot (5) \stackrel{?}{=} 40$$

$$15 + 25 \stackrel{?}{=} 40$$

$$40 = 40 \quad \text{True}$$

The solution is verified.

Check Yourself 6

Solve and check.

$$7x + 4x = -66$$

RECALL

In percent problems, A is the amount, B is the base, and R is the rate.

As with the addition property, many applications require the multiplication property. One of the most useful set of applications involves percent problems.

In Section 6.3, you learned to use the percent relationship

$$\frac{A}{B} = R$$

to solve percent problems. We did this by writing the percent relationship as a proportion in which $R = \frac{r}{100}$.

$$\frac{A}{B} = \frac{r}{100}$$

We then used the proportion rule to rewrite the equation.

$$100A = rB$$

RECALL

In Section 6.3, you learned to identify the base, rate, and amount in a percent problem.

So, for instance, if the question asked us to find 45% of 80, we would identify $r = 45$ and $B = 80$.

$$100A = (45)(80)$$

$$100A = 3{,}600$$

The next step was to divide both sides by the coefficient of the variable, just as we have been doing throughout this section.

$$\frac{100A}{100} = \frac{3,600}{100}$$

$$A = 36$$

So, 45% of 80 is 36.

We can simplify this process by using algebra. First, we can rewrite the percent relationship by multiplying both sides by the base.

NOTE

This says that the amount is equal to the product of the base and the rate.

$$\frac{A}{B} = R$$

$$\cancel{B}\left(\frac{A}{\cancel{B}}\right) = R \cdot B$$

$$A = RB$$

This form is especially useful if we write the rate as a decimal rather than a fraction.

Example 7 Using Algebra to Solve a Percent Problem

< Objective 3 >

(a) What is 75% of 360?

The rate is given as 75%, which we write as a decimal, $R = 0.75$.
Translating the question, we have

$$A = RB$$

$$A = (0.75)(360)$$

$$A = 270$$

75% of 360 is 270.
Alternatively, we can translate the question directly into an algebraic equation.

What	is	75%	of	360?
↑	↑	↑	↑	↑
A	=	0.75	Multiplication	360

Which gives,

$$A = 0.75 \cdot 360$$

$$= 270$$

(b) What percent of 246 is 342?

This time, we begin by translating the question directly into an algebraic equation.

What percent	of	246	is	342?
↑	↑	↑	↑	↑
R	Multiplication	246	=	342

NOTE

The base is 246 and the amount is 342.

This gives us

$$R \cdot 246 = 342$$

or $246R = 342$

We can solve this equation using the methods of this section.

RECALL

Because the amount is larger than the base, the rate is greater than 100%.

$$\frac{\cancel{246}R}{\cancel{246}} = \frac{342}{246}$$

$$R = \frac{342}{246}$$

$$= \frac{57}{41}$$ The GCF of 246 and 342 is 6.

$$\approx 1.39$$

$$= 139\%$$ Move the decimal two places to the right and attach the percent symbol.

Therefore, 342 is about 139% of 246.

Check Yourself 7

Use algebra to solve each percent application.

(a) 240 is what percent of 400?
(b) 57 is 30% of what number?

Of course, we can use algebra to solve applications that involve percents, as well.

Example 8 Solving a Percent Application with Algebra

A saleswoman earns a 5% commission on her sales. If she wants to earn $1,800 in commissions in one month, how much does she need to sell?

The question we are being asked is, "$1,800 is 5% of what number?" We can translate the question into an algebraic equation by writing the rate in decimal form, 5% = 0.05.

$$1{,}800 = 0.05x$$

$$\frac{1{,}800}{0.05} = \frac{0.05x}{0.05}$$

$$36{,}000 = x$$

She must sell $36,000 in order to earn $1,800 in commissions.

Check Yourself 8

Patrick pays $525 interest for a 1-year loan at 10.5%. What was the amount of his loan?

We can solve many types of problems with algebra besides percent problems.

Example 9 An Application Involving the Multiplication Property

On her first day on the job in a photography lab, Samantha processed all of the film given to her. The next day, her boss gave her four times as much film to process. Over the two days, she processed 60 rolls of film. How many rolls did she process on the first day?

Step 1 We want to find the number of rolls Samantha processed on the first day.

Step 2 Let x be the number of rolls Samantha processed on her first day and solve the equation $x + 4x = 60$ to answer the question.

RECALL

Always use a sentence to answer an application.

Step 3 $x + 4x = 60$

Step 4 $5x = 60$ Combine like terms first.

$\frac{1}{5}(5x) = \frac{1}{5}(60)$ Multiply by $\frac{1}{5}$ to isolate the variable.

$x = 12$

Step 5 Samantha processed 12 rolls of film on her first day.

Check: $4 \times 12 = 48$; $12 + 48 = 60$.

NOTE

The yen (¥) is the monetary unit of Japan.

Check Yourself 9

On a recent trip to Japan, Marilyn exchanged $1,200 and received 97,428 yen. What exchange rate did she receive?

Check Yourself ANSWERS

1. 4 **2.** −6 **3.** 6 **4. (a)** 21; **(b)** −32 **5.** 27 **6.** −6 **7. (a)** 60%; **(b)** 190 **8.** $5,000 **9.** She received 81.19 yen/dollar.

Reading Your Text

These fill-in-the-blank exercises will help you understand some of the key vocabulary used in this section. The answers to these exercises are in the Answers Appendix at the back of the text.

(a) Multiplying both sides of an equation by the same nonzero number yields an ______________ equation.

(b) Always return to the ______________ equation to check a solution.

(c) Dividing by 5 is the same as ______________ by $\frac{1}{5}$.

(d) The product of a nonzero number and its ______________ is 1.

11.5 exercises

Skills Calculator/Computer Career Applications Above and Beyond

< Objectives 1 and 2 >

Solve and check.

1. $5x = 20$ 

2. $6x = 30$

3. $9x = 54$

4. $6x = -42$

5. $63 = 9x$

6. $66 = 6x$

7. $4x = -16$

8. $-3x = 27$

9. $-9x = 72$ 

10. $10x = -100$

11. $6x = -54$

12. $-7x = 49$

13. $-4x = -12$

14. $52 = -4x$

15. $-42 = 6x$

16. $-7x = -35$

17. $-6x = -54$

18. $-4x = -24$

19. $\frac{x}{2} = 4$

20. $\frac{x}{3} = 2$

21. $\frac{x}{5} = 3$

22. $\frac{x}{8} = 5$

23. $6 = \frac{x}{7}$

24. $6 = \frac{x}{3}$

25. $\frac{x}{5} = -4$

26. $\frac{x}{7} = -5$

27. $-\frac{x}{3} = 8$

28. $-\frac{x}{8} = -3$

29. $\frac{2}{3}x = 6$ VIDEO

30. $\frac{4}{5}x = 8$

31. $\frac{3}{4}x = -15$

32. $\frac{7}{8}x = -21$

33. $-\frac{2}{5}x = 10$ VIDEO

34. $-\frac{5}{6}x = -15$

35. $5x + 4x = 36$

36. $8x - 3x = -50$

37. $16x - 9x = -42$

38. $5x + 7x = 60$

39. $4x - 2x + 7x = 36$

40. $6x + 7x - 5x = -48$

41. $8x = 5$

42. $12x = 8$

43. $15x = -9$

44. $-21 = 24x$

45. $\frac{4}{5}x = 10$

46. $\frac{3x}{2} = 8$

47. $\frac{3x}{4} = -16$

48. $-\frac{2}{3}x = -\frac{4}{9}$

49. $-x = 0$

50. $-x = 12$

51. $-x = -4$

52. $-x = -\frac{1}{2}$

We can also use the multiplication property to solve equations containing decimals. For instance, to solve $2.3x = 6.9$, we use the multiplication property to divide both sides of the equation by 2.3. This isolates x on the left, as desired. Use this idea to solve each equation.

53. $3.2x = 12.8$

54. $5.1x = -15.3$

55. $-4.5x = 13.5$

56. $-8.2x = -32.8$

57. $1.3x + 2.8x = 12.3$ VIDEO

58. $2.7x + 5.4x = -16.2$

59. $9.3x - 6.2x = 12.4$

60. $12.5x - 7.2x = -21.2$

< Objective 3 >

Solve each percent problem.

61. What is 65% of 300?

62. 15% of 140 is what number?

63. Find 80% of 80.

64. What is 6% of 550?

65. What percent of 220 is 66?

66. 104 is what percent of 260?

67. 102 is what percent of 85?

68. What percent of 130 is 299?

69. 15 is 4% of what number?

70. 16% of what number is 24?

71. Find the base if 240% of the base is 36.

72. Find the base if 375% of the base is 600.

Use x as the variable to write each statement as an algebraic equation.

73. 6 times a number is 72.

74. Twice a number is 36.

75. A number divided by 7 is equal to 6.

76. A number divided by 5 is equal to -4.

77. $\frac{1}{3}$ of a number is 8.

78. $\frac{1}{5}$ of a number is 10.

79. $\frac{3}{4}$ of a number is 18.

80. $\frac{2}{7}$ of a number is 8.

81. Twice a number, divided by 5, is 12.

82. 3 times a number, divided by 4, is 36.

Solve each application.

83. **Business and Finance** Roberto has 26% of his pay withheld for deductions. If he earns $850 per week, what amount is withheld? chapter 11 > Make the Connection

84. **Business and Finance** A real estate agent's commission rate is 3%. What is the amount of commission on the sale of a \$229,000 home?

85. **Social Science** Of the 60 people who started a training program, 45 were successful. What is the dropout rate?

86. **Business and Finance** In a shipment of 250 parts, 40 are found to be defective. What percent of the shipment is in good working order?

87. **Science and Medicine** There are 117 mL of acid in 900 mL of a solution (acid and water). What percent of the solution is water?

88. **Statistics** Marla needs to answer 70% of the questions correctly on her final exam in order to receive a C for the course. If the exam has 120 questions, how many can she miss?

89. **Business and Finance** Returning from Mexico City, Sung-A exchanged her remaining 450 pesos for \$34.20. What exchange rate did she receive?

90. **Business and Finance** Upon arrival in Portugal, Nicolas exchanged \$500 and received 382.15 euros (€). What exchange rate did he receive?

91. **Science and Technology** On Tuesday, there were twice as many patients in the clinic as on Monday. Over the 2-day period, 48 patients were treated. How many patients were treated on Monday?

92. **Number Problem** Two-thirds of a number is 46. Find the number.

93. **Statistics** Three-fourths of the theater audience left in disgust. If 87 angry patrons walked out, how many were there originally?

94. **Number Problem** When a number is divided by -6, the result is 3. Find the number.

95. **Geometry** Suppose that the circumference of a tree measures 9 ft 2 in., or 110 in. To find the diameter of the tree at that point, we must solve the equation

$$110 = 3.14d$$

Find the diameter of the tree to the nearest inch.

Note: 3.14 is an approximation for π.

96. **Geometry** Suppose that the circumference of a circular swimming pool is 88 ft. Find the diameter of the pool to the nearest foot by solving the equation.

$$88 = 3.14d$$

97. **Problem Solving** While traveling in Europe, Susan noticed that the distance to the city she was heading to was 200 kilometers (km). She knew that to estimate this distance in miles she could solve the equation

$$200 = \frac{8}{5}x$$

What was the equivalent distance in miles? 

98. **Problem Solving** Aaron was driving a rental car while traveling in France, and saw a sign indicating a speed limit of 95 km/hr. To approximate this speed in miles per hour, he used the equation

$$95 = \frac{8}{5}x$$

What is the corresponding speed, rounded to the nearest mile per hour?

Determine whether each statement is **true** *or* **false.**

99. To isolate x in the equation $\frac{3}{4}x = 9$, we can simply subtract $\frac{3}{4}$ from both sides.

100. Dividing both sides of an equation by 5 is the same as multiplying both sides by $\frac{1}{5}$.

Complete each statement with **always, sometimes,** *or* **never.**

101. To solve a linear equation, we _________ must use the multiplication property.

102. If we want to obtain an equivalent linear equation by multiplying both sides by a number, that number can _________ be zero.

Skills | **Calculator/Computer** | Career Applications | Above and Beyond

Use your calculator to solve each equation. Round your answers to the nearest hundredth.

103. $230x = 157$ **104.** $31x = -15$ **105.** $-29x = 432$

106. $-141x = -3{,}467$ **107.** $23.12x = 94.6$ **108.** $46.1x = -1$

Skills | Calculator/Computer | **Career Applications** | Above and Beyond

109. **Information Technology** A 500-GB hard drive contains 300 GB of used space. What percent of the hard drive is full?

110. **Information Technology** A compression program reduces the size of files and folders by 36%. If a folder contains 17.5 MB, how large will it be after it is compressed?

111. **Automotive Technology** It is estimated that 8% of rebuilt alternators do not last through the 90-day warranty period. If a parts store had six bad alternators returned during the year, how many did it sell?

chapter 11 > Make the Connection

112. **Agricultural Technology** A farmer sold 2,200 bushels of barley on the futures market. Because of a poor harvest, he was only able to make 94% of his bid. How many bushels did he actually harvest?

chapter 11 > Make the Connection

113. **Automotive Technology** One horsepower (hp) estimate of an engine is given by the formula

$$\text{hp} = \frac{d^2n}{2.5}$$

in which d is the diameter of the cylinder bore (in centimeters) and n is the number of cylinders.

Find the number of cylinders in a 194.4-hp engine if its cylinder bore has a 9-cm diameter.

114. Automotive Technology The horsepower of a diesel engine is calculated using the formula

$$\text{hp} = \frac{P \cdot L \cdot A \cdot N}{33{,}000}$$

in which P is the average pressure (in pounds per square inch), L is the length of the stroke (in feet), A is the area of the piston (in square inches), and N is the number of strokes per minute.

Determine the average pressure of a 144-hp diesel engine if its stroke length is $\frac{1}{3}$ ft, its piston area is 9 in.^{2}, and it completes 8,000 strokes per minute.

115. Manufacturing Technology The pitch of a gear is given by the number of teeth divided by the working diameter of the gear. Write an equation for the gear pitch p in terms of the number of teeth t and its diameter d.

116. Manufacturing Technology Use your answer to exercise 115 to determine the number of teeth needed for a gear with a working diameter of $6\frac{1}{4}$ in. to have a pitch of 4.

Skills | Calculator/Computer | Career Applications | **Above and Beyond**

117. Describe the difference between the multiplication property and the addition property for solving equations. Give examples of when to use each property.

118. Describe when you should add a quantity to or subtract a quantity from both sides of an equation as opposed to when you should multiply or divide both sides by the same quantity.

Answers

1. 4 **3.** 6 **5.** 7 **7.** -4 **9.** -8 **11.** -9 **13.** 3 **15.** -7 **17.** 9 **19.** 8 **21.** 15 **23.** 42 **25.** -20 **27.** -24 **29.** 9 **31.** -20 **33.** -25 **35.** 4 **37.** -6 **39.** 4 **41.** $\frac{5}{8}$ **43.** $-\frac{3}{5}$ **45.** $\frac{25}{2}$ **47.** $-\frac{64}{3}$ **49.** 0 **51.** 4 **53.** 4 **55.** -3 **57.** 3 **59.** 4 **61.** 195 **63.** 64 **65.** 30% **67.** 120% **69.** 375 **71.** 15 **73.** $6x = 72$ **75.** $\frac{x}{7} = 6$ **77.** $\frac{1}{3}x = 8$ **79.** $\frac{3}{4}x = 18$ **81.** $\frac{2x}{5} = 12$ **83.** \$221 **85.** 25% **87.** 87% **89.** 0.076 peso/dollar **91.** 16 patients **93.** 116 patrons **95.** 35 in. **97.** 125 mi **99.** False **101.** sometimes **103.** 0.68 **105.** -14.9 **107.** 4.09 **109.** 60% **111.** 75 alternators **113.** 6 cylinders **115.** $p = \frac{t}{d}$ **117.** Above and Beyond

11.6 Combining the Rules to Solve Equations

< 11.6 Objectives >

1 > Combine the properties of equality to solve equations

2 > Simplify and solve a linear equation

3 > Use linear equations to solve applications

NOTE

More generally, we reverse the order of operations to solve an equation so that we can *undo* each operation.

Up to this point, we solved equations by using the addition property or the multiplication property of equality. Most of the time, we need to use both properties, along with the order of operations, to solve an equation.

When an equation requires both properties, we **always** apply the addition property before the multiplication property. We begin with some examples.

Example 1 Solving Equations

< Objective 1 >

RECALL

Adding 5 *undoes* the subtraction on the left side of the equation.

The addition property requires us to add 5 to *both* sides of the equation.

(a) Solve

$4x - 5 = 7$

Our first goal is to get the variable term, $4x$, all alone on one side. We call this *isolating the variable term.* We accomplish this by adding 5 to both sides of the equation.

$$4x - 5 = 7$$

$$4x - 5 + 5 = 7 + 5 \quad \text{Add 5 to both sides.}$$

$$4x = 12 \quad \text{This isolates the variable term } 4x.$$

Now that the variable term is alone on the left side, we want to make its coefficient 1. This is called *isolating the variable*. In this case, we use the multiplication property to multiply both sides of the equation by $\frac{1}{4}$.

$$4x = 12$$

$$\frac{1}{4}(4x) = \frac{1}{4}(12) \quad \text{Multiply both sides by } \frac{1}{4}.$$

$$x = 3 \quad \frac{1}{4}(4x) = x \text{ because } \frac{1}{4} \cdot 4 = \frac{1}{4} \cdot \frac{4}{1} = 1.$$

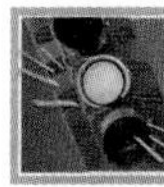

> CAUTION

Use the addition property before the multiplication property.

In this case, isolate the variable term by adding 5 before isolating the variable by dividing by 4.

The solution is 3. To check, replace x with 3 in the original equation. Be careful to follow the rules for the order of operations.

$$4(3) - 5 \stackrel{?}{=} 7$$

$$12 - 5 \stackrel{?}{=} 7$$

$$7 = 7 \quad \text{True}$$

(b) Solve

NOTES

Isolate the variable term.

$$3x + 8 = -4$$

$$3x + 8 - 8 = -4 - 8 \quad \text{Subtract 8 from both sides.}$$

$$3x = -12$$

Isolate the variable.

Now divide both sides by 3 to isolate x on the left.

$$\frac{3x}{3} = \frac{-12}{3}$$

$$x = -4$$

The solution is -4. We leave it to you to check this result.

Check Yourself 1

Solve and check.

(a) $6x + 9 = -15$ **(b)** $5x - 8 = 7$

The variable may appear in any position in an equation. Just apply the rules carefully as you try to write an equivalent equation, and you will find the solution.

Example 2 Solving Equations

Solve

$$3 - 2x = 9$$

$$3 - 3 - 2x = 9 - 3 \quad \text{First subtract 3 from both sides.}$$

$$-2x = 6$$

Now divide both sides by -2. This leaves x alone on the left.

$$\frac{-2x}{-2} = \frac{6}{-2}$$

$$x = -3$$

NOTE

$\frac{-2}{-2} = 1$, so we divide by -2 to isolate x.

The solution is -3. We leave it to you to check this result.

Check Yourself 2

Solve and check.

$10 - 3x = 1$

You may also have to combine multiplication with addition or subtraction to solve an equation. Consider Example 3.

Example 3 Solving Equations

(a) Solve

$$\frac{x}{5} - 3 = 4$$

To get the x-term alone, we first add 3 to both sides.

$$\frac{x}{5} - 3 + 3 = 4 + 3$$

$$\frac{x}{5} = 7$$

To undo the division, multiply both sides of the equation by 5.

$$5\left(\frac{x}{5}\right) = 5 \cdot 7$$

$$x = 35$$

The solution is 35. Always return to the original equation to check the result.

$$\frac{(35)}{5} - 3 \stackrel{?}{=} 4$$

$$7 - 3 \stackrel{?}{=} 4$$

$$4 = 4 \quad \text{True}$$

NOTE

We multiply both sides of the equation

$\frac{2}{3}x = 8$

by $\frac{3}{2}$ because

$\frac{3}{2} \cdot \frac{2}{3} = 1$

The product of a number and its reciprocal is always 1.

(b) Solve

$$\frac{2}{3}x + 5 = 13$$

$$\frac{2}{3}x + 5 - 5 = 13 - 5 \qquad \text{First subtract 5 from both sides.}$$

$$\frac{2}{3}x = 8$$

Now multiply both sides by $\frac{3}{2}$, the reciprocal of $\frac{2}{3}$.

$$\left(\frac{3}{2}\right)\left(\frac{2}{3}x\right) = \left(\frac{3}{2}\right)8$$

or

$$x = 12$$

The solution is 12. We leave it to you to check this result.

Check Yourself 3

Solve and check.

(a) $\frac{x}{6} + 5 = 3$ **(b)** $\frac{3}{4}x - 8 = 10$

In all of our examples, we used both the addition and multiplication properties. First, we used the addition property to *isolate the variable term.* Then, we used the multiplication property to *isolate the variable.*

Why did we do it in this order? Consider what happens to the problem in Example 1 when we multiply both sides by $\frac{1}{4}$ as the first step so that we have x instead of $4x$.

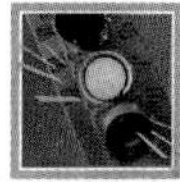

>CAUTION

When we think of dividing both sides by 4, we get

$\frac{4x - 5}{4} = \frac{7}{4}$

because we must divide the entire left side by 4.

Distribution requires us to divide 5 by 4 as well. The 4s do not cancel!

$$4x - 5 = 7$$

$$\frac{1}{4}(4x - 5) = \frac{1}{4}(7) \qquad \text{We must multiply the entire left side by } \frac{1}{4}.$$

$$\frac{1}{4}(4x) + \frac{1}{4}(-5) = \frac{7}{4} \qquad \text{Distribute } \frac{1}{4} \text{ on the left side.}$$

$$x - \frac{5}{4} = \frac{7}{4} \qquad \text{Now we are ready to add } \frac{5}{4} \text{ to both sides.}$$

Since $\frac{7}{4} + \frac{5}{4} = 3$, we have the same result. We consider this approach to be more difficult and do not use it. Instead, we first get the variable term all alone with the addition property. Then, we use the multiplication property to get the variable all alone.

In Section 11.4, you learned how to solve certain equations when the variable appeared on both sides. Example 4 shows you how to extend that work by using the multiplication property of equality.

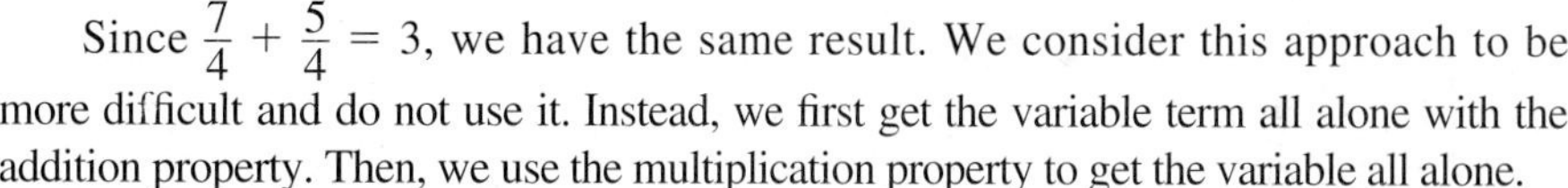

Example 4 Solving Equations

< Objective 2 >

Solve

$$5x - 11 = 2x - 7$$

Our goal is to finish with an equivalent equation that looks like $x = \square$. To do this, we need to bring all of the x-terms to one side and the constant terms to the other side of the equation. We use the addition property to subtract $2x$ from both sides. This brings the variable terms to the left side. We then add 11 to both sides to bring the numbers to the right side.

NOTE

Many students prefer to apply the addition rule vertically, rather than horizontally. Of course, the result is the same.

$$\begin{array}{rcl} 5x - 11 &=& 2x - 7 \\ -2x & & -2x \\ \hline 3x - 11 &=& -7 \\ +11 & & +11 \\ \hline 3x &=& 4 \end{array}$$

Subtract $2x$ from both sides.

Combine like terms to complete this step.

Add 11 to both sides.

Now that the variable term is isolated on the left side, we use the multiplication rule to divide both sides by 3. This isolates the variable.

$$\frac{\cancel{3}x}{\cancel{3}} = \frac{4}{3}$$ Divide both sides by 3 and simplify where possible.

$$x = \frac{4}{3}$$

To check our result, we always return to the original equation.

$$5x - 11 = 2x - 7$$ The original equation.

$$5\left(\frac{4}{3}\right) - 11 \stackrel{?}{=} 2\left(\frac{4}{3}\right) - 7$$ Substitute the proposed solution for the variable.

$$\frac{20}{3} - 11 \stackrel{?}{=} \frac{8}{3} - 7$$ Multiply first, as required by the order of operations.

$$\frac{20}{3} - \frac{33}{3} \stackrel{?}{=} \frac{8}{3} - \frac{21}{3}$$ Rewrite 11 as $\frac{33}{3}$ and 7 as $\frac{21}{3}$ to combine the fractions.

$$-\frac{13}{3} = -\frac{13}{3}$$ True

Check Yourself 4

Solve and check.

$7x - 5 = 3x + 5$

The basic idea is to use our two properties to form an equivalent equation with the x isolated. In Example 4, we subtracted $2x$ and then added 11. You can do these steps in either order. Try it for yourself the other way. In either case, the multiplication property is then used as the *last step* in finding the solution.

Here are two approaches to solving equations in which the coefficient of the variable on the right side is greater than the coefficient on the left side.

Example 5 **Solving Equations (Two Methods)**

Solve $4x - 8 = 7x + 7$.

Method 1

$$4x - 8 - 7x = 7x + 7 - 7x$$ Bring the variable terms to the same (left) side.

$$-3x - 8 = 7$$

$$-3x - 8 + 8 = 7 + 8$$ Isolate the variable term.

$$-3x = 15$$

$$\frac{-3x}{-3} = \frac{15}{-3}$$ Isolate the variable.

$$x = -5$$

We let you check this result.

To avoid a negative coefficient (-3, in this example), some students prefer a different approach.

This time we work toward having the number on the *left* and the x term on the *right*, or $\square = x$.

NOTE

It is usually easier to isolate the variable term on the side that results in a positive coefficient.

Method 2

$$4x - 8 = 7x + 7$$

$$4x - 8 - 4x = 7x + 7 - 4x$$ Bring the variable terms to the same (right) side.

$$-8 = 3x + 7$$

$$-8 - 7 = 3x + 7 - 7$$ Isolate the variable term.

$$-15 = 3x$$

$$\frac{-15}{3} = \frac{3x}{3}$$ Isolate the variable.

$$-5 = x$$

Because $-5 = x$ and $x = -5$ are equivalent equations, it really makes no difference; the solution is still -5! You may use whichever approach you prefer.

Check Yourself 5

Solve $5x + 3 = 9x - 21$ by finding equivalent equations of the form $x = \square$ and $\square = x$ to compare the two methods of finding the solution.

When possible, we start by combining like terms on each side of the equation.

Example 6 Simplifying an Equation

Solve.

$$7x - 3 + 5x + 4 = 6x + 25$$ Start by combining like terms.

$$12x + 1 = 6x + 25$$

$$12x + 1 - 6x = 6x + 25 - 6x$$ Bring the variables to one side.

$$6x + 1 = 25$$

$$6x + 1 - 1 = 25 - 1$$ Isolate the variable term.

$$6x = 24$$

$$\frac{6x}{6} = \frac{24}{6}$$ Isolate the variable.

$$x = 4$$

The solution is 4. We leave the check to you.

Check Yourself 6

Solve and check.

$$9x - 6 - 3x + 1 = 2x + 15$$

It may also be necessary to remove grouping symbols to solve an equation. Example 7 illustrates this property.

Example 7 Simplifying an Equation

RECALL

$5(x - 3) = 5[x + (-3)]$
$= 5x + 5(-3)$
$= 5x + (-15)$
$= 5x - 15$

Solve and check.

$$5(x - 3) - 2x = x + 7$$ Apply the distributive property.

$$5x - 15 - 2x = x + 7$$ Combine like terms.

$$3x - 15 = x + 7$$

We now have an equation that we can solve by the usual methods. First, bring the variable terms to one side, then isolate the variable term, and finally, isolate the variable.

$$3x - 15 - x = x + 7 - x$$ Subtract x to bring the variable terms to the same side.

$$2x - 15 = 7$$

$$2x - 15 + 15 = 7 + 15$$ Add 15 to isolate the variable term.

$$2x = 22$$

$$\frac{2x}{2} = \frac{22}{2}$$ Divide by 2 to isolate the variable.

$$x = 11$$

The solution is 11. To check, substitute 11 for x in the original equation. As always, we follow the rules for the order of operations.

$$5[(11) - 3] - 2(11) \stackrel{?}{=} (11) + 7$$

$$5 \cdot 8 - 2 \cdot 11 \stackrel{?}{=} 11 + 7$$

$$40 - 22 \stackrel{?}{=} 11 + 7$$

$$18 = 18$$ A true statement.

Check Yourself 7

Solve and check.

$$7(x + 5) - 3x = x - 7$$

We say that a linear equation is "solved" when we have an equivalent equation of the form

$x = \square$ or $\square = x$ in which $\square$ is some number

Step by Step

Solving a Linear Equation

Step 1 Use the distributive property to remove any grouping symbols.
Step 2 Combine like terms on each side of the equation.
Step 3 Add or subtract variable terms to bring the variable terms to one side of the equation.
Step 4 Add or subtract numbers to isolate the variable term.
Step 5 Multiply by the reciprocal of the coefficient to isolate the variable.
Step 6 Check your result.

There are a host of applications involving linear equations.

Example 8 **Applying Algebra**

< Objective 3 >

In an election, the winning candidate had 160 more votes than the loser did. If the total number of votes cast was 3,260, how many votes did each candidate receive?

We first set up the problem. Let x represent the number of votes received by the loser. Then the winner received $x + 160$ votes.

We can set up an equation by adding the number of votes the candidates received. This must total 3,260.

$$x + (x + 160) = 3{,}260 \quad \text{Remove the parentheses and combine like terms.}$$
$$2x + 160 = 3{,}260 \quad \text{Subtract 160 from both sides.}$$
$$2x = 3{,}100 \quad \text{Divide both sides by 2.}$$
$$x = 1{,}550$$

The loser received 1,550 votes. Therefore, the winner received $x + 160 = 1{,}550 + 160 = 1{,}710$ votes.

Check Yourself 8

The Randolphs used 12 more gallons (gal) of fuel oil in October than in September and twice as much oil in November as in September. If they used 132 gal for the 3 months, how much was used each month?

Percents are used in many applications. We conclude this section with one such application in Example 9.

Example 9 Solving a Percent Application

RECALL

26% = 0.26

Net pay is given by the difference between gross pay and the deductions.

Tony takes home $592 each week (net pay). Deductions for taxes, retirement, union dues, and a medical plan amount to 26% of his gross pay. What is his weekly gross pay?

We want to find Tony's gross pay.

Let x be Tony's gross pay.

Since 26% of his gross pay is deducted from his weekly salary, $0.26x$ is deducted from each paycheck.

$$x - 0.26x = 592 \quad \text{Gross pay} - \text{Deductions} = \text{Net pay}$$
$$0.74x = 592 \quad 1 - 0.26 = 0.74$$
$$\frac{0.74x}{0.74} = \frac{592}{0.74} \quad \text{Divide both sides by 0.74.}$$
$$x = 800 \quad \frac{592}{0.74} = 800$$

Check Yourself 9

Joan gives 10% of her take-home pay to charity. This amounts to $360 per month. In addition, her paycheck deductions are 25% of her gross monthly income. What is her gross monthly income?

Check Yourself ANSWERS

1. **(a)** -4; **(b)** 3 **2.** 3 **3.** **(a)** -12; **(b)** 24 **4.** $\frac{5}{2}$ **5.** 6 **6.** 5 **7.** -14
8. 30 gal in September, 42 gal in October, 60 gal in November **9.** $4,800

Reading Your Text

These fill-in-the-blank exercises will help you understand some of the key vocabulary used in this section. The answers to these exercises are in the Answers Appendix at the back of the text.

(a) The first goal for solving an equation is to ________ the variable term on one side of the equation.

(b) Apply the ________ property before applying the multiplication property.

(c) Always return to the ________ equation to check your result.

(d) An equation in the form $x = \square$ or $\square = x$ has been ________.

11.6 exercises

Skills | Calculator/Computer | Career Applications | Above and Beyond

< Objectives 1 and 2 >

Solve and check.

1. $2x + 1 = 9$

2. $3x - 1 = 17$

3. $3x - 2 = 7$

4. $5x + 3 = 23$

5. $4x + 7 = 35$

6. $7x - 8 = 13$

7. $2x + 9 = 5$

8. $6x + 25 = -5$

9. $4 - 7x = 18$

10. $8 - 5x = -7$

11. $3 - 4x = -9$

12. $5 - 4x = 25$

13. $\frac{x}{2} + 1 = 5$

14. $\frac{x}{3} - 2 = 3$

15. $\frac{x}{4} - 5 = 3$

16. $\frac{x}{5} + 3 = 8$

17. $\frac{2}{3}x + 5 = 17$

18. $\frac{3}{4}x - 5 = 4$

19. $\frac{4}{5}x - 3 = 13$

20. $\frac{5}{7}x + 4 = 14$

21. $5x = 2x + 9$

22. $7x = 18 - 2x$

23. $3x = 10 - 2x$

24. $11x = 7x + 20$

25. $9x + 2 = 3x + 38$

26. $8x - 3 = 4x + 17$

27. $4x - 8 = x - 14$

28. $6x - 5 = 3x - 29$

29. $5x + 7 = 2x - 3$

30. $9x + 7 = 5x - 3$

31. $7x - 3 = 9x + 5$

32. $5x - 2 = 8x - 11$

33. $5x + 4 = 7x - 8$

34. $2x + 23 = 6x - 5$

35. $2x - 3 + 5x = 7 + 4x + 2$

36. $8x - 7 - 2x = 2 + 4x - 5$

37. $6x + 7 - 4x = 8 + 7x - 26$

38. $7x - 2 - 3x = 5 + 8x + 13$

39. $9x - 2 + 7x + 13 = 10x - 13$

40. $5x + 3 + 6x - 11 = 8x + 25$

41. $8x - 7 + 5x - 10 = 10x - 12$

42. $10x - 9 + 2x - 3 = 8x - 18$

43. $7(2x - 1) - 5x = x + 25$

44. $9(3x + 2) - 10x = 12x - 7$

45. $3x + 2(4x - 3) = 6x - 9$

46. $7x + 3(2x + 5) = 10x + 17$

47. $\frac{8}{3}x - 3 = \frac{2}{3}x + 15$

48. $\frac{12}{5}x + 7 = 31 - \frac{3}{5}x$

49. $\frac{2x}{5} - 5 = \frac{12x}{5} + 8$

50. $\frac{3x}{7} - 5 = \frac{24x}{7} + 7$

51. $5.3x - 7 = 2.3x + 5$ 

52. $9.8x + 2 = 3.8x + 20$

< Objective 3 >

53. **Social Science** There were 55 more yes votes than no votes on an election measure. If 735 votes were cast in all, how many yes votes were there?

54. **Business and Finance** Juan worked twice as many hours as Jerry. Marcia worked 3 more hours than Jerry. If they worked a total of 31 hr, how many hours did each employee work?

55. **Business and Finance** Francine earns $120 per week more than Rob. If they earn a total of $2,680 per week, how much does Francine earn each week?

56. **Science and Medicine** To determine the upper limit for a person's heart rate during aerobic training, subtract the person's age from 220, and then multiply the result by $\frac{9}{10}$. Determine the age of a person if the person's upper limit heart rate is 153.

57. **Business and Finance** Jody earns $280 more per month than Frank. If their monthly salaries total $5,520, what amount does each earn?

58. **Business and Finance** A washer-dryer combination costs $950. If the washer costs $90 more than the dryer, what does each appliance cost?

59. **Problem Solving** Yan Ling is 1 year less than twice as old as his sister. If the sum of their ages is 14 years, how old is Yan Ling?

60. **Problem Solving** Diane is twice as old as her brother Dan. If the sum of their ages is 27 years, how old are Diane and her brother?

61. **Business and Finance** Tonya takes home $1,080 per week. If her deductions amount to 28% of her wages, what is her weekly pay before deductions?

62. **Business and Finance** Sam donates 5% of his net income to charity. This amounts to $190 per month. His payroll deductions are 24% of his gross monthly income. What is Sam's gross monthly income?

63. **Science and Medicine** While traveling in South America, Richard noted that temperatures were given in degrees Celsius. Wondering what the temperature 95°F would correspond to, he found that he could answer this if he could solve the equation

$$95 = \frac{9}{5}C + 32$$

What was the corresponding temperature?

64. **Science and Medicine** While traveling in England, Marissa noted an outdoor thermometer showing 20°C. To convert this to degrees Fahrenheit, she solved the equation

$$20 = \frac{5}{9}(F - 32)$$

What was the Fahrenheit temperature?

65. **Number Problem** The sum of twice a number and 16 is 24. What is the number?

66. **Number Problem** 3 times a number, increased by 8, is 50. Find the number.

67. **Number Problem** 5 times a number, minus 12, is 78. Find the number.

68. **Number Problem** 4 times a number, decreased by 20, is 44. What is the number?

Complete each statement with **always, sometimes,** *or* **never.**

69. To solve a linear equation, we ________ use both the addition and multiplication properties.

70. We should ________ check a possible solution by substituting it into the original equation.

71. **Agricultural Technology** The estimated yield Y of a field of corn (in bushels per acre) can be found by multiplying the rainfall r, in inches, during the growing season by 16 and then subtracting 15. This relationship can be modeled by the formula

$$Y = 16r - 15$$

If a farmer wants a yield of 159 bushels per acre, then we can write the equation shown to determine the amount of rainfall required.

$$159 = 16r - 15$$

How much rainfall is necessary to achieve a yield of 159 bushels of corn per acre?

72. **Construction Technology** The number of studs s required to build a wall (with studs spaced 16 in. on center) is equal to one more than $\frac{3}{4}$ times the length of the wall w, in feet. We model this with the formula

$$s = \frac{3}{4}w + 1$$

If a contractor uses 22 studs to build a wall, how long is the wall?

73. **Allied Health** The internal diameter D [in millimeters (mm)] of an endotracheal tube for a child is calculated using the formula

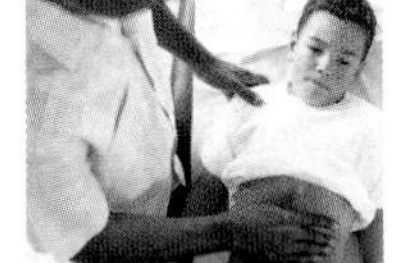

$$D = \frac{t + 16}{4}$$

in which t is the child's age (in years).

How old is a child who requires an endotracheal tube with an internal diameter of 7 mm?

74. **Mechanical Engineering** The number of BTUs required to heat a house is $2\frac{3}{4}$ times the volume of the air in the house (in cubic feet). What is the maximum air volume that can be heated with a 90,000-BTU furnace?

75. **Information Technology** A compression program reduces the size of files by 36%. If a compressed folder has a size of 11.2 MB, how large was it before compressing? VIDEO

76. **Agricultural Technology** A farmer harvested 2,068 bushels of barley. This amounted to 94% of his bid on the futures market. How many bushels did he bid to sell on the futures market?

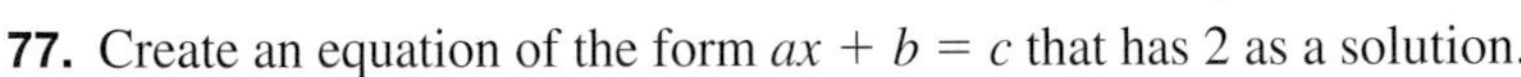

Skills | Calculator/Computer | Career Applications | **Above and Beyond**

77. Create an equation of the form $ax + b = c$ that has 2 as a solution.

78. Create an equation of the form $ax + b = c$ that has -6 as a solution.

79. The equation $3x = 3x + 5$ has no solution, whereas the equation $7x + 8 = 8$ has zero as a solution. Explain the difference between an equation that has zero as a solution and an equation that has no solution.

80. Construct an equation for which every real number is a solution.

81. Complete this statement in your own words: "You can tell that an equation is a linear equation when. . . ."

82. What is the common characteristic of equivalent equations?

83. What is meant by a *solution* to a linear equation?

84. Why does the multiplication property of equality not include multiplying both sides of the equation by 0?

85. Solve each equation. Express each solution as a fraction.

(a) $2x + 3 = 0$ (b) $4x + 7 = 0$ (c) $6x - 1 = 0$

(d) $5x - 2 = 0$ (e) $-3x + 8 = 0$ (f) $-5x - 9 = 0$

(g) Based on these problems, express the solution to the equation

$$ax + b = 0$$

where a and b represent real numbers and $a \neq 0$.

86. You are asked to solve an equation, but one number is missing. It reads

$$\frac{5x - ?}{4} = \frac{9}{2}$$

The solution to the equation is 4. What is the missing number?

Answers

1. 4 **3.** 3 **5.** 7 **7.** −2 **9.** −2 **11.** 3 **13.** 8 **15.** 32 **17.** 18 **19.** 20 **21.** 3 **23.** 2 **25.** 6 **27.** −2 **29.** $-\frac{10}{3}$ **31.** −4 **33.** 6 **35.** 4 **37.** 5 **39.** −4 **41.** $\frac{5}{3}$ **43.** 4 **45.** $-\frac{3}{5}$ **47.** 9 **49.** $-\frac{13}{2}$ **51.** 4 **53.** 395 votes **55.** $1,400 **57.** Jody: $2,900; Frank: $2,620. **59.** 9 years old **61.** $1,500 **63.** 35°C **65.** 4 **67.** 18 **69.** sometimes **71.** $10\frac{7}{8}$ in. **73.** 12 years old **75.** 17.5 MB **77.** Above and Beyond **79.** Above and Beyond **81.** Above and Beyond **83.** A value for which the original equation is true **85.** (a) $-\frac{3}{2}$; (b) $-\frac{7}{4}$; (c) $\frac{1}{6}$; (d) $\frac{2}{5}$; (e) $\frac{8}{3}$; (f) $-\frac{9}{5}$; (g) $-\frac{b}{a}$

summary :: chapter 11

Definition/Procedure	Example	Reference
From Arithmetic to Algebra		**Section 11.1**
Addition $x + y$ means the **sum** of x and y, or x **plus** y. Some other words indicating addition are *more than* and *increased by*.	The sum of x and 5 is $x + 5$. 7 more than a is $a + 7$. b increased by 3 is $b + 3$.	*p.* 621
Subtraction $x - y$ means the **difference** of x and y, or x **minus** y. Some other words indicating subtraction are *less than* and *decreased by*.	The difference of x and 3 is $x - 3$. 5 less than p is $p - 5$. a decreased by 4 is $a - 4$.	
Multiplication $x \cdot y$ $(x)(y)$ These all mean the **product** of x and y, or x **times** y. xy	The product of m and n is mn. The product of 2 and the sum of a and b is $2(a + b)$.	
Division $\frac{x}{y}$ means x **divided by** y, or the **quotient** when x is divided by y.	n divided by 5 is $\frac{n}{5}$. The sum of a and b, divided by 3, is $\frac{a + b}{3}$.	
Evaluating Algebraic Expressions		**Section 11.2**
Step 1 Replace each variable with the given number value. **Step 2** Compute, following the rules for the order of operations.	Evaluate $\frac{4a - b}{2c}$ if $a = -6$, $b = 8$, and $c = -4$. $\frac{4a - b}{2c} = \frac{4(-6) - (8)}{2(-4)}$ $= \frac{-24 - 8}{-8}$ $= \frac{-32}{-8} = 4$	*p.* 630
Simplifying Algebraic Expressions		**Section 11.3**
Term A number, or the product of a number and one or more variables and their exponents. **Like terms** Terms that contain exactly the same variables raised to the same powers.	$4a^2$ and $3a^2$ are like terms. $5x^2$ and $2xy^2$ are not like terms.	*p.* 639
Combining Like Terms		
Step 1 Add or subtract the numerical coefficients. **Step 2** Attach the common variables.	$5a + 3a = 8a$ $7xy - 3xy = 4xy$	*p.* 641
Solving Equations with the Addition Property		**Section 11.4**
Equation A statement that two expressions are *equal*. **Solution** Any value for the variable that makes an equation a true statement.	$3x - 5 = 7$ is an equation. 4 is a solution to the equation because $3(4) - 5 \stackrel{?}{=} 7$ $12 - 5 \stackrel{?}{=} 7$ $7 = 7$ True	*p.* 648
Equivalent equations Equations that have exactly the same set of solutions.	$3x - 5 = 7$ and $x = 4$ are equivalent equations.	*p.* 650

Continued

Definition/Procedure	Example	Reference
The addition property If $a = b$, then $a + c = b + c$. Adding (or subtracting) the same quantity to both sides of an equation yields an equivalent equation.	$x - 5 = 7$ $+5 \quad +5$ $x = 12$	p. 650
Solving Equations with the Multiplication Property		**Section 11.5**
The multiplication property If $a = b$ and $c \neq 0$, then $ac = bc$. Multiplying (or dividing) both sides of an equation by the same nonzero number yields an equivalent equation.	$5x = 20$ $\frac{5x}{5} = \frac{20}{5}$ $x = 4$	p. 664
To solve a percent problem algebraically, translate the problem into algebra (writing the rate as a decimal) and use the multiplication rule to solve.	30% of what number is 45? $0.3x = 45$ $\frac{0.3x}{0.3} = \frac{45}{0.3}$ $x = 150$	p. 669
Combining the Rules to Solve Equations		**Section 11.6**
Solving linear equations We say that an equation is solved when we have an equivalent equation of the form $x = \square$ or $\square = x$ in which $\square$ is some number. The steps for solving a linear equation follow. **Step 1** Use the distributive property to remove any grouping symbols. **Step 2** Combine like terms on each side of the equation. **Step 3** Add or subtract variable terms to bring the variable terms to one side of the equation. **Step 4** Add or subtract numbers to isolate the variable term. **Step 5** Multiply by the reciprocal of the coefficient to isolate the variable. **Step 6** Check your result.	Solve $3x - 6 + 4x = 3x + 14$ $7x - 6 = 3x + 14$ $7x - 6 - 3x = 3x + 14 - 3x$ $4x - 6 = 14$ $4x - 6 + 6 = 14 + 6$ $4x = 20$ $\frac{4x}{4} = \frac{20}{4}$ $x = 5$ Check $3(5) - 6 + 4(5) \stackrel{?}{=} 3(5) + 14$ $29 = 29$ True	p. 690

summary exercises :: chapter 11

This summary exercise set will help ensure that you have mastered each of the objectives of this chapter. The exercises are grouped by section. You should reread the material associated with any exercises that you find difficult. The answers to the odd-numbered exercises are in the Answers Appendix at the back of the text.

11.1 *Write each phrase symbolically.*

1. 5 more than y
2. c decreased by 10
3. The product of 8 and a
4. The quotient when y is divided by 3
5. 5 times the product of m and n
6. The product of a and 5 less than a
7. 3 more than the product of 17 and x
8. The quotient when a plus 2 is divided by a minus 2

11.2 *Evaluate each expression if $x = -3$, $y = 6$, $z = -4$, and $w = 2$.*

9. $3x + w$

10. $5y - 4z$

11. $x + y - 3z$

12. $5z^2$

13. $3x^2 - 2w^2$

14. $3x^3$

15. $5(x^2 - w^2)$

16. $\frac{6z}{2w}$

17. $\frac{2x - 4z}{y + (-z)}$

18. $\frac{3x - y}{w - x}$

19. $\frac{x(y^2 - z^2)}{(y + z)(y - z)}$

20. $\frac{y(x - w)^2}{x^2 - 2xw + w^2}$

11.3 *List the terms of each expression.*

21. $4a^3 - 3a^2$

22. $5x^2 - 7x + 3$

List the like terms.

23. $5m^2, -3m, -4m^2, 5m^3, m^2$

24. $4ab^2, 3b^2, -5a, ab^2, 7a^2, -3ab^2, 4a^2b$

Combine like terms.

25. $5c + 7c$

26. $2x + 5x$

27. $4a - 2a$

28. $6c - 3c$

29. $9xy - 6xy$

30. $5ab^2 + 2ab^2$

31. $7a + 3b + 12a - 2b$

32. $6x - 2x + 5y - 3x$

33. $5x^3 + 17x^2 - 2x^3 - 8x^2$

34. $3a^3 + 5a^2 + 4a - 2a^3 - 3a^2 - a$

35. Subtract $4a^3$ from the sum of $2a^3$ and $12a^3$.

36. Subtract the sum of $3x^2$ and $5x^2$ from $15x^2$.

11.4 *Tell whether the number shown in parentheses is a solution to the given equation.*

37. $7x + 2 = 16$ (2)

38. $5x - 8 = 3x + 2$ (4)

39. $7x - 2 = 2x + 8$ (2)

40. $4x + 3 = 2x - 11$ (−7)

41. $x + 5 + 3x = 2 + x + 23$ (6)

42. $\frac{2}{3}x - 2 = 10$ (21)

Solve each equation and check your result.

43. $x + 5 = 7$

44. $x - 9 = 3$

45. $5x = 4x - 5$

46. $3x - 9 = 2x$

47. $5x - 3 = 4x + 2$

48. $9x + 2 = 8x - 7$

49. $7x - 5 = 6x + (-4)$

50. $3 + 4x - 1 = x - 7 + 2x$

51. $4(2x + 3) = 7x + 5$

52. $5(5x - 3) = 6(4x + 1)$

11.5–11.6

53. $5x = 35$

54. $7x = -28$

55. $-6x = 24$

56. $-9x = -63$

57. $\frac{x}{4} = 8$

58. $-\frac{x}{5} = -3$

59. $\frac{2}{3}x = 18$

60. $\frac{3}{4}x = 24$

61. $5x - 3 = 12$

62. $4x + 3 = -13$

63. $7x + 8 = 3x$

64. $3 - 5x = -17$

65. $3x - 7 = x$

66. $2 - 4x = 5$

67. $\frac{x}{3} - 5 = 1$

68. $\frac{3}{4}x - 2 = 7$

69. $6x - 5 = 3x + 13$

70. $3x + 7 = x - 9$

71. $7x + 4 = 2x + 6$

72. $9x - 8 = 7x - 3$

73. $2x + 7 = 4x - 5$

74. $3x - 15 = 7x - 10$

75. $\frac{10}{3}x - 5 = \frac{4}{3}x + 7$

76. $\frac{11}{4}x - 15 = 5 - \frac{5}{4}x$

77. $3.7x + 8 = 1.7x + 16$

78. $5.4x - 3 = 8.4x + 9$

79. $3x - 2 + 5x = 7 + 2x + 21$

80. $8x + 3 - 2x + 5 = 3 - 4x$

81. $5(3x - 1) - 6x = 3x - 2$

82. $5x + 2(3x - 4) = 14x + 7$

Solve each application.

83. **Business and Finance** A mechanic charged \$75 an hour plus \$225 for parts to replace the ignition coil on a car. If the total bill was \$450, how many hours did the repair job take?

84. **Business and Finance** A call to Phoenix, Arizona, from Dubuque, Iowa, costs 55 cents for the first minute and 23 cents for each additional minute or portion of a minute. If Barry has \$6.30 in change, how long can he talk?

85. **Number Problem** The sum of 4 times a number and 14 is 34. Find the number.

86. **Number Problem** If 6 times a number is subtracted from 42, the result is 24. Find the number.

CHAPTER 11

chapter test 11

Use this chapter test to assess your progress and to review for your next exam. Allow yourself about an hour to take this test. The answers to these exercises are in the Answers Appendix at the back of the text.

Write each phrase symbolically.

1. 5 less than a

2. The product of 6 and m

3. 4 times the sum of m and n

4. The sum of 4 times m and n

Determine whether the number in parentheses is a solution to the given equation.

5. $7x - 3 = 25 \quad (5)$

6. $8x - 3 = 5x + 9 \quad (4)$

Solve each equation and check your result.

7. $x - 7 = 4$

8. $7x = 49$

9. $7x - 5 = 16$

10. $10 - 3x = -2$

11. $\frac{x}{4} = -3$

12. $7x - 12 = 6x$

13. $9x - 2 = 8x + 5$

14. $\frac{4}{5}x = 20$

15. $7x - 3 = 4x - 5$

16. $2x - 7 = 5x + 8$

Evaluate each expression if $a = -2$, $b = 6$, and $c = -4$.

17. $4a - c$

18. $5c^2$

19. $6(2b - 3c)$

20. $\frac{3a - 4b}{a + c}$

Simplify each expression.

21. $8a + 7a$

22. $8x^2y - 5x^2y$

23. $10x + 8y + 9x - 3y$

24. $3m^2 + 7m - 5 - 9m + m^2$

Solve the application.

25. A coffee shop earns a (marginal) profit of \$3.45 on each double latte it sells. However, the fixed costs associated with double lattes amount to \$200 per day. How many lattes must the shop sell in order to break even? *Hint:* The break-even point is the number of lattes that the shop needs to sell in order to avoid a loss.

cumulative review chapters 1–11

Use this exercise set to review concepts from earlier chapters. While it is not a comprehensive exam, it will help you identify any material that you need to review before moving on to your final exam. In addition to the answers, you will find section references for these exercises in the Answers Appendix in the back of the text.

Name the property illustrated.

1.2

1. $(7 + 3) + 8 = 7 + (3 + 8)$

1.5

2. $6 \times 7 = 7 \times 6$

3. $5 \cdot (2 + 4) = 5 \cdot 2 + 5 \cdot 4$

1.4

Round the numbers to the indicated place value.

4. 5,873 to the nearest hundred

5. 953,150 to the nearest ten thousand

1.7

6. Evaluate $2 + 8 \times 3 \div 4$.

2.2

7. Write the prime factorization of 264.

3.2

8. Find the least common multiple (LCM) of 6, 15, and 45.

2.3

9. Convert to a mixed number: $\frac{22}{7}$

10. Convert to an improper fraction: $6\frac{5}{8}$

Evaluate each expression.

2.5

11. $\frac{2}{3} \times 1\frac{4}{5} \times \frac{5}{8}$

2.6

12. $2\frac{2}{7} \div 1\frac{11}{21}$

3.4

13. $4\frac{7}{8} + 3\frac{1}{6}$

10.2

14. $9 + \left(-5\frac{3}{8}\right)$

15. **Construction** A $6\frac{1}{2}$-in. bolt is placed through a wall that is $5\frac{7}{8}$ in. thick. How far does the bolt extend beyond the wall?

4.2

16. Business and Finance You pay for purchases of \$13.99, \$18.75, \$9.20, and \$5 with a \$50 bill. How much cash do you have left?

8.3

17. Geometry Find the area of a circle whose diameter is 3.2 ft. Use 3.14 for π and round the result to the nearest hundredth.

4.4

18. Construction A 14-acre piece of land is being developed into home lots. If 2.8 acres of land will be used for roads, and each home site is to be 0.35 acre, how many lots can be formed?

4.5

19. Write the decimal equivalent of $\frac{8}{11}$. Use bar notation.

5.4

20. Solve for the unknown: $\frac{5}{m} = \frac{0.4}{9}$

21. Business and Finance You are using a photocopy machine to reduce an advertisement that is 14 in. wide by 21 in. long. If the new width is to be 8 in., what will the new length be?

6.2

Write each number as a percent.

22. 0.003

23. $\frac{5}{8}$

24. $3\frac{1}{2}$

6.3

25. 120% of what number is 180?

26. 72 is 12% of what number?

6.4

27. Business and Finance Luisa works on an 8% commission basis. If she wishes to earn \$2,200 in commissions in 1 month, how much must she sell during that period?

7.3

28. Complete the statement: 300 mg = _______ g

9.4

29. Business and Finance According to the line graph, what is the difference in benefits between 2010 and 2012?

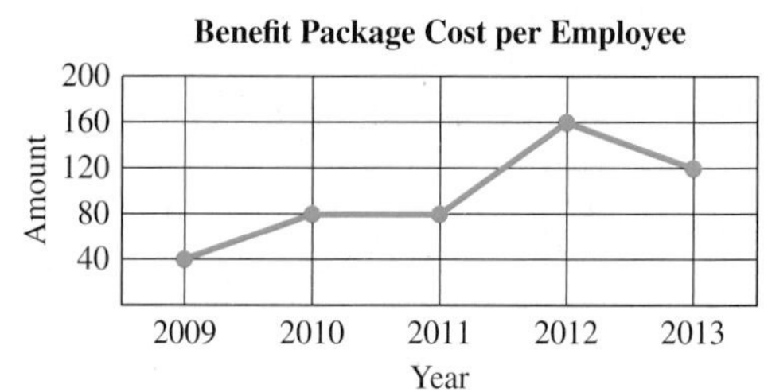

10.2

30. The opposite of 8 is _______.

10.1

31. The absolute value of -20 is _______.

Evaluate each expression.

32. $-(-12)$

33. $|-5|$

10.2

34. $-12 + (-6)$

10.3

35. $-8 - (-4)$

10.4

36. $(-6)(15)$

10.5

37. $48 \div (-12)$

11.1

Write each phrase symbolically.

38. 3 times the sum of x and y

39. The quotient when 5 less than n is divided by 3

11.2

Evaluate each expression if $a = 5$, $b = -3$, $c = 4$, and $d = -2$.

40. $6ad$

41. $3b^2$

42. $3(c - 2d)$

43. $\dfrac{2a - 7d}{a - b}$

Combine like terms.

11.3

44. $10a^2 + 5a + 2a^2 - 2a$

45. $5x - 3y + 2x + y - 7x$

Solve each equation, and check your results.

11.4

46. $9x - 5 = 8x$

11.5

47. $-\frac{3}{4}x = 18$

11.6

48. $2x + 3 = 7x + 5$

49. $\frac{4}{3}x - 6 = 4 - \frac{2}{3}x$

50. **Business and Finance** A grocery store adds a 25% markup to the wholesale price of goods to determine their retail price. Find the wholesale price of a gallon of milk that retails for \$2.59.

brief glossary

acute angle An angle of less than 90 degrees **(8.1)**

addends Numbers that are being added together **(1.2)**

additive inverse A number that is the same distance from zero on a number line, but in the opposite direction **(10.2)**

amount In a percent problem (denoted *A*), it is the part of the whole being compared to the base **(6.3)**

angle A geometric figure consisting of two line segments that share a common endpoint **(8.1)**

area A measure given to a surface. It is expressed in square units **(1.5)**

average A number that is typical of a larger group of numbers **(9.1)**

base 1. A number that is raised to a power **(1.7)**
2. The whole (denoted *B*) in a percent problem—it is the standard used for comparison **(6.3)**

circumference The distance around a circle **(8.3)**

common multiple A number that is exactly divisible by each number in a group of numbers **(3.2)**

complex fraction A fraction that has a fraction within the numerator or denominator **(2.6)**

composite figure A geometric figure that is formed by adjoining two or more basic shapes **(8.3)**

composite number Any whole number greater than one that is not prime **(2.1)**

convert Change units from one system to another—it occurs most frequently between the English system and the metric system **(7.4)**

deci- A prefix meaning ten **(1.1)**

decimal equivalent A decimal number that has the same value as a given fraction **(4.5)**

decimal fraction A fraction whose denominator is a power of ten **(4.1)**

denominate number A number that has a specific unit assigned to it **(5.1)**

denominator The bottom part of a fraction **(2.3)**

difference The result of subtraction **(1.3)**

digit Any of the numbers 0, 1, 2, 3, 4, 5, 6, 7, 8, 9 **(1.1)**

dividend A number that is being divided **(1.6)**

divisor A number divided into another number **(1.6)**

ellipsis Three dots that indicate that a pattern continues **(1.2)**

equation A statement that two expressions are equal **(5.4)**

equilateral triangle A triangle in which all three sides are the same length **(8.4)**

equivalent equations Two or more equations that have exactly the same solution(s) **(11.4)**

equivalent fractions Two fractions that name the same amount **(2.4)**

estimating The process of using rounding to find a reasoned guess for an answer **(1.4)**

evaluate Find the numeric equivalent **(11.2)**

exponent Also called a power. The number of times a factor is repeated in a multiplication **(1.7)**

expression 1. A combination of symbols that can be simplified to a single value **(3.5)**
2. A meaningful collection of numbers, variables, and signs of operation **(11.1)**

extrapolation Using an earlier trend to predict a future value **(9.4)**

factors 1. Numbers that are multiplied together **(1.5)**
2. Natural numbers that divide exactly into a given number **(2.1)**

fraction Whenever a unit or whole quantity is divided into parts, those parts are called *fractions* of the unit **(2.3)**

geo- A prefix meaning *earth* **(8.1)**

gram The basic unit of mass in the metric system—there are approximately 28 grams in one ounce **(7.3)**

graph A diagram that represents the connection between two or more things **(9.3)**

greatest common factor Also called GCF. Given a set of numbers, the GCF is the *largest* number that divides each of the numbers exactly **(2.2)**

hypotenuse The side opposite the right angle of a right triangle **(8.5)**

improper fraction A fraction in which the numerator is equal to or greater than the denominator **(2.3)**

integers The natural numbers, their negatives, and zero **(10.1)**

irrational number A number that cannot be written as a common fraction **(10.1)**

isosceles triangle A triangle in which exactly two sides are the same length **(8.4)**

least common denominator Also called LCD. Given a set of fractions, the LCD is the *smallest* number that is evenly divisible by every denominator in the set **(3.3)**

least common multiple Also called LCM. Given a set of numbers, the LCM is the *smallest* number that is evenly divisible by every number in the set **(3.2)**

like fractions Fractions with the same denominator **(3.1)**

like terms Terms with exactly the same variables to the same powers **(11.3)**

line A series of points that goes straight forever **(8.1)**

liter The basic unit of volume in the metric system—one liter is slightly more than one quart **(7.3)**

lowest terms A fraction is in lowest terms if the numerator and denominator have no common factors other than 1 **(2.4)**

maximum Often called the Max, it is the largest number in a data set **(9.2)**

mean An average found by adding together the numbers in a set then dividing by the number of items in the set **(9.1)**

median The number that is in the middle when a set of numbers is written in ascending order **(9.1)**

meter The basic unit of length in the metric system—it is slightly longer than one yard **(7.2)**

minimum Often called the Min, it is the smallest number in a data set **(9.2)**

minuend A number that has something subtracted from it **(1.3)**

mixed number The sum of a whole number and a proper fraction **(2.3)**

mode The item or number that occurs most frequently in a data set **(9.1)**

multiplicative inverse For any nonzero number a, it is the number $\frac{1}{a}$ **(10.4)**

natural numbers Counting numbers 1, 2, 3, ... **(1.2)**

negative number A number with a value that is less than zero **(10.1)**

numerator The top part of a fraction **(2.3)**

obtuse angle An angle of more than 90 degrees **(8.1)**

opposite A number that is the same distance from zero on a number line, but in the other direction **(10.1)**

origin A starting place. On a number line, it is zero **(1.2)**

parallel lines Two lines that never intersect—even if we extend the lines forever **(8.1)**

parallelogram A rectangle in which both pairs of opposite sides are parallel **(8.2)**

percent literally, "for each hundred" **(6.1)**

perimeter The distance around a closed figure **(1.2)**

perpendicular lines When two lines intersect, they form four angles. If the two lines cross such that they form four equal angles, we say the two lines are perpendicular **(8.1)**

pie chart A graph, drawn as a circle, in which each wedge represents the part of the whole that item makes up **(9.4)**

point A location with no size that covers no area **(8.1)**

polygon A simple closed figure with three or more sides in which each side is a line segment **(8.2)**

power See exponent **(1.7)**

prime number Any whole number that has exactly two factors, one and itself **(2.1)**

product The result of multiplication **(1.5)**

proper fraction A fraction in which the numerator is less than the denominator **(2.3)**

proportion A statement that two fractions, rates, or ratios are equal **(5.3)**

quartile The numbers that divide an ordered set into four equal parts **(9.2)**

quotient The result of division **(1.6)**

rate 1. A comparison of two denominate numbers with different types of units **(5.2)**

2. In a percent problem (denoted R), it is the ratio of the amount to the base. It is written as a percent **(6.3)**

rational number A number that can be written as a fraction **(10.1)**

ratio A comparison of two numbers or like quantities **(5.1)**

real numbers All positive and negative numbers, along with zero **(10.1)**

reciprocal 1. The inverted form or a nonzero fraction. The *reciprocal* of $\frac{a}{b}$ is $\frac{b}{a}$ **(2.6)**

2. Also called the *multiplicative inverse* **(10.4)**

rectangle A figure with four equal-sized corners **(1.2)**

rounding The process of expressing numbers to the nearest hundred, thousand, and so on **(1.4)**

scalene triangle A triangle in which no two sides are the same length **(8.4)**

similar triangles Two different triangles for which the measurements of corresponding angles are the same **(8.4)**

simplifying a fraction Rewriting a fraction as an equivalent fraction in which there is no common factor in the denominator and numerator **(2.4)**

solid A three-dimensional figure. It has length, width, and height **(1.5)**

solution (to an equation) A value for the variable that makes the equation a true statement **(11.4)**

square root (of a number) The special value that, when multiplied by itself, gives us the number **(8.5)**

subtrahend A number being subtracted from another number **(1.3)**

table A display of information in rows or columns **(9.3)**

term An expression that can be written as a number, or the product of a number and one or more variables and their exponents **(11.3)**

trapezoid A four-sided polygon with exactly one pair of parallel sides **(8.2)**

triangle A three-sided polygon **(8.2)**

unit price Relates a price to some common unit **(5.2)**

unit rate A *rate* (a comparison of two denominate numbers with different types of units) that is simplified so that it compares a denominate number with a *single unit* of a different denominate number **(5.2)**

unlike fractions Fractions with different denominators **(3.3)**

variable A letter used to represent a number **(11.1)**

volume A measure given to a solid. It is expressed in cubic units **(1.5)**

whole numbers The natural numbers with zero, so, 0, 1, 2, 3, ... **(1.2)**

Answers to Prerequisite Checks, Reading Your Text, Summary Exercises, Chapter Tests, and Cumulative Reviews

Reading Your Text for Chapter 1

Section 1.1 (a) decimal; (b) digit; (c) comma; (d) million
Section 1.2 (a) natural; (b) addition; (c) order; (d) read
Section 1.3 (a) difference; (b) less; (c) read; (d) borrowing
Section 1.4 (a) rounding; (b) place value; (c) rounding; (d) less
Section 1.5 (a) reasonable; (b) product; (c) powers; (d) rectangle
Section 1.6 (a) quotient; (b) remainder; (c) zero; (d) subtraction
Section 1.7 (a) first; (b) exponent; (c) parentheses; (d) one

Summary Exercises for Chapter 1

1. Hundreds **3.** Twenty-seven thousand, four hundred twenty-eight **5.** 37,583 **7.** Commutative property of addition **9.** 1,416 **11.** 4,801 **13.** (a) The total number of passengers. (b) There are five flights. There were 173, 212, 185, 197, and 202 passengers on these flights. **15.** 969 passengers **17.** 27 **19.** 15 **21.** 18,800 **23.** 2,574 **25.** \$536 **27.** 7,000 **29.** 550,000 **31.** < **33.** 18 ft **35.** Commutative property of multiplication **37.** Associative property of multiplication **39.** 1,856 **41.** 154,602 **43.** 24 ft^2 **45.** \$630 **47.** 0 **49.** 308 r5 **51.** 497 r1 **53.** 28 mi/gal **55.** 10 **57.** 40 **59.** 28 **61.** 36 **63.** 3 **65.** 24

Chapter Test for Chapter 1

1. Three hundred two thousand, five hundred twenty-five **2.** Hundred thousands **3.** 2,430,000 **4.** 21,696 **5.** 1,918 **6.** 13,103 **7.** 4,984 **8.** 55,414 **9.** 235 **10.** 40,555 **11.** 30,770 **12.** 12,220 **13.** 266 r7 **14.** 2,192 r6 **15.** 39 **16.** 15 **17.** Distributive property of multiplication over addition **18.** Associative property of addition **19.** Associative property of multiplication **20.** Commutative property of addition **21.** > **22.** < **23.** 7,700 **24.** 12 in. **25.** 12 in.^{2} **26.** 280 ft^3 **27.** \$558,750 **28.** \$892 **29.** 55,978 **30.** 72 lb

Prerequisite Check for Chapter 2

1. No **2.** No **3.** Yes **4.** No **5.** Yes **6.** 1, 3, 9 **7.** 1, 2, 5, 10 **8.** 1, 17 **9.** 1, 2, 3, 4, 6, 8, 12, 16, 24, 48 **10.** 1, 2, 3, 4, 5, 6, 10, 12, 15, 20, 30, 60

Reading Your Text for Chapter 2

Section 2.1 (a) One (1); (b) prime; (c) composite; (d) even
Section 2.2 (a) commutative; (b) prime; (c) common; (d) greatest
Section 2.3 (a) denominator; (b) proper; (c) mixed; (d) numerator
Section 2.4 (a) cross; (b) equivalent; (c) fundamental; (d) common
Section 2.5 (a) simplest; (b) simplify; (c) reasonableness; (d) reasonable
Section 2.6 (a) reciprocal; (b) complex; (c) divisor; (d) units

Summary Exercises for Chapter 2

1. 1, 2, 4, 13, 26, 52 **3.** Prime: 2, 5, 7, 11, 17, 23, 43; composite: 14, 21, 27, 39 **5.** None **7.** $2 \times 2 \times 3 \times 5 \times 7$ **9.** $2 \times 3 \times 3 \times 5 \times 5 \times 5$ **11.** 1 **13.** 13 **15.** 11 **17.** Numerator: 17; denominator: 23 **19.** Fraction: $\frac{5}{6}$; numerator: 5; denominator: 6 **21.** $6\frac{5}{6}$ **23.** $7\frac{2}{3}$ **25.** $\frac{61}{8}$ **27.** $\frac{37}{7}$ **29.** No **31.** $\frac{2}{3}$ **33.** $\frac{7}{9}$ **35.** Yes **37.** $\frac{1}{9}$ **39.** $1\frac{1}{2}$ **41.** $9\frac{3}{5}$ **43.** 8 **45.** \$204 **47.** \$45 **49.** 408 mi **51.** $\frac{5}{8}$ **53.** $\frac{2}{3}$ **55.** $\frac{3}{16}$ **57.** $\frac{3}{7}$ **59.** $\frac{3}{4}$ ft **61.** 56 mi/hr **63.** 48 lots

Chapter Test for Chapter 2

1. Prime: 5, 13, 17, 31; composite: 9, 22, 27, 45 **2.** $2 \times 2 \times 2 \times 3 \times 11$ **3.** 2 and 3 **4.** 12 **5.** 8 **6.** $\frac{5}{6}$; numerator: 5; denominator: 6 **7.** $\frac{5}{8}$; numerator: 5; denominator: 8 **8.** $\frac{3}{5}$; numerator: 3; denominator: 5 **9.** $4\frac{1}{4}$ **10.** $\frac{7}{9}$ **11.** $\frac{3}{7}$ **12.** $\frac{8}{23}$ **13.** Yes **14.** Yes **15.** No **16.** Proper fractions: $\frac{10}{11}, \frac{1}{8}$; Improper fractions: $\frac{9}{5}, \frac{7}{7}, \frac{8}{1}$; Mixed number: $2\frac{3}{5}$ **17.** $\frac{37}{7}$ **18.** $\frac{35}{8}$ **19.** $\frac{74}{9}$ **20.** $4\frac{1}{4}$ **21.** 15 **22.** $9\frac{1}{4}$ **23.** 3 **24.** $\frac{10}{21}$ **25.** 4 **26.** $\frac{5}{8}$ **27.** $3\frac{3}{7}$ **28.** $\frac{4}{15}$ **29.** $2\frac{2}{3}$ **30.** $\frac{9}{16}$ **31.** $1\frac{1}{7}$ **32.** $9\frac{1}{5}$ **33.** $1\frac{3}{11}$ **34.** \$1.32 **35.** 48 books **36.** 47 homes **37.** 20 yd^2 **38.** 190 mi

Cumulative Review Chapters 1–2

[1.1] 1. Hundred thousands **2.** Three hundred two thousand, five hundred twenty-five **3.** 2,430,000 **[1.2] 4.** Commutative property of addition **5.** Additive identity **6.** Associative property of addition **7.** 966 **8.** 23,351 **[1.4] 9.** 5,900 **10.** 950,000 **11.** 7,700 **12.** > **13.** < **[1.3] 14.** 3,861 **15.** 17,465 **[1.2] 16.** 905 **[1.3] 17.** \$17,579 **[1.5] 18.** Associative property of multiplication **19.** Commutative property of multiplication **20.** Distributive property **21.** 378,214 **22.** 686,000 **23.** \$1,008 **[1.6] 24.** 67 r43 **25.** 103 r176 **[1.7] 26.** 38 **27.** 56 **28.** 36 **29.** 8 **30.** \$58 **[2.1] 31.** Prime: 5, 13, 17, 31; composite: 9, 22, 27, 45 **32.** 2 and 3 **[2.2] 33.** $2 \times 2 \times 2 \times 3 \times 11$ **34.** 12 **35.** 8 **[2.3] 36.** Proper fractions: $\frac{7}{12}, \frac{3}{7}$; improper fractions: $\frac{10}{8}, \frac{9}{9}, \frac{7}{1}$; mixed numbers: $3\frac{1}{5}, 2\frac{2}{3}$ **[2.4] 37.** $2\frac{4}{5}$ **38.** 4 **39.** $\frac{13}{3}$ **40.** $\frac{63}{8}$ **41.** Yes **42.** No **43.** $\frac{2}{3}$ **44.** $\frac{3}{8}$ **[2.5] 45.** $\frac{8}{27}$ **46.** $\frac{4}{15}$ **47.** $5\frac{2}{5}$ **48.** $22\frac{2}{3}$ **49.** $\frac{3}{4}$ **[2.6] 50.** $1\frac{1}{3}$ **51.** $4\frac{1}{2}$ **52.** $\frac{5}{6}$ **53.** $1\frac{1}{2}$ **54.** \$540 **55.** 88 sheets

Prerequisite Check for Chapter 3

1. $2 \times 2 \times 2 \times 3$ **2.** $2 \times 2 \times 3 \times 3$ **3.** $2 \times 3 \times 3 \times 5$ **4.** $\frac{14}{3}$ **5.** $\frac{32}{5}$ **6.** $\frac{91}{10}$ **7.** $7\frac{1}{4}$ **8.** 7 **9.** 1 **10.** 27 **11.** $\frac{2}{3}$ **12.** $\frac{1}{4}$ **13.** $\frac{3}{5}$ **14.** $\frac{5}{8}$

Reading Your Text for Chapter 3

Section 3.1 (**a**) like; (**b**) numerators; (**c**) simplify; (**d**) difference
Section 3.2 (**a**) multiples; (**b**) smallest; (**c**) denominator; (**d**) greater
Section 3.3 (**a**) LCM; (**b**) unlike; (**c**) equivalent; (**d**) numerators
Section 3.4 (**a**) LCD; (**b**) regrouping; (**c**) like; (**d**) smaller
Section 3.5 (**a**) grouping; (**b**) exponents; (**c**) invert; (**d**) improper

Summary Exercises for Chapter 3

1. $\frac{2}{3}$ **3.** $\frac{15}{13} = 1\frac{2}{13}$ **5.** $\frac{4}{3} = 1\frac{1}{3}$ **7.** $\frac{11}{9} = 1\frac{2}{9}$ **9.** 12
11. 72 **13.** 60 **15.** 72 **17.** $\frac{7}{12}, \frac{5}{8}$ **19.** $>$ **21.** $<$
23. $\frac{10}{15}, \frac{12}{15}$ **25.** 36 **27.** 200 **29.** 132 **31.** 24
33. $\frac{19}{24}$ **35.** $\frac{7}{12}$ **37.** $\frac{107}{90} = 1\frac{17}{90}$ **39.** $\frac{7}{8}$
41. $\frac{85}{72} = 1\frac{13}{72}$ **43.** $\frac{5}{9}$ **45.** $\frac{1}{2}$ **47.** $\frac{5}{24}$ **49.** $\frac{7}{18}$
51. $\frac{11}{24}$ **53.** $\frac{13}{42}$ **55.** $\frac{1}{3}$ **57.** $10\frac{2}{7}$ **59.** $9\frac{37}{60}$ **61.** $3\frac{2}{3}$
63. $1\frac{43}{60}$ **65.** $8\frac{4}{5}$ **67.** $3\frac{1}{4}$ **69.** $9\frac{17}{24}$ **71.** $\frac{5}{12}$ cup
73. $19\frac{9}{16}$ in. **75.** $53\frac{9}{16}$ in. **77.** $3\frac{7}{12}$ gal **79.** $\frac{11}{12}$
81. $\frac{21}{4} = 5\frac{1}{4}$ **83.** $\frac{13}{54}$

Chapter Test for Chapter 3

1. 60 **2.** 36 **3.** $\frac{4}{5}$ **4.** $1\frac{3}{4}$ **5.** $1\frac{8}{15}$ **6.** $\frac{1}{9}$ **7.** $\frac{19}{24}$
8. $3\frac{23}{24}$ **9.** $\frac{1}{3}$ **10.** $\frac{63}{40} = 1\frac{23}{40}$ **11.** $7\frac{7}{10}$ **12.** $\frac{7}{12}$
13. $12\frac{3}{40}$ **14.** $10\frac{1}{4}$ **15.** $1\frac{11}{18}$ **16.** $\frac{9}{10}$ **17.** $\frac{25}{42}$
18. $\frac{23}{30}$ **19.** $9\frac{1}{7}$ **20.** $7\frac{11}{12}$ **21.** $\frac{2}{3}$ **22.** $13\frac{11}{20}$
23. $\frac{71}{60} = 1\frac{11}{60}$ **24.** $\frac{1}{12}$ **25.** $\frac{299}{24} = 12\frac{11}{24}$ **26.** 72
27. $\frac{1}{4}$ hr **28.** $1\frac{5}{12}$ cups **29.** 41,400 cups **30.** $5\frac{3}{4}$ hr

Cumulative Review Chapters 1–3

[1.2] 1. 7,173 **2.** 1,918 **3.** 2,731 **4.** 13,103
[1.3] 5. 235 **6.** 12,220 **7.** 429 **8.** 3,239
[1.5] 9. 174 **10.** 1,911 **11.** 4,984 **12.** 55,414
[1.6] 13. 24 r191 **14.** 22 r21 **15.** 209 r145
[1.7] 16. 5 **17.** 7 **18.** 3 **19.** 16 **20.** 20 **21.** 3
[2.3] 22. Proper fractions: $\frac{5}{7}, \frac{2}{5}$; improper fractions: $\frac{15}{9}, \frac{8}{8}, \frac{11}{1}$;
mixed numbers: $4\frac{5}{6}, 3\frac{5}{6}$ **23.** $1\frac{7}{9}$ **24.** $7\frac{1}{5}$ **25.** $\frac{23}{4}$
26. $\frac{55}{9}$ **[2.4] 27.** Yes **28.** No **[2.5] 29.** $\frac{1}{9}$ **30.** $\frac{1}{6}$
31. $\frac{3}{2} = 1\frac{1}{2}$ **32.** $\frac{17}{8} = 2\frac{1}{8}$ **33.** $\frac{48}{5} = 9\frac{3}{5}$ **34.** $\frac{34}{3} = 11\frac{1}{3}$
35. 8 **[2.6] 36.** $\frac{2}{3}$ **37.** $\frac{5}{6}$ **38.** $\frac{3}{16}$ **[3.1, 3.3] 39.** $\frac{4}{5}$
40. $\frac{61}{75}$ **41.** $\frac{71}{40} = 1\frac{31}{40}$ **42.** $\frac{1}{2}$ **43.** $\frac{5}{36}$ **44.** $\frac{5}{9}$
[3.4] 45. $6\frac{2}{7}$ **46.** $8\frac{1}{24}$ **47.** $4\frac{5}{9}$ **48.** $4\frac{1}{24}$ **49.** $3\frac{5}{8}$
50. $3\frac{13}{24}$ **51.** $14\frac{19}{30}$ hr **52.** $\frac{5}{8}$ in. **53.** $1\frac{11}{12}$ hr

Prerequisite Check for Chapter 4

1. Three and seven tenths **2.** Six and twenty-nine hundredths
3. Seventeen and eighty-nine thousandths **4.** 257 **5.** 164
6. 6,390 **7.** 79 r2 **8.** 7 r92 **9.** 136 ft^2 **10.** 225 in.^{2}
11. 23,000 **12.** 7,000 **13.** Ten thousands **14.** Tens

Reading Your Text for Chapter 4

Section 4.1 (**a**) decimal; (**b**) places; (**c**) exact; (**d**) ten-thousandths
Section 4.2 (**a**) decimal points; (**b**) value; (**c**) Perimeter; (**d**) following
Section 4.3 (**a**) add; (**b**) product; (**c**) zeros; (**d**) right
Section 4.4 (**a**) above; (**b**) past; (**c**) whole number; (**d**) left
Section 4.5 (**a**) divide; (**b**) bar; (**c**) terminating; (**d**) places

Summary Exercises for Chapter 4

1. Hundredths **3.** 0.37 **5.** Seventy-one thousandths
7. 4.5 **9.** $>$ **11.** $<$ **13.** 5.84 **15.** 4.876
17. $\frac{21}{25}$ **19.** 3.47 **21.** 37.728 **23.** 23.32 **25.** 1.075
27. 28.02 cm **29.** 6.15 cm **31.** 16.416 **33.** 69.44
35. 52 **37.** \$271.15 **39.** \$152.10 **41.** 4.65
43. 2.664 **45.** 1.273 **47.** 0.76 **49.** 0.0457
51. \$23.45 **53.** 54 lots **55.** 0.4375 **57.** $0.2\overline{6}$
59. $\frac{21}{100}$ **61.** $2\frac{3}{100}$

Chapter Test for Chapter 4

1. Ten-thousandths **2.** Two and fifty-three hundredths
3. 12.017 **4.** 12.803 **5.** 0.598 **6.** 23.57 **7.** 36,000
8. 0.049 **9.** 16.64 **10.** 10.54 **11.** 2.55 **12.** 1.4575
13. 2.35 **14.** 47.253 **15.** 17.437 **16.** 24.375
17. 0.004983 **18.** 0.00523 **19.** 735 **20.** 12,570
21. 3.888 **22.** 0.465 **23.** 0.02793 **24.** 2.385
25. 7.35 **26.** 0.051 **27.** 0.067 **28.** $<$ **29.** $>$
30. $>$ **31.** 0.4375 **32.** 0.429 **33.** $0.\overline{63}$ **34.** $\frac{9}{125}$
35. $4\frac{11}{25}$ **36.** \$543 **37.** 50.2 gal **38.** 32 lots
39. 7.525 in.^{2} **40.** \$573.40 **41.** \$3.06 **42.** $\frac{229}{500}$

Cumulative Review Chapters 1–4

[1.1] 1. Two hundred eighty-six thousand, five hundred forty-three
2. Hundreds **[1.2] 3.** 34,594 **[1.3] 4.** 48,888
[1.5] 5. 5,063 **6.** 70,455 **[1.6] 7.** 17 **8.** 35 r11
[1.7] 9. 29 **[1.4] 10.** 4,000 **[1.2, 1.5] 11.** *P*: 24 ft; *A*: 35 ft^2
[2.4] 12. $\frac{5}{17}$ **[2.5] 13.** $\frac{3}{4}$ **14.** $2\frac{6}{7}$ **[2.6] 15.** $\frac{9}{17}$
[3.1] 16. $\frac{5}{7}$ **[3.3] 17.** $\frac{7}{30}$ **[3.4] 18.** $3\frac{9}{10}$
[4.2] 19. 12.468 **[4.4] 20.** 3.9 **21.** 0.005238
[4.3] 22. 1.1385 **23.** 15,300 **[4.5] 24.** $\frac{43}{100}$
25. (**a**) 0.625; (**b**) 0.39 **[4.4] 26.** 0.429 **[4.4] 27.** 17.21
28. 39.829 **29.** 4.8 cm **30.** \$9.84 **31.** 230 lots

Prerequisite Check for Chapter 5

1. $\frac{3}{4}$ **2.** $\frac{8}{15}$ **3.** $\frac{3}{4}$ **4.** $\frac{28}{5}$ **5.** $\frac{3}{4}$ **6.** 525 **7.** 35
8. 5.56 **9.** 7.22 **10.** No **11.** Yes

Reading Your Text for Chapter 5

Section 5.1 (**a**) fraction; (**b**) like; (**c**) simplest; (**d**) mixed
Section 5.2 (**a**) like; (**b**) rate; (**c**) improper; (**d**) Unit
Section 5.3 (**a**) equal; (**b**) variable; (**c**) equal; (**d**) rates
Section 5.4 (**a**) equation; (**b**) coefficient; (**c**) read; (**d**) proportional

Summary Exercises for Chapter 5

1. $\frac{4}{17}$ **3.** $\frac{5}{8}$ **5.** $\frac{4}{9}$ **7.** $\frac{7}{36}$ **9.** $100\,\frac{\text{mi}}{\text{hr}}$ **11.** $50\,\frac{\text{cal}}{\text{oz}}$
13. $166\frac{2}{3}\,\frac{\text{ft}}{\text{s}}$ **15.** $6\frac{1}{2}\,\frac{\text{hits}}{\text{game}}$ **17.** Marisa
19. $0.09\,\frac{\text{dollar}}{\text{oz}}$ or $9\,\frac{\text{cents}}{\text{oz}}$ **21.** $0.095\,\frac{\text{dollar}}{\text{oz}}$ or $9.5\,\frac{\text{cents}}{\text{oz}}$
23. $14.95\,\frac{\text{dollars}}{\text{CD}}$ **25.** $\frac{4}{9} = \frac{20}{45}$ **27.** $\frac{110\text{ mi}}{2\text{ hr}} = \frac{385\text{ mi}}{7\text{ hr}}$
29. No **31.** Yes **33.** Yes **35.** Yes **37.** 2 **39.** 4
41. 180 **43.** 100 **45.** \$135 **47.** 15 in. **49.** 28 parts
51. 140 g

Chapter Test for Chapter 5

1. $\frac{7}{19}$ **2.** $\frac{5}{3}$ **3.** $\frac{2}{3}$ **4.** $\frac{1}{12}$ **5.** $4\frac{4}{5}\ \frac{\text{mi}}{\text{gal}}$ **6.** $8.25\ \frac{\text{dollars}}{\text{hr}}$
7. 20 **8.** 18 **9.** 3 **10.** 16 **11.** Yes **12.** No
13. Yes **14.** No **15.** $2.56\ \frac{\text{dollars}}{\text{gal}}$ **16.** $\frac{26}{33}, \frac{26}{7}$
17. $2.28 **18.** 576 mi **19.** 600 mufflers **20.** 24 tsp

Cumulative Review Chapters 1–5

[1.1] 1. Forty-five thousand, seven hundred eighty-nine
2. Ten thousands **[1.2] 3.** 26,304 **[1.3] 4.** 47,806
[1.5] 5. 4,408 **[1.6] 6.** 78 r67 **[1.3] 7.** $568
[1.7] 8. 3 **[1.2, 1.5] 9.** *P*: 16 ft; *A*: 12 ft^2 **10.** $1,104
[2.2] 11. $2 \times 2 \times 3 \times 7 \times 11$ **12.** 14 **[2.4] 13.** $\frac{1}{4}$
[2.5] 14. $\frac{6}{5} = 1\frac{1}{5}$ **15.** 10 **[2.6] 16.** $\frac{9}{2} = 4\frac{1}{2}$ **17.** $1\frac{9}{13}$
[3.3] 18. $\frac{25}{44}$ **[3.4] 19.** $7\frac{7}{12}$ **20.** $4\frac{1}{2}$ **[3.2] 21.** 180
[2.5] 22. 176 mi **[2.6] 23.** $48\ \frac{\text{mi}}{\text{hr}}$ **[4.2] 24.** 7.828
[4.4] 25. 1.23 **[4.3] 26.** 6.6015 **[4.5] 27.** $\frac{9}{25}$
28. 0.32 **[4.2] 29.** 14.06 m **[4.3] 30.** 50.24 cm^2
[5.1] 31. $\frac{6}{13}$ **32.** $\frac{10}{3}$ **33.** $\frac{4}{5}$ **[5.3] 34.** Yes **35.** No
[5.4] 36. 2 **37.** 15 **[5.2] 38.** $24.4\ \frac{\text{cents}}{\text{oz}}$
[5.4] 39. 600 km **40.** 50

Prerequisite Check for Chapter 6

1. $\frac{11}{25}$ **2.** $\frac{5}{4}$ **3.** 0.375 **4.** 5.5 **5.** $\frac{2}{25}$ **6.** $6\frac{1}{4}$
7. 41.6 **8.** 73.6875 **9.** 82.8 **10.** $63.\overline{3}$ **11.** 52.5
12. 0.96 **13.** 48 **14.** $58\frac{1}{3}$ **15.** 3.78 **16.** 8,000
17. The store's profit margin on the electric range.
18. The wholesale price is $674.96; the retail price is $899.95

Reading Your Text for Chapter 6

Section 6.1 **(a)** hundred; **(b)** fraction; **(c)** left; **(d)** greater
Section 6.2 **(a)** right; **(b)** decimal; **(c)** percent; **(d)** zeros
Section 6.3 **(a)** base; **(b)** rate; **(c)** rate; **(d)** amount
Section 6.4 **(a)** greater; **(b)** amount; **(c)** markup; **(d)** principal

Summary Exercises for Chapter 6

1. 75% **3.** $\frac{1}{50}$ **5.** $\frac{3}{8}$ **7.** $2\frac{1}{3}$ **9.** 0.75 **11.** 0.0625
13. 0.006 **15.** 6% **17.** 240% **19.** 3.5% **21.** 43%
23. 40% **25.** $266\frac{2}{3}\%$ **27.** 2,000 **29.** 330 **31.** 5,000
33. 66.5 **35.** 600 **37.** 3% **39.** $1,800 **41.** 7.5%
43. $102 **45.** 720 students **47.** $3,157.50
49. 500 s or 8 min 20 s

Chapter Test for Chapter 6

1. 80% **2.** 3% **3.** 4.2% **4.** 40% **5.** 62.5%
6. 0.42 **7.** 0.06 **8.** 1.6 **9.** $\frac{7}{100}$ **10.** $\frac{18}{25}$
11. (*A*) 50; (*R*) 25%; (*B*) 200 **12.** (*A*) Unknown; (*R*) 8%; (*B*) 500
13. (*A*) $30; (*R*) 6%; (*B*) amount of purchase **14.** 11.25
15. 20% **16.** 500 **17.** 750 **18.** 175% **19.** 800
20. 300 **21.** 7.5% **22.** $4.96 **23.** 60 questions
24. $70.20 **25.** 12% **26.** 8% **27.** $18,000
28. 6,400 students **29.** $18,500 **30.** 24%

Cumulative Review Chapters 1–6

[1.1] 1. Thousands **[1.5] 2.** 11,368 **[1.6] 3.** 89
[1.7] 4. 5 **5.** 9 **6.** 42 **[2.1] 7.** 53, 59, 61, 67
[2.2] 8. $2 \times 2 \times 5 \times 13$ **9.** 28 **[3.2] 10.** 180
[2.5] 11. $8\frac{1}{2}$ **[2.6] 12.** $1\frac{1}{3}$ **[3.4] 13.** $8\frac{7}{12}$ **14.** $4\frac{19}{24}$
[2.5] 15. $286 **[2.6] 16.** $54\ \frac{\text{mi}}{\text{hr}}$ **[3.4] 17.** $41\frac{1}{4}$ in.
[4.1] 18. Hundredths **19.** Ten-thousandths **20.** < **21.** =
[4.3] 22. 11.284 **[4.4] 23.** 17.04 **[4.3] 24.** $108.05
25. 6.29 m^2 **26.** 44.064 ft^3 **[4.5] 27.** $\frac{9}{25}$ **28.** $5\frac{1}{8}$
[5.1] 29. $\frac{2}{3}$ **30.** $\frac{17}{12}$ **[5.4] 31.** $18\frac{2}{3}$ **32.** 0.4
33. 350 mi **34.** $140 **[6.1] 35.** 0.34; $\frac{17}{50}$
[6.2] 36. 0.55; 55% **[6.3] 37.** 45 **38.** 500
[6.4] 39. 125 employees **40.** 8.5%

Prerequisite Check for Chapter 7

1. 428,400 **2.** 0.004284 **3.** $\frac{4}{3}$ **4.** $\frac{5}{12}$ **5.** $\frac{5}{2}$ **6.** $\frac{3}{5}$
7. $10.75/hr **8.** $50\ \frac{\text{mi}}{\text{hr}}$ **9.** 5.14 cents/fl oz
10. $0.45/DVD-R **11.** 54 in. **12.** 500 mm

Reading Your Text for Chapter 7

Section 7.1 **(a)** metric; **(b)** volume; **(c)** one; **(d)** like
Section 7.2 **(a)** metric; **(b)** yard; **(c)** centi-; **(d)** Kilometers
Section 7.3 **(a)** gram; **(b)** milligram; **(c)** liter; **(d)** milliliter
Section 7.4 **(a)** substitution; **(b)** unit-ratio; **(c)** gravity; **(d)** Celsius

Summary Exercises for Chapter 7

1. 132 **3.** 24 **5.** 64 **7.** 4 **9.** 4 ft 11 in.
11. 9 lb 3 oz **13.** 5 ft 7 in. **15.** 4 hr 15 min **17.** 5 lb 6 oz
19. 33 hr 35 min **21.** (a) **23.** (c) **25.** mm **27.** cm
29. 2,000 **31.** 3 **33.** 0.06 **35.** (a) **37.** (a) **39.** g
41. kg **43.** 5,000 **45.** 2 **47.** (c) **49.** (b) **51.** L
53. mL or cm^3 **55.** 5,000 **57.** 6 **59.** 326.77 **61.** 6.75
63. 5.51 **65.** 25.35 **67.** 1,737.36 or 1,729 **69.** 5,850
71. 0.8 **73.** 62.6 **75.** 37 **77.** 15 **79.** 41

Chapter Test for Chapter 7

1. 96 **2.** 48 **3.** 5,000 **4.** 3,000 **5.** 3 **6.** 101.6
7. 3.2 **8.** 55.1 **9.** 67.5 **10.** 30.5 **11.** 14.4
12. 75.2 **13.** 6 ft 9 in. **14.** 4 days 12 min **15.** 11 ft 5 in.
16. 2 lb 9 oz **17.** 15 hr 20 min **18.** 4 lb 6 oz **19.** (b)
20. (b) **21.** (b) **22.** (b) **23.** $1,050 **24.** 8.6 lb
25. $0.0013/mL

Cumulative Review Chapters 1–7

[1.5] 1. $896 **[1.6] 2.** $62 **[1.7] 3.** 16
[2.2] 4. $2 \times 2 \times 2 \times 3 \times 7$ **5.** 4 **[2.4] 6.** $\frac{3}{5}, \frac{5}{8}, \frac{2}{3}$
[2.5] 7. $\frac{3}{4}$ **[2.7] 8.** $\frac{5}{12}$ **[3.2] 9.** 60
[3.3] 10. $\frac{31}{30}$ or $1\frac{1}{30}$ **[3.4] 11.** $3\frac{13}{24}$ **[4.2] 12.** $18.30
[4.3] 13. 27.84 cm^2 **[4.2] 14.** 21.5 cm **[4.5] 15.** 0.5625
16. 0.538 **[5.4] 17.** 24 **18.** 400 mi **19.** 450 mi
[6.2] 20. 37.5% **[6.1] 21.** $\frac{1}{8}$ **[6.3] 22.** 3,526
23. 225% **24.** 150 **[6.4] 25.** $142,500 **[7.1] 26.** 120
27. 1 min 35 s **28.** 21 lb 11 oz **29.** 2 ft 9 in.
30. 21 hr 20 min **[7.2] 31.** 0.43 **[7.3] 32.** 62,000
[7.2] 33. 74 **[7.3] 34.** 14,000 **35.** 0.5 **36.** 0.375
[7.4] 37. 13.3 **38.** 149.6 **39.** 29.4 **40.** 48.2

Prerequisite Check for Chapter 8

1. 25 **2.** 169 **3.** 78.5 **4.** 36 **5.** 17.4 **6.** $\frac{45}{2}$
7. 19.63 **8.** 7.0 in. **9.** 4.4 m **10.** 118 ft
11. 144 mm^2 **12.** 173.25 in.^{2}

Reading Your Text for Chapter 8

Section 8.1 (**a**) earth; (**b**) perpendicular; (**c**) obtuse; (**d**) complementary
Section 8.2 (**a**) triangle; (**b**) regular; (**c**) sum; (**d**) nine
Section 8.3 (**a**) circumference; (**b**) radius; (**c**) composite; (**d**) area
Section 8.4 (**a**) equilateral; (**b**) isosceles; (**c**) similar; (**d**) corresponding
Section 8.5 (**a**) radical; (**b**) hypotenuse; (**c**) Pythagorean; (**d**) whole
Section 8.6 (**a**) three; (**b**) cubic; (**c**) sphere; (**d**) cube

Summary Exercises for Chapter 8

1. $\angle AOB$; acute; 70° **3.** $\angle AOC$; obtuse; 100°
5. $\angle XYZ$; straight; 180° **7.** 135° **9.** 47° **11.** 73°
13. 79° **15.** 67° **17.** 750 ft^2 **19.** $P = 60$ in.; $A = 216$ in.^{2}
21. $P = 93$ ft; $A = 457.5$ ft^2 **23.** 180 ft^2; 20 yd^2
25. $C = 37.7$ in.; $A = 113$ in.^{2} **27.** $P = 56$ ft; $A = 128$ ft^2
29. 60°; equilateral **31.** 45°; isosceles **33.** 14.3 **35.** 32 m
37. 18 **39.** 13.75 **41.** 55 **43.** (**a**) 40 in.^{2}; (**b**) 52 in.^{2}; (**c**) 24 in.^{3} **45.** (**a**) 289 in.^{2}; (**b**) $433\frac{1}{2}$ in.^{2}; (**c**) $614\frac{1}{8}$ in.^{3}
47. (**a**) 1,661.06 cm^2; (**b**) 6,367.4 cm^3 **49.** 20,736 in.^{3}
51. 491.61 cm^3

Chapter Test for Chapter 8

1. (**a**) Parallel; (**b**) neither; (**c**) perpendicular; (**d**) neither
2. (**a**) Straight; (**b**) obtuse; (**c**) right **3.** Obtuse **4.** Right
5. 50° **6.** 135° **7.** 137° **8.** 66° **9.** 67° **10.** 21
11. 327.74 cm^3 **12.** 216 ft^3 **13.** 60 mm **14.** 12 in.
15. 256 yd **16.** 62 m **17.** 94.2 ft **18.** 91.7 ft
19. 189 mm^2 **20.** 7 in.^{2} **21.** 2,800 yd^2 **22.** 54 m^2
23. 706.5 ft^2 **24.** 332.8 ft^2 **25.** (**a**) 160 cm^2; (**b**) 288 cm^2; (**c**) 320 cm^3 **26.** (**a**) 25.12 in.^{2}; (**b**) 31.4 in.^{2}; (**c**) 12.56 in.^{3}
27. (**a**) 12.56 cm^2; (**b**) 4.19 cm^3 **28.** 65 m **29.** 53.9
30. 2,376 BTUs

Cumulative Review Chapters 1–8

[1.1] 1. Ten thousands **[1.7] 2.** 64 **[2.1] 3.** $2 \cdot 3 \cdot 3 \cdot 5 \cdot 7$
4. 4 **5.** 120 **[2.5] 6.** $\frac{3}{4}$ **[2.6] 7.** $4\frac{1}{2}$ **8.** $578
9. $60\ \frac{\text{mi}}{\text{hr}}$ **[3.3] 10.** $\frac{43}{30}$ **[3.4] 11.** $3\frac{13}{24}$ **12.** $6\frac{7}{12}$ mi
[4.1] 13. Hundredths **14.** Ten-thousandths
[4.3] 15. $P = 24.2$ ft; $A = 35.04$ ft^2 **[4.5] 16.** $\frac{1}{8}$
17. 0.3125 **[5.2] 18.** $20,250/yr **[5.4] 19.** 4
20. 1,125 ft^2 **[6.1] 21.** 0.085 **22.** $\frac{3}{8}$ **[6.2] 23.** 67.5%
[6.3] 24. 51.2 **25.** 1,500 **[6.4] 26.** 1,350
[7.1–7.4] 27. 5,280 **28.** 0.25 **29.** 5,800 **[8.1] 30.** 45°
[8.2–8.3] 31. 49.6 ft **32.** 342 cm^2 **33.** 1,256 m^2
34. 60.63 ft^2 **[8.4] 35.** 73°; isosceles **[8.5] 36.** 40
37. 17 ft **[8.6] 38.** (**a**) 100 cm^2; (**b**) 150 cm^2; (**c**) 125 cm^3
39. (**a**) 314 cm^2; (**b**) 523.33 cm^3 **40.** 150 gal

Prerequisite Check for Chapter 9

1. $\frac{4}{5}$ **2.** $\frac{3}{4}$ **3.** 6 **4.** $\frac{17}{2}$ **5.** 27 **6.** 77 **7.** 90
8. 300 **9.** 62.5% **10.** $16.\overline{6}\%$ **11.** 0, 1, 2, 5, 7, 10, 11, 13
12. 7, 12, 21, 50, 55, 56, 81, 123 **13.** $1\frac{1}{8}$ in. **14.** 120°

Reading Your Text for Chapter 9

Section 9.1 (**a**) mean; (**b**) median; (**c**) mode; (**d**) bimodal
Section 9.2 (**a**) quartiles; (**b**) three; (**c**) summary; (**d**) box-and-whisker
Section 9.3 (**a**) rows; (**b**) cell; (**c**) graph; (**d**) legend
Section 9.4 (**a**) future; (**b**) extrapolation; (**c**) pie; (**d**) circle

Summary Exercises for Chapter 9

1. 6 **3.** 120 **5.** 85 **7.** 19; 20 **9.** 29; 28 **11.** 92
13. 21, 24, 28, 30, 35
15.

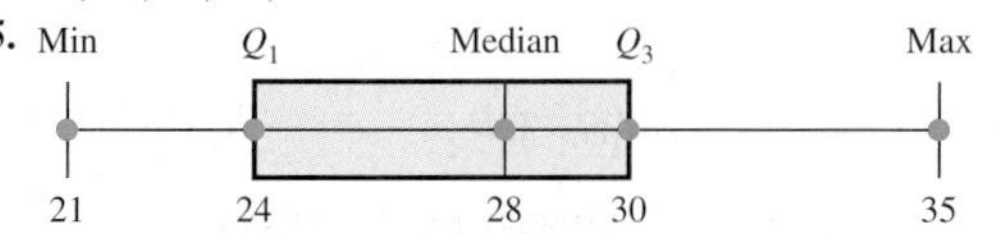

17. 482,000 vehicles; 9,629,000 vehicles **19.** 1.8% **21.** 12.4%
23. 77.6% **25.** 5,000
27.

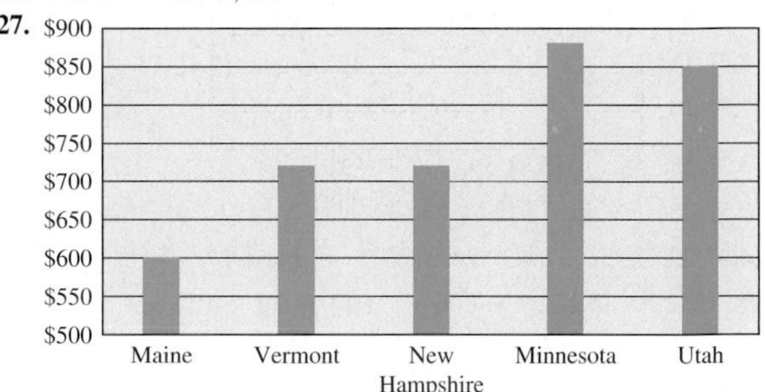

29. 250,000 **31.** 400,000 PCs
33.

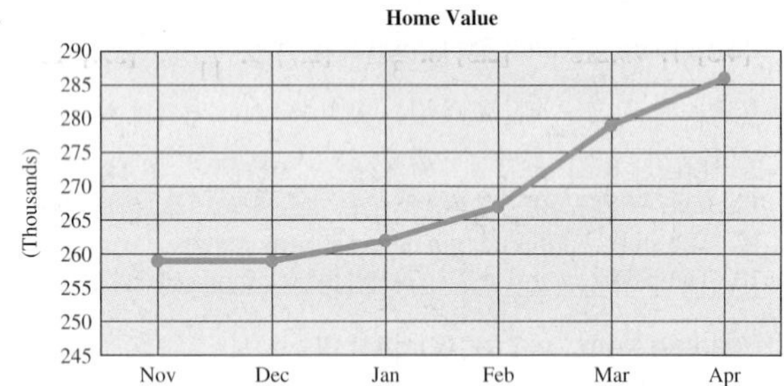

35. 10.4% **37.** Côte d'Ivoire; 30%
39.

Grade	Count	Percent	Degrees
A	7	17.5%	63°
B	12	30%	108°
C	13	32.5%	117°
D	5	12.5%	45°
F	3	7.5%	27°

Grade Distribution

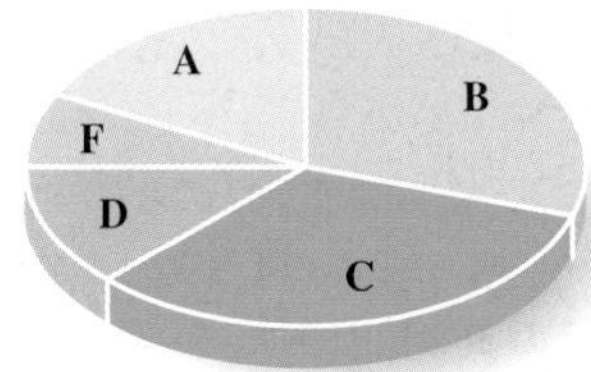

Chapter Test for Chapter 9

1. 15 **2.** 8 **3.** 11 **4.** 6 **5.** 23°F, 36.5°F, 51°F, 66.5°F, 76°F **6.** 204 passengers **7.** 93 **8.** Brown
9. 7,790,000,000 lb **10.** $9,460,000,000 **11.** 0.67%
12. 63%

13.

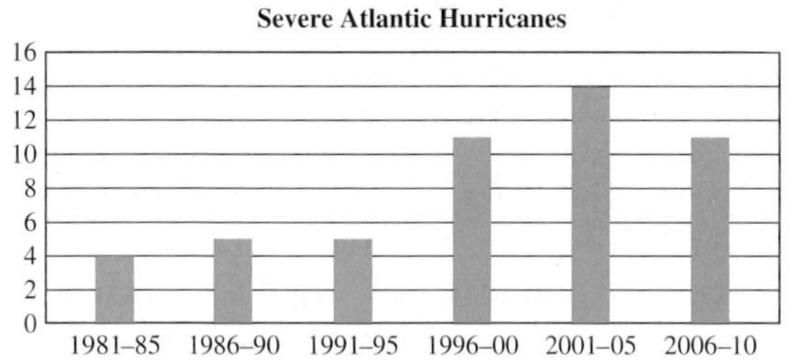
Severe Atlantic Hurricanes
16
14
12
10
8
6
4
2
0
1981–85 1986–90 1991–95 1996–00 2001–05 2006–10

14. 11 **15.** 27% **16.** 1996–2000 **17.** December
18. August and September
19.

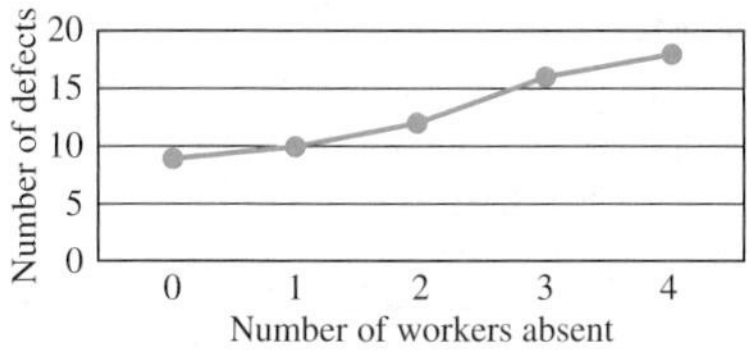
Number of defects
20
15
10
5
0
0 1 2 3 4
Number of workers absent

20. About 20 defects **21.** 45% **22.** 85% **23.** 180 items
24. 540 items **25.** $28,350

Cumulative Review Chapters 1–9

[1.1] 1. Thousands **[1.2] 2.** 32,278 **[1.3] 3.** 39,288
[1.5] 4. 26,230 **[1.6] 5.** 308 r6 **[1.3] 6.** 48,588
[4.3] 7. 75.215 **[2.5] 8.** $\frac{2}{3}$ **[2.7] 9.** $\frac{6}{11}$ **[3.4] 10.** $7\frac{1}{12}$
[5.4] 11. 14 **12.** 9 **[6.1] 13.** 0.18; $\frac{9}{50}$
[6.2] 14. 0.425; 42.5% **[7.1] 15.** 11 lb 5 oz **16.** 1 min 35 s
[7.2] 17. 8,000 **18.** 3 **19.** 5 **20.** 250
[9.4] 21. 2010 and 2011
[9.3] 22.

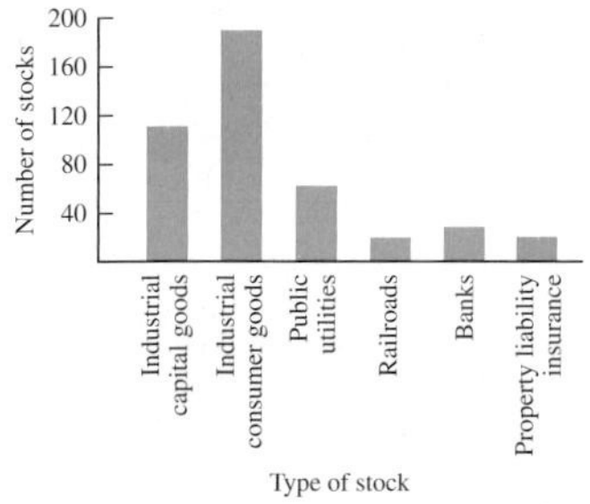
200
160
120
80
40
Number of stocks
Industrial capital goods
Industrial consumer goods
Public utilities
Railroads
Banks
Property liability insurance
Type of stock

[9.1] 23. Mean: 9; median: 9; mode: 11 **[5.2] 24.** 70 gal
25. $88\ \frac{\text{ft}}{\text{s}}$ **26.** $19\ \frac{\text{pitches}}{\text{inning}}$ **[8.2] 27.** $408 **[2.6] 28.** $48\ \frac{\text{mi}}{\text{hr}}$
[8.2] 29. $28\frac{3}{5}$ cm **30.** $55\frac{1}{3}$ ft **[6.3] 31.** 133 **32.** 0.2%
33. 185 **[2.5] 34.** $4\frac{1}{3}$ **[6.4] 35.** 2,800 students

Prerequisite Check for Chapter 10

1. 21 **2.** 79 **3.** 368 **4.** $15\frac{1}{4}$ **5.** 187 **6.** 32
7. Commutative property of addition
8. Distributive property of multiplication over addition
9. Associative property of multiplication
10. Distributive property of multiplication over addition
11. 33.91 mm **12.** 91.56 mm^2

Reading Your Text for Chapter 10

Section 10.1 (**a**) zero; (**b**) Negative; (**c**) ascending; (**d**) absolute value
Section 10.2 (**a**) negative; (**b**) negative; (**c**) absolute; (**d**) zero
Section 10.3 (**a**) addition; (**b**) subtraction; (**c**) opposite; (**d**) positive
Section 10.4 (**a**) negative; (**b**) positive; (**c**) identity; (**d**) reciprocal
Section 10.5 (**a**) negative; (**b**) positive; (**c**) Division; (**d**) positive

Summary Exercises for Chapter 10

1.

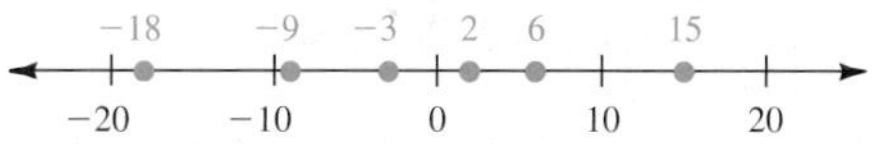
−18 −9 −3 2 6 15
−20 −10 0 10 20

3. −7, −3, −2, 0, 1, 4, 6 **5.** Max: 5; Min: −6 **7.** 9
9. −9 **11.** 4 **13.** −4 **15.** −11 **17.** 0 **19.** −18
21. −4 **23.** −5 **25.** 17 **27.** 0 **29.** 3
31. 4 **33.** 12 **35.** −1 **37.** −1 **39.** 100°F
41. −70 **43.** 45 **45.** 0 **47.** $-\frac{3}{2}$ **49.** 80 **51.** 10
53. 5 **55.** 9 **57.** −4 **59.** −1 **61.** −7 **63.** −21

Chapter Test for Chapter 10

1.

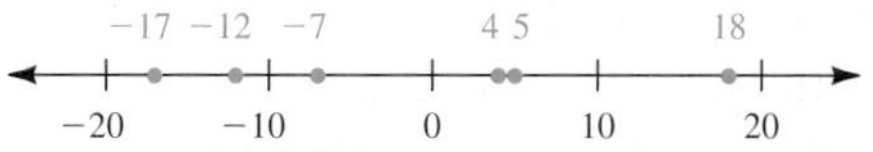
−17 −12 −7 4 5 18
−20 −10 0 10 20

2. −6, −3, −2, 0, $\frac{1}{2}$, $\frac{3}{4}$, 2, 4, 5 **3.** Max: 6; Min: −5
4. (**a**) 110°F; (**b**) 108°F **5.** 7 **6.** 7 **7.** −13 **8.** −3
9. −6 **10.** −24 **11.** −40 **12.** 63 **13.** −25
14. 3 **15.** 11 **16.** 11 **17.** −21 **18.** 9 **19.** 0
20. −27 **21.** −24 **22.** −5 **23.** Undefined
24. −24 **25.** −10

Cumulative Review Chapters 1–10

[1.1] 1. Ten thousands **[1.2] 2.** 142,231 **[1.3] 3.** 29,573
[1.5] 4. 53,445 **[1.6] 5.** 402 r28 **[4.2] 6.** 13.687
[4.3] 7. 1,837.353 **[2.5] 8.** $\frac{3}{7}$ **[2.7] 9.** 1
[3.4] 10. $8\frac{19}{54}$ **[5.4] 11.** 4 **[6.1] 12.** 0.58; $\frac{29}{50}$
[6.2] 13. 0.48; 48% **[7.1] 14.** 8 ft 10 in. **15.** 9 lb 4 oz
16. 12 ft 6 in. **17.** 2 lb 14 oz **18.** 10 hr 30 min
19. 2 min 9 s **20.** 14 ft 8 in.
[5.2] 21. The three smaller bottles **[4.3] 22.** 49.02 cm^2
23. $72\frac{1}{4}$ in.^{2} **[8.3] 24.** 25.7 ft **[6.3] 25.** 80 **[6.4] 26.** 7.5%
[7.3] 27. 0.017 **[7.2] 28.** 820 **[8.1] 29.** 160°
[9.3] 30. 5,000 students **[9.1] 31.** Mean: 17; median: 17; mode: 17 **[10.2] 32.** 4 **[10.3] 33.** 20 **34.** −32
35. −15 **36.** 31 **[10.4] 37.** 108 **[10.5] 38.** 4
39. 41 **40.** 72

Prerequisite Check for Chapter 11

1. 10 − 8 = 2 **2.** 3 + 5 × 6 = 33 **3.** $-\frac{1}{12}$ **4.** $\frac{8}{37}$
5. 1 **6.** 1 **7.** 1 **8.** 23 **9.** −64 **10.** 64
11. $14,117.65 **12.** $1.55

Reading Your Text for Chapter 11

Section 11.1 (**a**) variables; (**b**) sum; (**c**) multiplication; (**d**) expression
Section 11.2 (**a**) evaluating; (**b**) positive; (**c**) operations; (**d**) grouping
Section 11.3 (**a**) term; (**b**) number; (**c**) like; (**d**) distributive
Section 11.4 (**a**) equation; (**b**) true; (**c**) solution(s); (**d**) sentence
Section 11.5 (**a**) equivalent; (**b**) original; (**c**) multiplying; (**d**) reciprocal
Section 11.6 (**a**) isolate; (**b**) addition; (**c**) original; (**d**) solved

Summary Exercises for Chapter 11

1. $y + 5$ **3.** $8a$ **5.** $5mn$ **7.** $17x + 3$ **9.** -7

11. 15 **13.** 19 **15.** 25 **17.** 1 **19.** -3
21. $4a^3, -3a^2$ **23.** $5m^2, -4m^2, m^2$ **25.** $12c$ **27.** $2a$
29. $3xy$ **31.** $19a + b$ **33.** $3x^3 + 9x^2$ **35.** $10a^3$
37. Yes **39.** Yes **41.** No **43.** 2 **45.** -5 **47.** 5
49. 1 **51.** -7 **53.** 7 **55.** -4 **57.** 32 **59.** 27
61. 3 **63.** -2 **65.** $\frac{7}{2}$ **67.** 18 **69.** 6 **71.** $\frac{2}{5}$
73. 6 **75.** 6 **77.** 4 **79.** 5 **81.** $\frac{1}{2}$ **83.** 3 hr
85. 5

Chapter Test for Chapter 11

1. $a - 5$ **2.** $6m$ **3.** $4(m + n)$ **4.** $4m + n$ **5.** No
6. Yes **7.** 11 **8.** 7 **9.** 3 **10.** 4 **11.** -12
12. 12 **13.** 7 **14.** 25 **15.** $-\frac{2}{3}$ **16.** -5 **17.** -4
18. 80 **19.** 144 **20.** 5 **21.** $15a$ **22.** $3x^2y$
23. $19x + 5y$ **24.** $4m^2 - 2m - 5$ **25.** 58 double lattes

Cumulative Review Chapters 1–11

[1.2] 1. Associative property of addition
[1.5] 2. Commutative property of multiplication
3. Distributive property **[1.4] 4.** 5,900 **5.** 950,000
[1.7] 6. 8 **[2.2] 7.** $2 \times 2 \times 2 \times 3 \times 11$ **[3.2] 8.** 90
[2.3] 9. $3\frac{1}{7}$ **10.** $\frac{53}{8}$ **[2.5] 11.** $\frac{3}{4}$ **[2.6] 12.** $1\frac{1}{2}$
[3.4] 13. $8\frac{1}{24}$ **[10.2] 14.** $3\frac{5}{8}$ **15.** $\frac{5}{8}$ in. **[4.2] 16.** \$3.06
[8.3] 17. 8.04 ft^2 **[4.4] 18.** 32 lots **[4.5] 19.** $0.\overline{72}$
[5.4] 20. 112.5 **21.** 12 in. **[6.2] 22.** 0.3%
23. 62.5% **24.** 350% **[6.3] 25.** 150 **26.** 600
[6.4] 27. \$27,500 **[7.3] 28.** 0.3 **[9.4] 29.** 80
[10.2] 30. -8 **[10.1] 31.** 20 **32.** 12 **33.** 5
[10.2] 34. -18 **[10.3] 35.** -4 **[10.4] 36.** -90
[10.5] 37. -4 **[11.1] 38.** $3(x + y)$ **39.** $\frac{n-5}{3}$
[11.2] 40. -60 **41.** 27 **42.** 24 **43.** 3
[11.3] 44. $12a^2 + 3a$ **45.** $-2y$ **[11.4] 46.** 5
[11.5] 47. -24 **[11.6] 48.** $-\frac{2}{5}$ **49.** 5 **50.** \$2.07

Photo Credits

Chapter 1

Opener: © Corbis RF; p. 5: © Getty RF; p. 7(top): NASA & J. Hester/Arizona State University; p. 7(bottom): © Alamy RF; p. 8, p. 21: © Getty RF; p. 22(top): © Creatas Images/PictureQuest; p. 22(bottom): © Getty RF; p. 25: © Dynamic Graphics/Jupiter RF; p. 34: © Ingram Publishing RF; p. 38: © Corbis RF; p. 45: © Getty RF; p. 46: © PhotoDisc/Getty RF; p. 48: © The McGraw-Hill Companies Inc., Photo by Mark Steinmetz; p. 51: © Comstock Images/Jupiterimages RF; p. 60: © Ingram Publishing/SuperStock RF; p. 73: © Vol. 3/Corbis RF; p. 75: © Corbis RF; p. 77: © Ingram Publishing/Alamy RF; p. 87: © Blend Images LLC RF; p. 90: © Alamy RF; p. 91: © Blend Images LLC RF; p. 92: Brand X Pictures RF; p. 94: © Superstock RF.

Chapter 2

Opener: © BananaStock/PunchStock RF; p. 104, p. 111: © Photolink/Getty RF; p. 119 (top): © Getty RF; p. 119 (bottom): © Ingram Publishing/SuperStock; p. 121: © The McGraw-Hill Companies, Inc. Connie Mueller, photographer; p. 127 (top, bottom): © Getty RF; p. 131: © The McGraw-Hill Companies, Inc. Christopher Kerrigan, photographer; p. 144: © Vol. 16/PhotoDisc/Getty RF; p. 152 (top, bottom): © Ingram Publishing RF; p. 152 (bottom): © Ingram Publishing RF; p. 153: Getty RF; p. 155: © BananaStock/PunchStock RF.

Chapter 3

Opener: © Corbis RF; p. 172: © The McGraw-Hill Companies Inc., Photo by Mark Steinmetz; p. 180: © Getty RF; p. 184: © Blend Images LLC; p. 188(top, bottom): © Getty RF; p. 191: © Corbis RF; p. 197: © Blend Images LLC RF; p. 199: © Getty RF; p. 202: © IMS Communications Ltd./Capstone Design/FlatEarth Images RF; p. 206: © Getty RF; p. 207(top): © Alamy RF; p. 207(bottom): © Comstock RF; p. 208: © Ingram Publishing RF; p. 209: © Photolink/Getty RF.

Chapter 4

Opener: © Getty RF; p. 227: © Ingram Publishing RF; p. 246: © Getty RF; p. 247: © Brand X/Getty RF; p. 255, p. 257: © Ingram Publishing/SuperStock RF; p. 258: © Ingram Publishing/RF; p. 260: © Getty RF; p. 266: © Corbis RF; p. 269: © Comstock/Jupiter RF.

Chapter 5

Opener: © Getty RF; p. 281: © Getty RF; p. 284: © Purestock/SuperStock; p. 285, p. 287: © Getty RF; p. 290: © Corbis RF; p. 291: © Getty RF; p. 294(top): © The McGraw-Hill Companies, Inc.; p. 294(bottom): © Getty RF; p. 295: © BananaStock/Jupiter RF; p. 297, p. 302(top, bottom): © Getty RF; p. 310: © PictureQuest RF; p. 313: © BananaStock/Jupiterimages RF; p. 316: © Getty RF.

Chapter 6

Opener: © Digital Vision/Punchstock RF; p. 329: © Education/PhotoDisc/Getty RF; p. 330: © Digital Vision/Punchstock RF; p. 339: © Getty RF; p. 342: © Digital Vision/Getty RF; p. 349, p. 352: © Getty RF; p. 356: © Corbis RF; p. 359: © Ingram Publishing/SuperStock RF; p. 361(top, bottom): © Getty RF; p. 362: © Image Source RF; p. 364: © Getty RF; p. 365: © The McGraw-Hill Companies, Inc. Lars A. Niki, photographer; p. 368: © Masterfile RF; p. 370: © Alamy RF; p. 371: © The McGraw-Hill Companies, Inc.

Chapter 7

Opener: © Corbis RF; p. 385(top): © Design Pics RF; p. 385(bottom): © Blend Images LLC RF; p. 387: © Punchstock RF; p. 388: © Stockbyte/PunchStock RF; p. 396(top): © Lissa Harrison RF; p. 396(bottom): © Purestock/SuperStock RF; p. 397: © Getty RF; p. 401: © Corbis RF; p. 404: © Imagestate RF; p. 405: © Aaron Roeth Photography RF; p. 412: © Digital Vision/PunchStock RF; p. 414(top, bottom): © Getty RF; p. 416: © Corbis RF.

Chapter 8

Opener: © Corbis RF; p. 429: © Vol. 2/Corbis RF; p. 453: © Getty RF; p. 459: © Design Pics; p. 461: © Getty RF; p. 462, p. 463: © PhotoLink/Getty RF; p. 472: © Corbis RF; p. 478: © Corbis RF; p. 487: © Creaatas; p. 488: © Brand X Pictures/Jupiter Images RF; p. 498: © Alamy RF.

Chapter 9

Opener: NOAA Office of Marine and Aircraft Operations; p. 508: © Brand X/PunchStock RF; p. 510: © Digital Vision/Punchstock RF; p. 516: © PhotoLink/Getty RF; p. 523: © Corbis RF; p. 527: NOAA Office of Marine and Aircraft Operations; p. 528: © Purestock/SuperStock RF; p. 535: © Corbis RF; p. 537: © DesignPics RF; p. 551: © Punchstock RF; p. 556: © Corbis RF; p. 562: © PunchStock RF; p. 563: © Getty RF.

Chapter 10

Opener: © NOAA Photo Library, NOAA Central Library; OAR/ERL/National Severe Storms Laboratory (NSSL); p. 570(left): © Corbis RF; p. 576: © The McGraw-Hill Companies, Inc. Barry Barker, photographer; p. 586: © NOAA Photo Library, NOAA Central Library; OAR/ERL/National Severe Storms Laboratory (NSSL); p. 591: © Vol. 74/PhotoDisc/Getty RF; p. 594, p. 611: © Getty RF.

Chapter 11

Page 627: © Design Pics RF; p. 636: © Design Pics/Alamy RF; p. 638: © Getty RF; p. 647: © Brand X/SuperStock RF; p. 657: © Getty RF; p. 659: © Photo Alto RF; p. 663: © Digital Vision/Getty RF; p. 684(top): © Alamy RF; p. 684(middle): © Getty RF; p. 684(bottom): USDA.

index

Q

R

S